THE CLIMATE NEAR THE GROUND

THE CLIMATE NEAR THE GROUND

Rudolf Geiger

Translated by Scripta Technica, Inc.

HARVARD UNIVERSITY PRESS

Cambridge, Massachusetts

Distributed in Great Britain by Oxford University Press, London

Translated from the fourth German edition of

Das Klima der bodennahen Luftschicht,

published in 1961 by Friedrich Vieweg & Sohn,

Brunswick, Germany

Third Printing, 1971

Library of Congress Catalog Card Number 64-23191

SBN 674-13500-8

Printed in the United States of America

PREFACE

Anyone who merely turns the pages of this new edition will find that 48 percent of the figures are familiar to him from the third edition. But whoever reads it will discover that no three consecutive pages of text have been transferred unaltered. The enormous development that has taken place since 1950, particularly the surprising extension in the practical applications of micrometeorology, have made it necessary to rewrite the book. The rounding off of our knowledge also permitted the material to be arranged more clearly.

In producing this work, I had in mind two aims which were linked more closely to each other than I had at first dared to hope. The new edition was to be a clear and vivid textbook for those who were just taking up the study of microclimatology, and at the same time a reference work for those already familiar with the subject. For the first task I had in mind the students who would be regarding with horror the insurmountable barrier of an apparently unlimited and ever-increasing pile of literature, and thus were in need of real assistance. I was thinking also of colleagues working in related sciences, who have no time to study our literature; and finally, but not least important, I was thinking of all who work on the land, in forests and gardens, the architects, geographers, country planners, entomologists, doctors, transportation engineers, and others who—without having studied much physics—were anxious to acquire a knowledge of the rational physical principles governing the meteorologic laws that they have to put into practice. For the benefit of all these, I have at all times made renewed efforts to state the facts in the simplest and most uncomplicated manner possible. I have also tried at all times to improve the style so that the reader would be able to go on his way lightly, where the author had expended patient effort, tenacious industry, and much scrutiny of the material to be selected and the form of representation to be used. The extent to which I have been successful in providing genuine help to the reader remains to be seen.

The book should, in addition, everywhere lead up to the present-day status in research, and thus be of assistance to those already familiar with the subject. To stay within the required limits of space permitted only brief references to be made, in each section of microclimatology, to results that were useful and that pointed the way to the future. The novice will easily pass over these, whereas to the initiated they will provide access to the literature on the subject. This new edition is therefore at the same time a collection of data for the relevant problems. It was found possible, by careful selection, to limit the bibliography to about 1200 references. This was also made simpler by making use of the valuable collection of references in *Meteorological Abstracts and Bibliography*.

I owe a special debt of gratitude to my university colleague, Dr. Gustav Hofmann, who read the finished manuscript. The lively discussions that followed resulted in the introduction of many improvements in the book. Since experience has shown that many questions are raised concerning methods of observation and the use of instruments, I asked him to amplify this new edition by a chapter devoted to suggestions on the technique of measurement in microclimatology and micrometeorology. The brief yet copious exposition in Secs. 55-57 will certainly be welcomed by many users of the book.

I would like, further, to thank all those who assisted me by sending reprints of their works. Much came to my knowledge only by this means. Such assistance is indispensable in a technical field that has so many contacts with neighboring sciences. It is precisely the representatives of these adjacent fields of knowledge whom I would ask most urgently to bring to my notice any of the deficiencies and gaps in treatment, which are unavoidable with the present-day scope and tempo of research in microclimate.

During my work on the new edition, it became clear to me why, in spite of the ever-swelling flood of published papers, the number of usable textbooks is increasing much more slowly. In drafting almost every chapter, I could feel the spirit of a cherished colleague, at home or abroad, looking over my shoulder, and it seemed to me that he was more fitted to write just this chapter than I was. Then came the feeling of suffocation under the weight of good new literature, and the anxiety that the first chapter would be out of date before the last one could be finished. However, I was able to bear these tensions daily for three years, because without useful signposts no one can find his way any more in the labyrinth of science. If I have been able to point the way to a few, this will be my best reward for all the pains taken.

Munich-Pasing, November 1960
Perlschneiderstrasse, 18

RUDOLF GEIGER

PREFACE TO THE FIRST EDITION

I was introduced to microclimatology by Professor A. Schmauss. When he put me in charge of the organization and direction of the Bavarian special network for investigation of the air layer near the ground, and later when I had to make two extensive open-air investigations in forest meteorology, I had a good opportunity to get in closer touch with people dealing in forestry, moor cultivation, and agriculture. Thus I became acquainted with the difficulties that are met with everywhere in the practical application of the results of climatological research. This problem of application is indeed not new, and many valuable contributions have already been made, as, I hope, this book will demonstrate; but a systematic study has not been undertaken as yet, and it is just the practical man who has neither time nor opportunity to look for the available papers from the vast meteorological literature. When, therefore, I was invited to write a "Climate near the Ground," I was glad of the opportunity to attempt a first survey of microclimatological problems. This is the best way for me to thank the people referred to for the manifold suggestions that I have received from them, particularly those with a special interest in forestry. It is also a pleasure to express here my sincerest thanks to Professor Schmauss for his constant and unselfish furtherance of my work.

RUDOLF GEIGER

Munich, July 1927

CONTENTS

SOURCES OF ILLUSTRATIONS

1. *Archiv für Meteorologie, Geophysik und Bioklimatologie* [*B*] *6* (1955), 278, Fig. 4. Springer, Vienna.
4. *Forstwissenschaftliches Centralblatt 72* (1953), 178, Fig. 3. Parey, Berlin.
15. *Meteorologische Rundschau 10* (1957), 12, Fig. 2. Springer, Berlin.
16. *Ibid.*, p. 15, Fig. 7.
18. *Klima und Bioklima von Wien*, II (1957), 77, Fig. 12. *Wetter und Leben*, Vienna.
23. *Angewandte Meteorologie 1* (1948), 87, Fig. 1. Akademie Verlag, Berlin.
37. *Meteorologische Rundschau 2* (1949), 165, Fig. 4. Springer, Berlin.
44. *Ibid. 10* (1957), 3, Fig. 5.
45. *Ibid.*, p. 4, Fig. 6.
65. *Mitteilungen der Vereinigung der Grosskesselbesitzer* (1955), p. 56, Fig. 59. Stahleisen, Düsseldorf.
66. *Veröffentlichungen des Geophysikalischen Instituts der Universität Leipzig 12* (1940), 349, Fig. 21. Akademie Verlag, Berlin.
77. *Berichte des Deutschen Wetterdienstes in der US-Zone*, no. 28 (1951), p. 28/9, Fig. 5.
78. *Zeitschrift Flora 145* (1958), 526, Fig. 12. Fischer, Jena.
80. *Berichte des Deutschen Wetterdienstes in der US-Zone*, no. 28 (1951), p. 28/17, Fig. 12.
81. *Ibid.*, p. 28/18, Fig. 16.
83. *Abhandlungen des Meteorologischen und Hydrographischen Dienstes der Deutschen Demokratischen Republik 3* (1953), no. 19, supplement, Fig. 4. Akademie Verlag, Berlin.
84. *Berichte des Deutschen Wetterdienstes in der US-Zone*, no. 38 (1952), p. 38/198, Fig. 1.
85. *Mededelingen van de Landbrouwhogeschool te Wageningen Nederland 52* (1) (1952), 38, Fig. 8.
87. *Meteorologische Rundschau 9* (1956), 183, Fig. 1*a*-*e*. Springer, Berlin.
88. *Journal of the College of Agriculture, Tokio Imperial University, 13*, no. 5 (1936), Plate 30.
89. *Meteorologische Rundschau 7* (1954), 46, Fig. 1. Springer, Berlin.
93. *Orion* 4 (1949), 475. Lux, Murnau, Germany.
95. *Abhandlungen des Meteorologischen und Hydrographischen Dienstes der Deutschen Demokratischen Republik 4* (1954), no. 26, p. 43, Fig. 14. Akademie Verlag, Berlin.
99. *Annalen der Meteorologie* (1950), p. 95, Fig. 1. Seewetteramt, Hamburg.
103. *Archiv für Meteorologie, Geophysik und Bioklimatologie* [*A*] *8* (1955), 389, Fig. 3*a*-*d*. Springer, Vienna.
104. *Geografiska Annaler*, 1950, p. 201, Fig. 2.
105. *Ibid.*, p. 205, Fig. 7.
106. *Berichte des Deutschen Wetterdienstes in der US-Zone*, no. 38 (1952), p. 38/391, Fig. 3.
116. *Meteorologische Rundschau 6* (1953), 95, Fig. 1. Springer, Berlin.
128. *Sitzungsberichte der Deutschen Akademie der Landwirtschaftswissenschaften zu Berlin 5* (1956), pt. 5, p. 23, Fig. 7. Hirzel, Leipzig.
135. *Archiv für Meteorologie, Geophysik und Bioklimatologie* [*B*] 7 (1956), 253, Fig. 1. Springer, Vienna.
139. *Zeitschrift für Bienenforschung 4*, p. 93, Fig. 5.
149. *Veröffentlichungen des Reichsamts für Wetterdienst, Berlin, Wissenschaftlichen Abhandlungen 9*, no. 6, p. 26, Fig. 18.

153. *Időjárás 61* (1957), 274, Fig. 7.
158. *Forstwissenschaftliches Centralblatt 74* (1955), 61, Fig. 3. Parey, Berlin.
163. *Ibid. 71* (1952), 340, Fig. 2.
164. *Berichte des Deutschen Wetterdienstes 5* (1955), no. 28, p. 28/12, Fig. 7.
165. *Ibid.*, p. 28/13, Fig. 8.
166. *Ibid.*, p. 28/14, Fig. 9.
167. *Ibid.*, p. 28/43, Fig. 48.
171. *Ibid.*, p. 28/35, Fig. 42.
175. *Ibid.*, p. 28/16, Fig. 12.
176. *Ibid.*, p. 28/24, Fig. 25.
177. *Ibid.*, p. 28/22, Fig. 22.
178. *Ibid.*, p. 28/29, Fig. 35.
179. *Berichte der Deutschen Botanischen Gesellschaft 64* (1951), 217. Fischer, Stuttgart.
181. *Mitteilungen aus der niedersächsischen Landesforstverwaltung*, pt. 3 (1958), 78, Fig. 22.
193. *Forstwissenschaftliches Centralblatt 75* (1956), 232, Fig. 4. Parey, Berlin.
194. *Archiv für Forstwesen 3* (1954), 451, Fig. 8. Akademie Verlag, Berlin.
200. *Berichte des Deutschen Wetterdienstes 5* (1956), no. 29, p. 29/4, Fig. 1.
207. *Forstwissenschaftliches Centralblatt 77* (1958), 51, Fig. 4. Parey, Berlin.
215. *Gartenbauwissenschaft 23* (1958), 356, Fig. 10. Bayer. Landwirtsch. Verlag, Munich.
218. *Archiv für Meteorologie, Geophysik und Bioklimatologie* [*B*] *5* (1954), 310, Fig. 1. Springer, Vienna.
222. *Ibid. 4* (1953), 441, Fig. 7.
224. *Berichte des Deutschen Wetterdienstes in der US-Zone*, no. 12 (1950), p. 102, Fig. 3.
227. *Meteorologische Rundschau 6* (1953), 89, Fig. 5. Springer, Berlin.
231. *Berichte des Deutschen Wetterdienstes 7* (1959), no. 50, p. 50/23, Fig. 24.
232. *Ibid.*, p. 50/24, Fig. 26.
233. *Ibid.*, p. 50/35, Fig. 33.
234. *Meteorologische Rundschau 11* (1958), 94, Fig. 2. Springer, Berlin.
240. *Archiv für Meteorologie, Geophysik und Bioklimatologie* [*B*] *8* (1957), 222, Fig. 6. Springer, Vienna.
243. *Forstwissenschaftliches Centralblatt 78* (1959), 104, Fig. 5. Parey, Berlin.
245. *Ibid. 75* (1956), 300, Fig. 4. Parey, Berlin.
246. *Ibid. 77* (1958), 260, Fig. 3. Parey, Berlin.
247. *Planta 49* (1957), 625, Fig. 3. Springer, Berlin.
252. *Petermanns Mitteilungen 98* (1954), plate 13, Fig. 6. VEB Geogr.-Kartograph. Anstalt, Gotha.
258. *Archiv für Meteorologie, Geophysik und Bioklimatologie* [*B*] *5* (1954), 204, Fig. 4. Springer, Vienna.
261. *Wetter und Leben*, special section VI (1959), p. 170, Fig. 10.
266. *Mitteilungen der Schweizer Anstalt für des Forstliche Versuchswesen 29* (1953), pt. 2, p. 231, Fig. 7.
267. *Berichte des Deutschen Wetterdienstes 5* (1957), no. 32, p. 32/13, Fig. 16.
271. *Ibid. 7* (1959), no. 53, p. 53/7, Fig. 2.
272. *Mitteilungen des Deutschen Wetterdienstes 2* (1955), no. 12, p. 12/12, Fig. 3.
275. *Zeitschrift Umschau 59* (1959), 243, Fig. 6. Umschau, Frankfurt am Main.
278. T. Friedrichs, Hamburg.
279. Dipl.-Ing. Architekt Friedrich Tonne, Stuttgart N.
280. *Berichte des Deutschen Wetterdienstes 7* (1958), no. 48, p. 48/8, Fig. 7.
281. *Archiv für Meteorologie, Geophysik und Bioklimatologie* [*A*] *11* (1959), 236, Fig. 5. Springer, Vienna.

THE CLIMATE NEAR THE GROUND

INTRODUCTION

1. Microclimate and Research

Most people are familiar with the meteorological instrument shelters (screens) belonging to the observing stations of the national meteorological network. The instruments housed in these shelters are protected against radiation and precipitation, yet are well exposed to the air by means of numerous openings in the sides. The observer may be seen mounting a few steps to open the shelter, since the instruments are at a height of about 2 m above the level of the ground.

Positioning at this level was decided on after it had been established toward the end of the last century, by means of long series of observations, that at this height the chance influences of the position selected for making the observations had largely been eliminated. The nature and the state of the ground, and the vegetation growing on it, then no longer play a significant role. The meteorological station is thus "representative" of a wider surrounding area, and the climate thus measured is that experienced by a man walking upright, or human climate, as it is sometimes termed. Values measured at stations 20, 50, and often even a greater number of kilometers apart characterize the climate of the region in question, and this is often described today as large-scale climate or macroclimate. This is the kind of climate described in the meteorological yearbooks which are published regularly by geographers and meteorologists in all civilized countries, in works on the science of climatology, and in climatological atlases.

The air layer below this agreed level of about 2 m will be called the air layer close to the ground. In it substantially different conditions will, in general, be found. The closer the ground surface is approached, the more the speed of the wind is reduced by friction with the ground, and hence the less is the amount of mixing of the air, which might have reduced the differences among small areas. Then the ground surface has to be considered, absorbing radiation from the sun and sending out its own heat rays, now a source of warmth, now a cooling influence on the air in contact with it. It is also a source of water vapor, which escapes into the atmosphere by evaporation, of dust, and of gases emanating from the soil. The special conditions that develop, as a consequence, in the air layer near the ground are of great primary interest to the meteorologist, because these are the conditions prevailing in the boundary layer of earth and atmosphere, without a knowledge of which it is impossible to understand the processes taking place in the main body of the atmosphere. It is in this atmospheric layer near the ground that plants grow, especially young plants, which are still exceptionally sensitive to weather fluctuations; the climate of this air layer near the ground can therefore be called, quite simply, plant climate. Animals that are restricted to the proximity of the surface, or that live on the ground, must also be considered. The study of the climate near the ground is therefore more comprehensively called habitat climate, or, to use Greek terminology, ecoclimatology.

The immediate consequence of this is that the conditions of climate to which young growing plants are exposed cannot be deduced directly from the figures for climate published for the network of official stations. One example will be discussed in more detail in Sec. 42, where only one night of frost with -1.8°C was recorded during a particular month of May at the meteorological station in Munich, while in the same month 23 nights of frost with temperatures down to -14.4°C were measured in the air layer near the ground, 20 km outside the city. All who have to deal with plants—farmers, foresters, and gardeners—are interested in these facts. The ground also carries highways and railroads, however, and the architect uses it to provide foundations for buildings.

All the meteorological elements are subject to vertical changes because of the nearness of the ground; and in a similar way they also vary horizontally within short distances. These variations are brought about by changes in the nature and the moisture of the soil, even by minute differences in surface slopes, and by the type and height of vegetation growing on it. All these climates found within a small space are grouped together under the general description of microclimate, and are thus contrasted with the macroclimate experienced in the national network of meteorological stations.

In what follows, the internationally understood meaning of the term microclimate is adhered to. The term *Kleinklima* or *Kleinstklima*, which might be translated as "climate in a small space," and other frequently used terms such as local climate, position climate, miniature climate, piccoloclimate, and so forth, are not employed because of their ambiguity. For the moment we shall also leave out of consideration the question whether perhaps there are transitional stages between macroclimate and microclimate; we shall return to this in Sec. 47.

The difference between macroclimate and microclimate is well illustrated by taking, as an example, the measurements made by J. N. Wolfe, R. T. Wareham, and H. T. Scofield [968]* in attempting to determine the conditions of habitat climate for various plant communities in the Neotoma Valley in Ohio. The range of variation of a few meteorological quantities was worked out for the year 1942, first for the 88 normal observation stations, in representative

*The figures in brackets refer to the bibliography at the end of this book.

positions in the state of Ohio (area 113,000 km^2), and then for the 109 microclimate stations set up in the deep Neotoma Valley, over an area of 0.6 km^2. The results (in metric units) are shown in Table 1. The greatly varied conditions of microclimate observed in this one valley provide a contrast to the comparatively uniform general climate of the state of Ohio.

Table 1
Comparison of macroclimate and microclimate.

Meteorological variable	At the 88 meteorological stations in Ohio	At the 109 microclimate stations in the Neotoma Valley in Ohio
Highest annual temperature (°C)	33-39	24-45
Time of occurrence of the above	17-19 July	25 Apr.-19 Sept.
Lowest January temperature (°C)	-21 to -29	-10 to -32
Latest spring frost	11 Apr.-11 May	9 Mar.-24 May
Earliest autumn frost	25 Sept.-28 Oct.	25 Sept.-29 Nov.
Length of frost-free period (days)	138-197	124-276

"Climate" is an abstract concept. It can be grasped only by means of comprehensive calculations. Since climate includes the sum total of all individual meteorological occurrences, which we call weather processes, at a given place, it will comprise the average conditions and the regular sequences of weather, also repeatedly observed special phenomena such as tornadoes, dust storms, and late frosts. It is therefore not possible to understand climate, if one is unfamiliar with weather features.

Special weather phenomena can be found in the air layer near the ground, as shown by shallow ground fogs over forest meadows in the evening, or by sand sweeping above the ground over prairies. Normally, however, the special phenomena of the ground layer are not so obvious, but can be deduced only as special features of individual meteorological processes, through the use of instruments. Without an understanding of the processes involved it is not possible to understand microclimate. For this reason, a textbook on microclimatology must of necessity begin as a textbook on micrometeorology, as a glance at Chapters I to IV will show; these are devoted to the physical foundations. We shall therefore continually be faced with arguments that are at times micrometeorological and at times microclimatological; even the technique of measurement to be used, as will be shown in Sec. 55, differs from these two points of view. Those who are primarily interested in theoretical micrometeorology, which has been called the "mathematician's paradise" by O. G. Sutton [20], should consult his book, *Micrometeorology* [19].

The present book is not intended mainly for the theoretical meteorological physicist, but will serve the ever-increasing number of people who hope to learn about the average conditions of life near the ground, without having to use much mathematical physics. These are the people interested in plant life, the farmers, foresters, gardeners, vine growers, and botanists; those interested in animal life, the zoologists, entomologists, farmers, and breeders; those interested in the state of the ground, the traffic engineers, road builders, architects, water and soil experts, geographers and country planners, medical men, and bioclimatologists, even financial experts, who can no longer fail to take into account whether microclimate is favorable or unfavorable when assessing the just taxation of land. It is hoped that this book will lead all of them quickly and clearly into a realm of science whose wide connections with so many aspects of life have brought forth an abundance of literature, which even the specialist needs help in mastering.

The beginnings of microclimatology date back about half a century. The fact that in Finland in August 1893 Theodor Homén (1858-1923)—far ahead of his time—made comparative measurements of the heat budget in various types of soil seems to us today to be the real beginning of thinking in terms of microclimatology. Gregor Kraus (1841-1915), who published a book on soil and climate in very small volumes in 1911 [9], must be described as the father of microclimatology. He became aware of the extreme local conditions affecting the chalk countryside of the Main area near Karlstadt, and investigated them out of pure love of research. The fundamentals and practical applications of micrometeorology were then developed, mainly by Wilhelm Schmidt (1883-1936) in Vienna and August Schmauss (1877-1954) in Munich. Since that time this many-sided field of research has been built up by scientists of all faculties and in all countries. Micrometeorology provides, in fact, an uncommon and a fine example of scientific cooperation. While in other cases the breadth of a science can easily lead to a shallowness in research, and the depth of a science may lead to unhealthy specialization, micrometeorology seems fortunately to have avoided both of these dangers. As a specialized science it can and should extend in depth, and in its many ramified associations with a great number of neighboring fields of knowledge it has a stimulating and satisfying breadth.

CHAPTER I

HEAT BUDGET OF THE EARTH'S SURFACE AS THE BASIS OF MICROCLIMATOLOGY

2. Facts and Theories of the Earth's Radiation Economy

Radiation is undoubtedly the most important of all meteorologic elements. It is the source of power that drives the atmospheric circulation, the only means of exchange of energy between the earth and the rest of the universe, and the basis for organizing our daily lives. In spite of this paramount importance, it is only in the present century that meteorologists have turned their attention seriously to radiation problems. The reason for this has been largely the technical difficulties in the design and use of instruments for measuring radiation. These difficulties are much greater than those encountered with the coarser instruments for measuring temperature, air pressure, humidity, wind, and so forth. The question of cost is also of great importance, since the instruments have to be used all over the world.

The recent developments have resulted from an increasing recognition of the importance of radiation processes in all energy exchanges in the atmosphere. These processes must also be dealt with in detail in microclimatology. Experience has shown that the basic facts and the physical principles governing radiation processes are still all too little known. Even meteorologic textbooks have been falling behind the vigorous rate of advance. To provide the reader with a solid background for the later chapters, the most important laws of radiation and the basic theory of the earth's radiation balance are summarized below.

Heat can be transferred in the atmosphere in four different ways. If one end of an iron rod is heated, heat travels within the rod in a way not visible to the eye; the livelier motion of the heated molecules is transmitted by collision to the more slowly moving neighboring molecules. This process is called (true or physical) heat conduction.

When water is heated in a beaker, the same kind of conduction takes place; but the heating of the whole body of water is brought about much more effectively by the stream of heated particles rising from below; this motion can easily be observed in artificially colored water. Heat is transported with the moving water. This mass exchange, characteristic of both liquids and gases, is almost always present under atmospheric conditions, and plays a much greater part in minimizing temperature differences than does true heat conduction. The term "apparent heat conduction" is used occasionally and will be explained later.

Evaporating water extracts about 580 cal of heat from the earth's surface for every gram released as vapor. This heat becomes available to the air when the water vapor condenses in the upper atmosphere to form clouds. In this case heat is transported by means of the change of state of the water; there are other forms of the same phenomenon.

Finally, there is the process of thermal radiation which, in contrast to the previous three forms, does not require any material carrier, as is shown by the way solar radiation spreads through space. Thermal radiation is electromagnetic radiation in the range of wavelengths from 0.2 to 100 μ (1 μ = 0.001 mm = 1000 mμ = 10,000 Å). Radiation of longer wavelength is familiar in the form of VHF and other radio waves; radiations of shorter wavelength are known as x-rays, gamma-rays from radioactive substances, and cosmic rays.

The human eye perceives radiation between about 0.36 and 0.76 μ as light, and is most sensitive in the range 0.55–0.56 μ. Color perception is dependent on wavelength, as follows: 0.36 ← violet → 0.42 ← blue → 0.49 ← green → 0.54 ← yellow → 0.59 ← orange → 0.65 ← red → 0.76. Radiation of longer wavelength, which is not perceptible to the eye, is called infrared. Radiation below 0.36 μ is called ultraviolet, or UV for short. Visible light is therefore only a part of the radiation that concerns us in the atmosphere. Thermal radiation and light should not be confused with each other.

The earth's atmosphere reflects, scatters, and absorbs part of the solar radiation. Solar radiation which has not been weakened by these processes is termed extraterrestrial radiation. It varies by about 7 percent with the changing distance of the earth from the sun during the course of the year. For a mean earth-sun distance M of 150 × 10^6 km, the intensity of solar radiation, according to the best available observations, is 2.00 (±2 percent) calories per square centimeter per minute (cal cm^{-2} min^{-1}) when incident on a surface perpendicular to the direction of the rays. This quantity is called the solar constant, k. This insolation falls on an area of πR^2, where R is the earth's radius (6370 km).

According to the Stefan-Boltzmann law of radiation, $S = \sigma T^4$, all bodies radiate in such a way that the total intensity emitted S (summed over all wavelengths) is proportional to the fourth power of the absolute temperature T in degrees Kelvin (°K = 273 + °C). The constant of proportionality σ has the value 8.26 × 10^{-11} cal cm^{-2} min^{-1} $°K^{-4}$. Therefore

$$S = 8.26 \times 10^{-11} T^4 \qquad (\text{cal cm}^{-2}\ \text{min}^{-1}).$$

This law is valid only for so-called "black bodies," which absorb completely all radiation incident upon them. If the sun is considered to approximate a black body, its surface temperature T_s can be calculated from this law with the aid of the solar constant k. Since radiation decreases with the square of the distance, and the sun's radius is s = 695,560 km, while the earth's radius R is negligible in comparison with the distance M of the sun from the earth, $\sigma T_s^4/k = M^2/s^2$, from which the value of T_s is found to be 5793° K.

The Stefan-Boltzmann law also provides a conclusion about the earth's mean temperature, based on the assumption that the earth radiates like a black body. Since the earth's temperature is subject to variations in time but remains unchanged on the whole over thousands of years, the amount radiated by the surface of the sphere of area $4\pi R^2$ must be equal to the quantity received by the cross-sectional area πR^2 multiplied by the constant k (see above). The mean surface temperature of the earth T_E, calculated from the equation

$$\sigma T_E^4 \times 4\pi R^2 = k\pi R^2,$$

is found to be 278°K = 5°C. (The earth is thus at an ideal distance from the sun; Venus, which is closer, has an equilibrium temperature of 55° C, and more distant Mars has -47° C.) The surface temperature of the earth observed near the ground is higher (14° C) because of the protective effect of the atmosphere (see Sec. 4), which is correspondingly colder at higher levels (-50° to -80° C).

Even if the quantity of radiation received from the sun is equal to that radiated by the earth, as suggested by the maintenance of the present state of heat, the two types of radiation are fundamentally different in quality. The total intensity of solar radiation is spread over a wide range of wavelengths. According to Wien's displacement law, the product of the temperature T of a radiating body and the wavelength corresponding to maximum intensity of radiation, λ_{max}, is constant. With T in degrees Kelvin and λ_{max} in microns,

$$T\lambda_{max} = 2880\,(°K, \mu).$$

The higher the temperature of a body, the farther the radiation maximum is displaced toward shorter wavelengths. For the surface temperature of 5793° K, evaluated above for the sun, λ_{max} is 0.50 μ; the observed maximum is 0.47 μ, which means a higher sun temperature (the difference shows that the sun radiates only approximately as a black body). In either case, the most intense solar radiation occurs in the blue-green range of visible light. The wavelength of maximum intensity of radiation for the earth's actual surface temperature of 14° C or 287° K is about 10.0 μ, which is well into the invisible infrared.

Distribution of intensities over the spectrum is so asymmetric that 25 percent of the total radiation lies below λ_{max}, in the short-wavelength range, and 75 percent is above λ_{max}. It is therefore appropriate, according to G. Hofmann [44], to introduce a wavelength λ_s as a center of balance, such that 50 percent of the total intensity lies on either side of it; then $T\lambda_s = 4100$ (°K, μ). In Fig. 1, below the abscissa, which has a logarithmic wavelength scale, is shown a scale of temperature determined by this equation. For solar radiation, λ_s is 0.7 μ, in the visible red. Forty percent of solar radiation lies within the infrared part of the spectrum.

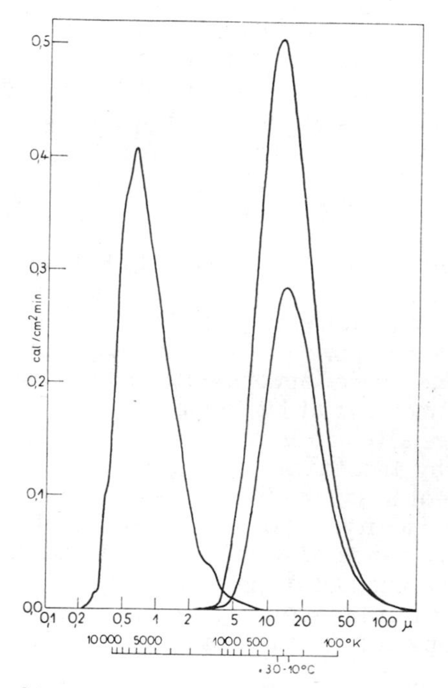

Fig. 1. Distribution of intensity in the two bands of atmospheric radiation, according to wavelength. (After G. Hofmann)

Figure 1 shows details of the distribution of intensity, according to Hofmann. On the left is the curve of solar radiation from observations. The area enclosed by the curve represents the total intensity, hence the solar constant, reduced to one quarter for the reasons mentioned above, for comparison with the earth's radiation. The two distribution curves on the right correspond to earth temperatures of +30° C and -10° C respectively (for details of the method of representation, see the original work). Figure 1 makes it clear that in meteorology it is correct to distinguish between two fundamentally different streams of radiation. Solar radiation and diffuse sky radiation are in the range from 0.3 to 2.2 μ; radiation emitted by the earth and its atmosphere lies between 6.8 and approximately 100 μ. The intervening range from 2.2 to 6.8 μ is used by both types of radiation to the extent of less than 5 percent. Hence there is a marked division between the two kinds of radiation which we shall refer to in the future as short-wavelength and long-wavelength radiations.

Attention will now be turned from this planetary consideration of radiation processes to conditions prevailing at the earth's surface, and to the question of heat balance.

3. Elements of the Heat Economy and Their Importance

Let us discuss the ideal case in which the earth's surface is extensive and entirely horizontal. In this case the boundary between ground and atmosphere is a plane, two-dimensional and massless. The plane contains no heat; but under normal circumstances a considerable exchange of heat occurs across it. The quantities that determine this heat exchange will now be established.

Radiation S is the major factor of heat exchange. Heat arrives at the earth's surface from the sun, the sky, and the atmosphere (insolation). Heat is sent back into space (outgoing or "terrestrial" radiation). Here, as in subsequent arguments, factors that add heat to the surface of the ground are considered positive; those that subtract heat from it are negative. The sum of insolation and outgoing radiation, that is, the balance, decides in individual cases whether S is positive or negative (unit: cal cm^{-2} min^{-1}, also called in English langley per minute, abbreviated ly min^{-1}).

The second factor B is determined by the flow of heat from the ground to the surface or in the reverse direction. During a cold winter night heat flows upward through the ground and B is therefore positive; on a summer afternoon, B is negative because heat is transported downward from the surface. The laws governing this transport of heat in the ground will be dealt with in more detail in Sec. 6.

Third, the air above the ground plays a part in the exchange of heat L. This factor, also, may be positive or negative. Transport of heat to or from the ground depends not only on physical heat conduction, as within the ground, but also on mass exchange (eddy diffusion) because of the great mobility of the air; more detailed explanations are to be found in Secs. 7-9.

Finally there is the effect of evaporation V. This is measured, like all the other heat-economy factors, in calories per square centimeter per minute. The quantity of heat in calories required to evaporate 1 g of water is called the latent heat of vaporization r_w and varies with temperature as follows:

t:	0	5	10	15	20	25	30	°C
r_w:	597	594	592	589	586	583	580	cal g^{-1}

When ice evaporates in winter, the energy requirement is increased by the latent heat of fusion of the ice. The latent heat of sublimation requires 677 to 678 cal g^{-1} at 0° C to -30° C. If a round figure of 600 cal g^{-1} is used for temperatures above 0° C, then V in cal cm^{-2} min^{-1} corresponds to the evaporation of a certain depth of water in millimeters per hour. This relation serves to illustrate V further. Normally V is negative, but here also positive values are possible, as when dew or hoarfrost form on the surface and heat of condensation or sublimation is released.

As stated above, since the earth's surface as a boundary surface can absorb no heat, according to the law of conservation of energy the following equation must be satisfied for all units of time:

$$S + B + L + V = 0.$$

This is the fundamental equation, containing the basic factors, that governs heat exchange at the earth's surface. Over oceans, lakes, and rivers the factor W, for the exchange of heat between water and its surface, is used instead of B. The equation then becomes

$$S + W + L + V = 0.$$

Conditions are not so simple in every case. From the surroundings of the area under consideration there can flow warmer or colder, moister or drier, air, a process that was not visualized in the introductory ideal situation but that often occurs under actual conditions. This advection process has an effect on the heat economy of the area and upsets the assumption on which the previous discussion was based, namely that horizontal counterinfluences are absent. Without verifying at this stage the effects it must have on the balance equation, we shall introduce this additional advection process, which is of considerable practical significance, into the equation in the usual way, as an extra factor Q. Advection processes will not be considered in more detail until Sec. 27.

Precipitation may entail a gain or a loss of heat for the ground, depending on its temperature, and this is given the symbol N. In a strict sense, this process also escapes our present method of analysis, since any precipitation brings about a change in the state of the ground. However, N may be taken as an input or a loss of heat through the surface to which the equation refers.

The complete equation for the heat exchange at a flat vegetation-free ground surface consequently becomes:

$$S + B\ (\text{or}\ W) + L + V + Q + N = 0.$$

If the ground is covered with vegetation, new factors are introduced (see Chapter V).

To illustrate the significance of the principal factors of this heat exchange, Table 2 gives the results of investigation made by F. Albrecht [435] at Potsdam, for selected times of day in 1903. During the day the heat exchange depends entirely on incoming radiation. The great difference between summer and winter is due principally to evaporation differences. During the night heat is lost both by outgoing radiation and by evaporation (the role of dew or hoarfrost is quantitatively insignificant). Heat is lost partly from the ground and partly from the air above it.

Table 2
Heat budget in Potsdam (cal cm^{-2} min^{-1}).

Typical period	Mean value (1903)	Factors				Total turnover
		S	B	L	V	
Summer day	June 12-13 hr	+0.407	-0.165	-0.094	-0.148	0.407
Winter day	Jan. 12-13 hr	+0.091	-0.082	-0.003	-0.006	0.091
Summer night	June 0-1 hr	-0.080	+0.070	+0.021	-0.011	0.091
Winter night	Jan. 0-1 hr	-0.065	+0.021	+0.060	-0.016	0.081

The relative importance of the four principal factors changes greatly when viewed over a long period of time. The annual average levels out, the heat accumulated in the ground and in the air by day and in the summer being lost at night and in the winter. The annual balance for 1903 in Potsdam was

$$S = +19{,}819\,;\ B = -181\,;\ L = +387\,;\ V = -20{,}025 \text{ cal cm}^{-2}\text{ yr}^{-1}.$$

Here B and L amount only to 0.9 and 1.9 percent respectively of the total exchange of 20,206 cal cm^{-2} yr^{-1}; they are not zero merely because air and ground temperatures on 31 December 1903 were by chance not the same as on 1 January 1903.

This shows the close relation between the heat balance and the water balance. It is incorrect to assume that the heat gained by day and in the summer is lost again by outgoing radiation during the night and in the winter. Day and night, summer and winter are by no means equally matched opponents, as A. Baumgartner [437] pointed out. Thus, in contrast to the equilibrium of the radiation balance of the earth as a planet (Sec. 2), the major part of the sun's heat, received in the form of radiation, is used for evaporation. This relation between the heat and water economies allows the results of investigations of heat and water balances to be cross-checked; an evaporation level calculated from the water economy must reappear in the heat economy with a corresponding amount of heat used up for evaporation (compare Secs. 25 *et seq.*).

4. Radiation Balance at the Earth's Surface

Of all the factors mentioned in Sec. 3 as taking part in the heat exchange at the surface of the earth, radiation is the most important. The symbol S, as explained on p. 9, means the radiation balance or net radiation. If insolation is greater than outgoing (terrestrial) radiation, the balance is positive; if it is less, the balance is negative. A negative balance is described as a net loss of radiation. In the literature on the subject, the term "outgoing radiation" is sometimes used to designate the Stefan-Boltzmann radiation loss, and sometimes for the negative radiation balance. In this book, only the unambiguous concept "effective outgoing radiation" is retained (see p. 21); otherwise, the ambiguous term "outgoing radiation" is avoided when possible.

The radiation balance consists of two radiation streams of different spectral ranges, as distinguished in Fig. 1. There is a short-wavelength part only as long as the sun shines, that is, during the daytime. Radiation reaching the surface of the earth consists of that part of direct solar radiation I that is not reflected by clouds, absorbed by the atmosphere, or scattered diffusely, and also that part of the nondirectional sky radiation H that represents diffusely scattered radiation that has reached the ground and provides "daylight" within the visible spectrum. The value of $I + H$ reaching a horizontal surface is called global radiation. Part of this radiation is reflected by the earth's surface. This short-wavelength reflected radiation R depends on the nature of the ground, in contrast to $I + H$. The reflection factor or albedo is the ratio of the reflected to the incident radiation, usually expressed as a percentage. More detailed information on albedo is given in Table 3, p. 15.

It was mentioned in Sec. 2 that incoming long-wavelength radiation is of no significance in the radiation balance of the earth as a planet. It is, however, of great importance for the radiation balance of the earth's surface. The atmosphere of the earth contains water vapor, carbon dioxide, and ozone, all of which absorb radiation and emit it according to Kirchnhoff's law. This long-wavelength atmospheric radiation G is termed counterradiation since it counteracts the terrestrial radiation loss. It occurs both by day and by night, and in fact is somewhat greater during the day, since it is dependent on temperature.

It might be expected that part of this long-wavelength radiation would also be lost through reflection by the ground. Laboratory investigations show, however, that the earth's natural surface cover can be considered to resemble a black body. Recently it has been established by J. T. Houghton [45], through measurements made from an airplane, that the earth radiates like a black body with an equivalent temperature between the shelter temperature and the grass minimum (see Sec. 14). G. Falckenberg [37] recorded an albedo between 0 and 8 percent in the 10-μ region; only light sand gave as much as 11 percent. D. M. Gates and W. Tantraporn [42] recorded for a number of plants and trees albedos from 0 to 6 percent over seven bands of wavelength between 3 and 25 μ, these values being exceeded only in isolated instances (lemon-tree leaves had an albedo of 17 percent at 10μ!). In general, the albedo of natural surfaces is less than 5 percent. Snow cover, which reflects so strongly within the visible spectrum that newly fallen snow produces a striking improvement in light conditions, in practically an ideal black body for long waves, reflecting at the most 0.5 percent of the incident radiation; as Falckenberg so aptly puts it, "snow can only be made more transparent to long waves by spreading soot on it." In the following discussion this reflection will be neglected.

According to the Stefan-Boltzmann law (see p. 6) the radiation emitted by the soil surface by day and by night would be exactly σT^4 (T is the surface temperature) if the ground were a black body. It has just been shown that this condition is largely fulfilled. To the extent that it is not fulfilled, the outgoing radiation will be reduced according to Kirchhoff's law; but at the same time the amount of outgoing long-wavelength reflected radiation would be increased, and it is not possible to distinguish it instrumentally from the terrestrial radiation.

The radiation balance S is therefore given by the equation

$$S = I + H + G - \sigma T^4 - R \qquad (\text{cal cm}^{-2}\,\text{min}^{-1}).$$

The last two factors depend on the nature of the ground surface, while the first three are independent of it.

Figure 2 shows the magnitude of these factors from 12:00 to 13:00 hr on a summer day (5 June 1954), and from 0:00 to 1:00 of the following night, from measurements made at the Meteorological Observatory in Hamburg and published by R. Fleischer and K. Gräfe [39]. The width of the arrow is proportional to the intensity of radiation. The remaining factors L, B, and V (Sec. 3) in the heat budget are included to make the diagram complete; they are taken from measurements made by E. Frankenberger [445] on clear days with light winds, near Hamburg in the same month.

It is clear from Fig. 2 how important radiation is in the total heat budget. Short-wavelength radiation is intense, but of short duration. Stefan-Boltzmann radiation lasts the whole 24 hr, and, since it is largely compensated for by counterradiation, the balance is small. At night, however, the heat balance is completely dominated by long-wavelength radiation, as shown in Fig. 2. This will be discussed in more detail in Sec. 5.

Figure 2 has been drawn for selected hours of the day. If it is desired to draw up a balance for a whole year, careful radiation-balance measurements made in Hamburg for the year 1953-54 yield, according to W. Collmann [32], the following values for the various factors:

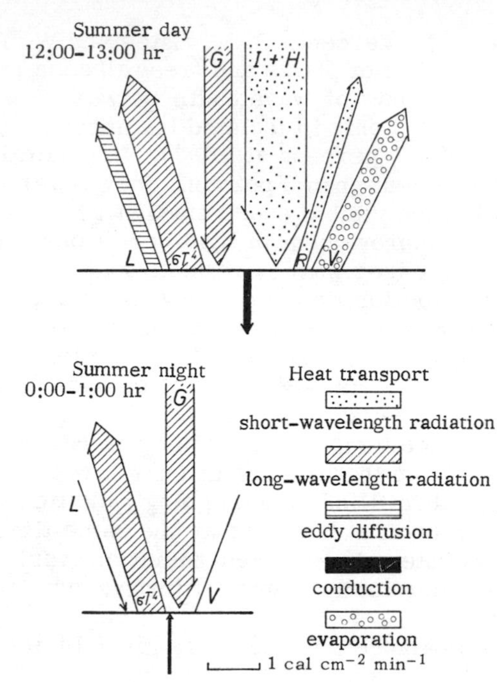

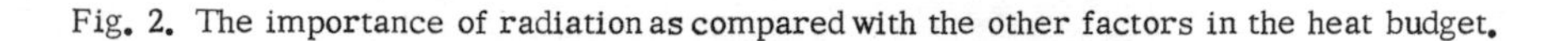
Fig. 2. The importance of radiation as compared with the other factors in the heat budget.

$$\begin{aligned} I &= +\ 34{,}153 \text{ cal cm}^{-2}\text{ yr}^{-1} \\ H &= +\ 43{,}444 \\ G &= +240{,}533 \\ \sigma T^4 &= -268{,}837 \\ R &= -\ 14{,}367 \\ \hline \text{Balance } S &= +\ 34{,}926 \text{ cal cm}^{-2}\text{ yr}^{-1} \end{aligned}$$

Out of this positive balance of 34,926 cal cm^{-2} yr^{-1}, 86 percent was used up in evaporation and 14 percent in heating the air; quantities of heat passing into and out of the ground canceled each other over the year.

In general it is costly and time-consuming to measure the individual components of short- and long-wavelength radiation separately. For most microclimatologic purposes it is sufficient to determine the net radiation, that is, the quantity S in the equation above.

After this brief survey, let us direct our attention once more to the reflection of radiation from natural surfaces. A distinction is made between diffuse and regular reflection. Reflection is described as diffuse when the incident rays are reflected in all directions, without preference for one particular direction. This is the type of reflection that normally occurs at the rough surfaces found in nature. Table 3 gives albedos for diffuse reflection for the total range of solar radiation. This table can be used only as a point of departure, when one thinks of the diversity of the surfaces found both in nature and in human settlements. S. Fritz and M. Rigby [41] have published a bibliography on albedo. Much observational material for microclimatologic requirements is found in the work of I. Dirmhirn [34], particularly in one of her recent studies on the spectral distribution of reflected waves [35]. A comparative list for 45 types of ground for the visible spectrum, which does not differ greatly from the values for total solar radiation, is given by R. Penndorf [52] on the basis of new measurements, including some from Russia.

Table 3
Albedo (percent) of various surfaces for total solar radiation, with diffuse reflection.

Fresh snow cover	75–95
Dense cloud cover	60–90
Old snow cover	40–70
Clean firn snow	50–65
Light sand dunes, surf	30–60
Clean glacier ice	30–46
Dirty firn snow	20–50
Dirty glacier ice	20–30
Sandy soil	15–40
Meadows and fields	12–30
Densely built-up areas	15–25
Woods	5–20
Dark cultivated soil	7–10
Water surfaces, sea	3–10

The figures given in Table 3 illustrate what can be observed directly from an airplane. The sea appears darkest, it is edged by white strips of surf and sand dunes. Over land, woods show up darker than fields, and only snow-covered areas are light. One can see here how the absorption capacity for incident solar radiation varies and how this affects the whole heat economy. This makes it possible to influence the heat budget of a locality, by artificially changing the structure of its surface (Sec. 20).

Albedo is influenced not only by the nature of a surface, but by its moisture content at any time. Experience shows that wet surfaces appear darker than dry surfaces. In this connection A. Ångström [24] observed that the albedo of gray sand decreased from 18 to 9 percent when it became wet; tall light-colored grass showed a decrease from 32 to 20 percent. In the Amrum Sand Dunes, K. Büttner and E. Sutter [31] noted a change from 37 to 24 percent. Table 4 shows the dependence on wavelength as found by F. Sauberer [55].

Table 4

Reduction of albedo when sand becomes wet, as a function of wavelength.

Wavelength	0.4	0.5	0.6	0.7	0.8 μ
Dry sand	20	23	29	30	30
Wet sand	10	12	15	16	19

Ångström gives the following explanation. When parts of the ground or of plants are covered by a layer of water, light rays may enter the layer from any direction, but they can escape again only if they reach the surface of the water from within the layer at an angle less than the critical angle for total reflection. Since the state of the surface undergoes daily and annual changes, there are corresponding periods of variation in albedo. Monthly values of albedo for a meadow and for the surface of the Danube near Vienna, for radiation from the sun and the sky, measured by F. Sauberer [57] and I. Dirmhirn [34], are shown in Table 5.

Table 5

Monthly average albedo (percent) near Vienna.

Month	I	II	III	IV	V	VI	VII	VIII	IX	X	XI	XII
Meadowland, depending on the development of vegetation	13	13	16	20	20	20	20	20	19	18	15	13
Meadowland, taking winter snow cover into account	44	39	27	20	20	20	20	20	19	18	21	36
Water surface of the Danube River	11.2	11.4	10.7	10.0	9.0	8.6	8.6	9.7	9.5	11.7	11.8	11.9

The effect of cloudiness and horizontal shielding may be found in additional data from Dirmhirn [34], and the daily variation in Sauberer and Dirmhirn [58]. The annual variation is determined by the condition of the ground, which has been charted for Vienna by F. Lauscher [49] in the form of frequency isopleths of snow cover, slush, wet and dry soils, and dry frost.

Albedos for the short-wavelength range of solar radiation, the ultraviolet below 0.36 μ, are lower throughout than the figures given in Table 5 for solar radiation as a whole. Only snow cover shows high reflection of ultraviolet (80 to 85 percent).

Regular or mirror reflection is directional, and therefore depends on the angle of incidence of the sun's rays. This type of reflection occurs at the surfaces of water and sand and is of importance to the climate of river banks and sea shores (see Sec. 22). The values shown in Fig. 3, for the relation between albedo and the elevation of the sun for the range of the solar spectrum, are the work of K. Büttner and E. Sutter [31]. From the zenith to an elevation of 40° there is little change from the values given in Table 3. The albedo then increases sharply, a familiar effect experienced when the dazzling rays of the setting sun are reflected by water. The surprising thing is that this is also true of a rough sandy surface.

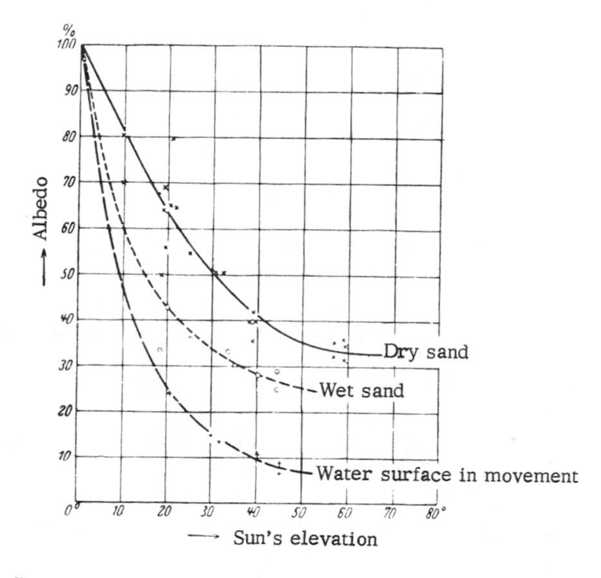

Fig. 3. Mirror reflection of sunlight at sand and water surfaces, for small angles of elevation of the sun. (After K. Büttner and E. Sutter)

The part of radiation not reflected by the surface is absorbed. This is obviously true in the case of a regular surface like a slab of rock or a concrete highway. Natural surfaces are mostly irregular in structure; they are made up of particles or grains between which traces of radiation can penetrate. As far as the heat budget is concerned, this is of no quantitative importance; but it is of great significance for the biology of the earth—the life of bacteria and algae and the germination of seeds—that requires the stimulation of light.

As distance below the surface increases, the long waves are more and more strongly represented. All previous measurements are in agreement on this point. Red light predominates within the earth because it is more strongly reflected than blue on each of the many occasions when there is reflection from a particle of soil.

F. Sauberer [55], who studied this question in detail in 1951, pointed out how great was the variation with different types of soil. The coarser the particles, the more can radiation penetrate. A. Baumgartner [25] investigated the distribution of light in the ground by using graded quartz sands, artificial light, and selenium cells. Figure 4 shows the results for grains with dimensions between 0.1 and 6 mm in a completely dry state. Light penetrates several

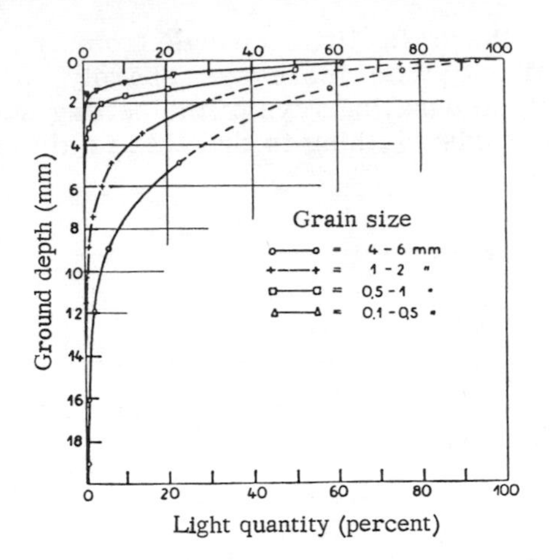

Fig. 4. Light penetration into coarse-grained sand. (After A. Baumgartner)

millimeters farther into such coarse-grained sands with appreciable intensity.

In the case of fine-grained sands, especially when mixed with a little loam, the penetrating light is reduced to one thousandth in the first millimeter. M. Köhn [46] has shown that only half the incident light penetrates to a depth of 0.015 mm in soil with texture as fine as dust.

5. Long-Wavelength Radiation at Night

At night, the terms I, H, and R in the equation of radiation balance are zero. The long-wavelength balance is $S = G - \sigma T^4$. The second term is always larger than the first, except in unusual circumstances. The negative balance prevailing at night may be shown more clearly by writing the equation in the form

$$-S = \sigma T^4 - G.$$

That is to say, the nocturnal radiation loss through emission is equal to the Stefan-Boltzmann radiation from the ground (as a black-body radiator) minus the counterradiation from the atmosphere above the place of observation. This counterradiation comes from water vapor, carbon dioxide, and ozone, in proportion to the quantities present and to the temperature. Figure 5, after F. Schnaidt [59], illustrates the selective absorption of water vapor (a) and carbon dioxide (b). The absorption coefficients, κ for water vapor (per 0.1 mm of precipitable liquid) and k for CO_2 (per 1 m of air under standard conditions), are shown above two different scales of wavelength λ. The same diagram also indicates the range

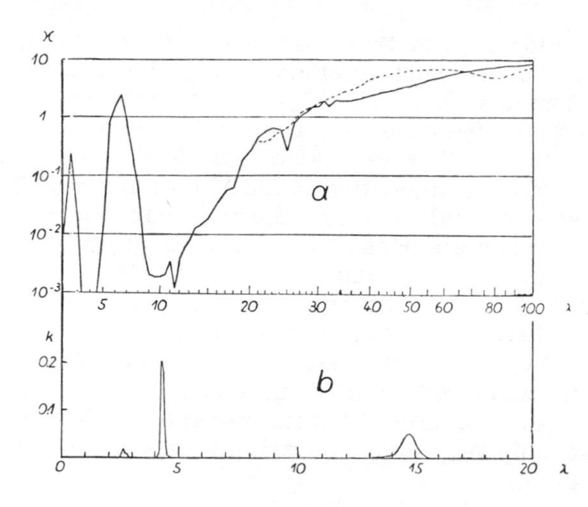

Fig. 5. Absorption spectra for (*a*) water vapor and (*b*) carbon dioxide. (After F. Schnaidt)

of wavelengths radiated by water vapor and carbon dioxide, since, according to Kirchhoff's law, the ratio of emission to absorption is constant for a given wavelength and temperature. Carbon dioxide has only a few well-marked bands, with centers at 2.8, 4.3, and 14.9 μ. It is present in the air in an almost constant proportion of 0.03 percent by volume, and hence its contribution to the counterradiation is also constant, but only to the extent of 1/6 of the total. The amount due to ozone, according to H. M. Bolz [26], is only about 2 percent of the total counterradiation. It follows, therefore, that the varying quantity and the changing temperature of water vapor in the atmosphere are mainly responsible for the changing amounts of counterradiation observed. Figure 5*a* shows that water vapor has a marked absorption band centered at 2.7 μ, and a very broad one with its maximum at 6.3 μ. As the wavelength increases beyond 12 μ, the atmosphere becomes increasingly opaque, until practically all radiation is absorbed. In the region between, there are two spectral ranges, one at about 4 μ and the other from 9 to 11 μ, that might be described as "windows" of the atmosphere. In the first window, absorption is practically nil, while in the second it is only about 10 percent. In these ranges, therefore, streams of radiation can escape into space. Long-wavelength radiation in the 4-μ band is of low intensity, as shown in Fig. 1; hence the leading role in the nocturnal radiation balance is played by the second window, from 9 to 11 μ. No matter how heavily the atmosphere is laden with water vapor, it can afford no protection.

Every layer of the atmosphere is involved in the counterradiation. Aerologic ascents are therefore needed to provide data on the water-vapor content and temperature for each layer. Since such ascents are only rarely available, a different method is used in micrometeorology. It is true that every layer takes part in counterradiation, but the contributions of individual layers are vastly different. Computations made by O. Czepa and H. Reuter [33] showed this to be the case with a normal atmosphere, that is, for average conditions. They estimated that, for a normal atmospheric water-vapor content amounting to 14.25 mm of precipitable water, the percentage of total counterradiation reaching the ground contributed by each of a series of layers (each of which contained 0.6 mm of precipitable water) was as shown in Table 6 for the six lowest layers. The very first layer, extending to only 87 m above the ground, contributes not less than 72 percent of the total counterradiation. This phenomenon depends partly on the low penetrating ability of long-wavelength radiation, a characteristic which in turn depends in a very complicated fashion on the wavelength. The upper layers make only very minor contributions, which decrease gradually with height.

Table 6
Contribution of various atmospheric layers to counterradiation.

Layer thickness (m)	87	89	93	99	102	108
Percent share of counterradiation	72.0	6.4	4.0	3.7	2.3	1.2

The procedure adopted is due to Ångström [23]. The air temperature T_L (°K) recorded in the normal meteorologic station and the water-vapor pressure e (mm Hg) are used to approximate the conditions prevailing in the part of the air that has an effect on counterradiation. From numerous measurements, Ångström deduced the following relation:

$$G = \sigma T_L{}^4 (a - b \cdot 10^{-ce}) \qquad (\text{cal cm}^{-2} \text{ min}^{-1}).$$

Here σ is the Stefan-Boltzmann constant (p. 6) and a, b, and c are other constants. Using comprehensive measurements, which they subjected to critical examination, H. M. Bolz and G. Falckenberg [28] calculated these constants to be: $a = 0.820$, $b = 0.250$, and $c = 0.126$. These figures are valid primarily for the German Baltic coast, where they were measured. They are somewhat low in comparison with other series of readings made in Central Europe, as pointed out by H. Hinzpeter [43]. Use of these figures in Fig. 6 at a later stage, however, demonstrates the relation among the quantities involved.

The long-wavelength night radiation balance of a body with temperature T_L is $-S_L = \sigma T_L{}^4 - G$, which, with the equation given above, becomes

$$-S_L = \sigma T_L{}^4 (0.180 + 0.250 \cdot 10^{-0.126e}) \quad (\text{cal cm}^{-2} \text{ min}^{-1}).$$

The value of $-S_L$ is measured by reading the temperature of the air and using an instrument for measuring radiation pointed in the direction of the sky. This is called "effective outgoing radiation."

Two considerations must be borne in mind. The first is that the rule, being statistical, is not capable of application to individual cases. We know, from investigations made by H. M. Bolz and H. Fritz [29], that the actual value of G deviates from the calculated value by ±5 percent, in extreme cases by ±10 percent; the 10-percent figure was exceeded in only 2 percent of all measurements made. This degree of accuracy is sufficient for practical purposes.

In the second place, the effective outgoing radiation $-S_L$ is different from the long-wavelength net radiation of the ground $-S_B$; this was first pointed out by F. Sauberer [54]. The reason is that the ground temperature T_B is usually a few degrees lower. The difference $\delta = T_L - T_B$ depends to a great extent on the type of ground, its state, and the weather situation. Much depends also on the distribution of water-vapor pressure in the layer between the shelter thermometer and the ground; this, however, may be neglected.

If it is necessary to calculate the radiation balance of the ground $-S_B$, $T_L - \delta$ is substituted for T_B in the equation $-S_B = \sigma T_B{}^4 - G$. Since the fraction δ/T_L is small in comparison with 1, the approximation $1 - 4\delta/T_L$ can be used for $(1 - \delta/T_L)^4$. Thus,

$$-S_B = \sigma T_L{}^4 \left(1 - \frac{4\delta}{T_L}\right) - G = -S_L - 4\sigma T_L{}^3\,\delta.$$

If the air temperature T_L is 273° or 288°K, that is, 0° or 15°C, the value of $4\sigma T_L{}^3$ is 0.0067 or 0.0079 cal cm^{-2} min^{-1} deg^{-1}, respectively. A difference of 1 C deg in ground temperature reduces long-wavelength radiation by 0.007 cal cm^{-2}min^{-1} on the average, which is not a negligible amount. The equation now becomes $-S_B = -S_L - 0.0007\ (T_L - T_B)$, or with the already computed value for $-S_L$:

$$-S_B = \sigma T_L{}^4\ (0.180 + 0.250 \cdot 10^{-0.126e}) - 0.007\ (T_L - T_B).$$

Figure 6 shows the values of $-S_L$ (abscissa) against air temperature T_L (ordinate, °C) and humidity at the level of the instrument shelter. Humidity may be shown as lines of equal vapor pressure e (mm Hg) or by curves of equal relative humidity (percent). The graph is for clear night skies on the German Baltic coast. The effect of a cloud cover will be discussed later.

The range of possible values for the net radiation (Fig. 6) is restricted in three main ways: in the lower right-hand corner by the zero line of water-vapor pressure, on the left-hand side by the 100-percent relative-humidity curve, and at the top left where the lines of increasing water-vapor pressure become progressively closer together.

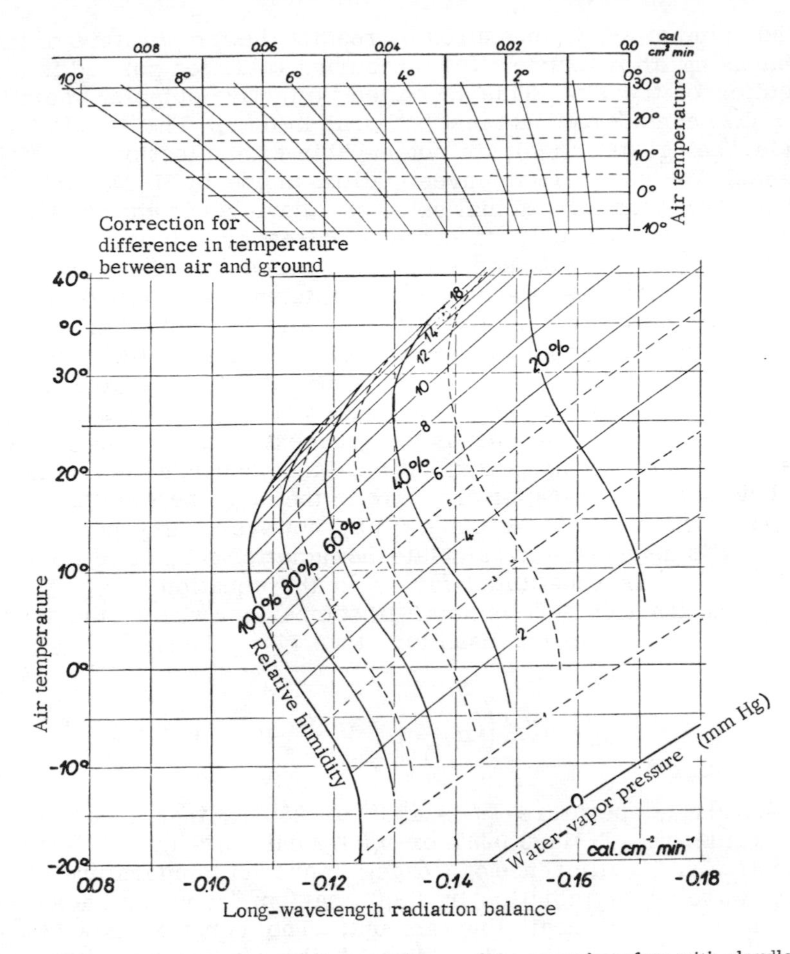

Fig. 6. Calculation of the night radiation balance of the ground surface with cloudless skies (explanation in the text).

An increase in water-vapor content of the atmosphere will not result in any decrease in loss by radiation, which might be desired, becuase the two "windows" toward the sky always remain open, as previously mentioned (Fig. 5).

For example, if the air temperature is 10° C and the relative humidity is 60 percent, the value of $-S_L$ from Fig. 6 is -0.122 cal cm^{-2} min^{-1}. If the ground temperature is 5 C deg lower, the correction for $\delta = 5$ C deg is obtained from the supplementary diagram at the top of the figure. If the ground is colder than the air at night, which is usually the case, the radiation loss from the surface is reduced. The correction is therefore subtracted from the value for the 10°, 60-percent point to give, in this case, a long-wavelength radiation balance of -0.085 cal cm^{-2} min^{-1} for the ground. If the ground is at a higher temperature than the air, as it may be when there is an influx of cold air, then the correction is added. This correction also permits the air temperature T_L to be taken into account in the equation on page 21, instead of using a mean value for $4\sigma T_L^3$. Figure 6 also shows how strongly night radiation losses are reduced because the ground temperature is lower. The "effective outgoing radiation," measured by a pyrgeometer directed toward the sky only, is substantially greater than that of the ground. If the pyrgeometer is also directed downward toward the ground, a radiation loss is obtained because $T_L > T_B$, and this is equivalent to the correction just mentioned. This difference must be taken into account in all considerations of microclimatologic heat budget.

So far it has been tacitly assumed that counterradiation came from the hemisphere of sky above the position in question. In microclimatology it is important to note that the streams of radiation received from different parts of the sky are substantially different. The thickness, and hence the mass, of air above any point on the ground is least in the direction of the zenith. The least amount of counterradiation is therefore received from the zenith direction, and the radiation loss is greatest in that direction. Falckenberg [38] showed that counterradiation from the zenith obeys an equation similar to that given on page 20 for counterradiation from the sky hemisphere, but with different values for the constants, becoming

$$G_{\text{zenith}} = \sigma T^4 \,(0.78 - 0.30 \cdot 10^{-0.065e})\ (\text{cal cm}^{-2}\ \text{min}^{-1}).$$

This zenith value is much smaller than that for the whole sky hemisphere. The thickness of the atmosphere, measured directly upward from a point on the ground, is less than the distance through the atmosphere in any other direction. This thickness increases as the angular zenith distance increases, becoming 1.5 times the zenith thickness at 48°, twice at 60°, and three times at 71°. The effective outgoing radiation S_L decreases proportionally as the angle of elevation from the horizon decreases. Table 7 gives relative values of S_L measured by P. Dubois [36] for a water-vapor pressure of 5.4 mm Hg, taking the effective radiation in the zenith direction as 100. These values are in agreement with recent measurements by Hinzpeter [43], who has made a detailed study of this subject.

Table 7
Effective outgoing radiation for different angles of elevation.

Angle of elevation	90 Zenith	80	70	60	50	40	30	20	10	0° Horizon
S_L (relative value)	100	100	98	96	93	89	81	69	51	0

Dubois's measurements made it possible for F. Lauscher [48] to determine effective radiation for a number of different types of topography. The results that follow are based on the Dubois value of 5.4 mm Hg water-vapor pressure. No measurements are available to show how these topographic features affect the radiation balance of the ground S_B. Nevertheless they are of considerable significance in many micrometeorologic problems. Five different features, shown in the sketches in Fig. 7, are considered.

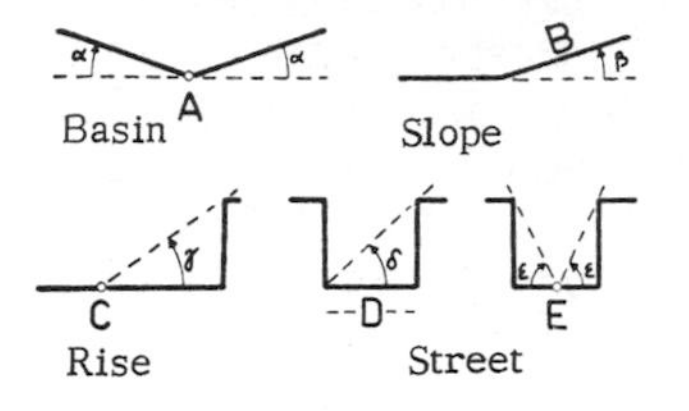

Fig. 7. Topographic features referred to in discussion of radiation.

A. When the horizon surrounding a place is uniformly obscured up to an angle α, as in a hollow in the ground, in the middle of a circular clearing in a wood, or in an amphitheater, the feature is termed a basin. The figures in row A of Table 8 give the effective outgoing radiation in parts per thousand of the unobstructed radiation that would be emitted by a level plane, for various angles of shielding. It may be seen that an angle of shielding of about 20°, which is quite noticeable on the ground, does not reduce the radiation by as much as 9 percent. This is an indication of just how small is the effect of obstructions close to the horizon on the outgoing radiation. Radiation measurements therefore are not less accurate if the horizon is not completely free of obstruction. The dome of the sky above 30° encloses half the solid angle of the arch of the heavens; almost 80 percent of the radiated energy is directed to this half of the sky.

Table 8
Ratio (parts per thousand) of the effective outgoing radiation from sheltered or inclined surfaces to the radiation from a completely open horizontal surface. (After F. Lauscher)

Angle (°)		0	5	10	15	20	30	45	60	75	90
Basin (α)	A	1000	996	982	955	915	793	549	282	79	0
Slope (β)	B	1000	996	986	970	951	900	796	667	528	396
Rise (γ)	C	1000	997	992	988	979	951	877	772	639	500
Surface of street (δ)	D	1000	930	862	797	737	622	452	296	143	0
Middle of street (ϵ)	E	1000	993	984	976	958	902	754	544	279	0

B. If a plane surface is tilted at an angle β from the horizontal, the effective outgoing radiation is reduced since no measurable quantity of radiation is directed below the horizon. (The assumption of a homogeneous temperature distribution for S_L is certainly not valid for S_B.) The decrease is negligible for small angles of tilt, and does not reach 10 percent even with the steepest hill pastures. A vertical wall ($\beta = 90°$), however, radiates only 40 percent of the emission from level ground.

C. In the vicinity of a sharp rise in the ground, such as a rock face, a man-made wall, a hedge, or the edge of a wood, thought of as extending indefinitely in either direction along the ground (perpendicular to the paper in Fig. 7), the effective radiation will be reduced on approaching the wall by an amount depending on the angle γ. This reduction is less, of course, than with the basin designated A. For $\gamma = 90°$, that is, at the foot of the wall, or at the edge of a stand of timber, radiation is 50 percent of that in open country, since half the sky is shielded.

D. It is useful to consider a theoretical street, infinite in length, with rows of houses of equal height on either side. This model is useful in town planning, for woodland clearings, and for long valleys. Line D in Table 8 gives the relative values for radiation from horizontal ground as a function of the angle δ (see Fig. 7). Hence, in an old-fashioned street, with houses of the same height as the width of the street on both sides, outgoing radiation was 45 percent of that in the open country ($\delta = 45°$).

E. Outgoing radiation varies with position in the street. Areas near the houses have more protection, and emission is at a maximum in the middle of the street. This maximum value is given in line E of Table 8, the angle ϵ being measured from the center of the street.

It should be mentioned here that the walls of the houses bordering the street also radiate, but to a markedly smaller extent than shown in line B for 90°, because each side of the street presents a barrier to the radiation of the other. Taken all together, the radiation emitted from the street surface and the two bordering walls must be as great as the radiation from a surface visualized as a lid covering the opening of the street to the sky. This is a factor of importance in a mountain climate.

When there is a cloud cover at night, there is, in addition to the atmospheric counterradiation just discussed, an additional radiation from water and ice particles in the undersides of clouds. From the previous discussion it is clear that the lowest layer of air is the most important for counterradiation. It therefore follows that low clouds reduce nocturnal radiation to a greater extent than do high clouds. In addition to type and height of the clouds, there is an effect from the extent of the cloud cover, which is measured in tenths of the sky (0.0 to 1.0) and indicated by the symbol w. Experience shows that when w is small the few clouds present are close to the horizon and are thus in a part of the sky that is of less importance for the radiation balance than the area near the zenith. Radiation from clouds therefore does not increase linearly with w, but almost as a quadratic function. This theory was proposed in

1928 by F. Lauscher [47], and has now been verified by H. M. Bolz [27].

Counterradiation G_W from a cloudy sky is expressed, according to Bolz, in terms of the value for a cloudless sky by the equation $G_W = G\,(1 + kw^2)$. The constant k is greater for lower cloud levels, and its values for different types of cloud are as follows: Ci, 0.04; Cs, 0.08; Ac, 0.17; As, 0.20; Cu, 0.20; St, 0.24. Effective radiation from a clear sky was (see p. 18)

$$-S_L = \sigma T_L{}^4 - G\,;$$

with cloudy skies it becomes

$$-S_W = \sigma T_L{}^4 - G_W,$$

or, if counterradiation is eliminated from the equation,

$$-S_W = -S_L - kw^2\,(S_L + \sigma T_L{}^4) \qquad (\text{cal cm}^{-2}\,\text{min}^{-1}).$$

The net radiation of ground surface can also be derived from the effective outgoing radiation when the sky is cloudy by the method described above.

During the night, both air and ground temperatures decline; as a result, radiation losses also decrease, but only to a small extent (provided there is no change in cloud cover during the night). The decrease is only about 10 percent during a night with clear skies, according to F. Sauberer [56], and about 15 percent on an overcast night. For rough calculations, such as the forecasting of night frosts, this variation with time can be neglected.

6. The Laws of Heat Transport in the Ground

The exchanges of radiation at the surface of the ground, discussed in Secs. 4 and 5, cause periodic (daily and yearly) variations of the surface temperature, while irregular changes in the weather produce irregular variations. These changes affect the temperatures of the soil below the surface and of the layer of air above it. In this section, the laws governing these effects will be discussed for the ground only, and the following section will deal with the influences on the air.

We must first consider whether or not some other influences might be at work in the ground. The high temperature of the interior of the earth immediately comes to mind. Since the rate at which temperature increases with depth in the ground is, on the average, 1 C deg every 33 m (a unit of geothermal depth), the upward flow of heat can be only about 0.0001 cal cm^{-2} min^{-1}, which can be neglected. This influence need be considered only in volcanic regions, near hot springs, or where there are underground fires.

The way in which air moves into and out of the ground can be taken into account along with radiation influences. M. Diem [62] has shown experimentally that with a sandy soil a volume of air was "breathed" through the surface of the ground in 1 day equal to a column 22 m high over the area concerned. This big turnover is of importance to air temperature, because air entering the ground quickly assumes its temperature. Consequently, air temperature near the ground is influenced by the nature of the soil by means of this breathing process. Since the density of soil is about 1000 times that of air, the reciprocal effect of air on ground temperature is always negligible.

One effect of importance, however, is that of the water vapor present in the air passing through the soil. Warm and moist air from outside may give up liquid water in contact with the colder soil ("internal dew"); every gram of condensed water vapor brings the soil a supply of about 600 cal of heat. G. Hofmann [63] has estimated that, even in favorable circumstances, which are effective for a limited time only, the amount of this condensation is less than 0.01 mm per hour. Nevertheless, this is equivalent to 0.01 cal cm^{-2} min^{-1}, which is of interest in building technology, and for water conservation in time of drought. In normal situations, such as those considered here, it can be neglected.

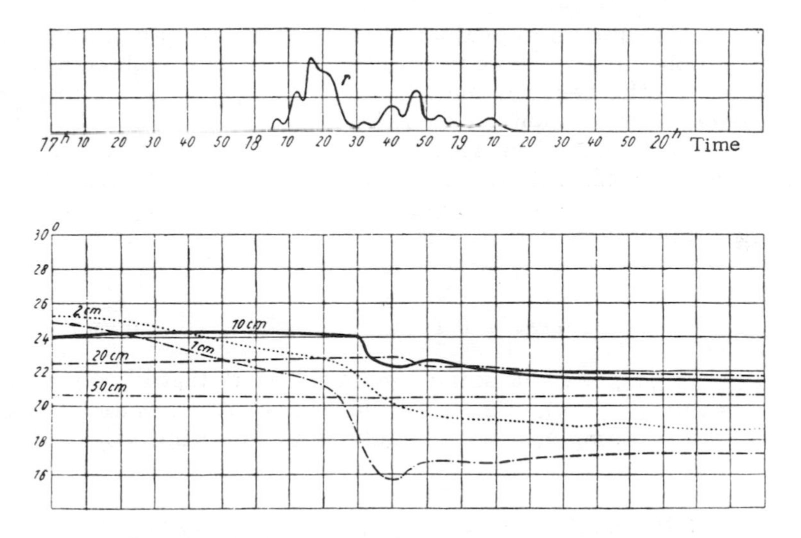

Fig. 8. Effect of cold rain on soil temperatures. (After F. Becker)

A more important external effect is that of the penetration of cold or warm rain into the ground, already mentioned on p. 10. Figure 8 shows how a 20.8-mm fall of cold rain affected the ground temperature; the variation with time is shown on the upper graph. These recordings were made by F. Becker [61] in Potsdam on

3 July 1936, using electric thermometers. Nineteen minutes after the rain started, the thermometer at a depth of 1 cm recorded the arrival of the cold water. After another 19 min it had penetrated to a depth of 20 cm. Only a heavy fall of very cold rain on permeable ground can give such excellent recordings as in Fig. 8. More usually the disturbance passes quickly. In the discussions below this exceptional case will not be taken into consideration.

All natural ground has three fundamentally different components: (1) the soil proper, consisting of chemical or organic substances, of density ρ_s (g cm^{-3}) and specific heat c_s (cal g^{-1} deg^{-1}); (2) free available water, not chemically bound to the soil; (3) the air occupying the spaces between the soil particles. The parts by volume occupied by these three elements, expressed as percentages, are v_s, v_w, and v_l, adding up to 100 percent. The values of these quantities, measured by S. Uhlig [330] on 29 August 1949, in an experimental plot at the Agricultural College of Hohenheim, were: $v_s = 50$, $v_w = 16$, and $v_l = 34$ at the surface, and $v_s = 59$, $v_w = 23$, and $v_l = 18$ at a depth of 0.5 m. The density of the soil ρ_m and its specific heat can be calculated from these figures. Air can be ignored because of its low density; therefore, taking the density of water as 1,

$$\rho_m = 0.01\,(v_s\,\rho_s + v_w) \qquad \text{(g cm}^{-3}\text{)}.$$

It is equally simple to evaluate the specific heat per unit volume (this is simpler and more convenient than the usual method of evaluating the specific heat per unit mass). The figure arrived at gives the amount of heat required to raise the temperature of 1 cm^3 of the soil by 1 C deg. The symbol $(\rho c)_m$ is used for this quantity in subsequent discussions, and it is termed the thermal capacity by volume. It is equal to the sum of ρ, c, and v; hence, using the approximation given above, and taking the specific heat of c_w of water as 1 cal g^{-1} deg^{-1}, we obtain

$$(\rho c)_m = 0.01\,(v_s\,\rho_s\,c_s + v_w) \qquad \text{(cal cm}^{-3}\text{ deg}^{-1}\text{)}.$$

This quantity $(\rho c)_m$ is a true soil constant for earth that is completely free of water ($v_w = 0$). Its value increases with increasing water content; for example, in a sandy soil, in which the volume

Table 9
Variation of density and thermal capacity by volume of sandy soil with varying water content.

Water content v_w	0	10	20	30	40	Vol.-%
Density ρ_m	1.50	1.60	1.70	1.80	1.90	g cm^{-3}
Thermal capacity by volume $(\rho c)_m$	0.30	0.40	0.50	0.60	0.70	cal cm^{-3} deg^{-1}

Table 10

Order of magnitude of some constants in the heat economy of the ground (arranged according to decreasing thermal conductivity).

Type of soil (or material)	Solid soil particles		Natural soil			
	Density, ρ_s	Specific heat, c_s	Density, ρ_m	Specific heat per unit volume $(\rho c)_m$	Thermal conductivity, 1000 λ	Thermal diffusivity, 1000 a
	($g\ cm^{-3}$)	($cal\ g^{-1}\ deg^{-1}$)	($g\ cm^{-3}$)	($cal\ cm^{-3}\ deg^{-1}$)	($cal\ cm^{-1}\ sec^{-1}\ deg^{-1}$)	($cm^2\ sec^{-1}$)
Silver	10.5	0.056	—	0.59	1000	1700
Iron	7.9	0.105	—	0.82	210	260
Concrete	2.2–2.5	0.21	—	0.5	11	20
Rock	2.5–2.9	0.17–0.20	2.5–2.9	0.43–0.58	4–10	6–23
Ice (see also Sec. 24)	0.92	0.505	1.7–2.3	0.46	5–7	11–15
Wet sand	2.6	0.20	—	0.2–0.6	2–6	4–10
Wet clay	2.3–2.7	0.17–0.20	1.7–2.2	0.3–0.4	2–5	6–16
Old snow (density 0.8)	—	—	0.8	0.37	3–5	8–14
Still water	1.0	1.0	—	1.0	1.3–1.5	1.3–1.5
Wet moorland	1.4–2.0	—	0.8–1.0	0.6–0.8	0.7–1.0	0.9–1.5
Dry clay	2.3–2.7	0.17–0.20	—	0.1–0.4	0.2–1.5	0.5–2.0
Dry sand	2.6	0.20	1.4–1.7	0.1–0.4	0.4–0.7	2–5
New snow (density 0.2)	—	—	0.2	0.09	0.2–0.3	2–4
Dry wood	1.5 (Wood fibers)	0.27	0.4–0.8	0.1–0.2	0.2–0.5	1–5
Dry moorland	1.4–2.0	—	0.3–0.6	0.1–0.2	0.1–0.3	1–3
Still air	0.0010–0.0014	0.24	—	0.00024–0.00034	0.05–0.06	150–250

v_s = 57 percent is occupied by grains of sand having a density ρ_s = 2.63 g cm^{-3} and a specific heat c_s = 0.20 cal g^{-1} deg^{-1}, the density and the thermal capacity by volume are as shown in Table 9. Further values are given in Sec. 21 for the case of ground frost, that is, when the water in the soil has changed to ice.

Values are given for ρ_s and c_s in Table 10. There is no difficulty in assessing these values for a homogeneous substance such as silver, because it is then a question of a true constant, which is only very slightly dependent on temperature. For substances described by such general terms as "sand," "clay," or "peat," only mean values can be given. The problem of specific heat by volume is much more difficult because of its dependence on water content. To indicate the correct order of magnitude of this quantity, a distinction is drawn between "dry" and "wet" soils. Values for this quantity are given in Table 10 because experience shows that they are a basic requirement for many problems discussed in the recent literature, although there are also reasons for not publishing them yet.

To use these values for calculating heat transport within the ground, we assume that the soil is homogeneous. All points at a depth of x cm will then have the same temperature t °C. The flow of heat B cal cm^{-2} sec^{-1} (see p. 9) is directed upward toward the surface because the temperature increases with depth, and it is proportional to the rate of change of temperature with depth:

$$B = \lambda\, dt/dx \qquad (\text{cal cm}^{-2}\ \text{sec}^{-1}).$$

The constant of proportionality λ (cal cm^{-1} sec^{-1} deg^{-1}) is termed the thermal conductivity, and is the amount of heat in calories that will flow through a 1-cm cube of the substance in 1 sec when the temperature difference between opposite faces is 1 C deg and there are no other variations in temperature.

For chemically pure substances, λ is a true constant; for instance, it is 1.00 cal cm^{-1} sec^{-1} deg^{-1} for silver. In natural soils, however, λ varies not only from place to place with the constitution of the soil, but also in one place as the water content of the soil changes. This is illustrated in Fig. 9, using measurements made by F. Albrecht [60] in sandy soil in Potsdam during the month of July 1937.

Precipitation is shown by the upper graph in Fig. 9, while in the lower half curves of thermal conductivity (cal cm^{-1} sec^{-1} deg^{-1}) are given for depths of 1, 10, and 50 cm. These are direct measurements made by an instrument for measuring conductivity, described in the same work. Since soil dries from the surface downward, there is normally an increase in the amount of water present farther down in the soil. The conductivity curve for 1-cm depth (solid line, Fig. 9) shows a rapid and marked response to rain. This effect is delayed and weakened as depth increases. The

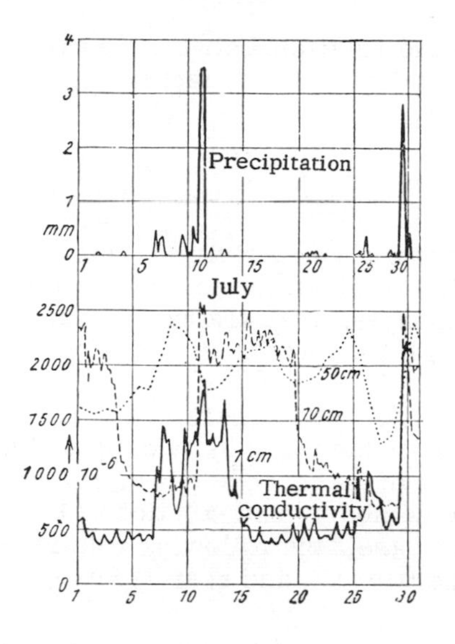

Fig. 9. Change in thermal conductivity of soil as a result of precipitation. (After F. Albrecht)

curves for depths of 10 cm (broken line) and 50 cm (dotted line) frequently run in opposite directions as a result of the process of percolation; while the conductivity at 10 cm is decreasing as the water seeps away, that at 50 cm is increasing as the water begins to arrive. This relation will be referred to again in Sec. 21. Considered as a whole, Fig. 9 shows that increasing water content alone is able to increase the value of λ three- or fourfold. Thus, the numerical values for λ given in Table 10 are valid only as a guide.

The temperature of a soil depends not only on the quantity of heat transported to it, but also on its ability to absorb this heat, that is, on the specific heat by volume $(\rho c)_m$. Temperature changes in an element of soil are brought about by alterations in the flow of heat B with depth x below the surface. The flow B is positive (see p. 8) when directed upward toward the surface. If B increases as x increases, that is, if the flow of heat is downward, then its value will decrease as the surface is approached. Consequently, heat will be retained within the soil, and the temperature of its elements will increase:

$$\frac{dB}{dx} = (\rho c)_m \frac{dt}{dz} \qquad (\text{cal cm}^{-3}\ \text{sec}^{-1}).$$

If the value of B from p. 30 is substituted, this becomes after a little rearrangement:

$$\frac{dt}{dz} = \frac{\lambda}{(\rho c)_m} \frac{d^2t}{dx^2} \qquad (\text{deg sec}^{-1}).$$

This equation gives the relation between the variations of temperature t with time z and depth x. The factor

$$a = \frac{\lambda}{(\rho c)_m} \qquad (\mathrm{cm^2\ sec^{-1}})$$

is called the temperature conductivity or thermal diffusivity. Numerical values for it are given in the last column of Table 10. It is well known, for example, that trapped air is a bad conductor of heat, but, because of its low density, a good transporter of temperature.

A harmonious rhythm of temperature variation is thus brought about at the earth's surface ($x = 0$) by daily and annual variations. The oscillation period of the heat cycle is therefore T = 86,400 sec, or 365-1/4 times that number. The solution of the differential equation gives, for this case, the following relation between the daily fluctuations of temperature s_1 and s_2 at depths x_1 and x_2:

$$s_2 = s_1 \exp\left[(x_1 - x_2)\sqrt{\frac{\pi}{aT}}\right] \qquad (\mathrm{deg}).$$

In this expression, "exp" has its usual meaning of exponential function ($e = 2.71828$). Consider, for example, a day when a variation s_1 of 38 deg in temperature was measured at the surface ($x_1 = 0$) of dry sandy soil ($a = 0.0013$ cm^2 sec^{-1}); the daily temperature variation at depth $x_2 = 8$ cm would work out as $s_2 = 10.0$ deg (or 9.97 deg to be precise).

The interval between the times of arrival of a temperature wave at two different depths is given by

$$z_2 - z_1 = (x_2 - x_1)\frac{T}{2\pi}\sqrt{\frac{\pi}{aT}} \qquad (\mathrm{sec}),$$

where z_1 is the time of arrival of an extreme value (maximum or minimum) at depth x_1, and z_2 is the time of arrival at x_2.

If the time of the maximum at the surface were 12:30 hr, for example, the maximum would arrive at a depth of 8 cm 18,398 sec later, that is, at 17:37. To simplify calculations, the square root appearing in the first equation is taken over into the second, which is then left without being further simplified.

The first of these two equations can be used to determine the depth of penetration of the daily or annual temperature fluctuation into various types of soil. Penetration depth is defined as the depth at which the fluctuation is reduced to 0.01 of its surface value. The daily fluctuation is calculated to penetrate to a depth of 764a cm, and the annual fluctuation to 19.1 times this distance. Table 11 shows the values for a few different types of ground.

Table 11
Penetration depth of temperature fluctuation.

Thermal diffusivity a ($cm^2 sec^{-1}$)	0.02	0.01	0.007	0.001
Type of soil	Rock	Wet sand	Snow cover	Dry sand
Daily fluctuation (cm)	108	76	64	24
Annual fluctuation (m)	20.6	14.5	12.2	4.6

When these equations are used in practice, the conditions on which the mathematical statement is based are not fulfilled. Conductivity and specific heat by volume both vary systematically with depth and change with time. The temperature wave originating at the surface, too, departs significantly from the sine curve assumed in the evaluation. Different values of a are therefore found in practice when it is evaluated from the rate of decrease of the temperature fluctuation with depth or its phase difference with depth.

H. Lettau [64] developed a more elaborate model than the simple one used above. He expressed the conduction of temperature as a function of depth in the ground, hence introducing another variable. The mathematical solution then becomes so difficult, however, that one can only approximate it, although with sufficient accuracy for practical evaluation. The original work should be referred to for further details.

The laws that have just been given, governing the transport of heat in the ground, allow a mean curve of temperature within the ground to be plotted; this will be discussed in Sec. 10. The influence of type of soil, ground cover, cultivation, moisture content, and ground freezing will be treated in Secs. 19–21.

7. Transport of Heat in the Atmosphere. Eddy Diffusion

Physical or true conduction of heat also takes place in the air. Since air possesses to a high degree the ability to transport temperature (see p. 32), it follows that the air temperature changes more quickly than the ground temperature. But there are other completely different factors at work as well.

The height to which the daily fluctuation of temperature would reach in the air through physical conduction is found from the equation given near the end of Sec. 6 to be about 3 m. This is about three times as far as it will penetrate into the best-conducting soil. Observations show, however, that at a height of at least 1000 m above the ground there is a marked difference of air temperature between day and night. Physical conduction therefore plays an insignificant, usually negligible, part in the actual movement of heat in the atmosphere.

The decisive factor in heat transport in the atmosphere is eddy diffusion, as already mentioned briefly on p. 9.

Observations made on the flow of fluids through tubes have shown that two types of flow have to be distinguished—laminar flow and turbulent flow. O. Reynolds made his famous experiment, now familiar to everyone, in 1883. By means of a thin capillary, a colored fluid is introduced into the middle of a glass tube, through which another fluid is streaming. When the rate of flow is small, a well-defined trace of color can be observed, and the current appears to consist of a bundle of threads running parallel to one another. As the speed of flow increases, a point is reached at which the thread of color is suddenly torn apart and, in the irregular movement that follows, the color is soon distributed uniformly throughout the tube. The first type of flow is called laminar, and the second turbulent. The sudden change from one type of flow to the other depends on the value of a constant that is directly proportional to the velocity of flow and the density of the fluid and inversely proportional to its viscosity (Reynolds number).

Laminar flow is much more rare in gases, which are more easily set in motion, than in liquids. Apart from some exceptional cases, motion in the atmosphere is always turbulent.

The turbulence of this motion is made visible when one tries to follow the path of a snowflake in the wind, when the smoke from a locomotive or from a chimney is observed closely, or when the movement of a winged seed can be followed on a sunny day. The individual parcel of air in random motion, sometimes called the "turbulence element" or "turbulence body," in these examples picks up a visible suspended particle (snowflake, soot particle, or seed) and deposits it in another chance position. These turbulence elements also transport invisible properties such as their content of heat, water vapor, kinetic energy, carbon dioxide, radon, and so forth. Sooner or later the elements disappear and share all these properties with the new environment. As the parcels of air undergo haphazard movement, all their properties move with them. This process is fundamental to the concept of eddy diffusion, and it is clear that it is of much wider significance than merely as a means of transporting heat.

Study of fluid flow also shows that, in places where the air comes into contact with a solid surface such as the ground or a wall, turbulence and hence eddy diffusion do not extend quite to the solid. A layer of air a few millimeters thick adheres with great tenacity to the wall or ground. This is termed the boundary layer. The laws of eddy diffusion are not valid in this layer, but the transition from the solid surface to turbulent air is completed within it, governed only by the laws of molecular physics. In this layer, heat is transported only by physical conduction, and water vapor by diffusion.

This boundary layer therefore constitutes a formidable barrier, as will be shown later.

Turbulence may be thought of as a supplementary motion in all directions, superimposed on the (horizontal) wind. The horizontal components of the turbulent motion may increase or decrease the speed of the wind, and cause it to deviate in direction from its mean path. This is the reason for gustiness in wind, which can be observed from any wind-speed record or swinging wind vane. As a rough evaluation, the lowest speed is about 0.2 times and the greatest about 1.9 times the mean wind speed. That is a measure of the magnitude of these horizontal components.

Although smaller, the vertical components of the turbulent motion are of more importance, since they provide a mechanism for the vertical transport of heat and in fact of all properties of the air.

Pioneering work in this field was carried out by Wilhelm Schmidt [82], and the following treatment, on which the basic equation of eddy diffusion is based, is borrowed from his work on the subject, published in 1925.

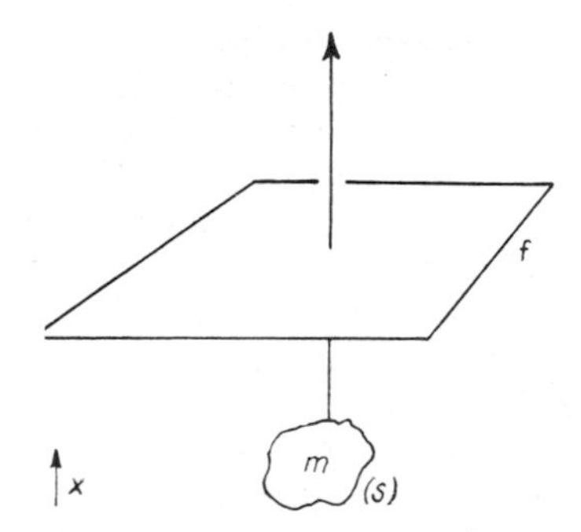

Fig. 10. Diagrammatic illustration of the basic equation of eddy diffusion.

A horizontal plane f (Fig. 10) is moving with the mean wind speed. It is therefore penetrated only by turbulent motion of the air, acting in all directions. In the following discussion only the vertical components of turbulence will be taken into account, as they alone effect the transport of the properties of the air to different levels. In order to include as many different cases as possible we shall consider an unspecified characteristic s, measured per unit mass and subject only to the stipulation that it is independent of the conditions in its environment. Such a characteristic might be the amount of dust, water vapor, or radon, or even a quantity of motion. Temperature cannot be independent of height, because of the vertical decrease of pressure. For the transport of heat, therefore, s is taken to be the heat contained in 1 g of air (cal g^{-1}), that is, $c_p T$, where c_p is the specific heat at constant pressure (0.241 cal g^{-1} deg^{-1}) and T (°K) is the potential temperature. The potential temperature is the temperature the air would have if it were brought down to a pressure level of 1000 mb. It therefore fulfills the condition of being independent of the environment. However, within a few meters of the ground—and only there—the difference between potential temperature and actual temperature is so insignificant that T can be considered to be the observed temperature without any noticeable error.

The question of the vertical transport of a characteristic s arises only when s varies with height. To simplify things, the zero of height is taken as the plane f. At that level $x = 0$ and the property s is s_0. The change of s with height can be expressed by a power series. This series will be taken only as far as the second-degree term, since experience shows that the true distribution of s in the vicinity of f can be represented with sufficient accuracy by a parabola. Therefore:

$$s = s_0 + \frac{ds}{dx}x + \frac{1}{2}\frac{d^2s}{dx^2}x^2 \qquad (\mathrm{P\ g^{-1}}),$$

the unspecified property being designated P in the statement of the unit. Turbulent motion brings particles from the layer at a distance $+x$ through the surface area f (cm^2). These will amount to a mass m_{+x} after a time z sec, and this mass will bring with it, from the layer $+x$, the average value of the characteristic s for that layer, that is, s_{+x}. The quantity of the characteristic flowing out of the layer $+x$ through f in time z is therefore $m_{+x}\,s_{+x}$. This will be taken as positive when directed downward toward the ground surface, in the same way as for the flow of heat L (see p. 9).

The transport of a characteristic such as this takes place not just from a distance $+x$ from the plane f but from many different distances. The total quantity carried in this direction is therefore the sum of all these separate amounts of the property transported: $\Sigma m_{+x}\,s_{+x}$. Similarly, the flow in the opposite direction during the same period is $\Sigma m_{-x}\,s_{-x}$. The net flow $\mathfrak{S}$ of the characteristic in question through unit area in unit time is therefore

$$\mathfrak{S} = \frac{1}{fz}\left(\Sigma m_{+x}\,s_{+x} - \Sigma m_{-x}\,s_{-x}\right) \qquad (\mathrm{P\ cm^{-2}\ sec^{-1}}).$$

If the expression for s given above is now substituted, the equation becomes:

$$\mathfrak{S} = \frac{1}{fz}\left\{s_0\left(\Sigma m_{+x} - \Sigma m_{-x}\right) + \frac{ds}{dx}\left(\Sigma m_{+x}\,x - \Sigma m_{-x}\,x\right) + \frac{1}{2}\frac{d^2s}{dx^2}\left(\Sigma m_{+x}\,x^2 - \Sigma m_{-x}\,x^2\right)\right\} \qquad (\mathrm{P\ cm^{-2}\ sec^{-1}}).$$

Since turbulent motion, in the final analysis, does not lead to changes in the total distribution of mass, the same quantity of mass must flow downward through f as upward; hence the expression in the first parenthesis is zero. It is assumed, in addition, that the exchange takes place symmetrically with respect to f: the mass m_{+x} coming from the distance $+x$ must be equal to m_{-x} coming from $-x$. This assumption has decreasing validity as the ground is approached. However, if we accept it, the third parenthesis will also become zero, and the equation will be simplified to

$$\mathfrak{S} = \frac{\Sigma m_{+x} x - \Sigma m_{-x} x}{fz} \frac{ds}{dx} \qquad (\mathrm{P\ cm^{-2}\ sec^{-1}}).$$

The coefficient of ds/dx is independent of the properties of the air mass in turbulent motion and is entirely descriptive of the vigor of the motion. If it is replaced by the symbol A, the equation becomes

$$\mathfrak{S} = A\, ds/dx \qquad (P\ \mathrm{cm^{-2}\ sec^{-1}}).$$

The quantity A ($\mathrm{g\ cm^{-1}\ sec^{-1}}$) is called the austausch coefficient. Its importance lies in the possibility it offers of expressing the apparently irregular movements of eddy diffusion in a numerical form. In the English literature the quantity $K = A/\rho$ is used instead of the austausch coefficient, and is called eddy diffusivity or volume transported through unit area in unit time. Its dimensions are $\mathrm{cm^2\ sec^{-1}}$, the same as those of thermal diffusivity.

If this general form is adapted to the treatment of a flow of heat, L ($\mathrm{cal\ cm^{-2}\ sec^{-1}}$) takes the place of $\mathfrak{S}$, the quantity $c_p T$ is used for s (see p. 35), and, since $dT = dt$, we have

$$L = A c_p \frac{dt}{dx} \qquad (\mathrm{cal\ cm^{-2}\ sec^{-1}}).$$

This equation is of the same form as that for thermal conductivity in Sec. 6. Here, in place of the thermal conductivity λ, the austausch coefficient is used (multiplied by c_p). Transfer of heat in the air therefore obeys the same type of law as heat transfer in the ground, but at different orders of magnitude because of the difference between λ and $A c_p$. Transport of heat by eddy diffusion is also termed pseudoconduction. It should be kept in mind that the factor A is subject to very great variations in space and time. In the boundary layer, where the principal mechanism is molecular conduction (p. 34), A is of the order of $10^{-4}\ \mathrm{g\ cm^{-1}\ sec^{-1}}$. The value increases with distance from the ground, very rapidly at first. At heights of 1 to 10 m it varies from 0.1 to 10 $\mathrm{g\ cm^{-1}\ sec^{-1}}$. In the atmosphere as a whole, many other orders of magnitude are possible.

8. Mixing Due to Friction and Convection

Eddy diffusion has two causes, frictional exchange and convection exchange.

Frictional exchange of mass may be called alternatively dynamic mixing, that is, mixing due to shear or forced convection; it is caused by variations in the roughness of natural surfaces and by changes (usually increases) in wind speed with height. The austausch coefficient A increases with increasing wind strength and height above the ground. The observation that follows, made by F. Katheder [71] at the Nuremberg airfield, illustrates this relation. On 23 September 1936, ground fog formed to a depth of 1 to 1.5 m shortly after 18:00. Visibility was still excellent above the fog blanket. Relative humidity in the instrument shelter was 86 percent, air temperature 15.2°C, and ground temperature about 12°C. At 18:40 a three-engined aircraft took off for a night flight to Munich. "During take-off a path completely free of fog was cut in the fog along the runway astern of the aircraft. After 4 or 5 minutes the sharp boundaries between the fog and the clear channel had dissolved, and a little later the previous situation was restored. The width of the fog-free pathway was approximately the same as that of the Junkers aircraft." The propellers had mixed the surface air with warmer, drier air aloft to dissipate the fog (perhaps with the help of hot exhaust gases).

At night the only type of mixing is frictional. E. Frankenberger [445] made detailed measurements of the heat economy of meadowland near Quickborn in Holstein from 1 September 1953 to 31 August 1954, by means of instruments attached to a radio mast at heights of 2, 13, 28, and 70 m. Figure 11 shows the austausch coefficient A, on an appropriate vertical logarithmic scale, against wind speed at a height of 10 m, for the levels 8, 15, and 30 m on clear summer days. The lower part of the diagram gives the data for clear nights. The increase of A with height is clearly shown by the relative position of the three isopleths, as is also the sharp rate of increase with increasing winds.

By day, however, when the earth's surface is heated as a result of the positive radiation balance, convective mixing is added to the frictional contribution. It is known, from considerations of general meteorology, that when the fall of temperature with height is exactly -1 C deg/100 m, which is termed the "adiabatic lapse rate," the air is in equilibrium. This means that any parcel of air which is lifted to a higher level, or brought down to a lower, will have the same temperature as the environment at each instant of its ascent or descent. If the lapse rate γ is greater than -1 deg/100 m, the stratification of the air is stable; this means that every vertical movement will be damped out, because ascending air enters a region in which it is colder than the surrounding air, while descending air will be warmer than its environment and therefore tends to return

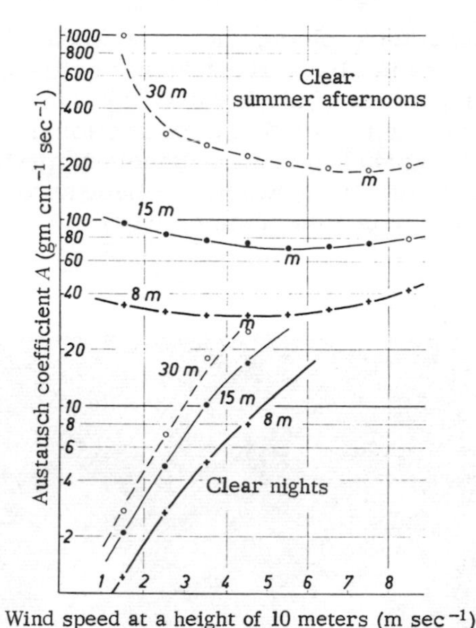

Fig. 11. Dependence of the austausch coefficient A on wind speed and height. (After measurements by E. Frankenberger)

to its original position. Thus the average lapse rate of -0.5 deg/100 m implies stable conditions. If temperature increases with height, that is, if γ is positive, stability is even more pronounced. Such a reversal of the normal vertical temperature gradient is called an inversion.

About noon there may be observed, in the free atmosphere, a lapse rate of more than 1 deg/100 m, especially in times of drought or in arid localities. This is a "superadiabatic lapse rate," which brings about unstable layering of the air, and is often the forerunner of a thunderstorm. In the vicinity of the ground, around midday, superadiabatic lapse rates like this are the rule rather than the exception. Overheated parcels of air rise from the ground into the atmosphere and colder air sinks downward to occupy the vacant space. This vertical circulation caused by heating is known as convection (in contrast to advection, the horizontal transport of other air masses).

The convection process gives rise to an irregular motion termed mixing by convection, thermal mixing, or free convection. It is obviously different in character from frictional mixing. The elements of air involved in the turbulent movements of convection are no longer directed at random, but usually tend to resolve themselves into ascending and descending currents. Figure 12 shows how this vertical motion was made visible by L. A. Ramdas and S. L. Malurkar [79], by spreading water over the surface of a heated plate. The lighter parts of the photograph show where the upward movement took place; the dark areas indicate descending currents. There is also a great deal of evidence to indicate that the elements or parcels of air involved in convection currents are larger than the turbulence elements of frictional mixing. At greater distances from the ground there are funnels of mixing air and large-scale areas of general uplift, which are made apparent by the development of cumulus clouds on summer days.

Fig. 12. The visible convection process. (After L. A. Ramdas and S. L. Malurkar)

Schmidt's statement of the law of transport referred to the random motion of frictional mixing, and was based on the assumption that the characteristic transported had no marked influence on the motion of the turbulence element. This is certainly true for moderate amounts of dust pollution, and for carbon dioxide content among other things. However, turbulence element that is warmer than its surroundings receives an upward thrust because of its lesser density. It therefore has an additional upward movement; similarly, a colder turbulence element has an added downward movement. As the contribution made by convection to the motion becomes greater, it gradually overwhelms the motion due to frictional mixing. When this occurs, the equation for eddy diffusion does not give an adequate description of what is happening. Its use must therefore lead to lack of precision and introduce obscurities into the results.

The measurements made by Frankenberger, shown in Fig. 11, for a clear summer day, indicate that in the upper part of the diagram there is no evidence of increase in the coefficient A with increasing wind strength. On the contrary, when winds are light its value is greater than for a strong wind. This is due to the influence of convection, because when winds are light the ground is strongly heated and the effect of this heating dominates the whole mixing process. As the wind strength increases, the value of A decreases to a minimum (m, Fig. 11), from which point onward frictional mixing again begins to predominate. The values of A for the strongest winds of 8 to 9 m/sec^{-1} are therefore situated approximately on the extrapolation of the night curves for frictional mixing.

Even today little is known about the interaction of frictional and convection mixing. It has been shown by C. H. B. Priestley [78], however, that the time during which the two processes are working together is only a transitional period of short duration (see also K. Brocks [112]). In the early morning, mixing is at first almost entirely frictional. As the elevation of the sun increases there is a sudden transition to convectional mixing. The more unstable the air is, the lower is the wind force at which this transition occurs; and the closer the observation point is to the ground, the earlier the change is observed in the morning. This may be expressed numerically (the Richardson number).

Both the surface wind speed, on which frictional mixing depends, and the intensity of convection have a maximum at midday and a minimum before sunrise. The austausch coefficient A, which depends on both these factors, shows a diurnal variation. Table 12 gives examples taken from four different investigations. The figures are averages for 2-hr periods. The first line gives the well-known series of results calculated by H. Lettau [76] from 5-yr measurements of the temperature in the lowest 100 m of the atmosphere, made by N. K. Johnson and G. S. P. Heywood (see Sec. 11). They apply to a height of 45 m above the ground, and show considerable variation in 2-hr values between 4 and 125 g cm^{-1} sec^{-1}. The curve of diurnal variation is smooth because of the extended period over which results were recorded. A. Baumgartner [437], using a new technique, measured the total heat exchange for the crown area of a young pine wood in the neighborhood of Munich on six clear days in July 1952. The values of A derived from his results are shown in the second line. R. Trappenberg [85] measured the way in which smoke spread from chimneys in a suburban area of Karlsruhe on eight windy days in the spring of 1954, using the method of O. G. Sutton mentioned in Sec. 9. These were by no means days of fair weather; the values of A derived (third line) are for a range of 12 to 24 m. The two latter sets differ by a factor of 100. This is attributable partly to the different conditions under which mixing took place, and partly to the methods of measurement. The Trappenberg figures are from measurements made of the motion of visible smoke particles, and are certainly not influenced by the additional upward and downward thermal motion described on p. 39, which is necessarily included in all values of A derived from heat measurements.

The basis of the last line was a practical method of making a quick and simple relative measurement of the amount of mixing near the ground. This method was developed by M. Halstead in Seabrook, New Jersey, to assist aircraft engaged in pest-control

Table 12

Diurnal variation in eddy diffusion.

Time of day (hour)	0	2	4	6	8	10	12	14	16	18	20	22	24
From temperature observations at a height of 45 m (After H. Lettau): A (g cm^{-1} sec^{-1})	21	5	4	10	48	125	94	34	28	31	46	40	
From calculations of heat budget (After A. Baumgartner): A (g cm^{-1} sec^{-1})	10	10	9	18	35	40	43	32	22	11	8	8	
From the spreading of smoke (After R. Trappenberg): A (10^{-2} g cm^{-1} sec^{-1})	31	22	23	71	84	86	91	75	75	34	19	16	
The smoke-puff method (After M. Halstead): z (sec)	-	-	-	29	17	14	13	13	15	25	-	-	-

operations in the use of the remaining daylight when atmospheric conditions had begun to settle down in the evening. A measured quantity (1.7 g) of a certain type of gunpowder was ignited on the ground under rigidly controlled conditions. The time z (sec) required for the dark cloud of smoke to become invisible, rising 1 to 2 m into the air and spreading out downwind, was measured by stopwatch. Contrary to expectations, measurements made by a number of observers agreed within ±3 percent. The times in seconds, obtained by this method in the "Great Plains Turbulent Field Program" (Secs. 25 and 26) at O'Neill, Nebraska, averaged for ten days in August and September 1953, are shown in the last line of Table 12 ([13], vol. I, p. 118). The short times for midday indicate vigorous mixing, while in the morning and evening it takes longer for the smoke puff to disperse. Unfortunately, this method can only give daytime values for A, since light is necessary for making the observation. (On diurnal variation see also Sec. 17.)

It has always been a matter of surprise that there is still a fair amount of mixing by night. At one time an explanation was put forward, based on research by A. Defant [67] and A. Schmauss [81], that this was attributable to convective descent or return convection, a term used to describe the vertical movement of dust particles. The dust is carried aloft by the wind during the day, and returns by night to the ground, descending partly because of its weight. The small solid particles of dust, thousands of which are often present in a liter of air, must cool the adjacent air by radiating, and this parcel of air, being strongly cooled, will sink (hence the paradoxical expression "cold convection"). However correct this may be in principle, quantitatively it is not of very great significance. Even at night frictional mixing is in operation, perhaps only as microturbulence, as H. M. Bolz [686] called it.

The study of turbulent mixing has today become a special branch of science. The simple and attractive basic ideas of Wilhelm Schmidt are certainly an excellent pictorial guide; but nature is seen here, as in many other fields, to be much more complicated than is apparent at first. All theoretical treatments of mixing are based on the idea that the motion involved is subject to the laws of probability, that it is molecular motion, as it were, on a larger scale, which is capable of numerical expression. But the very fact that the calculation of A for the same place and the same time can often produce different results, when different characteristics (heat, water vapor, dust content, mobility) are used in the computation, indicates that, in addition to the purely chance influences, there are others that are not subject to a probability law, and hence must be systematic influences. The turbulence elements often clearly acquire dimensions that place them beyond the scope of statistical treatment, and large-scale movement of layers controls the processes taking place in the surface layer as well.

On account of the lack of precision inherent in theories on mixing because of these facts (see p. 40), one welcomes the attempts now being made to measure the vertical flow of heat and motion directly, independently of any theory. This is being done in Quickborn near Hamburg by E. Frankenberger, and in Aspendale near Melbourne by C. H. B. Priestley. The trouble is that such investigations demand a very high degree of skill in observational technique and are extremely costly, since the data must be analyzed and evaluated by electronic computer. It is to be hoped that the results of these excellent investigations will soon be available to practical micrometeorology, which is obliged to use simpler and cheaper means of measurement.

9. Temperature Instability, Dissemination of Seeds, Smoke Pollution, and the Spreading of Gases as Problems of Eddy Diffusion

In the layer of air nearest the ground, several phenomena may be explained by eddy diffusion.

It should be remembered, in the first place, that all meteorologic elements are inherently unsteady, a fact that may be verified by observation when the measuring instruments are sufficiently free of lag and the time scale is sufficiently large. It is then possible to measure individual turbulence elements, and their special characteristics. Gustiness as a disturbance of wind speed and direction has already been mentioned (p. 35). It is possible to think of a gustiness of temperature in a similar way. Since there is little mixing close to the ground because of reduced wind speed and friction, the turbulence elements—when the surface is heated in the middle of the day—have greatly differing characteristics. It follows that temperature instability is particularly great in the air close to the ground.

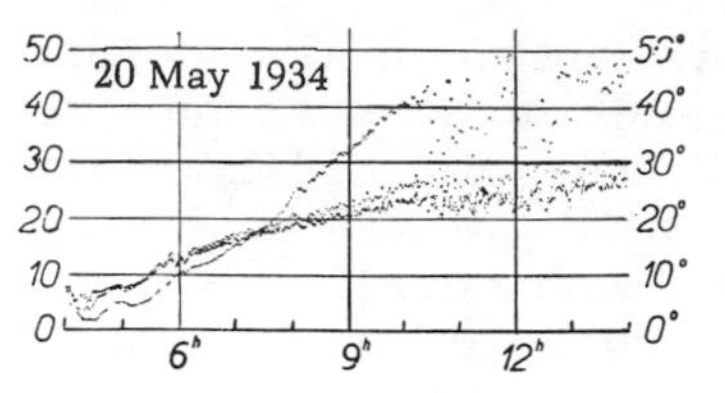

Fig. 13. Convection, setting in at 10:00 hr, leads to great temperature instability.

Figure 13 shows the temperature trace recorded over short grass on the Munich airport during a sunny forenoon in May 1934. Temperatures at heights of 200, 100, and 23 cm and from an electric thermometer lying in the grass were recorded every 20 sec with a point printer. At first the points join together to form a line; then about 10 a.m. convection sets in so vigorously that it looks as if there is a rising cloud of dots on the paper. This occurs because the variation at one position is greater than the difference between the calculable mean temperatures of two separate positions. The transition takes place suddenly, as shown by Priestley (see p. 41). It is possible to visualize the rising and descending currents mentioned on p. 39.

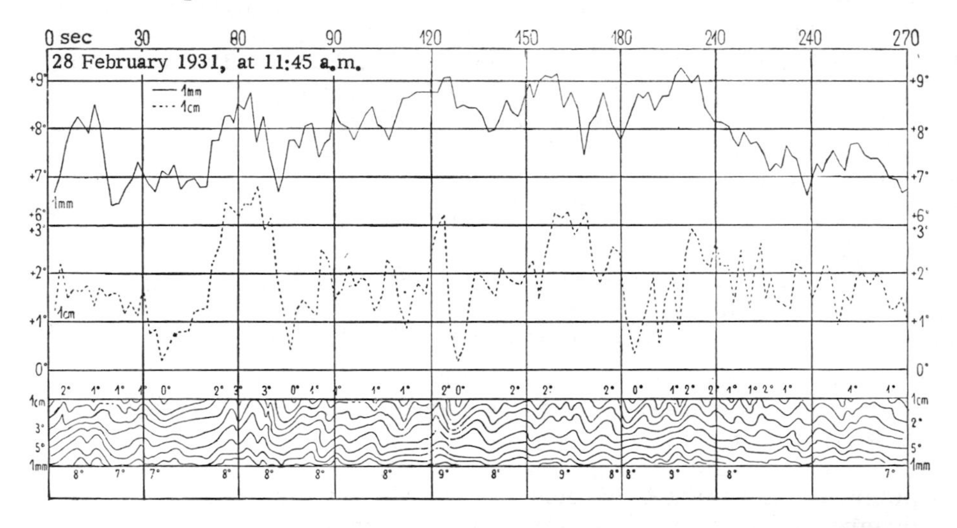

Fig. 14. Rapid recording of temperature over desert soil, using a single-strand platinum thermometer. (After W. Haude, in the Gobi Desert, 1931)

If recording is continuous, the variation can be followed directly. Figure 14 shows recordings made by W. Haude [70] over stony ground in the Gobi desert, with single wire platinum resistance thermometers. In an interval of 4-1/2 min, the air temperature at 1 mm (solid line) and at 1 cm (broken line) above the ground fluctuates up to 2 C deg on either side of the mean value, although the horizontal platinum wire, 10 cm in length, is recording the average of a large number of turbulence elements. The pattern of isotherms within the 9-mm layer of air is shown in the lower part of the diagram.

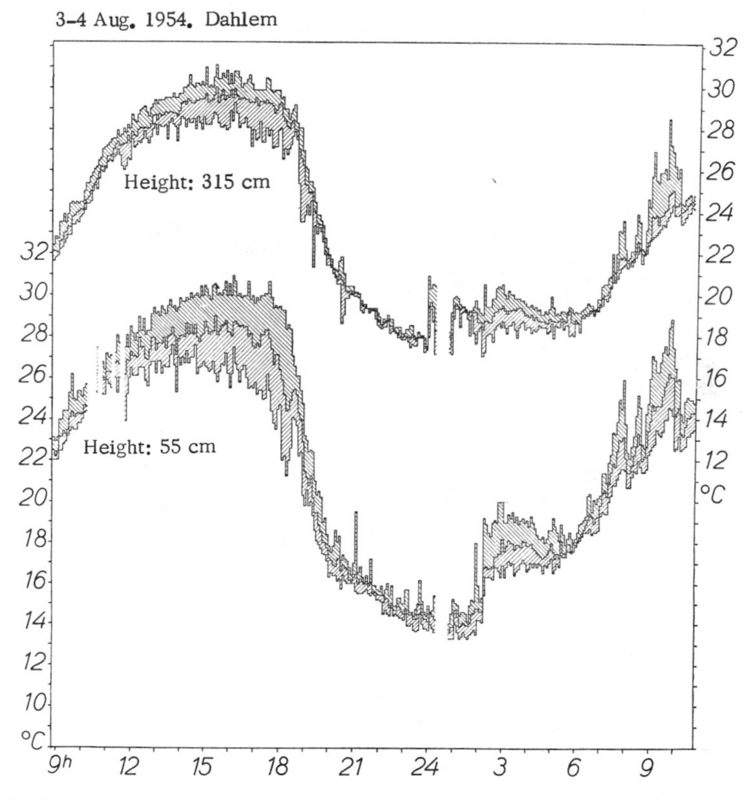

Fig. 15. Temperature instability during the day, at two heights above an alfalfa field. (After U. Berger-Landefeldt)

This instability of wind and temperature applies to other characteristics as well. W. C. Swinbank [84] has published data for four quantities, the vertical shear of wind, horizontal wind speed, air temperature, and specific humidity, during a 50-sec period under two different weather situations. Figures 15 and 16 show, by another method of representation, the instability of both temperature and water-vapor pressure on 3 and 4 August 1954. The measurements were made above a 50-cm crop of alfalfa in a garden in Berlin-Dahlem by U. Berger-Landefeldt, J. Kiendl, and H. Danneberg [66] by means of very rapid electrical recording of both elements with a remote-reading Askania measuring device. Its position at a height of 55 cm was just above the mean plant height. The average temperature (Fig. 15) and the average water-vapor pressure (Fig. 16) were determined by planimeter for 5-min intervals, and the highest and lowest instantaneous values within the 5 min were extracted. In this way the triple curves were obtained; the range of fluctuation above the mean value is shown by the area with hatching running downward from left to right, and the range below the mean by the area with opposite hatching.

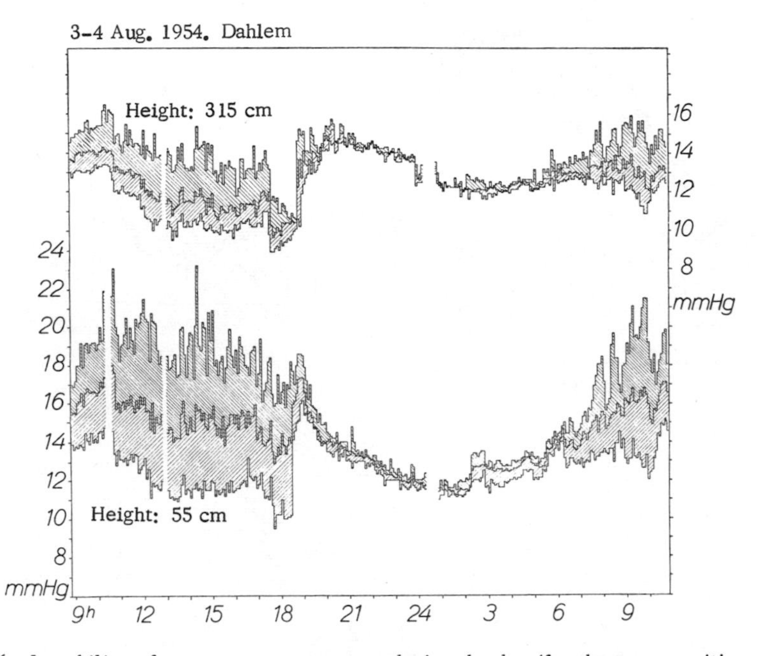

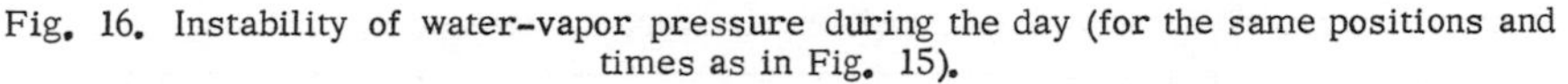

Fig. 16. Instability of water-vapor pressure during the day (for the same positions and times as in Fig. 15).

These illustrative diagrams show the increase of eddy diffusion and decrease of variability with height above the ground, which has already been mentioned. They also portray the marked diurnal variation, with great instability during the day and much more stable conditions at night. It is worth noting how the small increase in temperature at 2:10 hr was immediately accompanied by an increase in stability. The fluctuation in water-vapor content (note the scale) over the transpiring alfalfa crop is surprising. Within a 5-min interval it is greater than the total diurnal variation of the mean hourly values. The variation in humidity stops suddenly when the temperature falls quickly after 18:00 and, because less water vapor is given off to higher layers, the second daily maximum of water-vapor pressure is soon attained (cf. Sec. 15).

Another natural occurrence that can be explained only by eddy diffusion is the scattering of seeds, spores, pollen, and fruits. In a climate like that of Europe there is a greater lack of warmth than of water, and seeds are released when the air is dry enough for sufficient heat to be available, that is, when conditions are favorable for germination. This was found by L. Kohlermann [72] to be the case in the pine woods of two forests in Hesse, during a 7-week observation period. The scattering of seeds was found to be related to the relative humidity as shown in Table 13. The maximum fall of seeds in both cases was found to occur when the relative humidity was 55 to 65 percent. If the weather is more humid, the cone retains the seeds by swelling; drier days are rather rare.

Table 13
Seed dispersal and relative humidity (percent).

Relative humidity	100		75		65		55		45		35
Dudenhofen Forest Office		5		18		62		10		5	
Schlitz Forest Office		12		29		53		6		0	

If the only influences at work were the velocity of the wind u (cm sec^{-1}) and the rate of fall of the seed c (cm sec^{-1}), the trajectory would be a parabola similar to that followed by a bullet fired horizontally. However, the rate of fall varies from seed to seed. Kohlermann found, for example, that the rate of fall of *Populus nigra* seeds varied from 12 to 50 cm sec^{-1}, with a mean value of 26 cm sec^{-1}. This in itself provides a zone of scattering about the mean parabolic trajectory. No single seed, however, was able to exceed the height of its point of release.

Under a regime of eddy diffusion, each seed has an equal chance of being released into a rising or descending turbulence element. This leads to a broadening of the zone of scattering. The fact that each upward movement increases the time spent by the seed in the air means that, on the whole, upward movements have a greater effect than downward movements. S. Rombakis [80] was able to calculate a "probable flight path." This is defined by saying that a seed has an equal probability of being above this path or below it. Starting from the point of release with coordinates $x = 0$, $z = 0$ cm, the probable flight path rises to a height $z = Z$ (vertex) and descends to the level of release ($z = 0$) at a distance $x = X$ (range in the direction of the wind). The path is a parabola with the equation

$$z = 0.477\sqrt{\frac{4Ax}{\rho u}} - \frac{c}{u}x \qquad \text{(cm)},$$

in which A is the austausch coefficient (g cm^{-1} sec^{-1}) and ρ is the density of the air (g cm^{-3}).

If a normal air density of 0.0013 g cm^{-3} is assumed, the probable range (ignoring the solution $x = 0$, which gives the point of departure) is:

$$X = 700 \frac{Au}{c^2} \qquad \text{(cm)}.$$

The seed arrives there with a rate of fall of $\frac{1}{2}c$. The probable trajectory height is obtained by differentiating the equation of the parabola:

$$Z = 175 \frac{A}{c} \qquad \text{(cm)}.$$

The height of flight is therefore directly proportional to the austausch coefficient and inversely proportional to the rate of fall of the seed, but independent of the wind velocity.

The duration of flight τ is

$$\tau = 700 \frac{A}{c^2} \qquad \text{(sec)},$$

which is also independent of the wind speed, but more strongly dependent on the rate of fall. Table 14 gives numerical values for a number of seeds.

The shape of the trajectory is best seen from an examination of Fig. 17, which gives a few examples, for a moderate value of A (10 g cm^{-1} sec^{-1}) and a moderate wind speed (3 m sec^{-1}). For low plants, the height of release was taken as 0, and for trees z = 10 m. The heavy birch fruits do not get very far, but half the pollen grains of the Scotch pine travel as far as 1.5 km. The achenes of the dandelion remain airborne for over a minute under the conditions specified, and are strewn about 200 m away from the plant. The spores of mosses seem to wanter in the air at random. For example, the spores of ***Polytrichum*** will stay 1-1/2 days in the air under the conditions given in Fig. 17, rise higher than church steeples, and travel for 400 km. The spores of the puffball seem to be cosmopolitan, and must be present everywhere in the lower atmosphere.

The mathematical description of this scattering process has been proved experimentally by W. Schmidt. It is possible to do so, for example, by picking up airborne seeds over the sea at considerable distances from land, their distribution by kind being determined theoretically. Generally, observations confirm the theory, but many questions remain unanswered. Thus F. Firbas and H. Rempe [69] found to their surprise that pollen collected during flights at 2000 m and above did not agree with expected values for size and rate of fall. It is clear that the "chimneys" of the atmosphere suck up masses of pollen of all sizes, and that the rate of fall is of little significance. This is also the reason why sampling on night flights gives distributions by size that conform more closely to the theoretical. Another influence at work is the fact that a mass of pollen grains adhering to each other may later break up; and there are many other factors.

Table 14

Probable trajectory of airborne pollen, seeds, and fruits.

Plant	Rate of fall c (cm sec^{-1})	Range of flight X			Height of flight Z	Duration of flight τ
Austausch coefficient A (gm cm^{-1} sec^{-1})		50	20		20	20
Wind speed u (cm sec^{-1})		600	400	200	Independent	Independent
Puffball spores *(Lycoperdon)*	0.047	100,000 km	26,000 km	13,000 km	775 m	7-1/2 days
Clubmoss spores *(Lycopodium)*	1.76	70 km	19 km	9 km	21 m	68 min
Pollen of Scotch pine *(Pinus sylvestris)*	5.3	8 km	2 km	1 km	7 m	9 min
Achenes of dandelion *(Taraxacum officinale)*	10	2-1/4 km	580 m	290 m	360 cm	2-1/2 min
Fruit of birch *(Betula verrucosa)*	25	350 m	90 m	50 m	150 cm	23 sec
Fruit of spruce *(Picea excelsa)*	57	70 m	20 m	10 m	64 cm	4.5 sec
Fruit of fir *(Abies pectinata)*	106	20 m	6 m	3 m	34 cm	1.3 sec
Fruit of ash *(Fraxinus excelsior)*	200	5 m	2 m	1 m	18 cm	0.3 sec

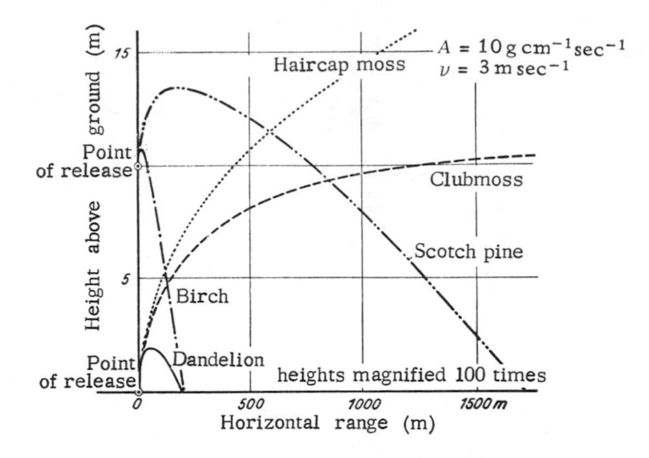

Fig. 17. Probable trajectories of seeds and pollen. (After the theory of S. Rombakis)

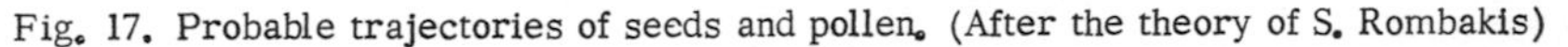

Dust consists of solid particles about 1 to 50 μ in diameter which make their way from the ground into the atmosphere. Gaseous products of natural radioactivity escaping from the ground and carbon dioxide are transferred from the ground layer by eddy diffusion into higher levels. These questions will be treated in Sec. 17.

Modern civilization has produced sources of atmospheric pollution situated at some height above the ground and continuously in operation. The most obvious of these are domestic chimneys and tall industrial stacks, which continually pour out great masses of ash and combustion products, causing a degree of smoke pollution injurious to health in thickly populated industrial areas (see also Sec. 52). The subject is well summed up by A. R. Meetham in his *Atmospheric Pollution* [77]. Investigations into the spread of dust from a large industrial undertaking have been made from a meteorologic point of view by M. Diem [68]. Methods of industrial production often make it impossible to prevent noxious gases or troublesome waste products from entering the atmosphere, although their concentration at the time of emission must be adjusted to the capacity of the atmosphere to disperse them. Disregard of this requirement has caused much damage to forests through the action of sulfur and fluorine compounds ([221]). Cement factories usually deposit large quantities of dust in their surroundings; recent figures have been published by W. Kreutz and W. Walter [74].

Finally, there is the danger of pollution by artificial radioactive contamination. This resulted from nuclear-weapon testing, and a potential threat is posed by the development of atomic-power stations. In every one of these cases the foreign particles reach the surface layer from above. There are two different processes involved. In one, it is a question of gases foreign to the atmosphere, or particles that have a negligible rate of descent and therefore spread in the same way as gases. In the other process, the bodies concerned are solid or liquid particles whose rate of descent cannot be ignored (as in the dispersal of seeds). For both of these, the rules developed by O. G. Sutton are the easiest to use in practice. In the following short treatment, we have essentially followed the excellent paper of H. Wexler [86] published in 1955 jointly by the U. S. Weather Bureau and the U. S. Atomic Energy Commission. This paper contains a bibliography of other literature on the subject.

A continuous source at height h (m) above the ground gives off a quantity Q of pollutant, measured in grams per second or, for radioactive sources, in curies per second (c $\sec^{-1}$). The distance x (m) from the source is measured in the direction of the wind, which has a mean velocity $\bar{u}$ (m $\sec^{-1}$). The coordinate y (m) is measured perpendicular to the line of drift, and hence gives the distance from the mean path. The concentration χ in the surface layer of air at the point (x, y) is given by the equation

$$\chi = \frac{2Q}{\pi C^2 \bar{u} x^{2-n}} \exp\left[-\frac{y^2 + h^2}{C^2 x^{2-n}}\right] \qquad (\text{g m}^{-3} \text{ or c m}^{-3}).$$

The state of the atmosphere at the time is indicated by the parameters C^2 and n. These may be defined by solving the equation; but it is more to the point if they are measured empirically, as has been done by E. M. Wilkins [87] by smoke-puff observations, using the Halstead method (p. 41). They depend mainly on the degree of stability or instability of the air, that is, they are functions of the vertical lapse rate of temperature γ (p. 38). For sources at a lower level (25 m) or a higher level (100 m), the figures given in Table 15 can be used as a point of departure.

R. Trappenberg [192] has made a statistical evaluation of several years of German wind records especially for use with the Sutton equation.

Table 15

Values of the parameters C^2 and n.

Atmospheric condition		Temperature gradient γ (approx.) (deg/100 m)	n	C^2	
				h = 25 m	h = 100 m
Unstable		-1.5°	0.20	0.043	0.015
Intermediate		-1.0°	0.25	0.014	0.005
Stable	inversion	+1.5°	0.33	0.006	0.002
Extremely stable	inversion	+3.0°	0.50	0.004	0.001

Figure 18 shows the different forms that the smoke plume from a chimney will take under different atmospheric conditions, as indicated on the left of the sketches by comparison of the actual lapse rate (solid line) with the dry adiabatic lapse rate (broken line). In unstable conditions, which normally prevail on bright summer afternoons, the smoke plume is strongly looped; high concentrations may reach down to ground level at times, and on the whole the smoke is broken up vigorously. Under average conditions, when the lapse rate is about −0.5 C deg/100 m, the smoke plume is in the form of a gradually widening cone. When temperature increases with height, as it normally does at night, there is very little scattering, and smoke from factory chimneys may be followed as far as 50 km. This is known as fanning. If the top of the stack reaches above the surface inversion at night, the smoke will settle at the level of the inversion, and the air near the ground will remain clear. This is called lofting. Should the top of the chimney be below the inversion level, however, all the smoke will be held in the surface layer, unable to escape through the inversion, and pollution will be at a maximum. This is fumigation.

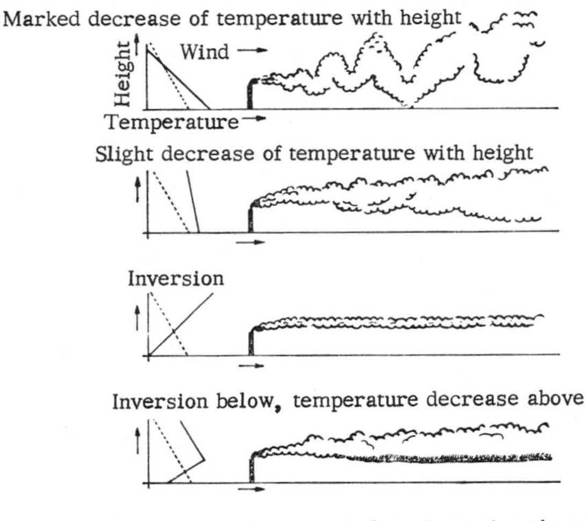

Fig. 18. Temperature stratification and the shape of smoke plumes. (After H. Wexler)

Sutton's equation is symmetric in y and h. The smoke plume can therefore be visualized as a cone of circular section with its apex at the source of smoke and its axis in the wind direction. The more stable the air, the more slender the cone. If the source is on the ground ($h = 0$) and only the most important concentration in the wind direction ($y = 0$) is considered, the exponent of e in the equation becomes zero, and the equation simplifies to

$$\chi = \frac{2Q}{\pi C^2 \bar{u} x^{2-n}}.$$

Figure 19 shows for these circumstances the dependence of the concentration χ on distance from the source. The heavy solid line is for the values of C^2 and n from line 1 of Table 15 for unstable conditions; the lighter continuous line is for the stable conditions of line 3 of the table. If the source is above 30 m, the distribution is shown by the two broken lines. From these curves, it can be seen that the height of the source is immaterial in unstable conditions beyond about 1 km, while in stable conditions it does not matter when the distance is greater than 6 km. The height of the source is of great importance, however, for its own immediate neighborhood. Thus it is possible to avoid completely having the highest concentration close to the ground (in this example, more than 10^{-4} g m^{-3}). This is the reason why it is right and necessary that smoke and gas should be allowed to escape only from the highest of chimney stacks. Figure 19 by no means shows the extremes of atmospheric conditions, which can in the one case cause much higher degrees of concentration, and in the other, effect much more rapid dispersal. The majority of atmospheric conditions will, however, lie within the two selected limits shown in the diagram.

If Q has a value different from 1 g sec^{-1} or 1 c sec^{-1}, as assumed in Fig. 19, the abscissa may be used in the ratio χ/Q (sec m^{-3}). The Sutton equation may be used for an instantaneous source as well as for continuous sources. Then the quantity Q becomes the

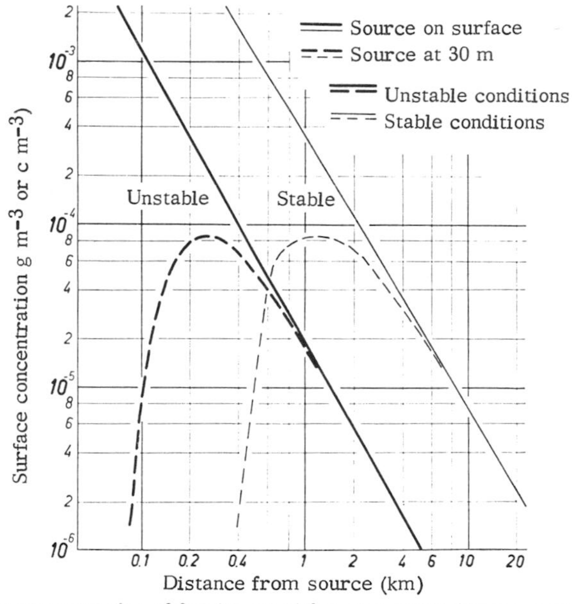

Fig. 19. Surface concentration of foreign particles at various distances downwind from a continuous source emitting 1 g sec^{-1} or 1 curie sec^{-1}, for two different atmospheric conditions. (From O. G. Sutton's equation)

quantity of smoke (g) or radioactive gases (curies) emitted. For example, in the case of an accident in an atomic reactor, where χ (g sec^{-1} m^{-3} or c sec^{-1} m^{-3}) is the dose rate at ground level, the ratio χ/Q is measured in seconds per cubic meter in this case also, and the diagram in Fig. 19 may be used.

When the emission is one of solid particles, such as soot, dust, or radioactive fission products, three new influences interact. First, the particles will tend to sink because of their own weight, at a rate dependent on their size, density, and shape (fallout), and this is the same for all particles only in the rarest of circumstances. Second, the particles combine with condensation nuclei and water droplets and arrive at the ground with precipitation before their expected time; this is called rainout. Finally, particles drifting through the atmosphere can be caught up in rain falling from clouds at a higher level; this is called washout. These processes are complicated, and difficult to assess numerically, but are already leading us away from our main interest, which is the air layer near the ground, and the reader is therefore directed to the bibliography [86] for further information.

CHAPTER II

THE AIR LAYER OVER LEVEL GROUND WITHOUT VEGETATION

10. Normal Temperature Stratification in the Underlying Surface (the Ground)

The intention in Chapter I was to describe the laws that govern processes in the air close to the ground. In this chapter attention is directed to these processes themselves. It is assumed, first, that the ground is completely horizontal, so that there are no topographic effects, and second, that the ground is free of vegetation, so that there are no plant effects. The development of temperature, humidity, wind, and other stratifications will be described for these conditions on the basis of available observations.

Because the ground surface, as pointed out in Chapter I, absorbs and emits radiation, evaporates and condenses water, and retards all air movement, the ground profoundly affects conditions in the overlying layer of air. The nature and state of the ground vary within wide limits from place to place and at different times. Moreover, the surface may be water, snow, or ice, instead of soil. Therefore the ground and its influence must be discussed more broadly, as the underlying surface. The influence of the underlying surface is so great that Chapter III will be devoted exclusively to this problem.

Without some knowledge of the processes that occur within the ground, the subsequent discussion of what takes place in the air layer near the ground could not be understood. For this reason, the temperature behavior in the ground itself will be considered first. Here, too, a simplification is made by assuming a "normal ground" under a homogeneous surface, having the same composition at all depths. The following results of observation fulfill this condition only approximately.

As was pointed out in Sec. 6, temperature in the ground is controlled by the temperature of its surface. This depends entirely on the prevailing weather conditions. The first and simplest example to be considered is a sunny summer day free of all disturbances. Figure 20 shows the first classical series of ground-temperature measurements made on the Wikkarais estate (60° 17′ N) in 1893, by the Finnish pioneer in microclimatology, T. Homén [98].

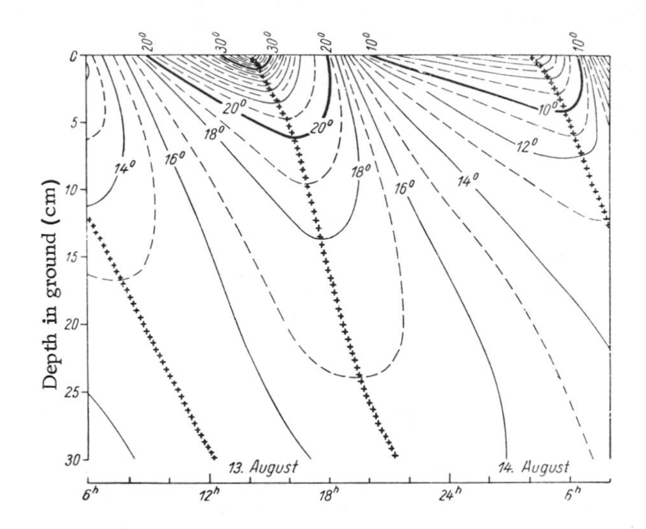

Fig. 20. Penetration of the daily heat wave into the ground on an undisturbed summer day. (From observations by T. Homén in Finland)

Three quantities must always be considered in the representation of ground temperature: depth in the ground (x), temperature (t), and time (z). The variation of temperature therefore corresponds to a surface in space, which may be represented on a plane in three different ways. One can draw isotherms in a coordinate system with z as abscissa and x as ordinate, x being vertical as in nature (see Figs. 20, 22, and 28). Isotherms within the ground are often called geotherms. Alternatively, one can make z the abscissa and t the ordinate and then draw the temperature pattern for selected depths (Figs. 23–25 and 27). Another method is to take t as abscissa and x as ordinate, using what are called tautochrones to show the variation of temperature with depth for a definite period of time (Fig. 21). Each of these three methods of representation has its particular advantages.

Figure 20 shows the typical temperature pattern on a summer day. In the early afternoon of 13 August 1893, surface temperatures above 34° C were recorded in sandy heathland; before sunrise the following day temperatures below 5° C were measured. This strong daily fluctuation decreases rapidly with depth. The heating to 20° C during the day penetrates only to a depth of 6 cm, and the night cooling below 10° C only to 4 cm. The deflection of the isotherms to the right with depth shows the lag in the time of penetration of the extreme temperatures into the ground, discussed on p. 32. The lines of crosses (+) join the points of highest and lowest temperature at the various depths, and therefore show the time lag quite clearly. In ideal homogeneous ground they are straight lines.

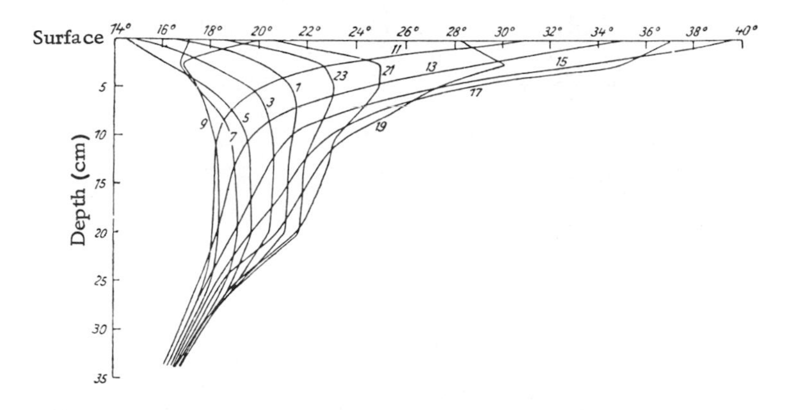

Fig. 21. Tautochrones of ground temperature on an undisturbed summer day. (After L. Herr)

Figure 21 shows the diurnal variations of ground temperature on a fair day in the form of tautochrones, using a more recent series of observations made by means of thermoelements. These were observed by L. Herr [97] in a plot of natural ground near the Geophysical Institute of the University of Leipzig. Readings were taken on 10 and 11 July 1934, for ten different depths in the ground; the temperature variation with depth shown here is for all the odd hours of the day. The tautochrones vary between two extremes, roughly defined by the 15- and 5-hr tautochrones. In the first case, with a strongly positive radiation balance, the maximum temperature of about 40°C is at the surface, and the temperature decreases at first quickly, then more slowly, with depth. This is called the incoming-radiation type. In the second case, with a negative radiation balance, the minimum is observed at the surface, and temperature increases with depth. This case is known as the outgoing-radiation type.

During the course of the day, the tautochrones move between those two extremes. Their pattern appears to be complicated by the fact that, in the intervening time, the heat at various depths in the ground may flow in different directions. For example, at 21:00 hr the highest temperature is recorded at a depth of 5 cm. Below this point the daytime heat still flows downward, but above it the upward heat flow already compensates to some extent for the loss of heat by the radiating soil surface. A contrasting picture is given by the tautochrones at 09:00 hr. The decrease of daily temperature fluctuations with depth is shown by the way the curves gradually converge. The fact that this bundle of curves below 20 cm runs upward from left to right only shows that this particular day occurred at a time when the weather was warming up.

It is worth mentioning here that the temperature of 40° C recorded in the upper layer of the ground is roughly that of a clear summer day in Central Europe. Under favorable conditions, as in poorly conducting soil and on southern slopes, a temperature of 60° C or more may be observed. (For more details see Sec. 20.)

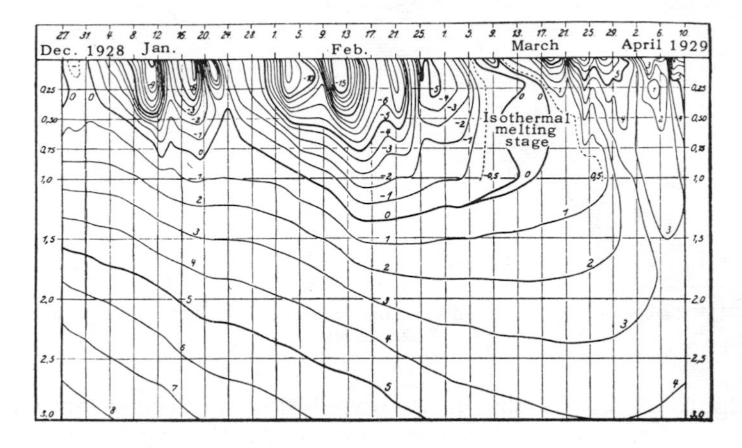

Fig. 22. Penetration of warm and cold waves into the ground in Potsdam in the winter of 1928-29.

The temperature of the ground surface responds more quickly than air temperature to changes in the weather. Even during the short break in solar radiation during the total eclipse of 30 June 1954, according to the observations of H. S. Paulsen [106] in Norway and K. Utaaker in Kleppe (60° 31′ N), the following drops in temperature were recorded at 13:33 hr during totality: 0.9 deg at 1 cm, 0.3 deg at 2 cm, and 0.1 deg at 5 cm below the surface. During the same eclipse, observed in the south of Sweden, B. Kullenberg [100], using thermistors shielded against radiation, observed a decrease in air temperature at a height of 10 mm above the ground from 24.6° C (1 hr before) to 15.8° C (at totality), and a rise again to 19.6° C (1 hr later). Figure 22 shows the unsteady behavior of ground temperature under the influence of changing warm and cold periods for the winter of 1928-29 in Potsdam, based on a sketch by J. Bartels. The fluctuations penetrate, at most, to a depth of 1 m. Below 1 m, as shown in Fig. 22, it is easy to see that the isotherms become steadier. It was for this reason that P. Lehmann [101] proposed that this first layer of the ground, subject to rapid temperature variations, should be called the base layer of the ground, in analogy to the first 1-1/2 km, or base layer, of the atmosphere. In these two base layers, all the diurnal changes and weather variations take place, with their repeated reversals of the lapse rate of temperature. The lives of most animals are spent in these two layers: above, birds, butterflies, and flying insects; below, mice, worms, and other insects. Because it is not waterlogged

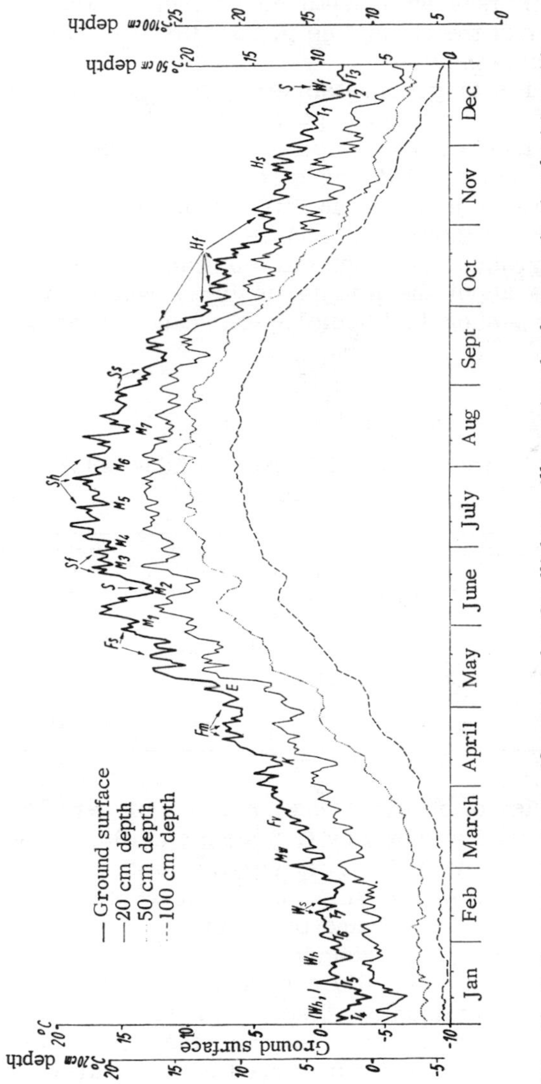

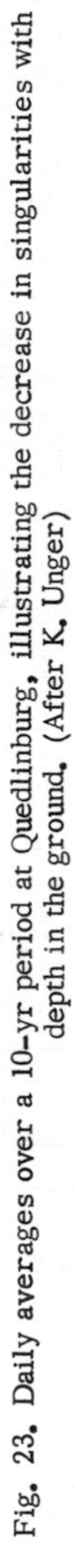

Fig. 23. Daily averages over a 10-yr period at Quedlinburg, illustrating the decrease in singularities with depth in the ground. (After K. Unger)

in its normal state, the base layer of the ground also contains the amount of air necessary to support life. Under the base layer lies the troposphere of the ground, which gives way, at a depth of about 20 m, to the stratosphere of the ground. In the ground stratosphere temperature increases with depth on account of the heat coming from the interior of the earth (see p. 26), just as it increases with height in the stratosphere.

A 10-yr series of ground-temperature observations made in Quedlinburg (51° 47′ N) for the period 1938-1947 has been evaluated by K. Unger [109]. The graph of the 10-yr daily averages for depths of 0, 20, 50, and 100 cm is shown in Fig. 23. This gives a very good picture of how singularities of the weather not only appear on the surface during the course of the year, but also penetrate into the ground as much as 50 cm. The cold spell that normally occurs about the middle of June, which in Germany is called the "sheep-shearing" cold spell, can be seen plainly even at a depth of 1 m.

Table 16

Soil temperatures at Potsdam, 1894-1948.
(After G. Hausmann)

Depth in ground (cm)	Average annual		Average annual fluctuation	Absolute annual fluctuation	Average time of year	
	Maximum	Minimum			Maximum	Minimum
100	20.7	1.0	19.6	25.4	30 July	11 Feb.
200	17.2	3.6	13.6	17.2	15 Aug.	4 Mar.
400	13.7	6.3	7.3	9.7	22 Sept.	3 Apr.
600	11.9	7.8	4.2	5.9	30 Oct.	4 May
1200	10.0	9.3	0.7	2.0	10 Feb.	10 Aug.

The long series of observations made in Potsdam from 1894 to 1948 have been used recently by G. Hausmann [95] to show the effects of large-scale weather irregularities and seasonal changes at greater depths below the surface. Table 16 shows the average yearly temperature extremes and the consequent yearly fluctuation. (Since all values have been rounded off to the nearest 0.1°C, the fluctuation may, at times, differ by 0.1 deg from the difference of maximum and minimum.) Also shown is the absolute yearly variation, which is the difference, over the whole 55-yr period, between the highest and lowest observed temperatures. This gives us some idea of the kind of temperature variation that is possible at such depths in Central Europe. The time of arrival of the extreme at these depths also shows the delay in penetrating the ground. The warmest time of the year at a depth of 12 m is exactly the same as that when the winter cold has penetrated to a depth of 1 m.

After considering the marked influence of changing weather on temperature in the ground, one may gather that the elegant graphs of the daily and yearly variations of ground temperature used in physics and meteorological textbooks to confirm the laws mentioned on pp. 31f are only the numerical results of long series of observations made for practical reasons in a piece of ground that had artificially been made more uniform. Despite this limitation, these well-balanced series of observations can be used to establish the relation between the yearly and the daily variations of temperature with the time of year and depth in the ground.

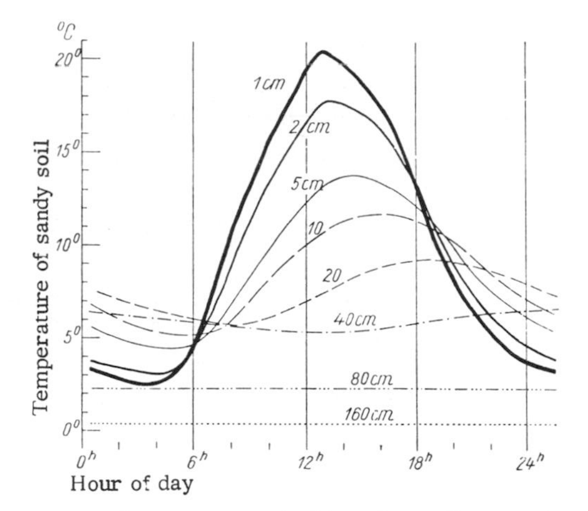

Fig. 24. Daily sequence of temperature in May, from 10-yr averages at Pavlovsk in a sandy soil. (After E. Leyst)

Figures 24 and 25 show the daily variation in two extreme months in sandy soil in Pavlovsk (59° 41′ N) for ten-year observations evaluated by E. Leyst [102]. The temperature at 1 cm (Fig. 24) responds to the sun's radiation much more quickly than the air temperature. Since the maximum of air temperature normally occurs between 14:00 and 15:00, it may be assumed that the ground-surface temperature follows the radiation curve almost without any delay. Although the ground is warming up in May from the heat of spring, the deeper layers at a depth of 80 and 160 cm are still cold. The reverse is the case in January (Fig. 25) when the ground at a depth of about 1 m is still unfrozen under a surface that has been kept free of snow.

The daily fluctuation of temperature varies with the time of year, as is shown by a comparison of the month of May with January. The daily fluctuation is unusually narrow at the high latitudes of Pavlovsk in winter, while in late spring it is very wide. The systematic influence of the time of year is shown most clearly in the uppermost layer of the ground, at a depth of 1 cm, and this is

Table 17
Extract of temperature measurements over 10 years at Pavlovsk.
(After E. Leyst)

Month	Temperature (°C) at a depth of 1 cm every 2 hr, for 2-hr periods												Diurnal fluctuation		Daily mean tempera- ture (°C)
	02	04	06	08	10	12	14	16	18	20	22	24	Periodic	Aperiodic	
January	-12.4	-12.4	-12.4	-12.5	-12.1	-11.1	-11.0	-11.5	-12.0	-12.2	-12.3	-12.3	1.7	5.8	-12.0
February	-13.2	-13.3	-13.3	-13.2	-12.0	-10.1	-9.4	-10.4	-12.0	-12.8	-13.1	-13.2	4.0	6.4	-12.2
March	-12.1	-12.6	-12.9	-12.4	-9.2	-5.8	-4.0	-5.0	-8.0	-9.4	-10.3	-11.0	8.9	10.2	-9.4
April	-0.8	-1.3	-0.7	2.8	7.8	12.7	12.9	10.3	5.9	2.3	0.9	-0.1	14.3	16.0	4.4
May	2.8	2.5	4.7	10.5	15.3	19.3	19.7	17.7	13.4	8.2	5.3	3.8	17.8	20.4	10.3
June	7.7	7.6	11.2	16.6	21.6	25.6	26.6	24.8	19.9	14.3	10.8	9.1	19.2	21.6	16.3
July	11.0	10.7	13.5	18.8	23.4	25.6	26.4	24.6	21.4	16.9	14.0	12.3	15.7	17.7	18.2
August	11.2	10.8	11.7	16.5	21.1	23.7	24.5	22.6	18.9	15.3	13.1	11.9	14.1	16.0	16.8
September	6.8	6.4	6.2	9.0	13.6	17.0	17.2	15.0	11.3	9.0	8.0	7.1	11.3	13.8	10.6
October	1.1	1.2	1.2	1.7	3.5	5.5	5.6	4.0	2.4	1.5	1.0	0.8	4.6	7.1	2.4
November	-2.9	-3.0	-2.9	-3.0	-2.0	-1.2	-1.2	-1.9	-2.5	-2.5	-2.7	-2.8	1.8	3.4	-2.4
December	-9.0	-9.0	-9.0	-9.1	-8.8	-8.3	-8.4	-8.8	-9.0	-9.2	-9.3	-9.4	0.9	4.8	-8.9

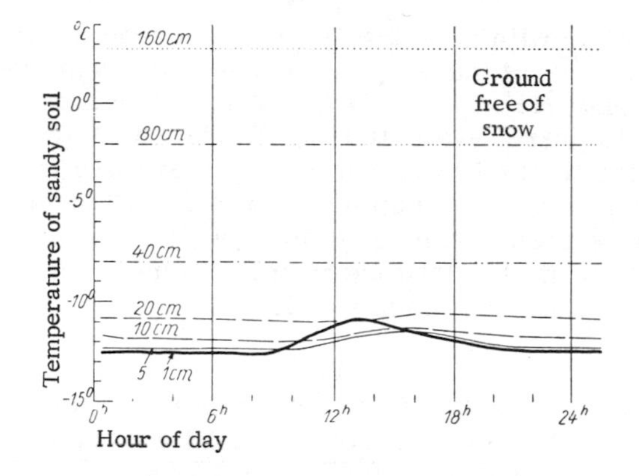

Fig. 25. Daily sequence of temperature in sandy soil in January, from 10-yr averages at Pavlovsk. (After E. Leyst)

illustrated by Table 17, which gives the temperatures every 2 hr for every month throughout the year. The difference between the warmest and coldest hourly averages is called the periodic daily fluctuation of temperature, and the difference between the average daily maximum and minimum is called the aperiodic fluctuation. Both of those are found in the table. At the same time, the daily averages shown in the last column of the table illustrate the yearly variation at the 1-cm depth.

Unfortunately, there are no sets of observations available for Central Europe extending over a long period of years. However, I. Dirmhirn [93] made a year of records with a remote-reading thermograph in 1951 in the garden of the Vienna Central Institute (48° 15′ N). These readings provide a valuable point of departure for practical discussion. The cylindrical thermometer holder was 17 cm in length and 2 cm in diameter, and was three-quarters buried in the ground. The readings were for the temperature at 1-cm depth, as verified by comparative measurements with thermoelements. The site was partly shaded by high trees, a fact which affected the results. Table 18 shows the monthly average of daily maxima and minima and the fluctuation deduced from these readings.

Table 18
Monthly average temperatures 1 cm below the surface, Vienna.

Month	I	II	III	IV	V	VI	VII	VIII	IX	X	XI	XII
Maximum	1.7	8.0	10.3	24.8	30.2	32.1	34.9	36.5	30.4	18.9	8.3	4.3
Minimum	-0.7	-0.3	0.2	6.4	11.2	13.0	15.2	15.0	12.6	6.0	2.6	0.0
Diurnal fluctuation	2.4	8.3	10.5	18.4	19.0	19.1	19.7	21.5	17.8	12.9	5.7	4.3

A comparison with an entirely different type of climate is given by the measurements made by W. Haude in the Gobi Desert in sandy soil at Ikengung (41° 54′ N, 107° 45′ E; elevation about 1500 m). These results were evaluated by F. Albrecht [88]. The figures (Table 19) give the daily variation of temperature in the uppermost layer of the ground to a depth of 0.5 m for 12 sunny August days. The boldface extreme values show how the extreme was delayed by depth. Even at this latitude in midsummer, the temperature at a depth of 2 mm did not reach 50°C.

Table 19
Daily sequence of ground temperature at Ikengung (Gobi Desert) from 1 to 12 August 1931.

Depth (cm)	0	2	4	6	8	10	12	14	16	18	20	22	24
0.2	18.5	17.4	16.9$^+$	18.9	30.3	40.7	46.5	45.0	37.0	28.5	23.0	20.1	18.4
0.6	18.7	17.6	17.2$^+$	19.1	30.3	38.9	42.7	44.2	37.0	28.9	23.5	20.4	18.6
5	21.4	20.0	19.6$^+$	19.8	23.9	30.1	34.7	35.4	33.6	31.0	26.9	23.7	21.3
10	24.9	23.7	22.8	22.4$^+$	23.4	25.4	28.2	30.5	31.0	29.9	28.4	26.3	24.8
25	26.3	25.5	25.0	24.4	24.0	23.9$^+$	24.4	25.4	26.5	27.4	27.7	27.1	26.2
50	24.7	24.7	24.7	24.6	24.4	24.2	24.1	24.0$^+$	24.1	24.3	24.5	24.6	24.7

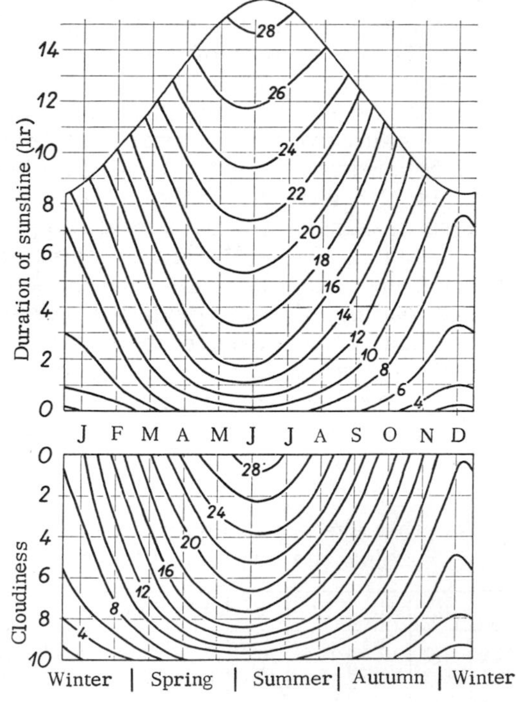

Fig. 26. Variation of diurnal fluctuation of temperature at a depth of 1 cm in the ground with duration of sunshine and with cloudiness in Vienna. (After I. Dirmhirn)

In addition to the seasonal influence, the daily range of temperature is also affected by cloudiness. This was evaluated by I. Dirmhirn [94] for the previously mentioned series of measurements made in Vienna, by supplementing them with a comparable series of measurements made in an unsheltered position. The results are shown in Fig. 26. The upper diagram shows the daily variation of temperature at a depth of 1 cm throughout the year as a function of the duration of sunlight. The lower diagram shows this variation as a function of cloudiness. Because of changes in the declination of the sun, the duration of sunlight possible at any time is a function of the season, and this forms the upper limit of Fig. 26. On a cloudless day in midsummer, the fluctuation of temperature in a day exceeds 28 deg, whereas on a cloudy winter day it does not even reach 2 deg. With decreasing cloudiness or increasing duration of sunshine, the fluctuation increases at first rapidly, and then more slowly. This is attributable to the moderating effect of convectional mixing (see p. 39).

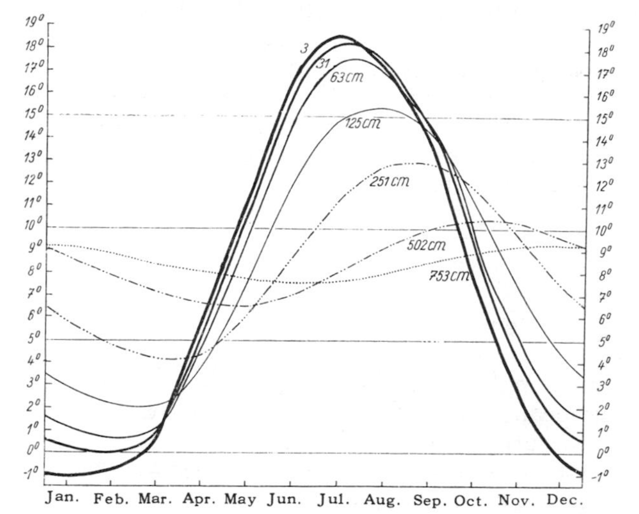

Fig. 27. Annual sequence of temperature at Königsberg. (After A. Schmidt and E. Leyst)

We now turn our attention from the daily pattern of temperature to the variations taking place throughout a year. Figure 27 shows the well-known series of observations made from 1873 to 1877 and from 1879 to 1886 at Königsberg (54° 43′ N) by A. Schmidt [107] and E. Leyst [103]. Although the curves show the damping of the annual wave of temperature with depth in such a classical form that they might seem to have been constructed from theory, they are only a reflection of actual observations. One might conclude from the positions of the curves that the mean annual temperature scarcely alters with depth. A study of the figures showed a small increase of temperature with depth, which corresponded approximately with that of the geothermal depth mentioned previously (p. 26). This, however, is not always the case. The 10-yr measurements made in Pavlovsk, examples of which have already been shown (Table 17), indicate that from 5 to 320 cm there is an increase of the annual mean temperature from 4.3 to 6.1° C, which gives a rate of change of 1 deg in less than 2 m. In contrast, the annual mean temperatures for Vienna (Table 18) give a decrease with depth.

At one time, much shrewd and penetrating thought was expended in the search for an explanation of this difference in the behavior of the annual mean temperature. Under stationary conditions, if one were to disregard the possibility of a systematic yearly change in the conductivity of the soil, the variation in the annual mean temperature would have to be a function of the geothermal depth. Otherwise, there would have to be a steady stream of heat downward or upward, and this cannot be reconciled with the heat economy. One might try to explain the observations by pointing out that the temperature of precipitation also influences ground temperature (see pp. 10 and 27), or that a conductivity is a function of the season. However, all these different results are easily explained since G. Hausmann [95] made his investigations into the effect of aperiodic fluctuations of weather on the ground temperature in Potsdam as already mentioned. Over the 54-yr period the annual mean temperatures varied as follows:

Depth (m)	1	2	4	12
Temperature variation (deg)	2.6	1.9	1.5	1.3

The graphs published by G. Hausmann of the variation of annual temperature between 1896 and 1949 for depths of 1 and 12 m may be used, for example, to deduce from a 3-yr range of observations the following change of mean temperature between 1 and 12 m: for 1907-1909, an increase of temperature with depth of 1 deg at 22 m, but a decrease of 1 deg at 12 m for 1947-1949. The first of these periods was a time of predominantly cold weather, especially in the cold winter of 1908-09. The years 1947 to 1949, however, were warmer, with a particularly dry summer in 1947. More recently, W. Müller [105] has demonstrated the parallel behavior of temperature in the air and in the ground under the influence of weather, by means of nine series of observations in Europe from 1911 to 1955. It is not permissible to draw conclusions about the conditions prevailing under the ground from measurements of the geothermal depth made over a limited period of time, as has already been attempted.

In conclusion, let us consider the measurements made at the Vienna Central Institute over the 33-yr period from 1911 to 1944, and recently published by M. Toperczer [108]. These are of considerable practical importance because, unlike earlier measurements, they were not made on an experimental plot kept free of snow and shaded from direct sunlight, but in the Institute gardens, which are filled with flowers and bushes and therefore subject to varying sunshine, as is generally the case in practice. Figure 28 gives the pattern of temperature for the year. The winter mean over the long period of years indicates only a slight degree of frost in the ground, protected as it was by the snow cover, and in midsummer the temperature never rose as high as 24° C. The relation to the temperature in the instrument shelter, approximately 2 m above the ground, is shown by the broken isotherms at the top of the diagram. It is recognized that the yearly variation of temperature in the air is more moderate than at the surface of the ground, but Fig. 28 tells nothing about what is happening in the 2-m layer of air immediately above the ground (see Sec. 12).

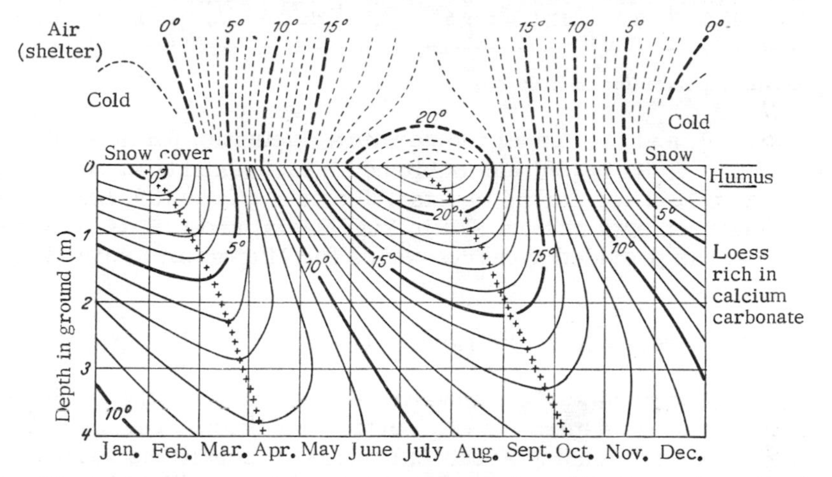

Fig. 28. Annual sequence of temperature in the gardens of the Vienna Central Institute. (After M. Toperczer)

Table 20 contains a few numerical values. In addition to the yearly mean and the yearly variation, temperatures shown for the extreme months of January and July—which were the warmest and the coldest mean monthly temperatures in the 33-yr series of observations—and the difference between them. The fact that in January the ground is protected by a snow cover is evident. Temperature increases with depth, but the difference between the extreme months decreases, and in fact does so the same way as in summer. The difference between the extreme values of the annual means are shown in the last column.

We are indebted to Jen-Hu Chang [92] for a comprehensive evaluation of ground-temperature measurements over the whole earth, and for having published world atlases of ground temperature at depths of 10, 30, and 120 cm for January, April, July, and

Table 20
Ground temperatures (°C) in Vienna. (After M. Toperczer)

Depth in ground (cm)	Annual mean temperature (1911-1940)	Annual fluctuation (1911-1940)	During the whole period of observation 1911-1944						
			January			July			Difference of extreme annual means
			Monthly mean Highest	Monthly mean Lowest	Difference	Monthly mean Highest	Monthly mean Lowest	Difference	
*	9.38	19.7	4.8	-9.3	14.1	21.6	16.1	5.5	3.6
About 1	10.89	23.3	3.4	-4.1	7.5	25.2	19.3	5.9	2.8
52	10.65	20.2	4.0	-0.5	4.5	22.7	18.3	4.4	2.4
98	10.51	15.7	5.7	2.9	2.8	19.6	16.5	3.1	2.5
193	10.36	9.8	8.5	6.1	2.2	15.2	12.8	2.4	1.3
288	10.30	6.2	10.2	8.4	1.8	12.5	10.9	1.6	1.1
377	10.26	4.1	11.1	9.6	1.5	11.0	9.7	1.3	1.3

*Shelter temperature (for comparison)

October [90], and the annual temperature fluctuation for those three depths [91]. E. Batta [89] has published measurements at Budapest for 1912 to 1941 for depths of 2 to 400 cm. P. Katić [99] has compiled measurements at Novi Sad in Yugoslavia, and J. S. G. McCulloch [104] has provided a yearly series for Muguga in Kenya (1° S).

11. Temperature in the Lowest 100 m of the Atmosphere

From the ground surface upward the temperature of the air layer near the ground is determined by surface conditions, regardless of effects of the surrounding area. In contrast to conditions within the ground, the role of molecular conduction is insignificant as compared with eddy diffusion, which has already been discussed on p. 34. There are other factors in addition to mixing, such as long-wavelength radiation, breathing of the ground (p. 27), the influence of the boundary layer (p. 34), and many others, all of which make atmospheric processes difficult to follow and difficult to evaluate. For this reason it is preferable to disregard momentarily what occurs within the air layer near the ground in the narrower sense of the term and to approach this problem, as it were, from above by first considering the processes taking place in the lowest 100 m of the atmosphere.

A great deal of information is available from measurements made over many years with electric resistance thermometers or thermoelectric devices. Many of these observations over the last 30 yr have been analyzed carefully and exhaustively. As long ago as 1896, A. Angot used temperatures recorded on the Eiffel Tower in Paris, which is 330 m high, to investigate the pattern of temperature at various heights above the ground. An important step forward in this much-discussed problem was taken in 1929 by N. K. Johnson [117] by his analysis of measurements made in England over the period 1923-1925 with modern electric measuring instruments. The original series of measurements made in England was extended to Egypt and India and to greater heights, and both instruments and methods of evaluation were improved. Table 21 contains a summary of these evaluated measurements extending for a period of at least 1 yr.

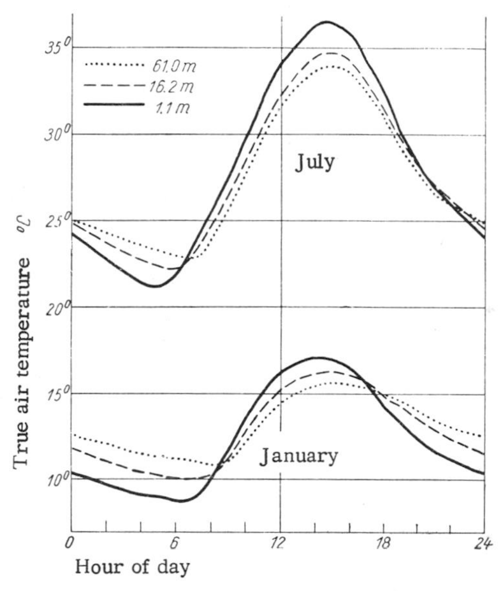

Fig. 29. Temperature sequence in an air layer 61 m thick over desert soil. (After W. D. Flower)

Figure 29 gives the daily pattern of temperature averaged over two extreme months, observed by W. D. Flower [115] at heights between 1 and 61 m above the ground in Egypt. The curves show how the daily fluctuation of temperature varies with distance from the ground, and how the extremes are delayed, an effect already seen in Fig. 24. One must always bear in mind the vastly different orders of magnitude involved. Between 1 and 61 m, the daily fluctuation of temperature in July decreases from 15.4 to 11.1 deg, as shown in Fig. 29. By careful analysis Flower established that the delay in the extreme caused a displacement of the time of maximum temperature from 14:55 hr at 1 m to 15:33 hr at 61 m. This measurement allows the temperature conductivity to be calculated from the equations given on p. 32, to obtain values of more than 10^4 cm^2 sec^{-1}. Table 10 shows, however, that the thermal diffusivity of air completely at rest—that is, an atmosphere which

Table 21
Analyzed observations of air temperature in the lowest 100 m of the atmosphere, for series of at least 1-yr duration, published since 1928 (arranged according to height).

Highest point of measurement (m)	Intermediate measurement levels (m)	Author, year of publication [reference no.]	Place of observation (surface)	Latitude, longitude, altitude (m)	Period analyzed	Elements analyzed	Method of evaluation (see note below)
106.7	47.2; 15.2 and 1.1	A. C. Best, E. Knighting, R. H. Pedlow K. Stormonth 1952 [111]	Rye, England (grassland)	50° 58' N 0° 48' E 4 m	3 years (1945–1948)	Temperature and humidity	YDT, YDG, FG, MI, Cloud, SF, SS, and calculation of absolute humidity
87.7	57.4; 30.5; 12.4 and 1.2	N. K. Johnson and G. S. P. Heywood 1938 [118]	Leafield, England (grassland)	51° 50' N 1° 34' W 186 m	5 years (1926–1930)	Temperature (wind at 94.5 and 12.7 m)	YDT, YDG, FG, MI, Cloud, NI, SF, WS
76.0	No fixed position of measurement, meteorograph raised and lowered every 15 min	J. Rink 1953 [121]	Lindenberg (meadow)	52° 13' N 13° 48' E 98 m	1 year (7842 hr of recording, 1950–51)	Temperature (and humidity)	YDG, FG, Mi (between 1 and 76 m), Cloud, relation between incoming radiative energy and air mass
70.0	28.0; 13.0 and 2.0	E. Frankenberger 1955 [445]	Quickborn, Holstein (damp meadow with hedges)	58° 44' N 9° 53' E 12 m	1 year (1953–54)	Temperature, humidity, and wind	YDT, YDG, FG, Cloud, humidity, wind and heat budget calculations
61.0	46.4; 16.2 and 1.1	W. D. Flower 1937 [115]	Ismailia, Egypt (desert almost without vegetation)	30° 36' N 32° 16' E 16 m	1 year (1931–32)	Temperature (wind at 62.6 and 15.2 m)	YDT, YDG, FG, MI, Cloud, NI, SF, WS

47.5	17.1 and 1.2	S. Mal, B. N Desai and S. P. Sircar 1942 [119]	Drigh Road, near Karachi (sandy desert)	24° 54' N 67° 08' E (about 20 m)	1 year (1930–31)	Temperature	YDT, YDG, FG, Cloud, NI, SF
17.1	7.1 and 1.2	N. K. Johnson 1929 [117]	Porton, Salisbury Plain (grassland)	51° 08' N 1° 44' W 111 m	3 years (1923–1925)	Temperature	YDT, YDG, FG, MI, Cloud, NI
1.2	0.3 and 0.025	A. C. Best, 1935 [124]			2 years (1931–1933)	Temperature (wind at 6 heights up to 2 m)	YDG, FG, Cloud, WS, investigations of heat economy and turbulence
10.0	5.0; 1.0; 0.5; 0.1 and 0.01	H. Henning 1957 [116]	Lindenberg (meadow)	52° 13' N 13° 48' E 98 m	1 year (1953–54)	Temperature (wind at 5 heights)	YDG, Cloud, NI, WS

Method of Evaluation:

- YDT — Tables of the yearly and daily sequences of temperature at all heights of measurement for every hour of every month (either directly or indirectly by giving tables for the lowest position of measurement and tables of lapse rates)
- YDG — Tables of the yearly and daily sequences of temperature lapse rate, for all layers and for every hour of every month.
- FG — Frequency statistics for all lapse rates found, for every month and for every layer.
- MI — Maximum values for inversions, mostly with explanations of individual cases.
- Cloud — Investigation of the influence of cloudiness on temperature lapse rates, mostly by analysis of clear and overcast days, and mostly by summer and winter examples.
- NI — Special investigations of formation and destruction of night inversions.
- SF — Special relations in foggy situations.
- SS — Special relations with snow cover.
- WS — Connection between temperature lapse rate and wind speed.

transports heat only by molecular conduction—is of the order of 10^{-1}, whereas that of solid earth is 10^{-3}. By using the given values for ρ and c, one can calculate the thermal conductivity λ (see p. 30). From this it follows that, for the place and time in question, as a result of mixing, the conductivity of heat and thermal diffusivity of the air have been increased 100,000-fold. If air undergoing mixing is compared with dry, sandy soil, its thermal conductivity is 10^5 times and its temperature conductivity even 10^7 times greater. This example illustrates the overwhelming importance of mixing.

The values shown here are averages for the 60-m layer of air. Within this layer, there must be substantial differences, because the amount of mixing is not constant with height. Hence, the small amount of mixing that takes place close to the ground gives rise to a greater temperature gradient, while at higher levels more lively mixing gives rise to a smaller temperature gradient. This also means that the range within which the temperature gradient can vary will become greater, the closer it approaches the ground.

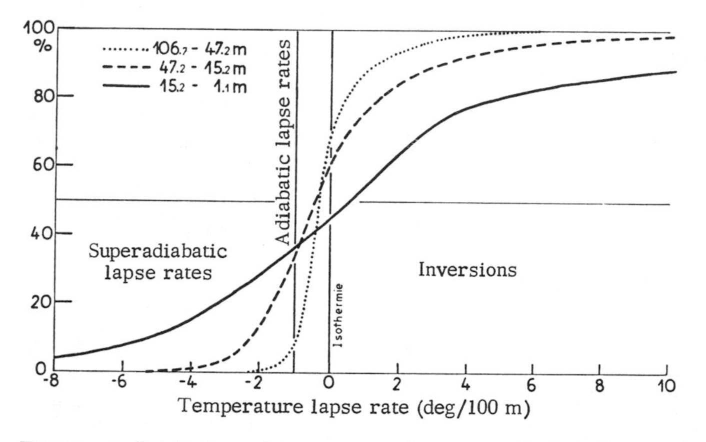

Fig. 30. Frequency distribution of temperature lapse rates in three layers of air below 107 m, from observations over three years in England. (After A. C. Best, E. Knighting, R. H. Pedlow, and K. Stormonth)

Figure 30 is based on the most recent and most comprehensive series of measurements made over 3 yr in Rye, Sussex, 5 km away from the south coast of England, by A. C. Best, E. Knighting, R. H. Pedlow, and K. Stormonth [111]. The abscissa is the temperature lapse rate γ (C deg/100 m; see p. 38). While the ordinate gives the frequency—that is, each ordinate gives the value of the summation curve or ogive of the percentage of all hours with a lapse rate below that of the corresponding abscissa value—the average value for γ in the free atmosphere is -0.5 deg/100 m. Thus in Fig. 30 the average γ lies between the adiabatic and the isothermal lapse rates. The uppermost layer of air (dotted curve) shows a superadiabatic lapse rate in only 8 percent of all cases, and these hardly ever exceed -2C deg/100 m. Inversions, on the other hand, are found in 32 percent of cases because of their stable stratification.

As the ground is approached, however, there is an enormous extension of the range within which the lapse rate may lie. In the highest layer observed, between 47 and 107 m as has just been shown, 60 percent of all gradients lie between the adiabatic and the isothermal, while between 15 and 47 m only 27 percent do so, and between 1 and 15 m only 9 percent.

During the 3-yr observation period the highest superadiabatic lapse rates occurred in the spring months of March to June. In the lowest layer (1 to 15 m) -20.5 deg/100 m was reached, always between 11:00 and 13:00 in clear weather. In the uppermost layer (47 to 107 m) the highest value was -4.2 deg/100 m. Such values were spread over a wider period of time and could also be observed in cloudy conditions. The strongest inversions were with clear skies, sometimes in conjunction with ground fog, at any hour during the night, and the maximum values in the lowest layer were 53.4 deg/100 m, and in the uppermost layer, 13.6 deg/100 m.

The magnitude of the temperature gradient increases as the positive or negative radiation balance becomes larger. J. Rink [121] has given in tabular form the relation during the day (positive balance) between mean hourly quantities of heat (recorded by a Moll-Gorczinski solarigraph) and the gradients. This effect may easily be observed from the daily and yearly variations of gradient.

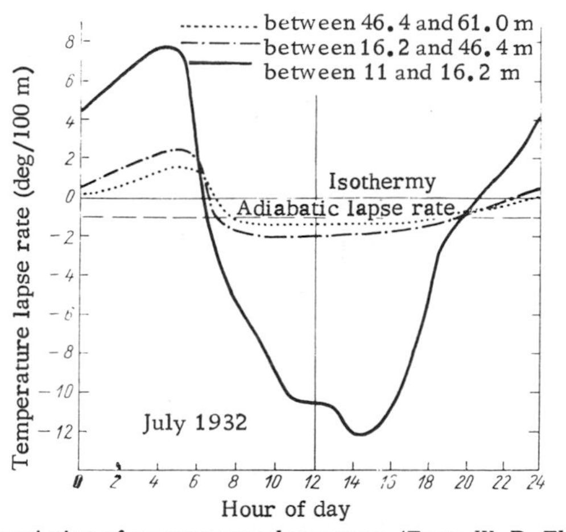

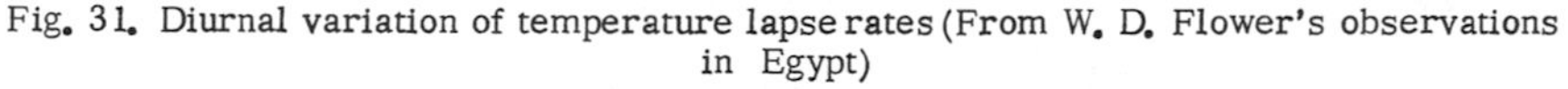
Fig. 31. Diurnal variation of temperature lapse rates (From W. D. Flower's observations in Egypt)

Figure 31 gives the daily pattern of temperature gradient in three layers at different heights, from the observations made by W. D. Flower [115] in Egypt. The values are for the cloudless month of July, and the rule that the closer it is to the ground, the greater the gradient, is fulfilled here. Before sunrise, which occurs shortly before 05:00 hr, the inversion is strongest. One hour after sunrise—and this is valid for all seasons of the year in Ismailia—the isothermal state begins to break down from below. In the lowest layer, superadiabatic gradients begin to increase substantially under the influence of increasing radiation; this is prevented in the middle and upper layers by the freer movement of the air. The initiation of convection allows the gradient curves shown in Fig. 31 to run almost horizontally above 16 m. Not until 1 hr, or at higher levels 2 hr, after sunset at about 19:00 hr is the isothermal state reached again. In contrast to conditions at sunrise, this time difference was seen to vary throughout the year, being greatest in summer and sometimes even negative in winter; that is, the isothermal state set in before sunset.

Table 22

Temperature lapse rate (C deg/100 m) for the air layer from 47.2 to 106.7 m, giving the diurnal and annual variation in the form of monthly averages for the three years 1945-1948 at Rye, England. (Corresponding humidity gradients are given in Table 31.)

Hour	2	4	6	8	10	12	14	16	18	20	22	24
January	0.26	0.31	0.17	0.18	0.03	-0.30	-0.38	-0.19	0.05	0.15	0.18	0.21
February	0.02	0.13	0.02	-0.06	-0.23	-0.53	-0.53	-0.47	-0.22	-0.16	-0.03	0.02
March	0.94	0.74	0.68	0.07	-0.68	-0.89	-0.74	-0.43	0.11	0.50	0.65	0.72
April	0.44	0.50	0.56	-0.21	-0.67	-0.91	-0.86	-0.76	-0.35	-0.11	0.24	0.26
May	0.74	0.64	0.23	-0.57	-0.68	-0.86	-0.72	-0.89	-0.42	0.27	0.64	0.90
June	0.72	0.64	0.15	-0.64	-0.64	-0.87	-0.52	-0.70	-0.41	0.21	0.66	0.74
July	0.94	1.00	0.63	-0.05	-0.11	-0.13	-0.03	0.02	0.18	0.49	1.00	1.02
August	0.90	0.89	0.70	-0.22	-0.50	-0.60	-0.59	-0.46	-0.25	0.15	0.57	0.81
September	0.61	0.58	0.67	0.21	-0.40	-0.48	-0.47	-0.39	-0.12	0.35	0.48	0.65
October	0.80	0.80	0.92	0.76	-0.17	-0.54	-0.53	-0.23	0.10	0.45	0.67	0.81
November	0.52	0.54	0.46	0.59	-0.05	-0.28	-0.27	-0.14	0.06	0.25	0.33	0.45
December	0.50	0.35	0.32	0.27	0.21	-0.15	-0.28	-0.09	0.07	0.18	0.22	0.35

In order to express numerically the variation of temperature gradient during the day and the year, we shall use figures from the investigation made by Best, Knighting, Pedlow, and Stormonth [111], since, at a later stage, we shall also use the humidity measurements made at the same time. Tables 22 and 23 give the mean

Table 23

Temperature lapse rate (C deg/100 m) for the air layer from 1.1 to 15.2 m, giving the diurnal and annual variation in the form of monthly averages for the three years 1945-1948 at Rye, England. (Corresponding humidity gradients are given in Table 32.)

Hour	2	4	6	8	10	12	14	16	18	20	22	24
January	6.1	6.6	5.6	5.2	1.6	0.1	0.4	3.7	5.7	5.2	6.0	5.5
February	3.9	3.2	2.9	1.8	-0.6	-0.6	-0.6	0.9	3.1	3.9	3.9	4.6
March	7.6	7.0	6.1	1.0	-2.6	-4.4	-3.9	-1.4	2.1	5.9	6.6	6.7
April	8.4	7.4	3.7	-2.2	-4.7	-5.5	-4.7	-2.5	0.2	5.2	7.3	8.3
May	6.2	6.9	0.2	-4.1	-5.6	-6.2	-5.6	-4.2	-1.6	3.3	6.0	6.5
June	7.2	6.1	-0.8	-3.1	-5.6	-5.6	-5.2	-3.1	-0.4	3.5	6.6	6.8
July	5.2	5.6	-0.7	-3.5	-5.4	-5.9	-5.5	-3.4	-0.4	3.2	5.7	6.2
August	5.6	5.2	0.3	-3.7	-5.6	-6.3	-5.5	-3.3	-0.1	4.6	6.5	5.8
September	6.8	6.3	4.4	-0.8	-3.7	-4.4	-4.1	-1.8	2.1	6.8	6.8	6.8
October	7.1	6.9	7.1	-0.1	-2.7	-3.4	-3.3	-0.6	5.0	8.1	8.4	7.0
November	3.7	2.8	3.2	1.8	-1.3	-1.5	-0.7	2.2	4.4	5.1	4.8	4.4
December	3.5	2.5	3.0	2.7	0.8	-0.8	-0.2	3.0	4.6	4.4	4.6	4.4

values of the gradient every 2 hr for each month of the year in the uppermost and lowest layers (47.2-106.7 m and 1.1-15.2 m). For the sake of uniformity and comparison, especially with later tables in Sec. 13, all gradients have been expressed in Celsius degrees per 100 m. For the small gradients at higher levels, the data are expressed in hundredths of a degree, corresponding to the accuracy of measurement, while for the larger gradients it is sufficient to express the data in tenths. Undoubtedly, even with 3-yr averaged values, chance effects of weather can still be noticed. Nevertheless, the order of magnitude of the gradients in both of these layers, and the change of sign from positive to negative, controlled by the time of year and the influence of increased radiation or reradiation can still be seen clearly. The influence of wind speed on the gradient will be given in Sec. 16 (pp. 120 to 124).

The graphs in Fig. 32, from the measurements of N. K. Johnson [117], show the effect of cloudiness in the lowest 17-m layer by contrasting bright and dull days of a summer and winter month, using a common temperature scale. In June the mean temperature in both situations is approximately the same, but the daily temperature

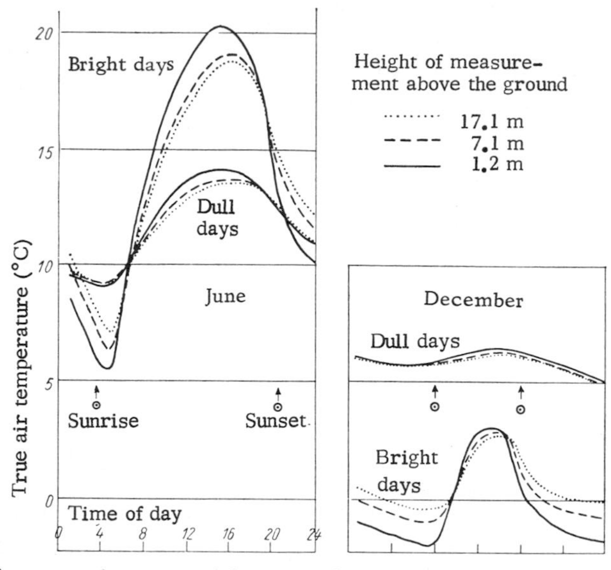

Fig. 32. Influence on the season of the year and the weather on the course of temperature near the ground. (After N. K. Johnson)

fluctuation, the temperature gradient, and the displacement of the maximum temperature with height are substantially greater on bright days than on dull days. In December the temperature level is higher and the gradients smaller on dull days than on bright days.

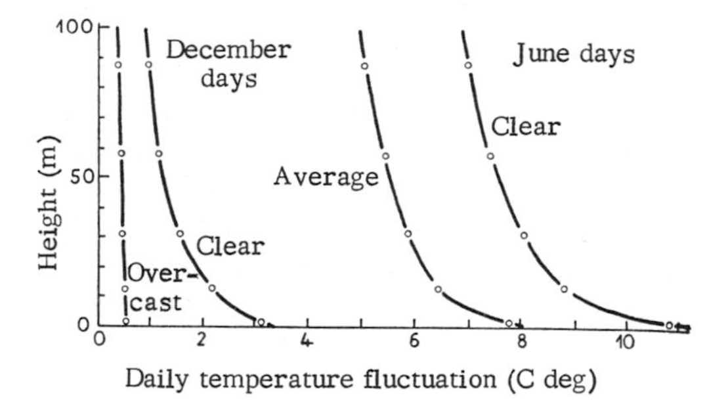

Fig. 33. Increase in temperature fluctuation as the ground is approached. (After N. K. Johnson and G. S. P. Heywood)

Table 24
Time (hours) of maximum temperature at various heights.

Height above ground (m)	1.2	12.4	30.5	57.4	87.7
December (mean value)	14.05	14.26	14.34	14.42	14.50
June (mean value)	14.55	15.35	15.30	16.06	16.20
June (clear days)	15.45	16.35	17.00	17.14	17.24

The increase of temperature fluctuation as the ground is approached is shown by Fig. 33 for a typical month and weather situation, from the results obtained by N. K. Johnson and G. S. P. Heywood [118] in Leafield. Table 24 gives the displacement of the time of maximum temperature with height, calculated from the averages over five years. The figures give the time in hours and minutes.

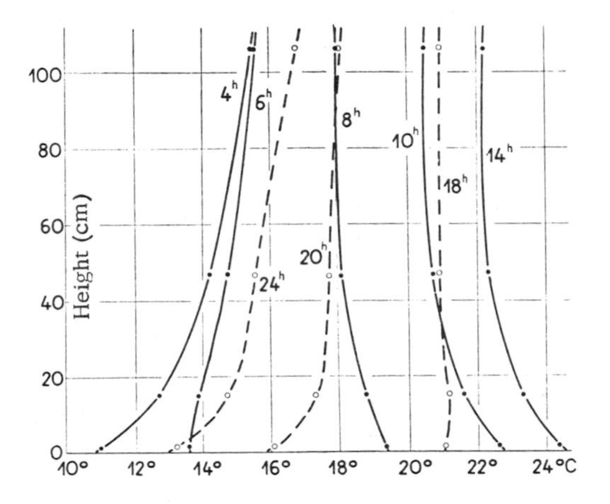

Fig. 34. Tautochrones of daily temperature sequence in the lowest 100 m of the atmosphere on clear summer days. (After A. C. Best *et al.*)

Finally, the tautochrones in Fig. 34 show the changeover from incoming-radiation to the outgoing-radiation type (see p. 57) for the first 100-m layer above the ground, once more using the measurements made in Rye, England. From the average values of 19 clear summer days, specially selected by the author from the data for three years, tautochrones have been drawn with full lines to show the change from the outgoing-radiation type at 04:00, to the steepening of the curve (06:00), through the isothermal stage and the quick setting-in of heating from below (08:00), to the incoming-radiation type at 14:00. The broken tautochrones show the transition in the opposite sense, which takes place during the afternoon and night. Corresponding values for relative humidity of the air are given in Sec. 15.

12. The Unstable Sublayer and the Inversion Sublayer

K. Brocks [112] has analyzed in detail all available observations made to 1938 in the lowest 100 m of the atmosphere. The method of representation is a double logarithmic coordinate system, as shown in Fig. 35. As before, t (° C) is the air temperature, z (m) the height above the ground, and γ (deg/100 m) the lapse rate. The abscissa is log y, and the ordinate log z. The hourly mean values of the lapse rate for every month, similar to those shown for a more recent investigation in Tables 22 and 23, were expressed in this coordinate system for all the measurements made in Central Europe and in Egypt. If the observed gradients had been assigned to the mean height of the layers of air in which they were measured, the values obtained near the ground would be too high on account of the rapid increase of gradient in that region. It was therefore necessary to introduce a suitable reduction. Although the hourly values were averages for all weather situations, nevertheless the times of strong incoming or outgoing radiation were outstanding because of the high values of gradient associated with them. The results are therefore indicative of the type of weather associated with strong radiation. They provide us with an insight into the normal or average situation prevailing in the air layer near the ground. In individual cases, of course, there may be substantial deviations from the normal rule.

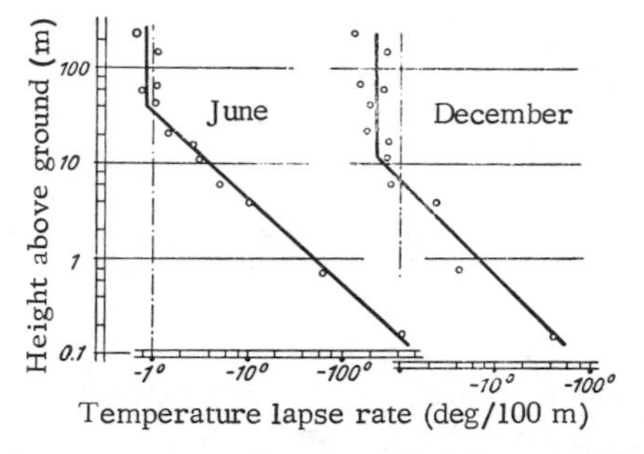

Fig. 35. Temperature lapse rate as a function of height, at midday. (After K. Brocks)

The example shown in Fig. 35 is for noon in two extreme months. These two observations, so different in time and place, unite to present an astonishingly uniform picture: that the layer of air can be divided into two distinct parts. Above the ground in summer (June) and in winter (December) there is a layer in which the alteration of temperature gradient with height may be represented by a straight line. Brocks called this the unstable sublayer. At noon in summer it is higher than in winter, and the slope of the straight line is somewhat greater in summer. Above this, there is a second layer in which the temperature gradient is constant with height, its value being less than the adiabatic gradient shown by a broken line in Fig. 35. This layer Brocks called the adiabatic intermediate layer. It is a few hundred meters in height.

The unstable sublayer is formed when the sun's elevation reaches about 10°. From then on, it increases in thickness, and so do the temperature gradients observed in it. During this period a substantial amount of the heat transferred from the ground is used to build up this sublayer. When the sun reaches an elevation of about 30°, the temperature gradient continues to increase, but the height of the unstable sublayer no longer increases, because by then the convective motion has become so vigorous that further supplies of heat from the ground are transferred mainly into layers higher up.

The height of the unstable sublayer is a minimum of 4 m at midday in December, increasing to 30 or 40 m by the time the sun changes direction again in summer. The yearly average is 21 m. The temperature gradient at a height of 1 m changes seasonally in a similar way from -6 deg/100 m to -45 deg/100 m, with a yearly average of -27 deg/100 m. It is a linear function of the sun's elevation.

The straight line in the unstable sublayer of Fig. 35 obeys the following equation: $\log \gamma = b \log z + \log a$, or

$$\frac{dt}{dz} = a\, z^{b} \qquad (\mathrm{deg\ cm^{-1}}),$$

in which a and b are constants. Since the temperature gradient decreases with height, b is negative. The equation governing the distribution of temperature with height is

$$t = a \int z^{b}\, dz + c \qquad (^{\circ}\mathrm{C}).$$

In order to determine the constant of integration, it is assumed that at height z_1 the temperature has the value t_1. Then the solution of the equation becomes, for $b = -1$,

$$t = t_1 + a \ln \frac{z}{z_1} \qquad (^{\circ}\mathrm{C}),$$

which gives a logarithmic distribution of temperature with height. When b is not equal to -1,

$$t = t_1 + \frac{a}{1+b}\left(z^{1+b} - z_1^{\,1+b}\right) \qquad (^{\circ}\mathrm{C}).$$

The broken curve in Fig. 36 shows the frequency distribution of the index $(1 + b)$ for the time when the unstable sublayer is present and the radiation balance is positive. The curves are derived from actual observations made by Brocks. They show a well-marked maximum for the value $1 + b = 0$. This, however, corresponds to a logarithmic distribution of temperature with height ($b = -1$).

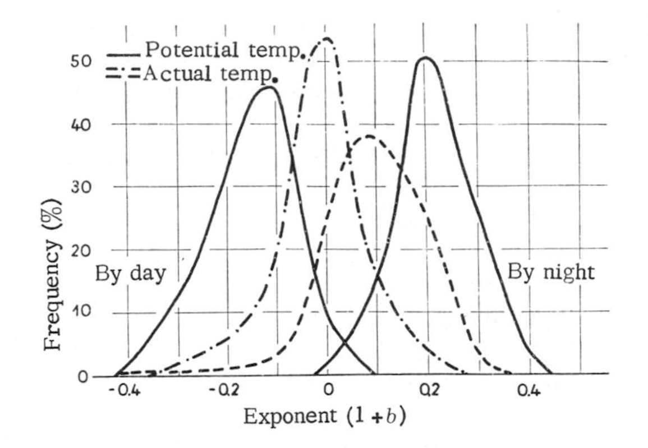

Fig. 36. Frequency distribution of the observed exponents in the temperature-height function, for observed and potential temperatures. (After K. Brocks)

The arrival of night brings similar conditions. An inversion is built up from below, because of the cooling of the ground through emission of radiation; it increases in strength and height until sunrise, often reaching 100 m in height and, in exceptional cases, extending beyond 1000 m. Within this well-known inversion lies an inversion sublayer in which temperature gradients develop similar to those shown in Fig. 35, but with a positive value of γ. Above this level the temperature gradient gradually decreases until it becomes isothermal at the upper limit of the inversion. The exponent $(1 + b)$ is mostly positive and its frequency distribution is shown in Fig. 36 by the dashed curve.

In contrast to the unstable sublayer during the day, the inversion sublayer does not give evidence of any marked yearly variation in its vertical extent. Its presence may even be observed on completely overcast nights. The temperature gradients observed at a

Table 25
Mean values for the inversion sublayer.

Period	Vertical thickness (m)		γ (deg/100 m)		Exponent $1 + b$
	Of the inversion	Of the inversion sublayer	At 1 m height	At upper boundary of inversion sublayer	
Yearly average	104	19	19.0	1.6	0.10
Seasons: spring	100	19	21.4	1.7	0.10
summer	90	18	16.9	1.9	0.20
autumn	130	15	16.6	1.8	0.10
winter	100	25	17.1	1.2	0.04
Clear nights: December	> 100	20	31.3	3.8	0.18
June	> 100	21	34.5	2.9	0.14

height of 1 m do not change during the course of the year by 7 or 8 times as they do during the day, but at the most by twice their value. This is best shown by the figures from Brocks in Table 25.

The air temperature as directly observed, usually termed the actual air temperature, is unsuitable for use when questions of eddy diffusion are considered. In a layer of air 100 m thick (see p. 35) the change of temperature resulting from the change of pressure with height is no longer negligible. Instead of the actual temperature, one must use the potential temperature, that is, the temperature that the air would have if reduced adiabatically to a pressure of 1000 mb.

Brocks [114] has also plotted the variation of potential temperature as a function of height; it is shown in Fig. 36 by the full line. This indicates that a logarithmic distribution of potential temperature with height is exceptional. By day negative values of -1/7, and by night positive values of 1/5, are the rule for the exponent $(1 + b)$.

Our discussion of temperature conditions in the neighborhood of the ground may therefore be summed up in the following way. Between the warm or cold surface of the ground and the free atmosphere, there is a layer of air, the temperature of which by day and by night is a uniform function of height. By day the exponents are negative; the unstable sublayer is built up during the course of the day to one in which the instability increases in proportion to the incoming radiation. The difference in temperature between the ground and the air is at first mainly retained within this layer; then, when the sun's elevation reaches 30°, it is passed on to layers of air at higher levels by a stream of heat. During the night the inversion sublayer has approximately the same strength. The exponents in the temperature-height function are positive. The inversion sublayer is also built up during the night, but the differences occurring over the year are small in contrast, because of the small differences in intensity of the outgoing radiation.

A change from one system to the other occurs twice daily. Figure 37 shows the change from day to night (left) at sunset, and the change from night to day (right) at sunrise. The graphs are based on observations collected by Brocks [113] for the periods from 2 hr before to 12 hr after sunset, and from 6 hr before to 2 hr after sunrise.

The curves at the top (d) show how the inversion intensifies during the night. On the average, the incoming-radiation type gives way to the outgoing-radiation type 1-3/4 hr before sunset in the neighborhood of the ground. While the vertical extent of the inversion increases steadily during the night, the inversion sublayer (b) is built up very quickly, and the temperature gradient (a) in the first meter of height above ground reaches its maximum 2 hr after sunset. The depth of the sublayer increases only slowly from 12 to 20 m, while the inversion itself increases from 20 to almost 200 m. The exponent of the height function (c) also remains

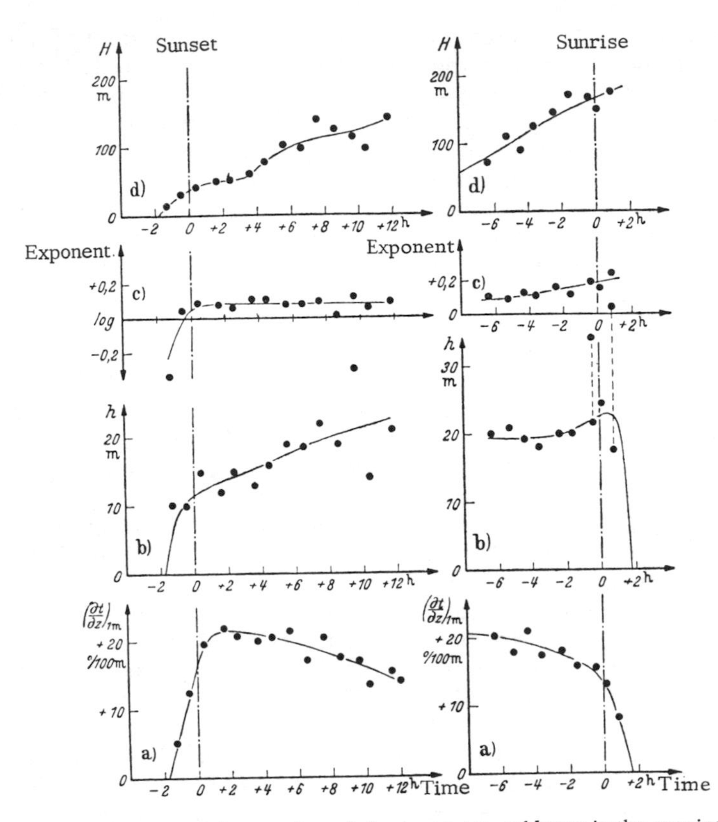

Fig. 37. Formation and destruction of the inversion sublayer in the evening and in the morning. (After K. Brocks)

almost constant during the night. The destruction of the inversion sublayer takes place even more suddenly during the morning. On the average, the incoming radiation type has been reestablished 1-2/3 hr after sunrise.

As already shown in Fig. 35, the unstable sublayer reaches down to a minimum of about 10 cm above the surface of the ground. The layer of air close to the ground in the proper sense, with its height of approximately 2 m, is therefore only a part of the Brock sublayer. The essential points of the temperature conditions prevailing in it are adequately described by the previous discussion. In the following discussion, a twofold task still remains: first, to demonstrate the practical consequences of the unstable sublayer and the inversion sublayer in the region of the ground as far as temperature and gradient values are concerned; and second, to point out special effects and anomalies due to the immediate proximity of the earth's surface, and to explain these effects.

13. Daytime Temperature of the Air Layer near the Ground

Whereas a good series of measurements made over many years for the first 100 m of the atmosphere is available, the situation is different for the first 2-m layer of air close to the ground. The increasing gradient of all characteristics near the surface of the ground demands instruments that are as small as possible. The shielding of these instruments against radiation from above and below presents considerable difficulty. In general, it is not possible to ventilate the instruments artificially, as is usually done, because this would alter the quantities to be measured. In addition to these mechanical and technical difficulties, there is the fact that chance influences of the type of soil, the condition of the soil, and the vegetation at the site of measurement make it extremely difficult to generalize the results obtained.

The only set of temperature records embracing all periods of the year for the layer of air close to the ground is that made by A. C. Best [124] at Porton, England, from August 1931 to July 1933. These were made in association with N. K. Johnson, and some of the data have already been given in Table 21. These carefully analyzed measurements were made with platinum resistance thermometers at heights of 2.5, 30, and 120 cm above grass, which was kept cut short. Tables 26 and 27 are based on these results.

A considerable number of shorter-term observations are available today for the air layer near the ground. The most valuable are those covering several characteristics, since from them a number of general questions on heat economy and variation of water-vapor content can be answered. The most comprehensive is certainly the program of observations carried out in August and September 1953 by the joint effort of sixteen universities and state institutions of the U.S.A. in O'Neill, Nebraska, on a uniform area of prairie grass [13].

To illustrate the temperature field in the neighborhood of the ground, we shall use the comprehensive series of observations made, with very good equipment, by the Climatological Laboratory in Seabrook, New Jersey (39° 34′ N, 75° 13′ E), by C. W. Thornthwaite and his assistants [141] since 1948. From this series of hourly electrical measurements of temperature at 10, 20, 40, 80, 160, 320, and 640 cm above the ground, five undisturbed spring days are selected from 1951 (17 and 18 March, 8-10 May) for which there was a minimum number of gaps in the observation of other characteristics. Frequent reference will be made to this series in what follows.

First of all, the daily sequence of temperature is shown in Fig. 38. The curves run at almost the same distance from one another, because the measurement heights are plotted logarithmically (Sec. 12). After the minimum before sunrise, which occurs about an hour later at higher levels, an isothermal state sets in

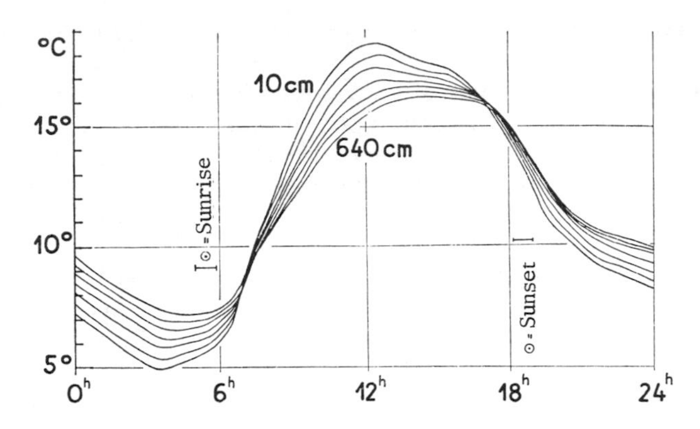

Fig. 38. Temperatures at 10, 20, 40, 80, 160, 320, and 640 cm in the air layer near the ground on bright spring days at Seabrook, New Jersey.

almost simultaneously at all levels 1 to 2 hr after sunrise. At a height of 10 cm maximum temperature is reached immediately after the midday maximum of radiation. By 6.4 m this is delayed until 14:30. The change from the incoming-radiation type to an outgoing-radiation type occurs here also, long before sunset (cf. p. 81) and the temperature gradient passes through the isothermal boundary, first at lower and then at higher levels. Since almost cloudless spring days are involved, the temperature at midnight is noticeably higher than at midnight 24 hr earlier.

Unfortunately, measurements between 10 cm and the surface of the ground are missing in this series too. If the validity of Fig. 38 extends into this level, as expected, then there should be an enormous temperature rise at midday and a temperature fall by night, before the temperature of the surface is reached. Some indication of this is given by the simultaneous measurements of surface temperature which at midday reached 23° C at a depth of 2.5 cm.

The tautochrones in Fig. 39 show the relation between temperatures in the lowest 1.5 m of the atmosphere and measurements made at depths of 2.5, 5, 10, 20, and 40 cm in the ground. Only a selection of the tautochrones is given, in order to provide a better picture of what is happening. Between +10 and -2.5 cm the course of the tautochrones can only be suggested. The tautochrone for 03:00 represents an outgoing-radiation type, that of 12:00, an incoming-radiation type. The beginning of diurnal heating is clearly seen at 07:00 by the flow of heat downward into the ground as far as 5 cm, and there is a weak indication of it in the air, as indicated by the temperature decrease with height. At 09:00, nocturnal conditions are found only below the 25-cm depth. The lighter, broken tautochrones for 19:00 and 23:00 show a similar return movement from incoming radiation to an outgoing-radiation type.

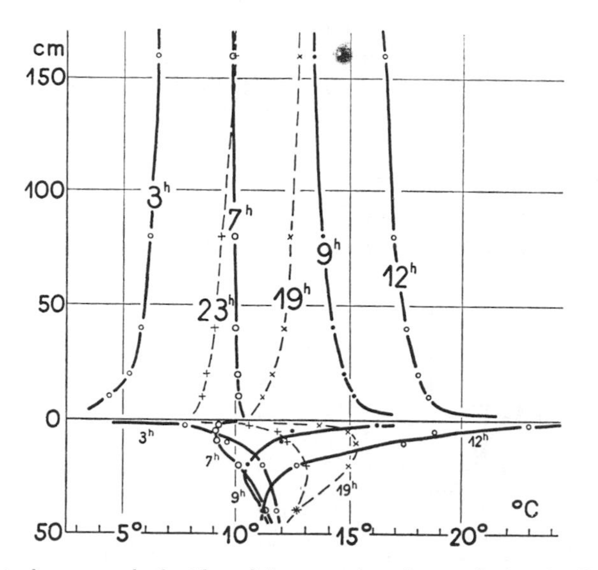

Fig. 39. Tautochrones on both sides of the ground surface at Seabrook, New Jersey.

Comparison of Fig. 39 and Fig. 34 shows considerable similarity. On a small scale, in the layer close to the ground, once again there is a substantial increase of temperature gradient toward the surface, as, on a larger scale, there is in the first 100 m of the atmosphere. It can be seen from Tables 26 and 27 that this is the case for all weather situations throughout the year. For purposes of comparison with Tables 22 and 23, which actually do not come from the same place but from the same area and climate, the temperature gradients obtained by A. C. Best [124] for the layers from 30 to 120 and 2.5 to 30 cm over short grass are repeated, after conversion to degrees per 100 m.

Even taking the mean monthly values, gradients in summer can be observed that are several hundred times the adiabatic rate. With $\gamma = -3.4$ deg/100 m, it is known that air density is constant with height. When gradients of this magnitude are reached, air density must increase in the lowest meter by about 2 percent. The laws applying to the free atmosphere show that this condition is highly unstable. The question is, however, whether this can still be said of the small volumes under consideration here.

W. A. Baum [123] adopted an idea from D. Brunt and showed, in 1951, that even with gradients of this strength it was still possible to have stable conditions. Lord Rayleigh had proved in 1916 that thin layers of fluid might still remain stable in spite of increasing density with height, if certain equations of condition were fulfilled. In addition to temperature, conductivity, variation of density, and thickness of the layer, these equations also contain the kinematic viscosity. By applying these results to the layer of air close to the

Table 26
Temperature gradient (deg/100 m) in the layer of air from 30 to 120 cm over short grass at Porton, England; 2-yr averages from 1 August 1931 to 31 July 1933.

Hour	2	4	6	8	10	12	14	16	18	20	22	24
January	25	33	21	17	-2	-7	-6	22	34	31	33	26
February	25	23	21	10	-15	-30	-20	7	31	32	31	30
March	46	41	39	-10	-45	-60	-49	-17	35	58	52	46
April	34	32	15	-27	-57	-62	-50	-29	7	38	39	38
May	30	27	-7	-42	-56	-64	-56	-34	-5	27	34	36
June	29	25	-9	-44	-66	-77	-60	-43	-10	25	36	34
July	23	18	-2	-26	-38	-48	-46	-28	-6	21	26	25
August	20	16	0	-27	-49	-59	-54	-36	1	33	34	31
September	29	26	15	-14	-36	-36	-28	-12	15	33	28	34
October	35	33	28	-4	-25	-32	-25	1	48	46	44	39
November	21	19	21	13	-14	-15	-7	22	30	26	26	23
December	18	18	20	19	-2	...	-1	23	30	27	31	25

Table 27
Temperature gradient (deg/100 m) in the layer of air from 2.5 to 30 cm over short grass at Porton, England; 2-yr averages from 1 August 1931 to 31 July 1933.

Hour	2	4	6	8	10	12	14	16	18	20	22	24
January	94	89	69	58	-46	-103	-36	105	143	121	115	109
February	119	117	101	61	-123	-220	-125	77	208	191	173	155
March	187	167	139	-99	-315	-397	-325	-46	195	258	228	197
April	135	113	48	-260	-426	-442	-359	-151	46	151	159	171
May	75	54	-115	-357	-460	-480	-408	-212	-12	101	101	89
June	97	85	-214	-513	-622	-682	-519	-327	-42	133	165	127
July	54	30	-71	-361	-456	-502	-420	-244	-28	105	109	83
August	63	42	-46	-252	-432	-492	-428	-202	42	127	111	101
September	61	73	32	-165	-321	-315	-199	-89	93	123	93	105
October	141	107	97	-65	-242	-262	-197	67	222	187	173	111
November	97	103	115	77	-97	-121	-36	127	147	121	139	135
December	99	105	107	111	-10	-65	0	147	143	133	155	161

ground, and by making certain probable numerical assumptions, Baum comes theoretically to the conclusion that gradients of -170 deg/100 m in the lowest 10 m of the atmosphere or of -2100 deg/100 m in the lowest 2 m may still represent stable conditions.

However, the distribution of temperature close to the ground cannot be entirely explained from the thermodynamic point of view alone. G. Falckenberg and his pupils [149-152] pointed out the significance of long-wavelength radiation very early, particularly its importance in the heat budget at night. A part of long-wavelength radiation has only a limited range, depending not only on water-vapor content and wavelength, but also on air pressure and temperature. For example, in an atmosphere with water-vapor pressure of 10 mm Hg, radiation in the band from 6.25 to 6.75 μ is half absorbed within a distance of 1 to 2 m, and over the whole range of the spectrum within a distance of 30 m. When temperature differences are large, this means that within the air layer near the ground there must be a radiation exchange that cannot be neglected. Here is a mutual influence of the earth's surface and the layer of air close to it, acting on each other to give the process which Falckenberg [149] has called wavelength transformation. The reason for it is that H_2O and CO_2 in the atmosphere emit radiation in bands, while the solid ground emits radiation in a continuous spectrum.

The discussion below on radiation conditions in the middle of the day is based on work by F. Möller [135] in 1955, dealing with radiation processes near the ground. If it is necessary to calculate the radiation balance of a single layer of air, then the radiation exchange with all other layers of air within the range of the radiation, some of them colder, some of them warmer, must be taken into account, and this makes the calculations very complicated. Where there is a linear increase of temperature with height and the absorption coefficient is constant with height, every layer of air would receive just as much heat from below as it emitted above; hence its own temperature would remain unaltered. With the gradients observed at midday, however, much more heat is received from below than is given off upward later. The consequent radiative heating is estimated to be of the order of magnitude of 1 to 50 deg/hr, and is therefore considerable.

The boundary layer at the ground (see p. 34) is a special case because, although heat is being radiated upward, no heat is being received from below. "Below" is the surface of the ground, with temperature equal to the air temperature at height 0 in the stationary state, and beyond there is no layer at higher temperature. Thus, the boundary layer cools very strongly through radiation and tries to make up for its temperature losses. This apparently paradoxical situation, based on theory, was first confirmed by F. Linke and F. Möller [134]. Using the Thornthwaite measurements mentioned earlier, and selecting from them three fair-weather days of the spring of 1952, Möller calculated the temperature change that would take place in the afternoon period from 13:00 to 14:00 through long-wavelength radiation and arrived at the results given in Table 28.

Table 28
Afternoon cooling at various heights by long-wavelength radiation.

Height above the ground (cm)	0	0.06	0.27	0.89	2.65	10	20	80	640
Observed (extrapolated) air temperature (deg)	(25.5)	(24.5)	(23.5)	(22.5)	(21.5)	20.3	19.8	18.9	18.1
Radiative heating (+) or cooling (-) (deg/hr)	-22.8	-5.1	+11.9	+16.5	+15.7	+10.6	+8.4	+8.6	+1.3

The surface temperature of the ground was calculated by extrapolating the measurements of ground temperature extending to within 25 mm of the surface by three different methods. These calculations gave the intermediate values for the air temperature close to the ground shown in parentheses in the table. At first sight the results looked surprising; the rate of cooling through radiation of the air in contact with the heated surface of the ground amounts to 23 deg/hr. But this cooling is limited to a boundary layer only 1.2 mm in height. In the layer of air close to the ground above this boundary layer, the air is strongly heated through long-wavelength radiation. In the case just quoted, the amount of heating in the first meter is between 8 and 17 deg/hr. Using a different method involving a radiosonde ascent in August, O. Czepa [125] calculated the gain of energy through radiation in the lowest meter of the atmosphere to be 1.7 cal/hr, which corresponds to a change of temperature of 60 deg/hr. This was based on an assumed surface temperature of 45°C, or 20 deg higher than Möller had extrapolated from Thornthwaite's measurements. There can be no doubt, however, that the amount of heating due to radiation is very substantial, when the incoming-radiation type is effective.

It is easy to explain why this radiative cooling of the boundary layer was never observed. The calculations are based on the assumption of a stationary state, in which the air at zero height has the temperature of the ground surface. For the case under discussion, Möller calculated that a temperature difference of 1.2 deg between the surface of the ground and the air would be sufficient to bring about the change of sign from cooling to heating in the boundary layer. In spite of the extremely small amount of eddy diffusion near the ground, with these strong gradients cold parcels of air reach the ground surface repeatedly. The alternative is that the 1.2-mm boundary layer will have to be done away with.

If the air temperature t increases with time z, the gain of energy per unit volume is $c_p \rho\, dt/dz$ in which c_p, the specific heat at constant pressure, is 0.241 cal deg^{-1} g^{-1} and ρ (g cm^{-3}) is the density of the air. This increase of temperature, ignoring advective influences, is made possible by the fact that the flow of heat caused by eddy diffusion (L) or by radiation (S) does not pass unaltered through the air at the level in question, that is, it is not independent of the height x (cm), but some of the heat is retained, and the flow becomes differentiated. The gain of energy in any position is brought about by the divergence of the heat flow, and makes itself felt through an increase in temperature. Therefore

$$c_p \rho \frac{dt}{dz} = \frac{dL}{dx} + \frac{dS}{dx} \qquad (\text{cal cm}^{-3}\ \text{sec}^{-1}).$$

Now, $L = Ac_p \cdot dt/dx$ (see p. 37). After substitution in this equation and differentiation we have

$$\frac{dt}{dz} = \frac{1}{\rho} \cdot \frac{dA}{dx} \cdot \frac{dt}{dx} + \frac{A}{\rho} \cdot \frac{d^2t}{dx^2} + \frac{1}{c_p \rho} \frac{dS}{dx} \qquad (\text{deg sec}^{-1}).$$

In the middle of the day the first term, depending on the vertical variation of the austausch coefficient, is negative, while the second austausch term and the radiation term are positive. This last term is already known numerically from the figures given in Table 28. The value of A must be known before the others can be calculated. So, after rearrangement and transformation, the first equation becomes

$$\frac{d}{dx} \cdot \left(\frac{A}{\rho} \cdot \frac{dt}{dx}\right) = \frac{dt}{dz} - \frac{1}{c_p \rho} \cdot \frac{dS}{dx} \qquad (\text{deg sec}^{-1}).$$

Integration gives:

$$\left.\frac{A}{\rho}\frac{dt}{dx}\right]_0^x = \int_0^x \left[\frac{dt}{dz} - \frac{1}{c_p \rho} \cdot \frac{dS}{dx}\right] dx$$

or

$$\frac{A}{\rho}\frac{dt}{dx} = \int_0^x \left[\frac{dt}{dz} - \frac{1}{c_p \rho} \cdot \frac{dS}{dx}\right] dx - \frac{A_0}{\rho_0}\left(\frac{dt}{dx}\right)_{z=0} \qquad (\text{deg cm sec}^{-1}).$$

In this equation all the quantities involved are known from observation with the exception of the Austausch coefficient A_0 at the surface of the ground. Next, substitute for this the value appropriate to the transport of heat by molecular conduction (about $2 \cdot 10^{-4}$ g cm^{-1} sec^{-1}). Möller evaluated A_0 for this case, using the heat-balance equation already mentioned on p. 10, and found the value $1.94 \cdot 10^{-3}$ g cm^{-1} sec^{-1}. This is 10 times the other value, because the surface of the ground in Seabrook is not a theoretical mathematical boundary surface but a real one, which produces some degree of microturbulence by its configuration.

Now A too can be determined, as well as the order of magnitude of the two austausch terms which govern the variation of temperature with time. The results for the selected hour of 13:00 to 14:00 in the middle of the day are shown in Table 29.

Table 29
Austausch coefficient at various heights.

Height above the ground (cm)	0.06	0.27	0.89	2.65	10	20	80	640
Austausch coefficient A (g cm^{-1} sec^{-1})	$5.1 \cdot 10^{-3}$	$1.5 \cdot 10^{-2}$	$4.5 \cdot 10^{-2}$	$1.3 \cdot 10^{-1}$	0.41	1.08	7.6	136
Order of magnitude of the two austausch terms (deg sec^{-1})	400	102	35	13	4	2.2	0.6	0.1

The layer of air close to the ground is thus characterized as the lowermost edge of the atmosphere, in which eddy diffusion increases rapidly with height (almost in a linear fashion according to theory) until it reaches the value obtained using observations from meteorological shelters, after which its order of magnitude changes only a little, even decreasing slightly later on. The flow of heat from the warm surface of the earth can make its way through this layer of rapidly changing eddy diffusion into the atmosphere with its strength more or less undiminished only provided the temperature gradient, too, changes rapidly with height, that is, if d^2t/dx^2 is large. Since dA/dx and d^2t/dx^2 must also have high values, the two austausch terms in the equation have magnitudes that make the radiation term appear insignificant. In Table 29 their value is shown in degrees per second, that is, in a unit that is 1/3600th of the unit of measurement of the radiation term in Table 28, in order to give figures that are easy to visualize. Adjacent to the ground, each of the austausch terms is 100,000 times the radiation term, and even at a height of 6 m each is still more than 100 times its value. It is true that the austausch terms are not independent of each other; the difference between them is small. However, it only requires an insignificant alteration of dA/dx to offset any radiation term, no matter how great it may be, positively or negatively. It does not follow, as we have recognized, that the absolute value of the radiation term is in any way negligible, but only that the temperature distribution in the layer of air close to the ground is determined principally by the enormous vertical falls in austausch and in temperature gradient.

W. A. Baum [123] has rightly shown that the temperature gradients in the neighborhood of the ground in the middle of the day could not be explained by the suggestion that they were always being regenerated by the enormous supply of heat, in spite of continual overturning of the layers of air. No such overturning is proved by observation. This may be recognized by a study of temperatures recorded above the ground in desert areas, such as those given in Fig. 14. In spite of the great unsteadiness of temperature, parcels of air from a height of 1 mm, which here have a temperature of 6° to 9° C, clearly do not reach a height of 1 cm, since the temperature at this level fluctuates only from 0° to 4° C. It must depend on a coupling of austausch and temperature gradient. As soon as the austausch decreases for any reason, the temperature gradient immediately increases and therefore maintains the flow of heat. Horizontal temperature contrasts within very small distances are also probably important here. Compensatory processes such as this are always in action, both in space and in time. The flow of heat, borne by eddy diffusion, oscillates to and fro. The facts we observe with our instruments are nearly all average values of the sum of individual events. Radiative heating is not, however, excluded from having a part in this massive short-term exchange.

The overturning on a grand scale of superheated layers of air close to the ground is an exceptional occurrence, accompanied by dust devils and sand devils. The upward swirling of the air drags with it dust, sand, leaves, twigs, paper, and so forth, and therefore becomes visible. Wandering off, usually slowly, it sucks other overheated layers into its orbit and so maintains itself. In arid zones of the earth, dust devils are a regular feature of the hot hours of the afternoon. They have been investigated rather closely by W. D. Flower [128] in Egypt, and by R. L. Ives [131] in the North American prairie. These are not to be confused with tornadoes and waterspouts, which come down from low-lying clouds toward the surface. B. D. Kyriazopoulos [133] found this striking micrometeorologic phenomenon portrayed even on the capitals of the Corinthian pillars of St. Sophia in Thessalonika, dating back to the fifth century B.C., where acanthus leaves are shown being picked up by a whirlwind.

Figure 40, based on Egyptian measurements from 1926 to 1932, shows the frequency distribution of dust devils during the day. The diagram also shows temperature gradients, measured for 1932 on days when dust devils were present. The connection between the two distributions is obvious. The dust devil is initiated by some chance happening, perhaps at a pile of crushed rock at the side of a highway, by an automobile, by a gust of wind at the edge of a wood, or, as described by K. Hartmann [130] in his report of a dust devil in Potsdam in 1953, by the confluence of two trails of dust at a street intersection. "In most cases," writes R. L. Ives, "the track of a small animal in the desert, perhaps a rabbit or a wolf, can be recognized from the trail of

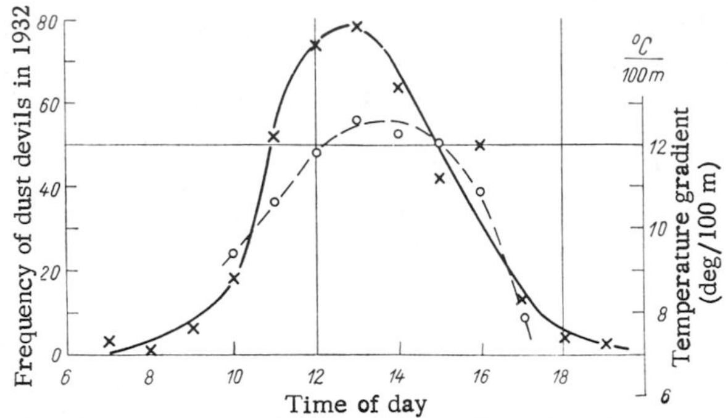

Fig. 40. Connection between the incidence of dust devils and vertical temperature gradients in Egypt. (After W. D. Flower)

small whirls of dust it leaves behind. Observations we have made in this territory suggest the possibility that most dust devils are started by small animals.''

By some as yet unexplained mechanism, the upward-swirling air starts to rotate. The direction of rotation appears to be decided by chance, since Flower reported 175 dust devils rotating in a clockwise direction, and 200 rotating in the opposite direction. Ives was able to observe that a dust devil which had almost come to the end of its existence might revive itself in contact with an obstacle, and rotate in the opposite direction. This is what happened to a Potsdam dust devil after collision with an apple tree, as observed by L. Klauser [132] in 1950.

In arid zones, dust devils can reach a height of 1000 m, and last for several minutes. In Utah a dust devil 800 m high was observed continuously for 7 hr along a 60-km path. Inside the dust devil there is a lowering of pressure and a rising current of air. One dust devil passed right over the weather station in Phoenix, Arizona, and the report made on it by H. L. De Mastus [127] states that the barograph trace fell by 1.3 mm-Hg and the disturbance lasted for 30 sec. Shortly before this, another dust devil had passed by at a distance of 23 m, producing in 6 sec a pressure drop of 1.2 and a rise of 0.8 mm-Hg. The true decrease of pressure in the interior of these dust devils is certainly much greater. This was established by Ives, who chased one in a Jeep, holding an instrument inside it on the end of a long stick. Upward currents of 10 to 15 m sec^{-1} have been estimated by seeing rats sucked up, and later measuring their actual rate of fall.

Klauser has given a good description of the Potsdam dust devil of 18 May 1950 already mentioned, showing the proportions of the phenomenon in Germany. "The light wind suddenly dropped during bright sunshine, and all at once a rustling sound [often alternatively described as a whistling] was heard; an observer thought a thunderstorm was about to break. Then a small whirl was observed in the form of a small brownish-black or gray-brown opaque dust devil. This rotated rapidly about its axis, which was vertical, reaching a height of about 3 m to begin with, and about 5 to 8 m at its maximum intensity. Its diameter was about 1 to 2 m. There was a pulsing variation in its thickness as it wandered on its way. Around the whirlwind proper a rotating ring of dust could be seen, about 1.5 m higher and with an estimated diameter of 6 to 10 m" (summary). This dust devil traveled 180 m and lasted 1.5 min.

Dust devils can cause damage from time to time. The one just described broke several hotbed windows and threw some over barriers 4 to 6 m high for a distance of 20 m. H. Schlichtling [139] writes of one observed on 19 May 1934 in Lübeck. "It was interesting to see how three people were caught up in the whirlwind. A woman managed to get out of it quickly, but two men were in it for some time. Their clothes were flapping about vigorously and they had to hold on tightly to their hats. They had great difficulty in keeping their feet."

The optical effects associated with superheated layers of air close to the ground will be reported in Sec. 18.

14. Night Temperature of the Air Layer near the Ground

The type of temperature distribution that normally sets in above ground cooling by radiation is already familiar to us under the name of the outgoing-radiation type. It has been shown in Fig. 34 for the 100-m layer by the 4-hr tautochrone, and for the air layer near the ground in the strict sense as well as the uppermost layer of the soil, by the 3-hr tautochrone in Fig. 39.

S. Siegel [164] made a more detailed study of the changing nocturnal temperature pattern above the ground at the Meteorological Institute of the University of Hamburg by placing 23 thermoelements between a height of 0 and 4 m. Figure 41 shows the results obtained for a typical cloudless radiation night in summer. The falling-off of wind during the night is shown at the top of the diagram. The night inversion, which builds up slowly, reaches its maximum height just before sunrise. After sunrise it is destroyed quickly, as already shown in Fig. 37.

On any individual night the temperature behaves much more irregularly, of course, because of variations in wind speed. Figure 42 shows a typical example: this is a record of the isotherms on the night of 9-10 July 1935. An increase of wind shortly before 02:00, shown at the top of the diagram, is followed by a quick breakdown of the inversion, which forms again as soon as the wind drops once more. The wind speed here gives only an indication of the amount of eddy diffusion which carries the cooling of the ground surface (because of long-wavelength radiation) up into the air.

It was pointed out as early as 1932 by L. A. Ramdas and S. Athmanathan [159] that in India, especially in the area of the

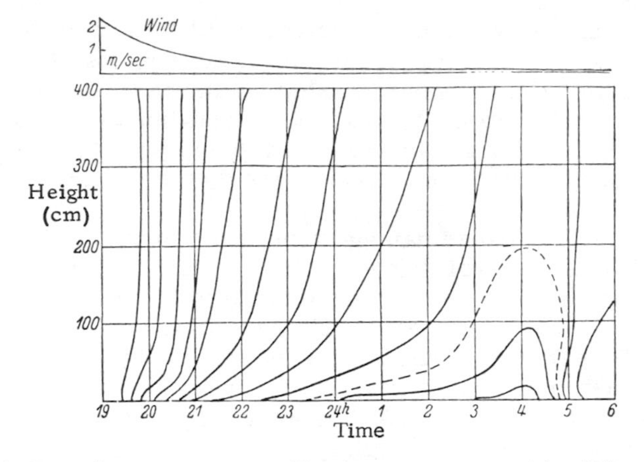

Fig. 41. Typical temperature stratification on a summer night. (After S. Siegel)

dark cotton-growing soil, the lowest temperatures at night frequently were observed not at the ground surface but a few centimeters, and sometimes even as much as 1 m, above it. This was confirmed in India on many occasions. As an example, Fig. 43 gives the record of measurements made by K. R. Ramanathan and L. A. Ramdas [158] in Poona on a January night in 1933. The arrowhead pointing upward shows the temperature of the ground surface; the circle at zero height gives the temperature of the air immediately at the ground. Surprisingly, it is much lower than the ground temperature and continues to decrease with height to give a minimum, in this particular case, between 10 and 30 cm, depending on the time the observation was made. These

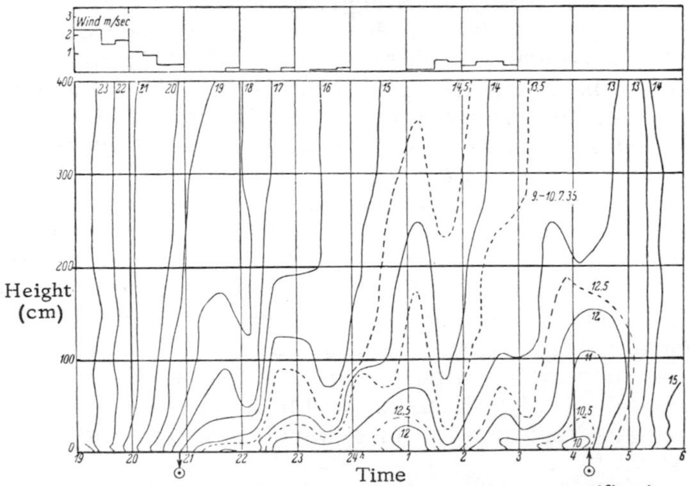

Fig. 42. Connection between air movement and temperature stratification on a night in July. (After S. Siegel)

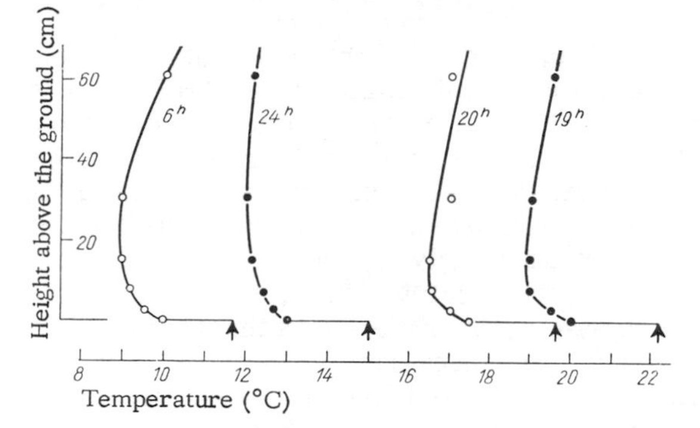

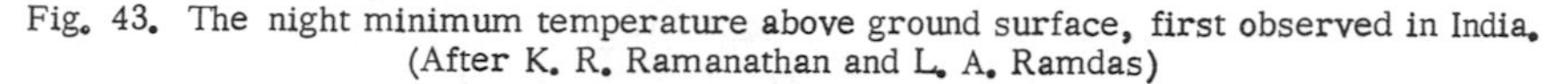
Fig. 43. The night minimum temperature above ground surface, first observed in India. (After K. R. Ramanathan and L. A. Ramdas)

results were at first accepted with some reservations, since similar conditions might be arrived at if cold air had flowed in from the environs, as from the radiative cooling of the surface of a plant. In the meantime, however, so many confirmatory observations have been recorded—by F. Albani [142] in Argentina, R. A. Fleagle [153] in the U.S.A., J. V. Lake and J. L. Monteith [157] in England, J. Szakály [166] in Hungary, and above all by K. Raschke [161]—that there can no longer be any doubt that a temperature minimum above the surface of the ground, as a special case, is often encountered. It is now only a question of the correct explanation.

K. Raschke, in a series of investigations in India for the purpose of elucidating this question, eliminated the possibility of advective influences, by taking his radiation-compensated thermoelements to the top of a table mountain, Chatturshringi Hill, near Poona. Figure 44 shows the tautochrones for the night of 7–8 January 1955, using a logarithmic height scale. Although the maximum was at the surface, the temperature distribution shows clearly the development

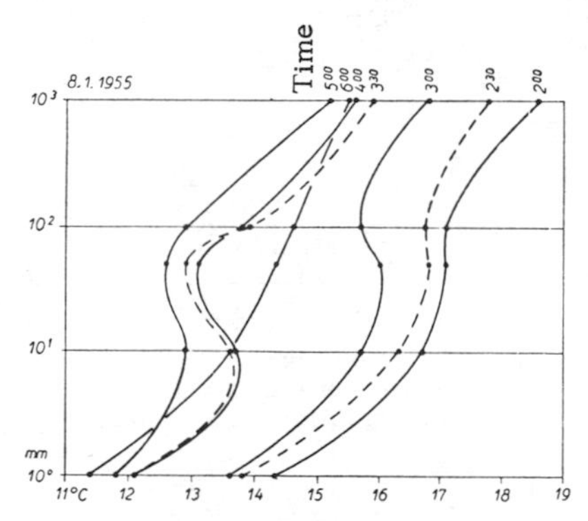

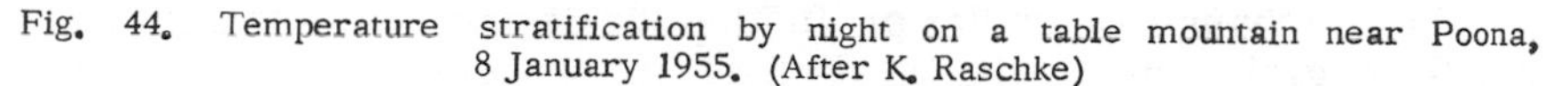
Fig. 44. Temperature stratification by night on a table mountain near Poona, 8 January 1955. (After K. Raschke)

of a second minimum at a height of about 100 mm. That this did not become the principal minimum was due to the weather conditions on that particular night, as was proved beyond doubt by observations on 183 nights near Poona. Figure 45 gives the temperature in the early morning hours of 29 December 1954 at 0.1, 1, 5, 10, 100, and 1000 cm above the ground. The trace added at the bottom is of gusts recorded nearby, giving an indication of the magnitude of eddy diffusion. From 01:30 to 02:30, when wind and eddy diffusion were at their greatest, the trace shows that temperature distribution was of the normal outgoing-radiation type, with the lowest temperature in the lowest position of measurement. The inversion between 1 mm and 10 m is 4 to 5 deg. But as soon as the wind speed 20 cm above the ground remained below 0.5 m sec^{-1}, the minimum became established at a height of 1 to 5 cm, as may easily be seen at the start of the trace. This could be broken down even by artificial ventilation (waving a wooden lath). By way of contrast, during nights when the wind was strong and the normal outgoing-radiation type of distribution was the rule, this anomaly once more came into being whenever the wind speed suddenly dropped below a critical value. This anomaly is caused by long-wavelength radiation.

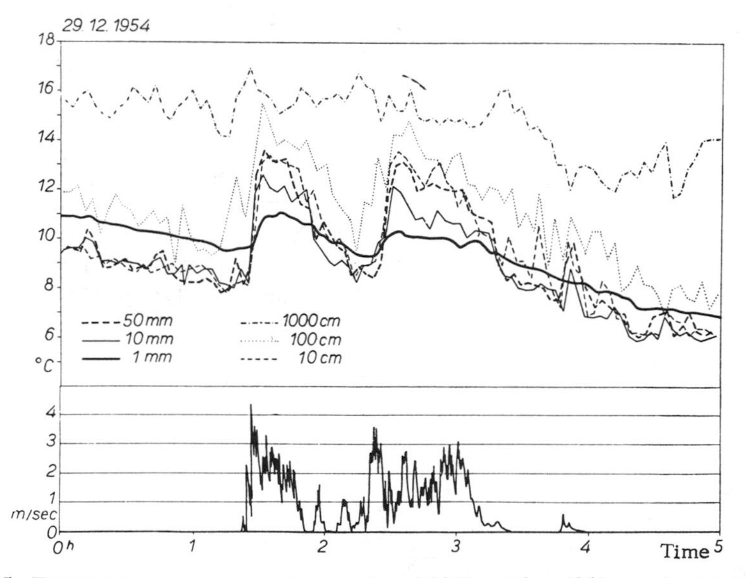

Fig. 45. Temperature sequence in the morning of 29 December 1954 over bare ground in Poona, showing the change in temperature distribution controlled by radiation and eddy diffusion. (After K. Raschke)

The exchange of radiation in the air layer near the ground is not negligible, particularly at night (G. Falckenberg). As in the state prevailing during the day, illustrated by Möller's results (see pp. 87ff), there is again found, at a time of strong outgoing radiation at night, the apparently paradoxical state of radiative heating of a millimeter-thick boundary layer (in the absence of a sudden temperature discontinuity between the surface of the ground and the air). Above this, the layers close to the ground are cooled throughout by long-wavelength radiation. This has been evaluated by Möller [135] for Thornthwaite's records from Seabrook to give the following average values for the period from 03:00 to 06:00:

Height above the ground (cm)	0	10	40	160	640
Radiative cooling (-) (deg/hr)	+5.4	-2.0	-1.3	-1.2	-1.4

It was calculated theoretically by R. G. Fleagle [153], during his studies in 1953 on the formation of fog, that above a cold and dark surface long-wavelength radiative exchanges would give rise to considerable heating close to the surface, and that the greatest cooling would take place at a height of about 1 m. In 1956 he succeeded, by making optical measurements of light refraction, in measuring temperature gradients near the gound with a degree of accuracy not previously attained. Theory and observation agreed that there was a radiative heating at night immediately at the radiating surface (water in this case) of 10 deg/hr, a maximum of radiative cooling of -6 deg/hr at a height of 10 cm, and from 30 cm up to 1.5 m a practically constant rate of cooling of -3 deg/hr.

Nocturnal distribution of temperature is therefore a function of both long-wavelength radiation and eddy diffusion. If mixing is particularly low, which is usually the case at night close to the ground surface, the influence of radiation predominates, and the minimum is found about 10 cm above the ground. With light air movement, convective mixing, which must be brought about by sinking of the coldest air at 10 cm, is reinforced to the extent that the minimum descends to the surface or remains only as a secondary minimum such as is seen in Fig. 44, finally disappearing completely.

In view of these facts, it is of interest to obtain quantitative values of the austausch coefficient A at night, as a function of height above the ground. The calculation of A from observations is always beset with great uncertainty, and is determined to a great extent by the theoretical assumptions made. In spite of all this, at the present time there is a set of values that can be used. The fact that they are based on numerous different methods of measurement and calculation only adds to their value on the whole. Figure 46 summarizes the matter.

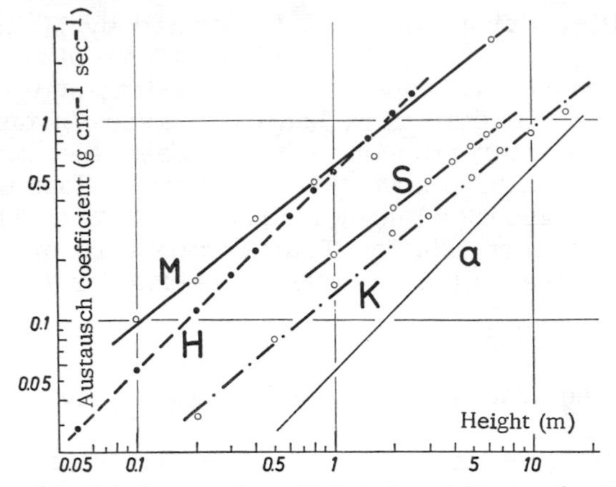

Fig. 46. The value of the austausch coefficient A at night, as a function of height.

The line M gives the values of A from midnight to 04:00, calculated by Möller [135] from the observations made by Thornthwaite at Seabrook in the spring of 1952. The values obtained by H. Hoinkes and N. Untersteiner [449] in August 1950 over the Vernagtferner, averaged for the hours from 18:00 to 07:00 when there was a considerable amount of wind (1 to 3 m sec^{-1} at a height of 28 cm), are shown by the line H. Sverdrup's measurements [165] on the Isachsens Plateau in Spitsbergen over a period of seven summer weeks in 1934 made it possible to compute A from the distribution of temperature and wind. The group average for the lowest wind speed (0.44 m sec^{-1}), corresponding most closely to night conditions, is shown by the line S. H. Kraus [451] has been kind enough to permit the use of the observations he made on the cloudless outgoing radiation night of 23-24 September 1954, from 18:30 to 01:00 at the Munich-Riem airport; these A-values are shown by K. The average wind speed at a height of 2 m was 0.74 m sec^{-1} and the inversion between 0.2 and 15.0 m amounted to 2.7 deg.

As might be expected, there is an appreciable scatter in the values obtained, roughly by a factor of 10. This is chiefly due to the influence of wind speed. It can be seen from the graphs that higher wind speeds lead to higher A-values. In all cases, however, the computed values lie along a straight line in each set. If these lines had the slope of line a, it would mean that there was a linear increase of austausch with height, such as adiabatic layering of the atmosphere would require in theory. With stable layering at night, the exponent of z is less, and lies between 0.75 and 0.99.

The austausch will not continue to increase the way it does in Fig. 46. At higher levels these lines must bend downward; and if we extrapolate backward toward the ground, the values at 1 mm above the surface would be 0.5 to $2.4 \cdot 10^{-3}$ g cm^{-1} sec^{-1}, which is about 5 to 10 times the amount of molecular conduction ($2 \cdot 10^{-4}$ g cm^{-1} sec^{-1}). This is quite likely and proves that there is a gradual change from still powerful eddy diffusion at a height of 1 to 2 m to molecular transport of heat in the vicinity of the surface.

K. Brocks [114] has advanced independent evidence of this transition, from observations made over snow by A. Nyberg [421] in Uppsala. If heat is being carried by molecular conduction, the temperature will alter linearly with height. The exponent $(1 + b)$ in the normal height distribution of temperature given on p. 79 must then have the value 1. Brocks worked out the distribution curve for these indices from layer to layer, in the way shown in Fig. 36, and found maximum frequencies for the following values of the index $(1 + b)$:

Layer (mm)	25-20	15-12	10-8	6-5	3-2	2-1
Maximum frequency	0.1-0.2	0.2-0.3	0.4-0.5	0.5-0.6	0.6-0.7	0.8-0.9

At a height of 2 to 3 cm above the snow, $(1 + b)$ is therefore seen to have the small positive value already shown in Fig. 36. As we approach 1 mm, however, the value of A given by this report also seems to approach that of molecular heat conduction, since $(1 + b)$ approaches unity.

With normal temperature distribution at night, the ground is colder than the temperature indicated by the thermometers in the instrument shelter. This is of great economic importance in the spring and autumn periods of frost hazard. For this reason, weather stations that have the responsbility of advising farmers, nurserymen, market gardeners, fruit growers, builders engaged in concrete work, and other commercial interests in preparing frost warnings do not just use the shelter minimum temperature but also take into account the minimum of a thermometer supported horizontally 5 cm above the surface. This is called for short, if not quite correctly, the "ground minimum" or, since it is usually on a grass plot, the "grass minimum." The difference between shelter and ground minima has been investigated in Germany by G. Schwalbe [163]. The frequency distribution of this temperature difference, averaged over the years 1937 to 1944 for the Agricultural and Meteorological Experimental and Advisory Center in Geisenheim on the Rhine, is shown in Fig. 47, and has been taken from a paper by F. Witterstein [168]. In extreme cases, the temperature at 5 cm was 6.5 deg colder than in the shelter. On dull nights with advection of cold air, the reverse can be the case, the 5-cm temperature remaining higher than the shelter temperature. Figure 47 is for nights when the wind strength was less than Beaufort force 3, since only then is the temperature difference of significance. Naturally, the difference is greater in clear radiation weather

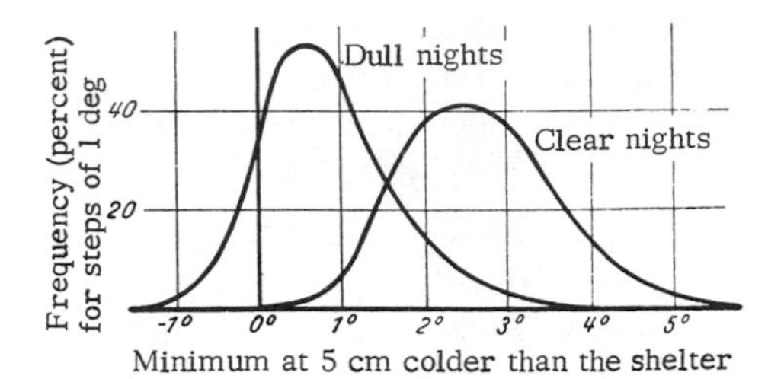

Fig. 47. Frequency distribution of the difference in temperature between the shelter and the ground minima at Geisenheim. (After F. Witterstein)

than in dull weather. There was a maximum in the spring months of April and May, which is exactly the most critical time for frost danger. A second, less pronounced maximum appears in November. F. Hader [155] also found a sharp maximum in spring, in an investigation of minima he made, covering one year, in St. Pölten-Viehofen in Austria, the monthly average differences for March, April, and May being 3.6, 4.1, and 3.6 deg, respectively. J. van Eimern and E. Kaps [148] recorded a distribution similar to that of Fig. 47 at five stations in the lowlands of the Elbe basin. L. Dimitz [146] has shown that, for the Austrian meteorological network, frost sets in at 5 cm on the average 10 days earlier and finishes 14 days later than at shelter level.

These low night temperatures, taken along with the extremely high midday temperatures mentioned in Sec. 13, produce a microclimate of extreme temperature fluctuation at the ground. This peculiarity of microclimate close to the ground may be recognized immediately from Fig. 48. It is a photograph of the sandstone balustrade of the city hall of Winterthur, Switzerland, from a paper by F. de Quervain and M. Gschwind [144]. It shows how the weathering of the stone becomes more pronounced near the ground. The effect of large temperature fluctuations is enhanced by the influence of water. Splashing water and falling snow make the stone nearer the ground more subject to frequent changes from wet to dry. In winter there is also the splitting effect of frost to be considered, since water on freezing at 0° C expands by about 9 percent of its volume.

It is difficult to support this increase of fluctuation by numerical evidence, because of the difficulty of measuring temperature close to the ground, particularly during the day. Some help is obtained from Thornthwaite's Seabrook measurements on certain selected spring days, already mentioned on p. 83 and used in Fig. 39. Taking the values for the warmest and coldest hourly observations, the following increase in daily fluctuation is found as the ground is approached:

Height (cm)	640	320	160	80	40	20	10
Temperature fluctuation (deg)	9.1	9.5	10.1	10.8	11.7	12.7	14.4

Fig. 48. Weathering of Berne sandstone on the city hall in Winterthur. (From F. de Quervain and M. Gschwind)

From earth temperatures at four different depths, the corresponding surface-temperature fluctuation can be calculated to be 16 to 17 deg. The weathering processes in the soil and on the surface of rocks give rise to the "bloom" of salts contained in the earth in prairie lands; the natural preparation of soil for seeds and other phenomena is dependent on this large daily fluctuation of temperature.

A decisive factor in this respect is the frequency of frost changes. By a frost change is meant the passage of the temperature through 0° C, regardless of the direction in which it takes place, whether from + to - or vice versa. A frost-change day is a day on which one or more movements through 0° C occur. The number of frost changes occurring on a frost-change day is called the density of frost change. This number must be greater than 1, and in our climate it is usually 1.5 to 2. At high altitudes in the tropics, where general temperatures do not alter with the time of year, the temperature is always above 0° by day and below 0° by night, giving a frost-change density of exactly 2.

The importance of this number as one of the significant elements of macroclimate was recognized by C. Troll [167], who expressed numerically the great difference between high latitudes where frost changes are restricted to spring and autumn, and high altitudes in the tropics which experience up to 337 frost changes per year (El Misti in southern Peru). It hardly needs to be mentioned that the density of frost change increases with proximity to the ground, as shown by a few figures from L. Dimitz [146] for six Austrian stations. It can reach very high values close to the surface because of the high fluctuation of temperature there, but without any great practical consequences. The frequency of frost change is greatest at the surface itself, and decreases rapidly with depth. E. Heyer [156] gives the following values for Potsdam observations from 1895 to 1917:

Depth in the ground (cm)	0	2	5	10	50	100
Number of frost changes per year	119	78	47	24	3.5	0.3
Average density of frost change	1.8	1.8	1.7	1.5	1.1	1.0

In the instrument shelter (at 1.9 m) the yearly figure was 131; on the observation tower (34 m) it was 95; and at both positions the frost-change density was 1.8.

15. Distribution of Water Vapor above the Ground

The ground surface is as important in the water budget of the atmosphere as it is in the radiation economy. Evaporation takes place either from the ground surface or from its vegetation cover. This stream of vapor is directed upward. Water returns to earth in quite a different fashion, being precipitated in liquid or solid form. Only at night, in special situations, can there be a downward transport of water in the form of vapor, that is, when there is a "fall" of dew or of hoar frost. This exceptional case will be termed "humidity inversion," by analogy with nocturnal temperature inversion.

The annual precipitation in the German Federal Republic amounts to 800 mm, almost half of which is returned to the atmosphere in the upward stream of vapor. The amount of dew per year is of the order of 30 to 40 mm. It follows that the downward stream of vapor is only about one-tenth of the upward stream. In contrast to the reciprocal streaming of heat, therefore, the flow of water vapor is distinctly one-sided.

Just as eddy diffusion plays a preponderant role in the transport of heat as compared with molecular conduction, as soon as the surface boundary layer is passed, so the influence of mass exchange overshadows that of diffusion in the physical sense of the term, as far as transport of water vapor is concerned. In the eddy-diffusion equation (see pp. 35ff) the quantity of a characteristic per unit mass was discussed. In the problem now at hand, this becomes q, the mass of water contained in 1 g of moist air. Aside from differences of air pressure within the air layer near the ground, the water-vapor pressure e can also be used, since it is proportional to the specific humidity for a given atmospheric pressure. In the following discussion e is measured in millimeters-of-mercury, and therefore results quoted from other writers have been converted from millibars to these units. The absolute humidity a (g m^{-3}), which is used occasionally, cannot be converted directly into vapor pressure, since it is dependent on temperature. The conversion factor is 1.06 at 0°C and 0.99 at 20°C, that is, it is not greatly different from 1, so that for all practical purposes a may be taken as vapor pressure (see Fig. 49).

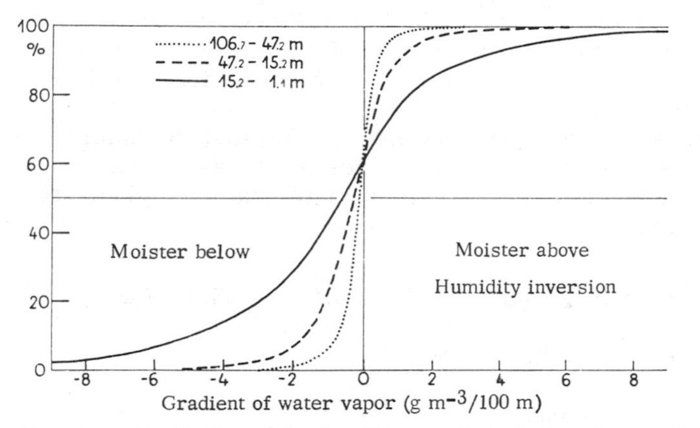

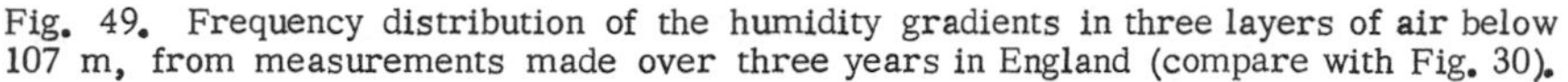
Fig. 49. Frequency distribution of the humidity gradients in three layers of air below 107 m, from measurements made over three years in England (compare with Fig. 30).

All discussions about the direction and magnitude of water-vapor transport must be based on these quantities. In nature, however, the quantity of greatest significance is relative humidity, since the process of swelling depends on it, a fact familiar from the behavior of human hair. In biological studies, attempts are made to investigate the layered structure of relative humidity close to the ground. Therefore, the distribution of vapor pressure will be discussed first, and relative humidity afterward.

As with temperature, the discussion will start with the problem on a large scale, considering the first 100 m of the atmosphere.

The three-year series of observations made in England (see pp. 74f) by A. C. Best, E. Knighting, R. H. Pedlow, and K. Stormonth [111] included humidity measurements made with a Gregory hygrometer, the action of which is based on changes in the electrical resistance of lithium chloride with variations in atmospheric humidity. Figure 49 shows humidity gradients in a way similar to that used in Fig. 30 to show temperature gradients. The quantity shown is absolute humidity. The abscissa gives the variation of water-vapor content in grams per cubic meter per 100 m. Here, too, positive numbers mean an increase of humidity with temperature (humidity inversion), while negative figures indicate the normal situation in which humidity decreases with height.

It comes as a considerable surprise that, while these two possibilities do not occur with equal frequency, they do occur comparably often. The flow of moisture is directed upward in only 60 percent of cases, and this applies at all levels, while on 40 percent of the occasions it is drier below than above. At first sight this appears to contradict the fact that water vapor is transported upward for the most part. However, the total movement of water vapor in the x-direction (vertically) is proportional to the product $A\,dq/dx$ or approximately proportional to $A\,de/dx$. The positive values of de/dx in Fig. 49 are all for hours during the night, when A is very small, while the negative values are associated with values of the austausch coefficient ten times greater, during the day.

Figure 49 also shows, as did Fig. 30, that the humidity gradient increases near the ground surface in a way similar to that of the temperature gradient. The maximum gradients observed over the three years were:

Height in air layer (m):	1.1-15.2	15.2-47.2	47.2-106.7
Decrease in humidity ($g\ m^{-3}/100\ m$):	-40.8	-13.1	-7.5
Humidity inversion ($g\ m^{-3}/100\ m$):	+22.2	+11.3	+5.5

Not only did the inversion have these record values in all layers, but the highest value of the inversion for the month was about two-thirds of the greatest decrease of humidity with height. In the layer nearest the ground its appearance is closely related to the time of day, that is, about 08:00 to 10:00 in the forenoon for humidity decrease, and about midnight for the humidity inversion. As distance from the ground increases, other factors become involved, causing a certain spread in the time of appearance of these phenomena. The maximum lapse of humidity is by no means restricted to fair-weather situations, but was found once, for example, with a thunderstorm.

Figure 50 shows water-vapor pressure tautochrones for 19 clear summer days (calculated from the mean temperature and absolute humidity in this case) for the same hours as were used for the temperature tautochrones in Fig. 34. Before sunrise there is a flow of water vapor from a height of about 40 m toward the ground, to form dew. The height of this layer is in good agreement with the theoretical considerations presented by G. Hofmann [484]. Vigorous evaporation is initiated by heating after sunrise, as may be seen from the increase in water-vapor pressure

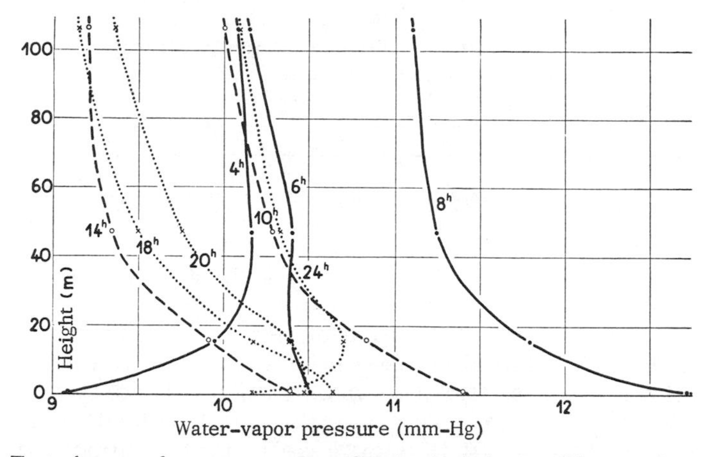

Fig. 50. Tautochrones of water-vapor stratification in the lowest 100 m on clear summer days (compare with Fig. 34).

in the layer nearest the ground until 06:00. Since eddy diffusion is still restricted, this supply of water vapor is trapped near the ground, and the daily maximum of vapor pressure is reached about 08:00 with powerful humidity gradients (full line). Without any marked change of the gradient, the tautochrones (broken) then become displaced toward the region of lower vapor pressure, because of increasing eddy diffusion. This transport out of the layer close to the ground brings about a minimum value at 14:00. At 18:00 the decrease of water-vapor pressure with height is still normal, but by 20:00 the decreasing eddy diffusion and increasing water-vapor content of the air have reestablished the humidity inversion at the ground, and as time goes on it gradually increases in height, in a similar way to the temperature inversion.

These results were best demonstrated by E. Frankenberger [445] at Quickborn in Holstein for the 70-m layer of air. From observations made on clear July days in 1954 it was calculated, taking all the same periods together, that the vertical distribution of water-vapor pressure was as shown in Table 30. Here, too, the flow of water vapor is directed downward at night toward the ground surface (covered with dew), and the fall is remarkably large in the hours before sunrise. During the forenoon, vapor pressure rises close to the ground, and falls at the 70-m level. The transition from day to night conditions can be recognized easily in the average values for the period 20:00 to 22:00.

The daily pattern is shown in Fig. 51 for heights of 2, 13, and 70 m. The well-known double wave of vapor pressure is easily recognized. This appears at all levels, but the amplitude of fluctuation increases with approach to the ground. In all these layers, the evening maximum is higher than the morning value.

Table 30
Water-vapor pressure (mm-Hg) at Quickborn on clear July days.

Height (m)	Time of day					
	22-2	2-6	6-8	8-14	14-20	20-22
70	8.7	9.0	8.9*	8.6*	9.2*	9.2*
28	8.7	8.7	9.0	8.8	9.3	9.3
13	8.6	8.5	9.1	9.1	9.5	9.3
2	8.2*	7.8*	9.4	9.8	10.0	9.2

In other months, for example, on clear days in August and September, it dominates to the extent of becoming a single maximum, so that the double daily wave is simplified to a single wave. British observations show a forenoon maximum which is stronger than that in the evening, as seen from Fig. 50, but the double wave is maintained at the ground, with a secondary minimum at 15:00. The night minimum close to the ground is substantially deeper than that at midday, both here and in Britain; but the difference decreases with height, so that in Quickborn they are equal at a height of 70 m. In England, the late-afternoon minimum is deepest at a height of 107 m, while there is only an indication of the morning minimum, so that here also it is a matter of a single daily wave, but with an early maximum about 08:00. The picture of the daily trend of water-vapor pressure, therefore, is, as we might expect, dependent on the season of the year, the weather situation, and the geographic location of the place of observation.

If the average of all days is taken, the picture obtained of the layered structure of water-vapor distribution is similar, but a rather deflated version of the picture for a clear day. Tables 31 and 32 give the figures for two hourly periods for each month from the observations made in Rye by A. C. Best, E. Knighting, R. H. Pedlow, and K. Stormonth [111], similar to the tables of temperatures from the same source (Tables 22 and 23); gradients given here are of the absolute humidity a (g m^{-3}) calculated for differences of 100 m. The algebraic sign preceding the figures does not have the same meaning as the direction of transport of water vapor, which is determined by the direction of the gradient of the specific humidity q (g g^{-1}). Since $a = \rho q$ (ρ is the density of air), and because, in proximity to the ground, there are marked density differences arising from large temperature gradients, the signs of da/dx and dq/dx may be different near 0. For example, in the lowest layer from 1.1 to 15.2 m on June nights, it might be thought that evaporation was taking place (negative da/dx), instead of which the flow of water vapor was directed downward. The alternation of signs, now positive, now negative, in the layer from 47.2 to 106.7 m is for this same reason, and not due only to irregularities in weather conditions, which nevertheless are still to be seen in the three-year averages. Larger gradients remain unaffected.

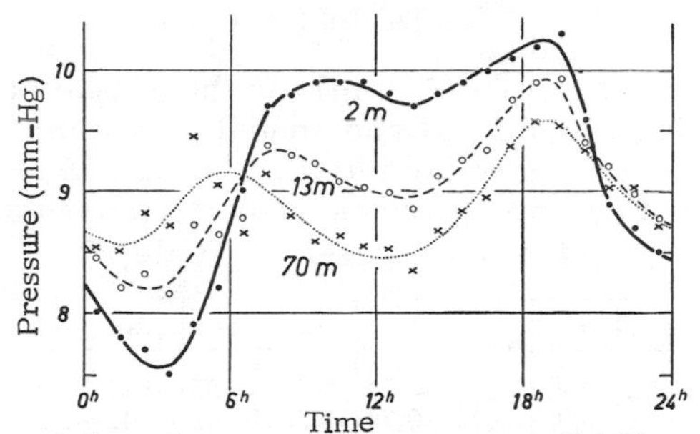

Fig. 51. Diurnal variation of water-vapor pressure at Quickborn on clear July days. (After E. Frankenberger)

If conditions in the 100-m layer show a clear-cut relation, the same cannot be said for the first few meters close to the ground. In the majority of cases there is a repetition, on a small scale, of what we found for the first 100 m. This is true particularly in regions of dry climate, rich in radiation. Observations made by L. A. Ramdas [174] in Poona (18.5° N) during clear winter weather in the years 1933 to 1937 were used to evaluate the mean distribution of water-vapor pressure (mm-Hg):

Height (cm)	305	122	91	61	30	15	7.5	2.5	0.8
At sunrise	8.8	8.2	8.0	7.8	7.6	7.5	7.4	7.4	7.5
At midday	8.3	8.5	8.6	8.7	8.9	9.0	9.4	9.6	10.0

The night inversion is well marked, and the daily fluctuation of 2.5 mm-Hg at 0.8 cm is five times its value at 305 cm. Observations made at Pretoria, South Africa (26°S), by E. Vowinckel

Table 31
Humidity gradient (g m^{-3}/100 m) for the air layer from 47.2 to 106.7 m, giving the diurnal and annual variation in the form of monthly averages for the three years 1945-1948 at Rye, England. (Corresponding temperature gradients are given in Table 22.)

Hour	2	4	6	8	10	12	14	16	18	20	22	24
January	-0.07	-0.08	-0.08	-0.08	-0.08	-0.20	-0.22	-0.20	-0.17	-0.12	-0.07	-0.12
February	0.00	0.03	0.00	-0.02	-0.10	-0.08	-0.08	-0.10	-0.03	-0.05	-0.03	-0.02
March	0.00	0.00	0.07	-0.07	-0.20	-0.03	-0.08	-0.10	-0.08	-0.03	-0.03	0.00
April	0.05	0.05	0.12	-0.08	-0.07	0.02	-0.05	-0.05	-0.08	-0.05	-0.07	-0.02
May	-0.02	0.07	-0.07	-0.12	0.08	0.05	0.05	0.12	0.15	0.12	0.10	0.03
June	-0.07	-0.10	-0.02	-0.05	0.12	-0.05	-0.08	0.07	-0.07	0.03	0.02	0.03
July	-0.20	-0.05	-0.24	-0.15	-0.07	-0.12	-0.20	-0.10	-0.18	-0.28	-0.28	-0.25
August	-0.25	-0.20	-0.22	-0.44	-0.34	-0.42	-0.32	-0.40	-0.47	-0.47	-0.60	-0.30
September	-0.22	-0.15	-0.17	-0.42	-0.42	-0.44	-0.49	-0.45	-0.45	-0.54	-0.50	-0.47
October	-0.39	-0.34	-0.12	-0.15	-0.60	-0.72	-0.72	-0.55	-0.44	-0.47	-0.32	-0.35
November	-0.00	-0.05	-0.07	-0.02	-0.18	-0.17	-0.08	-0.10	-0.08	-0.07	-0.08	-0.07
December	0.08	0.03	0.02	0.02	-0.03	-0.18	-0.13	-0.03	0.00	0.00	0.02	0.03

Table 32

Humidity gradient (g m^{-3}/100 m) for the air layer from 1.1 to 15.2 m, giving the diurnal and annual variation in the form of monthly averages for the three years 1945-1948 at Rye, England. (Corresponding temperature gradients are given in Table 23.)

Hour	2	4	6	8	10	12	14	16	18	20	22	24
January	1.27	1.55	0.99	0.71	-0.21	-0.71	-0.57	0.21	0.78	0.57	0.92	1.06
February	0.28	0.21	0.07	-0.14	-0.71	-0.50	-0.42	-0.42	-0.28	-0.71	0.14	0.28
March	1.91	1.77	1.70	0.00	-0.63	-0.50	-0.42	0.71	0.92	1.55	1.84	1.77
April	1.34	1.41	0.21	-2.69	-2.19	-2.05	-1.91	-1.34	-0.71	0.35	0.99	1.34
May	0.71	0.78	-2.12	-4.32	-3.75	-3.75	-4.39	-3.47	-2.05	-0.35	0.57	0.78
June	-0.14	-0.00	-4.25	-6.09	-5.53	-5.45	-5.53	-5.03	-4.32	-2.41	-0.64	-0.28
July	2.55	2.90	-0.35	-3.26	-3.97	-4.18	-4.04	-3.82	-2.69	-0.64	0.99	2.05
August	0.85	0.92	-1.77	-5.17	-3.75	-3.75	-3.26	-2.83	-3.82	-1.70	0.00	0.42
September	1.48	1.63	0.63	-1.55	-3.19	-3.40	-2.90	-2.69	-1.91	0.35	0.71	1.20
October	1.84	1.84	2.26	-1.48	-2.69	-2.41	-1.98	-1.55	0.14	1.77	1.91	1.55
November	0.28	0.00	0.42	-0.07	-1.84	-1.27	-0.85	-0.07	0.28	0.57	0.57	0.57
December	0.35	0.21	0.28	0.35	-0.35	-0.99	-0.92	-0.21	0.50	0.42	0.64	0.57

[178], using an aspiration psychrometer, gave the average results for cloudless days from 21 to 27 June 1950 shown in the table below. In this dry climate, the inversion is on the average substantially better marked than the normal decrease of water-vapor pressure.

Twelve-hr averages of water-vapor pressure (mm-Hg).

Height (cm)	08:00-19:00	20:00-07:00	Diurnal fluctuation
130	5.31	5.38	1.4
5	5.36	4.61	2.3

The position is quite different in the damper climates of higher latitudes. Figure 52 shows, in its lower half, the average daily variation of water-vapor pressure for three dry August days in 1934 in the parish of Pälkäne in Finland (about 61° N). The readings were made at three different heights above ground level by M. Franssila [307], using platinum resistance thermometers for temperature and an aspiration psychrometer for air humidity. The night inversion is only weakly marked, but in contrast the fall during the day is very great. The double wave of water-vapor pressure becomes a single wave close to the ground, with a maximum in the middle of the day; this is because, in these northern latitudes, convection is not vigorous enough to carry away upward, sufficiently rapidly, the supply of water vapor, which increases with rising temperature. Finnish observations of V. Rossi [138] made above fallow land in Lauttakylä, using psychrometers with thermoelements, even gave a night maximum of water-vapor

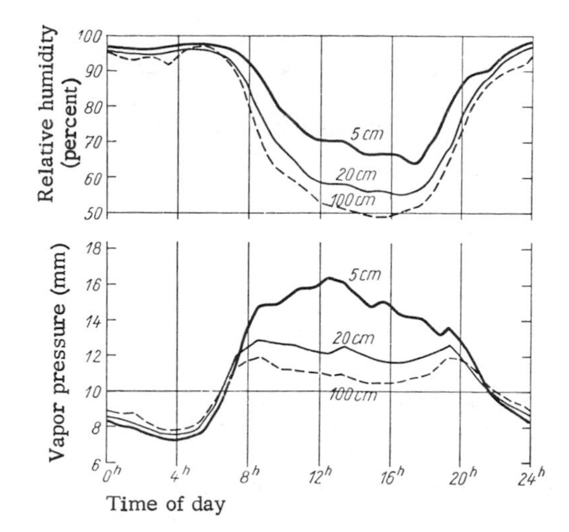

Fig. 52. Diurnal pattern of air humidity in the lowest meter. (From observations by M. Franssila in Finland)

pressure at a height of 1 to 2 m. This corresponds to the humidity inversion (excluding the possibility of the advection of moister air from a neighboring meadow). The decrease of water-vapor pressure by day is so strongly preponderant here that it continues to be vigorous in the middle of the day, as may be seen from Fig. 53.

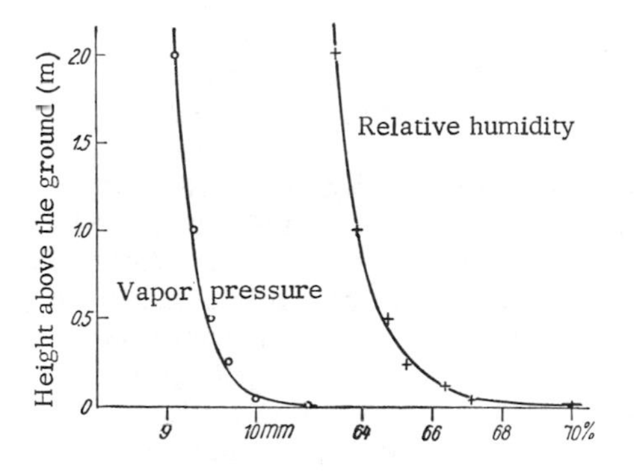

Fig. 53. Mean daily values of vapor pressure and relative humidity. (From observations by V. Rossi in Finland)

There seem, however, to be cases in which a humidity inversion is no longer established near the ground at night. This is shown by a careful set of observations made by C. W. Thornthwaite in Seabrook (36.6° N), even during the clear spring days, for which the temperature conditions were mentioned earlier (pp. 83f). It is a pity that the observations between 23:00 and 05:00 are missing, but from 05:00 to 23:00 the water-vapor pressure decreases continuously from 5 cm upward, and during the night the decrease is noticeably lower than by day, as may be seen from the following mean values for the days in question:

Height (cm)	640-80	40	20	10	5	2.5
14:00-17:00 (mm-Hg)	5.7	5.8	5.9	6.0	6.3	6.4
20:00-23:00 (mm-Hg)	6.9	7.0	7.0	7.0	7.1	7.1

Occasional deviations below 5 cm can be left out of consideration as local peculiarities. The water-vapor-pressure graphs published in the interim reports for Seabrook also show that the humidity inversion is an exceptional feature. Even in the Nebraska project (see Sec. 25), the measurements made in August 1953 in O'Neill (42.5° N), which have been published in detail, show almost without exception a decrease of water-vapor pressure with height. Occasional measurements made by W. Vieser [177] over sandy soil at Karlsruhe with a Diem small-hair hygrometer produced similar results, but these still need to be confirmed.

This is surprising, because dew forms at night in these places too. The conclusion must be that water present in the atmosphere is hardly used at all in the formation of dew, as enough water comes out of the damp ground for this purpose. This is supported, for example, by the fact that in Germany more dew is observed in the autumn with its warmer earth than in the spring when the ground is colder, under circumstances that are otherwise similar. It may serve us better, however, to remember that P. Lehmann and H. Schanderl [173] proved as long ago as 1942 that evaporation and the deposition of dew can take place simultaneously.

It is well known that the depositon of dew takes place when a surface is at a temperature lower than that of the saturation temperature of the air with which it is in contact. Since solid surfaces cool more quickly than air, the first dew forms when the relative humidity reaches 80-90 percent; and in warmer climates, where radiation is stronger, it may even form when humidity is only 60 percent. Evaporation depends on a difference between saturation vapor pressure corresponding to the temperature of the evaporating surface and the momentary vapor pressure of the air, and is proportional to this difference; hence it is still taking place. Obviously, the same surface cannot receive a deposit of dew and have evaporation taking place from it simultaneously. But it may nevertheless happen that the tips of plants are having dew deposited on them at the same time as the ground, or lower parts of the plant, are still giving off water vapor. Even with ground free of vegetation, poorly conducting parts can be receiving a covering of dew while better-conducting parts are still evaporating. This may be observed directly when the dew forms as rime. It was thus that Lehmann and Schanderl were able to register a marked fall of dew while at the same time and in the same place the lysimeter still showed a loss of weight. If in the evening a surface feels covered with dew, it does not follow that the evaporation process, in the sense of applying to the water cycle as a whole, is necessarily over for the day.

Since the formation of dew depends on outgoing radiation at the particular site (horizontal shielding, cloudiness), temperature, the heat budget of the particular part of the surface, humidity of the air, and local air currents, it is not a continuous process, even if the quantity of dew present usually increases with the passage of time during the night. There is a short-term changing rhythm of evaporation and deposition of dew, depending on the type of plant cover and structure of the ground. Lehmann and Schanderl have proved this by using a sensitive device by means of which the change can be detected quickly in any object being tested.

Plants are not the only recipients of dew but, as becomes more and more clear, the hygroscopic particles present in the atmosphere can also extract considerable quantities of water from it, and this may play an important part in the water economy of the atmosphere. This was recognized earlier in India's cotton-growing soils and is being investigated by K. Büttner [171] in the U.S.A.

Relative humidity is calculated from the water-vapor pressure and the saturation vapor pressure, which is a function of the temperature. Its daily behavior is something like a mirror image of the daily variation of temperature. Therefore in the early afternoon it is relatively low in the air layer near the ground, and relatively high about sunrise.

Within the air layer near the ground there are two extreme possibilities which are probably influenced by the macroclimate. In cold, damp climates it may remain relatively moister at the ground, both by day and by night, than at a height of 1 m. This is seen from the observations of M. Franssila [307] shown in Fig. 52 above. Even with measurements made above bare humus there was a marked increase of humidity during the day, down to a height of 5 cm, and a weaker increase by night. This is shown by the observations of M. Diem [172] in Karlsruhe in July 1947, using a hair hygrometer. Only below 5 cm was the relative humidity somewhat lower.

The reverse situation holds for dry climates. An investigation was carried out by L. A. Ramdas and M. S. Katti [175] from 4 to 8 January 1933, using aspiration psychrometers over a water-absorbing soil in India. Figure 54*a* shows the isopleths of air and ground temperatures in order to give a complete picture. The distribution of water-vapor pressure (Fig. 54*b*) shows the usual noon maximum with over 9 mm-Hg at the ground, not coinciding with the temperature maximum but occurring earlier, about the time of the radiation maximum. At night there is a strong radiation inversion. Relative humidity, shown in Fig. 54*c*, is lower, both by day and by night, at the ground than it is at 1 to 2 m.

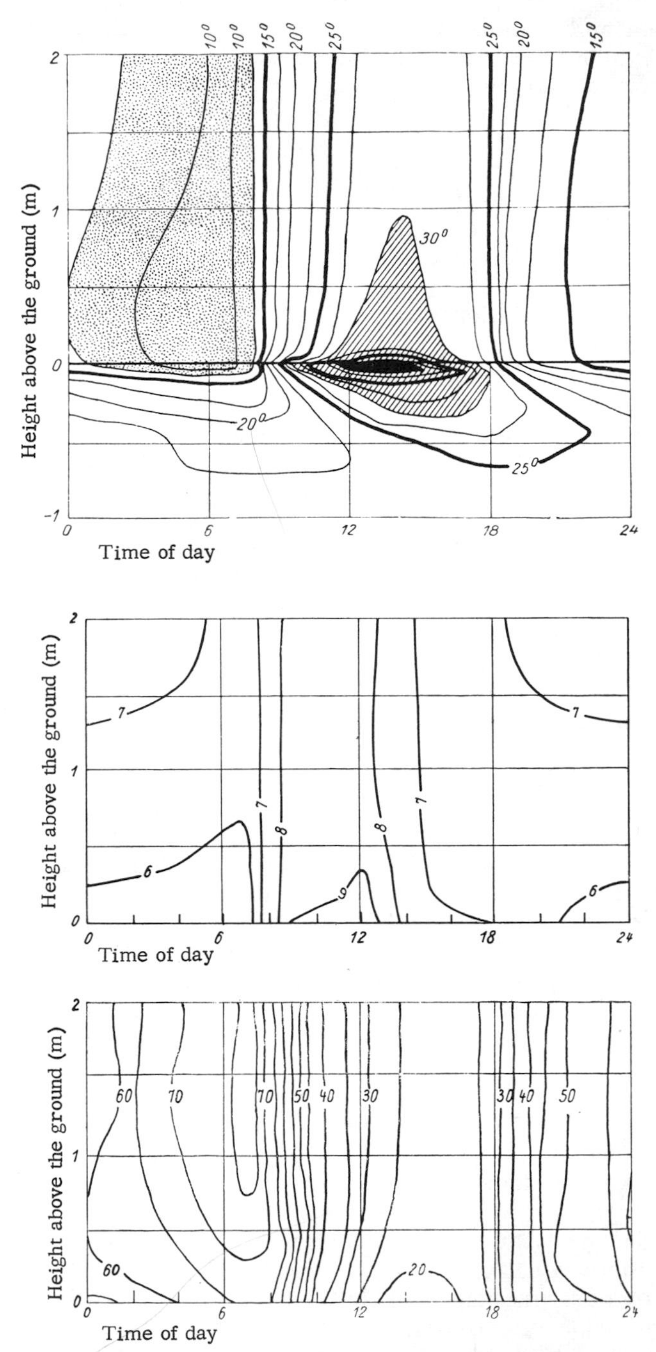

Fig. 54. Measurements of (*a*) temperature, (*b*) water-vapor pressure, and (*c*) relative humidity during a day in the arid climate of India. (After L. A. Ramdas and M. S. Katti)

An increase in the fluctuation of water-vapor content is found, as the ground is approached, similar to the increase in fluctuation of diurnal temperature. The difference between the values for the moistest and driest hours for 19 clear summer days in England was:

Height (m):	1.1	15.2	47.2	106.7
Diurnal fluctuation (g m^{-3}):	3.31	2.13	2.04	2.07

Other examples have already been given (Fig. 52 and pp. 105f). The unsteadiness of water-vapor pressure in the layer near the ground was mentioned on p. 46 (Fig. 16). Some new figures have been published by U. Berger-Landefeldt, J. Kiendl, and H. Danneberg [169] for ground covered with vegetation; these show that, within the 2.6-m layer with which they were concerned, there could be temporary pressure differences in the evening of up to 6 mm-Hg. It is also very instructive to watch the behavior of a hair hygrometer, with the hair horizontal over the ground surface on a day of dry summer heat, and observe how its rapid fluctuations indicate the quick change from moister to drier parcels of air, as described by A. Büdel [170].

16. The Wind Field and the Influence of Wind near the Ground

The horizontal transport of an air mass is a consequence of large-scale differences in air pressure, that is, of the horizontal pressure gradient. The wind is a "gradient wind," and as such can be observed in its undisturbed state only at higher levels. The surface of the earth acts as a brake on the movement of air. In general, the friction layer is 1000 to 1500 m thick; only above this height can the true gradient wind be observed. In the following discussion attention will be directed first to the disturbed wind field near the ground, and then to the great influences exerted on the characteristics of climate close to the ground by the more or less powerful gradient wind.

Statistical analysis of wind observations shows that wind speed usually increases with height; and in the Northern Hemisphere there is a veer of wind with height. For the purposes of this book, which is concerned only with the lowest 100 m of the atmosphere, this shift of wind direction will be left out of consideration. The increase of wind strength with height is, however, a factor of some considerable importance for the layer near the ground. This may be observed directly by looking at the rime in Fig. 55. Rime is formed when supercooled water droplets, carried by the wind in fog or cloud, come in contact with a solid surface and freeze immediately. If an even distribution of such droplets in the lowest

Fig. 55. The banner of rime shows the increase of wind with height. (Photograph from Mount Washington)

meter is assumed, then the amount of rime deposited on the fence post seen on the right of the picture will be greater, the greater the number of drops transported in a given time, that is, the greater the wind strength. The length of the banner of rime is therefore a measure of the strength of the wind, blowing from the left.

Figure 56 gives the diurnal variation of wind speed at various heights above the ground, for observations made over a year (mean values from September 1953 to August 1954) by E. Frankenberger [445] on a radio mast in Quickborn, Holstein. At lower levels there is a daily sequence with a well-marked maximum in the hours around noon and a minimum by night; the curves lose this distinct form with increasing height, until by 70 m the trace is approximately uniform. The movement of the air at higher levels, where the wind speed depends on the position of the regions of high and low pressure and not on daily influences, is carried downward to lower levels by eddy diffusion. The maximum of wind strength at the ground near midday is due to the existence of the maximum of the austausch coefficient at this time (see p. 41). This corresponds to a minimum of speed at heights over 100 m, because at this time there is more energy carried away downward. There is, then, a maximum of wind by night at these heights, which may be seen in Fig. 56 for a height of 70 m. The change from the type of daily variation near the ground to that of the inverse form at higher levels takes place in the so-called "intermediate layer." This is relatively high when either convection or frictional mixing is great, that is, in summer, in cloudless

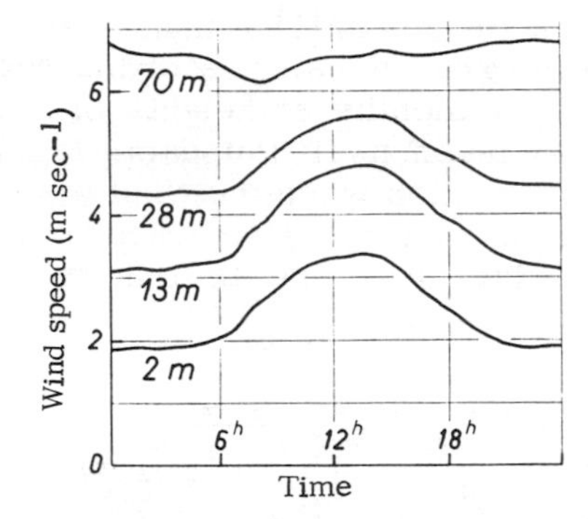

Fig. 56. Diurnal variation of wind speed in Quickborn, at four different heights above ground. (After E. Frankenberger)

weather, in low-pressure troughs, or in gales; on the other hand it is low in winter, with overcast skies, in high-pressure areas, and with light winds. The height of this intermediate layer varies between 50 and 100 m; but these limits may be exceeded in both directions. M. H. Halstead [184] has made a new theoretical evaluation of the daily variation in the lowest 250 m, using new ideas about frictional mixing, and basing his results on new observations from New Jersey, Texas, and Ohio.

It should be mentioned that this picture of a daily pattern is limited to latitudes affected by migrating high- and low-pressure areas. In the tropics there is a double wave of variation corresponding to the double variation in diurnal pressure. At sea there is very little diurnal variation of wind.

The well-marked daily period of wind speed is followed by an accumulation of times of calm, particularly by night. H. Henning [116] measured the number of calms, expressed as a percentage of a year's observations made from April 1953 to March 1954 at the Lindenberg Observatory, Berlin (Table 33). The increased frequency of calms by night and close to the ground can be recognized easily. The significance of this feature will be discussed later on.

Table 33

Number of calms as percentage of a year's observations, at Lindenberg Observatory.

Height of measurement (m)	Hour of day											
	2	4	6	8	10	12	14	16	18	20	22	24
10	4.1	5.0	2.8	2.2	0.8	0.6	0.6	1.4	2.8	6.6	6.1	5.7
5	9.1	9.7	8.0	3.3	1.7	1.1	1.7	3.0	4.4	10.2	10.0	9.4
1	15.7	16.3	10.2	4.4	2.5	1.9	1.9	3.0	8.8	17.9	16.8	17.6

As early as 1918, G. Hellmann [185] made a series of measurements even closer to the ground, using recording cup anemometers, over a period of several months at heights of 5, 25, 50, 100, and 200 cm on the Nuthe-Wiesen near Potsdam. Figure 57 shows the number of hours of observation when the wind was calm, expressed as a percentage, from his results. The dark hatching shows clearly the tendency of the layers close to the ground to become an "air sump."

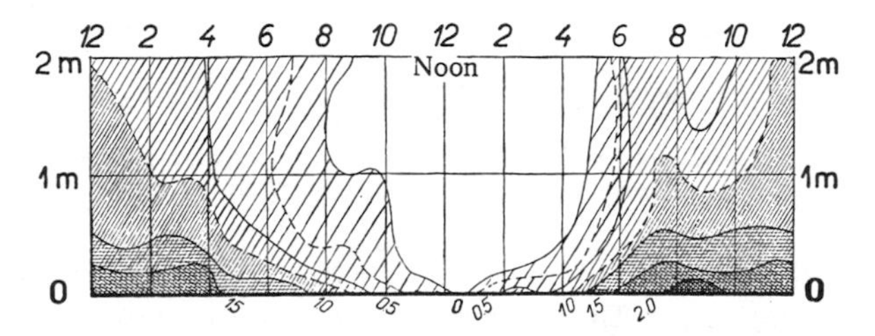

Fig. 57. Frequency of hours of calm in the lowest 2 m. (After G. Hellmann)

The features of the vertical wind profile, and knowledge of the laws that govern the increase of wind speed with vertical height, are of great interest, both for the theory of air streaming and turbulent mass exchange and for the heat and water economy of the atmosphere, as well as for many other practical problems. The following discussion will be restricted to the principal features that are capable of practical evaluation.

Once more, we turn to the observations made by C. W. Thornthwaite in Seabrook (see pp. 83 and 109). Four selected hourly average values from the five undisturbed spring days in question are shown in Fig. 58; the number on the line is the hour for which these mean values apply. On the left, the wind profiles are shown with linear scales. It can be seen that the rate of change of wind speed with height increases as the ground is approached, in a way similar to the gradients of temperature and water-vapor pressure. The simplest way in which the relation between wind speed

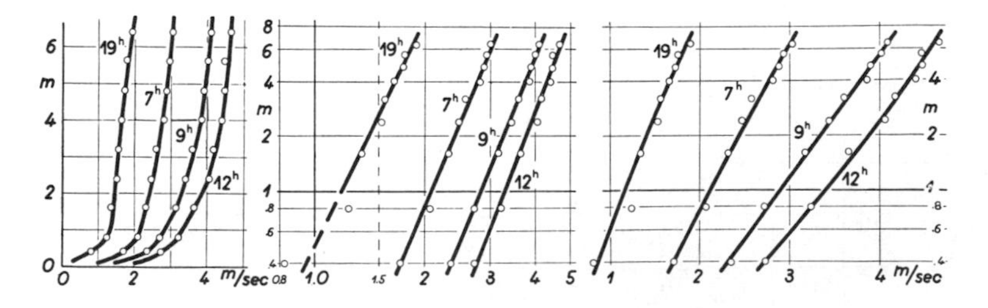

Fig. 58. Wind profiles at Seabrook, in three different methods of representation. (Corresponding temperature profiles are shown in Fig. 39.)

u (m sec^{-1}) and height z (m) can be expressed is by a power law:

$$u = u_1 z^a \qquad (\text{m sec}^{-1}),$$

in which u is the wind speed at a height of 1 m. With the aid of logarithms, the equation becomes

$$\log u - \log u_1 = a \log z,$$

from which it follows that, if a double logarithmic coordinate system is used, the relation between wind speed and height will become a straight-line graph. This can be seen in the center diagram of Fig. 58 for the Seabrook measurements. Although 5-hr averages are used here, the observations fulfill these conditions approximately.

The differing slopes of these four straight lines make it immediately clear that the exponent a in the foregoing law is not a constant; but, since a is equal to the tangent of the angle the line makes with the log z axis, it must become smaller as wind speed increases. Since we have seen that there is a dirunal variation of wind speed, it follows tha a must alter with the time of day. In addition to this, a is a function of height, and increases with proximity to the ground. O. G. Sutton [190] used the observations made by G. S. P. Heywood [186] for the lowest 100 m to compute the daily variation of this exponent, subdivided for summer and winter readings. From April to September a varied between 0.07 at noon and 0.17 at night, and from October to March between 0.08 and 0.13. H. Henning [116] found seasonal averages deduced from the measurements made in Lindenberg to be 0.32, 0.39, 0.53, and 0.28 during the day in spring, summer, autumn, and winter, respectively, while by night the values were 0.38, 0.49, 0.59, and 0.28. These values were for the lowest 10 m; but the autumn values were for the first 5 m only, and are rather high for that reason. The actual value of a also depends on the temperature structure and on the roughness of the ground surface.

A quick comparison of wind readings taken at different heights in the air layer near the ground can be made most simply by using the power law. If a basic value for a of 0.25 is used, then the wind speeds are in the same ratio as the fourth roots of the heights at which they were measured.

A realistic view of the connection, however, can be obtained only if the results of laboratory experiments on streaming are used as a point of departure. As a preliminary simplification, the assumption is first made that there is no effect due to the layered structure of the temperature distribution, that is, that there is no instability which might favor vertical transfer of wind motion, nor is there a stable layering which might suppress it. For neutral or adiabatic equilibrium we have L. Prandtl's [189] logarithmic law:

$$u = \frac{u_*}{k} \ln\left(\frac{z}{z_0}\right) \qquad (\text{m sec}^{-1}).$$

In this equation u_* (m sec^{-1}) is the velocity of shear, which is indicative of the amount of turbulence, and its value is independent of height for a given wind profile; k is the von Karman constant, which has been evaluated in various ways as 0.4 (a dimensionless number) for the layer of air close to the ground; and z_0 is the so-called roughness parameter, which has the dimensions of length and allows a numerical value to be put on the roughness of the ground. This parameter will be dealt with in more detail when the time comes to discuss the influence of vegetation (see Sec. 30). Changing to common logarithms and collecting all the constants together, we arrive at the law of wind change with height in its simplest form:

$$u = c \log\left(\frac{z}{z_0}\right) \qquad (\text{m sec}^{-1}).$$

This equation is a straight line in a system with coordinates u and $\log z$. Thornthwaite's observations are plotted in this manner on the right of Fig. 58. When Fig. 39 is considered, the temperature distribution at 07:00 and 09:00 on these spring mornings, after the night inversion has been broken down and before diurnal heating has properly set in, corresponds most closely to the conditions of neutral or indifferent stability that were presupposed in formulating the foregoing law. The layered structure of the wind for both these hours, seen on the right of Fig. 58, is therefore in close conformity with a linear law.

Many writers have tried to extend Prandtl's law to other temperature structures. This they have done by improving the technique of measurement, by eliminating the influences of environment, and by making a very large number of observations. The mutual relation between wind and temperature profiles may be seen readily from a study of Fig. 59. The observations used this time are those made by W. D. Flower [115] in the winter of 1931-32 in the desert near Ismailia on the Suez Canal, already cited. The abscissa of Fig. 59 is the wind speed at 62.6 m, at the top of the observation mast. The ordinate is the wind increase between 15.2 and 62.5 m, which itself increases as the gradient wind increases. However, the amount of the increase is strongly influenced by the temperature structure. The four curves shown in the fig. are for four different temperature gradients observed over the same range of height, the value of the gradient, in degrees per 100 m, being given alongside the curve. The four curves tend to converge toward the origin of coordinates, since a calm aloft is accompanied by still air below. If the temperature gradient is negative, then an increase in the upper wind makes only a small difference to the rates of change of wind with height. On the other hand, when there is a stable inversion, the cooler air will remain near the ground, and an increase in the gradient wind will only increase the difference between the winds above and below but will not be able to extend its influence down to the ground.

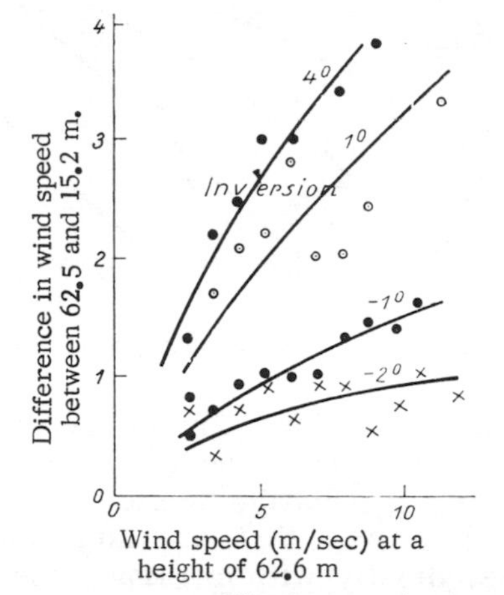

Fig. 59. Influence of temperature stratification on the increase of wind with height. (After measurements by W. D. Flower in Egypt)

If the simple form of Prandtl's equation is differentiated we get:

$$\frac{du}{dz} = c z^{-1}.$$

E. L. Deacon [181], who carried out a new series of measurements during the last war at Porton, England, where N. K. Johnson (see p. 75) determined the temperature profile, was able to show, by making a critical analysis of all previous work, that the simplest form of the law was

$$\frac{du}{dz} = c z^{-\beta}.$$

The factor β depends on the temperature structure. If this is adiabatic, $\beta = 1$ and the equation reduces to the Prandtl equation. When the atmosphere is stably layered, the increase of wind strength with height is greater than in neutral stability, and β becomes less than 1. In these circumstances the $u = \log z$ line develops a concave curvature toward the u-axis. This can be seen on the right of Fig. 58 in the Seabrook curve for 19:00. When the atmosphere has an unstable structure, β is greater than 1 and the curve is convex toward the u-axis. The 12:00 Seabrook data show this configuration. From results available up to now, the value for the lowest 10 m of the atmosphere is $0.7 < \beta < 1.2$. Wind increase with height, therefore, according to E. L. Deacon, follows a law that is completely analogous in form to the law of variation of temperature with height derived by Brocks (see p. 78).

Figure 60 shows results of observations made by Deacon over long grass in the summer of 1941. To allow for the effect of vegetation cover, 25 cm is subtracted from all heights before

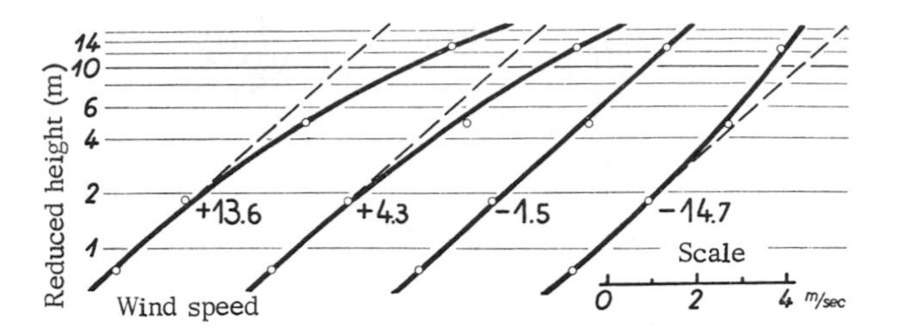

Fig. 60. Dependence of the wind profile on temperature stratification. (After E. L. Deacon)

they are represented logarithmically in the graph. More will be said about this in Sec. 30. The reduced heights are used in Fig. 60. Four examples, using a relative scale for u, are shown beside each other in such a way that on the left are the stable temperature structures, mostly with low true wind speeds, and on the right the unstable temperature structures. The temperature gradients measured between 17.1 and 1.2 m are written alongside, converted into degrees per 100 m.

Figure 60 also shows with great clarity the dependence of wind increase on the temperature structure and that this dependence is greater in the higher layers than it is in the lowest 2 m, where it almost becomes nonexistent. The assisting or restricting influences of stable or unstable structures must depend, naturally, on a certain mobility of the air, if they are to be effective. This is lacking, as we know, precisely in these layers near the ground. In this layer, in the narrower sense of the term, the simplified Prandtl equation can be used, even with temperature profiles that are not adiabatic, without involving too large errors.

In practical climatology it must always be borne in mind that laws governing wind increase with height apply to mean values only over a long period of time, or, if applied for short terms, it must only be with observations from which all other influences have been eliminated. G. Hellmann [185] established, in the above-mentioned measurements, that "there often are threads of stronger wind among lesser air movements." Figure 61 shows wind profiles measured by W. Schmidt [1130] near Vienna in 1928 for May nights with light winds. These show the very great irregularities that can arise in individual cases.

While the wind profile is subject to correction from the coexisting temperature structure, this correction is negligible in the lowest 2 m; by contrast, the influence of wind speed on the temperature structure is extraordinarily great, and is greatest in the lowest 2 m.

Figure 62 shows temperatures recorded on three consecutive nights in Kentfield, California, by A. G. McAdie [187]. The wind increase in the middle of the night swept away the cold inversion layer. For the layers near the ground, mixing means heating,

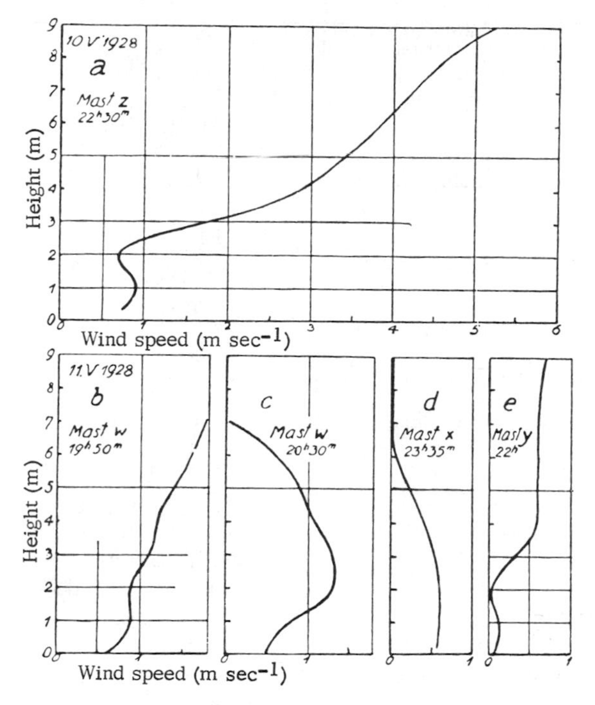

Fig. 61. Irregular structure of wind above the ground. (After Wilh. Schmidt)

which in this extreme case amounted to about 10 deg. The farmer and the fruitgrower have, for this reason, no fear of night frosts in spring, when the wind does not die away in the evening.

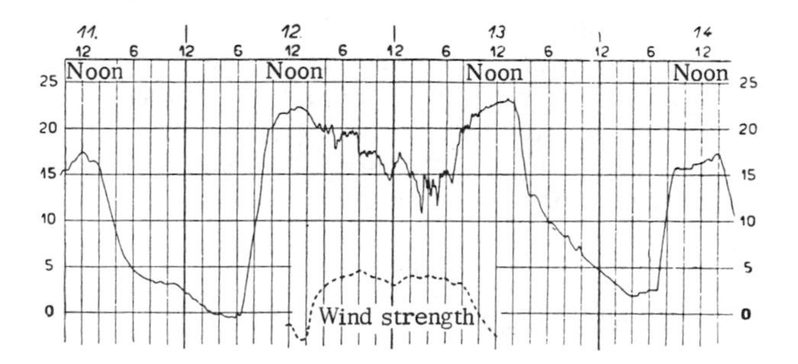

Fig. 62. Night temperatures at Kentfield, California, on 11-14 December 1911. (After A. G. McAdie)

Figure 63 is designed to give a spatial picture of what Fig. 62 showed in a time reference. The Los Angeles area is bounded on the north by the San Gabriel and the San Bernardino mountains, between which lies Cajon Pass. During a night of severe frost

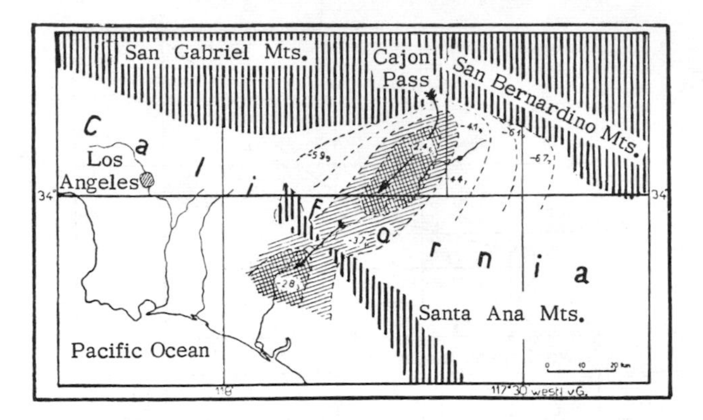

Fig. 63. Frost protection by wind. (After measurements by F. D. Young in California)

damage on 19-20 January 1922, F. D. Young [193] was making night-minimum-temperature observations in the threatened orange groves. Their distribution is shown by the broken isotherms (°C), while the influence of the northeast wind coming down Cajon Pass is shown by the higher values in the shaded areas. There may also have been some heating due to foehn effect.

While, therefore, lack of mobility of air allows the characteristics of the climate near the ground to be maintained through differences in the utilization of incoming and outgoing radiation, stormy conditions cause all differences to vanish by mixing the air completely. Those who wish to carry out microclimatological investigations in the field, therefore, must try to find weather situations with light winds or still air, in which all influences are clearly seen, and must avoid windy days and bad weather situations which confuse all issues.

Two orders of dependence are found between wind strength and temperature structure for a given wind speed, when this relation is investigated quantitatively: (1) the dependence is substantially greater by night than by day, and (2) the dependence increases with proximity to the ground.

Table 34

Mean temperature gradients (deg/100 m) at Leafield.

Hour of day	Wind (m sec-1)	Height (m) 87.7		57.4		30.5		12.4		1.2
Noon (12:00)	2.5		-1.2		-1.2		-1.6		-8.0	
	7.4		-1.1		-1.1		-1.5		-6.4	
Night (02:00)	2.5		-0.0		0.8		1.8		8.0	
	7.4		-0.3		0.1		0.4		2.4	

N. K. Johnson and G. S. P. Heywood [118] reported the mean temperature gradients shown in Table 34, evaluated for two groups of wind speeds from their 5-yr records of observation in Leafield (see p. 77). At noon above 10 m there is virtually no effect of wind speed, or it is overshadowed by other processes (for example, cloudiness) and in the lowest layer it is barely detectable. At night the influence of wind is easily recognized, and in the lowest 10 m it is even strongly marked.

The following relation between the wind measured at a height of 2.25 m and the temperature gradient was found by S. Siegel [164] in 77 individual observations for the layer from 220 to 6 cm at night in Hamburg:

Wind speed (m sec^{-1}):	0.3	1	2	3	4
Temperature gradient (deg/100 m):	14.5	10.3	7.5	5.6	4.2

For bright days (from 10:00 to 14:00) M. Franssila [129] found that there was "no noticeable effect" of wind speed on temperature gradient for the layer from 240 to 5 cm in Finland.

A. C. Best's [124] 2-yr series of observations in Porton, which have already been mentioned (pp. 83f) went even nearer the ground. Below 30 cm the observations were arranged according to five wind-speed groups for the midday hours 11:00 to 13:00, giving the following table when the temperature gradients are converted into degrees per 100 m:

Wind speed (m sec^{-1})	0	1	3	5	7	>7
30-2.5 cm, March	-412	-501	-522	-364	-327	
30-2.5 cm, June	—	-707	-599	-568	-452	

In June, a decrease in lapse rate with increasing wind can be recognized. In the layer from 30 to 120 cm, however, no effect could be shown for the middle of the day.

At night the situation is different. Between 23:00 and 01:00 the temperature gradients were as shown in Table 35. Spring nights show the influence of wind more clearly than do summer nights, because of the greater daily fluctuation of temperature earlier in the year.

Table 35
Temperature gradients (deg/100 m) at night in Porton.

Wind speed (m sec-1)	0	1	3	5	7	>7
120-30 cm, March	112	74	38	31	27	
120-30 cm, June	98	48	38	10	—	
30-2.5 cm, March	396	203	183	167	160	
30-2.5 cm, June	315	196	197	92	—	

Finally, in Fig. 64 we have a diagrammatic picture of both effects, using E. Frankenberger's new observations [445] from Quickborn. The graphs are for clear summer midday periods (dashes) and clear nights (solid lines). The temperature gradient, used as the ordinate, is negative during the day, and positive at night. The influence of wind speed on the temperature field is practically negligible on bright days such as these, when there is brisk convectional exchange. At night, in contrast, the dependence is better seen, the closer the range of measurement approaches the ground. Initiation of wind always brings about a substantial reduction of the temperature gradient, while a further increase in its speed has a lesser significance. Since it was possible to make reasonably accurate readings of the surface temperature by night (in contrast to daytime) the lowest 2-m layer is also included in the diagram; there is a special ordinate scale on the right of the diagram for this curve.

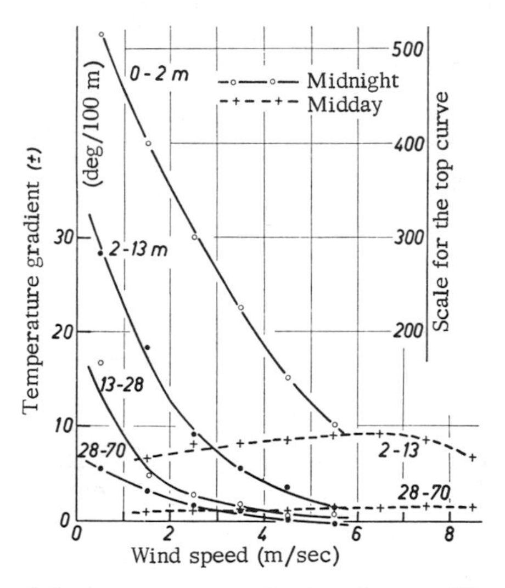

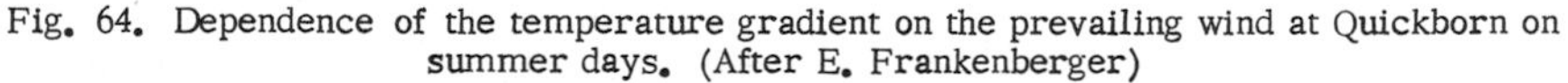
Fig. 64. Dependence of the temperature gradient on the prevailing wind at Quickborn on summer days. (After E. Frankenberger)

It is worth mentioning at this point that the laws of wind effect on temperature near the ground just discussed are valid only as long as there is no advection of air at other temperatures. Undoubtedly, a powerful influx of cold air can still lead to cooling, with freshening winds, even after the night inversion has been removed. In small-scale features, too, a similar sequence of events may take place. C. Hallenbeck [183] mentions one case when a few gusts of wind caused a fall of temperature in the Roswell fruit-growing area, because they advected air from a colder neighborhood.

17. Air Plankton and Foreign Gases

Besides heat, water vapor, and wind, the distribution of which in the layers close to the ground has already been discussed, there is still another range of elements whose behavior is of interest. These are solid particles suspended in the atmosphere or sinking slowly through it (air plankton) and gases, other than those, such as nitrogen, oxygen, and argon, which are constituent parts of air.

The particles and gases under discussion are those whose distribution is controlled from the ground in a way similar to that of water vapor or heat. The dust content of the air, the carbon dioxide, and the radioactive gases are derived mainly from the ground or out of it. Their vertical distribution is determined by this source, the state of the ground at the time, and the conditions prevailing in the air near the ground.

Particles and gases have also other origins. Dust and sand are transported by storms from prairies and deserts. Industrial plants pour out smoke, waste gases, cement dust, and other products; carbon dioxide is liberated in combustion, and the dissemination of radioactive particles and gases from the testing of military nuclear weapons has become a source of anxiety to man. In addition to the influence of the wind blowing at the time, the spread of these foreign materials is mainly a function of eddy diffusion, which has been treated in Sec. 9. This problem will arise again from time to time in discussing climatological aspects of topography (Sec. 47) and the climate of cities (Sec. 52). It remains to show here that this form of advective dissemination is also affected by the periodic fluctuation of atmospheric processes close to the ground. This can best be seen by observing the concentration of industrial smoke and sulfur dioxide (SO_2), the daily distribution of which is given in Fig. 65, from observations made in Leicester, England, by A. R. Meetham [77, 220].

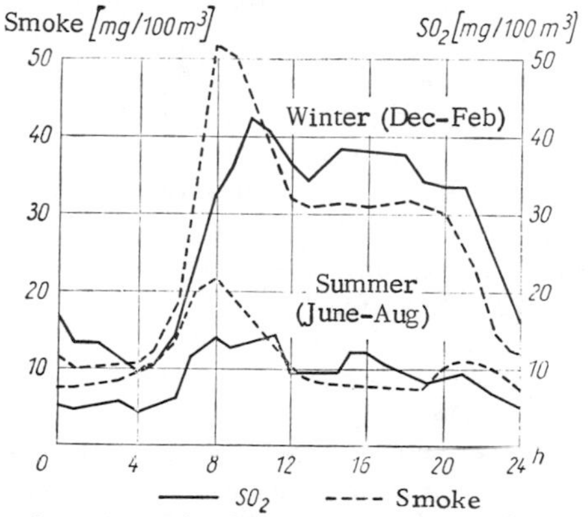

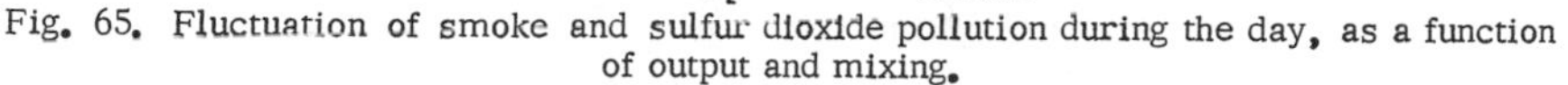
Fig. 65. Fluctuation of smoke and sulfur dioxide pollution during the day, as a function of output and mixing.

Domestic consumption of fuel is considerably greater in winter than in summer, and therefore the concentration of products from it is markedly higher. When fires are lit in the morning, there is a maximum, first seen in the form of smoke (the result of incomplete combustion) and then in the concentration of sulfur dioxide, which is roughly proportional to the quantity of fuel consumed. Concentration falls off rapidly after 08:00 and remains almost constant during the day. Some other observations even show a definite minimum in the early hours of the afternoon. This is not a reflection of the quantity of soot particles and sulfur dioxide from fires, but is due to the daily fluctuation of eddy diffusion, which carries much of the pollution aloft in the middle of the day and therefore reduces its concentration near the ground.

A change from advective processes to others controlled from the surface of the ground is illustrated by the sweeping and driving of sand and snow in the layers close to the ground, and by the driving of spray in the layer close to the water surface in heavy seas. In these cases a strong wind is able to remove particles of sand, snow, or water from the surface and carry them into the air. The concentration of the particles in the air depends on the wind speed, and on the size and shape of the particles and their specific gravity. The term "sweeping" is used when the horizontal visibility is not affected, while "driving" is used for cases in which the particles from the surface are lifted high enough to cause a marked reduction in visibility. These processes are of some considerable importance, since, along with the particles, the properties of the underlying surface are also transported into the layer of air near it. In desert sandstorms, the particles of sand, at temperatures much above that of the air, give up heat to it, and in addition absorb short-wavelength radiation, thus acquiring more heat.

First we shall consider the sweeping and driving of sand. As wind speed increases, some grains of sand begin to roll along the surface. Through collision they set other grains in motion. When the wind speed reaches about 5 m sec^{-1}, the grains that have been hit tend to fly into the air, rising at an angle between 30° and 70° and falling again in a flat curve, meeting the ground at an angle of 2° to 15°. The range of flight is about six times the maximum height reached. If the wind speed becomes very high, the particles of sand remain suspended in the air.

Within the ground layer both the size of the grains and the quantity of sand decrease rapidly with height. The following distribution was measured during a sandstorm at Colomb-Bechar on 14 April 1956 by L. Demon, P. de Felice, H. Gondet, L. Pontier, and Y. Kast [197]; the wind speed was 11 m sec^{-1} at 1.4 m and the figures give the number of grains of sand in 1 m^3 multiplied by 100; the maximum frequency is displaced toward grains of smaller size as height above the ground increases; the quantity of sand transported was 89 mg cm^{-3} at 60 cm and 32 at 140 cm:

Diameter of grains (μ)	40	50	60	70	80	90	100	120	140	160
At height of 60 cm	25	24	30	45	57	59	52	25	8	2
At height of 140 cm	50	77	65	53	46	41	33	10	2	1

Investigations made by K. H. Sindowski [217] on the island of Norderney have shown that grains are selected according to shape, and spherical grains are carried highest. Under the conditions obtaining in Germany, grains were not lifted higher than 1 m. The normal distribution of sand within the air layer near the ground can be taken from Table 37, the figures in which were measured at the upper edge of the steep beach in Norderney with a wind of 11 m/sec^{-1}at a height of 10 cm in the condition called "smoking sand dunes." The mass of sand, rounded off to the nearest gram for each section and unit of time, is expressed as a percentage of the maximum value at the ground surface. The main quantity of sand is seen to be carried into the first few centimeters above the ground. If we take the wind distribution to be the simplified version given on p. 117 for $a = 0.25$, the wind profile shown in the next line of the table is obtained. The amount of energy transported is proportional to the mass of the sand and the square of its velocity (= mass flow × speed). Since the mass of sand decreases with height but the wind speed increases, the maximum of energy was found to be at about 15 cm. If this value is equated to 100, the energy distribution of the sweeping sand grains is obtained, as shown in the last line. Heights about 15 cm above the ground therefore constitute the danger zone in cases like this. This result is in good agreement with the observations of E. Blissenbach, mentioned in the last edition of this book, that in the Egyptian desert telegraph poles were subjected to the greatest wear at a height of about 10 cm.

Table 36

Effect of wind on sand, Norderney Island.

Height above ground (cm)	0	5	10	15	20	25	30	40	85
Quantity of sand carried ($g\ cm^{-2}\ hr^{-1}$)	117	34	33	31	27	22	18	11	1
As percentage of maximum value	100	30	28	26	23	19	15	9	0
Wind speed ($m\ sec^{-1}$)	0.0	9.2	11.0	12.2	13.1	13.8	14.5	15.6	18.8
Relative energy (percent)	0	85	97	100	95	84	65	45	4

The incidence of sweeping sand has been shown by W. Haude [206] for the Gobi desert to be a function of the season of the year, the time of day, and the relative humidity of the ground. Sand and dust devils have already been discussed (pp. 91f).

This question becomes a matter of importance in the climate of Central Europe, when the problem of soil erosion by wind arises. Fertile topsoil is blown away and the exposed seed grows up defective or not at all, young shoots suffer mechanical damage, and young plants are smothered in sand. This problem in Lower Saxony has been studied in detail by R. von Gehren [203]. His investigation showed that "medium sand," with grains 0.1 to 0.5 mm in diameter, was most liable to be blown away. It was more difficult for the wind to loosen smaller grains from the soil and heavier grains offered a greater resistance because of their weight.

Sweeping and driving of snow occur in a similar way in the winter landscape. The local significance of this will be dealt with in Sec. 24; it can also affect the water budget. The most striking example of this phenomenon comes from F. Loewe [416], who found that in the Antarctic at least 20,000 tons of snow were swept over each meter of the coast of Adélie Land into the sea each year, which is an amount equal to half the precipitation falling in the 200-km coastal strip in the icecap. From 1086 measurements of blowing snow he obtained the following mean vertical distribution of the snow transported:

Height above snow cover (m)	10	20	50	100	200	300
Quantity of driving snow (10^{-7} g cm^{-3})	12	6.4	2.6	1.3	0.56	0.22

In severe gales (35 m sec^{-1} at anemometer level) 1 m^3 of air at a height of 1.5 m contained 6 to 10 g of snow when the visibility was under 10 m. I have no knowledge of any similar measurements made in Central European latitudes.

In the following pages, horizontal transport will be left out of consideration, and the discussion will be limited to air plankton and gases that come from the soil or from within the soil. First to be considered is the dust content of the air.

The term dust is used to cover the dry, rough, microscopic yet visible components of air plankton. Diameters of dust particles are between 1 and 50 μ, which would give terminal velocities between 0.1 and 200 mm sec^{-1}. These dust particles, which are unusually very fine pulverized soil particles, can therefore be lifted up into the atmosphere by wind and eddy diffusion, and will fall back to the ground very slowly. Only precipitation is able to cleanse the atmosphere efficiently (hence the process of "washing" the air by forcing it through a veil of spray in an air-conditioning plant). Particles larger than 5 μ are forced out of the body by coughing, but smaller particles can make their way into the breathing passages. Many measurements have therefore been made of the dust content of the air for health purposes, and most of these have been for heights of 1.5 to 2 m above the ground [77, 202].

Naturally, the amount of dust in the atmosphere varies greatly from place to place and from time to time. Normally the number of particles in 1 cm^3 is between 1 and 100. The smallest quantities are found over the open sea or over a winter snow cover, and the greatest quantities in industrial areas and large cities. According to A. Niemann [213], the absolute dust content lies between 0.5 and 5 mg m^{-3}.

The decrease in dust content with height can be measured either directly by a dust counter, such as the Zeiss konimeter, or indirectly by optical means. In the latter case it is not easy to allow for the additional influence of water droplets in mist. H. Goldschmidt [205] was able to prove, for example, by using a searchlight, that the atmospheric turbidity near the ground was considerably greater than the calculated value. From direct dust counts made in aircraft flights over Jena, H. Siedentopf [216] showed that 10 to 60 percent as much dust as was present in the air near the ground was still to be encountered at a height of 1 km. The percentage is naturally greater in summer, with its strong convection, than in winter. It is hardly ever possible to detect a preponderance of a particular size of particle, since particles of all sizes are carried aloft, and the differences in terminal velocity hardly have time to come into effect.

Unfortunately, there are virtually no systematic observations available of the decrease in dust content in the first few meters of the atmosphere, although such measurements would be of great use in the study of eddy diffusion. However, a few conclusions may be drawn from variation of dust content close to the ground. Figure 66 shows records made over the observation field of the geophysical observatory of Leipzig in Collmberg, from 28 to 30 July 1939, from records made by E. F. Effenberger [198]. The instrument used was a Zeiss recording konimeter, which could pick up dust particles between 0.1 and 30 μ.

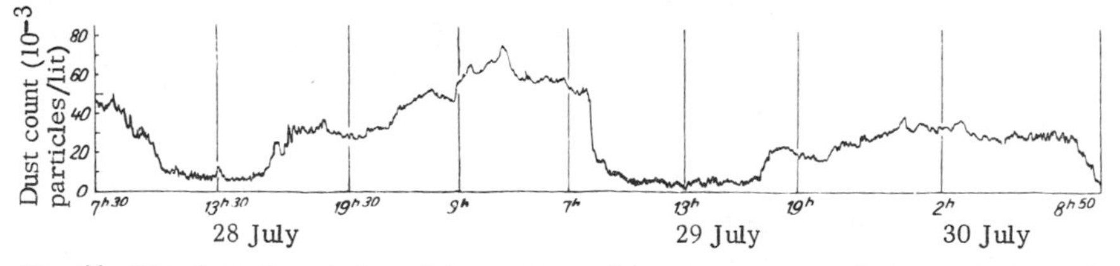

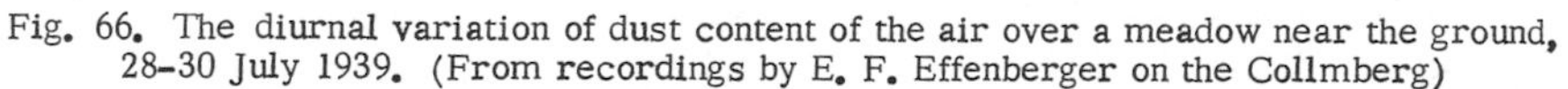

Fig. 66. The diurnal variation of dust content of the air over a meadow near the ground, 28–30 July 1939. (From recordings by E. F. Effenberger on the Collmberg)

These were days of fine summer weather, a fact confirmed for other places by other writers, with a marked diurnal fluctuation, showing a maximum at night and a minimum at midday. When the night inversion has formed, all dust is retained within it, having for the most part come down again near the ground through cooling in the evening (see p. 42). As soon as convectional mixing sets in with insolation in the morning, the dust is carried upward to higher layers, and there is a corresponding decline in the content at the ground. This is shown particularly well by the curve at 07:00 on 29 July. This agrees with the fact that the dust content of the air on high mountains is greatest about noon, because only at that time is it transported from the dust-filled low-lying country to sufficiently high levels.

Carbon dioxide (CO_2) is present in dry air in an average quantity of 0.3 parts per thousand by volume. A slow increase has been observed over the past 50 yr, corresponding roughly to the quantity of carbon dioxide produced by world industry. In Europe, the annual average fluctuates between 0.27 and 0.36 parts per thousand. The quantity of carbon dioxide absorbed by vegetation has reduced the supply available in the atmosphere over the past 40 yr by about $2 \cdot 10^{11}$ tons, which is equivalent to the amount provided by bacteria, by nocturnal exhalation by plants and animals, and by volcanoes and industry. It follows that there must be great variations in the carbon dioxide content of the air, both in locality and in time.

According to E. Glückauf's comprehensive study of the matter [204], sea water contains about 150 times as much carbon dioxide in dissolved form as the same volume of air. The carbon dioxide content of air above the sea is therefore mainly determined by the state of equilibrium with the water, that is, it depends on the temperature of the water. The smallest values found so far were 0.15 parts per thousand in the polar seas. The value is high over warm water, and also where cold water is welling up, presumably because it is particularly rich in carbon dioxide from decaying organic matter. The carbon dioxide gradient over the sea is small, as shown by K. Buch [196], because of the long period available for the establishment of equilibrium and because the sea is uniform over wide areas.

Since continental plant life requires carbon dioxide for respiration, the amount present is of immediate practical interest. New measurements, made by B. Huber and J. Pommer [209], have shown that there is no appreciable annual fluctuation of carbon dioxide content. This fact proves that the large consumption by plants in summer, in contrast to winter, disturbs the equilibrium hardly at all. By way of contrast, it is possible to observe a marked diurnal variation near the ground.

Measurements made over the past decade generally give substantially higher values by night than by day. Effenberger [199] has collated the results. Using an infrared absorption recorder (called URAS) belonging to the Baden Aniline and Soda Factory at Ludwigshafen, which is capable of detecting a change in carbon dioxide concentration in the air of 0.0001 percent, B. Huber [207] was able recently to chart the daily variation as a function of height above the ground. Figure 67 shows the average

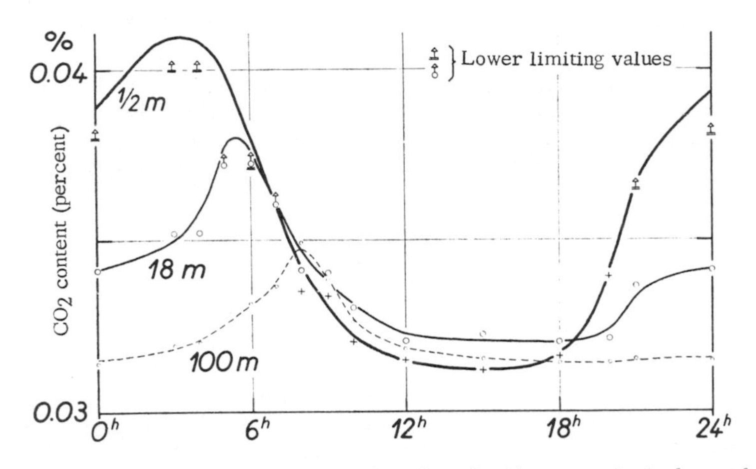

Fig. 67. The marked diurnal variation of carbon dioxide content in the lowest layer of the air. (From observations by B. Huber near Munich in summer)

of ten investigations made between 20 June and 1 August 1952 above a potato field near Munich. The suction tubes for taking air samples were raised to heights of 100 m by means of captive balloons.

While there is a nocturnal inversion, the proportion of carbon dioxide in the air increases. I. Iizuka [210] has demonstrated that there is a rapid increase, easily recognized, in the evening. Close to the ground it is so much steeper than at higher levels that there are surprisingly large vertical differences at sunrise. When the wind increases during bad weather at night, eddy diffusion increases and the process of enrichment in carbon dioxide is interrupted. Carbon dioxide content decreases when the wind increases, as proved by B. Huber [208] through simultaneous measurements of wind and carbon dioxide content. When heating starts in the morning and eddy diffusion sets in, there is a sharp decrease in concentration. Vertical distribution is similar from late forenoon until evening, there being almost complete mixing throughout the lowest 100 m. Distribution of carbon dioxide therefore follows a pattern similar to that of dust (Fig. 66) or smoke (Fig. 65). It is not yet possible to decide to what extent the decrease during the day is affected by the quantity of carbon dioxide consumed through plant respiration.

In the distribution of radioactive gases, a new factor is introduced—the time-dependent process of radioactive decay. The decay products of the radium and thorium series are of concern here. Other things being equal, a gas with a long half-life will be found at higher levels than one with a short half-life. According to R. Mühleisen [212] the air above land usually contains 20-400 · 10^{-18} curie m^{-3}, while over the sea, where no supply is available from below, the value is only about $1 \cdot 10^{-18}$c m^{-3}. If the austausch coefficient A is taken as 23 g cm^{-1} sec^{-1} at 100 m, and the power law for the increase of wind with height applies (see p. 117), where $a = 0.14$, then the decrease in content of radioactive gases with height, expressed as a percentage of its value at a height of 1 cm above the ground, will be as shown in Table 37, according to J. Priebsch [215].

Table 37
Content (percent of value at 1 cm) of radioactive gases in air.

Height (m)	0.1	1	10	100	1000	13,000
Radon	91	80	66	44	18	3
Thoron	62	19	1	0	—	—
Thorium B	89	74	51	28	4	0

The amount of emanation in the layer close to the ground depends on the state of the ground. Wet and frozen ground give off less than dry ground; and a few centimeters of snow can block the escape of emanation completely. The state of the weather and the type of soil (pore volume) are of some influence. It is also a fact that the distribution of the parent material depends on the geologic nature of the rocks below. The quantity of emanation is subject to great variation in place and time. Long periods of observation are therefore essential if the natural radioactivity of any locality is to be determined prior to possible construction of an atomic reactor to provide power.

The amount of radium emanation (radon) in the air layer near the ground has been measured near Frankfurt am Main by F. Becker [195]. Figure 68 shows the diurnal variation for 4–5 April 1943 at heights of 1 m and 13 m above the ground. On the average, as might be expected, the amount is greater at the lower than at the higher level. The similarity of this graph to that of the daily variation of carbon dioxide content (Fig. 67) is obvious. Here too one finds a lag of about 3 hr between maxima and the minima at

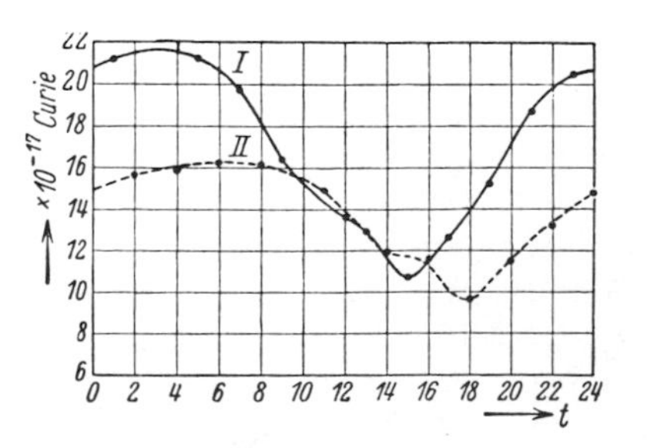

Fig. 68. Diurnal variation of radon content of the air at 1 m (I) and 13 m (II) above the ground. (After F. Becker)

the 1-m level and the 13-m level, as well as an increase in concentration with the small amount of mixing at night, and the uniform distribution effected by convection during the day. Only further measurements will show whether this is random or systematic.

The reverse is the case with ozone (O_3), since its source is in the upper regions and it can be brought downward to the layer near the ground only by mixing. As in now well known, ozone is formed by solar radiation at heights of 15 to 30 km. In the lower levels of the atmosphere, where it decomposes, it is found to the extent of only 2 to $4 \cdot 10^{-6}$ volume percent on the average. If the ozone content is measured in micrograms (μg or γ) per cubic meter, it varies between 10 and 60. Where there is a surface inversion or very still air without supply from above, the quantity can sink temporarily to zero.

Ozone, for these reasons, exhibits the opposite kind of diurnal variation from that of gases coming out of the earth. It was shown by R. Auer [194] and later by A. and H. Ehmert [201] that ozone content is greatest near the ground in the early hours of the afternoon, and least at night. In mountains, which lie closer to the source of ozone, there is a noticeable decrease in the daily fluctuation. Local fluctuations in a health resort have been investigated by G. Zimmermann [222].

Comparative measurements for various heights above the ground have been made at Lindenberg Observatory, Berlin, by F. Teichert [219]. During the summer of 1954, 222 separate observations were made at a height of 80 m on a tower, and at table height at the foot of the tower. The mean value at the top, that is, nearer the source of ozone, was higher throughout than the value below. For example, in the summer month of August, with vigorous mixing, it was 37 above and 32 $\gamma\ m^{-3}$ below; in September the figures were 23 and 22 $\gamma\ m^{-3}$. Figure 69 illustrates a typical daily fluctuation, that of 7 July 1954, when the weather became more settled in a ridge of high pressure that formed after there had been an incursion of polar air. In spite of the unavoidable scatter of the readings made on a single day, it can be seen how extraordinarily low were the values of ozone content at the end of the night, and how, when convection set in, the increase occurred at the top of the tower before it was felt on the ground. The maximum is in the late afternoon, when vertical differences are smoothed out by good mixing. The increase occurring before midnight, clearly associated with a wind increase (it started to rain at 01:00), was turned downward from above. The curve in Fig. 69 is like a mirror image of those in Figs. 65 to 68, which were for sources on the ground.

It must be mentioned here that ozone may originate in the layer of air close to the ground, in certain circumstances. The urban area of Los Angeles, whose sheltered setting has already been shown in Fig. 63, is often isolated from above by temperature inversions. In addition to nearly 3 million vehicles emitting exhaust

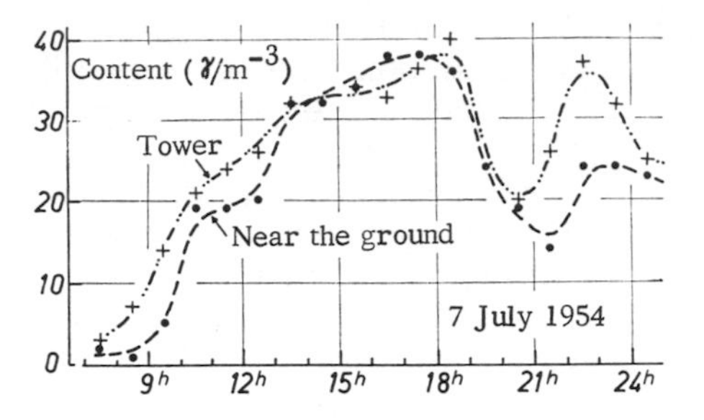

Fig. 69. Diurnal variation of ozone content at two different heights above the ground. (After E. Teichert)

fumes, the petroleum industry pours 55,000 tons of gas and sulfur products into the air daily. H. K. Paetzold [214] has shown that, as a result of a complicated chemical reaction in the presence of sunlight, ozone will be produced. The maximum concentration of this anthropogenic ozone hence occurs about noon, when the air of the city contains about $50 \cdot 10^{-6}$ percent or ten times the normal amount. A thick haze develops, which is called "smog"—a combination of the words smoke and fog. Ozone is about 1000 times more poisonous than the much-feared carbon monoxide. The irritation set up in the mucous membranes of the eyes and nasal passages becomes so great that the entire vehicular traffic of this great city must be brought to a standstill for hours. Even if the Los Angeles smogs are rare today, they remind us how watchful we must be about foreign gases in the lower atmosphere.

18. Optical and Acoustical Phenomena

The density of the air layer near the ground is not uniform because of large contrasts of temperature and water-vapor content, and the small amount of eddy diffusion. Hence it is also optically inhomogeneous. Parts of the air that are of greater or lesser density—called optical schlieren—deflect rays of light, but, since the schlieren are disposed irregularly both spatially and in time, the direction and amount of deflection are both subject to variation. This gives rise to an optical unsteadiness, which is called terrestrial scintillation.

If, on a warm summer day, one looks along a hot country road, over a sandy surface, or along the embankment of a railroad, the lower parts of distant objects appear to flicker, and the edges of houses seem to waver. Usually, optical phenomena like this are observed from a height of about 1.5 m, or eye level. It is well worth while, however, when visibility is good, to bend down slowly, bringing the eyes closer to the ground. Under no other circumstances will one appreciate so well the simultaneous increase in the temperature gradient and the decrease in eddy diffusion as by this experience of the rapid increase in the flickering movement. Then it becomes clear that the special characteristic of the layer close to the ground is its unsteadiness.

When the lapse rate is $\gamma = -3.42$ deg/100 m, the density of air is constant with height (see p. 85). Since, however, the lapse rate is usually considerably lower than this, density generally decreases with height. Since this is the case, the light ray AB in the top left of Fig. 70 is bent toward the denser layer. This bending is in the same sense as the curvature of the earth, and is said to be positive. An observer stationed at B therefore sees the object A (a star, the sun) in the direction A', that is, higher than its true position. This bending of light rays is called refraction.

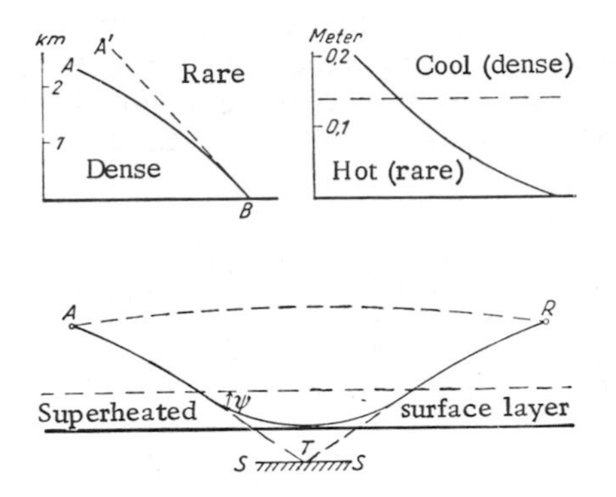

Fig. 70. Sketch of path followed by rays in forming a mirage.

The coefficient of refraction κ is the ratio of the radius of curvature of the observed ray to the radius of curvature of the earth. According to K. Brocks [227] κ usually lies between $+0.10$ and $+0.25$, that is, the curvature of the ray is about 1/10 to 1/4 the curvature of the earth. The relation between the local coefficient of refraction and the temperature gradient is given by the equation

$$\kappa = c\,(3.42 + \gamma) \sin \zeta,$$

in which ζ is the zenith distance of the ray of observation and c is a constant that depends on air pressure, temperature, refractive index of air, and local conditions of place and time.

In the surface layer, larger negative values of γ than -3.42 are found during the day (see Sec. 13). The bending of the light ray can therefore become negative, as shown on the top right of Fig. 70, which occurs as soon as the density of the air increases with height. The number of times this happens may be deduced from Tables 26 and 27 by looking for values of γ in excess of -3.42. A rule from K. Brocks [226] is that on clear summer days in Central

Europe the density of air increases up to 10 or 15 m, and on overcast days up to about 8 m; even on overcast December days it may increase to a height of over 1 m. In prairie and desert areas the corresponding heights are about 20, 15, and 8 m. Figure 71 shows the diurnal variation in κ for the Central European type of climate, for a horizontal line of sight ($\zeta = 90°$), from tables published by K. Brocks [227] for heights (on a logarithmic scale) up to 3 m, for the month of June. Anamalous curvature of the light path observed reaches 50 times the curvature of the earth shortly before noon on clear days at a height of 10 m. The change to normal curvature does not take place until about 12 m above the ground. Even with overcast skies, the curvature is negative during the whole of the day, and only changes sign at a height of 10 m, but the degree of curvature is naturally less. Where precise leveling measurements have to be made, or exact trigonometric height finding is to be carried out over flat territory, or the depression of the horizon has to be measured at sea, this optical peculiarity of the surface layer acquires considerable practical importance.

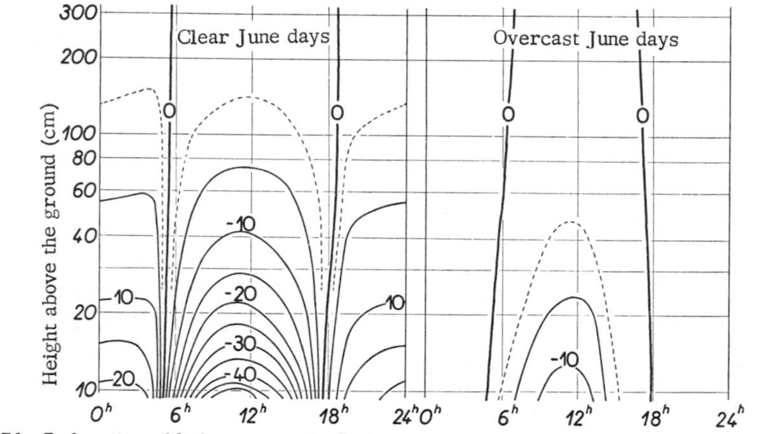

Fig. 71. Refraction of light rays in the layer near the ground on clear and overcast June days. (After K. Brocks)

The technique of measurement in this zone of mutual interest to the sciences of geodesy and microclimatology, used since 1939 by K. Brocks [225-228] and R. G. Fleagle [232], and in Greenland by R. A. Hamilton [233], provides a unique opportunity to reverse the procedure and to use optical methods to determine the layered structure of temperature differences near the ground. The advantage of this approach lies in the fact that the results obtained are not, as in all other methods, purely local values, but are averages over a wide range of distance. Their variation with time can also be followed with great precision. Care must be taken, all the same, that the distribution of water vapor should not be too strongly layered, as this has not been allowed for in the equation for κ given above.

In extreme cases, when there is an unusually large fall of temperature near the surface, it may happen that rays that enter the superheated surface layer at an angle ψ close to the horizontal will be bent upward again. In this case there is total reflection (Fig. 70, lower diagram). It is as if the incoming ray were reflected at a mirror, *SS*. The angle ψ is of the order of a few minutes of arc. An observer in position *A* sees the object *R* twice, that is, directly, via the light path shown with positive curvature *AR*, and also reflected from below, along the path *ATR*. This phenomenon is called inverted or inferior mirage. This was mentioned by W. Köppen as one of the special characteristics of the surface layer of air. This type of mirage may be observed in prairie lands and deserts, but most often on the coast. At the coast there is always the required range of sight that is needed if the angle ψ is to be small enough, and when the land is heated in the morning the necessary layered temperature structure is also present. The sight of a coastal strip "floating on the water" is an everyday occurrence for seamen and those who live on the coast.

Figure 72 illustrates exactly how these mirages are formed. Let *A* be the position of the observer's eye; *AF* is therefore the height of his eye above the sea. The direction of the astronomical horizon is *AH*, since the angle *HAF* is a right angle. The limit of vision over the sea is determined by the ray *AKW*, which touches the surface of the water at *K*, and which is slightly curved by normal refraction. The limit of horizontal vision is *AK*. The angle ν is the depression or dip of the horizon. Under conditions of normal refraction, the dip of the horizon and its distance depend on the height of the observer, as given in Table 38.

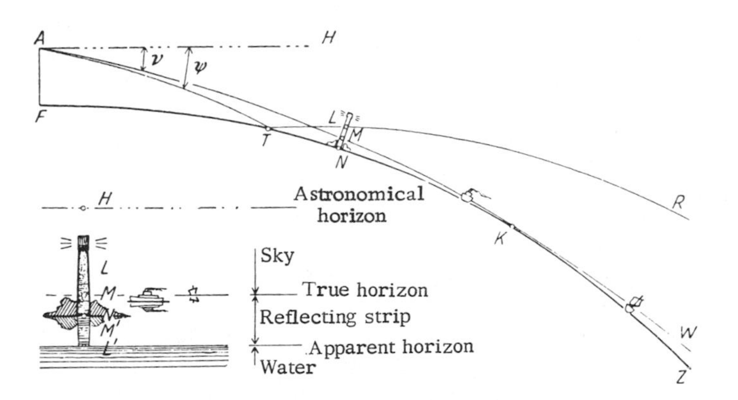

Fig. 72. The formation of an inverted mirage.

Table 38
Range of vision as a function of height.

Height of eye above sea level (m)	1	2	5	10	30	50
Depression of horizon (minutes of arc)	1.8	2.5	4.0	5.6	9.7	12.6
Range of vision (nautical miles)	2.1	3.0	4.7	6.6	11.4	14.8
Range of vision (km)	3.8	5.5	8.6	12.2	21.2	27.3

Now consider what happens when mirage occurs. If the angle ψ is the greatest angle for which there will be total reflection, then all light rays entering the observer's eye in the angle $\psi - \nu$ will have been totally reflected. Hence *TK* is the width of the reflecting strip on the surface of the sea. Since this reflecting surface is convex, all objects reflected will appear compressed, that is, shortened vertically. Everything within the angle *WKZ* is invisible; everything in the space *RTKW* can be seen both directly and reflected. The sketch illustrates the resulting picture. The lighthouse is reflected only below the level *L*, where *L* is the height of intersection of the limiting reflected ray *RT* with the lighthouse. The whole of the steamship on this side of the horizon, and the upper part of the sailing ship beyond the horizon, are visible directly and seen in reflection. The line in which the direct picture and the mirage meet moves upward as the distance of the object increases within the reflecting strip. At *T*, it coincides with the line of contact of sea and reflected sky, that is, with the apparent horizon, while from *K* onward it coincides with the true horizon.

The theory of mirage is discussed by A. Wegener [244]. It has been pointed out by R. Meyer [238] that most observers grossly overestimate the angular height of mirages; floating and compressed images, caused by variability in the air structure, become exaggerated, particularly in association with the fantasies of those wandering in deserts. Photographs of them are therefore often disappointing. W. E. Schiele collected all known temperature readings made at the same time as mirages were observed and in 1935 analyzed them, proving that the superheated layer of air in which reflection took place was only a few centimeters thick. This is why, as we know by experience, wind and traffic do not disturb a mirage. This is particularly noticeable on the German autobahns; the "highway mirage," as it is called, gives the impression that there are great pools of water on the road. This is caused by total reflection of the sky, and the effect is the same as that observed by travelers across deserts. Thus, the prophet Isaiah (35:7), speaking of the blessed time to come, says literally, "The mirage shall become a real pool of water." An excellent photograph of this phenomenon was published in 1932 by L. A. Ramdas and S. L. Malurkar [241]. The presence of this kind of reflection does not depend on the actual temperature, but only on its rate of decrease at the ground; hence it was possible for J. Fényi [230] to observe it in Hungary even with temperatures below the freezing point. Mirages have been observed regularly in northern Wales by S. E. Ashmore [223] between mid-February and the end of October, unless prevented by snow cover, precipitation, or fog. He gives a height of 18 mm for the layer of air that causes it. Observations of the phenomenon in the area of the Great Salt Lake have been published (with photographs) by R. L. Ives [236], and by W. Heybrock [234] for occasional observations in Morocco. Figure 73 is a reproduction of a photograph taken by W. Findeisen [231] of the coast near Cuxhaven from a distance of 12.2 km. A normal view from the same point is given above for comparison. In the picture of inferior mirage (below) both the seamark on the left, a 30-m conical beacon, and the coast on the right are mirrored below. The angle of total reflection $(\psi-\nu)$ in this case was $5'$.

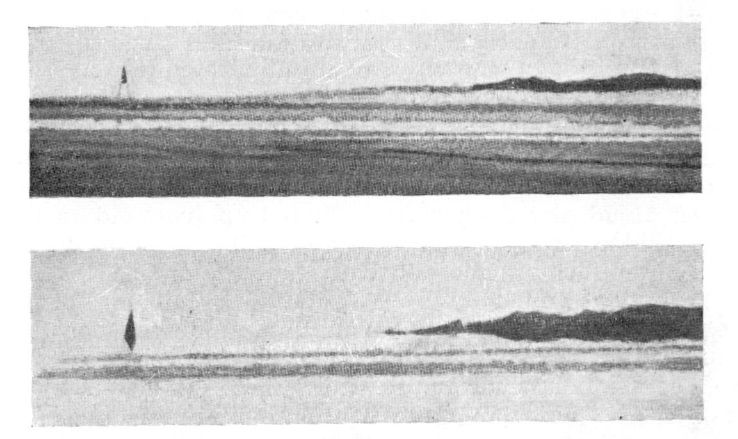

Fig. 73. (Above) The coast near Cuxhaven from a distance of 12.2 km.
(Below) The same coast with an inverted mirage. (Photographs by W. Findeisen)

To give an approximate idea of the orders of magnitude involved, the following example from A. Wegener is given: height of the eye, 10 m; temperature discontinuity of 5 deg at the surface of the water; depression of the horizon $\nu = 5.6'$; the limiting angle ψ up to which total reflection occurred was $12.2'$, and the width of the reflecting strip *TK* (Fig. 72), about 12 km.

The best proof that this type of reflection is caused by a superheated skin of air is that it can also take place at a sunny wall. A photograph published by J. M. Pertner and F. M. Exner [239] shows a boy leaning against a heated wall, both directly and doubled by reflection. The lens of the camera in this case was only 16 cm away from the wall. The line of sight was at grazing incidence on the wall and could therefore be subject to reflection in the same way as when a person looks in the direction of a distant horizon.

The rainbow may also be seen near the ground, brought into existence by conditions there, as in the spray from fountains or other forms of spray. These mechanically produced drops of water are larger by far than any found naturally in showers, and so the colors of these artificial rainbows are extraordinarily vivid. Rainbows are often observed appearing and disappearing in the spray borne up momentarily in the air from breaking wave crests. What can be more entrancing than to sail through the still heaving seas in the tail of a storm, when the clouds are being ripped apart, and the sun breaks through astern, to produce before the bow rainbow after rainbow, flashing and sparkling as if enchanted in the silvery spray tossed up against the dark background of the sea.

Halo too, caused by reflection and refraction by ice crystals, may sometimes be observed near the ground. W. Portig [240] reports that a halo was seen by H. Seilkopf on 6 March 1931, in hoarfrost crystals shaken out of a tree by gusts of wind. Portig himself saw both parhelia, part of the 22° halo and the upper tangential arc in an ice fog caused by the evaporation of water during the dousing of gas coke in the cold (-13°C) saturated atmosphere of the port of Hamburg. W. Marquardt [237] observed a halo in a valley of the Erzgebirge near Altensburg, on 8 February 1959, partially visible in front of the mountain slope. It arose in a driving cloud of ice crystals, whirled up from old snow on the crest of a ridge, and was clearly reinforced by evaporation of water droplets in the foehn wall of clouds.

From optical phenomena we now turn to acoustical phenomena in the air layer near the ground. It is generally known that the propagation of sound is entirely dependent on the weather. A powerful rumble of thunder is at best heard a distance of 15 km because of the peculiar stratification of temperature found in association with thunderstorms. The sounds of heavy artillery fire or catastrophic explosions, on the other hand, are audible at distances of hundreds of kilometers. The path of sound waves through the atmosphere is determined by the wind and temperature structure. From this fact it can be concluded that the surface layer, with its strong gradients, will have its own peculiar sound climate.

My attention was drawn by H. Ertel in a new publication [229] to the fact that Alexander von Humboldt correctly attributed the diurnal changes in audibility to the lack of homogeneity arising from vigorous turbulence in the medium of transmission, a phenomenon known to the ancient Greeks. He remarked that "the rumbling of the volcanoes Cotopaxi and Guacamayo sounds much louder at night than during the day, and the increase in the sound heard over the sea is less than over a flat stretch of land, but this in turn is greater than on the slopes of the Andes, at a height of

3000 feet. He made similar observations in the plains around the Atures Mission, where the sound of a huge waterfall on the Orinoco, more than a French mile distant, was heard three times louder at night than during the day." There are therefore two influences affecting the climate of sound in the surface layer. H. Ertel, who has given an exhaustive theoretical solution of the problem, differentiates the "geometric effect," arising from atmospheric stratification, and the "Humboldt effect," brought about by turbulance. Both of these produce maximum audibility by night and a minimum during the day. The extinction coefficient of sound intensity is directly proportional to the distribution of convective temperature fluctuations.

Improved audibility in the evening and during the night can frequently be observed. A. Schmauss pointed out that, in the streets of a city at night, the footsteps of a distant pedestrian or a whispered conversation can be heard. H. Wagemann noticed, on the coast, that the sound of the engine of a fishing boat is best heard in the spring, when there is a layer of warm air over the cold sea. In a polar climate, with its exceptionally strong and persistent inversions, unusual audibility is a striking phenomenon. In the diary of Captain Scott [243], on 1 August 1911, just such a weather situation in the Antarctic is described: "The light was especially good today, the sun was directly reflected by a single twisted iridescent cloud in the north, a brilliant and most beautiful object. The air was still, and it was very pleasant to hear the crisp sounds of our workers abroad. The tones of voices, the swish of ski, or the chipping of an ice pick carry two or three miles on such days—more than once today we could hear the notes of some blithe singer—happily signalling the coming of spring and the sun."

The practical importance of the way sound behaves near the surface is brought to mind by the sounding buoys and other devices used to safeguard shipping. In sound ranging for military purposes it was necessary, as stated by L. Aujesky [224], to select the most suitable terrain from the point of view of audibility, and it was found that there were "often surprisingly large differences in the conditions of sound climate between two closely neighboring positions." In this present age of noise abatement the problem has acquired new meaning. Sound waves are bent around hills, buildings, and woods, but unfortunately the sound protection available in the lee of obstructions applies more to the deeper tones than to piercing, irritating, high-pitched noises. Attention should be paid to this point when selecting sites for rest clinics and convalescent homes. The acoustics of large cities are dealt with by B. Hrudička [235].

CHAPTER III

INFLUENCE OF THE UNDERLYING SURFACE ON THE ADJACENT AIR LAYER

19. Type of Soil, Soil Mixtures, Tilling of Soil

The extreme diversity of microclimate can be attributed to the varied nature of the surfaces underlying the air layer near the ground. The surface may be solid ground with or without vegetation, a snow cover, a water surface, or ice in the form of frozen lakes or glaciers. In Secs. 19 to 21 the discussion will be limited to solid ground without vegetation. The laws governing conduction of heat in the soil have been dealt with in Sec. 6, and the average temperature variation in the soil in Sec. 10. The wide limits within which the heat economy of the ground may vary have already been shown in Table 10.

In spite of the great contrasts in the distribution of heat, water vapor, wind, air plankton, and so forth, pointed out in Chapter II, it is very much simpler to achieve a balance in air than in solid ground. The ground consists of close particles of different types of soil, differing in density, grain size, heat conductivity, and water content, which also varies with time. Embedded stones, tree roots, dead organic matter, earthworms and other animals, and water passages, all combine to transform the soil into a veritable mosaic. A great many samples must be taken before it is possible to determine the type of soil, or its representative moisture content over an area of any size. The same is true for soil temperature. In contrast to methods used to measure air temperature, simultaneous readings must be taken at several neighboring points, to make sure that the distribution obtained is not a local random distribution. Where plants are growing, "the whole volume of the ground," as G. Winter [266] writes, "becomes split up into smaller areas each centered about a root. Every plant residue that penetrates into the soil creates about itself, by the inhibiting and activating substances it contains and other specific components, a special set of conditions for the existence of soil microflora, an effect that becomes further complicated by the interaction of colonizing bacteria."

This finer soil structure will have to be ignored in the discussion that follows. Only typical and homogeneous types of soil will be dealt with, because even from the meteorologic point of view, the influence of the ground on the air overlying it is extremely complicated. All factors involved in the heat balance of the earth's surface (see Sec. 3) play a part in shaping the heat conditions in the soil and in the air layer near the ground.

It is best to use as a point of departure the heat-balance equation $S + B + L + V = 0$ (p. 10), in which $S = I + H + G - \sigma T^4 - R$ (p. 13). Of all these quantities only the short-wavelength global radiation $I + H$ and the long-wavelength counterradiation of the atmosphere G are independent of the nature of the earth's surface. All other factors come under the influence of the nature and state of the soil. Reflected radiation R depends on the reflection index, or albedo, which is only one of the properties of the ground surface. This determines what share of global radiation is to be made available to the ground. Because of the great importance of soil-surface characteristics, they will be dealt with in greater detail in Sec. 20.

The quantities B and V depend on the properties of the ground as well as on its temperature. The thermal conductivity λ and the specific heat by volume $(\rho c)_m$ will be dealt with in the present section in so far as they are determined by the nature and state of the ground. The water content of the soil, which also affects the quantity B and, above all, the evaporation factor V by virtue of the amount present and by the way in which it is bound to the soil, will be treated separately in Sec. 21.

The quantities σT^4 and L of the heat balance depend only on soil temperature and not on any of its other characteristics. For a given long-wavelength counterradiation G, the nocturnal radiation balance S is consequently greater and the loss of heat less when the soil surface temperature T is lower. For example, it was found by F. Sauberer [261], from a series of measurements made in 1934-35, that the radiation balance S for summer nights with three-tenths cloud cover at the most had the following values, reduced for equal horizontal screening (see Sec. 55): hard roadway, 0.136 cal cm^{-2} min^{-1}; sandy soil, 0.103; bare earth, 0.098; meadow, 0.072. The first two well-conducting (and possibly also drier) surfaces give off more of their large heat supply by irradiation because of their higher surface temperature than the poorly conducting, cooler (and probably also wetter) surfaces.

A similar role is played by the factor L. The hotter the surface becomes in the middle of the day, the greater is the temperature contrast of ground and air, and therefore the stronger the convective mixing, and the greater will be the quantity of heat transferred from the ground to the adjacent air. Similarly, by night, the colder parts of the ground surface will have transferred to them more heat from the air, through radiation and mixing, than the warmer parts.

This survey shows how difficult it is to obtain the pure influence of the soil itself, and how carefully all these quantities mentioned must be assessed in trying to judge temperature measurements. Attempts are therefore always being made to arrive at a theory of heat transport in the ground, including the water factor, in spite of all the obstacles. Great advances have recently been made in Holland by D. A. de Vries [264, 299, 335]. On the other hand, these studies open up a wide field of possibilities, still too little used today, of influencing artificially the microclimate of the soil, and altering it to useful ends.

As already indicated in Sec. 6, every soil consists of three elements: the soil substance, water, and air. The water content, which varies with rainfall, and the air enclosed in the soil, which, besides depending on the water content, varies with the soil structure and the state of cultivation, lead to continual changes in soil properties. In agriculture, the productivity of a soil is determined by its grain size and composition, the calcium carbonate content, and the quantity of humus in it. A soil is described as sandy when coarser grains predominate and there are not more than 25 percent of particles present that can be turned into mud (diameter less than 0.01 mm). It is called loamy when this percentage lies between 25 and 65 percent, and clay when more than 65 percent of mud-forming particles are present. Since these soils also vary in the amount of work required for fillage, a sandy soil is often spoken of as a light soil, and clay as heavy soil.

The point to be considered now is the influence the type of soil exerts on the heat balance of the ground, and hence on that of the adjacent air layer.

How surprisingly temperature conditions vary in different soils can be illustrated by considering a series of observations made daily in 1925 by N. K. Johnson and E. L. Davies [250] in various types of soil at six different places in Salisbury Plain, England. A 15-cm layer of soil was placed in boxes 1 m square. Temperatures were measured at a depth of 1 cm below surface, as shown in Table 39. The last row of the table gives the air temperature for purposes of comparison, measured in the Stevenson shelter at a height of 1.2 m above the ground.

Table 39
Variation of subsurface temperature (° C) with type of soil.

Type of ground	June			January		
	Max.	Min.	Variation	Max.	Min.	Variation
Tar macadam	35.4	10.0	32.6	8.1	1.2	6.9
Earth	35.1	10.4	25.0	7.1	1.7	5.4
Sandy soil	31.1	9.1	26.0	7.4	2.0	5.4
Gravelly soil	42.6	9.9	21.2	7.1	1.4	5.7
Under grass	29.3	13.3	16.0	6.2	2.9	3.3
Loam	24.6	13.1	11.5	6.7	1.7	5.0
(Air temperature)	21.8	7.6	14.2	8.3	1.7	6.6

The greatest contrasts occur at the soil surface, but they are too difficult to measure. However, even at a depth of 1 cm, the monthly average temperature maxima differs by up to 18 deg and the minima, by up to 4 deg. Even if field investigations were influenced by many other factors, these figures serve to indicate the order of magnitude of the temperature differences that occur, and they are likely to be still greater in other climates than that of England, which is relatively low in radiation and rich in precipitation.

To get an idea of the influence of the type of soil, apart from all other factors, a comparison may be made between soils with good and bad conductivity, by assuming that both absorb equal quantities of heat through the surface during the day, and give off equal quantitites of radiation by night (equal values of S). According to Sec. 6, then, the daily and the annual heat variation must penetrate deeper into the soil with good conductivity (see Table 11) than into the bad conductor. Correspondingly, at the surface the daily maxima will be lower for a good conductor, and the daily minima higher. The surface of the soil with good conductivity has more moderate temperatures; by analogy with macroclimatologic processes, one might say it had an oceanic surface climate. This acquires great practical importance during the critical spring weeks in our temperate latitudes. Plants in a soil with good conductivity are not stimulated to premature growth by too much heat during the day, and the higher night temperatures experienced for the same reasons reduce the danger of late frosts. A further consequence is that soils with good conductivity give off less heat during the day to the adjacent air, and therefore the L-term in the heat economy is smaller. More of the heat received in radiation is retained within the soil; the average temperatures of well-conducting soils are higher. Diurnal temperature fluctuations penetrate deeper into such soils, however, than into poorly conducting soil.

These statements are supported by the comparison of granite and sandy heath illustrated in Fig. 74. This is a classical experiment carried out as long ago as 1893 in Finland by Theodor Homen [249], a pioneer far in advance of the times in which he lived. Simultaneous temperature measurements were made in three different kinds of soil. For an average of three clear August days, the tautochrones (see p. 56) of the warmest and the coldest hours are shown, and the curves are extended beyond the surface to the temperature extremes measured at the level of the top of the grass cover. The air-temperature maxima and minima at 2 m in the shelter are shown by small circles.

Granite, which is an excellent conductor, shows marked diurnal variation at a depth of 60 cm. It is warmer, on the average, than the other soils, but in the surface layer the maximum is nearly 10 deg lower, and the minimum 5 deg higher, than in the poorly conducting sandy soil. The bog soil conforms only partially to the rules mentioned. It is true that it could be recognized as the worst conductor of these three soils by its generally lower temperature, by the small penetration of diurnal temperature fluctuations, which hardly get beyond 20 cm, and by the steep increase of diurnal fluctuations in the top few centimeters. But the surface maximum was 15 deg lower than that of the sandy soil. Bog soil, which is rich in moisture, loses too much heat by evaporation; the V-term in the heat economy upsets this comparison of soils.

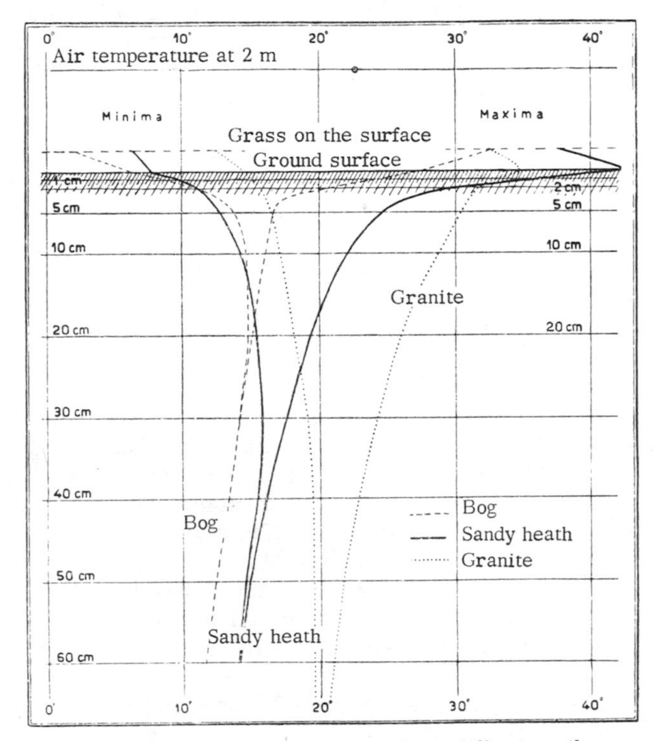

Fig. 74. Range of temperature fluctuation in three different soils on a summer day in Finland. (After T. Homén)

In Fig. 75 some results of modern investigations are illustrated by a method already familiar from Fig. 20. Comparative temperature measurements were made in Sapporo, Japan (43° 08′ N), by R. Yakuwa [267] during the summers of 1927 to 1929 at the surface and depths of 5, 10, 20, 30, and 40 cm in a sandy soil, a loam, and a clay soil (and two other types of soil). The readings for 15 August 1929 were published by him in detail, to provide a typical example of a clear day, and these are shown in Fig. 75. The spaces between the 5-deg isotherms are alternately clear and shaded to make comparison simpler. The dashed intermediate isotherm of 22.5° is added to show what the position is at deeper levels.

In this case we are presented with a two-layer soil, the temperatures of which at first sight appear to contradict the rules illustrated by Fig. 74. In the topmost 5-cm layer the thermal

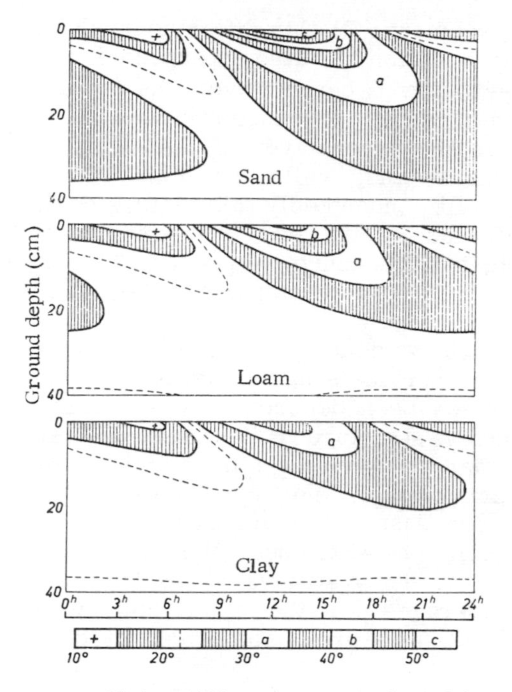

Fig. 75. Temperatures on 15 Aug 1929 in three soils that exhibit greater differences at some depth than at the surface. (After R. Yakuwa)

diffusivity increases from sand to loam and then to clay, as shown in Table 40. Thermal conductivity, for which we have no data, almost certainly is in the same order. The temperature maxima at

Table 40

Temperatures in four types of soil (see Fig. 75).

Type of soil		Sand	Loam	Clay	Bog
Thermal diffusivity (10^{-3} cm^2 sec^{-1})	0 - 5 cm	2.2	3.2	3.6	2.8
	5 - 30 cm	4.4	2.8	2.6	1.1
Maximum surface temperature (°C)		53.5	46.7	35.9	39.0
Minimum surface temperature (°C)		13.5	13.1	14.4	15.8
Diurnal temperature fluctuation (deg)	0 cm	40.0	33.6	21.5	23.2
	5 cm	19.4	18.5	13.7	13.9
	10 cm	12.3	10.7	7.7	5.4
	30 cm	1.6	0.7	0.6	0.3
Depth (cm) with diurnal temperature fluctuation of 0.1 deg		57	47	47	40

noon are in agreement with this differing conductivity; the minima and the diurnal variation are also as expected.

In the layer below 5 cm, however, the order of the three soils is reversed with respect to conductivity. It follows that the high temperatures in the upper layer of the sandy soil are conducted down to greater depths, while the loam and clay below remain comparatively cold. Undoubtedly this picture of the behavior of the soils is reinforced by the effect of their moisture content and by evaporation, but this cannot be proved. The bog soil, which has been added to the table for comparison, behaves as in the Finnish observations.

Observations in various types of soil, over many years, at the German Weather Service's agrometeorologic station at Giessen, which have already been described by W. Kreutz [252], indicate a similar two-layer structure in a sandy soil and a loess. Mean values of thermal diffusivity a ($cm^2\ sec^{-1}$) were calculated by H. Wächtershäuser [265] for the three summer months of 1947 and 1948, from the decrease in the diurnal temperature fluctuation with depth (Table 41). The extreme values are given as percentages of the mean value. They still vary by ± 20 percent, although they are monthly averages for the same soil, which should have a smoothing effect. This gives some idea of the effect of weather and soil moisture.

Table 41

Thermal diffusivity for three types of soil at different depths during summer.

Type of soil	Depth (cm)	Mean ($10^{-3}\ cm^2\ sec^{-1}$)	Extreme values (percent of mean)
Humus	2.5 - 5	2.58	81 - 120
	10 - 20	2.29	78 - 110
Sand	2.5 - 5	4.33	65 - 169
	10 - 20	11.47	88 - 110
Loess	2.5 - 5	8.60	55 - 152
	10 - 20	6.59	85 - 120

The influence of the type of soil on the temperature variation is clearly illustrated in the results obtained by L. Nidetzky [254] in the experimental garden of the Institute of Soil Studies in Vienna. He used boxes with a surface of 100 × 70 cm filled to a depth of 20 cm with each type of soil, and found for 29 August 1950, to take one example, the values shown in Table 42. Each

Table 42
Maximum temperatures and temperature variation in four materials at different depths on a summer day.

Type of ground	Maximum temperature (°C)			Diurnal temperature variation (deg)		
	1 cm	10 cm	20 cm	1 cm	10 cm	20 cm
Garden soil	38.0	29.3	27.4	19.0	5.8	2.8
Brick chips (3-cm diameter)	27.2	23.8	22.7	9.1	2.6	1.2
Brick chips (6-cm diameter)	26.2	23.8	22.7	7.6	2.6	0.8
Sphagnum (moss)	24.1	22.8	22.4	6.6	2.4	0.4

soil was artificially moistened to the same degree during the experiments.

Figure 75 shows that hardly any temperature differences arise at night in the upper layer of the soil, in contrast to the middle of the day. They amounted to only 1.3 deg in the minimum, according to the figures in Table 42. Nevertheless, these small differences are of great importance, because the presence or absence of late frosts during critical spring nights may depend on them. For this reason, there have been many attempts to elucidate theoretically what effect the type of soil has on the temperature decrease at night. One theoretical approach, by H. Philipps [259], has recently been amplified by H. Reuter [260]. This theory takes account of the interaction of exchange, air-temperature stratification, and the nature of the soil on the modification of ground-surface temperature at night. As before, A (g cm^{-1} sec^{-1}) is the exchange coefficient and γ (deg/100 m) the lapse rate, positive in inversions, both of which had to be taken as constant with height in order to carry out the calculations; ρ (g cm^{-3}) is the air density, and c_p (0.24 cal g^{-1} deg^{-1}) its specific heat at constant pressure. The quantities mentioned in Sec. 6, λ (cal cm^{-1} sec^{-1} deg^{-1}) and $(\rho c)_m$ (cal cm^{-3} deg^{-1}), were used for the ground. If S (cal cm^{-2} min^{-1}) is the radiation balance of the ground surface, negative at night, the change of temperature dt in time z, according to H. Reuter, would be given by the characteristic equation (cf. Sec. 55):

$$dt = \frac{2}{\sqrt{\pi}} \frac{(S + \gamma \cdot c_p \cdot A)}{(\sqrt{\lambda \cdot (\rho c)_m} + c_p \sqrt{A \cdot \rho})} \sqrt{z} \quad \text{(deg)}.$$

If the units employed throughout this book are used, and ρ is given the value 0.001293 g cm^{-3}, this transforms, when z is more conveniently expressed in hours, into the numerical equation

$$dt = \frac{1.128\,(1000\,S + 1.44\,\gamma\,A)\,\sqrt{z}}{1000\,\sqrt{\lambda\,(\rho c)_m} + 8.65\sqrt{A}} \quad \text{(deg)}.$$

For the special case where $\gamma = 0$, this reduces to Philipps's equation.

At night, S is negative, and therefore dt is normally also negative. It may happen, however, that with large values of A and γ (strong inversion) the expression in parentheses in the numerator may become zero, and therefore also $dt = 0$. In this situation, there is a flow of heat from the air above the ground down to the surface, brought about by intense mixing, which is equal to the heat loss from the surface by radiation. The decrease in surface temperature is proportional to the square root of time, thus, after 4 hr of night conditions dt is twice and 8 hr later, three times as high as at the end of the first hour. To determine the influence of the remaining factors, the writer calculated the drop in temperature during a 10-hr night for four types of soil, four temperature gradients, and two different exchange coefficients. The results are given in Table 43. The overwhelming importance of the type of soil can be seen immediately. It is clear that the differences calculated from theory are greater than those actually observed. Exchange weakens the contrasts and can do this effectively only if mixing is intense and the temperature inversion strong. These two conditions are seldom found together in nature. C. H. Grasnick [248] has extended Reuter's investigations to the temperature decrease within the air layer near the ground, and has published nomograms for heights of 2, 15, 30, and 76 m above the surface. These give the drop in temperature during 2- and 8-hr periods as functions of γ, A, and S, but only for a few types of soil (similar to wet sand).

Table 43

Calculated nocturnal drop of temperature (deg) at ground surface in 10 night hours, after H. Reuter.

Exchange coefficient A			1				10			
Temperature gradient γ			-1	0	+1	+3	-1	0	+1	+3
Type of ground	λ	$(\rho c)_m$								
Rock	0.011	0.52	5.1	5.1	5.0	4.9	4.7	4.2	3.7	2.7
Wet sand	0.004	0.40	8.9	8.8	8.7	8.5	7.1	6.4	5.6	4.1
Dry sand	0.0004	0.28	22.5	22.3	22.0	21.5	12.3	11.0	9.7	7.0
Bog soil	0.00015	0.09	35.2	34.8	34.4	33.5	15.5	13.8	12.2	8.8

From the previous discussions on the different properties of soil it is concluded that improvements can be brought about by mixing soils, in cases where existing conditions are disadvantageous. Everyone who owns a garden has already made use of this process when he has mixed peat into a heavy soil or added sand to a rather compact loam.

This process of mixing soils is used practically in improving bog soils more than anywhere else. This is not a question of putting a layer of sand on top (cf. Sec. 20), but of mixing it in. In Japan, R. Yakuwa [268] mixed peat and loam in five different proportions, placed the mixtures in 16-m^2 boxes, and measured the temperature variation to a depth of 20 cm during two summer months. In 1951 in the Donaumoos, H. Kern [251] compared an old cultivated bog with a neighboring soil which was a half-mixture with sand. Y. Pessi [256] made a detailed comparison of four mixtures containing different proportions of sand and peat soil, each with an area of 100 m^2, over the years 1952-1954 in central Finland (Pelsonsuo, 64° N). Table 44 has been extracted from these rich sources. The thermal conditons in the mixed soils were improved both by improved in temperature diffusivity and by reduced evaporation. The difference was greatest in spring and early summer.

Table 44

Effect of adding sand to bog soil.

Quantity measured	Depth (cm)	Quantity of sand added to bog soil (m^3 ha^{-1})			
		0	200	400	800
Thermal conductivity λ (10^{-3} cal cm^{-1} sec^{-1} deg^{-1}) (2 test days)	0-10	0.81	1.08	1.27	1.75
	0-20	0.91	1.15	1.29	1.90
Thermal diffusivity a (10^{-3} cm^2 sec^{-1}) (2 test days)	0-10	1.17	1.45	1.78	2.58
	0-20	1.30	1.60	1.80	2.68
Heat absorbed by ground in 24 hr (cal cm^{-2}) (9 June 1953)		79	84	86	95
Mean temperature (°C) June 1953	5	15.7	16.0	16.0	17.0
	20	7.1	10.7	11.1	12.9
July 1953	5	16.0	16.6	16.5	17.3
	20	12.5	13.8	14.2	15.3
Temperature minima (°C) within plant cover (oats), mean of					
14 frost nights in May 1951		-3.6	-3.2	-2.9	-2.7
10 frost nights in June 1951		-3.9	-2.7	-2.6	-2.0
Mean snow cover (cm) in 1954					
January		17	24	21	18
February		26	32	30	19
March		27	32	28	28
17 April		20	20	15	10

W. Baden and R. Eggelsmann [245] made valuable microclimatic observations in the air layer over an uncultivated bog and a meadow reclaimed from this bog soil by 40 yr of cultivation, in the extensive Königsmoor near Lüneburg. Table 45 gives monthly means from readings taken at 14:00 daily at a height of 5 cm above the ground with an Assmann psychrometer, and of the night minimum with an alcohol thermometer that had been shielded during the day, between 8 May and 17 August 1951. As might be expected from the previous results, temperatures above the virgin bog were more extreme, warmer by day and colder by night. The air was also drier, both relatively and absolutely, than over the meadow. From continuous measurements of the water content of the soil it appeared that the uncultivated bog soil contained 80-90 percent of water, but experienced less evaporation than from grass cover of the meadow, where the water content varied much more with changing weather and, in the hot days of high summer, dropped as much as 50 percent by volume. The air above the meadow is therefore richer in water vapor and more humid, because of evaporation, than above the original bog. Continuous sampling at random up to a height of 2 m on two days in August confirmed this fact.

Table 45
Monthly mean temperatures (°C) and humidities 5 cm above an uncultivated bog (B) and a reclaimed meadow (M).

Month	Air temperature						Air humidity			
	Maxima		Minima		Diurnal variation		Relative (percent)		Water-vapor pressure (mm-Hg)	
	B	M	B	M	B	M	B	M	B	M
May	23.6	21.4	2.9	3.4	13.3	12.4	65	79	8.4	10.3
June	22.4	20.8	5.6	6.2	14.0	13.5	63	77	12.8	14.2
July	23.5	20.5	8.0	8.2	15.8	14.4	71	84	14.8	15.3

Cultivation also changes the nature of the soil, since it increases the proportion of air in it, and therefore reduces its ability to conduct heat. Gardeners and farmers therefore avoid loosening the soil in spring when there is danger of frost in order not to increase the diurnal temperature fluctuations at the surface, which would lower the night minima. K. Bender [246] writes, "I will always remember how, after a night frost in a potato field, the plants in the part that had been weeded the day before suffered frost damage without exception, while the part that, fortunately in this case, had not been worked showed no signs of damage." In a forestry report, I read the advice that weeds in the ground around young oaks should not be cleared in the autumn, since this would increase the risk of damage to the shoots of the sensitive young plants. In farming, when there is a danger of spring frost, grain should not be harrowed, nor should potato plants be hilled up.

Temperature readings confirm these facts. W. Schmidt [262] found as long ago as 1924 that, at the end of an August night, the surface temperature of a freshly plowed field was 2 deg colder, and at a depth of 10 cm 1 deg colder, than in an untilled field. During the day, the surface temperature of the loosened soil at 15:00 was 5.5 deg warmer, because of its reduced conductivity. It follows that increasing the density of the surface layer will help in guarding against frost. This is why soils are rolled, especially loose bog soils. The effect of several types of roller has been investigated in Sweden by A. Olsson [255], by taking temperature readings 5 cm above and 10 cm below the surface of three different soils. The comparative measurements of Y. Pessi [257] in Finland showed that in addition to an increase of soil temperature by 1.5 deg there was also an increase in the yield of oats.

The proportion of air in the soil increases with grain size. D. A. de Vries [264] showed that the dependence of thermal conductivity on grain size can be arrived at theoretically, by assuming that soil particles are ellipsoidal in shape. Using his results, R. H. A. van Duin [247] calculated to what extent the thickness of the loosened layer of the surface soil would affect the diurnal temperature fluctuation, on the assumption that the flow of heat in the ground remains unchanged. The fluctuation rises sharply at first and then reaches a constant value when the depth of loosening reaches 10 cm. Bearing in mind the danger of night frosts, including the alteration in heat flow, we can estimate that loosening of the upper soil layer to a depth of 2 cm, in a European climate, would reduce the night minimum by 2 deg for moist soil and by 3 deg for dry soil. During the course of a year, tilling of the soil in autumn and winter seems to lower the mean temperature of the plowed layer, but to raise it in spring and summer.

The heat budget of the soil, finally, can be improved by heating it artificially. Research carried out in 1956 by A. Morgen [253] in the agrometeorologic experiment station in Trier, with central heating at a depth of 20 cm, showed how difficult it is to distribute heat evenly in soil, which is always inhomogeneous, if an economical number of heating ducts is to be used. It was not possible to know exactly the increase in yield that resulted from raising the temperature at root level, but it was of the order of 10 percent for a 1-deg increase in temperature. At night, however, only the layer of air next to the surface is kept warmer, while plants cannot be protected against frost damage at a height of 0.5 m.

20. Ground Color, Surface Temperature, and Ground Cover (Mulching)

The albedos of natural surfaces have already been listed in Table 3. Together with soil moisture, which controls the amount of evaporation (Sec. 21), the albedo determines how much of the heat that reaches the ground in the form of radiation will remain available. If the radiation balance is measured over different types of surface, under the same weather conditions, substantial differences are found.

F. Sauberer [293] found, for example, the following figures for radiation balance on a fair September day at 13:30. The highest values were found over water surfaces which reflect only slightly, namely, 0.612 cal cm^{-2} min^{-1} above the Lower Lunz Lake; while a tributary with colder water, which consequently emitted less long-wavelength radiation, had a value as high as 0.645. Further values were: 0.526 over a potato field, 0.484 over Alpine sorrel, and 0.430 over a mossy meadow. Light-colored earth free of vegetation had the lowest value; a gravel path had 0.402, and a hard road 0.368 cal cm^{-2} min^{-1}.

Since reflected radiation is visible to the human eye, R also determines the color of the ground, an effect that can often be observed. Figure 76 reproduces the results of field researches by L. A. Ramdas and R. K. Dravid [286] in India. On the left is the course of temperature in 40 days at the test surfaces, and on the right for an unaltered control surface. Five days after observations were started (time A) the dark cotton-growing soil of the test patch was covered with a thin layer of white powdered lime. The immediate cooling effect can be observed in the sharp rise of the isotherms. Before wind and water were able to weaken the effect of the white coloration, the difference in temperature between the surfaces reached 15 deg and the effect was felt at the greatest depths for which readings were taken. After the powder was removed (time B), another week elapsed before the test surfaces reached the same temperatures as the control surface.

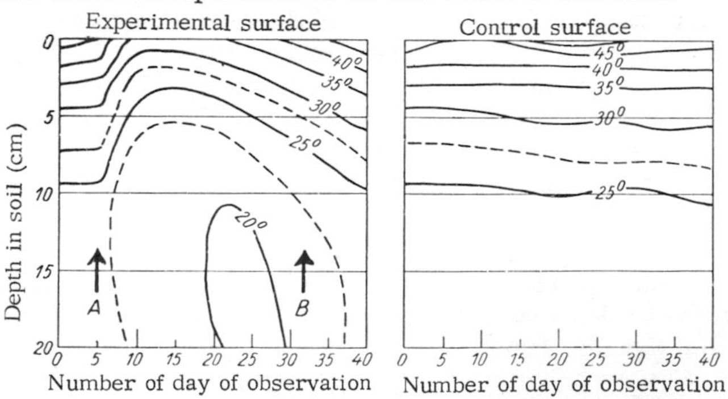

Fig. 76. Changes in soil temperature caused by whitening the surface, from an experiment in India.

In Geisenheim on the Rhine, N. Weger [300] covered two surface areas of 34 m in a beanfield, one with light powdered china clay and the other with black coal dust, and followed the temperature readings at 5-cm depths during the summer of 1949. Figure 77 records the values he obtained, expressed in pentade means (5-day periods) for the 15:00 observation. The continuous curve is for the natural soil, the broken line for the dark surface, and the dotted line for the light surface. As expected, the temperature below the dark surface is higher, by 9.8 deg for the 5-day average, and as much as 13.0 deg on an individual day. During the night it was somewhat cooler, so that the dark coloring has the effect of increasing the diurnal fluctuation. The amount of radiation returning from the light surface was measured by a Lange photometer at a height of 70 cm and found to be six times that from the black surface. The effect in June was first to produce a lower temperature than that of the dark surface. This difference, however, decreased gradually until it became zero by the second half of July, and then it acquired the opposite algebraic sign. This was not due to the destruction by weather of the white powder, which was renewed when necessary. Perhaps it was a result of the progressive development of the bean foliage. It was not found possible to compare the yield from the three fields, since cold June nights had caused considerable damage to the beans which thrive on heat. Figure 77 can be used, above all, as a warning of the care that must be taken, before making practical use of such measures, not to use too short a series of tests, since substantial differences can accumulate with the passage of time.

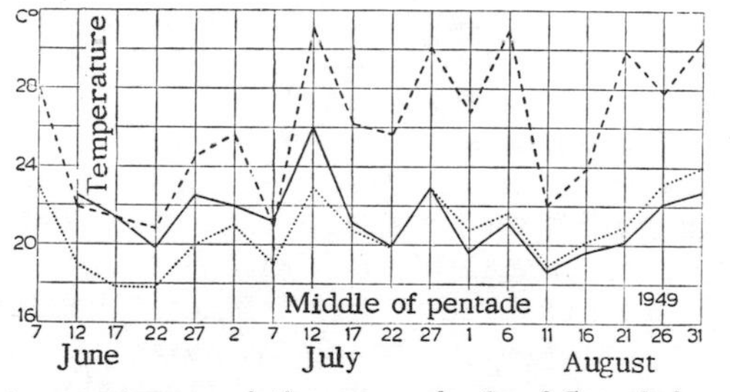

Fig. 77. Summer temperature variation at a depth of 5 cm below natural (full line), blackened (dashes), and whitened surfaces (dotted). (After N. Weger)

Comparative measurements of air temperature 50 cm above natural and artificially darkened and lightened surfaces were made from 1948 to 1951 in the neighborhood of Murmansk and Moscow by P. G. Aderikhin [269]. On windless days the air temperature was 3 to 5 deg higher over the darkened soil than over the light-colored surface, and 2 to 3 deg higher than over the natural surface. The difference increased as the water content of the soil decreased.

Temperature differences of up to 10 deg were measured 1 cm below the surfaces of dry soil that had been made white and black by dusting, in Mendoza (32° 53′ S), by W. W. Ehrenberg [274]. He tested the possibility of making practical use of soil coloration on a grand scale from both theoretical and technical points of view. It is of particular interest that this investigation combined a color test with a humidity test, which has already been mentioned. A white and a black field were moistened to the same extent. Since at first the greater evaporation from the black field was in opposition to the influence of color, the temperature difference was only 5 deg. After a few weeks without rain, the black field was dried out, but the white field was still giving off water, and the difference rose to 14 deg.

In models of water tanks for irrigation, of dimensions 40 × 40 × 10 cm, R. Yakuwa and F. Yamabuki [302] measured water temperature as a function of the color of the paint. While the night minimum temperature, as was to be expected, was independent of the color, the maximum temperature in the white-painted tank was 31.8° C, in the dark red 35.2° C, and in the black 36.3° C. In one tank, which was protected against evaporation by a thin floating zinc plate, also painted black, a temperature as high as 44.9° C was reached.

It is of interest to note the findings of S. Sato [291] and Y. Funahashi [292], that when a black powder is scattered over the water surface in rice fields the amount of heat in the ground and in the air close to it is reduced in comparison with a field with a clear water surface. Here also the radiation by night is independent of color. The turnover of heat during the day may be adduced from the example of 8 August 1955, from 12:00 to 13:00. With a global radiation $I + H$ of 81.0 cal cm^{-2} hr^{-1}, more heat went into the darkened water, 8.1 compared with 5.7. But the greater amount of evaporation V brought about a numerically greater loss of 47.1 against 41.8 cal cm^{-2} hr^{-1}. The heat loss due to conduction and long-wavelength radiation was also greater, being 19.7 against 17.2. The soil of the rice field, under its flat water layer therefore absorbed only 6.1 cal cm^{-2} hr^{-1} below the black surface, in comparison with 13.8 below the clear water surface. In this way the actual state of affairs, which is at first sight surprising, can be explained.

The influence exerted by paint on the temperature of wood has been investigated by C. Dorno [272]. For this purpose he set up four cylindrical pieces of wood, 3 cm in height and 2.5 cm in diameter, on a balcony with a southern exposure in Davos. He found that the incoming solar radiation of 1 cal cm^{-2} min^{-1} produced the following increase in temperature of the wood over that of the surroundings: blue-white paint, 10.8 deg; rose paint (white zinc and Dammar varnish), 11.0 deg; yellow-ochre paint, 14.4 deg; red oil paint, 15.7 deg; and finally for a soot-coated piece of wood, 16.9 deg. K. Schropp [295] made a series of measurements for various surfaces used in technology. The surfaces in question were mounted on an insulating cork plate 5 cm thick and their temperatures measured by a thermoelement in sunshine and still air. He found that under similar conditions, black paper or black enamel rose to 45° or 55° C, white surfaces to 15° to 20° C, while a polished aluminum foil reached only 15° C. At night all the surfaces had a temperature deficit of 2 to 4 deg. I. Dirmhirn [271] has recently made carefully analyzed surface observations of blackened and whitened natural cork and natural stone surfaces.

The importance of the color of vertical surfaces has also been recognized, and has been evaluated practically. H. Schanderl and N. Weger [294] investigated the effect of coloring trellised walls in Geisenheim over a period of two years. These were left partially with their natural light-brown color, and other parts were painted white or black. The growth and yield of tomatoes planted in front of a wall 3-m high, orientated in a SW direction, were followed continuously. At a distance of 10 cm, differences in air temperature could no longer be detected, but radiation conditions had altered. On a sunny June day the amount of reflected short-wavelength radiation in front of the white wall was 56 percent higher than in front of the black; this increased the yield of tomatoes. The long-wavelength radiation of the black wall, which was warmer, allowed the plants to grow more rapidly, but their yield remained lower. The higher yield in front of the white wall justifies the expenditure on paint.

These tests were repeated by N. Weger [300] in 1943 with peach trees and in 1947 and 1948 with vines. With the peach trees also, the growth of wood was greater in front of the black wall, in fact, 31 percent more than in front of the white and 24 percent more than in front of the natural color. The number of buds on a 1-m length of the young shoots was 75 percent and 190 percent more, respectively, in front of the white wall in two tests than in front of the black. With vines the best growth of wood was found in front of the white wall and there was (not yet verified) an increase in the weight of grapejuice.

Previous reports on the effect of the surface color of soil showed, naturally, that an ideal black surface would have the highest temperature. What is sought, from the practical point of view, is the maxima for natural surfaces; there are many reasons why this should be interesting, when one thinks of the numerous other factors that influence surface temperature. Apart from the intensity of incident radiation, which is controlled by latitude, elevation of the sun, state of the sky, and slope of the land, it depends on the water content and the thermal conductivity of the underlying soil, as well as on the movement of air over the surface. It is still a matter of debate what values the maxima can have in our climate. A. F. Dufton and H. E. Beckett [273] demonstrated that surface temperatures could exceed that of a black-bulb thermometer (a blackened sphere in a vacuum). Roofing paper on top of a heat-insulating base, for example, heated to 65.5° C, while the black-bulb thermometer stood at 56.1° C. When the temperature was measured in a box, heat-insulated in the best possible way, with blackened sides, and closed by a glass plate, it was found to be 120° C.

From observations made by B. Huber [279] it appears that surface temperatures up to 70° C have been recorded under natural conditions in Germany. Hardly any higher values will be established here, assuming that methods of measurement are free of faults, which in surface-temperature measurement presents great difficulties. When taking careful account of all possible errors in measurement, O. Vaartaja [297] found surface temperatures from 50° to 60° C, with a maximum of 63° C, in south Finland, which is in agreement with the above. If the temperature measurements made at depths of 2 cm and below in sandy soil under extreme local conditions by M. Andó [270] and A. Kiss [281] in Hungary are extrapolated to the surface, the same order of magnitude is arrived at. I. Dirmhirn [903] made observations of the surface temperature of rocks in the high Alps, which will be mentioned later in another connection (Sec. 46).

F. Firbas [275] has shown that the temperature may be very high by day in oak and beech woods in Central Europe, in spring before the trees break into leaf, because of extremely poor heat conduction of litter of dead leaves on the ground. When the spring sunshine was able to penetrate unhindered between the bare stems and twigs, while there was some protection offered against outgoing radiation by night, he was able to measure in the first few days of May temperatures up to 43° C within the litter, where anemones and hepatica grow.

Surface temperature must be taken into account in the construction of highways and railroads, but unfortunately no measurements are known as yet. K. R. Ramanathan [258] published observations of the temperature of railway lines in India, where only 61° C was reached.

It is to be expected that in dry subtropical regions even higher surface temperatures will be attained. At the same time as temperature rises, the amount of convection increases, which in turn has a regulating influence on the temperature. There are many indications that, even in these areas, substantially higher surface temperatures may not in fact be reached.

Figure 78 shows a record made by A. Vaupel [298] in Palermo, on 19 July 1955, using thermoelements. The cloudless, sunny nature of the day is shown by the upper graph, which is a recording made by a Robitzsch actinograph, giving the global radiation. The position was on Monte Pellegrino (38° 10′ N), 425 m above sea level, and above the brownish-red soil there were only the dried-up remains of vegetation, including a few *Opuntia ficus indica*. The temperature was measured in three positions on a fixed horizontal flax thread at a height of 8 cm above the ground (see also Sec. 29). Although this was an extreme position, under

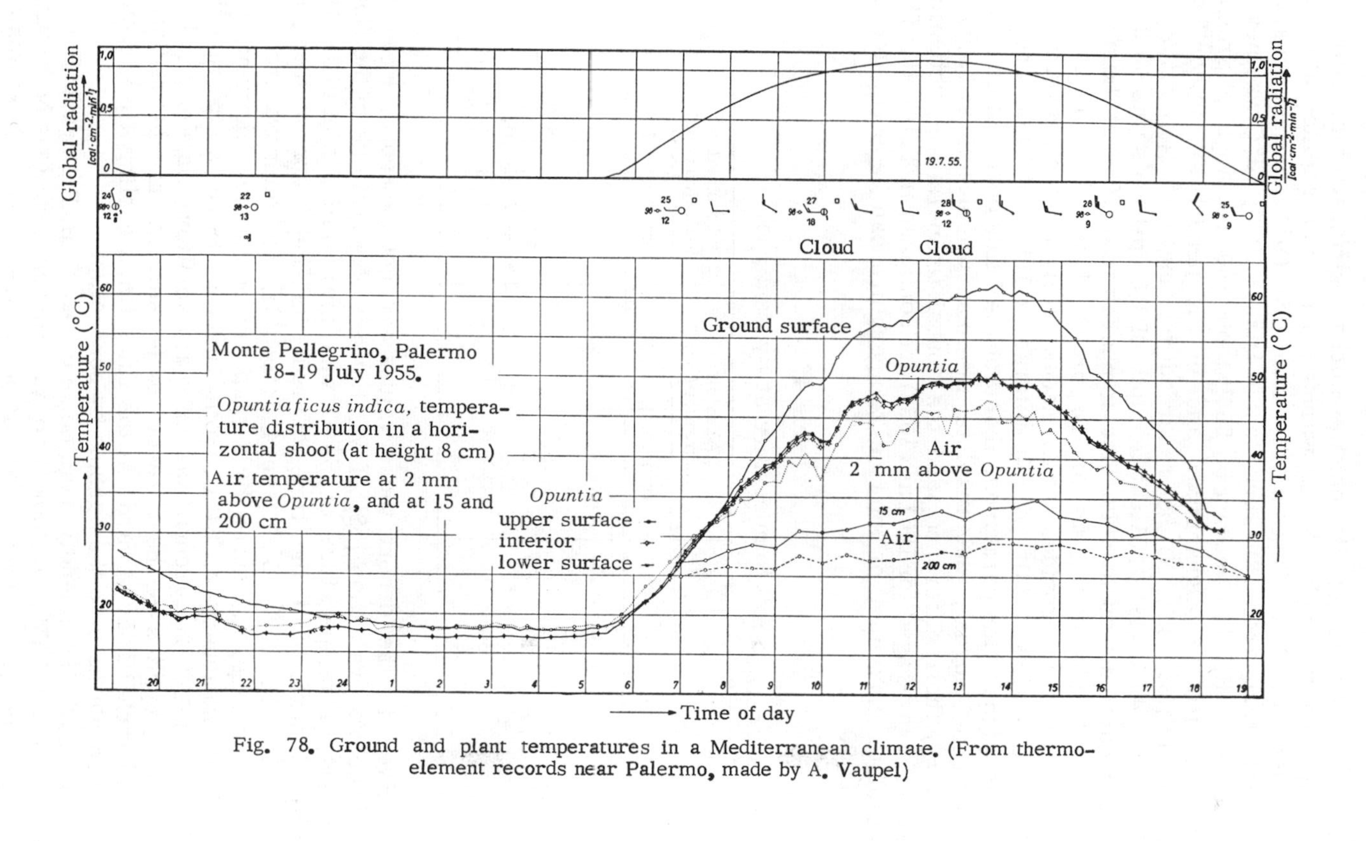

Fig. 78. Ground and plant temperatures in a Mediterranean climate. (From thermo-element records near Palermo, made by A. Vaupel)

extreme weather conditions, the midday surface temperature of the ground was only 33 deg warmer than the air temperature at a height of 2 m, and hardly exceeded 60°C. In the period of investigation, from 21 June to 20 August 1955, 70°C was never exceeded. H. F. Neubauer [283] took 800 readings of ground temperature in Afghanistan, and in the desert to the north of Jalalabad (34° 27′ N) he recorded the highest temperature at 14:20 on 4 August 1951, at a depth of 1.5 cm. This was 60.2°C, which corresponds to a surface temperature of 70°C at the most.

Useful information as to the order of magnitude of surface temperature is provided by A. Hadas [278], who took readings on the Lydda airport (31° 57′ N) in Israel in 1951 and 1952. The absolute highest temperatures, those of the 14:00 observation in the month of August, the absolute minima, which fell at 08:00 in January, their difference, and the annual temperature fluctuation at 14:00 (averaged for 1951 and 1952) are contained in Table 46. The absolute maximum of surface temperature can be estimated at 65° to 70°C.

Table 46

Soil temperature at Lydda airport.

Depth in soil (cm)	Absolute extremes (°C)		Annual variation (deg)	
	Maximum	Minimum	Absolute	14:00
1	61.8	2.3	59.5	50.6
2	57.8	2.8	55.0	46.6
10	40.5	7.5	33.0	28.6
20	34.5	10.3	24.2	22.8
40	33.3	12.6	20.7	20.0
60	32.0	14.4	17.6	17.0
80	31.2	15.0	16.2	15.5

Even in the desert these values are not exceeded. From F. Albrecht's evaluation [88] of W. Haude's observations in the Gobi Desert, a temperature of 55°C was measured 2 mm under the ground surface at the time of radiation maximum at Ikengung (41° 54′ N, 107° 45′ E) on 4 June 1931. The temperature 2 cm above surface, measured by a platinum resistance filament, remained stationary at 34°C from 10:00 to 16:00, a sign that the layer nearest the ground was only a transitional layer, by which further supplies of heat reaching the surface through radiation were carried away upward. At the same time, cooling air currents are brought down to the surface by convection.

Through the friendly assistance of Dr. N. B. Richter in Sonneberg, I was able to look into the new observations he brought back from the Sahara Expedition of the Berlin Geographical Society. These were ground-temperature readings made by Dr.

Richter and his wife in the spring of 1955 in Wadi Faregh (24° 47′ N, 16° 53′ E; 400 m above sea level) in the sand. Figure 79 summarizes the mean temperatures of the almost cloudless and completely rain-free days from 7 to 14 April. Ground temperatures were measured by mercury thermometers, air temperature and humidity by a sling psychrometer at a height of about 95 cm. The 8-day average maximum temperature at a depth of 5 mm in the ground was 45°C, reaching 53.5°C on the hottest day. Extrapolation to the surface of the ground leads to the conclusion that the surface did not reach 60°C. The mean maximum air temperature at a height of 95 cm was about 29°C, and 30° to 31°C at 37 cm. The midday flattening of the temperature curve can still be recognized at 95 cm. The range of relative humidity shows the extreme aridity of the observation site, and hence the reduction of the part played by evaporation in the heat budget.

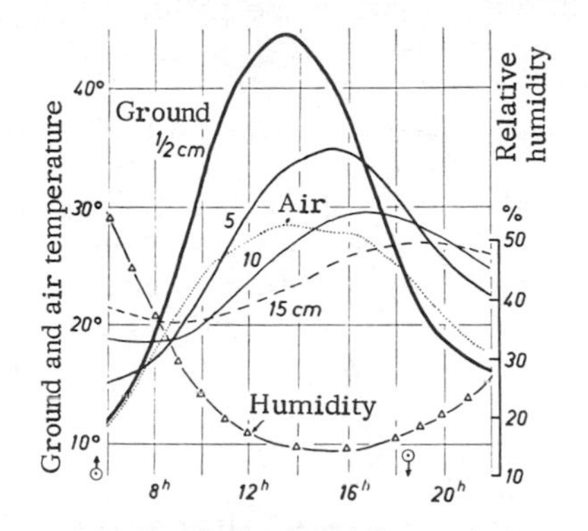

Fig. 79. Diurnal temperature and relative-humidity variations in the Sahara; averages for 7-14 April 1955. (From observations by N. B. and L. Richter in Wadi Faregh)

Surface temperatures are high even in a polar climate. The German Antarctic Expedition of 1938-1939 under A. Ritscher was surprised to find a group of unfrozen lakes in the ice continent at 70° 41′ S. The meteorologist of the expedition, H. Regula [288], showed that volcanic warming is not necessary to explain this phenomenon. The unfrozen water is due to the warming of the dark red-brown rock by solar radiation at an air temperature of about 0° C.

Unfavorable properties of ground surface can be improved by simple means. Besides changing its color, the following methods can be used as well, and will be the subject of further discussion. Either a surface layer with the required qualities can be superimposed, or an unsatisfactory surface can be rendered harmless by removing it or mixing it with a better subsoil; or the surface can be covered with some foreign material such as brushwood, straw, stone chips, roofing slates, or the like, which is called mulching the ground.

The temperature variation in a soil with an artificial covering has been investigated theoretically by D. A. de Vries and C. T. de Wit [299] on bog soils that had been sanded over to reduce the frost hazard. The experience was that out of this two-layer soil there soon developed a soil with three layers, namely, dry sand, moist sand, and bog soil, all of which differed in their ability to conduct heat and in heat capacity. Calculations showed that a sand cover 10 to 15 cm thick is sufficient to reduce the very high night frost danger of bog soil to that of pure sand. If the sand dries out, which is usually the case in the first 3 cm with Western European spring weather, then even less than 10 cm of sand will be enough to make the disadvantages of a bog soil ineffective. This top-sanding procedure is, of course, more successful than if the sand is mixed into the soil, which is often done for other reasons (see p. 151).

N. Weger [300] investigated the influence of a layer 5 cm thick of peat or sawdust spread on top of a mineral soil to protect plants germinating on it. There was a large rise in the temperature of the covering layer itself in the middle of the day, because of its poor conductivity; for example, on a fair summer day (24 July 1947) it rose to 44° C at 2 cm below the surface of the sawdust, and to 53° C under the dark peat. The maximum on the same day, at a distance of 2 cm within the natural soil, was only 30° C in the peat and 29° C in the sawdust, against 44° C in the unprotected earth. The delicate parts of the plants emerging from the soil, and the improving grounds for vine plants, were protected from overheating and drying. The ordinary "hilling-up" process makes use of this principle.

Von Ramin [287] recommends, after making the drills in sugar-beet cultivation, the spreading of crude potash salts, when there is a tendency for wind erosion. After the first rain has fallen, even after the first night with dew, the upper layer of soil becomes moister because of the hygroscopic salts, and hence becomes firmer. Thus the mechanical properties of the uppermost layer are altered.

K. Keil [280] gives an impressive example of the removal of an unfavorable top layer. The ground in Yakutsk in northern Siberia is permanently frozen at depth, only the surface layer thawing out in summer. A thick, poorly conducting layer of moss prevents the heat received from radiation, in so far as it is not used up for evaporation, from reaching the lower levels. During the Second World War this layer of moss was removed from a number of experimental areas. This led to heating, which became more noticeable from year to year. The yield from plants was also so good that the experiment was extended to wide areas. Thus there was developed in this region, barren up to 1939, a flourishing agriculture, which in its turn helped the exploitation of the rich ore deposits.

Such radical action is not possible as a rule. When good soil lies close below the surface, methods of tilling can bring success. This involves a combination of cultivation methods and soil mixing, which has already been mentioned in Sec. 19. R. Geiger and G. Fritzsche [277] carried out an enquiry to this end in a laboratory district of the Forestry College in Eberswalde. In a weed-covered, frost-damaged pine plantation one part of the ground had been deep plowed as if for planting a crop. This "volume break-up" had torn apart the dead, poorly conducting surface soil and mixed it with the mineral soil lying deeper down. In the spring of 1937 and up to the end of the 1939 winter, different parts of the plantation were treated in this way. The night minimum temperatures were measured at a height of 10 cm above the surface of two worked areas and one natural area in the spring and summer of 1939. Four observation stations, all at exactly the same elevation above sea level, gave the results shown in Table 47.

Table 47
Night minimum temperatures in a pine plantation.

Type of ground	Nights with late frost 1939			Average of 30 cold nights	Number of frost nights	Date of last frost
	13-14 May	29-30 May	10-11 June			
Weed-covered frost pocket	- 3.5	- 8.1	- 4.1	- 0.6	17	12 July
Weed-covered area of level cultivated soil	- 1.6	- 6.5	- 3.7	+ 0.2	15	28 June
Soil completely turned over in 1937	- 0.3	- 3.9	- 1.0	+ 3.3	9	15 June
Soil completely turned over in 1939	+ 0.2	- 3.4	+ 0.5	+ 4.0	6	15 June

The influence of the type of soil is clearly seen in all the figures, and always in the same sequence. The change in the soil brought about by deep plowing favorably changed the heat economy of the air layer near the ground and lessened the danger of frost damage to the crops. The more recent deep tillage proved itself to be more effective than the older. The land plowed in 1937 settled in time and, even to the eye, was seen to develop a weed cover on the consolidated surface. On the one hand, this provides a new proof that a transformation of the type of soil is responsible for the changes in night temperature, and on the other hand it gives an indication that the frost protection arising from deep tillage will be lost in proportion as the vegetation on it becomes denser and more frost-resistant.

As far as mulching of the soil is concerned, its effectiveness in increasing crop yield has been tried and proved repeatedly. The change it brings about in microclimate has, however, been measured properly only on rare occasions.

N. Weger [300], who has made the most valuable contributions in this field of study, has rightly pointed out how many-sided the problem is. The existing climate of a place and the target aimed at will decide the success and the profitability of the scheme. In Germany we try to obtain a surface climate that is generally cool by day and warm by night, an improvement in the water balance of the soil, and better use of radiation by increasing the radiation from the covering substance. Since no single material can provide the answer to all the requirements, the selection depends on which factor of the climate is at a minimum level, and therefore must be given first consideration. A few examples will illustrate possible solutions.

In Israel, the second planting of potatoes usually takes place in September. It is desirable to advance this date, but this would involve a risk of damage from heat, owing to the high temperatures prevailing in August. During the summer of 1949 J. Neumann [284] tested the effect of a 5-cm layer of straw over the soil. The air temperature at a height of 10 cm showed an increase in maxima of 1 to 6 deg, caused by the higher reflective power of the straw, which also means greater heating of the plants. The minima were 1 to 3 deg lower than for the untreated areas of the ground. The climate of the ground under the straw was greatly moderated, however, particularly as long as the development of the plant foliage only overshadowed the ground to a slight extent. The average of two series of readings taken between 23 August and 31 October showed that the temperature fluctuation at a depth of 15 cm below the ridges was only 57 percent of that found in the untreated fields (5.1 deg against 9 deg for a daily fluctuation of 13.9 deg in shelter temperature). F. Fuss [276] has published the results of his experiences over many years in the use of straw coverings in orchards in Germany.

S. Sato [290] was faced with a similar task in looking for a method of reducing the high temperatures to which the young plants were exposed in the water-covered rice fields in Japan. He put rice straw or grass mowings on the surface of the water, and observed the temperature maxima given in Table 48, with 12 August 1953 as an example.

Table 48
Temperature maxima in a rice field.

Measurement site	Open rice field	Mulch	
		Grass	Rice straw
In water	41.4	38.5	38.2
Ground surface	42.3	37.5	36.6
5 cm in ground	38.0	34.5	33.0
20 cm in ground	30.3	29.3	29.0

Mulching with inorganic material was tested in our own climate by N. Weger [300] at the agrometeorological experiment station in Geisenheim. Gherkin and tomato plots were covered with aluminum foil 0.035 mm thick, which best withstood the effects of the weather and reflected light three times as strongly as the natural ground. Figure 80 shows the daily air-temperature pattern at a height of 10 cm for the average of two sunny July days, each illustrated by two curves between which lie the widely scattered readings of the platinum resistance thermometer, which has a low rate of lag and is insensitive to radiation. The solid lines are for a normal gherkin plot, the broken lines for the plot with a covering of aluminum foil. As it did in Israel, the strongly reflecting aluminum foil gave rise to higher air temperatures when the temperature was increasing during the forenoon. In the afternoon and at night the radiation from the foil, which was lying loosely on the ground, gave rise to lower temperatures. The only exception was provided by the late hours of the night from 02:00 to 06:00, which Weger attributed to heat released by condensation, because the lower side of the aluminum sheets, next to the ground, was coated with water droplets, while the uncovered ground appeared to be dry. Figure 81 gives the diurnal temperature variation at a depth of 5 cm and shows, by contrast, how the climate of the ground is made milder, here too, by mulching. The water content of the soil was higher throughout under the aluminum foil, especially in the uppermost soil layer.

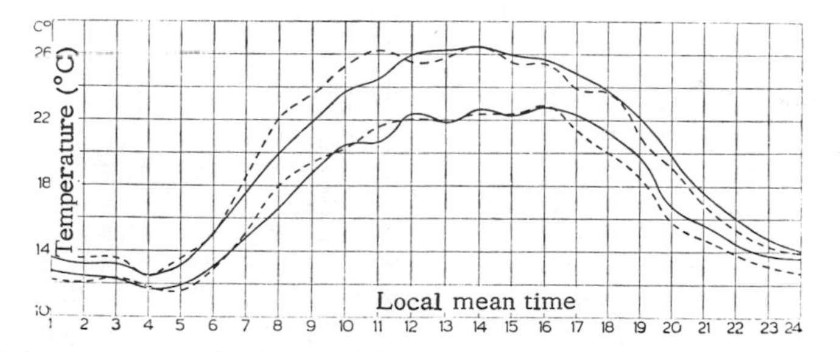

Fig. 80. Scatter of ground temperatures in gherkin plots with (broken line) and without (solid line) a covering of aluminum foil on the ground. (After N. Weger)

Similar experiments were carried out with glass-fiber matting, and with dark gray and white slate. It is also worth mentioning here that an experiment was made by N. Weger [301] using glass panes employed by gardeners to cover hotbeds. These panes were laid on either side of rows of tomato plants, inclined at an angle of 5° so that the water would run in, and radiation reflected by the glass could be used by the plants. Below the glass, in a free space of 3 cm, the air temperature reached 40° to 55° C at midday; this is due to the well-known "greenhouse

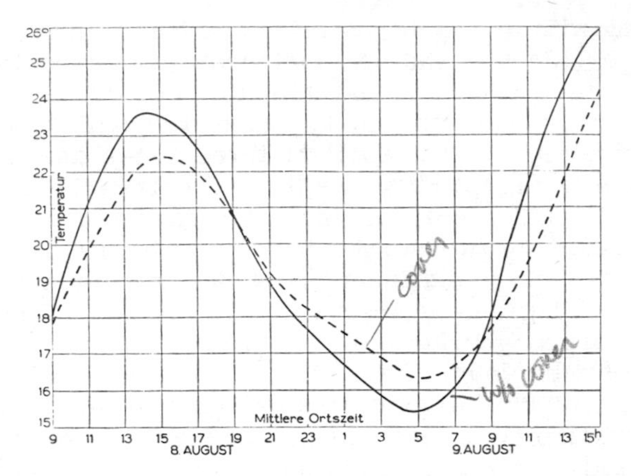

Fig. 81. Effect of aluminum foil on the temperature in the ground. (After N. Weger)

effect," where the short-wavelength radiation is allowed to pass through, but long-wavelength radiation is not allowed to escape. Even at night the enclosed air remains 4 to 6 deg warmer than the open air. This procedure had a very useful side effect in that only one weed *(Portulaca oleracea)* was able to survive this "murderous microclimate." On 3 June 1952, the soil temperatures at a depth of 5 cm as given in Table 49.

Table 49
Effect of glass covering on soil temperature 5 cm below surface.

Temperature (°C)	Bare soil	Rows of plants between glass sheets	Below glass
Highest	24.4	25.7	38.1
Diurnal mean	19.2	21.1	27.3
Lowest	15.6	18.5	20.3

Covering with glass is therefore a good method of obtaining a substantial gain in heat. Artificial watering was used to better advantage by the glass-covered plots, while the rain water was better used by the uncovered beds. The water content of the soil was greatest in the rows of plants. In the glass-covered parts the length of the side roots of the tomato plants was 27 percent greater; the yield of ripe fruit was 17 percent and of unripe harvested fruit 104 percent greater than in the uncovered field.

Surface characteristics have acquired new importance today in all problems of traffic and transport. The problem of trafficability of paths and fields for automobiles, tractors, tracked vehicles, and so forth, and the influence of the season, weather, melting snow, the nature of the soil, and many other factors have virtually developed into a separate science. More can be read on this subject in C. W. Thornthwaite [296].

21. Soil Moisture and Ground Frost

The water content of soil may be expressed as a percentage either by weight or by volume. The first of these is defined as the ratio of the weight of water to the weight of dry soil, expressed as a percentage; the second is the ratio (percent) of the volume of water absorbed to the volume of moist soil in its natural state. It is possible to convert from one to the other only if the volume of air spaces in the soil and the density of the pure substance of the soil are known. A guide to the determination and conversion of these quantities has been given by S. Uhlig [330]. The term "field moisture capacity" means the water content of the soil, as established after precipitation, when the seepage water has had time to run off after 2 to 3 days.

The mosaic structure of soil (see p. 142) is also applicable to its water content. For example, when S. Uhlig [330] took 25 equally distributed samples from an area 0.4 × 0.4 m of a loamy soil near Bad Kissingen, he found that the water content varied at depths between 10 and 20 cm from 19.9 to 21.9 percent by weight. Figure 82 shows how the tracks of cartwheels in dry sandy soil near Konin were deeply impressed and remained visible over a period through the growth of *Panicum lineare*. The soil below the cartwheels had become firmer, thus providing a better base for germination of the seeds pressed into it, and the channel so formed was able to collect not only more seeds blown in by the

Fig. 82. The tracks of a wagon are permanently marked by a dense growth of weeds in them. (Photograph by R. Tüxen)

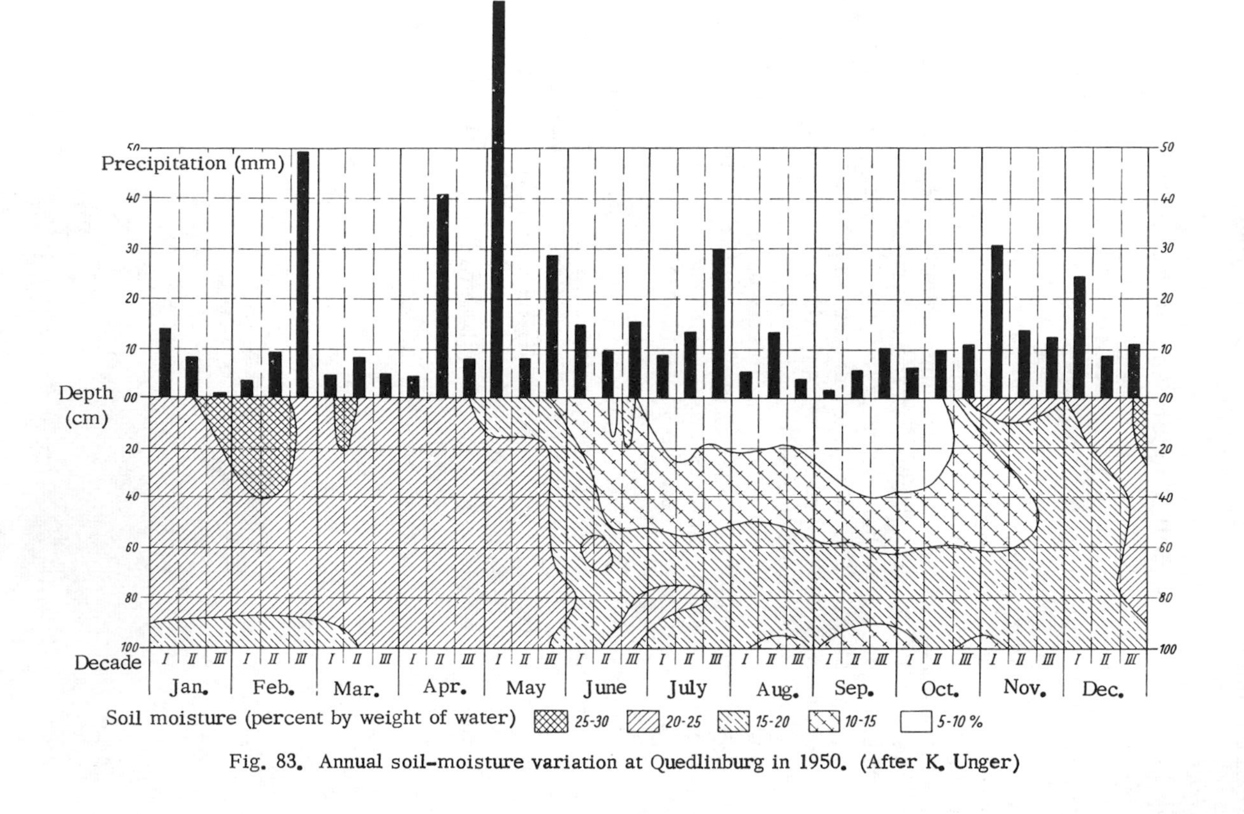

Fig. 83. Annual soil-moisture variation at Quedlinburg in 1950. (After K. Unger)

wind but rainwater as well. Thus it was made possible to photograph a local difference in humidity. Dr. R. Tüxen, who has observed this feature on many occasions, kindly permitted the use of his photograph.

Near Hiddensee, on the island of Rügen in the Baltic, one summer forenoon I saw the spoor of a fox, the paw marks projecting 1 to 2 cm above the sandy surface of the dunes. It seems that the animal, in pressing down on the still damp sand, had consolidated it. When the morning sun dried out the sand, and the wind set it in motion, these more solid and sheltered parts gradually stood out in relief, so that on the windward side of the slope (and only there) a negative of the spoor appeared.

The amount of water in the soil can be represented, like its heat content, by lines of equal soil moisture. Figure 83 gives measurements made by K. Unger [333] in Quedlinburg in 1950. The soil in question was loam with a light layer of humus, and grass was growing on it. Beyond a depth of 80 cm there was a gradual change to a gravel base at 120 cm. The water content in weight percent was measured from core samples every 20 cm of depth, three times a month. The amount of precipitation is shown for 10-day periods. The normal annual variation shows up very clearly for this particular year, with a soil-moisture maximum at the end of the winter, and a minimum in late summer. S. Uhlig [331] has published a similar graph for readings taken twice a week down to a depth of 50 cm from autumn 1949 to spring 1950, and also, for comparison, graphs for six West German agrometeorologic stations for the calendar year 1950 [330]. Further summarized results are to be found in R. Pfau [321], H. Rettig [325], and S. Uhlig [332].

Annual soil-moisture variations also depend on the type of soil. From 1948 to 1951, E. Unglaube [334] averaged over the first 50 cm the weekly water content in volume percent for the loess soil of an orchard and the stony soil of a vineyard near Geisenheim. The annual variation, based on these weekly values, is shown in Fig. 84. These are well-balanced curves, since they give the averages of a 4-yr period that contained both a dry year and a wet year. The time of least water content for both of these

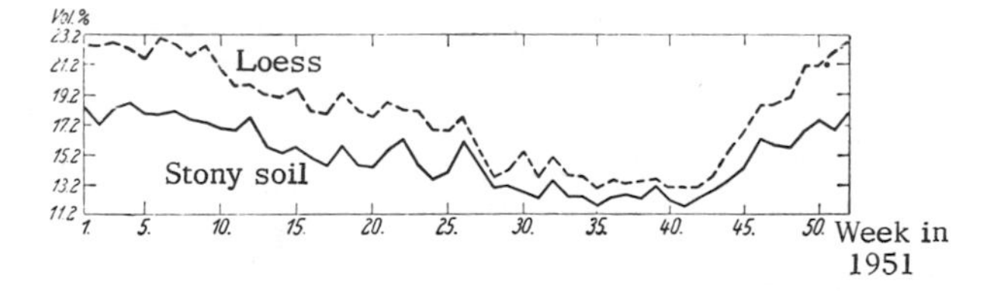

Fig. 84. Annual moisture variation in two different types of soil at Geisenheim. (After E. Unglaube)

soils falls between the 35th and 40th weeks, or the period from the end of August to the end of September, and the maximum is in February, which agrees well with Fig. 83. Not only is the water content of the loess always higher than that of the stony soil but also the difference between the two values is greater in winter than in summer. The annual variation is smaller in the stony ground with its poorer water-holding capacity than in loess.

The question now is, what is the effect of varying water content on the heat economy of the soil? The color of the ground, and therefore its albedo, are altered by soil moisture; a few figures illustrating this difference were given earlier in Table 3. Of greater importance is the change in soil constants brought about by air in the pores of the soil being driven out by water. As already shown in Sec. 6, both the density and the heat capacity of the natural soil will increase with rising water content. The same applies to the thermal conductivity λ.

D. A. de Vries [264] has shown that the heat conductivity of a moist granular soil can be calculated approximately if the assumption is made that the grains have the shape of an ellipsoid of revolution. The results of his comprehensive calculations agree well with his laboratory experiments. For a fine sand, 89 percent quartz and 11 percent feldspar, the grains occupying 57.3 percent of the volume, the relation between the heat conductivity λ and the water content in weight percent is shown in Fig. 85. At normal temperatures (20° C), λ first increases very quickly, because even a very thin film of water on the grains makes a substantial increase in the surface areas in contact. From about 7 percent water content λ increases more slowly. At very high temperatures (75° C) a maximum of λ is found when water content is about 6 percent.

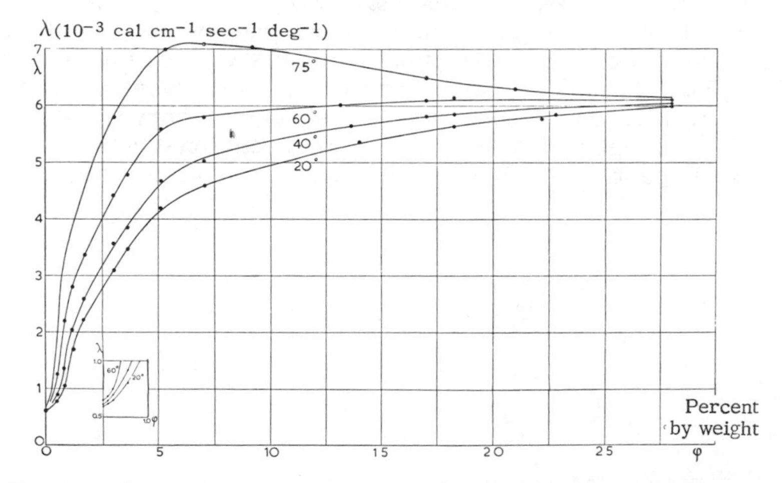

Fig. 85. Dependence of thermal conductivity λ on water content (weight percent) of a sandy soil, at different temperatures. (After D. A. de Vries)

In a more recent work de Vries [335] used the relation between the water content v_w (volume percent) of a clay on which grass was growing and the heat conducitivity λ, with the idea of using it in reverse to find v_w from λ. The increase of λ with increasing v_w is naturally similar to its increase with water content expressed in weight percent. This is also shown by the figures below. If the values for heat capacity $(\rho c)_m$ as a function of v_w (given in Table 9), which are also approximately valid for this soil, are used, then the thermal diffusivity a of the soil can also be found:

Water content v_w (volume percent):	0	10	20	30	40
Thermal conductivity λ (cal cm^{-1} sec^{-1} deg^{-1}):	0.0006	0.0024	0.0036	0.0040	0.0043
Thermal diffusivity a (cm^2 sec^{-1}):	0.0020	0.0060	0.0072	0.0066	0.0061

While λ continues to increase with soil moisture, a shows a maximum for medium moisture values. Dry soil conducts temperature more slowly because λ is small, and wetter soil more slowly because $(\rho c)_m$ is large. This has been confirmed recently in laboratory studies by E. Maruyama [319] in Japan.

In a way similar to that used in Fig. 9, showing the variation of heat conductivity with changing water content of soil for July 1937, J. Bracht [303] published the variation of λ at depths of 2, 10, and 30 cm in a loam, the top 10-cm layer of which was mixed with an equal quantity of humus, for the period June to August 1940 in the neighborhood of Leipzig. De Vries [335] using the method described above, published the value of λ at depths of 4, 8, and 16 cm in a clay, for the whole of the growing period in 1951, giving at the same time the variation of water content in volume percent for the three depths.

A daily variation in λ is to be expected, since the upper layer of the soil is dried to some extent during the day in fair weather, and takes up water again at night. M. Franssila [307] was able to measure a maximum value of λ in the topmost 2 cm of the soil toward the end of the night, which was 16 percent higher than the minimum in the late afternoon. This diurnal variation can be recognized also from the pulsation of the curve (Fig. 9) for 1-cm depth during the precipitation-free period of fair weather at the beginning, and in the second half of July. This variation of v_w and λ has also been confirmed by de Vries in his random measurements in May, mentioned earlier.

Figure 86 shows the influence of artificial watering on temperature distribution in the soil. L. A. Ramdas and R. K. Dravid [323] observed soil temperatures at two identical surfaces at 14:00 under Indian radiation conditions, then at the time W (Fig. 86) one of the surfaces was watered. The steep change in the slope of the isotherms shows that the water added had the effect of a "cold shower." This, however, is not the result of the lower temperature of the water, which soon reaches soil temperature, but of the heat released on evaporation.

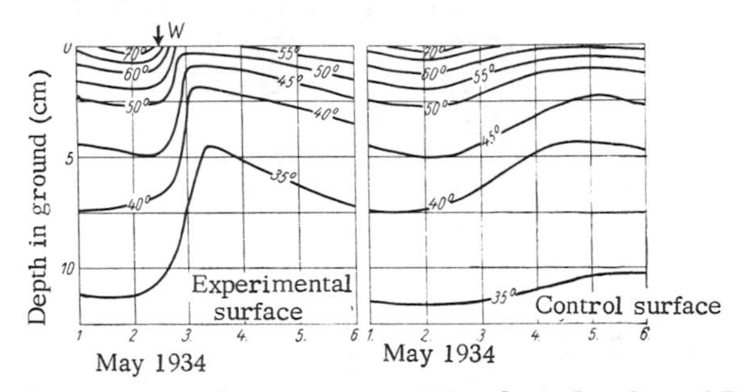

Fig. 86. Effect of watering on soil temperatures. (After L. A. Ramdas and R. K. Dravid)

It should be borne in mind that the flow of heat in wet soil is much greater than in dry soil. The albedo of a moist surface is lower in the first place; but the main influence is the increase in thermal conductivity λ. F. A. Brooks and D. G. Rhoades [304] carried out an experiment in a California pear orchard at 39° 04′ N, making a comparison by direct measurement of the flow of heat in two experimental areas of 2 hectares each, one watered and the other unwatered. It was 2 to 3 times as much in the watered as in the unwatered plot. The highest noon value in the first case was 0.36 cal cm^{-2} min^{-1}, against 0.17 in the second. At a depth of 3 mm, however, the noon temperature in the dry soil was 54.4° C, in comparison with only 33.9° C in the wet.

From Fig. 86 it can be seen that the watered soil does in fact begin to warm up after a few days, but retains its special features for the duration of the experimental period. Brooks and Rhoades found in their experiments that it took 1 to 2 weeks before the two soils reached the same temperature again (see also Sec. 51 on artificial sprinkling).

Not only does the variable water content of the soil determine its temperature, but, inversely, the temperature stratification in the soil affects the water distribution in it. Water is transported first by gravity and capillary action. When, however, the air pores in the soil are not filled with water and the air can circulate between soil particles coated with a film of water, a transport of water vapor takes place in the soil, directed toward the region of lower water-vapor pressure. Since saturation pressure for water vapor increases with temperature, temperatures decrease in the same direction as does water-vapor pressure. Other things being equal, water moves in the ground from regions of higher toward regions of lower temperature, through its evaporation in the one region and condensation in the other.

Generally, this method of water transport is masked by the downward seepage of rain water and the capillary rise of water. There are times, however, when this relation between soil temperature and moisture alone is able to account correctly for the observed data. This occurs, as shown by H. Rettig [324], when precipitation is low but there are many changes in temperature. Figure 87 shows measurements made at the German Weather Service's experiment station in Neustadt an der Weinstrasse. The upper curve (*a*) shows the variation of the daily mean air temperature in comparison with the long-period temperature variation; the hatched areas show warm periods. Section (*b*) shows its penetration into the soil in the form of geotherms. Precipitation, shown in (*c*), was not followed by any seepage but was taken up by the top 25 cm of the soil, except in one instance, that of 13.3 mm on 24 May 1956. The daily evaporation from bare soil was measured with a small lysimeter, of the Popoff type, and this is shown by (*d*) together with the saturation deficit of the air at 14:00. At the bottom of the diagram (*e*) the seepage water (stippled area) and the water transported upward are shown. It can be seen that seepage is restricted to the warm periods. It follows that the small amounts of precipitaion, which usually fall in cool periods, do not immediately flow downward. This takes place only subsequently in warmer intervals, when the temperature and water-vapor-pressure gradients are in a downward direction. (See also R. Pfau [322].)

The movement of water vapor to colder parts of the soil becomes significant in magnitude and of considerable practical importance when the ground freezes.

Soil freezing in winter is welcomed by the farmer to the extent to which it helps, by repeated freezing and thawing, to break up the coarse blocky soil in bare areas into a more crumbly structure, which might be called "tilling by frost." On the other hand, he fears its action on the planted land, since it draws the young plants out of the ground or breaks their roots. Ground freezing is also a source of great worry in the maintenance of highways, because of the heavy demands placed on them by increasing density of traffic and loading of trucks. After freezing, the movements of the ground may break the surface of a highway, and after a thaw from the surface there forms over the remaining ice a water-filled roadway with a greatly reduced load-bearing capacity. Frost danger in road construction has been studied in detail, because of its great economic importance, in recently published works by A. Dücker [305], R. Ruckli [326], J. Schmid [328], and L. Schaible [327].

Water in the soil does not freeze at 0° C. The depression of the freezing point, investigated closely by Ruckli, is the greater the lower the water content of the soil and the more fine-grained the soil is. When the temperature decreases, the mobile water in the soil pores freezes first, then the looser water films round

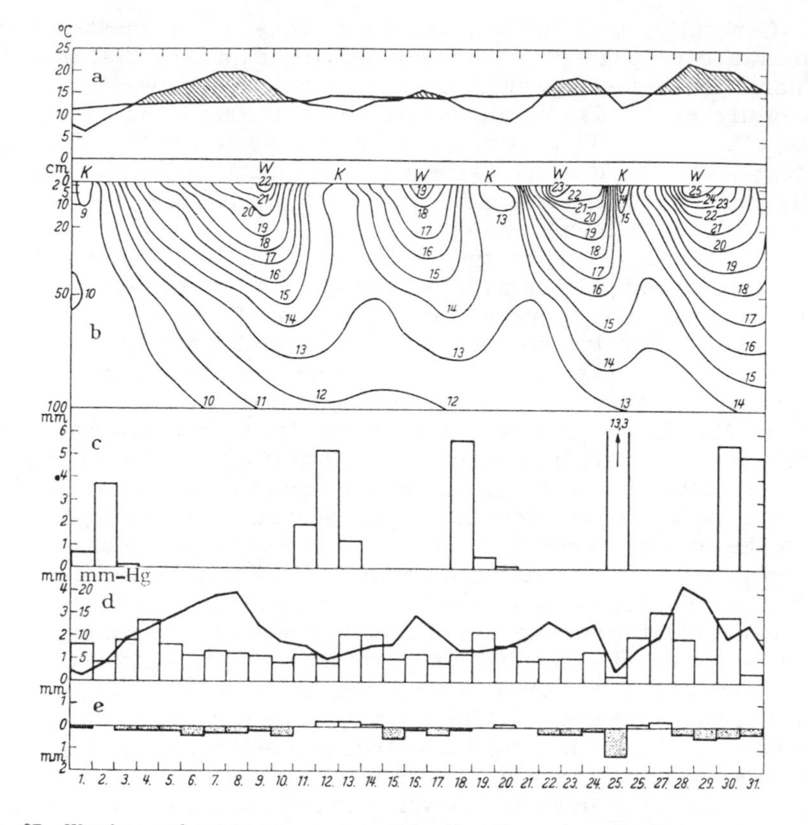

Fig. 87. Weather and water transport in the soil at Neustadt an der Weinstrasse in the relatively dry month of May 1956. (After H. Rettig)

the particles, and lastly, or sometimes not at all, the strongly adsorbed films in the smallest pores. In freshly poured concrete, for example, the water in the pores freezes only at -3° to -4° C. In many meteorologic problems it is desirable to know whether the ground was in fact frozen during the night or not, and this is not easy to determine by means of a minimum thermometer. P. Lehmann [318] has devised a frost indicator, called the geloscope, for this purpose. It is based on the traces of frost that form on the surface of fine soil through the growth of ice needles, which remain after thaw and should not be confused with dessication cracks, which they often resemble. Geloscope measurements show good agreement with frost observations in the area where they are installed. The use of geloscopes on a large scale in the field—the material for 50 geloscopes only costs one-tenth as much as a minimum thermometer—gave results in conformity with experience of frost hazard and frost damage collected over a period of years in the area in question.

When the air is below 0° C but the ground is not yet frozen, a striking phenomenon can be seen occasionally: the formation of mush frost, also called stalk-ice, hair-frost, ice needles, or ice fibers, known in Sweden as "Pipkrake." This is formed in loose damp soil, mostly loam or sandy loam, mainly on drainage slopes, at road cuttings, often on pine needles in forests, on decaying wood, and so on. Ice needles about 1 mm in diameter grow perpendicularly, or in a slightly bent form, out of the ground, close together in such great numbers that they look like a comb. In growing, they lift straw, particles of soil, and stones (sometimes as big as a fist) to heights of several centimeters. Literature on this subject has been collected by J. Schmid [328] and the formation of mush frost has been studied in detail, mostly experimentally, by H. Fukuda [308].

Mush frost forms most easily when (warm) rain is suddenly followed by a sharp frost without further precipitation. The plentiful supply of soil water feeds the ice needles from below, while crystal formation and the decrease in temperature assist the water supply. The needles grow most quickly at the beginning. Occasionally it is possible, as seen in Fakuda's photograph in Fig. 88, to see growth on successive days. The first 15 cm below the surface of the soil play the greatest part in producing this effect; the water content of the soil below 30 cm has no effect. Mush frost may also form under a snow cover. Usually the ice needles are a few centimeters in height. Heights of 15 cm have often been observed, and much greater heights (50 cm) in rare cases.

Fig. 88. Mush frost, showing diurnal layers. (Photograph by H. Fukuda)

After a time, as winter progresses, the ground freezes, bringing about a number of changes in its physical state. The first of these is the change in specific volume from 1.00 for water at 0° C to 1.09 $cm^3 g^{-1}$ for ice at 0° C. This increase of almost 10 percent in volume causes heaving of the ground, about which more details are given below. The specific heat of ice at 0° C is 0.505 cal $g^{-1} deg^{-1}$, which is only about half that of water. The thermal capacity of the soil (see p. 28) therefore decreases when freezing occurs, to an extent dependent on the amount of water present. Since $v_s \rho_s = v_e \rho_e$, the thermal capacity of the frozen soil becomes

$$(\rho c)_{m\ (ice)} = 0.01\,(v_s \rho_s c_s + 0.505\, v_w) \qquad (cal\ cm^{-3}\ deg^{-1}).$$

In sandy soil with $\rho_s = 2.63$ g cm^{-3} and $c_s = 0.20$ cal cm^{-1} deg^{-1}, which has already been used in the example on p. 28, we have the values given in Table 50.

Table 50

Thermal capacity of unfrozen and frozen soil.

Water content, v_w (volume percent)	0	10	20	30	40
Ice content, v_e (volume percent)	0	11	22	33	44
$(\rho c)_m$ unfrozen (cal cm^{-3} deg-1)	0.30	0.40	0.50	0.60	0.70
$(\rho c)_m$ frozen (cal cm^3 deg-1)	0.30	0.35*	0.40*	0.45*	0.50*

*A recent amendment.

From Table 10 it can be seen that the thermal conductivity of ice is roughly four times that of water. The thermal diffusivity is therefore also increased; and since the thermal capacity is reduced in the frozen state, this gives an additional increase to the thermal diffusivity. An increase of 20 to 50 percent in the thermal diffusivity can therefore be expected when the soil freezes.

Of great importance in the total exchange of heat is the latent heat of 79.5 cal gm^{-1} released when a gram of water freezes. It was calculated, for example, by J. Keränen [312] that during the winter of 1915-16 in Sodankylä (67° 22′ N), the heat loss, totaling 1605 cal for every column of cross-sectional area 1 cm^2 and depth 1.1 m in the frozen soil layer, was offset to the extent of 69 percent (= 1104 cal) by latent heat. Of this 24 percent (= 389 cal) came out of deeper layers, from stored summer heat, and only 7 percent (= 112 cal) caused a lowering of temperature of the frozen soil layer during the winter (the difference between 168 cal for cooling and 56 cal for reheating between November and April).

Another effect of soil freezing is that the water-vapor-pressure gradient in the unfrozen ground below is always directed upward and is independent of the diurnal temperature variation at the surface. Therefore both the water rising under capillary action and the water vapor still moving through the soil pores come up against a barrier at the frozen layer. Water arriving at this barrier freezes to an extent dependent on how much of the heat it has brought with it and the latent heat released by every freezing gram can be conducted away upward. Experience shows that, on the average, the temperature decreases linearly with height in a homogeneous frozen layer. Freezing in winter changes a uniform soil into a two-layered soil.

The enrichment in water of the frozen layer can be seen in Fig. 89 from observations made by N. Weger [336] in Geisenheim on the Rhine in January 1953. The water content in the fine sandy-loam soil of an orchard was measured continuously for 10-cm layers. The average value for the topmost layer was assigned to the depth 5 cm with which the diagram begins. The rectangles show the depth of freezing. There were only 2.0 mm of precipitation during the month, which had no influence worth discussing on water movement in the soil, as in Fig. 87. The soil had been well moistened by rain in the second half of December. The lines of equal water content in percent by weight show the increase in the frozen layer at the expense of the unfrozen layer below. If the change is worked out from 30 December 1952 to 16 January 1953 for a column of area 1 cm^2, a gain in water of 2.14 g is found for the frozen layer, and a loss of 2.08 g for the layer below down to a depth of 50 cm. Transport of water in the first half meter of the soil in January was therefore ten times the precipitation falling on the surface. When the ground thaws at the end of the month, the lines of equal moisture diverge; some of the water seeps downward (20-percent line), but most evaporates from the open soil into the atmosphere. Repeated freezing and thawing thus dries up the ground and the process is therefore feared by the farmer.

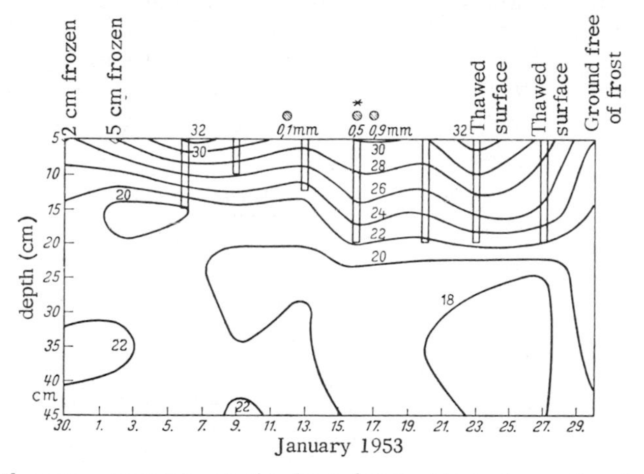

Fig. 89. Soil-water migration in the frost layer. (From observations by N. Weger in Geisenheim)

Homogeneous freezing of the ground is to be distinguished from the formation of ice lenses. Homogeneous freezing occurs in noncohesive soils down to a grain size of 0.05 mm where the grains are surrounded by a crust of ice and become linked together. It may happen, however, for a number of reasons, that the soil gets displaced from its original position by the formation of mainly horizontal ice lenses, a matter being investigated by road-construction experts. These lenses consist of clear ice, generally with a vertical fibrous structure, occasionally containing air inclusions. They are found in all sizes from microscopic up to 20 cm (35 cm exceptionally). The reasons why they form have been explained recently by G. Kretschmer [314].

The frequency and depth of soil freezing in winter depend on the weather. The mean frequency of freezing at depths of 0, 20, and 50 cm has been computed by K. Unger [109] for each day of the 10-yr series of observations at Quedlinburg. The surface freezes there from the beginning of November until the middle of April, on more than 60 percent of the days in midwinter. At 50-cm depth it hardly reaches 10 percent, beginning on 22 December at the earliest, and ending on 1 April.

The greatest depth of freezing in winter is a point of practical importance for the laying of pipes that might be affected by it, such as water mains. There are great differences from year to year. W. Kreutz [316] has compared the time and depth of freezing of the same soil during four mild and four severe winters between 1939 and 1949. The deciding factor is the depth and duration of snow cover; winters with small amounts of snow are the most dangerous with respect to freezing. The experimental area of clean gravelly sand used in the long series of observations at Potsdam from 1895 to 1948 was deliberately kept clear of snow. During this period, as shown by G. Hausmann [95], the temperature at a depth of 2 m never dropped below 0° C. At 1 m frost was observed 8 times in the 54 years, shown chronologically in Table 51. These results give, at the same time, limiting values for ground without snow protection.

Table 51
Duration of frost and minimum temperature at two depths.

Winter	Frost duration (days)		Lowest temperature (°C)	
	1 m	0.5 m	1 m	0.5 m
1894-95	6	32	- 0.2	- 5.1
1900-01	8	21	- 0.3	- 6.5
1916-17	2	38	- 0.1	- 4.7
1921-22	10	33	- 0.3	- 5.2
1928-29	36	63	- 2.7	- 9.6
1939-40	41	58	- 1.6	- 7.4
1941-42	40	64	- 1.1	- 7.6
1946-47	51	80	- 1.7	- 8.1

Observations in Vienna from 1911 to 1944, analyzed by M. Toperczer [108], were made in a loess soil rich in limestone with a covering of 30 cm of humus, the winter snow covering remaining undisturbed. The thermometer at 98 cm never fell below the freezing point. The frost depth exceeded 80 cm once during the 34 years, 70 cm four times, 60 cm nine times, and 50 cm twelve times. Frost therefore did not penetrate beyond half a meter in two-thirds of all these years.

Soil freezing is naturally much influenced by the type of soil. Damp soil freezes more slowly because of the latent heat released, and to a lesser depth than dry soil, but it also thaws more slowly in the spring. J. Keränen [311] has published the following figures for average frost depth (cm) in Finland:

Type of ground	Sand and gravel	Plowed land	Clay	Bog
North Finland	126	100	90	88
South Finland	72	47	50	42

Figure 90 shows the duration and depth of ground freezing in four different types of soil at the agrometeorologic experiment station at Giessen, for the cold winter of 1939-40, after W. Kreutz [315]. The lighter-shaded areas indicate where temperature was negative, and the darker areas where it was less than -4° C. In the dry basalt gravel the frost penetrates at a rate of 2 cm per day on the average, and reaches 67 cm; but the whole soil is thawed by 25 February. The rate of frost penetration into moist humus is only 0.6 cm per day, and reaches only 32 cm, but it lingers until 22 March under a surface that is already thawed out. Attempts have been made to calculate the depth of frost penetration theoretically, for purposes of road construction, taking all the various factors into account and including both the case of homogeneous soil freezing and that of ice-lens formation. The bases of the arguments are to be found in R. Ruckli [326].

The 9-percent expansion of water on freezing, and ice-lens formation, cause soil heaving. Hard-surfaced roads are subject to heaving, develop frost fissures, buckle, and suffer other forms of frost damage. Heaving can amount to some tens of centimeters. In natural soil, which is the main concern here, the amount of movement is much smaller, as shown by the following frequency table for various amounts of displacement evaluated by R. Fleischmann [306] from observations made in Hungary from 1931 to 1935:

Heaving (mm/24 hr)	0-5	5-10	10-15	15-20	20-25
Number of cases	57	38	4	3	2

G. Kretschmer [313] and J. Schmid [328], in a new investigation

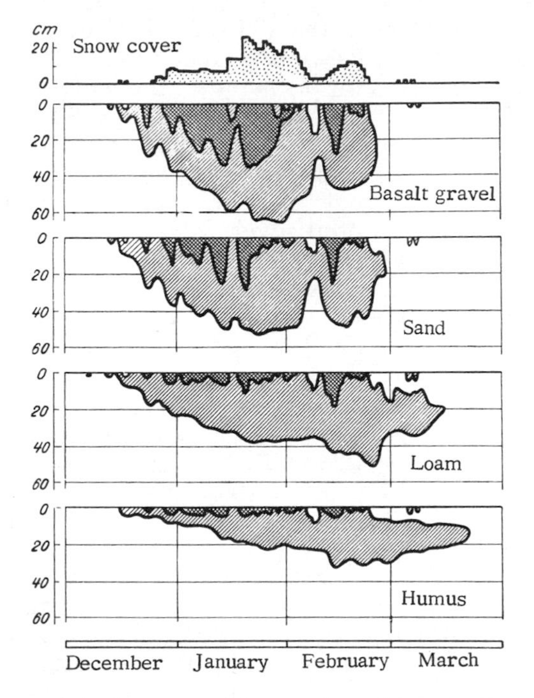

Fig. 90. Duration and depth of ground frost in four different soils in winter of 1939-40 at Giessen. (After W. Kreutz)

with improved techniques, found that the movement of arable land did not exceed 24 mm a day, and reached a maximum of 36 mm for the whole winter.

Figure 91 gives three typical periods taken from a continuous series of observations made by G. Kretschmer in loose plowed soil near Jena during the winter of 1954-55. At the turn of the year, when there was a dry frost, there occurred the first strong heaving, typical of the start of winter. The curves for two recording instruments differ only in the magnitude and not in the type of heaving. On 1 January there was a fall of snow, and the snow cover reached a depth of 12 cm by 3 January. Frost heaving came to an end. There was freezing weather from 22 to 25 January, with a short thaw in the middle of the day. Each time it thawed the earth settled again a little. During a period of thaw from 28 to 31 January, with short night frosts, the surface of the ground was subjected to marked periodic movements. All these movements can damage the young shoots of grain or woodland plants by pulling the roots out of the soil so that they dry up or freeze, or by tearing roots at greater depths, mostly of 2 to 6 cm.

Before concluding the discussion of Secs. 19-21 on soil temperature and moisture, three meteorologic processes should be mentioned that show the connection between soil characteristics and heat exchange. These are the melting of freshly fallen snow, the formation of hoarfrost, and the formation of glaze.

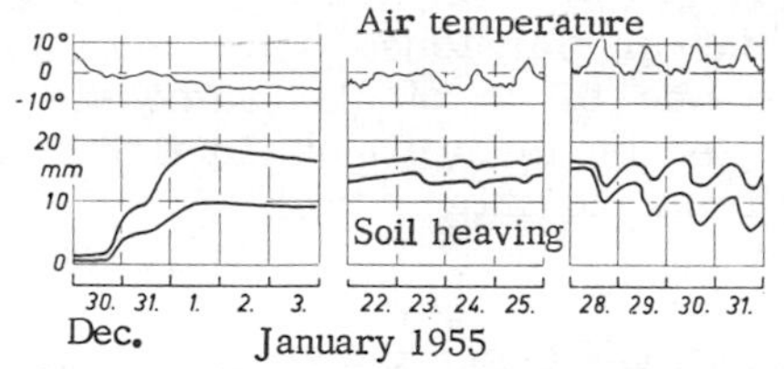

Fig. 91. Typical frost heaving of the surface of plowed land. (From measurements by G. Kretschmer near Jena)

Snow is formed in the upper levels of the atmosphere, and is therefore independent of microclimate. In mountain areas, with temperatures a little above 0° C, while wet snow is falling, it can be seen on looking over wide stretches of countryside that the lower limit of the snow cover coincides with a contour. But as soon as the snow settles, the influences of radiation, wind, and heat of the ground are excluded, microclimatic differences quickly make themselves felt, and all the more quickly, the thinner the snow cover. The snow starts to melt at its lower boundary. Over soil of good conductivity the flow of heat from below quickly makes the snow edge climb to higher levels, and the same thing happens on slopes where radiation brings heat down onto the surface of the snow from above, thus removing it. The melting processes have been illustrated in maps with explanatory text by H. Slanar [329], published in 1942.

A very fine example is shown in Fig. 92, from a photograph by H. Mayer [320]. Members of the Frankfurt Meteorological Institute, on a visit in April 1943 to the Research Station on the Jungfraujoch, had gone into the Lauterbrunnen Valley. Mayer writes: "Continuous snow cover about 10 cm deep and 0°C lay over warm ground above 0° C. Air temperature was also 0° C. The cloud ceiling was very low and the only light was from weak diffused sky radiation, so that the snow was melted mainly by heat coming from the ground. It melted first on rocks with growth on them, then on grassland and overgrown debris fallen from the slopes. It had all disappeared completely when we saw in front of the Staubach Valley a part still completely snow-covered, among the green slopes. The Staubach Waterfall, tumbling down 300 m of perpendicular wall left by an old glacier trough, had carved a small erosion valley in the old consolidated and overgrown debris from the cliffs. The debris in the new valley still had a complete snow cover. The rock forming this debris cone, brought out by the water in comparatively recent times, almost certainly had the same high thermal conductivity as the soil. The conductivity of the entire debris cone, however, was determined by the air present in the hollow spaces between the rocks, and was therefore much less than that of the solidified ground. The snow began to melt only when the weather cleared up and the sun broke through."

J. L. Monteith [320a] compared two photographs, one of an area paved with uneven flagstones with small plants growing between the stones, the other of the same view after a snowfall. The snow had melted on the well-conducting stones but remained on the plants, outlining the pattern with white bands.

Fig. 92. Snow melts last on a heap of stones with poor conductivity. (Photograph by H. Mayer)

Equally conclusive observations can be made when there is hoarfrost. Whereas snow falls everywhere indiscriminately and microclimatic differences become visible only when it melts, the formation of hoarfrost depends on the characteristics of the place where it forms. The time at which hoarfrost melts should also be observed. Piles of wood are still white in the morning long after the earth around them, which is a good conductor, has become dark. Water pipes, which prevent the movement of heat, become visible through white hoarfrost in an otherwise uniform street. Once in Bad Kissingen I saw the complete structure of the smooth tin roof of a large shed outlined in snow-white lines a few minutes after sunrise. Where there was only a thin foundation below the tin, the hoarfrost melted quickly in the morning sun; but where there were joists and cross-ties underneath, these had taken up some of the sun's heat, so that the hoarfrost did not melt until a little later.

A few trees had been removed, and their stumps completely uprooted, from an avenue in the Hofgarten in Munich in the autumn, and after the holes had been filled in and smoothed over their position had become quite invisible. The following spring A. Schmauss noticed that, after cold nights, the whole area that had been dug out was white with frost. The earth in the holes was looser than elsewhere, and therefore had lower thermal conductivity.

Figure 93 is an air photograph taken by the Royal Air Force in Lincolnshire, England, at the suggestion of O. G. S. Crawford. It shows the land around the medieval village of Gainsthorpe during hoarfrost. This photograph, published in *Orion* by J. Herdmenger [310], has been placed at my disposal by the kindness of the publishers of the periodical. The walls of the old village, which had been lost to sight since 1610, show up in hoarfrost because of differences in conductivity of the ground below the surface. Historical researches have frequently made use of such processes, also outside Europe.

Fig. 93. Traces of the foundations of an old village are sketched by hoarfrost. (Air photograph by R.A.F.)

Glaze, however, is the element most sensitive of all to changes in soil characteristics. Glaze forms in two ways, either by the freezing of supercooled rain drops on the (warm) ground, or by the freezing of rain drops (above 0°C) on very cold ground. Anyone who walks around with open eyes during severe glaze will not cease to be surprised, to ask questions, and to investigate. Every street, every wall, every surface, every type of stone has its own style of glaze. Houses with central heating have an effect on the sidewalk outside. Old street excavations show up vividly again. The roughness of the surface, the thickness and the type of stone surfacing, the slope of the ground, all these have their effects. Truly, if one wants to set oneself an examination in microclimate, a walk during glaze would answer all the questions set by nature.

22. The Layer of Air above Small Water Surfaces

If the lower boundary of the atmosphere rests on water rather than earth, its behavior is conditioned by the changed state of the surface. In water there is an exchange of mass that is not present in soil. Earth and water react differently toward short-wavelength radiation. Actual or effective evaporation can be reduced below the possible or potential level only in solid ground in the absence of an adequate supply of water. The potential rate of evaporation is determined by the temperature of the evaporating surface and the state of the air above it. Even a small roughness in water surfaces modifies the wind field over water as compared with that in the air layer near the ground, and hence produces changes in exchange values.

The heat economy in solid ground was controlled almost exclusively by molecular heat transport, that is, poor conduction. In water, which is mobile, there is mass exchange. The action of wind on the surface causes a certain amount of mixing in the uppermost layers. To this frictional exchange there is added, as in air, convectional exchange. Cooler parcels of water sink downward while warmer parcels rise. Water therefore has the properties of extremely conductive soil. Following the rules established for good conductors (see p. 145), the daily temperature fluctuation at the surface will be small, amounting in the open sea, for example, to only a few tenth of a degree. In contrast, much thicker layers take part in the process of heat exchange. This applies also to the annual turnover of heat, giving rise to the differences between maritime and continental climates.

It must be borne in mind, however, that in certain circumstances mass exchange may be totally lacking or very small in amount. During cooling in autumn, denser water above +4° C will sink until the temperature becomes uniform and all convectional mixing ceases. Spontaneous freezing can set in during the calm periods often found in winter nights, since frictional mixing is then also absent. According to K. Keil [342], a Swedish invention is able to eliminate the difficulties this causes to shipping; long pipes with small holes in them were laid out in the lanes to be kept free. When there was a danger of the surface freezing, compressed air was fed into the pipes. It escaped through the holes, causing an artificial mixing of the surface layers with water at 4° C at lower levels, and delayed ice formation for a short time until the ships still at sea could reach their berths. If there is sufficient frost to inhibit mass exchange completely, as over the polar seas, climate adopts continental traits. The layer of air above water then becomes a layer over snow or ice.

Under circumstances of interest to microclimatology the effect of mixing in water can often be substantially reduced. There may be vigorous mixing taking place in a small body of water, but the quantity of water taking part in it is too small to act as a heat reservoir, which tends to create uniformity. Vegetation growing in the water, such as reeds and creeping plants, may restrict mixing partially or totally. We shall return to this point later.

In making a comparison with earth, there is, in addition, the different behavior of water to long- and short-wavelength radiation. Water behaves much the same way as earth toward long-wavelength radiation; both are subject to the same conditions of absorption and emission by night. The reflection of polarized and unpolarized light as a function of the sun's position (angle of incidence), cloudiness, and state of the water surface have been investigated many times, and recently very cogently summarized by F. Lauscher [343]. Table 3 showed that diffuse reflection from water surfaces was less than that of all other natural surfaces. Theory shows that a sphere of water will reflect only 6.6 percent of solar radiation. It follows that solar and sky radiation falling on water is utilized almost completely. Only when the sun is low is there much mirror reflection (Fig. 3). This is of practical importance at the seashore and on the banks of lakes and rivers. The microclimate of vineyard terraces may be influenced by reflected light from the river below. In February, when the sun is low at midday, O. H. Volk [354] measured light intensity on the Steinberg near Wurzburg by photoelectric cells and found that sunlight, skylight (light from above), and reflection from the river Main (light from below) were in the ratios 42:11:41; the average of five sample tests made in March showed that light from below amounted to 65 percent of light from above. "The best situations for vine terraces," writes Volk, "are those that enjoy this extra light. To the east and west of the Main Valley there are within a few kilometers from the river many slopes that have the same exposure, the same inclination and geologic structure, the same soil characteristics, which in short are not recognizably different in any respect from the south and west slopes of the Main Valley. The latter, however, no longer grow vines, or produce inferior wines. There is no question of there being microclimatic differences between the Main and Wern Valleys. I was unable to account for the difference in wine production until I became aware of the difference in the amounts of light from above and below." This difference can be seen clearly also in the distribution of wild flowers.

In lakes, the western shores receive a detectable amount of additional reflected radiation from the morning sun, and the eastern shores from the evening sun. The westward-facing shore, as H. Frey [339] pointed out is overrated subjectively, since the average person does not like to get up early, and therefore rarely sees the morning sun.

The principal difference, however, in the behavior of water toward radiation is the depth to which radiation penetrates. Every swimmer knows that objects can be recognized deep down and that visible light must therefore penetrate there. The intensity of solar radiation at various depths in clear water is given by Wilh. Schmidt [351] for three selected ranges of wavelength in Table 52, the intensity being expressed as a percentage of the radiation reaching the surface. In the range from ultraviolet to orange (0.2-0.6 μ) almost three-fourths of the radiation penetrated as far as 10 m, and about 6 percent reached 100 m. From red to infrared (0.6-0.9 μ) penetration is poorer. Longer wavelengths are absorbed in the topmost layer of the water.

Table 52
Percentage of incident solar radiation reaching various depths in water.

Wavelength (μ)	Depth					
	1 mm	1 cm	10 cm	1 m	10 m	100 m
0.2-0.6	100.0	100.0	99.7	96.8	72.6	5.9
0.6-0.9	99.8	98.2	84.8	35.8	2.6	0.0
0.9-3.0	65.3	34.7	2.0	0.0	0.0	0.0

Differences in the transmission of different wavelengths in the spectrum may be seen more clearly from Fig. 94. The top curve, after W. R. Sawyer and I. R. Collins [taken from 353], shows the percentage of radiation of various wavelengths penetrating a layer of clear water 1 m deep. In the infrared range, penetration decreases sharply beyond 850 mμ, where the diagram ends. Water surfaces may therefore be considered as behaving in the same way as solid ground toward long-wavelength radiation.

These values for clear water are not valid for natural bodies of water such as lakes, ponds, or even pools. Dissolved and suspended substances not only color the water but affect its transparency. In comparison with clear water it is usually lower as a whole, and the optimum penetration of rays is shifted toward longer wavelengths. Figure 94 contains two curves from I. Dirmhirn [338], typical of the Lunzer Untersee in Austria in two different years. The greater or smaller quantity of suspended matter has a pronounced effect on penetration, while the spectral distribution for one and the same body of water remains substantially unaltered (see also Fig. 107).

Examined more closely, the radiation exchange taking place in water is much more complicated, since, in addition to absorption by pure water, the amount of absorption by dissolved substances and foreign particles in suspension has to be taken into account. The effect of light refraction is such that, viewed from inside the body of water, all light from outside comes from within a cone of semivertical angle 48.6°, the critical angle for total reflection. Besides being absorbed, light undergoes scattering. Scattered radiation comes first from the surface of the water. Light comes also from below into the water, and may be increased considerably by reflection from the bottom, provided this is not too far down. Weakening of radiation intensity with depth is therefore not merely a result of absorption; it is described, taking all factors into account, as extinction. More details can be found in a book by F. Sauberer and F. Ruttner [349], and in a new summary by F. Lauscher [343].

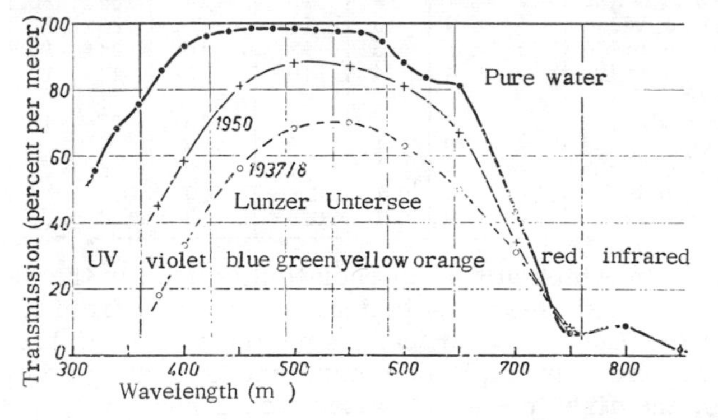

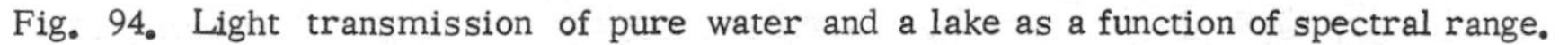
Fig. 94. Light transmission of pure water and a lake as a function of spectral range.

Table 53 shows the wide range of variation of the penetration of light in natural waters, within the visible spectrum, particularly if the small water surfaces and streams, which are of importance to microclimatology, are included. Values already given in Fig. 94 for pure water are included for comparison, on the first line. The results for Austrian waters are from F. Sauberer [348, 349] and those in Brandenburg from O. Czepa [337]. Among the lakes, the Neusiedlersee southeast of Vienna occupies a special position, because it is only 40 to 80 cm deep, with a surface area varying between 250 and 300 km^2. A layer of mud about 0.5 m thick on the bottom is stirred up by even moderate winds; in the wide shore area, covered with reeds which hinder the mixing process, penetration is at these times not so extremely low as it is in the open water. The small lakes Kalksee and Flakensee, in Brandenburg, have thick growths of algae, and the Flakensee is made even more turbid by the dirty water of the Löcknitz, which flows into it. In rivers, the principal role is played by suspended matter carried along in the water. It is hoped that observations of this type will be made in other areas, as they are of great importance for many practical problems such as pollution by effluence, assimilation by aquatic flora, vision of predatory fish, and so forth.

Table 53

Light transmission by water (percent per meter).
(After F. Sauberer and O. Czepa)

Wavelength (mμ)	375	400	450	500	550	600	650	700	750
Range of spectrum:	Violet		Blue	Green	Yellow	Orange		Red	
Pure water	84	93	98	98	97	87	81	43	7
Achensee	51	65	80	85	82	73	57	33	8
Lunzer Untersee	18	33	56	68	70	63	50	31	7
Lunzer Obersee	2	9	26	39	46	47	41	27	6
Müggelsee near Berlin	-	-	8	23	34	36	36	28	5
Arm of Danube, clear	3	8	15	21	26	25	21	16	4
Danube, slightly turbid	0	1	5	11	16	20	15	8	2
Kalksee I, near Berlin	-	-	4	10	15	17	15	13	3
Neusiedlersee									
Area with reeds	-	0	2	8	13	17	17	12	-
Open water	-	0	1	3	6	7	6	4	-
Flakensee, near Berlin	-	-	1	3	6	5	5	5	1
Danube, very turbid	-	0.0	0.1	0.1	0.3	0.7	0.8	0.5	0.1
Flat moorland pool	-	0.0	0.1	0.2	0.8	1.8	2.8	4.8	1.3

The way in which absorbed radiation is used in the heat budget in water is best seen by taking an example from I. Dirmhirn [338], for the Lunzer Untersee. Assuming a daily global radiation of 600 cal cm^{-2} in high summer, and excluding other types of heat loss, the first meter of water would heat up by 4.3 deg, the second by 0.6 deg, and the third meter by 0.3 deg. The temperature maximum lies below the surface if, in addition to penetrating short-wavelength radiation, only the long-wavelength radiation of the water surface is taken into account, that is, if evaporation and mixing below the surface are ruled out. H. Reuter [427] was able to demonstrate this fact, and the question will be raised again in Sec. 24, since a snow cover conforms to similar radiation laws.

To gain insight into the state of the air layer close to a water surface, a distinction must first be made between small water surfaces, lakes, and the sea. Besides these stationary bodies of water, which have time to reach some form of thermal equlibrium with their environment, we have to consider flowing water, which bears its heat content into unfamiliar surroundings. The following classification, due to W. Pichler and amplified by W. Höhne [341] will be used to help in presenting the following discussions:

I. Stationary waters
   1. Small bodies of water (dealt with in Sec. 22).

(*a*) Puddles: small, flat, mostly temporary collections of water, the temperature of which is determined by the ground, so that there is no temperature stratification within the water (depth about 10 cm).
(*b*) Pools: permanent or temporary collections of water, which show heating from the ground, but in which there is thermal stratification subject to daily variations (depth 10 to 70 cm).
(*c*) Ponds: usually a permanent body of water, in which there may be diurnal variation of stratification but in which there is no development of a summer discontinuity layer (depth up to several meters).

2. Lakes, in which a summer discontinuity layer develops (see Sec. 23). The term discontinuity layer or thermocline is used to describe a zone at some depth in the water through which there is a particularly great change in temperature. It separates the surface layer, the temperature of which is determined by daily changes and by the weather, from the thermally stable deep waters. Since its position changes considerably as the weather changes, it does not usually give evidence of its existence in mean values for long periods of time.
3. The open sea.

II. Flowing water (see I, 2 and 3, and Sec. 23)

In winter, puddles always, and pools often, lose their characteristics as bodies of water, since they freeze. In the German climate, this happens to ponds and large volumes of flowing water only at times of exceptional cold. O. Pesta [345] has shown how the lives of various creatures develop in high alpine pools according to whether they freeze solid in winter, or whether life can be maintained under an ice cover.

Theoretical and experimental investigations have recently been made into the microclimate of small bodies of water by W. Höhne [341], at the Agrometeorological Institute of Halle University. The lower half of Fig. 95 gives tautochrones for a 2.5-cm deep puddle of water on a sunny day. The whole layer of water is practically isothermal. Its temperature is controlled by the ground below, which in its turn is affected directly by the radiation absorbed, and indirectly by the water-free ground in the neighborhood as well. Cooling begins at the water surface and quickly affects the parts below. Loss of heat by evaporation has the effect that, even in the middle of the day, the top layer of the water is cooler, and acts in opposition to the process of heating from below. The microclimate of the puddle is "maritime" in comparison with the solid ground surrounding it. Because of the nearness of the "shores," the water in the middle will be coolest on cloudy days, if the shores are flat. The shore effect amounts to about 0.5 deg, according to measurements made by Höhne.

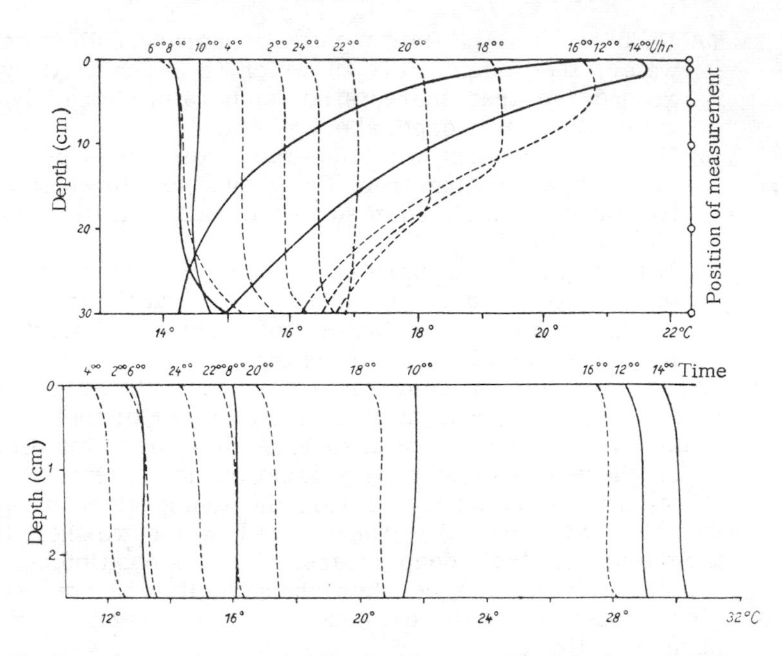

Fig. 95. Tautochrones in a pool (above) and in a puddle (below) on a bright summer day. (After W. Höhne)

When the sun is shining and the banks are steep, the horizontal temperature distribution will be determined by the amount of shielding; then it is not the middle that is coldest, but the water in the shade of the southern bank.

The shape of the isotherms is very different in pools, an example of which is given in the upper part of Fig. 95. The diurnal temperature variation of the water surface is considerable, being 8 deg on bright days. But this heating during the day penetrates only slowly into the 30-cm deep mass of water, so that the bottom remains almost 8 deg colder than the surface. It is not possible to identify any heating due to absorption of residual radiation at the bottom of the pool in this case. The form of the tautochrones at night allows us to conclude that heating from below is taking place here, and also that the flow of heat in the neighboring solid ground is directed upward at night.

By way of contrast, Fig. 96 illustrates a series of observations made by W. Pichler [346] in an Obersteiermark pool of surface area 12 m^2 and depth 40 cm, with a growth of horsetail rushes. Heating from the bottom of the pool can be seen here too in the 14:00 tautochrone, and this may be attributed to the clearer water. It is useful to make a comparison between the tautochrones of the top section of Figs. 95 and 96 with tautochrones plotted as in Fig. 21 and to observe the great difference between water and solid earth as an underlying surface.

Comparative measurements by Höhne in ponds and pools of depths 2.5, 7, 25, and 35 cm gave the same mean daily temperatures under similar weather conditions. The fact that these are independent of depth is because all small bodies of water of this size require only a short time to reach a state of equilibrium. The daily temperature fluctuation, observed at the surface, decreases with depth in the water, the rate of decrease being inversely proportional to the fourth root of the depth.

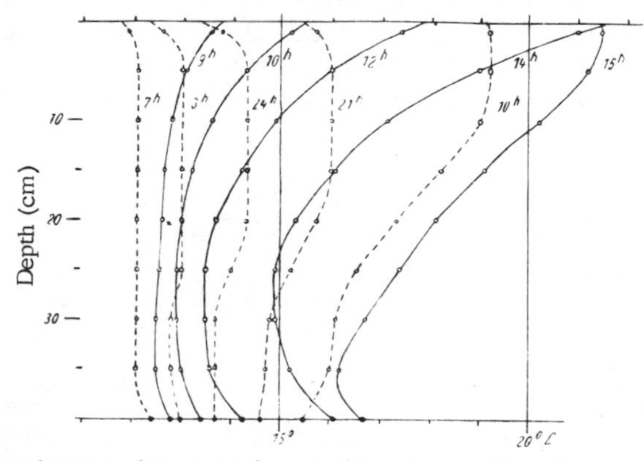

Fig. 96. Tautochrones for a pool containing horsetail rushes, on a bright summer day. (From W. Pichler)

In the biggest of these small waters, that is, in ponds, inertia of the water mass causes the fluctuation of the surface temperature to be very small. To illustrate this, measurements made by J. Herzog [340] in the Kirchenteich near Leipzig can be used. This is a stretch of water 1.1 km long, averaging 200 m in breadth and 2 m in depth. Figure 97 shows temperature measurements made at seven different depths, by electrical resistance thermometers, on a bright summer day with light winds (17 July 1934). Remarks on the type of weather are at the top of the diagram. The isotherms, which are largely horizontal below 1 m, indicate that deeper water hardly takes any part in daily variations. Even at the surface, the difference between day and night is only 2 deg.

Temperatures in the air close to the water are determined by water-surface temperatures, in much the same way as those in the layer near the ground are controlled by the temperature of the ground surface. With such small water surfaces, however, advective influences from surrounding areas are not negligible. When there is a positive radiation balance, air temperature close to the water is very low, because of the influence of the solid ground surrounding it. The difference between it and the shore areas increases with the size of the body of water, which responds more slowly to changes in radiation, and depends on the speed of the wind that brings air out over the water. The air layer close to the water in small bodies of water therefore properly belongs to the transitional microclimate of boundary zones (see Sec. 27).

Ordinarily, an increase of humidity might be expected over the surface of small areas of water. However, at the time of most intense evaporation, in the middle of the day, the water is cooler than the surrounding land, and therefore gives off less moisture. W. Höhne consequently found that water-vapor pressure was somewhat lower in the air layer near the water than at the neighboring shore. This agrees with readings taken by A. Willer [355] in the reed zone of the Müggelsee on 31 May 1944. Between 12:00 and 18:00, in spite of higher air temperature, the relative humidity was 45 to 65 percent in comparison with 40 to 50 percent over the open water.

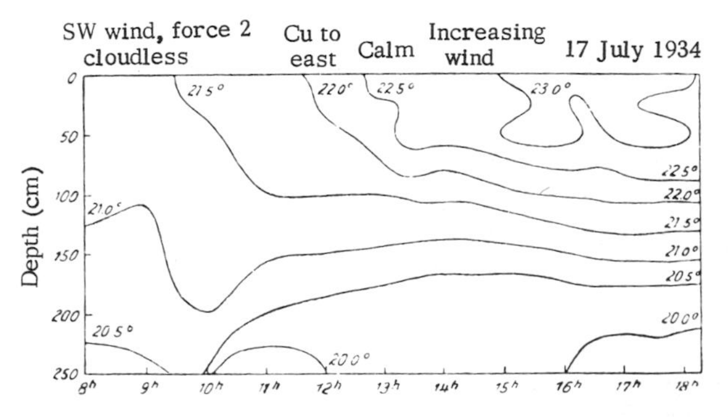

Fig. 97. Diurnal temperature variation in a pond, the Kirchenteich near Leipzig (2 m deep). (After J. Herzog)

At the Zeppelin factory on Lake Constance, engineers took their tools, which were liable to rust, out to the water in the floating dock at night. Even if the warmer lake water evaporated more than over the land, the relative humidity increased much more over land because of the sharp fall in temperature there, and it was therefore relatively drier over the water and less damaging to the tools.

The general opinion about the influence of small bodies of water is correct all the same, but the reason is not to be found in an increased supply of water vapor from water surfaces, but rather in the rich sources of water in the vegetation on the shores. This evaporates more than the drier parts of the shore area and more than the surface of the water itself. The only water surfaces that provide rich supplies of vapor are warm waters (thermal springs or gaps in polar ice) and the spray of waterfalls, as shown by E. Rathschuler [347] for the great Krummler waterfall in Salzburg area.

Ice formation in small bodies of water does not begin at the banks but on small solid objects at the surface. The surface water, to a depth of 1 cm at most, becomes supercooled to from -0.5 to -0.9° C, then freezes spontaneously with a rise of temperature to 0° C. Melting begins from below. Valuable observations on both of these processes have been published by W. Höhne [341] for the first 5 cm above and below the water surface.

If a small body of water has a growth of reeds, water plants, algae, and so forth, they absorb the short-wavelength radiation that penetrates the water, and rise to a higher temperature than the water itself. H. Schanderl [350], using thermocouples, was able to measure excess temperatures of up to 6.3 deg in plants compared with the surrounding water in areas of thickly growing algae and pondweed in Lake Geneva. Strong radiation, calm, and transparent water are prerequisites for such a difference. Normally the excess temperatures are 1 or 2 deg because the heat is carried away quickly by the water. Owing to this process the water is warmer in areas that are silting up than where there is no vegetation.

Figure 98 shows W. Schmidt's [352] results for the shore zone of the Lunzer Untersee on a calm, sunny, and warm autumn day, 13 November 1926. The positions above and below the surface at which readings were taken, making it possible to draw the tautochrones, are marked on the right-hand side of the diagram. The 10° line is everywhere drawn in full for ease of comparison, and temperatures are only indicated by short strokes for each degree. The letters *a* to *d* indicate successive times of day, *a* being for incoming radiation about noon, and *d* outgoing radiation in the evening.

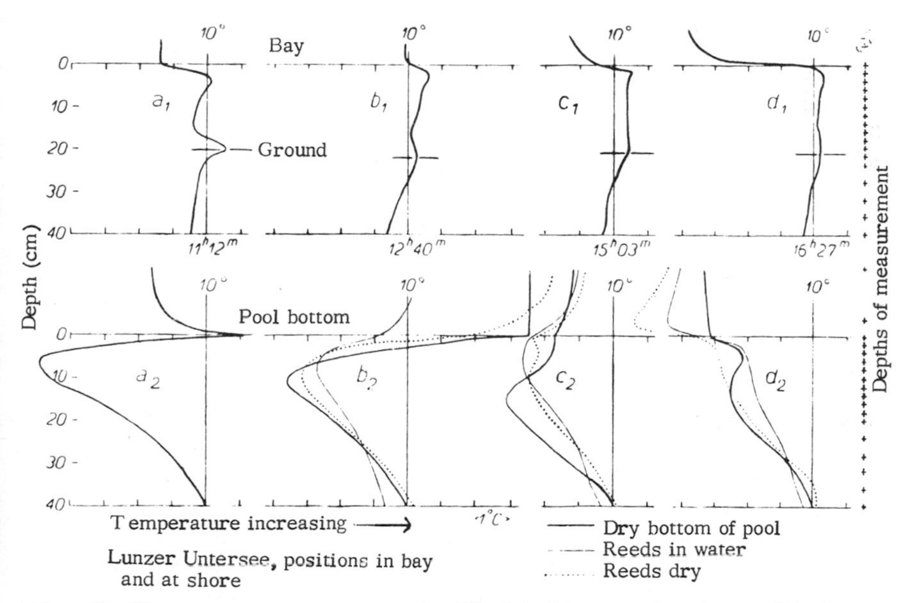

Fig. 98. Temperature measurements by W. Schmidt near the shores of the Lunzer Untersee (explanation in text).

The upper row of curves gives the measurements made in a shallow bay. The bottom, which absorbs radiation, shows a temperature maximum (a_1) at midday, and there is a second maximum just below the evaporating water surface. With reduction (b_1) and cessation (d_1) of short-wavelength radiation, water temperatures become balanced. The temperature of water near the shore drops sharply under the influence of the land.

In the lower row of diagrams, the heavy solid line refers to a pool which had been dried out a few weeks before. The tautochrone a_2 shows the incoming-radiation type on both sides of the solid surface with a reservoir of cold stored from nighttime below 6 cm, presenting a fine contrast to the mild maritime microclimate of the bay, shown in a_1 above. At 15:03 (c_2) the beginning of outgoing radiation from the surface can already be seen. Toward evening (d_2), there can be seen plainly the chronological sequence of the cold previous night, the warm autumn day following it, and the next night setting, arranged in layers one on top of the other. The cold zone near the ground here corresponds to the advected cold air over the neighboring water in d_1.

The dotted lines in Fig. 98 show the temperature distribution in a dried-out area covered with reeds that were mostly dead. Shielding of the surface by the reeds (b_2) makes it slightly cooler than the vegetation-free soil. However, a short distance away from the ground the capacity of the reeds to absorb radiation becomes evident. Night cooling (d_2) is considerable, perhaps because there is still an evaporation influence. The fine solid lines are for an area where reeds are growing in 10 cm of water, which behaves more like dry land than like free water.

23. The Air Layer near the Water Surface of Lakes, Seas, and Rivers

The first noticeable difference found in large masses of water, such as lakes or the sea, is that the daily temperature fluctuation becomes smaller and smaller in comparison with the annual variation.

V. Conrad [359, 360] found that the daily range of temperatures in the surface waters of Alpine lakes was only 1 or 2 C deg on the average, even in midsummer. The June average for the Wörther Lake was 2.6 deg. In winter it decreases to a few tenths of a degree. This small diurnal variation has superimposed on it the irregular changes in water temperature caused by currents in the water, the welling up of colder water from the depths, wind influence, especially in spring and early summer, unequal vertical mixing caused by changes in the wind field in space and time, cold precipitation, and other processes. More detailed descriptions have been given for Lake Constance by W. Peppler [374]. The day-to-day temperature change in the surface waters averaged 0.18 deg for the quietest month (January or February) and 1.22 deg for the most unsettled month (April or May), according to Peppler. Conrad [359] recorded 0.23 and 0.99 deg for Lake Gmund, and 0.07 and 1.12 deg for Lake Pressegger in Austria. Similar results are given in a new collection of data published by O. Eckel [363] for 14 Austrian lakes. It seems, therefore, that the irregular change from one day to another is at least as great as the regular fluctuation in the course of a single day.

In the open sea, the daily temperature variation is reduced to a few tenths of a degree. Figure 99 has been constructed by E. Wahl [385] from a large amount of observational data, to show the daily temperature variation in surface waters. Temperatures were measured by taking samples of water with a sea-bucket made of sailcloth or tin, hence giving only mean values for a surface layer about 1 m deep. The solid line (*K*) gives the results obtained on the German Meteor Expedition in the South Atlantic Ocean by E. Kuhlbrodt and J. Reger [370]; the dotted curve (*W*) is for the North Sea, from measurements by E. Wahl [384]. In spite of the vast geographic differences between the two areas, the temperature in both fluctuates only about 0.1 deg on either side of the mean value during the day.

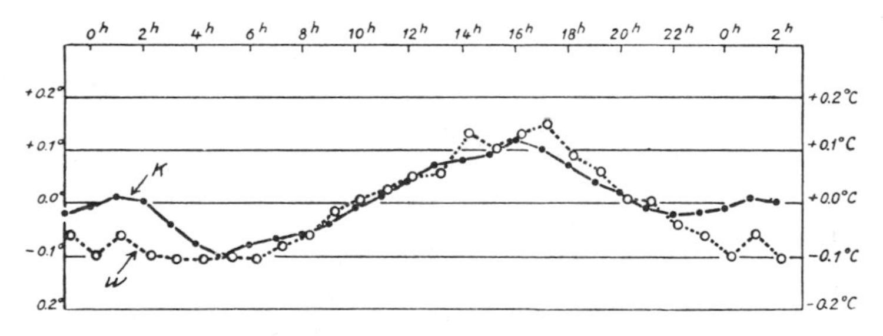

Fig. 99. Diurnal temperature variation in the surface water of the sea.

It should be noted that the daily temperature fluctuation in the air near the surface is greater—perhaps under the influence of direct absorption of radiation, or, according to M. Koizumi [369], as a result of pressure variations. From the Meteor observations the air maximum, between 12:00 and 13:00, was 0.25 to 0.45 deg higher than the minimum, while the maximum for the surface water was only 0.26 deg higher, and was not reached until about 18:00. U. Roll [376] showed, in observations made near Heligoland, that the diurnal fluctuation increased with height within the air layer near the water. He found in August, when there was a range of 0.4 deg in the surface-water temperature, that in the air it increased from 0.6 deg at 20 m to 0.9 deg at 150 m.

So far as the influence of the underlying surface is concerned, therefore, air temperatures near the water in great lakes and at sea are determined by the annual temperature variation. Figure 100 illustrates the results of a 23-yr series of water-temperature observations (from 1927 to 1950) in the Hallstatt Lake, in the form of tautochrones of monthly average temperature down to a depth of 40 m, evaluated by O. Eckel [364]. This lake is 8 km long, 1 to 2 km wide, and 125 m deep; it lies in the Salzkammergut district of Austria, and has been investigated, from the thermal point of view, in more detail than any other lake in the country. Warming up above the winter temperature that corresponds to that of maximum water density, or 4° C at all depths, takes place only from the surface downward. August is the warmest month, down to a depth of about 5 m; at a depth of 30 m it is September; and at still greater depths the effect of summer heat is not felt fully until October. While, during the period of warming up, even the monthly-average figures show substantial temperature gradients in the top few meters, convective mixing in the water spreads the cooling process more evenly throughout all layers. Other lakes show similar typical temperature variations.

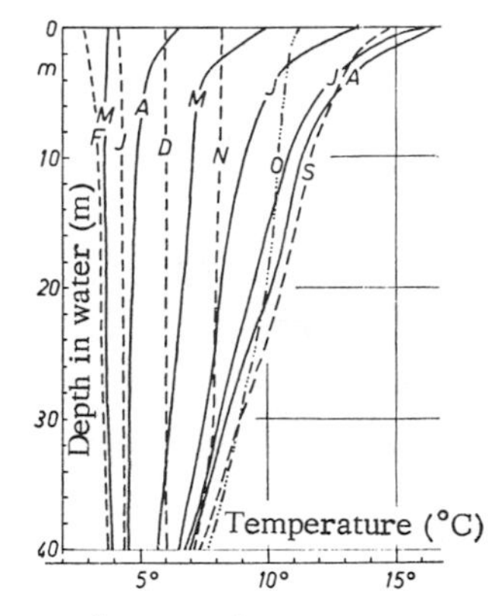

Fig. 100. Mean monthly tautochrones of water temperature in the Halstatt Lake in Austria. (After O. Eckel)

The temperature difference between surface water and the air over it is no longer determined by the regular temperature variation of the underlying surface of such large inert masses of water, but by the irregular change of air masses. If the water is warmer than the air, the air near the surface behaves in much the same way as the air near the ground at a time of incoming radiation. An unstable sublayer is built up, as described in Sec. 12, in which gradients are superadiabatic. K. Brocks [357] investigated this layer in the summers of 1949 to 1951 by optical methods (see Sec. 18) at the German North Sea coast, between the "Roter Sand" lighthouse and Büsum. The thickness h of the superadiabatic layer increased with increasing temperature difference Δt between the water surface and the air at a height of 5 m, in the following way:

Δt (C deg)	0.5	1.0	1.5	2.0	2.5	3.0
h (m)	6	11	15	19	21	23

If, on the other hand, water is colder than air, the layer close to the surface behaves in a way similar to the outgoing radiation situation above the ground.

Figure 101 gives temperature measurements made by thermocouples on both sides of the water surface on a clear summer day, at the positions and times shown by the small circles. These were made by H. Bruch [358] in the Baltic Sea near the Greifswalder Oie. Heights above and below the surface are shown on a scale that contracts with distance, to make the processes near the surface clearer.

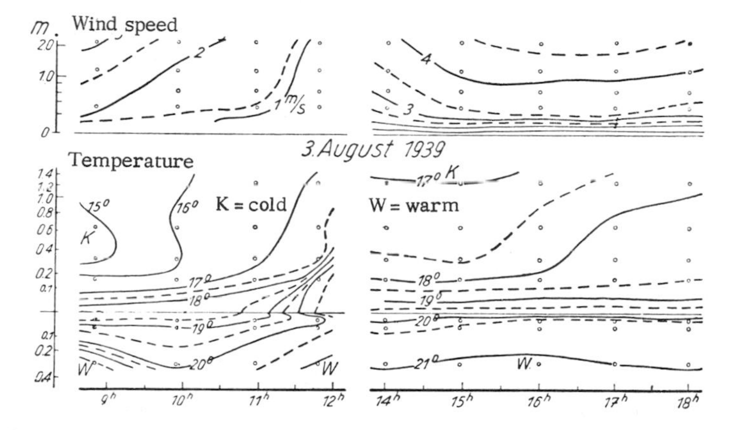

Fig. 101. Wind and temperature stratification on a summer day above the sea near Greifswald. (From observations by H. Bruch)

The right-hand half of the diagram shows that the warmer water gave off heat to the cooler air, in conditions of normal wind stratification (top section, right), so that the largest gradients were found in the narrow boundary layer in which there is little mixing, the isotherms being separated only for purposes of illustration. A significant decrease in wind can be observed in the left-hand half of the diagram after 11:00. This caused, on the one hand, a marked rise in the temperature of the surface layer of air, because of reduced mixing, and on the other hand the formation of a cold film on the water. As shown elsewhere (pp. 188 and 212), this follows of necessity from radiative exchange, and is almost certainly reinforced by evaporation heat losses.

Figure 102 shows this cold layer very clearly. This diagram shows the situation on two July days, the figures being from H. Bruch. The temperature differences between air and water was small and in the opposite sense from Fig. 101, the water being colder than the air. The isotherms show the cold surface layer of water and the way it influences the air close to it. The effect is very marked in the calm conditions of the left-hand part of the diagram, while with winds of 2 to 4 m sec^{-1}, as on the right, it still exists but is clearly recognized only when isotherms are drawn at intervals of 0.2 deg.

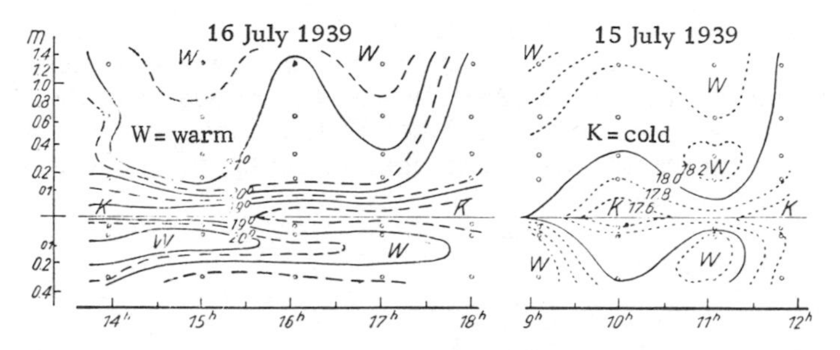

Fig. 102. The cold layer at the boundary surface of air and water. (From observations by H. Bruch)

H. Bruch used his thermocouples only in the 4 cm of water under the moving water surface. H. U. Roll [380] employed "knitting-needle thermometers" to measure the temperature profile in salt-water pools left behind at ebb tide in the Neuwerker Wattenmeer, from +8 cm to −6 cm, under changing conditions in August 1950. In practically all sets of observations the same surface cooling was found in a layer 2 to 10 mm thick. The degree of cooling may be seen from the following example on 10 August 1950, between 15:30 and 16:05:

	Water			Air			
Height (cm)	−6	−2	−0.5	0	0.2	1	10
Temperature (°C)	26.1	26.2	26.1	25.8	26.4	26.5	26.8

Roll [378] demonstrated the existence of this cold-water film indirectly by quite another method. Evaluation of 165 temperature profiles, taken from three sources in different places, gave, when extrapolated, a water-surface temperature 0.5 to 1.5 deg lower than that obtained simultaneously by the usual method (see above). Since this figure is valid over a wide area, the quantity in question must represent the amount of cooling due to evaporation in the boundary layer between water and air.

The temperature, water-vapor, and wind profiles are extremely closely linked together over the sea. One of the basic problems of oceanography arises from a study of the air layer close to the water surface, particularly with regard to the quantities of water evaporating from the surface; but we cannot go into that further here. For information about water-vapor stratification in the air over the sea, the works of G. Wüst [387], R. B. Montgomery [372], H. U. Sverdrup [382], T. Takahashi [383], and K. Brocks [356] should be consulted. Wind stratification which is influenced by friction at the surface, that is, by the state of the sea, is treated by H. U. Roll [377-379], J. S. Hay [368], G. Neumann [373], E. L. Deacon [361], T. Takahashi [383], and recently by H. Lettau [371]. Textbooks on oceanography should also be consulted.

Attention is now turned from conditions above stationary water to consideration of the special microclimatic features of flowing water in brooks, streams, and rivers.

All brooks start off with the temperature of the spring water at the source. As a rule, particularly in summer, this is lower than the temperature the water would have in equilibrium with the environment. In winter, spring water may be at a higher temperature, so that it is cooled when it first emerges, especially if the source is in mountain country. This applies even more to warm springs.

In midsummer, therefore, the flowing water is heated from day to day, but also has its own diurnal temperature pattern under the influence of radiation. O. Eckel [366], who made a detailed study of Austrian river temperatures, established that the rivers of the eastern Alps in Austria did not generally achieve the equilibrium temperature even after running 100 to 400 km through the country. In its early stages, the water may certainly rise in temperature by several degrees on a summer day; but by the time it has grown in size, 30 to 80 km from the source, the increase is only about 0.6 deg and at 150 to 350 km only 0.15 deg for every 10 km of the river bed. The modification of the river temperature with time along its course is inversely proportional to the distance from the source. This rule, which was derived from observations of the Austrian hydrographic network, is valid for the early morning water-temperature reading, which is approximately the daily minimum.

The heat economy of flowing water has been studied in detail by O. Eckel and H. Reuter [367]. In addition to the general conditions affecting stationary water, there now enter into the problem the amount of shading by the river bank and the influence of heat exchange with the river bed, which depends on the mean temperature difference between the flowing water and the bottom, and which has a component dependent on the time of day. Another point to be taken into account is the drag of the flowing water on the neighboring air above it.

This drag effect has been investigated on the Inn River by E. R. Reiter [375], and on the Main near Frankfurt and on the Rhine near Bonn and Cologne by K. O. Wegner [386]. Wind speed must be less than 4 m sec^{-1}, otherwise frictional mixing by the gradient wind reaches down to the surface. Reiter therefore carried out his research during calm in the early morning, while Wegner waited until the hours of late evening and night.

Reiter dropped smoke bombs into the Inn River from a suspension footbridge crossing it at Stams, where it was 75 m wide, and filmed the track of the smoke. Figure 103 shows the smoke trails from 4 of the 63 experiments made at intervals of 1 sec. In experiment *a* (top left) the air was stationary at 3.3 m. Below this level it was entrained by the river, with increasing speed as the surface was approached. A few turbulence elements can be seen disturbing the movement of the smoke, and above the stationary level there is evidence of a light wind blowing up the valley. Although experiments *b* to *d* followed within a few seconds, the height of the stationary level shifted, moving down to 2.5 m. In all measurements the water surface had the same degree of roughness with such a slack wind (roughness parameter z_0 = 109 cm;

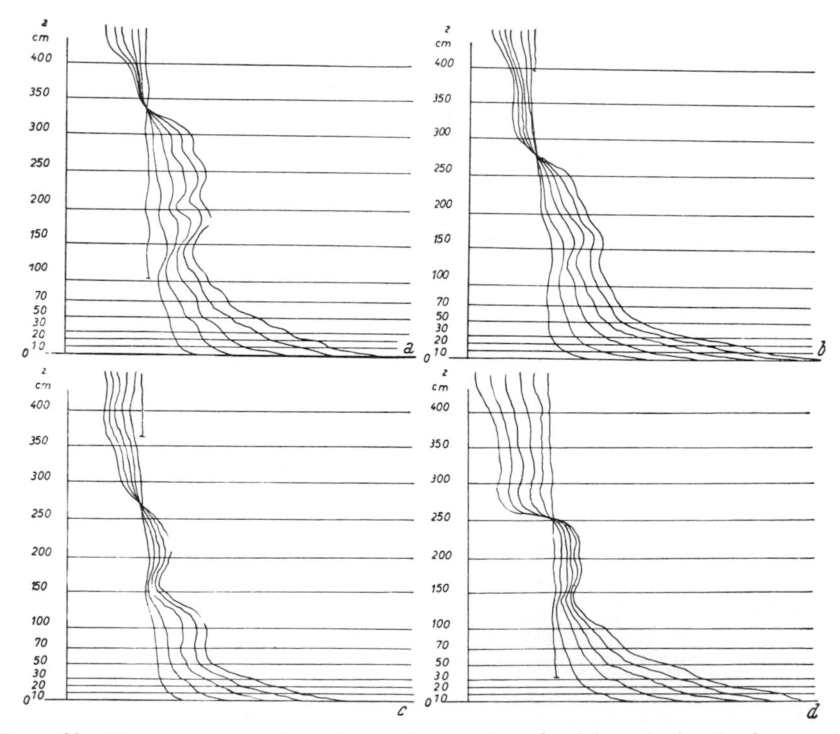

Fig. 103. The movement of smoke trails at intervals of 1 sec in the air close to the surface of a river, showing the drag effect of the flowing water and the influence of a slight wind up the valley. (After E. R. Reiter)

see p. 118). When the winds up the valley became stronger, there was a sudden jump in roughness and capillary waves created by the wind came into being.

In Wegner's observations on the Main and Rhine, made from a float towed behind a boat, no decrease with height was found in the air dragged along with the water, but there was a "surface layer" a few tens of centimeters thick (150 cm at the maximum), which was swept along bodily by the river. Above this layer there was a sudden change in wind direction and a transition to gradient wind, speed increasing with height as over the ground. This unexpected state of affairs is explained by Wegner as follows: increasing frictional exchange accompanying stronger winds makes it possible for impulses received from the moving water to remain effective up to higher levels. The substantial differences in wind profile within this layer of air, accompanying the flowing water, may have their origin both in differences in width of the rivers and the type of surrounding country and in the time of day at which the measurements were made, because Wegner always had warm air over cold water, while Reiter found very little or a negative difference. It is hoped that further observations will clarify this matter.

In comparison with areas of stationary water, large in extent or surrounded by flat countryside, rivers, which cut into the land or, as often happens, are bordered by trees, have their radiation quota reduced by horizontal shielding (see Secs. 5 and 55). O. Eckel and H. Reuter [367] established, for a river 30 m wide, the relative amount of radiation received, as a percentage of unhindered radiation (shielding angle 0°) for the March and September equinox (declination of the sun 0°) and for midsummer (declination +20°). Table 53 gives the results. The decrease of outgoing radiation by night has already been dealt with (p. 25).

The diurnal temperature variation in the water of a river flowing in a completely level plain has been evaluated for a sunny July day by Eckel and Reuter, using an iterative graphic method for solving the differential equations, to illustrate the influence of initial temperature and depth of the water. The results may be taken from Fig. 104. The daily water-temperature variation is shown for four initial water temperatures (Tw_A): 10°, 15°, 20°, and 25°C, and for four depths: 30, 60, 100, and 300 cm. It is easy to recognize the equalizing effect of large water masses, and the more rapid gain of heat associated with low initial temperatures. In each case the water begins the second day with a different initial temperature, and starts to adapt itself to its environment once again, reaching an equilibrium temperature with time, unless the environment itself has changed in the meantime (changing weather, new countryside).

Figure 104 portrays the events recorded by an observer moving with the stream. The situation is quite different when we inquire

Table 53
Relative radiation received by rivers as percentage of unimpeded incoming radiation.

Greatest angle of shielding		6°	11°	17°	22°	26°
Corresponding to tree heights (m)		3	6	9	12	15
	Equinox					
	N-S	86	75	67	59	54
Direction of valley	NW - SE or NE - SW	88	77	68	60	54
	W-E	91	80	71	62	54
	Midsummer					
	N-S	88	78	70	63	57
Direction of valley	NW - SE or NE - SW	91	83	76	70	64
	W-E	93	88	84	79	76

into the variations of temperature at a fixed position in the river. For example, let us suppose that water is flowing out of a reservoir with a constant temperature of 5° C in sunny summer weather; the daily temperature variation observed downstream will be a function of the distance from the dam overflow, and of the speed u of the water. Initial time is taken as 04:00 in Fig. 104 and the depth of water as 30 cm. Figure 105 shows the characteristic temperature variation as a function of distance, this being expressed as a multiple of the rate of flow u in kilometers per hour.

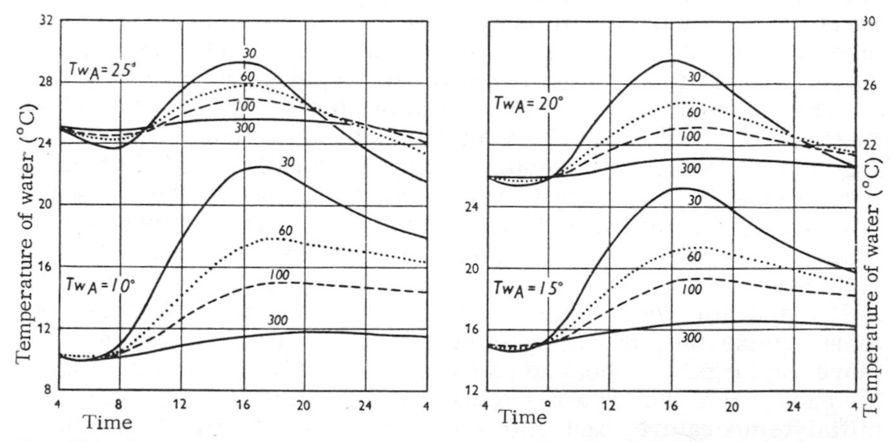

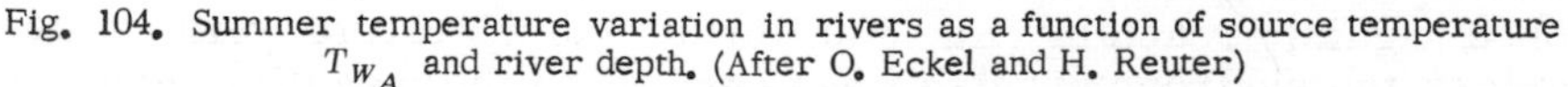
Fig. 104. Summer temperature variation in rivers as a function of source temperature T_{W_A} and river depth. (After O. Eckel and H. Reuter)

The dashed curve shows the gradual adaptation of the mean daily temperature, from its starting value of 5° C, to summer radiation conditions. The solid curve gives the daily temperature fluctuation of the river water. At the outflow point it is assumed to be zero. It increases steeply up to the distance reached in flowing for 12 hr, but still decreases at night to 5° C. At greater distances water temperature no longer cools quite so much at night, while the maximum temperature remains constant. There is, therefore, a decrease in the fluctuation up to the 24-hr distance. At this point the night temperature no longer falls below 16° C, and the maximum is a little above 20° C. Further on, the maximum rises again, and so the daily fluctuation increases once more. This interesting pattern in daily water-temperature fluctuation with distance from the dam is amplified in Fig. 105, in which the times of temperature extremes are shown. The small triangles show when the temperature was a maximum, and the crosses when it was a minimum at the position in question, against the time scale along the right-hand margin of the diagram.

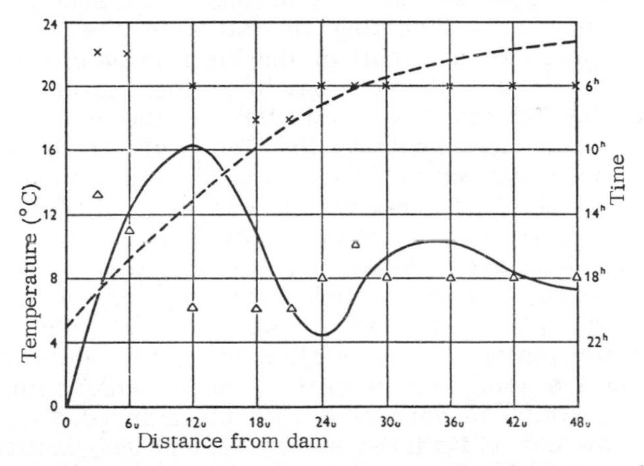

Fig. 105. Temperature relations in flowing water, with reference to a fixed system, assuming a constant initial temperature of 5°C. (After O. Eckel and H. Reuter)

These astonishing computations by Eckel and Reuter throw new light on the conditions that affect the ecology of plants and animals in flowing water. Comparison with observed river temperatures gave a satisfying confirmation of these theoretical calculations.

24. The Air Layer near Snow and Ice

Snow cover is as varied an underlying surface as the ground. Snowflakes can be deposited so loosely that density of the snow cover is only 0.01 g cm^{-3}. In general, new snow has a density of 0.1 to 0.2, and old snow 0.3 to 0.5 g cm^{-3}. Rain falling into snow cover increases its water content and therefore its density. The density of permanent snow cover on high mountains increases as it changes to firn snow and then firn ice to 0.8 g cm^{-3} and glacier ice has a density of 0.9 g cm^{-3} and more. Sea ice, with a salt content of 3 percent, and with 3 percent by volume of trapped air, has been found to have a density of 0.91, and with 9 percent by volume of air, 0.86 g cm^{-3}.

International organizations are trying to establish a unified classification for snow [410]. Those who wish to learn something about the many different types of snow cover to be found should read a stimulating book by W. Paulcke [422] on snow and avalanches. In it are described the dozen different types of snow that an observant walker or skier should be able to distinguish. One will learn from it the many changes the snow undergoes when it is transported and redeposited by wind and storms, and when it settles on aging, and as a result of water movements, melting and freezing, and hoarfrost and firn formation ("diagenesis" of snow). W. Paulcke was the first to construct a snow profile from actual measurements depicting the state of the snow cover at a given moment as a result of the alternation of new snowfalls and aging processes, rather similar to geologic profiles.

Figure 106 shows the accumulation, the melting, and the structure of the snow cover on the Hohenpeissenberg in the Alps from measurements made by J. Grunow [405] in the winter of 1951-52. The field of observation was on a ridge of the 977-m high mountain, trending westward. Readings made daily in the main observation area, close by, for the Weather Service are slightly different, and are included for comparison with average snow density (above) and snow depth (below). The upper half of the diagram shows the new-fallen snow between November and March, and the increase in average snow density for the whole snow cover, which is connected with the aging of the snow. The depth of snow cover, its inner structure, and the quantity of water seeping out of it are given in the lower part of the diagram. After a new snowfall the older snow settles under the weight above it. Density therefore increases with depth. Experience shows, however, that the maximum density is not at the bottom, because the water formed when snow at the top melts under the influence of radiation or advected warm air does not penetrate as far as the bottom of the snow, or at least does not do so until the snow cover begins to melt in spring. The decrease of temperature in winter from the warmer deeper layers of snow to the colder snow surface produces, in addition to the capillary rise of water, an upward transport of water vapor by diffusion, similar to that described for frozen ground in Sec. 21. Grunow observed that, when a colored snow surface was covered by new snow, more colored water rose upward into the new snow than trickled downward. According to H. Hoinkes [409], this density stratification is maintained over several years in the firn areas of glaciers in the Eastern Alps, with a maximum about the middle of the snow cover, which can therefore be used to calculate the limits of the annual snow balance.

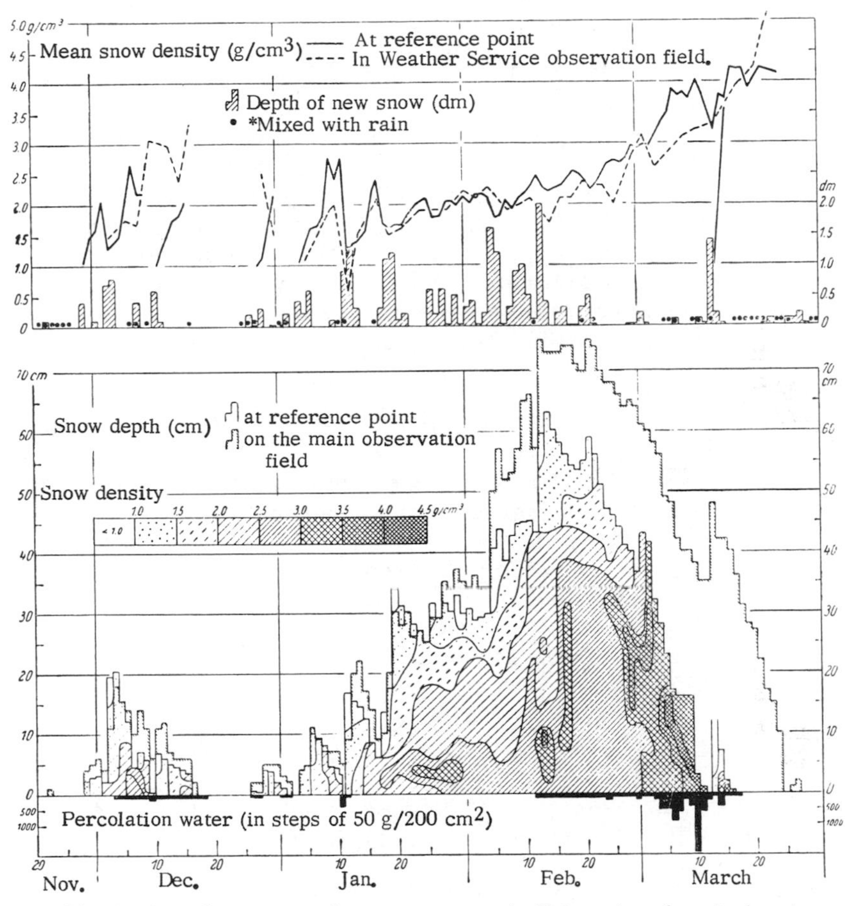

Fig. 106. Height and structure of snow cover on the Hohenpeissenberg in the winter of 1951-52. (After J. Grunow)

It is not possible to mention here the infinitely varied forms the snow surface may assume; this is always one of the most important subjects of research in Arctic and Antarctic expeditions. The property that concerns this discussion most, in exploring the nature of the boundary surface of snow and air, is the high albedo of snow for short-wavelength global radiation. Table 4 shows that new snow occupies the leading position, with an albedo of 75 to 95 percent. According to F. W. P. Götz [403], a figure of 100 percent was measured on a number of occasions at Arosa, at 1800 m elevation. In such cases the total global radiation simply does not enter into the heat budget of the snow cover. Light conditions are greatly improved every year by the arrival of snow in the dark winter months. As the snow becomes older and its surface becomes dirty, the albedo decreases quickly, and may fall as low as 40 percent.

H. Hoinkes [407] gives a value of 40 to 60 percent for the albedo of firn snow. This decreases still further as firn gradually changes to ice. Glacier ice has an albedo of 20 to 40 percent, depending on how dirty it is. F. Sauberer [430] found a value of 7 to 8 percent for clear ice from the Lunzer Untersee. The reflected light seen by an observer consists not only of radiation reflected from the surface, but also of spectral light filtering through the surface from below (cf. p. 187). As I. Dirmhirn has shown [397], this makes up a significant, sometimes even the preponderant, part of the counterradiation measured by instruments. Numerical values for ice, firn, and old snow are:

Albedo (percent):	10	20	30	40	60	80
Fraction (percent) of light from below:	31	48	80	92	88	64

As already shown by Sauberer, reflected radiation is only slightly dependent on wavelength, and then only to the extent that the middle range of the visible spectrum is reflected most strongly. This has been confirmed for Antarctic snow fields by G. H. Liljequist [414] at Maudheim station, where reflection on overcast days varied between 92 percent (red) and 97 percent (green, yellow). Dirmhirn found the following albedo values for Alpine glaciers in 1950:

Wavelength (μ):	0.4	0.5	0.6	0.7	0.8
For clean ice (percent):	44	54	56	48	32
For dirty ice (percent):	24	53	36	31	19

This shows that no serious error will be involved if a single albedo value is used for the whole range of short-wavelength radiation.

Solar and sky radiation may also penetrate into snow, but not so easily as into water, of course (see Sec. 22). While in the latter case penetration was measured by the percentage penetrating each meter, it is more useful here to use the extinction coefficient ν (cm^{-1}), which is defined by the relation

$$I_x = I_o e^{-\nu x} \qquad (\text{cal cm}^{-2}\ \text{min}^{-1}),$$

in which I_o is the radiation received at the surface and I_x is the radiation penetrating to the depth x cm. The relation between the transmissivity D and the coefficient ν is given by $D = 100e^{-100\nu}$; for example:

Extinction coefficient ν (cm^{-1}):	0.005	0.01	0.02	0.03	0.04	0.05
Transmissivity D (percent m^{-1}):	60.7	36.8	13.5	5.0	1.8	0.7

In snow, with its similarity to water of low transmissivity, extinction coefficients of about 0.07 and 0.23 cm^{-1} have been recorded. It therefore varies within wide limits; however, according to F. Löhle [415], this is not related in any ascertainable way to the amount of water in the snow. It can easily be seen from Fig. 107 what percentage of the radiation penetrating the snow surface reaches down to various depths, for values of ν within the range just mentioned. The diagram has been extended to include ice, which will be discussed further on, and also pure (distilled) water for comparison with Secs. 22 and 23. The value of ν for distilled water is about 0.0003 cm^{-1} in the green-yellow range. The curve for $\nu = 0.005$ corresponds roughly to the water of the Lunzer Untersee in 1937-38 for the same spectral range (cf. Fig. 94).

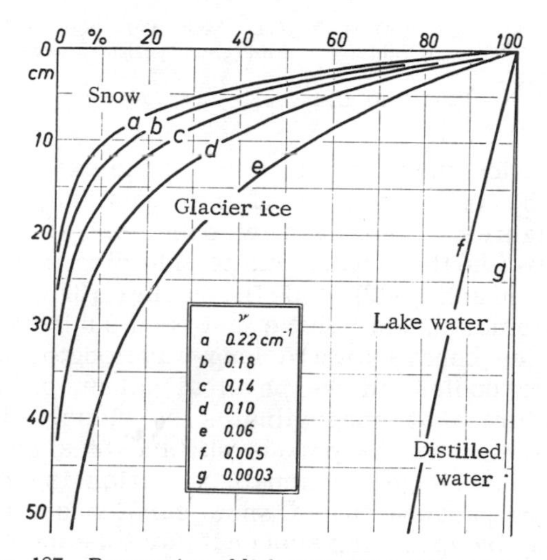

Fig. 107. Penetration of light into snow, ice, and water.

F. Sauberer [430] demonstrated that there were considerable differences between individual cases in the penetration of light in the range 0.38 to 0.76 μ into snow, but no systematic relation with the wavelength could be found. This seems to apply to loose snow everywhere. If, however, the amount of water contained in the snow increases, then the extinction coefficient increases, as in water, with increasing wavelength. G. H. Liljequist [414] found, at the Antarctic station of Maudheim mentioned above, that with snow of density 0.4 g cm^{-3} extinction coefficients measured by photoelectric cells and filters were:

Wavelength (μ):	0.42	0.52	0.59	0.65
Color of light:	blue	green	orange	red
Extinction coefficient, ν (cm^{-1}):	0.066	0.083	0.114	0.172

Ice is naturally more penetrable than snow for short-wavelength radiation. F. Sauberer [430] found an average extinction coefficient of 0.03 cm^{-1} for snow plates 4 cm thick, with a negligible dependence on wavelength. New and careful measurements made by W. Ambach [390] on the Hintereis Glacier in the Ötztal Alps gave, as a mean of 228 individual measurements, a value of $\nu = 0.057$ cm^{-1}. This value corresponds approximately to the e-line in Fig. 107. At a depth of 20 cm, therefore, we still find 30 percent and at 40 cm, almost 10 percent of the short-wavelength radiation penetrating the ice surface. Dependence of the extinction coefficient on wavelength is much the same for ice as for water, as shown by Sauberer [431]. Ice from the Lunzer Untersee had the following extinction coefficients:

Wavelength ($m\mu$):	313	350	400	450	500	550	600	650	700	750	800
Color of light:		violet		blue	green	yellow	orange		red		
Extinction coefficient, ν (10^{-3} cm^{-1})	1.0	0.5	0.4	0.5	0.8	1.3	2.0	3.4	6.0	10.6	17.7

There is a sudden jump in the value of the extinction coefficient on reaching the infrared part of the spectrum, as may be seen from these figures. Using a new spherical type of instrument [391], which avoids the errors due to shielding that are otherwise present, W. Ambach [392] and H. Mocker [393] have made new sets of measurements, which have just been published.

Snow and ice behave toward long-wavelength radiation practically as black bodies (in the physical sense, p. 7). There is no surface in nature that approximates so closely the ideal of the hollow-box radiator as the porous surface of a snow cover. The albedo of snow for long-wavelength radiation is only 0.5. Hence the Falckenberg paradox that a snow surface can be made to reflect better only by spreading soot on it (for this spectral range). In nocturnal radiative exchanges, therefore, both snow and ice behave in the same way as the ground. During the day, however, the internal radiation balance is always positive, since long waves are radiated by the surface. It decreases to 0 at most when the weather situation or the depth inside the snow or ice cover is such that there is no more short-wavelength radiation.

The thermal conductivity λ of snow is extremely low, as we have already seen from Table 10. This is the main reason for the sudden cold that usually follows the first snowfall, because it reduces the flow of heat from the soil, or stops it altogether.

Thermal conductivity is a function of the snow density ρ. This relation has been expressed by several writers in a number of different equations. A simple and useful relation is that discovered by H. Abels [388]:

$$\lambda = c\rho^2.$$

If ρ is in grams per cubic centimeter, then λ will be in calories per centimeter second degree. The constant c has the value 0.00677 according to Abels, while F. Loewe [416] found that there was good justification for the value of 0.0066 he obtained from measurements made in Adélie Land, and J. Bracht [303] obtained a value of 0.0049 in measurements with snow of density 0.19 to 0.51 g cm^{-3}. Thermal conductivity may be taken from Table 54 for three different values of c and various snow densities, the larger values being more probably correct than the lower. As the density increases, λ gradually approaches the value for ice, given in Table 10 as 5 to 7×10^{-3} cal cm^{-1} sec^{-1} deg^{-1}.

Table 54
Thermal conductivity λ (10^{-3} cal cm^{-1} sec^{-1} deg^{-1}) for various values of the constant c.

c	Snow density (g cm-3)							
	0.1	0.2	0.3	0.4	0.5	0.6	0.7	0.8
0.005	0.05	0.20	0.45	0.80	1.25	1.80	2.45	3.20
0.006	0.06	0.24	0.54	0.96	1.50	2.16	2.94	3.84
0.007	0.07	0.28	0.63	1.12	1.75	2.52	3.43	4.48

H. Reuter [426] has adduced proof that in less dense snow there is a distinct mixing process of a convective nature at work in the air spaces, which increases thermal conductivity in the upper layers to 7 or 8 times the value of molecular conduction. If computations are made according to the rules laid down in Sec. 6 for the ground, which do not take an exchange of this type into account, incorrect values are obtained for the decrease of diurnal temperature fluctuation with depth, and to a smaller extent for the phase lag. This results in an appreciable uncertainty in the values of λ given in Table 54.

The thermal conductivity of ice increases as temperature decreases. The values for ice from physical tables are:

Temperature (°C)	0°C	-20°C	-40°C	-60°C
Temperature conductivity λ (10^{-3} cal cm^{-1} sec^{-1} deg^{-1}):	5.3	5.8	6.4	7.0
Specific heat c (cal gm^{-1} deg^{-1}):	0.50	0.47	0.43	0.40

In saline sea ice the specific heat increases very quickly as the salt content increases, becoming, for example, 10.8 cal gm^{-1} deg with 1 percent of salt and a temperature of -2°C. During the Maud Expedition, λ for Arctic ice was found to be between 1.7 at the surface and $5.0 \cdot 10^{-3}$ cal cm^{-1} sec^{-1} deg^{-1} at a depth of 2 m.

The low thermal conductivity of snow gives rise to large vertical temperature gradients within the snow covering. Figure 108 shows temperature variations in snow varying in depth between 32 and 65 cm at Davos in the winter of 1937-38, measured by O. Eckel and C. Thams [398]. It should be noted that the isotherms near the surface are drawn for much greater intervals than deeper down in the snow. When surface temperatures are -33° C, the temperature of ground surface seldom falls below 0° C. This shows the high degree of protection afforded by snow cover.

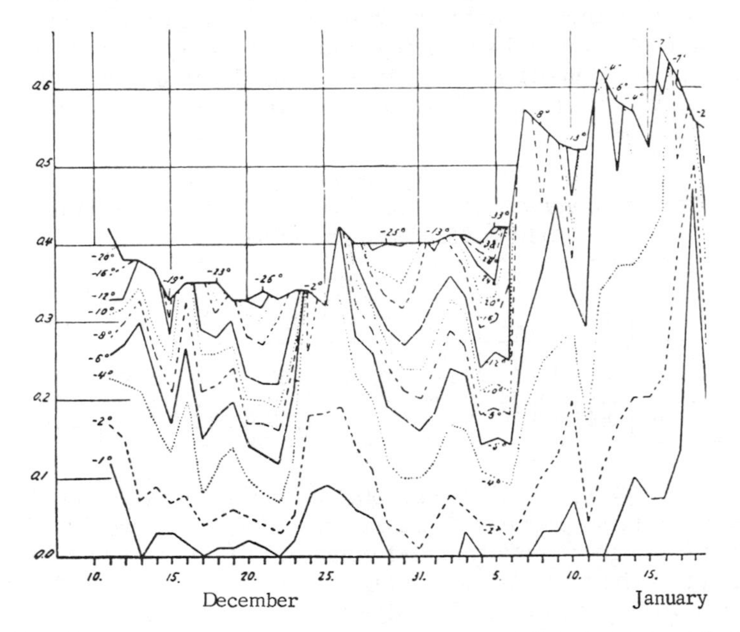

Fig. 108. Temperatures in winter snow cover at Davos. (After O. Eckel and C. Thams)

The daily temperature variation in snow may be studied from the tautochrones in Fig. 109, which were drawn from observations made by E. Niederdorfer [455] in Carinthia, Austria. The diagram is for 16 January 1932. The reading for 09:45 shows a temperature difference of 9 deg between the surface and a depth of 20 cm, and is still indicative of outgoing night radiation. Warming during the forenoon brings the snow surface up to the melting point. It should be noted that the maximum lies 1 to 2 cm below the snow surface. This follows, of necessity, from the penetrability of snow to radiation, proved theoretically by H. Reuter [427]. He

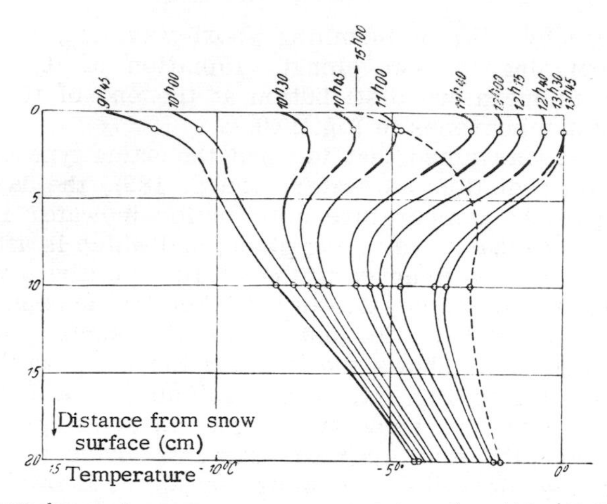

Fig. 109. Tautochrones in snow cover on a sunny winter day. (After E. Niederdorfer)

evaluated the temperature variation that would obtain in a penetrable medium, on the assumption that isothermal conditions existed to begin with, that only short-wavelength radiation was able to penetrate to some extent into the medium, that a constant amount of long-wavelength radiation was emitted from the surface only, and that no other influences were at work. The results are shown for one computed example in Fig. 110.

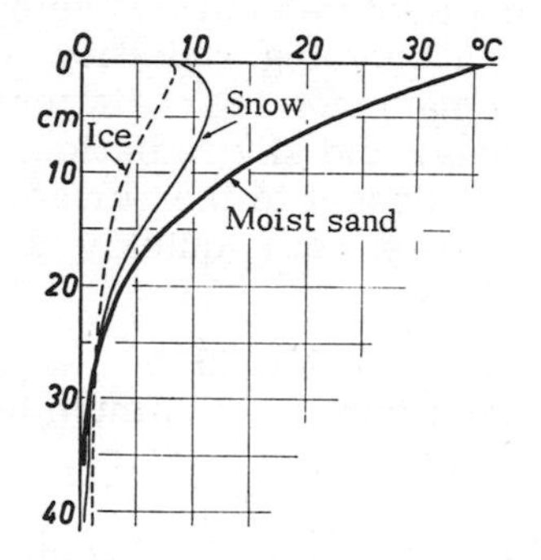

Fig. 110. Temperature distribution in ice, snow, and ground under the influence of radiation only. (According to the theory of H. Reuter)

Three different underlying surfaces are compared here; moist sand: $R = 10$ percent, $(\rho c)_m = 0.34$ cal cm^{-3} deg^{-1}, $\lambda = 0.0032$ cal cm^{-1} sec^{-1} deg^{-1}; old snow: $R = 60$ percent, $\rho = 0.4$ g cm^{-3}, $c = 0.505$ cal g^{-1} deg^{-1}, $\lambda = 0.00109$ cal cm^{-1} sec^{-1} deg^{-1}; and ice: $R = 0$, $\rho = 0.917$, $c = 0.505$, and $\lambda = 0.0051$. In each of these three cases it was assumed that initially isothermal conditions

were followed by 3 hr of incoming short-wavelength radiation of 0.79, and outgoing long-wavelength radiation of 0.10 cal cm^{-2} min^{-1}. The temperature distribution at the end of this period is shown by the tautochrones in Fig. 110.

Since it was assumed that ice had the same type of selective absorption of radiation as water (see p. 186), the dashed curve will also give the temperature distribution in water in which no form of mass exchange is taking place, and which is affected only by radiation. It would appear, therefore, as already shown on p. 188, that, even if evaporation is not taken into account, the temperature maximum will lie just under the surface of water as well. The sandy soil shows the familiar incoming-radiation type, somewhat similar to Fig. 21. The maximum, at a depth of a few centimeters under the snow surface, is in good agreement with Fig. 109, and with similar measurements made by J. Keränen [411]. This also provides an explanation for F. Loewe's observation [416] that the annual average temperature in Antarctic snow in Adélie Land was 0.6 deg higher at a depth of 5 cm than on the surface. According to Reuter's theory, it follows that the position of the temperature maximum in the stationary state does not depend on the thermal conductivity or the thermal diffusivity of the snow but on the intensity of the incoming and outgoing radiation, and on the extinction coefficient ν, to which the depth x_m of the maximum is inversely proportional. Table 55 gives the theoretical depth of the maximum temperature in centimeters below the snow surface, for the four values of the extinction coefficient used in Fig. 107. The values selected for effective short-wavelength incoming radiation and long-wavelength outgoing radiation are probable values, taking account of the fact that incoming radiation means here only the unreflected portion of solar and sky radiation. The maximum temperature difference with respect to the surface depends on thermal conductivity, and increases very quickly as the density of the snow decreases.

Table 55
Theoretical depth (cm) of maximum temperature below snow surface.

Radiation (cal cm^{-2} min^{-1})		Extinction coefficient ν (cm^{-1})			
Short-wavelength incoming	Long-wavelength outgoing	0.10	0.14	0.18	0.22
0.4	0.15	4.7	3.4	2.6	2.1
0.4	0.10	2.9	2.1	1.6	1.3
0.8	0.15	2.1	1.5	1.2	0.9
0.8	0.10	1.3	1.0	0.7	0.6

Temperature profiles in a snow cover have been measured at night, using minute thermistors, by Y. Takahashi, S. Soma,

and S. Nemoto [433] in Japan. It was also found that the night minimum, like the day maximum, did not fall at the surface, but was 7 mm below it. This could be explained theoretically by assuming that the porous structure of the snow allowed outgoing radiation to be emitted down to a certain depth under the surface. It was only when the amount of mixing was very small, and therefore the transport of heat *L* from the layer of air close to the snow was very small, that the minimum was found at the surface.

Measurements made by W. Ambach [389] in the ice of Vernagt Glacier in the Ötztal Alps showed that the latent heat of fusion of ice (79.5 cal g^{-1}) played an important role in determining the behavior of night temperatures. When cold penetrates at night into the glacier ice, which has been warmed up to 0° C during the day, the latent heat released on freezing retards the drop in temperature. Calculation of the thermal balance on three nights in July 1952, when observations were made, gave the following average values for the period from 18:00 to 06:00:

heat given out:	
through outgoing radiation	46 cal cm^{-2}
through evaporation	1
heat received:	
from eddy diffusion in the air	17
from latent heat of fusion of ice	16
amount lost on cooling, therefore	14

The heat released on freezing, which makes practically as much available as is received from the air near the ice, is therefore included in the quantity *B* in the thermal balance equation of Sec. 4.

A further consequence is that the night temperature field is limited below by the 0° C isotherm, when night frost is invading, and this boundary is displaced downward to the extent that the latent heat released is able to flow away toward the radiating upper surface of the ice. Between the frost boundary and the surface there is therefore established (in sharp contrast to unfrozen solid ground) a linear decrease of temperature, which is clearly recognizable in the nocturnal ice-temperature records of Ambach. This role of latent heat of fusion has already been noted in the discussion on ground freezing (p. 176). It also determines the temperature variation in snow cover, but here it is much more difficult to verify because of the inhomogeneous structure of snow cover.

Let us now turn our attention to temperatures in the layer of air close to snow or ice. To begin with, Fig. 111 shows qualitatively the temperature pattern for three different types of weather situations. These temperatures are for Munich airport, recorded by rod-shaped electrical-resistance thermometers, which in spite of their nickel-plated upper surfaces respond well to radiation ("test-element" temperatures).

The upper record is for 9 January 1935, when a fresh snowfall during the night covered the thermometer lying on the ground. Its temperature increased while the air temperatures near the snow continued to decrease until morning. The middle record, for 20 January 1935, in clear frosty winter weather, shows the temperature variation within the 9-cm snow cover; this is completely lacking in short-period fluctuations as well as in the air near the snow, where there is a slight temperature decrease with height during the day, and a strong inversion at night. The traces in the lowest part of the diagram, for 12 February 1935, are for melting snow that is still 6 cm deep. Snow temperature is therefore constant at 0° C, but 10° C is exceeded in the layer of air close to it, because of the effects of solar radiation and the advection of warm air. Between 14:00 and 15:00 the incoming-radiation type is well marked, changing at sunset with the onset of frost to the outgoing-radiation type.

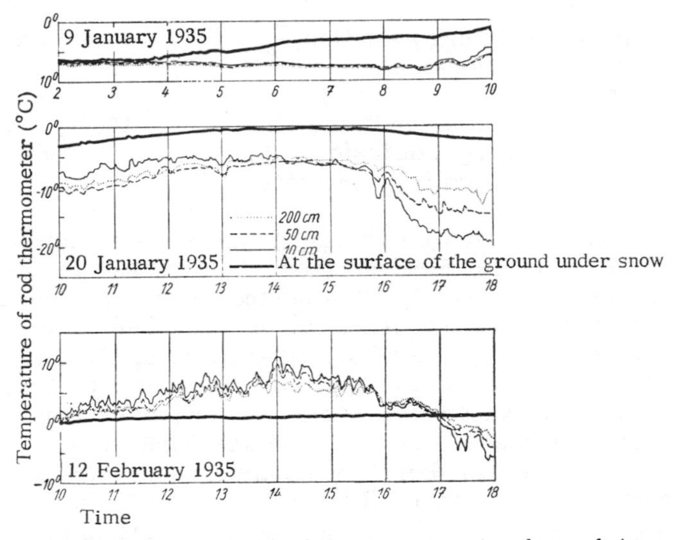

Fig. 111. Temperature measurements with rod thermomenters in a layer of air near snow in Munich area, for fresh snow, winter frost, and thaw.

It is difficult to make accurate measurements of true air temperature close to snow during the day, on account of the strong reflected radiation from the surface. Even small polished thermoelements require some form of radiation shielding, as shown in recent experiments by G. Band [394]. Things are simpler at night. A. Nyberg [421] has made very careful measurements of the night inversion in the lowest 25 mm above snow cover in

Sweden. These observations, made with single-strand resistance thermometers, gave the relation between wind speed and the increase of temperature with height shown in Table 56. These figures give a striking demonstration of the decrease in the lapse rate and the rise in temperature associated with increasing wind speed.

Table 56

Temperature (° C) as a function of wind speed and height above snow surface.

Wind speed (m/sec)	Number of observa-tions	Height above snow surface (mm)						
		1	5	10	15	20	25	1400[a]
Absolute calm	37	-17.6	-17.0	-16.4	-16.1	-15.9	-15.7	-12.1
0.3-0.6	30	-11.5	-10.7	-10.1	- 9.8	- 9.4	- 9.2	- 6.7
0.9-1.2	21	- 9.3	- 8.7	- 8.4	- 8.2	- 8.1	- 8.0	- 6.4
1.8	20	- 4.1	- 3.7	- 3.5	- 3.4	- 3.3	- 3.3	- 2.7

[a] Extrapolated.

Both Nyberg and Hoinkes [406] found a logarithmic distribution of temperature with height (see Sec. 12). H. Berg [395] confirmed this type of distribution down to within 1 mm of the snow surface in measurements he made at an elevation of 1240 m in the Allgäu Alps. Differences of up to 6 deg were recorded between the temperatures at 1 mm and at the snow surface. A new and rich source of material, dealing with the first 3 m above the snow cover of the inland icecap of Greenland, has been provided by the addition of the observations of the French Polar Expedition, Mission Paul-Émile Victor, to the original observations of J. Georgi [402]. The comprehensive temperature profiles were published by P. Pluvinage and G. Taylor [424].

The inversion found at night over the air layer near snow can also appear during the day in the polar climate in winter, and in the European climate in summer over glaciers. This feature has been studied by H. Hoinkes and N. Untersteiner [449] on the Vernagt Glacier in the Ötztal Alps, and by Hoinkes [406] on the Hornkees in the Zillertal Alps and on the Hintereis Glacier [448] in the Ötztal Alps. Table 57 gives a few values extracted from the observations made on the Hornkees in September 1951 at an elevation of 2262 m by means of an Assmann aspiration psychrometer. The increase of water-vapor pressure with height followed, like temperature, a logarithmic rule, leading to a value of 4.76 mm-Hg for the boundary layer at the ice, which is approximately the saturation vapor pressure at 0° C.

W. Vieser [177] has published the daily variation at heights of 118, 28, and 0 cm above a snow cover and 22 cm under it, in the form of averages for a few undisturbed days in February and March 1950. This gives only an indication of G. H. Liljequist's discovery [414] in Maudheim, Antarctica, that on 23 clear southern winter afternoons in July 1950 there was a linear increase of temperature in the lowest 10 m of the air layer above the snow.

Table 57

Air temperature and water-vapor pressure above glacier ice.

3 - 9 September 1951	08:00	12:00	15:00	18:00	Absolute		Average 09:00-16:00
					Maximum	Minimum	
Air temperature (°C) 130 cm	6.3	7.6	8.1	7.1	12.0	—	7.47
10 cm	4.2	5.9	6.2	4.6	—	1.2	5.70
Water-vapor pressure 130 cm	5.1	5.4	5.6	5.7	—	—	5.47
(mm-Hg) 10 cm	4.9	5.1	5.3	5.4	—	—	5.19

In the plant world, a snow cover means both protection and danger. The temperatures given in Fig. 108, for the interior of the snow, show that plants within the cover, or to the extent that they are partly below the snow surface, are protected against severe winter cold, and are completely shielded from wind within their moist packing. As long ago as 1902 W. Bührer [396] propounded the rule, based on measurement of the minimum temperature at the snow surface and at the ground surface below it, that even 1 cm of snow would afford some protection, and that 5 cm would give effective protection. At 20 cm the maximum is reached and further snow is useless from this point of view.

Plants within the snow cover run the risk, however, that icing up of the surface will stop the circulation of air within the snow and smother the plants (winter survival of seeds). There are, unfortunately, hardly any measurements of the permeability of snow to air. Laboratory investigation by O. Gabran [401] showed that this permeability is similar to that of an equally deep layer of sawdust free of wood splinters. The contention that the amount of carbon dioxide emanating from the ground and exhaled by plants at night might reach dangerous proportions under snow has been disproved by F. Pichler [423]. He measured the carbon dioxide content in a rye crop under snow from 35 to 135 cm deep, lasting from 17 to 92 days, and only once obtained a value of 0.21 percent by volume. He ascribed this low value to the permeability of the snow cover, to low plant respiration, and to absorption of carbon dioxide by the snow.

The situation is much more dangerous for the parts of the plants projecting above the snow. They are exposed during the day to both direct and indirect solar and sky radiation, which are reflected from the snow surface to a considerable extent.

They are therefore heated strongly, a fact that is immediately evident from the melt holes developing around every projecting twig or stem in sunny weather. Figure 112 is a photograph taken by A. Baumgartner in March 1954 on the southeast slopes of the Gross Falkenstein in the Bavarian Forest to illustrate this feature. The parts of plants near the snow are stimulated to evaporate, but the supply of moisture fails at temperatures below freezing, or at least becomes difficult. At night, outgoing radiation from the snow surface produces very low temperatures. Besides, when winds are strong, physical damage can be caused by the hard crystals of drifting snow (Sec. 17). The parts of plants that suffer are those that are exposed only because of the accidental depth of the snow at the time, and are therefore not specially adapted to offer resistance.

Fig. 112. The melted areas around each trunk are the result of radiation. (Photograph by A. Baumgartner)

Figure 113 is a reproduction of a photograph taken by P. Michaelis [417] of a pine tree at the tree line in the Allgäu Alps. The broken line shows the level of winter snow cover. The configuration of the tree is controlled by the double surface. Above the winter snow, lower branches are missing from the right-hand (north) side as a result of the abrasive action of the wind-driven snow. This part of the trunk is also lacking in lichens which are prevalent elsewhere; often the bark is deeply scored, since ice crystals are as hard as glass at very low temperature. On the south side, however, on the left of the picture, the branches are withered but still present as dead wood and are overgrown with lichens. The very large temperature fluctuations over the strongly reflecting snow surface are to blame for this damage.

Excellent color photographs of similar damage on a smaller scale have been published by A. Niemann [420]. One is of a cherry laurel of which the leaves above the snow have a reddish-brown dead appearance, while those below the snow are a fresh green color. Others are pictures of frost damage at the edges of a path cleared by a snow plow, and of a yellowish-brown track left in a field of red *Erica carnea* flowers, caused by the passage of a skier when it was covered by snow.

Fig. 113. Damage to a fir tree on a mountain above the level of winter snow (broken line) through drying-out (left) and drifting snow (right). (After P. Michaelis)

In the climate of Central Europe the principal cause of the decrease of snow cover when the temperature is lower than 0° C is evaporation due to radiation that is not reflected. This evaporation is possible only when the vapor pressure of the air is less than that of the snow surface; if the reverse is the case, then hoarfrost will be deposited on the snow. Since the latent heat of vaporization has the high value of 690 cal g^{-1}, only about 0.2 to 1.0 mm can be evaporated (converted into the depth of the equivalent layer of water) on sunny days without precipitation. Even on days with extremely favorable conditions, the amount did not exceed 6 mm.

The snow will disappear more quickly if a similar supply of heat energy is used to melt it, since the latent heat of fusion,

80 cal gm^{-1}, is less than one-eighth of that required for evaporation. Melting always takes place at 0°C; H. G. Müller [419] has demonstrated that the humidity of the air plays an important role in the melting process. The saturation vapor pressure over melting snow is 4.58 mm-Hg. If the vapor pressure of the air is lower than this, only evaporation will take place. If, however, it is higher than this value, water will condense from the air onto the snow surface, and by releasing 600 cal of heat for every gram condensed, is able to melt 7 to 8 times the quantity of water condensed. The moister the air is, therefore, the more quickly the snow melts. If radiation is discounted and heat supply from the ground below is taken to be negligible, the snow surface will behave in the same way as a wet-bulb thermometer, as long as none of the snow melts. The heat used up in evaporation is equal to the stream of heat from the air above 0° C. For a wet-bulb temperature of 0° C, psychrometric tables give the following connection:

relative humidity (percent):	100	80	60	40	20	0
air temperature (° C):	0.0	1.2	2.5	4.2	6.3	9.4

Snow therefore melts the sooner with increasing air temperature the higher the relative humidity is; and under otherwise similar conditions, the amount melted increases as the relative humidity increases.

The quantity of water produced on melting can also be calculated, if certain assumptions are made as to the amount of eddy diffusion taking place in the air layer close to the snow and the quantity of heat exchanged between air and snow, using Müller's theory. For example, making plausible assumptions, it can be shown that an air temperature of 2.5° C along with 60 percent relative humidity will not lead to melting, and that with 100 percent humidity 7 mm per day may be melted, which, for snow of density 0.1 g cm^{-3}, would correspond to a decrease of 7 cm in snow level in 24 hr. With the unusually high temperatures that are found in association with foehn winds in the foothills of the Alps, the snow may disappear very suddenly. J. Grunow [405] reports that on 13 March 1951 the snow cover on the Hohenpeissenberg decreased 24 cm in depth and lost 60 mm of melt water.

These well-known facts in general climatology also apply to microclimatology, since there are substantial differences in radiation, temperature, and humidity, even on a small scale. An example has already been given in the photograph in Fig. 92, which shows that snow melts at different rates. On that occasion melting was caused by differences in the thermal conductivity of the ground. In Japan, T. Fukutomi [400], in analyzing 144 observations made in January and February 1950, while searching for warm springs, established that there was such a good correlation between the temperature of the ground and the thickness of snow cover, for temperatures ranging from 1° to 10° C and snow cover from 70 to 0 cm, that the extremes of these observations could be deduced from each other. J. L. Monteith [418] observed at Harpenden, England, on 7 January 1955 that snow melted more quickly over the short grass on a cricket pitch than on the longer grass in the neighborhood, because the thermal conductivity was better, but it melted later on 14 February, in the same year, because this time melting was caused by warm air, to which the longer grass responded more readily.

It is possible for this reason to indicate the microclimatic peculiarities of a restricted area by charting the state of its snow cover, in a similar way to that in which special charts (reported in the international meteorological code) are used in the Weather Bureau to plot the state of the ground to give a picture of large-scale differences in climate at a given time. H. Slanar [432] was probably the first to plot maps of the melting process, which he has published for the area around an institute in Vienna. K. Kreeb [413] plotted the area around Plochingen on 22 February 1952 and showed that the mapped state of snow cover was in good agreement with phenologic charts for the same area. G. Waldmann [891] used snow maps to delineate the boundaries of similar plant growths in the Bavarian Forest, in the interests of forestry. Figure 243 is an example of such a map. The melting of snow cover in the Alps at the end of winter follows the same pattern every year, as shown by H. Friedel [905]. The same relative pattern of snow and ground appears every year in the same sequence, even if at different times in different years, depending on the weather. Friedel writes, "We have only to remember the figure of the 'Falconer', which appears about the middle of May on the slopes of the north chain, below the Rumerspitze, often seen from Innsbruck, and the shape called the 'white scythe' which appears on the Monte Veneto toward the end of June, and which is a sign to the farmers in the Innster district that it is time to begin mowing hay." The surprising thing is that this melt pattern is so strongly influenced by microclimate that photographs or charts of the same piece of ground present the same appearance year after year as the snow melts, even if the date on the calendar may be different. A discontinuity in large-scale climate, however, exerts a strong influence on microclimate. This can be studied from the work of M. Roller [428] on snow conditions on the Grossglockner Alpine highway on either side of the mountain range.

As the snow cover melts, ice plates are often formed. The process by which they form is illustrated in Fig. 114. Freshly fallen snow is piled up in a mound over the tufts of grass (1). During the following day there is a certain amount of melting on the "southern slope," and at night the water so produced freezes again, forming ice (2). The next day, solar radiation penetrates almost without hindrance through the thin ice plate and continues the melting process at the heated grass (3). The ice plates remain in position above the disappearing snow, often like piles of broken glass on top of the grass (4). High temperatures can often be produced in the hollow spaces under the ice plates because, supported on the blades of grass, they have no link that will conduct heat from the warmer ground.

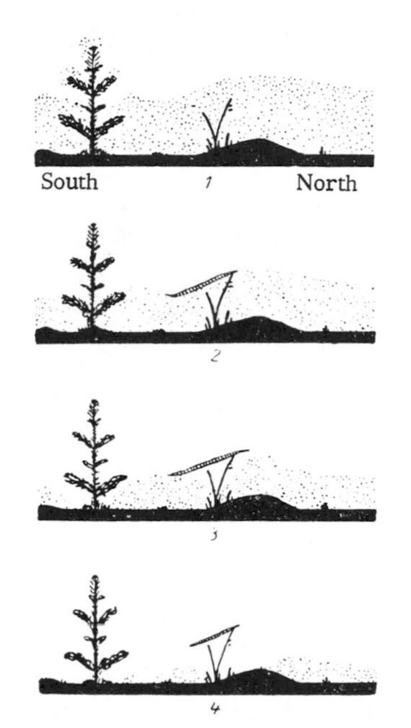

Fig. 114. The formation of ice plates when snow melts.

Once large open spaces form, snow melting continues rapidly because of intense heating of the exposed earth. Figure 115 records temperature measurements made by M. Köhn [412] on 29 May 1937 in the Black Forest. Conditions of winter cold still prevail under the patch of snow, which is a few meters in diameter, while 2 cm away from its edge soil temperatures reach 15° C.

In situations like this, "snow smoke" may be observed. This has been investigated more closely by F. Rossmann [429]. If air temperature and humidity are high and the wind is very light, a very fine misty veil becomes visible above the snow. When there is a breath of wind this misty veil forms on the windward side of the snow patch and dissolves again not far beyond the limit of the snow on the lee side. F. Rossmann even managed to record data for this microclimatologic phenomenon by means of an Assmann aspiration psychrometer on the summit of the 1497-m high Feldberg peak in the Black Forest on 26 May 1931. The average of a number of windward readings gave 18.1° C and 82-percent relative humidity for the air, while in the lee it was 15.2° C and 89 percent. The phenomenon is therefore caused by cooling of the warm flow of air near the ground by the snow surface.

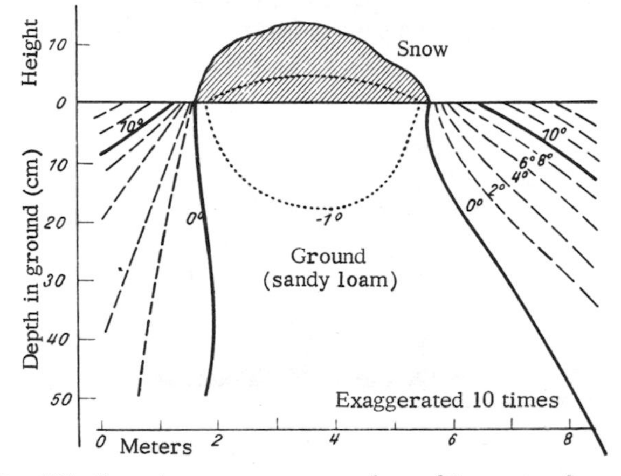

Fig. 115. Ground temperature around a melting mass of snow.

An unusual, rare, but striking occurrence within the layer of air close to a snow surface is the formation of snow rollers or cylinders, with which we will conclude this section. When there is a strong wind, which must reach at least 20 m sec $^{-1}$ in gusts, snow rollers or cylinders may be formed in a continuous snow cover on smooth fields and meadows, shown diagrammatically in Fig. 116, by W. Gressel [404]. The term roller is used when there is a hole through the axis of the cylinder, so that one can see through it, and it is called a cylinder when completely filled. The width of the rollers varies between 15 and 80 cm and the diameter between 8 and 50 cm; sometimes they form in hundreds and then look like molehills when seen from a distance. The tracks that they leave behind them during the course of formation (Fig. 116) may be up to 1 cm deep and are usually very clearly marked; their length is 10 to 20 m, the highest value recorded being 34 m.

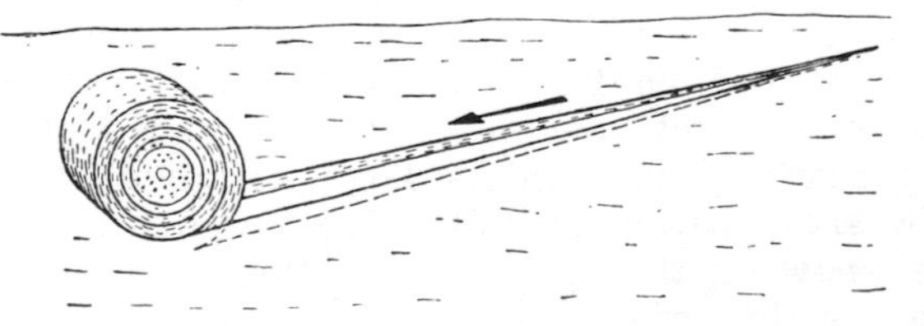

Fig. 116. The formation of snow rollers illustrated diagramatically.

These are not due to avalanche formation in miniature, since they may be observed to move upward on gently inclined slopes. W. Friedrich demonstrated [399] that temperature conditions played an important part in their formation, in addition to the preponderant influence of wind. Powder snow does not have the power of cohesion necessary for the formation of snow rollers, while heavy soggy snow cannot be whirled up and have its form altered by wind action. The unusual conditions necessary for their formation are only a transitional stage, therefore, which are present only for short periods. This is confirmed by meteorologic observations. A warm wind will make the loose snow, whisked up into the air, moist and gluey enough to be able to increase in size with further rolling. Topography also has some effect, since small irregularities disturb the wind field and lead to lee eddies that initiate the process. It seems likely that they all start as rollers and are changed by pressure and subsidence into cylinders. More details, references, descriptions, and photographs can be found in W. Gressel [404].

Snow cover will be mentioned again later when we describe the microclimate of high mountains in Sec. 46.

CHAPTER IV

QUANTATIVE DETERMINATION OF HEAT-BALANCE FACTORS

25. Basis and Methods of Evaluation

Earth's radiation equilibrium with the surrounding universe was portrayed by a few rough strokes of the brush in Sec. 2. The various elements playing various parts in determining the heat budget at the base of the atmosphere, that is, at the ground, were mentioned in Sec. 3, and the laws governing these elements were disucssed in Secs. 4 to 9. Knowledge of these matters is a prerequisite for understanding the state of the air near the ground, the changes taking place in it, and those in the underlying surface, to which Chapters II and III were devoted.

In the discussion of heat balance, questions that so far have been described qualitatively can now be subjected to quantitative examination, in order to arrive at a true understanding of the factors involved. This is not simply a step forward, but is the real aim of all study. It ought to be a fundamental rule of climatology that real understanding of microclimatological phenomena can be achieved only when all the factors involved in heat balance are followed quantitatively throughout their sphere of influence. This is still very difficult to accomplish today, and it demands, at the least, a very big outlay on instrumental equipment, great observational effort, and much hard work. Even where this aim cannot be fully realized, the object of all measurements should be to approach it as closely as possible.

The same requirements must be applied to the study of the water budget. It is true that this subject can be treated separately, but the heat required for evaporation forges the closest of links with heat balance. It is unfortunate that, for restricted areas, the water budget is less reliably known than the heat budget, with present methods of measurement. Neither the water content of the ground, which involves numerous problems of physics, chemistry, and plant physiology, nor evaporation canbe determined satisfactorily. Measurement techniques that are good enough in themselves place too much constraint on the natural state, and are not truly representative; if, on the other hand, natural states are not interfered with, the possible means of measurement lack precision. It is only to be hoped that great advances will soon be made in the arid zones of the earth, where this problem is being studied intensively.

For these reasons the discussion will be restricted for the moment to the heat balance. A few pointers will be given in Sec. 28 on evaporation, which is common to both budgets.

As in Chapter I, the problem will be considered first in a broad sense. The three horizontal bands in Fig. 117 show, on the same scale, the heat budget of the earth as a whole (above), of the atmosphere (middle), and of the surface of the ground (below). Incoming solar radiation is taken as 100 (equivalent to about 700 cal cm^{-2} day^{-1}). All other quantities are expressed as fractions of this amount. These were first ascertained by W. Trabert, in the form of annual balances for the earth, and have been improved many times since. Figure 117 is based on F. Lauscher's figures [452]. Heat gained is shown to the left of the center line, heat lost to the right; for the annual average, therefore, the bands are of equal lengths to the right and to the left.

Nineteen percent of solar radiation penetrates through to the surface of the earth; 37 percent is scattered diffusely in the atmosphere, of which 26 parts reach the ground as sky radiation, and 11 are lost to outer space. This 11 percent, together with 28 percent reflected by clouds, make up the reflection index or albedo of the earth, which is 39 percent (Jupiter 41 percent,

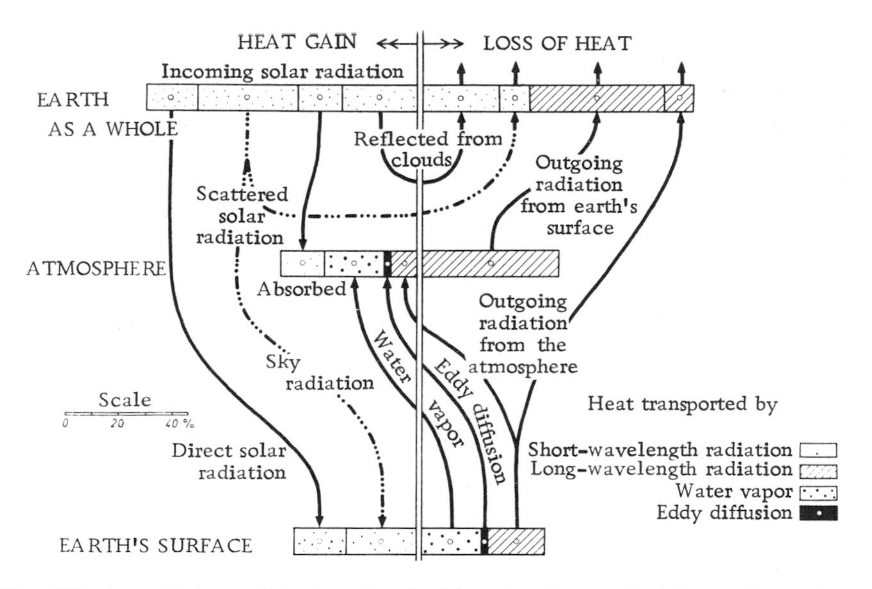

Fig. 117. Annual balance of the planet Earth, of the atmosphere, and of the earth's surface.

Venus 49 percent). Sixteen percent of the solar radiation is absorbed by water vapor, carbon dioxide, and ozone in the atmosphere, and is therefore converted into heat which can be felt. While incoming radiation lies within the short-wavelength range only, outgoing radiation is 61 percent long-wavelength; 50 parts of these are made up of the radiation given off by the atmosphere and the other 11 escape from the ground through a "window" in the atmosphere (p. 19) into outer space.

The heat exchange in the atmosphere is exactly half as great as that of the Earth as a planet. Against the radiation loss of 50 percent, there is a 16-percent gain from absorption of solar radiation, already mentioned. Then there is 10 percent retained of the outgoing long-wavelength radiation from the ground. To this there is also added on the positive side of the heat of condensation and sublimation (22 percent), which is used up on the earth's surface to evaporate water, and is later released when water droplets and ice crystals form in clouds. Eddy diffusion also transfers more heat from the ground to the atmosphere (2 percent) than in the reverse direction.

In the total heat exchange at the earth's surface, amounting to 45 percent, only short-wavelength incoming radiation appears on the positive side. This is because, in the annual balance, the long-wavelength radiation reaching the ground (the factor G in the heat budget) is by far outweighed by the long-wavelength outgoing radiation; this is represented by 21 percent, of which 10 parts are retained by the atmosphere, and the other 11 escape into space.

Now let us consider the heat balance of the ground surface. For every instant of time, and for every time interval, the complete heat-balance equation of p. 10 is valid for the ground-air boundary surface:

$$S + B\ (\text{or } W) + L + V + Q + N = 0\ .$$

Since the present concern is with solid ground, the factor W can be discarded. It is always very difficult, even impossible, to evaluate the gain or loss of heat caused by transport of warmer or colder air from surrounding areas. Research is therefore so organized that Q will become 0. This assumes that the environment has the same characteristics as the actual point of measurement, which means equal roughness and the same type of soil and vegetation cover; and orographic effects (for example, downslope winds) can be ruled out only in a plain. The first quantitative measurements of heat balance, such as those of H. U. Sverdrup [461] or E. Niederdorfer [455], were carried out over level snow surfaces, which satisfy these requirements ideally. If it is known how long the air will take to adapt itself to the underlying surface, then for a given wind speed it can be shown how far downwind conditions will become homogeneous. In the "Great

Plains Turbulent Field Program," complete homogeneity of the environment of the observation area was achieved; O'Neill, Nebraska, was especially selected from this point of view. According to H. Lettau and B. Davidson [13], the ground surface was homogeneous 1300 m downwind, and the observation site was surrounded by a circle of radius 16 km that was completely uninterrupted except by a few trees along a river bank, and these were 8 km distant. As a rule, however, the condition $Q = 0$ is not fulfilled, and for this reason Sec. 27 is devoted to the problems of advective effects, which are of great practical significance.

The quantity N, representing the gain or loss of heat due to precipitation, has little direct influence on the heat budget. Precipitation, which usually falls from higher levels, mostly has a cooling effect (Fig. 8). Its indirect effects are great, by way of contrast, because it changes the state of the ground by increasing its water content, and by altering B, this in turn affects all the other elements in the heat budget. If precipitation occurs in solid form, the resulting snow cover transforms completely all heat relations in the former reference surface (the air-ground boundary). An attempt is therefore made to have $N = 0$ as well. Fair-weather periods free of precipitation are therefore sought, but these bring into the analysis all the inherent disadvantages of a deliberate selection of observations. These disadvantages are not serious, since the main object of the operation is to achieve an understanding of numerical relations, which is easiest to obtain under conditions of strong radiation.

After having taken all these special precautions, the task still remains to determine the four heat-balance factors, which are connected by the equation

$$S + B + L + V = 0 .$$

Two of these factors, namely S and B, can be measured directly. In Sec. 4 it was shown that the radiation balance S consisted of several amounts of short- and long-wavelength radiations. Figure 2 showed the proportion of these quantities for two selected times. Recordings of all components in the radiation balance, made over a long period of years by Lupole instruments, after R. Schulze, are available from the Hamburg Meteorological Observatory of the German Weather Service. They provide a good insight into the variation of all these components throughout the year. Four-year monthly averages from 1 March 1954 to 28 February 1958, made available through the kindness of Dr. R. Fleischer, are shown in Fig. 118, with the symbols used in this book. Analyses of the figures for two individual years have been published by R. Fleischer [442, 443].

The top curve in Fig. 118 gives the long-wavelength radiation emitted by the ground. In addition to the small amount of reflected long-wavelength radiation (see p. 13), this consists principally of radiation emitted by the ground (σT_B^4) according to the Stefan-Boltzmann law, and therefore increases and decreases as the temperature of the ground rises and falls. This radiation would cause a tremendous loss of heat, in comparison with incoming solar radiation, even in high summer, if it were not compensated to a considerable extent by the counterradiation G. The vertically hatched area between the two curves therefore represents the (negative) long-wavelength radiation balance; the loss of heat is greater in spring and summer than in winter. On individual days it is a maximum with clear skies; with a low overcast or fog it can approach zero.

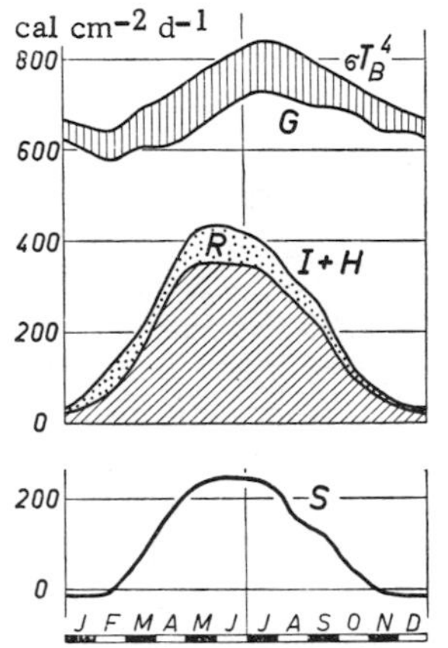

Fig. 118. Individual components of the radiation balance throughout the year, from measurements made at the Hamburg Meteorological Observatory. (After R. Fleischer)

In the same section of the diagram, the annual variation of global radiation $I + H$ is shown, from which is to be subtracted the quantity R reflected by the earth's surface (shown by the dotted area). The area with oblique hatching indicates the value of the short-wavelength radiation balance, varying with the meridian altitude of the sun during the year. The albedo of the Hamburg observation field averages 16 percent but varies during the year under the influence of changing weather. In winter, when snow is present, it may rise to 85 percent. The influence of snow cover in February may be detected even in the average values for the 4-yr period.

The difference between this area with oblique hatching and the vertically hatched area at the top of Fig. 118, that is, the sum of the short- and long-wavelength radiation balances, gives the total radiation balance S, which is shown separately at the foot of the diagram. Only in winter can the loss of heat be greater than the quantity received in incoming radiation, that is, the balance becomes negative. This curve shows the value of the radiation balance obtained by direct measurement; this is only slightly different from the values obtained by computation from the individual components. In the following discussion only the radiation balance S will be used.

In order to determine the heat exchange in the soil, hence the factor B, a set of earth thermometers are used, which must be able to reach to the depth where temperature becomes isothermal. Unfortunately, measurement of the temperature of the boundary surface between earth and air, which is very important, presents the greatest difficulties. Until accurate methods of measuring the long-wavelength radiation emanating from the surface are ready for use in the field, only extrapolation of temperature to the surface can be used. If the rules given in Sec. 6 and Secs. 19-21 are used to calculate the eddy diffusivity of the soil, which varies with time, and this is then used to evaluate the vertical flow of heat B, a homogeneous soil without horizontal temperature differences is presupposed. It is difficult to find a representative site at which to make these measurements, and it is often appropriate to install earth thermometers at two or three different points. Attempts are made at times to measure the vertical flow of heat directly by appropriate instruments. R. C. Staley and J. R. Gerhardt [13] have discussed the various methods of computation and the degree of precision possible in the measurement of B from the O'Neill results. Special care must be taken when there is ground frost, because then the latent heat released introduces a new factor and the temperature decrease in (homogeneous) soil will become linear, in contrast to its normal distribution, because the frost will penetrate into the soil to the extent that the latent heat released can be transported away (see p. 176). Similar care has to be exercised when the frozen soil thaws. It will also be shown that B is one of the smaller quantities involved in the heat budget, so that a degree of accuracy that falls short of desirable precision will not weigh very heavily in the problem.

If the ground is covered with vegetation, it becomes important to use, as a reference surface in discussions on heat balance, not the ground surface but one at a higher level called the "outer effective surface" (see p. 272). For example, when A. Baumgartner [437] investigated the heat balance of a pine thicket near Munich, he selected the average height of the crowns of young pines at 5 m as the reference surface. The factor B then acquires a slightly altered meaning and comprises the flow of heat entering or leaving the ground, the vegetation on it, and the air layer below 5 m.

The two quantities S and B may therefore be measured directly. The total value of the two remaining quantities is therefore known, $L + V = -(S + B)$. From Sec. 7,

$$L = c_p A \Theta \frac{d\Theta}{dx} \qquad (\text{cal cm}^{-2} \text{ min}^{-1}),$$

in which c_p (0.241 cal g^{-1} deg^{-1}) is the specific heat of the air for constant pressure Θ the potential temperature (pp. 35f), x (cm) the height above the ground, and A_Θ (g cm^{-1} sec^{-1}) the appropriate austausch coefficient for the thermal pseudoconductivity. In a similar way, the heat required for evaporation V is given as:

$$V = r_w A_q \frac{dq}{dx} \qquad (\text{cal cm}^{-2} \text{ min}^{-1}),$$

in which r_w is the latent heat of vaporization of water, which varies with temperature (being 597 cal g^{-1} at 0°C and 586 at +20°C, for example), q is the specific humidity or the mass of water vapor per gram of moist air (see p. 102), and A_q is the austausch coefficient for transport of water vapor.

In making his first calculations of the heat balance, H. U. Sverdrup [461] made use of the assumption first postulated by Wilhelm Schmidt that the same mixing process would transport all characteristic properties in the same way, so that $A_\Theta = A_q$. Using this assumption, the ratio

$$\frac{L}{V} = \frac{c_p \, d\Theta/dx}{r_w \, dq/dx}$$

can be calculated from the temperature gradient and water-vapor gradient of the air layer near the ground (this ratio is often called Bowen's ratio in English publications). From this it follows that

$$V = -\frac{S + B}{1 + L/V} \qquad \text{and} \qquad L = -\frac{L/V\,(S + B)}{1 + L/V}.$$

Then all the heat-balance factors can be calculated. If L or V is known, the austausch coefficient itself can be calculated.

Two simplifications are considered permissible in evaluating observations. If height differences within the air layer near the ground are small, the observed temperature t is used instead of the potential temperature Θ. Also, if differences of air pressure with height are neglected, the vapor pressure e may be used instead of the specific humidity q, the two quantities being related by the equation

$$q = \frac{0.622e}{p - 0.378e} = ce.$$

For a normal air pressure p of 750 mm-Hg, the value of the constant c is between 0.832 and 0.840×10^{-3} mm^{-1} for water-vapor pressures between 1 and 20 mm-Hg. With a value of 0.836×10^{-3} mm^{-1}, a mean r_w of 592 cal g^{-1}, and c_p = 0.241 cal g^{-1} $deg^{-1}$the ratio becomes

$$\frac{L}{V} = 0.49 \frac{dt/dx}{de/dx},$$

where t is in degrees Celsius and e in millimeters-of-mercury. If the differences in temperature and water-vapor pressure are taken for the two positions at the same level,

$$L/V = 0.49 \frac{\Delta t}{\Delta e}.$$

The quantity $1 + L/V$, which is used in all further calculations, has been tabulated by G. Horney [450] for the usual range of values of Δt and Δe. Instantaneous values should not be used for gradients, but rather values average over an interval of time sufficiently long to smooth out the random effects of individual turbulence elements.

This procedure does not apply when $S + B = 0$ and when $L + V$ also becomes zero. Frankenberger [446] has shown that this is always the case when the gradient of the equivalent potential temperature is zero, for then the surface of the earth will behave as an ideal psychrometer. Even when $S + B$ has a very small value, as happens mostly at night, and during the day also in cloudy weather, errors in measurement have too great an effect. An attempt must then be made to separate L from V by some means, and to measure them separately. For example, it is possible to use the vertical wind profile du/dx to calculate an austausch coefficient A_u when the lapse rate is adiabatic, and with other lapse rates it is possible to make an estimation. If the assumption that $A_u = A_\Theta = A_q$ is considered valid, it will be possible to use the austausch coefficient and the temperature gradient to calculate L, or to obtain a value for V from the gradient of water-vapor pressure.

In Sec. 8 attention was directed to the fact that, in any theoretical calculation of mixing by turbulent streams in gases, the point of departure was the presence of internal friction. The usual method of deriving the austausch equation (see p. 36) presupposes that turbulent mixing, because of the size and number of turbulence elements involved, permits the use of the probability law (for a large sample). It is further assumed that the characteristic transported has no reciprocal influence on the process of transport. There is justifiable doubt as to the validity of these assumptions. Warm air, for example, experiences an uplift in cooler surroundings; therefore, in convective mixing at least, systematic deviations cannot be ruled out. One might well suppose that mixing of other properties would not be disturbed to quite the same extent. Many measurements support this surmise.

The premise that $A_\Theta/A_q = 1$, on which the Sverdrup method is based, is therefore open to question. Comprehensive writings on the subject (summarized in [446]) suggest a value between 0.5 and 1.7 for the ratio A_Θ/A_q, on the basis of independent measurements of the austausch coefficients. In any case, the best and most numerous measurements give a value of about 1. M. Halstead [447] used a value of 0.86 in his heat-budget calculations for O'Neill. Frankenberger [446] has recently worked out the ratio experimentally for the German type of climate by making a double evaluation of evaporation, first via the heat budget by the Sverdrup method, then a second time by lysimeter measurements, for which the humid meadowlands near Quickborn provided favorable conditions. For selected September days he found a value of 1.0 (see also [445]).

In this state of affairs, and in consideration of the degree of accuracy that can be achieved, the Sverdrup method must still be considered a useful practical approach today. It is always applicable at times of the day when there is a large heat exchange, and these are also the most important periods. At the same time, this ensures that the law of conservation of energy is maintained as the most certain law; it requires only a suitable separation of L and V. In addition, a considerable amount of progress can be made today if the three austausch coefficients are not considered equal, but all proved measurement techniques are used together, and there are tremendous possibilities in this respect, depending on climate and the site of measurement. Only by testing the results obtained can the difference be shown between actual reality and assumptions. M. H. Halstead has made an evaluation using such a process [447], and H. Lettau has developed a new model. These complicated interrelations arising from the struggle toward discovering the best description of heat balance of the earth's surface, capable of being used in equations, are best described in the short exposition that H. Lettau (in [13]) published as a result of the great O'Neill undertaking.

Until now, the heat balance of the ground surface only has been considered. Now we can proceed to a further stage, and inquire into the heat balance of the air layer near the surface.

In the stationary state, roughly true of measurements in the noon hours, the vertical streams of heat S, L, and V are constant with height. This means that exactly the same amount of energy is transported through the upper boundary x_2 of a layer of air as through its lower boundary x_1. In the nonstationary state, a divergence of the thermal stream is established. Then a part of the radiation stream between x_1 and x_2, called ΔS, is retained in the layer or is lost to it as a negative quantity. The same applies to the conduction stream (ΔL) and the water-vapor stream (ΔV). The energy gain $\Delta S + \Delta L + \Delta V$ must be equal to the increase in the same layer of sensible and latent heat. The increase in sensible heat is $\int_{x_1}^{x_2} \rho c_p \frac{d\Theta}{dz}\, dx$, where again z is time, x (cm) the height, Θ the potential temperature, ρ (g cm^{-3}) the air density, and $c_p = 0.241$ cal g^{-1} deg^{-1}. The increase in latent heat is $\int_{x_1}^{x_2} \rho r_w \frac{dq}{dz}\, dx$, where q is the specific humidity and r_w the latent heat of vaporization of water. The heat-balance equation of an air layer near the ground therefore is $\Delta S + \Delta L + \Delta V = \int_{x_1}^{x_2} \rho c_p \frac{d\Theta}{dz} dx + \int_{x_1}^{x_2} \rho r_w \frac{dq}{dz} dx$ (cal cm^{-2} min^{-1}). Where there is no change in the total quantity of water vapor present in the layer, the term ΔV and the second integral have the same value. In this case the change in temperature of the layer corresponds to the divergence of the radiation and conduction streams.

In either case the amounts of energy involved are small, and difficult to measure. Their determination can, however, lead to valuable conclusions, as may be exemplified by research carried out by H. Kraus [451] into radiation fog in a woodland clearing near Munich. He evaluated the heat budget of the layer of air between 50 and 600 cm above the grass-covered soil on the evening of 12 October 1956. The latent heat released on fog formation was found to be negligible: "the fog, which appears to the observer to be the most significant feature turns out, from the energy point of view, to be merely a side effect of cooling in the evening." The heat-balance factors varied from hour to hour, as may be seen from Table 58.

Table 58

Heat-balance factors (10^{-3} cal cm^{-2} min^{-1}) in the formation of radiation fog in a woodland clearing.

Time	ΔS	ΔL	$\int_{50}^{600} \rho c_p \frac{d\Theta}{dz} dx$	ΔV
17:00	-3.9	-12.7	-16.6	2.6
18:00	-9.2	0.0	- 9.2	-5.2
19:00	-8.2	3.5	- 4.7	-4.5
20:00	-7.2	4.5	- 2.7	-2.6
21:00	-7.1	5.0	- 2.1	-1.2
22:00	-6.5	6.5	0.0	-0.0
23:00	-5.2	6.9	1.7	-0.5

Heat losses through radiation from the air layer near the ground are greatest after sunset, and decrease during the night, as we already know from different investigations (see Sec. 5). At 17:00 the conduction stream still drew a good deal of heat from this layer, and hence made a significant contribution to the cooling process and to the formation of "radiation fog" which was seen to begin forming at 17:30. However, ΔL became 0 by 18:00 and, as the night inversion strengthened, the flow of sensible heat offset the loss by radiation from the air layer to an increasing extent. When the fog was swept away by stronger winds after 22:00, this became even greater and effected a temperature increase in the air layer near the ground. Only at 17:00 did evaporation from the grass surface transfer more water vapor into the layer near the ground than the latter gave off upward. This layer then became drier in spite of fog formation (heavy deposit of dew on the grass).

26. Results of Previous Heat-Balance Measurements

Beginning with the diurnal variation of the factors involved in the heat balance for the climate of Central Europe. E. Frankenberger [445] measured and evaluated temperature, water vapor, and wind from 1 September 1953 to 31 August 1954, up to heights of 70 m above a flat meadow, not much above the water table, near Quickborn, Holstein. Figure 119 gives the results for clear days with light

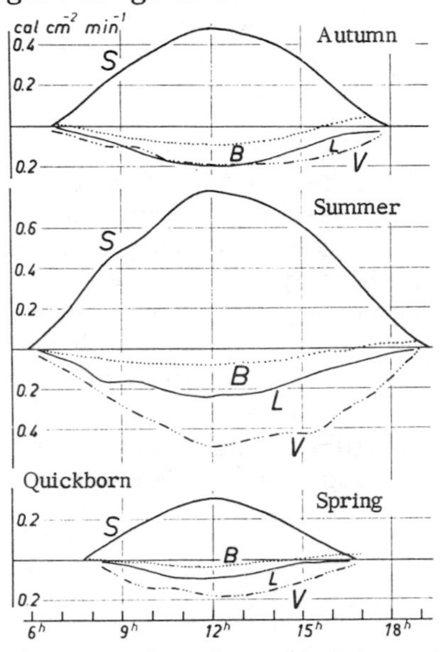

Fig. 119. Daily heat exchange on clear days with light winds at Quickborn, Holstein. (After E. Frankenberger)

winds in three seasons of the year. The figures for the small heat exchange at night (and also in winter) were not sufficiently reliable, so that the heat balance is given only for the hours between sunrise and sunset. Since only clear days were considered, the radiation balance S is a smooth, balanced curve, symmetric about the 12:00 line. During the day it is used up to heat the ground (B) and the air (L), and for evaporation (V). These three elements are therefore always shown below the line when they take heat away from the surface, in this diagram and in Figs. 120 to 122, in agreement with their algebraic sign in the heat-balance equation (p. 9). In Central European climate most of the heat is used up in evaporation, particularly in summer. If daily observations for a longer period are plotted, as was done for Potsdam in June 1903 by F. Albrecht [435], they produce a uniform, flattened picture as a result of bad-weather effects. This justifies the selection of fair-weather periods, as long as they are used only to promote understanding of the processes involved, and never when it is a question of climatologic mean values.

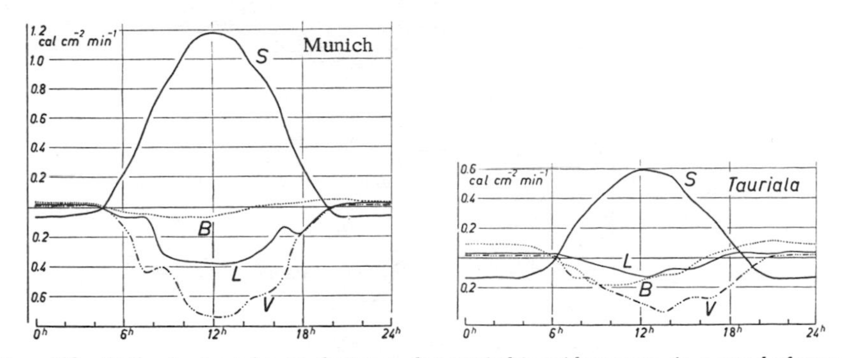

Fig. 120. Daily heat exchange during a dry period in midsummer, in a stand of young firs near Munich. (After A. Baumgartner)

Fig. 121. Daily heat exchange on bright summer days in Finland. (After M. Franssila)

The factors B, L, and V exhibit singularities in their diurnal variation, more clearly seen in Fig. 120 which is for a period of drought in midsummer, as may be deduced from the value of S. A. Baumgartner made the calculations—including the night period in this case—for a spruce thicket 5 to 6 m in height in the neighborhood of Munich. The quantity B therefore comprises, as mentioned on p. 229, the heat exchange not only in the woodland soil, but also in its protective covering of trees. The value of B is comparatively small; the way in which it constitutes a number of individual elements will be discussed later (Fig. 167). A chance variation in the weather during the recording period from 29 June to 7 July 1952, causes the curves in Fig. 120 to appear somewhat more disturbed than those in Fig. 119. For the most part, however, fluctuations are systematic in nature. When the morning sun touches the moist grass, which may be wet with dew, or the treetops in a wood, it will first dry these places. If this is over by 08:00, then V will decrease in spite of rising temperature, and the air, which up to this time had little heat left for it, suddenly receives a much greater share. And the ground, which is dried gradually during the day to greater depths, begins to give off heat too, from about 15:00 on. It follows that B is asymmetric about the noon line, and this may be recognized in Fig. 119 as well, although less distinctly.

The same result is evident in measurements made in northern latitudes by M. Franssila [307]. Figure 121 is the mean of three August days in 1934, for a meadow near Tauriala, Finland. The ground, which had stored a surprising amount of heat during the forenoon, released heat after 16:00. The nocturnal exchange of heat was large, perhaps on account of the longer radiation in high latitudes, or insufficient accuracy of the radiation-measuring instruments available at that time.

The situation is different in areas low in precipitation. Two examples are given in Fig. 122. The upper graphs show the heat budget averaged for 4 August and 2 September 1953 by H. Lettau

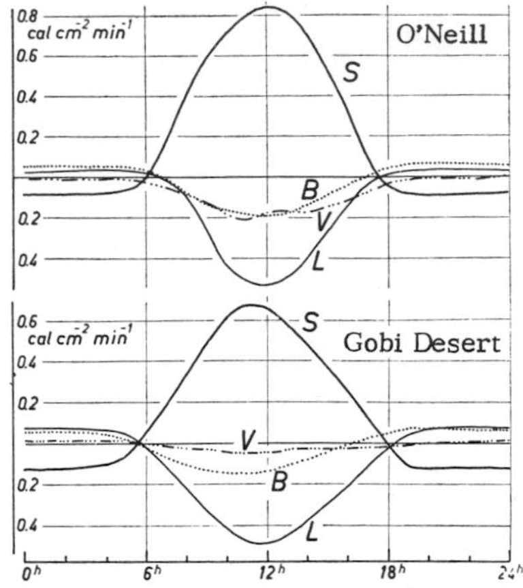

Fig. 122. Daily heat exchange in arid areas: O'Neill, Nebraska (above), and the Gobi Desert (below). (After F. Albrecht and W. Haude)

[13], from the "Great Plains Turbulence Field Program" at O'Neill, Nebraska. Below are results from the Gobi Desert for 11-20 May 1931, from W. Haude's observations in Ikengung during Sven Hedin's China Expedition, evaluated by F. Albrecht [436]. The smoothness of the curves results from the circumstance that only 2-hr averages were published. The agreement between the two is striking when it is considered how varied they were, both from the point of view of instruments and personnel. Evaporation naturally plays a minor part; in the Central Asian desert steppe of the Gobi it is even lower than in the North American prairies,

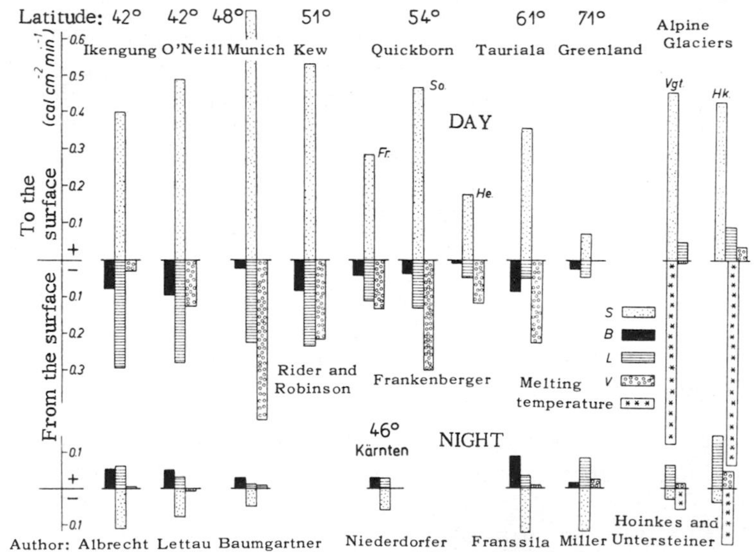

Fig. 123. Results of all microclimatologic heat-balance measurements, separated into night (below) and day (above).

which are almost in the same latitude. In O'Neill at night there is no condensation, but a small amount of evaporation, in contrast to Figs. 119-121. In Ikengung the accuracy of measurements is not good enough to permit conclusions to be drawn.

Diurnal variations will now be treated in two time intervals, day and night. The boundary between the two may be selected in a number of ways, and is also affected by the way in which some writers have published their material. In constructing the composite picture in Fig. 123, the criterion that was used as far as possible was the change in algebraic sign of the radiation balance. At the same time this summary allows the range of material to be extended, details being given in Table 59.

The upper half of Fig. 123 gives the values of the various factors by day, while the lower half shows the smaller night values on the same scale. Since the figures illustrated are the quantities per minute, averaged over the whole day or night period, the absolute values are small, even if the whole period was one of fair weather. The primary factors are determined during the day by the season of the year and the kind of weather. Except for the Alpine glaciers, on the extreme right, the places where the observations were made have been arranged in order of increasing latitude. Haude's measurements in Ikengung were in May, those at O'Neill in August and September. Munich gives the highest radiation value, because the observations were made practically at the turn of the summer sun and also during mid-summer drought. The new observations introduced here, those made by N. E. Rider and G. D. Robinson [458] at Kew Observatory, are averages of seven individual sets of readings taken between 20 and 24 June 1949 about noon (11-15). The Quickborn figures are arranged according to the season, as in Fig. 119. Since these contained no night figures, there were inserted here the results of E. Niederdorfer's nocturnal heat-balance measurements [455] over a snow surface in 1932. Greenland values are the work of the French and German Expeditions to the "Eismitte" station (70°54′N, 40°42′W, 3030 m), taken from D. H. Miller [454]; they refer to the 14 hr of sunshine during the polar day, and to the polar night. Between the latitudes of the Greenland and Finland stations, there are heat-balance figures from Point Barrow, Alaska, made by R. A. Bryson [438], on which a preliminary report is available. The calculations for 17 and 18 June 1953, however, gave such improbably high values for L and B, in spite of a thawing soil, and such an unusual diurnal variation, that they have not been included in the diagram.

Taken as a whole, Fig. 123 provides us with a clear picture of the heat exchange at the earth's surface. In a temperate climate most of the heat is used up in evaporation, both in the hottest periods and in times of drought, provided there is sufficient vegetation (Munich). In dry climates at lower latitudes the water available is insufficient while at higher latitudes heat is insufficient.

Table 59
Details of the data used to illustrate the heat balance in Fig. 123.

Country	Observation site				Observer and analyst	Data used in Fig. 123		Year of publication
	Latitude	Longitude	Height (m)	Situation		Period of recording	Weather conditions	
China	41.9°N	107.8°E	1500	Gobi Desert near Ikengung	F. Albrecht, from W. Haude's observations [436]	11 to 20 May 1931	Fair weather	1941
USA	42.5°N	98.5°W	603	Prairie grassland near O'Neill, Nebraska	H. Lettau, from the observations of several teams [13]	9, 13, 19, 25 August; 1, 8 September 1953	Fair weather	1957
South Germany	47.9°N	11.7°E	645	Fir thicket 5-1/2 m high, 30 km SE of Munich	A. Baumgartner [437]	29 June to 7 July 1952	Summer drought	1956
England	51.5°N	0.3°W	5	Grass plot at Kew Observatory	N. E. Rider and G. D. Robinson [458]	20 to 24 June 1949, 11:00 to 15:00	Fair weather	1951
North Germany	53.7°N	9.9°E	12	Flat meadow, 1 m above ground water at Quickborn, Holstein	E. Frankenberger [445]	1 September to 31 November 1953, and 1 March to 31 August 1954	Clear days with light winds only	1955
Austria	46.5°N	14.6°E	560	Snow field near Eisenkappel, Carinthia	E. Niederdorfer [455]	14 to 15 January 1932	Clear and calm nights	1933
Finland	61.2°N	24.4°E	Low	Meadow on a plain near Tauriala	M. Franssila [307]	Three August days in 1934	Fair weather	1936
Greenland	70.9°N	40.8°W	3000	Ice cap at "Eismitte" station	D. H. Miller, from observations by the A. Wegener and R. E. Victor expeditions [454]		Clear summer day (14 daylight hours) and winter night	1956
Austria	46.9°N	10.8°E	2973	Glacier ice on the Vernagt Glacier in the Ötztal Alps	H. Hoinkes and N. Untersteiner [449]	21 to 31 August 1950	Period of fair weather	1952
Austria	47.0°N	11.8°E	2262	Glacier ice on the Hornkees Glacier in the Zillertal Alps	H. Hoinkes [448]	3 to 9 September 1951	Period of fair weather in midsummer	1953

The heat balance of Alpine glaciers in summer was determined by H. Hoinkes and N. Untersteiner [449] for the Vernagt Glacier in the Ötztal, and by Hoinkes [448] for the Hornkees Glacier in Zillertal; it was tested by making simultaneous measurements of the ablation or quantity of water produced by melting. The amount of energy required to melt the ice appears here as factor B. It follows from the great temperature contrast between the glacier and its environment in mid-summer that L remains positive both day and night, and even V is nearly always positive, which means that heat is fed to the glacier by condensation. The main source of supply of the heat used up is radiation. Untersteiner [462] found similar results for the Chogo-Lungma Glacier in the northwest Karakorum in latitude 37° N at an elevation of 4000 to 4300 m.

Considering now the annual variation in the heat budget, we have at our disposal the pioneering work of F. Albrecht [435] from 1940, and M. I. Budyko's attempt [439], unfortunately with a text in Russian only, to show the components of heat balance in a world atlas. The discussion is thus extended into tropical regions, and includes the heat balance of seas in every latitude. It helps considerably toward a fuller understanding of conditions in the air over land and sea to have a true conception of the quantities involved in this global problem.

The factors S, L, B or W, and V for the 12 months of the year, subdivided into two half-years by a vertical line, are shown in Fig. 124 for six land stations, and in Fig. 125 for five ocean stations. In both diagrams latitude increases from left to right. Thus S is seen to change from being positive throughout the year, to become negative in winter in higher latitudes. Land and sea do not differ greatly in this respect since cloudiness has a zonal distribution and

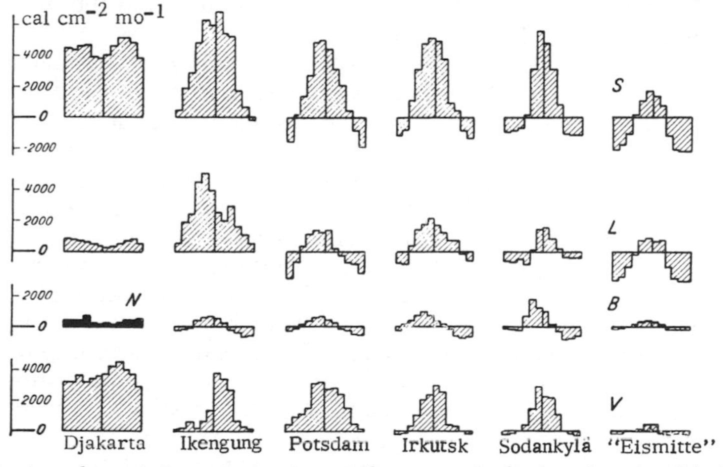

Fig. 124. Annual variation of the heat balance at six land stations in all latitudes. (After F. Albrecht)

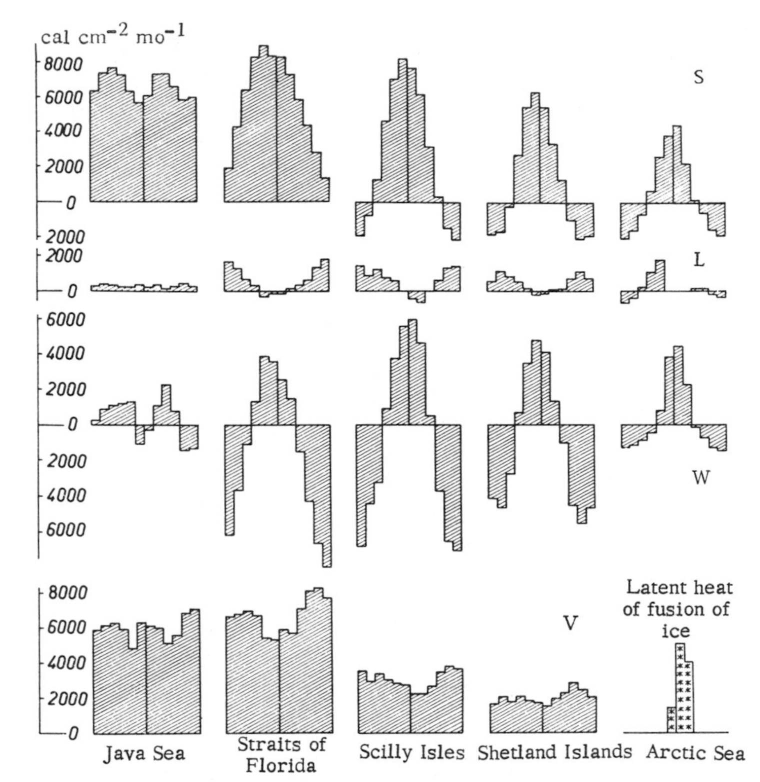

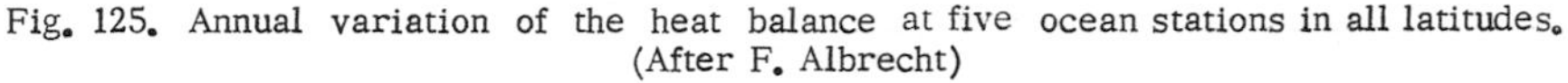

Fig. 125. Annual variation of the heat balance at five ocean stations in all latitudes. (After F. Albrecht)

the influence of temperature and other properties of the surface does not carry much weight. However, the ways in which the sun's heat is utilized are very different indeed.

At land stations the heat exchange in the ground is comparatively small. At Djakarta (Batavia) on Java, where there is practically no difference from month to month, it can be neglected. In the diagram, the quantity N of heat used to warm the cold precipitation water is substituted for B, since rain is always cold in the tropics. In Central European climate, with its changing air masses, N may be neglected for a yearly period, since precipitation may be either warm or cold. The greater the contrast between the seasons, the greater becomes the factor B. The maximum in Fig. 124 is for Sodankylä in Finland (67° N), which is in good agreement with the particularly large value of B for Tauriala in Fig. 123. The factor B is unimportant in tropical climates, which are always warm, and in polar climates.

The greater part of S is used up in evaporation V, and is large in the tropics and low at the poles. The actual amount of evaporation, however, depends also on the quantity of water available; therefore V is lower at Ikengung (42° N)—apart from the short summer monsoon period—than it is at Potsdam (52° N). In the far north (Eismitte) condensation may even be greater than evaporation, as we saw from Fig. 123 during the polar night.

The quantity L tells us the amount of heat gained or lost by the atmosphere. The figures for Ikengung, where very little is used in evaporation, show that the arid parts of the earth act as intense sources of heat for the atmosphere. Such places provide good fields for research into the surface layer of air, as we may see from the work of the Australian school of microclimatology. In higher latitudes large quantities of heat are withdrawn from the air in winter, and can be replaced only by advective means. The heat exchange in a given position, and large-scale horizontal exchanges of heat, may be seen to be closely linked with each other.

The magnitude of the heat exchange W with the changing seasons at the ocean stations in Fig. 125 is most striking. It is a maximum in the Straits of Florida (25° N) and near the Scilly Isles (50° N), decreasing toward the tropics, which are always warm, and also toward the poles, where the total heat exchange decreases. A comparison between B and W in Figs. 124 and 125 shows the difference between continental and maritime climates. At sea the actual evaporation is equal to the potential evaporation and therefore decreases in a fairly regular manner toward higher latitudes. It can be neglected in polar seas because of the small amount involved (results of the Maud Expedition). In its place the quantity of heat used to melt ice in the summer months has been inserted in the diagram, when $L = 0$ with temperatures about the freezing point.

Finally, Fig. 126 shows the annual balance of the heat budget for all geographic latitudes, for both land and sea, based on the

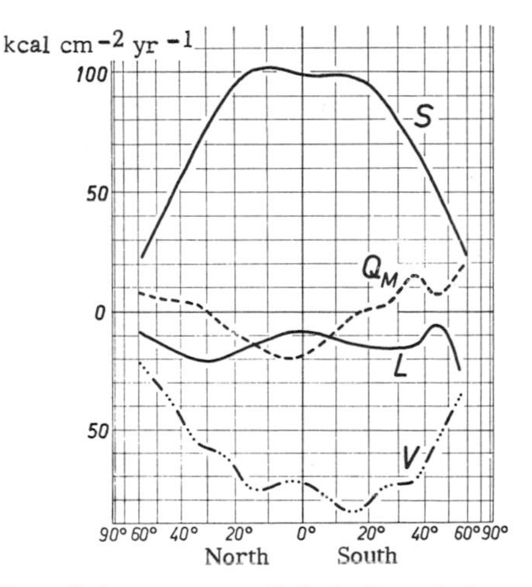

Fig. 126. Annual heat balance over all latitudes of the earth. (After M. I. Budyko)

results of M. I. Budyko [439] from a report published by H. Flohn. The scale of latitude is contracted toward the poles in order to give a proper idea of the area between respective circles of latitude.

The factor B (or W) does not appear in this annual balance since, over the year as a whole and discounting long-range temperature fluctuations, the amount gained balances the amount lost, as may be deduced directly from study of Figs. 124 and 125. The radiation balance S is always positive within the range of latitude from 60° N to 60° S, for which it has been calculated. Between +15° and -15°, that is, within the range in which the sun can be found at the zenith, it has an almost constant value of 10^5 cal cm^{-2} for the year. In all latitudes heat is transferred from the ground surface to the air (L), which is in conformity with the experience that where a positive radiation balance exists the annual ground temperature is always higher than that of the air. This has also been confirmed by Fig. 117. The two arid zones in roughly 30° latitude provide most of the heat for the air layer near the ground, as we have already seen from a study of Fig. 122. The heat consumption of the evaporation process (V) exhibits its usual dependence on latitude.

The balance can no longer be explained here, however, without reference to the advective transport of heat. The quantity of heat carried away by air masses, with their low thermal capacity, is very small in comparison with the quantities transported by ocean currents. From 30° N and 20° S toward the poles, there are added to radiation the amounts of heat supplied per square centimeter, designated Q_M in Fig. 126, which also make a contribution toward V and L. In the equatorial regions, which lie in between, substantial quantities of heat are given off. Advection to some extent equalizes the effects of the differential radiation of different latitudes. The zonal heat flow crossing various parallels of latitude was worked out in 1950 by P. Raethjen [456], which work should be consulted for further information.

27. Advective Influences. Transitional Climate and Dependent Microclimates

The heat balance of the ground surface was computed in Secs. 25 and 26 from the vertical flow of energy toward or away from the boundary surface between air and ground. Besides vertical gradients there are also horizontal gradients present; therefore the method of computation used up to now is no longer permissible, since it was based on the assumption that such horizontal gradients were nonexistent.

Advective processes, at first sight, do not concern the ground surface, but apply to the layer of air close to it. Since the ground surface, for which the heat economy equation $S + B + L + V = 0$ is valid, is a horizontal surface of reference it cannot be affected by horizontal, that is by advective, heat flow. However, we may postulate a volume of air bounded by the ground surface, a cover surface parallel to it, and sides formed of vertical surfaces and then consider the heat balance of this volume. G. Hofmann [467] was able to prove, for such a case, that the factor *L* of the heat budget—apart from changes in the heat content of the volume of air—composed of five separate components, which are then described in more detail. The equations thus derived offer the possibility of distinguishing advection proper from the vertical flow of heat occasioned by frictional and convective exchange (see Sec. 8).

Just how much heat finds its way by advection into the volume of air being considered depends on the magnitude of the horizontal heat flow and, naturally, on the height selected for the cover surface. If this surface is put high enough, the flow of heat through it will become negligible; then the heat transported advectively into the volume of air will be equal to the flow of heat from the air to the ground. This was the method used by F. Möller [472] for Europe, and by F. Täumer [478] for the area of the Spree River, to calculate the quantity of evaporation for the region, that is, the factor *V* of the heat balance, using figures for humidity and wind distribution measured in aerologic ascents from a number of stations. The procedure may be used to calculate *V* in a similar way to *L*.

It is necessary to avoid advection when investigating microclimate; hence the observation site has to be selected so that there will be no horizontal gradients. This ideal state was almost completely realized in the Great Plains Turbulent Field Program, which has already been mentioned. But there are no places like O'Neill in ordinary climatology, which must therefore always contend with some kind of advective influence; so this problem will be investigated thoroughly.

There are many research data available today, from which we can assess the significance of advective processes, and the nature and magnitude of their effects. These are most easily appreciated where two different kinds of surfaces abut on each other (land, water, snow), or where the ground is put to different uses (road, meadow, plowed land, wood).

At all such boundaries there develops a transitional or boundary climate. This feature was also taken into account in classical large-scale climatology at an early stage. A coastal climate is one example of such transitional climate, extending its influence inland for 30 km at the most, since after traveling this distance air coming from the sea has completed the process of adaptation to the land. R. A. Craig [465] has investigated the reciprocal process by which air passing from land to sea undergoes transformation, in a detailed study in Massachusetts Bay up to a height of 300 m. There are descriptions of coastal climate from every country in the world. Turning to microclimate it becomes evident that, for the layer of the atmosphere used by humans, the contrast between land and water exerts its influence only within a few hundred meters of the boundary, and consequently the coastal climate should be subdivided to give a river-bank or seashore climate. This has been analyzed recently by H. Berg [463] to demonstrate its importance for health resorts. The local climate of the health resort of Heringsdorf on the island of Usedom has been described by H. Zenker [480].

Consider now the dimensions of microclimate; the small amount of mixing near the surface ensures that those characteristics of the air that depend on the underlying surface (see Chapter III) are at first retained *in situ*. The extraordinary contrasts that may thus arise within a very small space will be illustrated for the boundary of land and water, to begin with.

It was mentioned previously, in the discussion on small bodies of water (Sec. 22) that water and land surfaces exerted a mutual influence on each other. The boundary zone on either side of the line of discontinuity has been investigated by H. Berg [464] and A. Mäde [471], and much new information is available. Figure 127 shows the temperature and water-vapor profiles over a flat sandy part of the bank of the Rhine, near Rodenkirchen, from 09:00 to 10:30

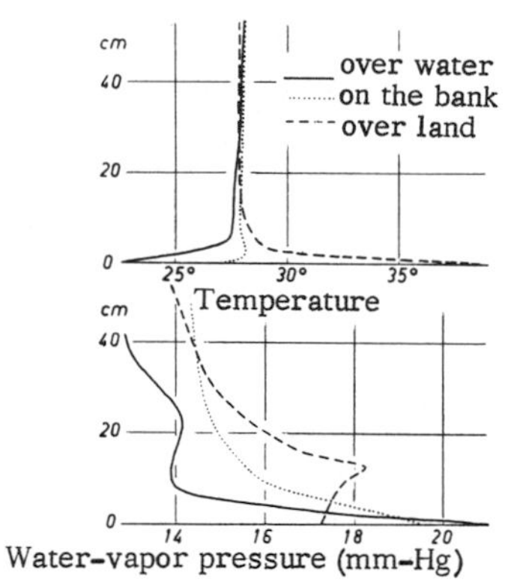

Fig. 127. Temperature and water-vapor-pressure profiles on the banks of the Rhine on a morning in July. (After H. Berg)

on 2 July 1952, when winds were light. The observations were made by H. Berg, using an aspiration psychrometer. In the lowest 10 cm there are in juxtaposition a cool moist layer of air over the water, and a layer 13 deg warmer over the sand only 8 m away. A little higher, between 10 and 40 cm, temperatures are practically in equilibrium, but at this height over the water the water-vapor pressure is slightly less than it is over the warmer bank. Further measurements and observations have led Berg to conjecture that, during the day, cool moist parcels of air from over the water rise over the hotter land air, but during the night cold air descends toward the water. Without progressing quite so far as the establishment of a miniature land and sea breeze, a zone of weak mixing is set up. This is able to blunt the contrast near the surface but not to prevent it from existing. By making temperature measurements when sea smoke made its appearance, W. Vieser [177] was able to show that the cold air flowing out from the coast was heated and enriched with moisture in its lowest 10 cm only.

Many valuable data were provided by the long series of thermoelectric records made by A. Mäde [471] on Lake Süssen near

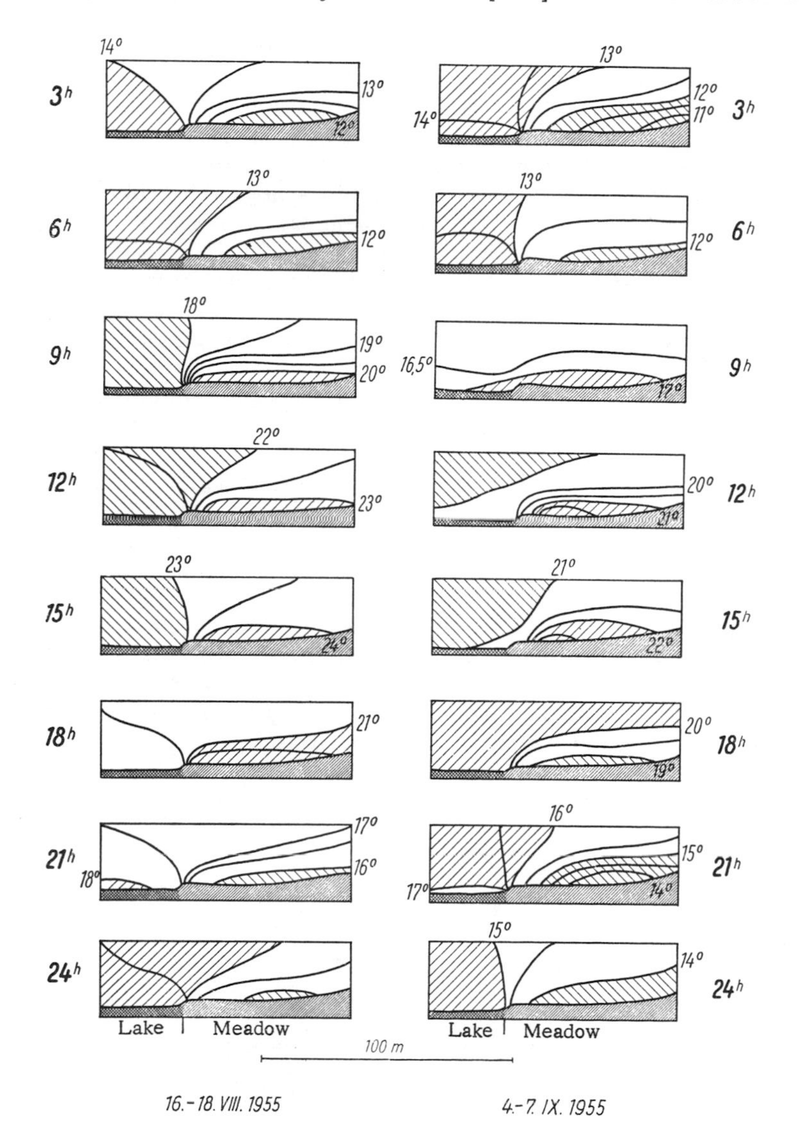

Fig. 128. Temperature field near the banks of Lake Süssen near Eisleben during the day. (After A. Mäde)

Eisleben. This lake lies in a shallow saucer of land, between 2 and 3 km in area, and its vegetation-covered shores are only about 5 cm above water level. Recordings were made up to a height of 1 m and for a distance of about 100 m horizontally on both sides of the water's edge. Figure 128 shows the different distributions of isotherms at 3-hr intervals for two series of observations, from 16 to 18 August 1955 (left), and 4 to 7 September 1955 (right). Relatively warm areas are hatched with lines sloping up to the right, and relatively cold areas with lines perpendicular to these, sloping up to the left. Although there is no fringe of reeds here that might hinder the equalization of temperature, in the lowest half meter contrasting temperatures are found right up against each other. There is very little sign of the moderating influence of water surfaces on their immediate neighborhood at night that is often mentioned in the literature. What compensatory process takes place above this lowest layer is not known.

A. Nyberg and L. Raab [475] demonstrated some features of the eddy-diffusion process in air flowing from sea to land, in a number of clever experiments. Four masts were erected on a plain with sparse vegetation on the west coast of Öland Island, Sweden. Thermistors were fixed on these masts at the heights shown by small circles in Fig. 129. Temperature and wind speed and direction were then recorded. The distance traveled by the wind coming from the sea on 1 and 2 July 1954 between 11:30 and 12:30 up to the time it reached the masts is shown in the diagram as the "effective coastal distance." On these two cloudless days the ground surface was heated to more than 35°C by radiation. Isotherms in the first 20 m of the atmosphere are shown in full lines (published values) and the deduced austausch coefficients A (g cm^{-1} sec^{-1}) in broken lines.

The greater the distance inland from the sea, the larger becomes the austausch coefficient A. That part of eddy diffusion due to convective processes, initiated by vigorous heating by the ground, comes into play only after a certain time in the air mass coming

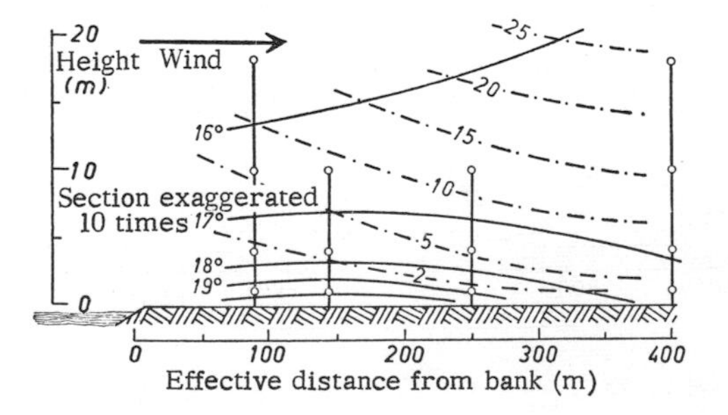

Fig. 129. Temperature and eddy diffusion in an air mass passing from a lake over heated land. (After A. Nyberg and L. Raab; section exaggerated 10 times)

from the sea. Since the ground is heated everywhere roughly to the same extent, throughout the area shown, the heat transferred to the air is restricted to a shallow layer in the part near the shore, where the austausch coefficient is small. This explains the observed result, at first surprising, that the temperature at a height of 1 m inland from the coast first increases, then decreases.

A. Nyberg and L. Raab [474] were able to show in another case that a certain amount of time is required for adjustment when air passes from one type of underlying surface to another. They measured the temperature profile above the unfrozen Indal River near Östersund, central Sweden, by means of thermistors fixed on a 6-m bamboo pole lowered from a bridge over the water. The following readings were obtained on 12 February 1953, when the air temperature was -18°C:

Height (cm):	0	1/2	1	2	20	1200
Temperature (°C):	0.0	-6.5	-11.0	-14.2	-16.3	-18.0

Although air stratification above the warm water of the river was highly unstable, the calculated value of the austausch coefficient was appropriate to extreme stability. The authors write: "Convection cells will not develop in a very small space, they must have adequate room. Observations by the Swedish glider pilot, Mr. Söderholm, lead one to the conclusion that a bay with warm water must have an extent of at least 1/2 km to provide the 'thermals' used by gliders" (translated). These Swedish measurements provide a new aspect of transitional climate. Similar temperature profiles were recorded by A. H. Woodcock and H. Stommel [479] in cold air at 0°C flowing at night over a pool with warm water at 15°C, near Cape Cod.

The strong contrast between snow cover and bare patches of soil has already been made clear in Fig. 115. This feature may often be observed at its best in mountainous regions where radiation is strong, and in the spring fields of flowers may be seen blooming next to deep snow. There is a fine colored photograph in W. G. Kendrew's *Climatology* (3rd ed., 1949, Plate 12) where this contrast has been brilliantly captured.

Over land itself there are many kinds of transitional climates to be found, owing to differences in types of soil and in types and heights of vegetation growing on it.

Figure 130 shows measurements made at four different heights above the ground on a clear, calm night after 22:00, when a slight dew began to be deposited. These readings were taken by W. Knochenhauer [469] over the concrete landing strip (left) and the grass beside it (right) on Hanover airport. The first thing noticed in this figure is that the lines of equal temperature and equal moisture run more vertically than horizontally. Then it is clear that the air close to the building is warm and dry, while it is cool and moist out over the grass.

In the layer nearest the ground the isopleths then bend toward the horizontal in such a way that the contrasts appear to have increased.

Over the grass there is a film of cool moist air, and over the concrete there is warm and dry air. The influence of the latter is stronger (perhaps a wind effect), because it is only after a distance of 30 to 40 m over the grass from the boundary of the two surfaces that equilibrium is established at a given instant between the two influences; there the lines of equal temperature and equal humidity are vertical. There is a region of maximum temperature at a height of about 1 m over the grass, which is also a dry zone, if only weakly identifiable, in the humidity field. It looks as if the warm dry air that has built up over the concrete is flowing out over the grass at this height, and sliding over the cold, moist surface air in the process. Temperature over airfields is of considerable importance in deciding take-off conditions for jet aircraft; therefore it is hoped that this will soon be investigated more closely.

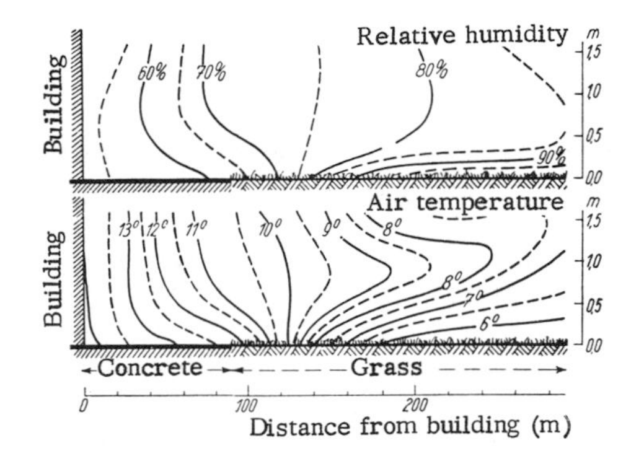

Fig. 130. The transitional climate of a landing strip and meadow on an airfield. (After W. Knochenhauer)

Another instance of this kind of thing, as pointed out by H. Runge [477], is the fact that in thick fog there is usually a shallow layer of air that is clear, or only slightly turbid, over a highway, rising to a height of about 35 cm. Visibility can therefore be improved substantially, when driving at night in fog, if an extra headlamp is fitted low down on an automobile in such a way that the beam points downward at an angle such that scattering of the light upward is avoided. Here also the effect of the dry road surface, contrasting with the surrounding cultivated land, extends upward to about 35 cm.

Every change in type and height of vegetation produces similar transitions in climate. This subject will be returned to in Chapter V. On the edge of woods the peculiarities of local climate are used to advantage in nursing the young plants, in practical forestry. Section 37 deals especially with this kind of transitional climate typical of the boundaries of a stand of timber.

Very often the contrast of microclimate can be seen directly. The following observation was made on 7 March 1950 near Heidelberg by J. Landeck and S. Uhlig [470]. "A field of young grain in which the ground was partly shaded by the small plants and covered with heavy dew, and a piece of fallow land that had been plowed but not yet harrowed, lay under the same strong morning sunshine, the latter warming up more quickly than the former. About 13:00 a cold NW wind set in. The vertical temperature decrease, and therefore also the amount of eddy diffusion, became greater over the unsown land than over the field with crops. Now there formed over the open unplanted soil, because of the vigorous transport of water vapor, a shallow ground fog like sea smoke, which was whirled up into streamers and veils by the turbulent motion of the air, rising from the ground, swaying, swirling, and dancing to a fast tempo, providing a fascinating spectacle for almost an hour." Further details, particularly why this fog never crossed the boundary of the cultivated field, in spite of the wind blowing in that direction, can be found in the original work.

The term transitional climate or boundary climate may appropriately be restricted to places near the boundary line between the various underlying surfaces. Advective elements, originating some distance away, may also have their effects. This is particularly true in areas where topographic inequalities give rise to local winds which may transport the characteristics of the locality for some considerable distance.

I myself directed attention, as long ago as 1929, to the need for inquiring whether in judging microclimate its peculiarities could be explained from the conditions obtained in the locality, or whether foreign influences from the close or more distant surroundings might have to be taken into account as well. In the first case one might speak of independent microclimate, in the second case of dependent microclimate [466]. D. Zoltán [481], who attempted to make a theoretical distinction between these two, called them quite simply nonadvective and advective microclimates. Nowadays the terms native (autochthonous) and foreign have come into use. In any case, it is useful to make a distinction between the two, although we still know very little quantitatively about the advective elements. The distinction is independent of a particular weather situation.

28. Remarks on Evaporation

It was recognized in Sec. 26 that evaporation was the most important of all the factors in the heat budget, after radiation, in the climate of Europe at least. It will now be shown that evaporation is also one of the most important factors in the water economy. Since a certain amount of evaporation also demands a definite amount of heat, it provides a link between the water budget and the heat budget, allowing a cross check of one against the other.

The term evaporation includes two processes of a different nature, which it is often necessary to differentiate. A moist sandy soil or a wet concrete roadway will evaporate, a process that is completely controlled by physical laws; this is the customary interpretation of the term evaporation. When a living plant evaporates water, however, the processes of plant physiology play a part in addition to the purely physical laws. In such cases it is termed transpiration. When water is in short supply, the plant can decrease the amount of transpiration by closing the stomata. At the same time the plant must give off water in order to live, since its food is carried from the soil in the stream of sap. For every gram of solid nourishment obtained, several hundred times the quantity of water must be absorbed, then given off into the air again. For this reason transpiration is sometimes called productive evaporation, and physical evaporation is termed unproductive evaporation by way of contrast. A soil covered by natural vegetation mostly gives off water both from the dead soil and from the living plants. C. W. Thornthwaite suggests that the term evapotranspiration be used when both kinds of evaporation are taking place.

As with heat balance, this study will be started with a short survey of the water budget for the world as a whole. The water supply of the world has been estimated as 1,370,000,000 km^3 in the oceans, 23,000,000 km^3 in the form of ice in the Arctic and Antarctic, 250,000 km^3 in lakes and rivers, and 13,000 km^3 in the atmosphere. This water circulates, as shown by W. Meinardus [491], in a layer hardly extending beyond 15 km above and below the surface of the earth. During the course of a year the exchange of water in this shallow layer amounts to 415,000 km^3, that is, evaporated from the surfaces of the sea and land and returned in the various forms of precipitation; this quantity is equivalent to a mean water depth of 81 cm. If this amount is compared with the water content of the atmosphere, 13,000 km^3, it is seen that all the water in it is turned over every 11 to 12 days (in contrast, a drop of water in the oceans—apart from social intervention—will only be evaporated once every 4100 years).

For the atmosphere as a whole, therefore, the evaporation V will be equal to the precipitation N. From land areas, distinguished by the suffix L, an outflow A is included, due to ground water, springs, and rivers. If the amount of ground water returning is discounted, which will be true if averages are taken for fairly long periods of time, then A_L will equal $N_L - V_L$; and A_L must also be equal to $V_M - N_M$ (the suffix M referring to the sea), otherwise the general level of the sea would change. From these two equations it follows that $N_L + N_M = V_L + V_M$, a result that is easily understood since it means that the water supply of the earth will remain constant only if total evaporation and total precipitation are in equilibrium, and this is in agreement with the observations.

From the work of E. Reichel [494] the water budget of the earth is known today with an accuracy of 3 percent at least, thanks to the possibility of checking heat and water budgets against each other by

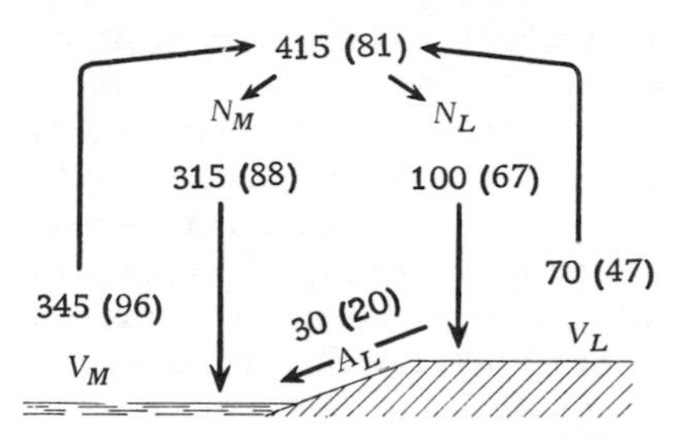

Fig. 131. The hydrologic cycle of the earth in 10^3 km^3 (cm of precipitable water). (After E. Reichel)

means of the evaporation factor V. Figure 131 is based on his results. The figures are mean totals for the year expressed in thousands of cubic kilometers, the numbers in brackets being the more easily visualized depth of water in centimeters. When making comparisons, it should be remembered that the ratio of sea to land surface on earth is 25:11.

These figures are much easier to grasp if one realizes that the average annual precipitation in Germany is roughly the same as

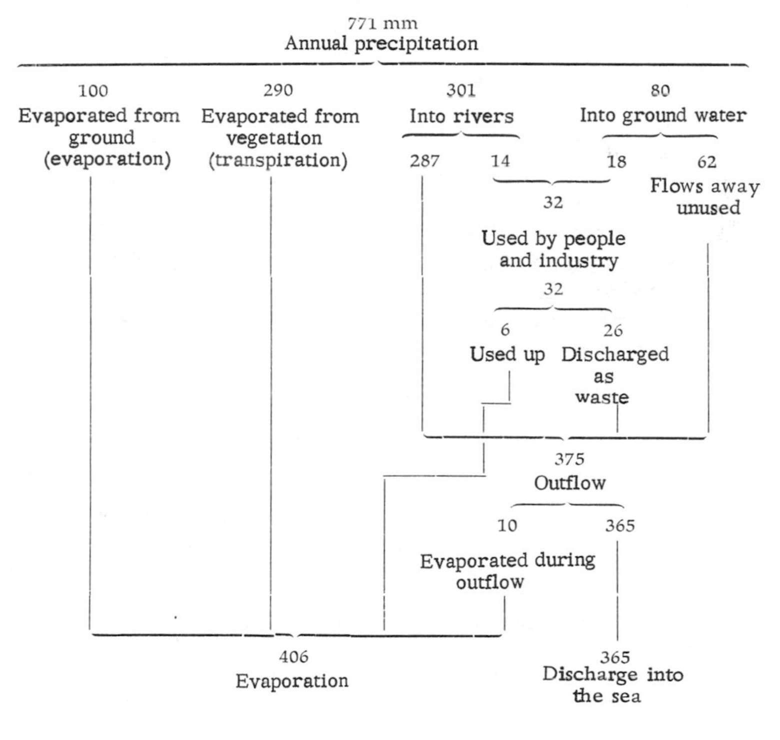

that for the earth as a whole. The diagram on page 251 illustrates the water budget of the German Federal Republic as summarized by R. Keller and S. Clodius [488]. The figures are in millimeters, for 1951, but do in fact illustrate the average situation.

These facts, showing the importance of the part played by evaporation in the water budget, form the foundation for the study of evaporation in its physical aspects, as a factor in microclimate.

Both water and energy are needed for evaporation. If there is a liberal supply of water available, one can speak of potential evaporation. If its supply is restricted, there is an effective evaporation which is less than or, at most, equal to the potential evaporation. Where there are free water surfaces, these two quantities are generally equal. It is essential to realize the difference between these two concepts; it is fundamental, and knowledge of it might have led to the avoidance of much confusion in earlier literature.

The energy required to evaporate 1 g of water is called the latent heat of vaporization r_w (cal g^{-1}). Its dependence on temperature has already been indicated in Sec. 3. With snow and ice the term evaporation is also used, but in place of latent heat we now have the heat of sublimation of 677 cal g^{-1}.

Insight into the process is greatly assisted by following G. Hofmann [484, 485] and subdividing it, according to the source of energy supply, into a radiation fraction V_S and a ventilation fraction V_V. We then have

$$V = V_S + V_V = -r_w \omega_s (S + B) - \frac{r_w \omega_v}{E} a_L (E - e) \qquad (\text{cal cm}^{-2} \text{ min}^{-1}).$$

Here V (cal cm^{-2} min^{-1}) is the heat used in evaporation which corresponds approximately (see p. 10) to a rate of evaporation in millimeters per hour, e (mm-Hg) is the water-vapor pressure, E (mm-Hg) is the saturation vapor pressure of the air, a_L is the heat-transfer coefficient which expresses the amount of heat flowing from the surface to the air per unit area and unit time for a 1-deg temperature difference between them; ω_s and ω_v are coefficients that are functions of the air temperature t only; they

Table 60

Coefficients in equation for the heat used in evaporation.

Coefficient	Air temperature (°C)									
	-20	-10	0	0	5	10	15	20	25	30
	Evaporation from ice			Evaporation from water						
$r_w \omega_s$	0.24	0.31	0.49	0.43	0.51	0.58	0.65	0.70	0.75	0.80
$\frac{r_w \omega_v}{E}$ (deg/mm-Hg)	2.16	1.76	1.30	1.28	1.11	0.94	0.78	0.65	0.53	0.44

have, according to Hofmann, the values given in Table 60. The advantage of the Hofmann method is that the temperature of the evaporating surface does not enter into the discussion. Although this has a great effect on the amount of evaporation, as we shall see later, it is very difficult to measure and therefore remains practically unknown.

Let us assume first that the air is saturated with moisture, so that the saturation deficit $E - e = 0$. Then the fraction $V_V = 0$, and evaporation is determined by $S + B$ only. When the radiation balance is positive, or heat is transferred out of the ground, evaporation will take place in spite of the saturated state of the air, which will lead to the formation of mist or fog in the layers near the ground. As shown in Sec. 26, S is far in excess of B during the day, in general, so that V_S is called for brevity the radiation fraction. At times, however, B plays an important part as well. When the streets are "smoking" after a cold thundershower, this evaporation is

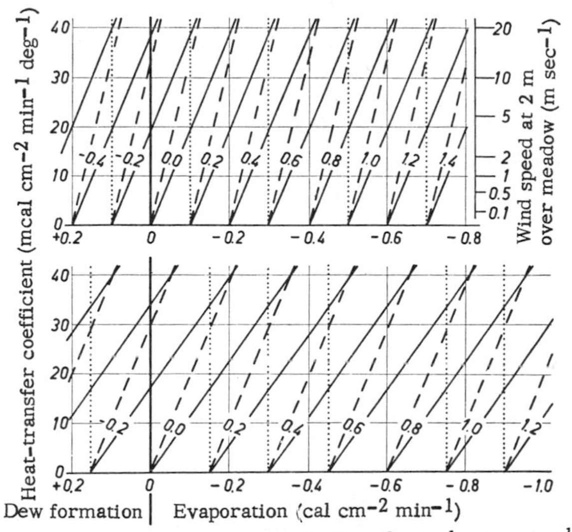

Fig. 132. Dependence of potential evaporation (cal cm^{-2} min^{-1} or mm hr^{-1}) on the balance $S + B$ and the heat transport number a_L for 100-percent (dots), 60-percent (dashes), and 30-percent (full line) humidity and an air temperature of 5°C (upper half) and 25°C (lower half). (After G. Hofmann)

caused by the heat stored in the material of the roadway (B) and the factor of sunshine breaking through again (S) although the air may still be saturated from the thundershower. When it is a question of very warm water in the presence of cold air, the factor B may play the major role, as for example, with hot springs which "steam" or with open sea in the Arctic ice, "sea smoke," or when warm ocean currents flow into cold zones.

Figure 132 is designed to help in understanding the connection between the quantities involved. Based on the Hofmann equation, it allows the amount of evaporation (abscissa) to be determined when the heat-transfer coefficient a_L (ordinate), the balance $S+B$ (italic numbers within the figure), and the relative humidity are given. The upper half of the diagram is for air temperature 5°C and the lower for 25°C. The range selected for a_L is the values it usually has over a level surface. The coefficient a_L increases approximately as the square root of windspeed u; it also depends to a considerable extent on the form and size of the evaporating body, increasing as the size diminishes (see Sec. 29). On the right-hand margin of Fig. 132 a wind scale has been added (m sec^{-1}) to show approximately the connection between a_L for the surface of a meadow and the wind speed measured at a height of 2 m above the field.

If we consider the case of saturated air once again, $E-e=0$ and $V_V=0$. This corresponds in the diagram to $a_L=0$ (although the actual value of $a_L=0$ can never be zero). The total evaporation V then depends only on $(S+B)$. Therefore the 100-percent humidity line (dotted) corresponding to each $(S + B)$ number (italic) runs parallel to the ordinate (a gap is left where figures coincide with the coordinate lines). The factor $r_{w}\omega_s$ indicates the extent to which the radiation balance may be utilized by evaporation. As the figures in Table 60 have already shown, this increases with rising temperature. This is why in the lower diagram, for 25°C, the isopleths of equal $(S+B)$ are farther apart than in the upper part, for 5°C.

If the air is not completely saturated, the fraction V_V increases the amount of evaporation to an extent that is proportional to the extent of ventilation, and hence to a_L, and that will be greater the drier the air is. This is shown by the broken lines for 60 percent relative humidity, and the continuous lines for 30 percent relative humidity in the diagram, which of course intersect the dotted 100-percent line on the horizontal axis, for $a_L = 0$.

The quantity V may also become positive; then there will be condensation and the formation of dew. The thickened vertical line for $V=0$ separates the two possibilities. From Fig. 132, for example, with a nocturnal radiation balance $S=-0.2$ cal cm^{-2} min^{-1} $(B=0)$ for temperature 25°C and humidity 60 percent dew will first begin to form when a_L falls below 30 mcal cm^{-2} min^{-1} deg^{-1}. Figure 132 therefore indicates the conditions necessary for the deposition of dew, since the two processes of evaporation and condensation differ only in sign.

A mere glance at the relation displayed in Fig. 132 shows the overwhelming importance of radiation for these two processes. If a value for V that is proportional to the saturation deficit $E-e$ is substituted in most evaporation formulas without taking radiation into account in any way, it will lead to values that are only partially useful, the constants having been established empirically, since in meteorology there is a close correlation among all factors involved.

If the temperature of the evaporating surface is known, and hence the saturation vapor pressure for this temperature, which will be given the symbol E', we can proceed by using the old Dalton evaporation formula, without knowing the value of the individual factors in the heat balance:

$$-V = c\,(E' - e) \qquad (\text{mm hr}^{-1}).$$

The quantity $E' - e$ is also often called the saturation deficit, although as a rule it is substantially different from $E - e$, which has caused some confusion. The constant c was determined empirically from measurements of the potential evaporation by means of evaporimeters (atmometers). It has the following values, depending on the wind speed u:

u (m sec^{-1}):	0.1	0.5	1	2	5	10
c (mm hr^{-1} mm-Hg^{-1}):	0.007	0.016	0.023	0.032	0.050	0.071

The Dalton formula helps us to assess the importance of the temperature difference between the ground and the air. Table 61 shows the

Table 61

Rate of evaporation (mm hr^{-1}) from a surface at different temperatures, as a function of air temperature and relative humidity.

Air temperature, t (°C)	Temperature (°C) of evaporating surface is—							
	$t-3$				t			
	100%	80%	60%	40%	100%	80%	60%	40%
0	Condensation	0.00	0.03	0.06	0.00	0.03	0.06	0.09
10		0.01	0.06	0.12	0.00	0.06	0.12	0.18
20		0.02	0.13	0.24	0.00	0.11	0.22	0.34
30		0.04	0.24	0.45	0.00	0.20	0.41	0.61
	$t+3$				$t+6$			
0	0.04	0.06	0.09	0.12	0.08	0.11	0.14	0.17
10	0.07	0.12	0.18	0.24	0.14	0.20	0.26	0.32
20	0.11	0.23	0.34	0.45	0.25	0.36	0.47	0.58
30	0.19	0.39	0.60	0.80	0.41	0.61	0.82	1.02

rate of evaporation in millimeters per hour for four different temperature differences, as a function of air temperature and relative humidity. The figures show how much the potential evaporation increases when the evaporating surface becomes warmer than the layer of air around it, whether this heat has been caused by the absorption of radiation or had its origin in the water which is evaporating.

It is not easy to determine the value of the potential evaporation, and the difficulties multiply when an attempt is made to measure the effective evaporation. The problem has been approached from several different angles, but it is not possible to discuss them in detail here. Measurements made by lysimeters, in so far as they are of interest to microclimatology, will be given in Sec. 30. Those who wish to become informed on this problem should read H. L. Penman [493] and I. C. McIlroy [490].

Methods have been developed in Australia recently for reducing the amount of evaporation from open water surfaces by artificial means. These methods, which promise to effect great savings of water in reservoirs, depend on covering the surface of the water with a thin layer of a substance that will hinder, to a great extent, the passage of water into the air. The substance requires a high viscosity so that it can resist the action of wind and waves, yet must not interfere with water biology and be economically practical. Films of oil reduce evaporation but are too easily broken up. The most effective is cetylalcohol, which forms a film one molecule thick on the surface of the water.

According to W. W. Mansfield [489], this monomolecular layer offers such resistance to the escape of water molecules from the surface into the air that evaporation is reduced by 75 percent. However, the reduction of V brings about changes in the heat budget of the water surface such that its temperature must rise substantially. It follows that $E' - e$ and V will increase, so that the influence of the protective layer will be broken down to some extent again. In experiments with flat basins of water at 30°C surface temperature, and with relative humidity of the air at 23 percent, and a wind of 3 m sec^{-1}, the expected reduction of evaporation by about 75 percent was attained in the absence of sunshine. In the presence of midday sunshine the reduction was only about 30 percent. If, however, the radiation term in evaporation was the major one, then, with a wind of only 0.6 m sec^{-1}, the reduction was reduced again to only a few percent. More favorable results were obtained with deeper water, when the extra heat could spread over a great mass of water. In the arid climate of Australia, an average reduction of about 50 percent could be substantial. According to a more recent report by J. L. Monteith [492], 37 percent of water was saved in large-scale investigations over a 14-week period in Australian reservoirs, at a cost of less than 1 cent per hectoliter of water saved.

So that a monomolecular layer may be maintained over a large surface in constant movement as a result of wind and waves, there must be a small excess of the substance, and this is also required because of continual losses at the shores. Altogether the requirement is about 1 g of cetylalcohol for every 3 m^2 of surface, contained in small spheres of 2 to 4 mm diameter in fine-mesh baskets over the surface. The small baskets near the shore are so arranged that they can be lifted out of the water when there are onshore winds, to avoid unnecessary loss of material.

This procedure, which is now being tested in all arid areas in the world, has been mentioned here since it shows that it is possible to alter the heat and water budgets of a surface under natural conditions by making an artificial alteration in the factor *V*. There are very few such opportunities in microclimatology.

CHAPTER V

THE AIR LAYER NEAR PLANT-COVERED GROUND

Vegetation occupies the space between the earth's surface and the atmosphere. A continuous plant cover not only takes up space, but by virtue of its properties forms a transition zone, because the individual organs of the plants, such as leaves, needles, twigs, and branches, behave like solid ground, absorbing and emitting radiation, evaporating, and playing their part in the exchange of heat with the surrounding air. However, the air is still able to circulate within the plant cover more or less freely. Thus vegetation forms a new component part of the air layer near the ground.

The total mass of plant components taking part in radiative, conductive, and moisture exchanges with the environment has a remarkably small thermal capacity. The thick, 5-m high stand of pine trees for which A. Baumgartner worked out the heat budget (Fig. 120) was shown by K. Mauerer [604], after a close survey of the stand, to contain a plant mass equivalent to a layer of wood only 19 mm thick. The heat capacity by volume was given in Table 10 as $(\rho c)_m$ = 0.1-0.2 cal cm^{-3} deg^{-1} for dry wood. With double this value, 0.4, to allow for the water content of the living wood, a figure of 0.8 cal cm^{-2} deg^{-1} is obtained for the thermal capacity, expressed per unit area. The thermal capacity of the air within the stand is about 0.2 cal cm^{-2} deg^{-1}. It follows that the thermal capacity of a pine forest is of about the same order of magnitude as that of air itself, in spite of the amount of space it occupies.

Plant parts differ from solid ground, however, in that short-wavelength radiation passes through them. Below a closed canopy of foliage in a wood there is a subdued greenish light. Besides, plant organs are not parts of a dead physical system; the process of living means playing an active part in relations with the environment, as in the orientation of leaves and flowers toward the direction of incident light, or the closing of leaf stomata to reduce evaporation.

Plants, being anchored to the place where they grow, are dependent on the climatic conditions of the locality; these may be advantageous or injurious to the life of the plant. On the other hand, the effect of plants on the microclimate of the locality increases as the plants grow, with increasing development of neighboring plants. The sum total of a number of influences, at first small, gradually acquires increasing importance. The justifiable question whether an earth completely covered with forests would have a different climate from that of an earth without any woods shows how extensively these reciprocal influences may extend beyond the locality proper.

This chapter, dealing with the influence of vegetation cover on the climate near the ground, contains a number of problems of great scientific interest, and is of considerable practical importance. Horticulture, farming, and forestry are all concerned with the growth of plants that are particularly sensitive in their early stages to the influence of climate and weather. In highly cultivated countries the "level ground, free of vegetation" that has been discussed up to now is indeed exceptional. This new discussion will therefore provide closer contact with things as they actually are in nature.

Before discussing the mutual influence of plants and local climate, we must first consider how plant organs enter into the heat budget, and how plant temperatures become adjusted to the temperatures of the ground and the air.

29. Heat Balance and Temperature of Plant Components

When radiation falls on a leaf or a needle, part of it is reflected from the surface and is expressed as a percentage of the incident radiation by the reflection index or albedo R (see p. 12). Another part passes through the leaf or needle, the quantity emerging again being expressed as a percentage D known as the penetrability coefficient. The remainder, A percent, is absorbed and therefore converted into heat which raises the temperature of the leaf. In all circumstances $R + D + A = 100$ percent.

From Table 3 it will be seen that for vegetation R lies between 5 and 30 percent in the range of short-wavelength solar and sky radiation. Exceptionally, the albedo may rise to 60 percent at the lighter surfaces of variegated leaves. The value of R is smaller in the ultraviolet range; K. Büttner and E. Sutter [31] found only 2 percent for heather on sand dunes.

If photographs of the countryside are taken on film that is sensitive to infrared radiation, trees appear very light, almost white, which was demonstrated by E. v. Angerer [497] as early as 1930. In the long-wavelength range, therefore, leaves and needles reflect more strongly, with an albedo of 35 to 50 percent, in contrast with solid ground. Figure 133 shows the results of F. Sauberer's measurements [517] of the amount of reflection from a concrete surface (dashes) and from a meadow (solid line). With the first surface, reflection depends very little on wavelength, but the meadow has a weak maximum in green light (500 mμ), then a high value of 40 percent and more in the range from 750 to 1000 mμ. For the long-wavelength ranges that have to be considered in nocturnal radiation, *R* decreases sharply (see Fig. 1). With 2.4μ K. Egle [501] found 5 to 16 percent for the green leaves of five different types of plant; with 10 μ—approximately that of the maximum of nocturnal radiation—G. Falckenberg [502] measured 4 percent for pine needles and 5 percent for green pelargonium leaves. It follows that a vegetation cover, like solid ground, behaves almost as a black body at night.

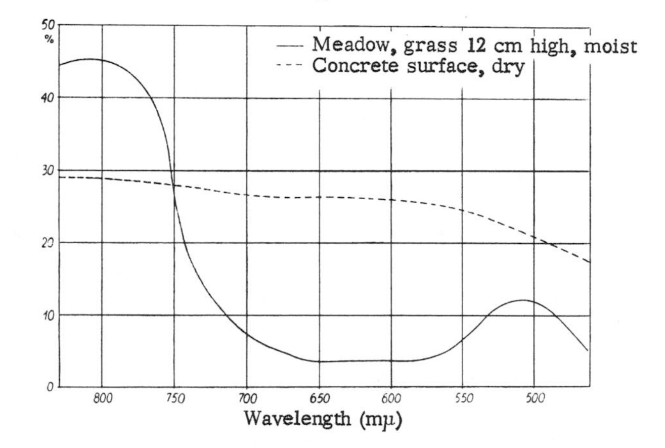

Fig. 133. Change in albedo *R* with decreasing wavelength for meadow and concrete surfaces. (After F. Sauberer)

While the reflection is usually diffuse, there may also be specular reflection at times, as is evident from the bright appearance of the leaves of some Mediterranean evergreens. The radiation that penetrates through the leaves is, however, always diffuse. The penetrability *D*, often called transmissivity by physicists and meterologists, and also known to botanists as diathermance, varies with the type of plant within wide limits, as has been shown by H. Schanderl and W. Kaempfert [520] in a comprehensive work, and is also affected by the movement of chloroplasts. It also depends on wavelength. Figure 134 illustrates the measurements made by F. Sauberer [517, 518] with a young red beech leaf (full line), a primrose (broken line), and a hellebore (dot, dash). There is a very striking increase in penetrability from about 10 percent to four or six times that value when the wavelength passes beyond 700 mμ. The marked maximum is maintained until about 1500 mμ, after which it declines slowly. We would therefore describe the light within woods as preponderantly infrared if our eyes were as sensitive to it as they are to the weak maximum in the yellow-green range (550 to 580 mμ). Between 1000 and 2400 mμ K. Egle [501] obtained D values of 25 to 47 percent, in the investigation mentioned above.

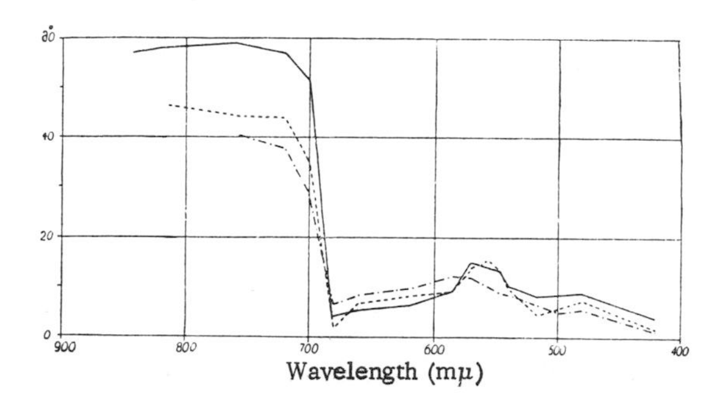

Fig. 134. Change in penetrability D with decreasing wavelength, for various types of leaf. (After F. Sauberer)

According to Sauberer, it makes a difference whether the upper or the lower sides of the leaves are exposed to radiation. A leaf of white poplar, for example, gave D = 22 percent for the upper surface and D = 15 percent for the lower. This difference changes with changing wavelength. There are also big differences from leaf to leaf of the same plant, arising from the many differences in the separate lives of individual parts. K. Raschke [514] found a mean value of D = 10 percent for the leaves of *Alocasia indica*, with values for individual leaves varying from 4 percent to 29 percent.

Penetrability produces a spectrally filtered green light in the woods, which is spoken of as green shade. According to A. Seybold [521], this must be distinguished from the blue shadow found perhaps on the north side of a wall, in which the light is the blue light that is most easily scattered. The distinction between green and blue shadows is important botanically.

In the short-wavelength radiation budget, therefore, $A = 100 - R - D$ percent is retained inside the leaf and used to heat the leaf, or utilized in some other way such as evaporation. The leaf behaves practically as a black body toward long-wavelength radiation, and therefore absorbs both the counterradiation of the sky and terrestrial radiation that falls on it from below, and in turn it emits the long-wavelength radiation appropriate to its own temperature. To obtain a correct quantitative appreciation of this radiation exchange of leaves, the following example is taken from the researches of K. Raschke [514], which was obtained by calculation and confirmed experimentally. The following radiation exchange was determined, using a horizontally placed *Alocasia indica* leaf with R = 21 percent and D = 10 percent, under Indian climate (all figures except temperature are in cal cm^{-2} min^{-1}, in which cm^2 refers to the upper and lower surfaces of the leaf):

	By Day	By Night
Temperature of ground surface	65°C	35°C
Temperature of leaf	46°C	34°C
Short-wavelength radiation budget		
From direct solar radiation (I)	+0,80	—
From sky radiation (H)	+0,15	—
From radiation reflected by the ground (R)	+0,15	—
There remains in the leaf	+1,10	—
Long-wavelength radiation budget		
Incoming radiation from the atmosphere (G)	+0.63	+0.54
Incoming radiation from the ground	+0,98	+0.67
Radiation emitted by the leaf	−1.61	−1.37
Hence the long-wavelength balance is	0,00	−0.18
Total calculated radiation balance	+ 1,10	−0.18
Measured values differed from this by	-6%	5%

A plant can influence the radiation balance of its own leaves. This is not merely a question of the structure of the plant, by which it avoids excessive heating by corrugations on the leaves, by a field of thorns in cacti, or by pubescence. Much more can be read about this in the work of the botanist B. Huber [503]. It is also a question of the ability of leaves to reduce the amount of radiation falling on them by their own movement. There are leaves that take up a position of "daytime sleeping" when irradiated for 15 min by a sun lamp. This is done by movement of the peculiar joints of the leaf so as to reduce the area exposed to the sun. O. W. Kessler and H. Schanderl [506] have published photographs of white melilot with its leaves in various attitudes, which are appropriate to this study. In dry Mediterranean areas and in hot prairies there is "a peculiar change in the physiognomy of the landscape at certain hours of the day" in which such reorientation can be recognized. The compass plant illustrates this method of regulating the radiation balance (see Sec. 41).

The factor B corresponds in leaves to the heat supplied through the stalk of the leaf and the storage of heat, dependent on the thermal capacity of the leaf. These can be neglected.

The thermal capacity of a normal leaf is very small; for example, in the fairly fleshy leaves of *Alocasia* just mentioned it is only 0.02 cal cm^{-2} deg^{-1}. Leaves therefore adjust themselves within a few seconds to changes in their surroundings. The amount by which the temperature of a leaf rises above that of the air when there is a positive radiation balance, or the amount of cooling with a negative radiation balance, depends on the other two factors in the heat economy, namely L and V.

The exchange of heat with the surrounding air (L) is determined, as we have already seen in Sec. 28, by the heat-transfer coefficient a_L. It is $L = -a_L(t_l - t_o)$, where t_l is the temperature of the leaf surface, and t_o is the temperature of its environment. A boundary layer a few millimeters thick is formed at the surface of the leaf (see pp. 34f), in which t_l gradually changes to t_o. When the air is completely still, a_L has a value of about 10 mcal cm^{-2} min^{-1} deg^{-1}. It increases in a way roughly proportional to the square root of the wind speed, and depends on the size and shape of the leaf and the position on the leaf at which measurements are made. The peak values, which are reached with small leaves and high winds at exposed parts of the leaves, amount to about 300 mcal cm^{-2} min^{-1} deg^{-1}.

In assessing the factor V, the leaf will first be considered as a physical object that takes part in evaporation but not in transpiration. The leaf is thought of as having its entire surface wet, and the potential evaporation is evaluated as in Sec. 28. The physiologic processes are accounted for by the introduction of a water-coverage factor, following the method of K. Raschke [514]. This allows for the fact that transpiration takes place only through the open stomata, which are present in varying numbers in different types of leaves and which can be regulated in size. The water-coverage factor is the fraction to which evaporation from the whole of a theoretical leaf is reduced in a natural leaf. It varies not only with the type of plant, but with the time of day and weather conditions. Values ascertained vary between 0.004 and 0.6. In a more recent work, K. Raschke [515] has made some improvements in his leaf model.

The temperature of a plant is therefore usually determined by the temperature of the air surrounding it. It depends on the three factors S, L, and V.

The diurnal variations of the elements involved in the heat budget of leaves are shown in Fig. 135, from measurements by K. Raschke [513, 514], who has made the best investigations available so far. Heat taken in (above) and heat given out (below) are placed in juxtaposition, and are of equal magnitude on either side, since the thermal capacity of a leaf is negligible. The measurements are for a horizontally placed leaf of *Alocasia indica* in Poona (18°31′N) on 7 and 8 May 1954, a clear day. The curves are drawn through half-hour average values, so that short-term variations are eliminated. The legend on the diagram explains what is happening, after all that has been said about it so far.

It is more interesting to note the differences between the upper and lower surfaces, when the much greater transpiration from the lower surface is easily recognized.

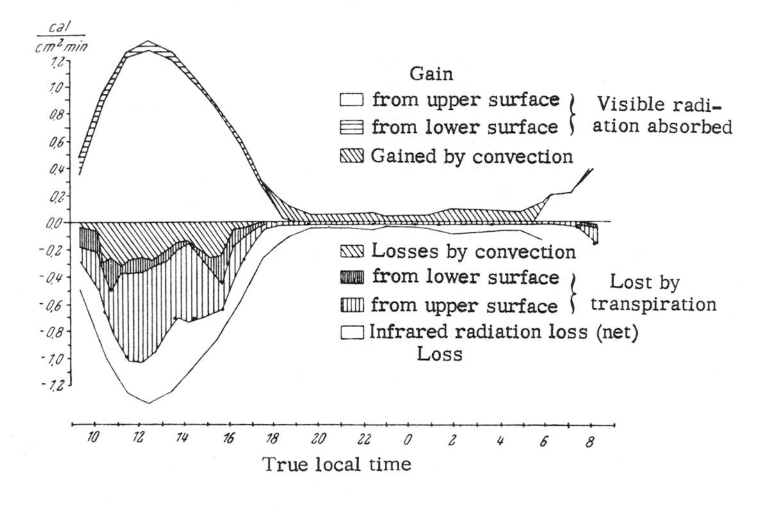

Fig. 135. Diurnal variation of the heat budget of a leaf of *Alocasia indica*. (From research K. Raschke)

Since L, which is labeled as convection in Fig. 135, appears as a heat loss during the day and a heat gain at night, it follows that during the day the temperature of the leaf must be higher, and at night lower, than the air temperature. Figure 136 gives an idea of how much the temperature difference may amount to between parts of a plant and the surrounding air in the climate of Central Europe. This also is from the work of K. Raschke [514] and gives the excess (+) or deficit (-) of leaf temperature as a function of α_L (shown on a logarithmic scale) and of the radiation balance S. The day figures above the zero line are based on an air temperature of 22°C, 61-percent relative humidity, water-vapor pressure 12 mm-Hg (summer day), and a water-coverage factor of 0.03. The night figures are based on 13.2°C, 95 percent, 11.4 mm-Hg, and 0.01.

Even during the day, if there is a brisk wind (large α_L) and a small amount of incoming radiation (cloudy skies), transpiration may reduce the leaf temperature a few tenths of a degree below air temperature. On the other hand, with strong radiation and a calm wind, the temperature excess may be more than 10 deg, although the rise in temperature will cause increased transpiration which in turn will produce a noticeable cooling. Since the amount of radiation exchange is small during the night, the fall of leaf temperature below that of the air seldom exceeds 2 deg.

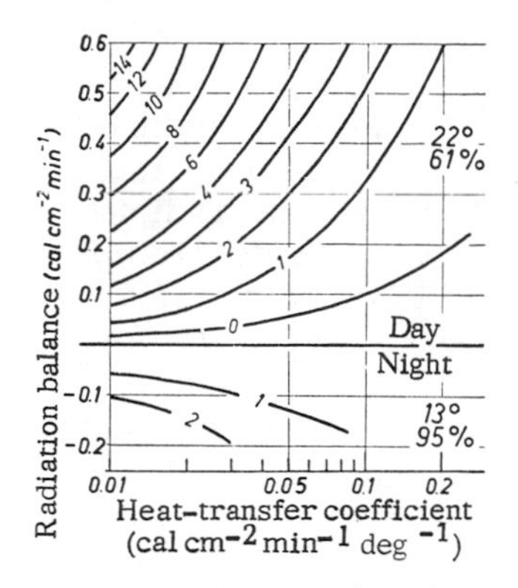

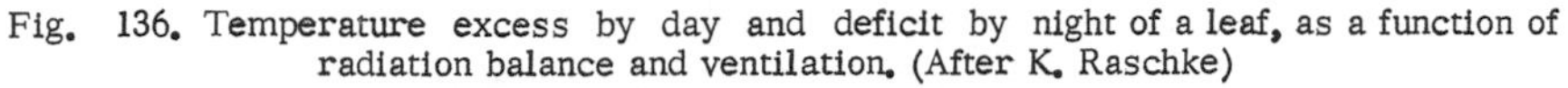
Fig. 136. Temperature excess by day and deficit by night of a leaf, as a function of radiation balance and ventilation. (After K. Raschke)

A good rule, due to B. Huber [503], states that the temperature of the leaf or needle will lie between that of the surrounding air and that of the ground surface, which is either receiving or emitting radiation. The larger, fleshier, and drier the leaf, the stronger the radiation; and the smaller the exchanges with the air are, the more closely will the plant approach the temperature of the ground surface. This can be seen directly from the magnitude of the diurnal temperature variation. K. Raschke found this to be 7 to 8 deg higher in the *Alocasia* leaf than in the air. The twigs of cherry blossoms were found by K. Kunii [507] to fluctuate by 2 to 22 deg during the day in Japan, whereas the variation in the air was only 1 to 13 deg. In Fig. 78 it was shown that, on a particular July day near Palermo, A. Vaupel recorded air temperatures that did not exceed 30°C while the ground temperature rose above 60°C. Between these two are found the temperatures of the *Opuntia ficus indica*, which was fixed in a horizontal position 8 cm above the ground. Recordings made in a flax sprout are shown in two neighboring curves; the full line with shaded circles is for the upper surface, while the broken curve with open circles and crosses is for the interior, which heats up more slowly in the morning but retains its heat longer than the upper surface in the afternoon. The work of B. Huber [504] contains many similar temperature recordings for plants. K. Takasu [524] has published recently, in Japan, a series of rapid recordings which give details of individual processes. An example is shown in Fig. 137, which is the record of observations made on the 2720-m summit of Mount Shirouma during the forenoon of 16 July 1943.

The high elevation, and the presence of light fog patches around the summit at the time the readings were taken, account for the large variations and the high peak values of radiation intensity, shown in the lowest curve *e*. The temperature of the lower surface of a *Lagotis glauca* leaf, a perennial plant with leaves near the ground, was measured at a height of 5 cm above the ground by means of a radiation-compensated thermoneedle, the recording being shown in curve *b*. Close analysis of the two curves shows that the leaf temperature responds more rapidly to a sudden increase than to a decrease in solar radiation. The air immediately below the leaf (*c*) shows a greater temperature instability than does the leaf itself, but follows, rather surprisingly, the fluctuations of leaf temperature very closely, whereas air temperature at a height of 1 m (*d*) in all its short-term unsteady movement balances to a large extent the variations in radiation. The temperature of the ground surface (*a*) is highest and shows the greatest lag in responding, as was expected.

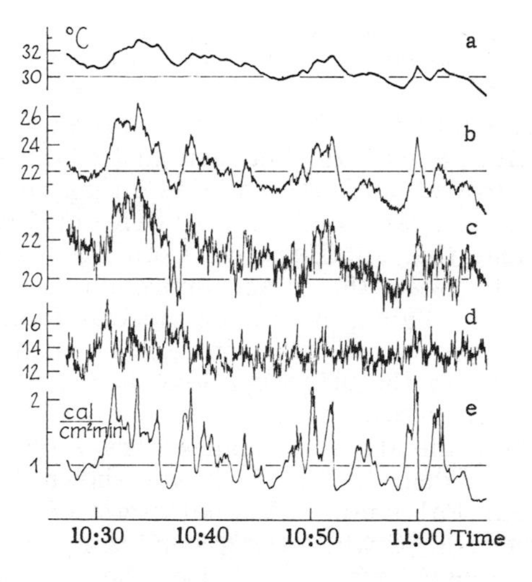

Fig. 137. Temperature at ground surface (a), in a leaf (b), and in the surrounding air (c, d) under radiation conditions on a high mountain (e) in Japan. (After K. Takasu)

Lichens, which lie directly on the ground or on stones and can dry out, may reach temperatures of 70°C. This was investigated experimentally by O. L. Lange [508, 509]. On the Kaiserstuhl (350 m elevation) at Freiburg in the Black Forest, for example, he obtained readings in a *Cladonia furcata,* for three branches at different heights, of 42°C at 4 cm from the ground, 46°C at 1 cm, and 66°C on the thallus covering the ground. The color of these lichens, which may vary considerably, plays an important part. In greenhouses, when the air becomes saturated with water vapor and *V* is consequently low, plants may rise to lethally high temperatures.

The temperatures shown in Fig. 138 were not measured in a leaf, but in the bark of a twig of green alder by G. and P. Michaelis [511]. There was still snow in the tiny Walser Valley in the Allgäu, at a height of 1670 m on 16 March 1933. The ordinate 0 indicates the position at which the twig emerged from the snow. With air temperatures between -2° and +4°C, the temperature of the bark reached 30°C; however, this maximum was not at the snow surface but some 15 cm above it, as a result of the cooling effect of the melting snow. At night the minimum was always found at the snow surface, which was then emitting radiation. Details of temperature distribution in tree trunks are given in Sec. 41.

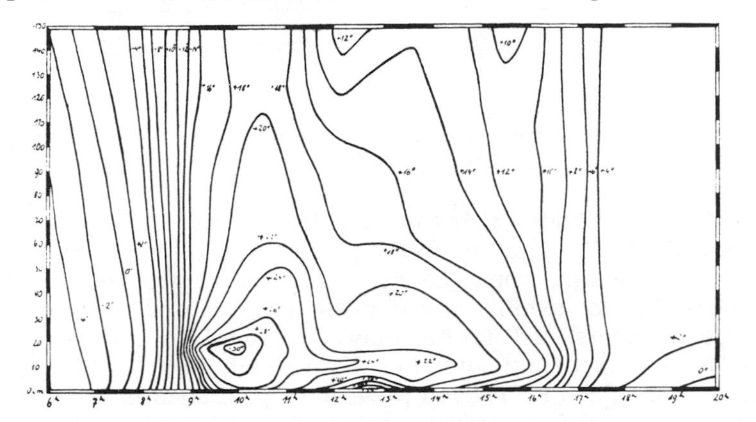

Fig. 138. Temperature variations in a twig of green alder, projecting through snow, on a March day. (After G. and P. Michaelis)

The heat economy of spherical fruits (apples, tomatoes) was explored by Y. Nakagawa [512], who established that out of the solar and sky radiation absorbed only 3 to 7 percent was stored up, while 6 to 12 percent was given off in long-wavelength radiation and 82 to 90 percent was transferred to the surrounding air (this last figure must include the factor *V*, which has not been mentioned).

A. Büdel [499] investigated the temperature variation in the blooms of spring flowers, for the benefit of beekeepers, since the secretion of nectar depends on the temperature of the flower. Figure 139 gives the following information for a sunny day: the dotted line is the temperature recorded by a thermoelement inserted between the petals of the dandelion flower (*Taraxacum officinale*); the crosses give the air temperature in the immediate neighborhood of the flower; and the broken line records the air temperature at a height of 1 m. The rapid response of this comparatively large flower is striking; the large fluctuations are due to shielding by the neighboring trees. With midday temperatures of 17°C, the flower temperature rises to 30°C.

Plant temperatures are of particular interest when air temperature approaches the frost point and there is danger of freezing. It is not possible to go into botanical research on frost resistance here, but there are a few significant meteriologic aspects to which attention should be directed.

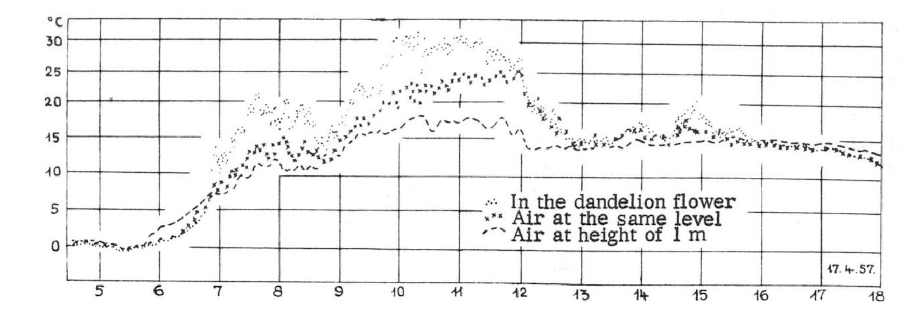

Fig. 139. Temperature variations in a flower of *Taraxacum officinale* on a spring day. (After A. Büdel)

Living winter buds of ***Anemone pulsatilla*** and ***Helleborus niger*** were found by K. Hummel [505] to be up to 10 deg warmer than the surrounding air, when the air temperature dropped to -12°C. Dead buds did not exhibit any excess temperature. In addition to this effect, plants and fruits may supercool to a considerable extent before freezing sets in. H. Ullrich and A. Mäde [526], S. Suzuki [523], and K. Takasu [524] have published measurements of the temperature variation inside leaves, which show clearly the sudden increase in temperature when latent heat is released at the onset of freezing. In both of the Japanese papers, the increase is subdivided on occasion into two parts, with an interval of 2 to 4 min between. The first increase results from freezing of the dew on the leaf, and the second from freezing of the sap in the cells.

When, in breeding experiments, the flowers are covered with protective pollenproof bags, they are subjected to quite a different kind of microclimate. This was first investigated by N. Weger [527]. Figure 140 shows the average daily temperature variation for five sunny days in May and June 1937 inside four different types of bag. Compared with air temperature (heavy continuous curve), the interior temperatures reach an excess of 15 deg. The fact that the flowers respond to radiation in the same way as the plants in a greenhouse is shown by the maximum being reached earlier than in the open air. This is in agreement with the observation by N. Weger, W. Herbst, and C. F. Rudloff [528] that enclosed buds on a pear tree blossomed 2 to 4 days earlier. At night the temperature falls 1 to 2 deg below air temperature.

This problem has been considered recently from every aspect by E. Rohmeder and G. Eisenhut [516]. They also found excess temperatures, at midday, of up to 18 deg, and deficits of even more than 5 deg at night. Inside the coverings themselves there was a marked thermal stratification; in the middle of the day the side receiving radiation was 6 deg warmer than the opposite side,

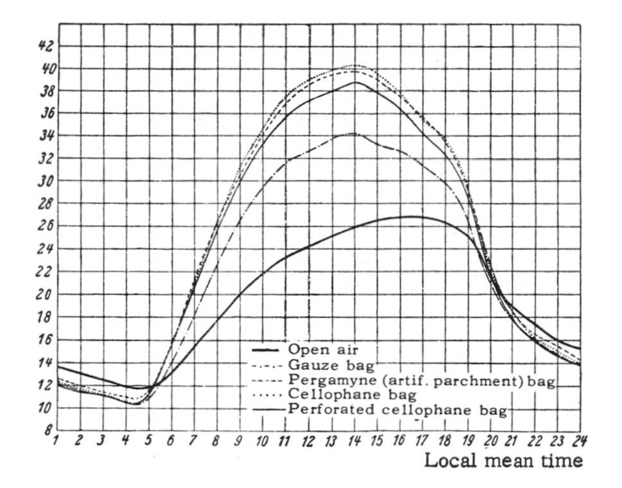

Fig. 140. Extreme temperature variation in flowers enclosed in dusting bags. (After N. Weger)

and at night the side radiating was 2 deg colder, which explains the uneven nature of the damage often found inside the protective covering. Inside the bags during the day the air is saturated with water vapor, as in a forcing house, with a relative humidity about 35 percent higher than outside.

The size of the dusting bag is of very little influence. The intensity of light in the interior varies between 95 percent (polyethylene) and 54 percent (silk) of the light outside, but is always sufficient to satisfy requirements. The material of which it is made, on the other hand, has a marked influence, as may be seen from Fig. 140. Polyethylene bags are cheap and the flowers can be seen easily inside them, but they prevent any diffusion of gases; condensation water collects inside them, up to 800 g in 14 days, and nocturnal cooling is excessive. Woven bags are more resistant to wear and tear, the mesh of the weave can be adjusted for pollen size, and gases are able to diffuse through the texture; variation of temperature is thus moderated, and an inspection window can be fitted. Imitation parchment bags are cheap, but easily damaged, and show extremes of microclimate in the absence of any gaseous diffusion.

There are means whereby the often lethal microclimate within these bags can be made less extreme. It may be seen from Fig. 140 for the two cellophane bags that small perforations help, provided the experiment allows this to be done. A well-tried method, due to Rohmeder and Eisenhut, is to attach the protective bag rigidly to a wire basket, 10 to 15 cm larger in diameter, with a shade over the side toward the sun. Excess temperatures are then reduced by half and the deficit raised somewhat. W. Tranquillini [525] found, in experiments with glass bowls in high Alpine radiation conditions, that the temperature excess inside was 25 deg (maximum temperature of the leaves 52°C). These conditions could be alleviated by the introduction of a Schott heat filter BG 21 without interfering with the natural processes of the plant, or by artificial ventilation with a current of air of speed at least 0.5 cm sec^{-1}.

The danger of frost in the bags can be delayed for half an hour by wetting them, since the latent heat released when the bag freezes delays freezing inside the bag for about that time. Rohmeder and Eisenhut were able to save some valuable breeding material by introducing tiny electric heaters (weight 30 g) into the bags. The fins of these heaters had a maximum temperature of 40°C and were able to protect flowers, without any difficulty, down to - 8°C outside temperature.

30. Radiation, Diffusion, and Evaporation in a Low Plant Cover

Having gained some knowledge of the heat balance of various plant organs, we can consider some general features that determine the microclimate of plant colonies, after which individual types of growth can be discussed.

The total surface area of vegetation growing on a meadow is 20 to 40 times the area of the ground on which it grows. This magnification of surface area by vegetation has no influence on the radiation budget. In a given weather situation, bare ground and vegetation-covered soil receive the same radiation $I + H + G$ per square meter (see p. 13). Both types of surface also give off, per square meter, the same amount of radiation by day as by night, provided their surface temperatures are equal. Only differences in albedo for long- and short-wavelength radiation, and differences in surface temperature, not the increase in surface area as such, may bring about a difference in the radiation balance between bare and plant-covered soil.

There is, however, a basic difference in the vertical distribution of the radiation gained or lost. The amounts of solar radiation (as shown in Fig. 141) received by the bare ground and the plant surfaces are naturally equal. At half the height of the growth on a meadow of lush grass and *Dactylis glomerata,* in which measurements were made by A. Ångström [529], the value of the radiation was somewhat reduced, because the tips of the grass blades had absorbed a small part of the incident radiation. The amount of absorption increased farther down where the grass became denser and only one-fifth of the original radiation reached the ground.

A similar distribution is found with outgoing nocturnal radiation. Since the soil has a grass cover which, to a large extent, has the same temperature and therefore emits the same radiation as the soil, the amount lost by radiation is low near the ground.

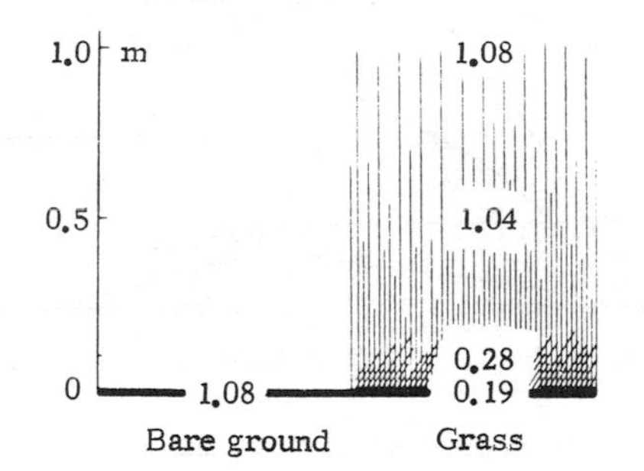

Fig. 141. Absorption of radiation in a meadow. (After A. Ångström)

The greater the height within the grass cover, the less is the position shielded from the open night sky by other parts of plants. Therefore the outgoing radiation increases until finally, at the upper surface of the growth, it reaches the value equivalent to that of bare ground at the same temperature.

It appears, then, that individual parts of plants receive more or less solar radiation, and lose greater or smaller quantities of heat by radiation, depending on the accidental amount of shielding that results from the density and form of the vegetation. There must arise, in consequence, substantial differences in temperature between individual parts of plants within the vegetation. In the calm that normally prevails inside a vegetation cover, an attempt is made to reduce these temperature differences by a lively exchange of long-wavelength radiation, in addition to turbulent mixing.

The quantity of light penetrating to the ground depends also on the density and structure of the plant cover. Figure 142, from F. Sauberer [517], shows the diurnal variation of visible light as the soil underneath various types of plants, in comparison with bare ground. It is easy to distinguish the different degrees of shielding provided by summer barley, only 12 to 15 cm high, winter rye nearly 1 m high, and clover with its dense umbrella of leaves. The quantity of light reaching the ground is of importance for the growth or the suppression of weeds. In pea fields of all kinds K. Unger [540] measured, from 23 to 25 June 1950, for 650 mμ, the wavelength of maximum assimilation in red, degrees of illumination inside the stand of plants varying between 38 and 84 percent of the value outside. Since it is practically impossible to combat weeds in the field in such cases, this ability of different kinds of plants to provide a screen against radiation is an important property from the grower's point of view.

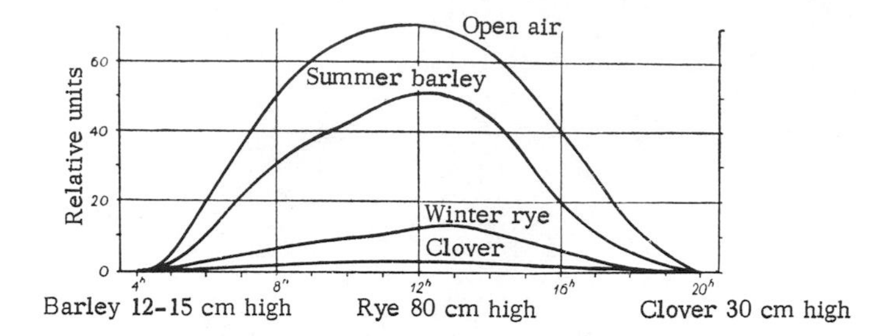

Fig. 142. Diurnal variation of light at the ground below different types of vegetation on a day in May. (After F. Sauberer)

The extent to which illumination can be reduced is illustrated by the figures (Table 62), also from Sauberer (in percent of outside illumination), measured in a 3-m-high thicket of young elms with dense undergrowth, and overgrown with clematis.

Table 62

Percentage of outside illumination within an elm thicket.

Date	Height above ground (cm)			
	1	10	25	100
5 July 1936	0.01	0.06	0.13	2.1
19 July 1936	0.03	—	20.17	2.2
15 November 1936	0.50	22	30	59

When the low plants, which were at first sprouting, became fully developed by the beginning of July, only 0.01 percent penetrated to within 1 cm of the ground, and only 2 percent came through the first 2 m of the growth. It became a little lighter when the undergrowth died, but it was only when the leaves fell (15 November) that the situation changed; in winter 6 to 7 percent reached a height of 1 cm above the ground.

With vegetation of such thickness, the ground surface loses entirely its function as a boundary surface with the atmosphere and set of heat exchange. A. Woeikof discusses an "outer effective surface" corresponding approximately to that of the upper surface of the vegetation, which in many respects assumes the role of the ground surface. This is like the surface of tree crowns in an old close-growing wood, which will be discussed in Sec. 33.

The described vertical distribution of radiation received and emitted by vegetation means that the temperature extremes cannot be so great as when the ground is bare. The temperatures of a vegetation cover are lower by day and higher by night. Vegetation makes the climate milder, more maritime in type.

Besides the changed radiation conditions, changed conditions of eddy diffusion determine the kind of microclimate in growing plants.

Any careful observer is bound to have noticed the extraordinary absence of wind movement within vegetation. If one lies down in heather above which stormy winds are raging one will soon appreciate this phenomenon. O. Stocker [539] found through a large number of measurements that the winds in vegetation were so weak that plants of the herbaceous type would hardly ever be exposed in our climate (not in deserts!) to a wind of more than 1 m sec^{-1}; normally it is about one-tenth of this value.

When strong winds blow over the tops of plants that are easily swayed, as in a lush meadow or a field of grain, a wave motion develops as at the boundary surfaces of water and air (sea), shallow water and sand (seas that dry out at low tide), and sand and air (deserts). Less pliable or more irregular plant surfaces, on the other hand, facilitate the development of large turbulence elements. Figure 143 illustrates the wind structure over a field of wheat stubble (above) and over a beet field with a growth of leaves reaching 40 to 50 cm (below). The measurements were made in the Rhineland by W. Schmidt [82] by filming the movement of very light pieces of gauze which responded quickly to changes in wind. The area was completely flat, and the wind had a run of sufficient distance to have adjusted itself to the roughness of the surface.

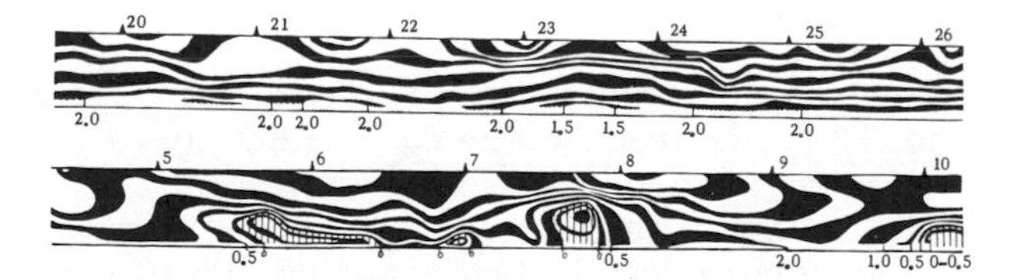

Fig. 143. The different wind structure over a stubble field (above) and a beet field (below). (After W. Schmidt)

The diagram is for a vertical section 1.5 m high above both the stubble and the beet fields. There is a time scale in seconds on top, so that rapid turbulent variations can be captured. Lines of equal wind speed are in stages of 25 cm sec^{-1} and the spaces between are alternately black and white. If the wind blows in a direction opposite to that of the main wind direction, this is made to stand out by vertical hatching. While the wind increases fairly regularly with height over the stubble, major disturbances appear over the beet field. Three times within 6 sec there is a reversal of wind direction immediately above the leaves, while in between there are intervals of relatively quiet conditions.

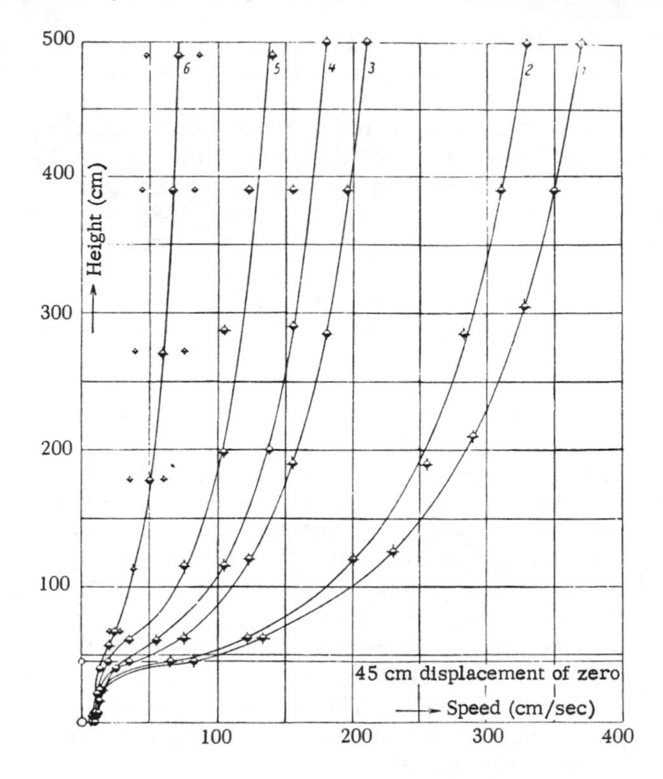

Fig. 144. Wind profiles in and above a beet field. (After W. Paeschke)

In discussing the increase of wind speed over vegetation-free ground in Sec. 16 we derived for the case of an adiabatic temperature distribution the simple relationship $u = c \log(z/z_0)$ (p. 118). Figure 144 is for wind profiles measured by W. Paeschke [188, 538] in the neighborhood of Göttingen, during six different periods of the day, arranged in order of increasing wind speed. With the linear form of representation selected, the profile above a height of about 0.5 m resembles that for bare ground on the left in Fig. 58. It is clear that the same rule can be used if, instead of the true height z at which the wind-recording instruments were placed, a reduced height $(z - d)$ is inserted in the equation. The quantity d is not the average height of the vegetation surface; only by examination of the wind profile can it be decided what value of d will best satisfy the equation. For the case shown in Fig. 144 the value is $d = 45$ cm. With the recordings made by E. L. Deacon, on which Fig. 60 was based, a value $d = 25$ cm was used, although the vegetation in question was grass with flower stalks 60 to 70 cm high. The high degree of pliability of grass under the pressure of wind is responsible for the difference.

Then the equation for wind increase with height under an adiabatic temperature lapse rate and with vegetation cover becomes

$$u = c \log \frac{z - d}{z_o}.$$

The constant c was explained on p. 118; z_o (cm) was called the roughness parameter, since the determination of z_o from wind-profile measurements can allow direct conclusions to be reached as to the roughness in the aerodynamic sense. (The quantity z_o in Paeschke's work has a different meaning.) Attention having been directed in Fig. 143 to variations in roughness of different types of natural surfaces, Fig. 145 now gives a review of values already established as summarized by Deacon [181] and other writers.

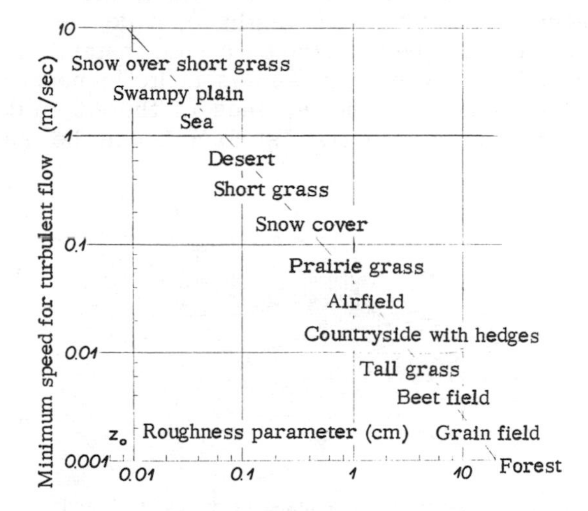

Fig. 145. The roughness of various natural surfaces. (After E. L. Deacon *et al.*)

As the abscissa on a logarithmic scale shows, z_o is spread over four powers of 10. The ordinate gives the wind speed in meters per second that must exist at a height of 1 m at least over the surface, so that the surface should become completely rough in the aerodynanic sense, that is, for flow to become turbulent. Since natural surfaces are extremely varied, the diagram can only indicate approximate mean values for the quantity z_o. This is indicated by the position in the diagram where the type of surface is entered. Starting with snow and water surfaces (top left), which offer little frictional resistance, the scale runs down to wooded country (bottom right). Individual measurements may differ considerably. For grass, for example, Deacon found the following relation:

Height of the grass (cm):	1	2	3	4	5
Roughness parameter (cm):	0.1	0.3	0.7	1.6	2.7

As the ordinate values show, flow over vegetation-covered surfaces is practically exclusively turbulent.

Calm within the vegetation cover means that there can be little eddy diffusion. Relative humidity is high among the transpiring organs of the plants, the turbulent fluctuations of all factors is low (compare Sec. 9), and a quiet, moist, protective climate prevails. R. Kanitscheider [536] measured the air temperature electrically at intervals of 2 sec in a stand of dwarf pines 2 to 3 m high at an elevation of 1600 m on a southern slope near Innsbruck. Figure 146 is a record of the mean difference of two consecutive readings (in tenths of a degree) as a function of time of the day and height above ground. Instability is greatest with respect to time when the sun is at its highest and greatest with respect to space at the surface of the 2.5-m dwarf pines. Here also the "outer effective surface" can be seen clearly.

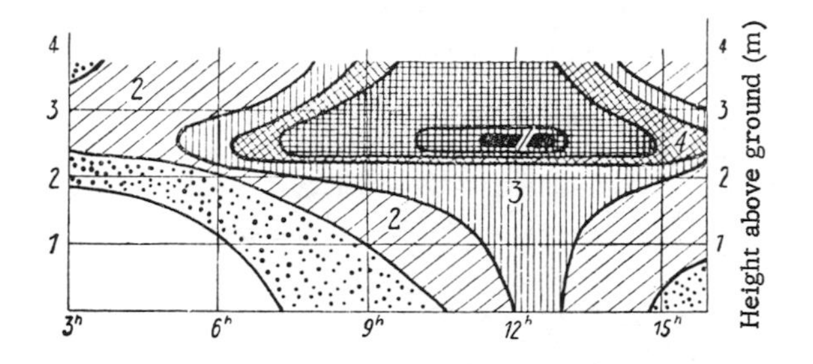

Fig. 146. Temperature instability in a stand of dwarf firs near Innsbruck for 28 July 1931, a cloudless day. (After R. Kanitscheider)

A third distinction in microclimate between bare and vegetation-covered ground arises from differences in water content.

In order to feed themselves, and thus live, plants must absorb large volumes of water through their roots and return it again to the atmosphere by transpiration. The transpiration ratio expresses the number of grains (or liters) of water a plant must use in order to produce 1 g (or kg) of dry substance. This quantity depends on the type of plant; values for trees range between 170 (pine) and 400 (beech), according to J. N. Köstler [537]. For agricultural plants it is between 400 and 600 on the average, from H. Walter [541], with extremes of 300 (millet) and 900 (flax). Plant-covered ground therefore evaporates more than bare ground.

The same conclusion can be reached, without taking account of the living process, by merely considering the surface areas involved. The leaf surface of a sparse wood is from 4 to 12 times the area of the ground on which it is growing. P. Filzer [531] gives a value of 20 to 40 times for grassland, and suggests, on the basis of experiments, that the quantity of water given off would increase proportionally as the surface area increases. This is true as long as eddy diffusion and wind are able to remove the moist boundary layer of the leaves fast enough. A natural limit is also soon reached because evaporation requires a considerable supply of energy.

The increased evaporation from vegetation-covered ground can be expressed numerically through lysimeter measurements. The results given here are the work of J. Bartels and W. Friedrich [530, 533] at the Forestry College in Eberswalde. Three similar boxes of capacity 1.5 m^3 with surface area 1 m^2 were placed in the ground, level with the surface and 3 m apart. Their weight could be measured from a cellar with an accuracy of 100 g, which is equivalent to 0.1 mm of precipitation; percolation water could be collected and measured, or the boxes could be sealed from below to give a ground-water level at any desired height in the boxes. Table 63 gives annual averages for evaporation and percolation at a depth of 1.5 m taken from comprehensive evaluations made by K. Göhre [534]. The readings comprise different sets of annual measurements, details of which are given in the table. To facilitate comparison, evaporation and percolation are both expressed in percent of the annual precipitation in the same period. The two figures add up only approximately to 100, since the amount retained in the soil enters in as a third, though insignificant, factor. Since the figures are averaged over a number of years in every case, the variation in the quantity retained by the soil is never more than $\pm$ 5 percent of the annual precipitation. There are two observation series for ground that has been artificially kept free of vegetation by the removal of all weeds. The second observation series was begun in April 1949 at the time when oaks were planted in a neighboring lysimeter. The results for 1950 to 1953 were evaluated in a form comparable to that published in the report by Göhre [535] after World War II (table in the appendix).

In the first period, which was rich in precipitation, the bare ground lost a quarter of its rainfall in unproductive evaporation, and in the second, drier period, half of it. At least half of the precipitation is always retained as ground water. Variations with time are not surprising, since the amount of evaporation from bare soil changes quickly as the weather changes, while vegetation, by means of its roots, is able to overcome the barrier to water conduction erected by the dried out upper layer of the soil. The plant covers, arranged in order of increasing requirement,

Table 63

Annual average evaporation and percolation at depth of 1.5 m.

Vegetation cover	Period	Average yearly totals				
		Precipitation (mm)	Evaporation (mm)	Percolation (mm)	Evaporation (%)	Percolation (%)
Bare ground	1929–1932	674	178	484	26	72
	1950–1953	533	270	266	51	50
Short grass	1929–1937	615	356	259	58	42
Pines (3 yr old in 1932)	1932–1937	576	450	149	78	26
Oaks (3 yr old in 1949)	1950–1953	533	454	117	85	20

take up an ever-increasing proportion of precipitated water. If the growing period from May to September, instead of the whole year, is considered, the share of precipitation taken up is seen to be even greater; for example, the young oaks consumed 145 percent of the precipitation during this period, which was possible only by drawing on the water stored in the soil.

From 1933 to 1937 the ground-water level in a lysimeter on which grass was growing was kept artificially at 40 to 50 cm below the surface. With an annual precipitation of 706 mm, evaporation amounted to 121 percent. This is approximately the value of the potential evaporation, and shows not only the requirements of vegetation but also the value of watering. W. Friedrich [533] has reduced this result into a simple formula; the annual amounts of evaporation from bare ground, short grass, young pines, and meadow above a high water table are in the ratios 2 : 4 : 5 : 8.

An even better idea is obtained from the daily evaporation values computed by J. Bartels [530] for well-defined meteorologic conditions, in the months of May to August from 1930 to 1932 (Table 64). The bare sandy soil restricts its evaporation considerably in dry periods; the grass area can do so to a limited extent only, and shows the results of water deprivation by yellowing and dying. Water surfaces, which, because of the slower temperature increase after rain, evaporate less than the other surfaces, give off much more in dry periods because of rising temperature.

Table 64

Evaporation (mm d^{-1}) from soil under various conditions.

Meteorologic condition	From bare ground	From short grass	From a water surface
On a day after rain	2,38	2,80	2,24
On bright days	0,47	2,15	3,61
On days of drought	0,26	1,14	3,80

In conclusion it must be pointed out that temperature and moisture distributions within the soil change considerably when the soil has a vegetation cover. This will be discussed in a more appropriate context at the end of the following sections.

31. The Microclimate of Meadows and Grain Fields

In the summer of 1953 W. R. Müller-Stoll and H. Freitag [553] measured temperatures in the ground and in the vegetation of six different types of meadow communities in the Spree Forest, which were near enough to each other in the undulating countryside for comparisons to be made. From these measurements we can recognize clearly how the microclimate becomes milder as the covering of the soil increases (Sec. 30).

The measurements used are those for 23 July 1953, a sunny day, and were made in three communities of green plants. The grasses on a sandy soil above a low water table were loosely arranged in thickets about 15 cm high, and consisted of *Nardus stricta* and *Festuca rubra* with rosette plants in between, so that 25 percent of the ground remained bare. A moist pipe-grass meadow was growing on humic sand with the water table at 55 cm; it was a thick growth 20 cm high with individual grasses of *Molinia* towering to 40 cm. A wet sedge was only 35 cm above the water table on partially decomposed peat and was a dense growth 80 cm high. Comparative measurements produced the results shown in Table 65.

Table 65

Temperatures (°C) and daily temperature fluctuation (deg) in three types of vegetation.

Type of growth (arranged in order of shielding the ground)	Maximum at the—		Daily temperature fluctuation			
	Ground surface	Vegetation surface	-10 cm	0 cm	5 cm	100 cm
Grass thickets	43	35	4	**26**	10	14
Pipegrass	27	31	1½	9	**13**	12
Sedge	24	31	½	7	10	**14**

The increasing amount of shade reduces the day maximum at the ground surface. At the same time the zone of greatest temperature fluctuation (in the absence of minimum temperatures, the reading shortly after 06:00 was used) moved from the ground up to the outer active surface (boldface figures). The change in type of ground from dry to moist, which is closely associated with the change in the type of plant on the meadow, exerts the same kind of influence on temperature.

If we postulate a continuous plant community completely covering the ground, its plant structure will have a far-reaching effect on microclimate. Individual plants in the meadow are tall relative to their horizontal extent. This applies even more to grain fields. This section is devoted to plants with this type of structure.

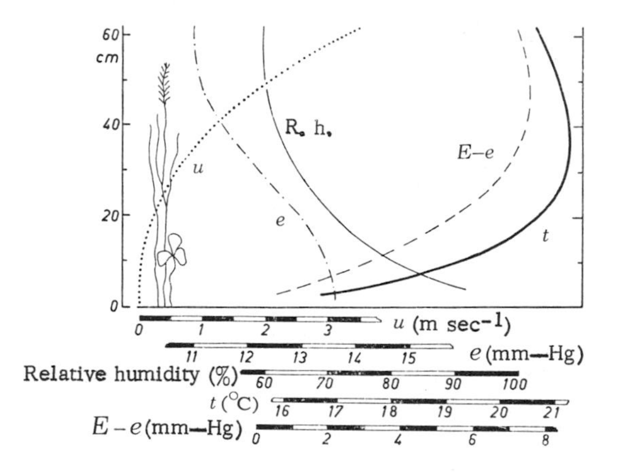

Fig. 147. Temperature, humidity, and wind in a meadow on a summer afternoon. (After F. L. Waterhouse)

Figure 147 illustrates the typical distribution of all meteorologic elements within grass 50 cm tall, which F. L. Waterhouse [569] recorded in Scotland on a sunny June day between 15:00 and 16:00 following precipitation. According to the temperature profile (t), the outer active surface is at a height of 30 cm; in the grass layer nearest the ground where dead grasses are also found mixed together, it is remarkably cool, and insects that are sensitive to radiation find complete shelter there. The relative humidity is also close to saturation point, water-vapor pressure (e) is a maximum, and the air is calm. The curve for the saturation deficit $E - e$ follows that of temperature. The temperature difference between this protective zone close to the surface and the air above the grass therefore shows considerable variations with time, confirmed by rapid gradient measurements made by K. Heigel [546].

Temperature recordings were made in winter by J. T. Norman, A. W. Kemp, and J. E. Tayler [554] in uncut grass and grass that had been cut short, near the Thames valley, to find out its importance for the growth of grass, winter cattle fodder, and for the process by which the grass becomes green again in the spring. The two meadows were only about 5 m apart. The grass that had been cut a number of times in the previous growing period was 2 to 3 cm tall; the other meadow had reached a height of 30 to 45 cm. The temperature at 2.5 cm above the ground was recorded continuously during the winters of 1953-54 and 1954-55, in both of these fields. Table 66 gives the temperature differences recorded for four-week periods of the first winter 1953-54, the difference being (+) when the thermometer in the long grass was warmer than the one in the short grass, and (-) when colder.

Table 66
Temperature difference produced by tall grass.

Winter of 1953-54	Temperature difference (deg)			Number of nights with frost	
	Average maximum	Daily average	Average minimum	Tall grass	Short grass
4 November—1 December	- 0.9	1.0	2.2	0	3
2 December—29 December	- 0.1	1.1	2.1	0	6
30 December—26 January	- 0.3	1.1	2.1	11	20
27 January—23 February	- 2.7	0.8	2.4	18	20
24 February—23 March	- 4.6	- 0.3	1.6	7	13
Mean (or sum)	- 1.8	0.7	2.1	36	62

The most striking result is the degree of heat protection afforded by the tall grass. The number of nights with temperatures below 0°C (nights with frost) is only half that in the short grass. During winter afternoons it is only a little cooler under the tall grass; but toward spring, with the quick growth of grass in the mild English climate, the difference increases so rapidly that in March even the daily average temperature in the tall grass is lower than that in the short grass. These results show how the microclimate becomes milder as the grass grows taller.

The temperature variation in a field of winter rye near Munich was measured by R. Geiger [5]. The upper part of Fig. 148 contains the midday temperature profile, from which it is possible to see the upward movement of the zone of greatest heating as the rye grows. At first it was substantially below the upper surface of the grain, since both solar radiation and wind are able to penetrate into the slender vertical structure. As the grain ripens and becomes lighter in color, the maximum falls again toward the ground and reaches the surface after reaping. At night (lower part of Fig. 148) the pool of cold air forming at the upper surface of the grain sinks only slowly between the thickening stalks, so that just before reaping the lowest temperatures are found at a height of 1 m above the ground.

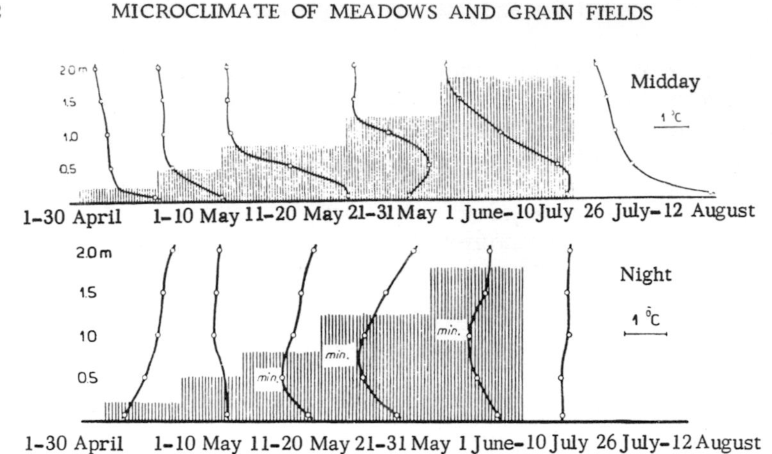

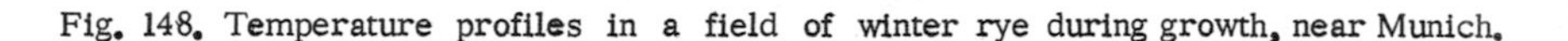

Fig. 148. Temperature profiles in a field of winter rye during growth, near Munich.

Random samples in the tropical savannas of French Guinea and on the Ivory Coast by B. Kullenberg [549] give reason to believe that conditions similar to those in mid-European latitudes prevail there at times of incoming radiation. Table 67 gives values for air temperature and relative humidity found on 12 August 1954 in grass consisting mainly of *Imperata cylindrica* on Mont Nimba (8°N, 500 m elevation) with a half cloud cover.

Table 67
Air temperature and relative humidity in a tropical savanna.

Height above ground (cm)	Position of instrument in the vegetation	10:30		11:30	
		Temp. (°C)	R. H. (%)	Temp. (°C)	R. H. (%)
100	Above the vegetation	23.6	82	26.2	75
60	Between the tips of leaves	23.9	85	26.8	75
17	In the lowest third of the mass of leaves	28.0	—	29.5	—
0.5	Between the stalks on the ground	26.7	—	27.3	—

Naturally, it is desirable to compare the temperatures within a plant stand with those over bare soil. Figure 149 shows, on the top left, the temperature variation recorded by platinum resistance thermometers shielded against radiation, the measurements being made by A. Mäde [551] at Müncheberg (Brandenburg). The readings plotted are the deviations of temperature from its value at a height of 2 m above grass during sunny weather in August 1936, and, as was expected, it is warmer near the ground by day and colder at night than in the meteorological screen.

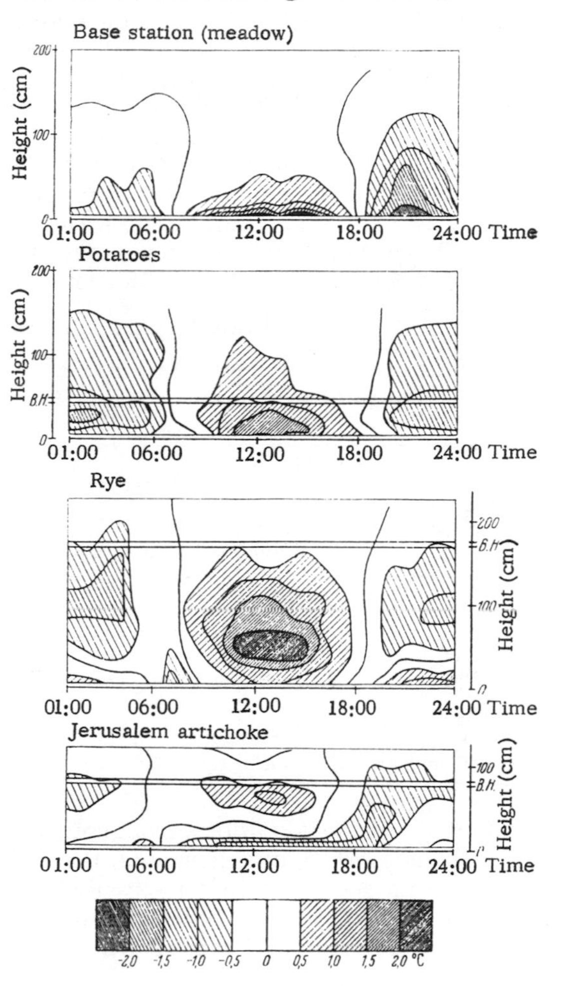

Fig. 149. Temperature deviations from the base-station values with various types of crops. (After A. Mäde)

The difference between temperatures in a field of rye and over a meadow are shown in the top right diagram of Fig. 149. The height of the stand is marked B.H. The zone of diurnal heating extends much higher than over the meadow, and at night the lowest temperatures occur at a height of 1 m, while at the foot of the grain stalks it is warmer than in the grass at the same height. Both of these results are in good agreement with Fig. 148. This method of making comparisons with a base station is to a large extent independent of the actual temperature and changes in the weather, which therefore allows it to give a good picture of the special character of climate for the crop in question.

M. J. Delany [544] measured the temperature and humidity profiles in heather 15 cm high and neighboring open land near Exeter in southwest England, while studying insect life. Y. Daigo and E. Maruyama [543] in Japan calculated the regression equations for air temperature in open land and a wheat field for clear and cloudy days, and compared the temperature variations at the two places. Comparative measurements were made in Hungary by G. Szász [563] at two different heights in a field of winter barley and neighboring bare land, at various periods of growth. The temperature differences shown in Table 68 were recorded.

Table 68

Temperature difference (deg) between a barley field and bare soil; +, barley field warmer; -, barley field cooler.

Period of obser- vation	Development stage of barley	Height of obser- vation (cm)	Temperature difference at—							
			04:00	06:00	08:00	10:00	12:00	14:00	16:00	18:00
22–27	Start of lateral	5			-0.8	-2.4	-4.0	-3.0		
October 1953	growth	10			0.0	0.2	0.2	0.8		
17–26	Period of sprouting	5			-0.6	-2.4	-4.4	-5.0	-5.2	-4.0
April 1951	Barley 35–55 cm high	10			0.2	0.4	1.2	-0.1	-1.4	-2.2
12–13	Maturity, barley	5	-2.2	-2,2	-3.0	-5.2	-8.5	-9.8	-8.2	-7.0
June 1951	130–140 cm high	10	-1.0	0,6	2.0	2.6	1.8	1,0	-1,4	-1.6

At a height of 5 cm it is colder throughout in the barley, with its sheltering influence. At 10 cm, however, the radiation absorbed by the stalks is retained to such an extent, in spite of the loose structure of the barley, that it has higher temperatures, which are lost again in the early hours of the afternoon (effect of the daily wind variation?).

On an August day in 1953 E. Tamm and H. Funke [565] measured the temperature in a field of corn 2.10 m tall and adjoining bare land, at 16 and 9 different heights above the ground respectively. During the period of incoming radiation the corn at 40 to 60 cm above the ground was 0.5 to 2.5 deg cooler, and above this up to 210 cm about 0.5 to 1.5 deg warmer than the bare land. The outer active surface was not very well marked, between 80 and 180 cm, and changed its appearance with different amounts of radiation and weather conditions. At the period of outgoing radiation the corn field was 0.5 to 1.0 deg warmer to a height of 80 cm and above this 0.5 deg cooler; the outer active surface was better marked than by day, although differences remained small. The writers consider themselves justified, in spite of the small amount of observational material, in concluding that a general relation between temperature over bare land and in crops cannot be established. Only measurements in the crops themselves can give useful information.

Our knowledge of the humidity of the air in meadows and grain fields is still very sparse. Numerous random samples show an enrichment in water vapor in the still air of the transpiring plant cover, which was already shown in Fig. 147. This is particularly evident in the relative-humidity values found with the cooler temperatures within the plant cover during the day. F. L. Waterhouse [568] distinguished between three zones affecting the lives of insects, in measurements made in tall grass in June afternoons near Dundee (56°30′N):

Height (cm)	Air temperature (°C)	Relative humidity (%)
50–25	24–25	46–47
25–10	25–26	47–55
10– 0	17–18	90–100

The density of growth is also a factor of importance. P. Filzer [531] experimented with planting corn of different density. He specified density by the area in square centimeters of leaf surface per cubic centimeter of air space. The average values he found for a series of measurements were: in the densest corn (1.81 $cm^2\ cm^{-3}$), 73 percent; a medium dense crop (0.82), 64 percent; in a thin crop (0.38), 51 percent; in the open the relative humidity was 40 percent.

In fields that are watered artificially, the distribution of moisture and temperature are substantially different. L. A. Ramdas, R. J. Kalamkar, and K. M. Gadre [556] were the first who investigated this in artificially watered sugar-cane fields near Poona (18°N). Their measurements were extended in 1951 by K. M. Gadre [545].

Figure 150 compares the temperature profiles, during the dry period of the Indian northeast monsoon, of bare land, a field of millet 150 to 180 cm high, and an artificially watered field of sugar cane 2.5 m high. The sugar cane uses the incoming radiation around noon mainly for evaporation; the air temperature over the water is therefore 14 deg lower than over the vegetation-free ground, and 8 deg lower than under the shade of the millet. The air flowing among the canes therefore owes its heat to the radiation taken up by the canes above the water, and of course also to the surrounding areas that have not been watered. At night there is the situation with a minimum above the ground, already discussed on pp. 94ff. The water in the cane field acts as a heat reservoir, storing the daytime heat and producing from the ground up to

1.2 m a temperature somewhat higher than over bare land, while in the loose-growing millet field, grom a height of 0.5 m, the radiating plants lead to lower temperatures than over the bare land.

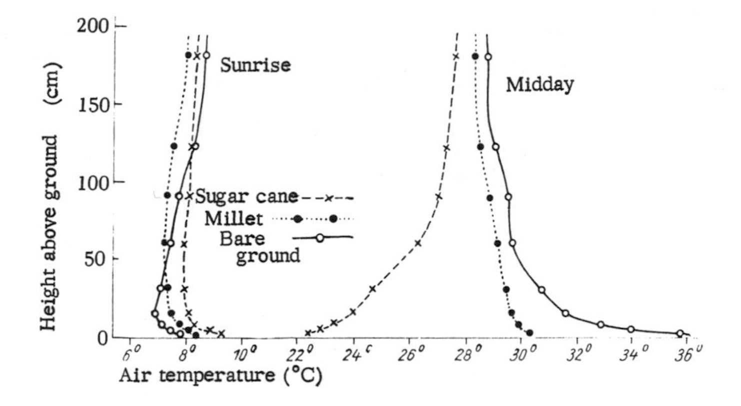

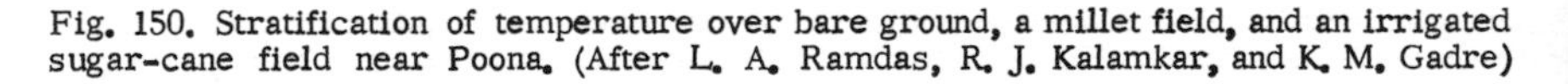
Fig. 150. Stratification of temperature over bare ground, a millet field, and an irrigated sugar-cane field near Poona. (After L. A. Ramdas, R. J. Kalamkar, and K. M. Gadre)

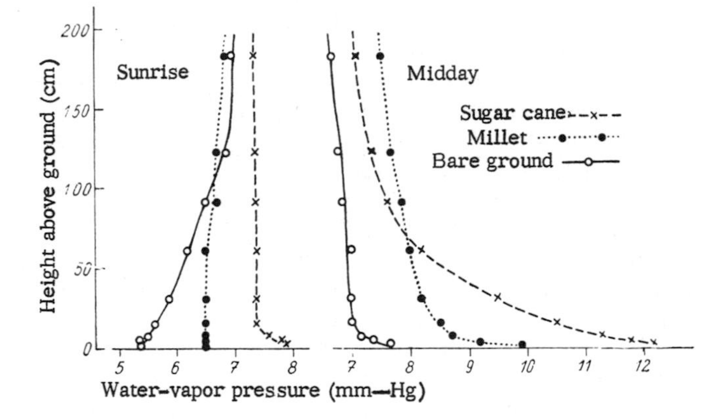

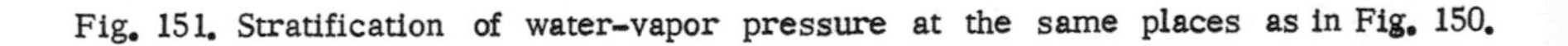
Fig. 151. Stratification of water-vapor pressure at the same places as in Fig. 150.

The corresponding moisture distribution is given in Fig. 151 (vapor pressure) and Fig. 152 (relative humidity). The highest humidity, both absolute and relative, is found over the watered field, followed by the millet, and then the bare land. The gradient of water-vapor pressure in the sugar-cane field is surprisingly high around noon.

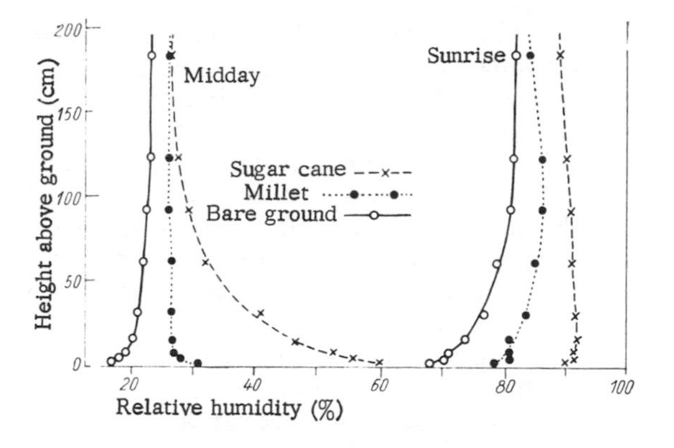

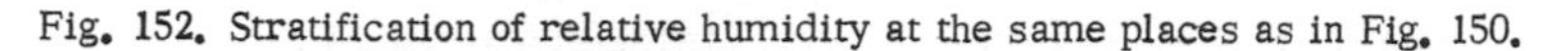
Fig. 152. Stratification of relative humidity at the same places as in Fig. 150.

K. M. Gadre [545] used the following method to investigate the microclimate of growing sugar cane, particularly the boundary influences, in an experimental field near Poona. Thirty-three rows of canes running east to west, 1.2 m apart, were planted, and psychrometer readings were taken in them once a week at noon at six different heights, the observers standing in rows 6, 13, and 21. The same readings were taken simultaneously in a neighboring unwatered field, free of growth. Noticeable differences began to appear only 3 to 4 months after planting, when the canes had reached a height of 70 to 90 cm. In November, when the canes were nearly 5 m high, air temperature, water-vapor pressure, and relative humidity were measured in the middle of the field (Table 69); the boundary effect extended to row 4.

Table 69

Temperature t (°C), water-vapor pressure e (mm-Hg), and relative humidity r (percent) over bare ground and a sugar-cane field.

Height (m)	Bare ground			Sugar-cane field			Difference		
	t	e	r	t	e	r	t	e	r
6	30.2	14.0	43	29.1	15.0	49	-1.1	1.0	6
5	30.3	14.0	43	29.6	16.5	53	-0.7	2.5	10
4	30.5	14.0	42	30.2	18.4	57	-0.3*	4.4	15
3	30.7	14.0	42	29.7	17.2	55	-1.0	3.2	13
2	31.0	14.0	41	29.0	16.7	55	-2.0	2.7	14
1	31.6	14.0	40	28.0	17.9	63	-3.6	3.9	23
0	38.0	18.5	37	26.1	21.0	83	-11.9	2.5	46

Similar relations are found in watered rice fields. The microclimate of paddy fields has been investigated by the Japanese from every possible point of view, and results of their investigations appear regularly in the *Journal of Agricultural Meteorology*. The influence of water temperature on the growth and yield of rice was studied by S. Sato and others, and suitable methods of irrigation were proposed. E. Inoue and his collaborators investigated wind and turbulence in rice fields; temperature measurements were made in 1943 by S. Suzuki [561] and more recently by H. Satone [558], Y. Tsuboi and Y. Nakagawa [566], and S. Sato [557]. The temperature pattern in the double underlying surface of water and soil is dealt with theoretically by Suzuki [562], and the absorption of solar radiation as a function of water depth by K. Yabuki [570]. A complete exposition of all the factors in the heat balance of rice fields was attempted by M. Kumai and T. Chiba [550]. Unfortunately, lack of knowledge of the Japanese language prevents me from dealing with their findings.

R. Wagner [567] investigated temperatures in a rice field near Szeged (46°N) in Hungary recently, during a sunny August day in 1956; the results are shown in Fig. 153. The rice was 45 to 50 cm high and the water 14 cm deep. Details of wind and cloudiness are given at the top of the diagram, and information on dew deposits, morning mist, and sunshine at the bottom. The temperature variation is explained by heating from the outer effective surface (perhaps also from the surroundings) and by the double surface underlying the air.

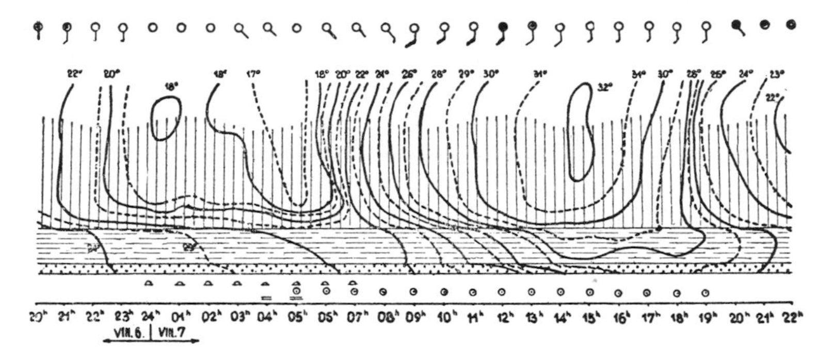

Fig. 153. Temperature variation in an irrigated rice field. (After R. Wagner)

A vegetation cover exerts a strong influence on the temperature and water content of the soil. The data are not yet sufficient to give answers to all the problems arising here. The existing measurements do, however, give us some insight into the order of magnitude of the differences brought about.

J. L. Monteith [552] has thoroughly investigated this problem, under climatic conditions in England. By making direct measurements of heat flow just under the ground surface, he determined the quantity of heat absorbed during a time of positive radiation balance and emitted during times of negative radiation balance.

Measurements of this kind were made for a meadow, a wheat field, and a potato field; the quantities of heat involved on a cloudless day in July 1956 are shown in Table 70. During the day, $-B$ is between 13 percent (wheat) and 16 percent (potatoes) of the radiation balance S. The average of all observation days in June and July gave a higher mean value; nine-tenths of all values lay between 18 and 23 percent. At night, however, the corresponding figures were 39 to 46 percent; there was an extraordinary scatter of individual values with changing weather (30 to 140 percent!). The three fields were little different in this respect. On the meadow the ratio $-B:S$ decreased as the wind speed increased, because the stronger wind delivered more heat through turbulence to compensate for radiation losses. This relation did not exist with the wheat and potatoes. The decisive factor here was soil moisture, so that very large values of the ratio $-B:S$ were found after a rainfall; this may be attributed either to improved thermal diffusivity of the soil or to increased evaporation from its rain-wet surface.

Table 70

Heat exchange (cal cm^{-2}) in fields with different crops.

Period of time	Radiation balance, S	Heat exchanged in the soil, $-B$		
		Meadow (2 cm)	Wheat (100 cm)	Potatoes (75 cm)
Day	+340	+52	+45	+54
Night	-46	-20	-18	-21
24 hours	+294	+32	+27	+33

E. Tamm [564] made temperature recordings in five adjacent fields at a uniform depth of 5 cm from April to October 1936 near Berlin. The values for the warmest hours in the afternoon, 12:00 to 16:00, when differences are most strongly pronounced, have been published for grass, winter rye, potato, hemp, and soya fields. Comparison of the hemp and soya fields gave the following annual ground temperature variation. For example, the two fields had the same temperature at first in May; then the soil of the soya field, which was less shaded than that under the hemp, became steadily warmer. By the end of May the difference was 2.5 deg at the depth of 5 cm, and increased to 9.2 deg in the second half of June. When the soya plants then reached maturity, the difference decreased, became zero at the beginning of August, and then changed in algebraic sign.

In Upper Austria, R. Biebl [542] compared seven different locations in summer 1950, using the method of Wilh. Schmidt by measuring ground temperature by means of thermoelements and small celluloid tubes which had been previously introduced into the ground. Table 71 gives values of diurnal temperature variation obtained for comparable sunny days in August.

Table 71

Daily temperature variation (deg) in different kinds of ground.

Kind of ground	Depth in ground (cm)				
	1	5	10	20	30
High moorland knoll	17.3	14.1	8.5	2.6	0.6
Meadow (cut)	15.7	6.8	4.3	1.7	0.6
Moor with turf removed	13.1	8.6	4.0	1.4	0.6
Moorland meadow	7.5	5.5	2.9	0.8	0.4

Temperatures at a depth of 10 cm under three typical natural plant communities in a Mediterranean climate were measured by R. Knapp and H. F. Linskens [547] on the Italian Riviera near Bordighera. They found surprisingly large differences which determined the life of the plants. On a cloudless July day in 1952, with air temperatures between 22° and 28°C (at a height of 2 m), they found a variation between 23° and 42°C under a 90-cm-high Cistus-Macchie (evergreens) covering 75 percent of the ground, and 29° to 38°C under 40-cm-tall dry *Andropogon* grass, while under a 7-m-high stand of cluster pines (*Pinus pinaster*) the soil temperature varied only between 26° and 30°C.

The water content of the soil as well as its temperature is determined by the type of plant cover. The dryness of soil in the neighborhood of the tap roots of trees is well known. Comparative measurements of soil moisture were made in the botanical gardens at Cologne between April and July 1951, by R. Knapp, H. F. Linskens, H. Lieth, and F. Wolf [548] under *Lolium perenne* and in black fallow using the plaster-block method. Records published for four depths in both areas show a considerably greater fluctuation of moisture under the lawn than under the bare plowed surface.

H. Schrödter [559] took 12 random samples every 3 days in June 1950 at two depths in soil under four differently treated areas, 20 by 13 m, of an experimental raspberry field at Aschersleben, in order to determine the temporal and spatial soil-moisture variation. The published results show that, for the untreated section, fluctuations at depths between 10 and 50 cm varied from less than 18 to more than 20 weight percent. In the two sections, which were either fertilized with manure, or watered every 2 days, the moisture content varied between 19 and 21 percent, and in the part both watered and fertilized with manure between 21 and 23 percent. The mosaic character of the soil (see pp. 142, 167) could also be seen clearly from this investigation, but there was also a boundary influence from the neighboring areas as the plots selected were very small.

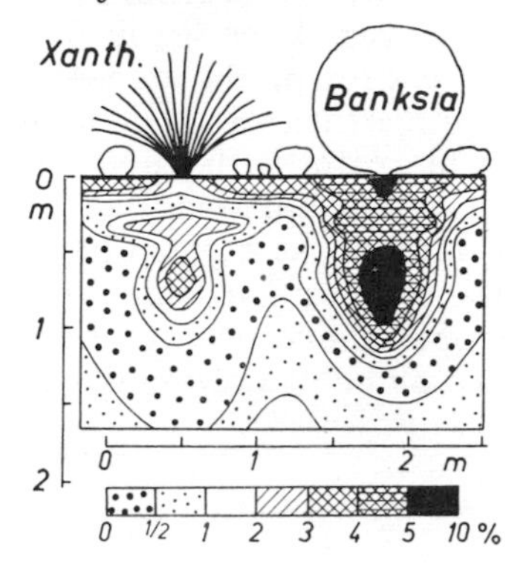

Fig. 154. Humidity of soil below steppe vegetation in Australia after a rainfall. (After R. L. Specht)

The Australian botanist R. L. Specht [560] discussed an extreme case of the influence of plants on soil moisture in 1956 at the Canberra Symposium on the Microclimatology of Arid Zones. Figure 154 illustrates the distribution of soil moisture (percent) on a summer day in 1955 under desert vegetation, from measurements made at six depths 30 cm apart horizontally. The previous day, 24 mm of rain had fallen after a long dry spell. This dry spell can still be recognized in the part at middle depths shown with large dots, beyond which moisture increases a little with depth normally. The plants, whose very different structure is shown in the sketch, first use up the water immediately in evaporation, but then direct it along their surfaces or deposit it in drops in places important to them for their continued existence. It is therefore surprising to find, the day after the rain, that the highest moisture is not found under the spaces between the plants where it can fall unhindered, but that the largest quantities are collected in the root zone. If this type of vegetation is left to itself for 50 years, only the large types will be found, while the smaller plants fail to survive the struggle for existence.

32. The Microclimate of Gardens, Potato Fields, and Vineyards

Having completed the study of plant communities with a more or less vertical structure found in meadows and grain fields, we now turn our attention to plant communities where the individual members, in their growth and form of leaves, present a more horizontal structure.

The first example to consider is a flower bed planted with antirrhinum in a Munich garden [5]. Figure 155 shows the temperature profile. In July, when the plants are small and form an open type of cover, the midday temperature profile (upper diagram) is still similar to that over bare ground. In August, however, when the plants are fully grown, the dense leaf structure raises the zone of maximum temperatures to a higher level much more markedly than happens in grain, with its vertical type of structure (Fig. 148). The active outer surface lies just under the upper surface of the stand of plants. At night, when the upper surface of the plants radiates, the cooler air can sink to the ground more easily than in grain crops; therefore the minimum always lies at the ground surface in flower beds (lower diagram).

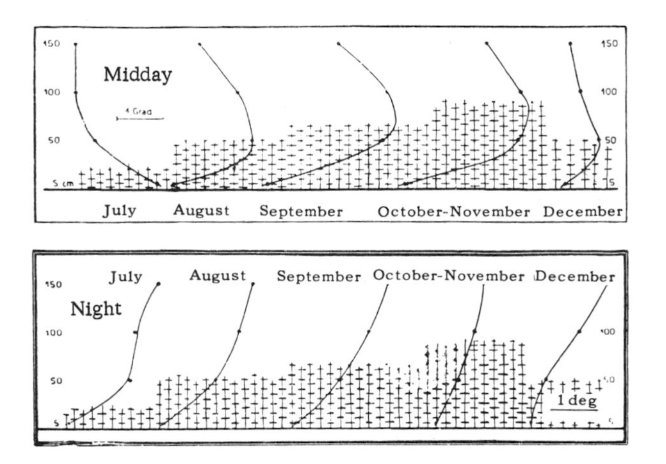

Fig. 155. Temperature profiles above a flower bed near Munich, during growth.

Leaf structures are not always as dense and close together as in flower beds. Where rows of vegetables, potatoes, beets, and the like, are planted in drills in a field, circumstances are rather different. L. Broadbent [571] was able to show, in investigations on a potato crop at Rothamsted, England, that the midday maximum temperature was at a height of 30 cm in a thickly planted crop 60 cm high, and at 10 cm in the sparsely planted. In contrast to Fig. 155, when the ground was dry the minimum was observed to be at the radiating upper surface of the crop. This is similar to the temperature variation observed by A. Mäde [551] in a Berlin potato field in 1936; the lower left-hand diagram in Fig. 149 shows the temperature difference observed during the day in comparison with the grassland base station on the top left. In

Rothamsted the temperature minimum was displaced only to the 10-cm level when the ground was wet. This was also the level of highest air humidity when the ground was wet, while with dry ground strong evaporation from the leaves raised its level to 30 cm. Short-term wind fluctuations may cause rapid changes in the temperature profile, as shown in the figures published by Broadbent.

J. M. Hirst, I. F. Long, and H. L. Penman [574] measured temperature and humidity relations at night at six different heights between 10 and 320 cm over a potato field at the same place. In a normal night, with deposition of dew, the gradient of water-vapor pressure was directed both from above and from below toward the upper surface of the crop. Investigations like those of A. Raeuber [579], with five different varieties of potatoes near Rostock, make it possible to discover the essential conditions for life of various potato parasites.

J. Justyák [576] analyzed a 5-year set of observations made in tomato fields at Debrecen in Hungary. There were plantings of various densities, with the plants arranged in rows in east-west and north-south directions. In these, soil temperature and moisture, air temperature, and vertical temperature gradients were investigated.

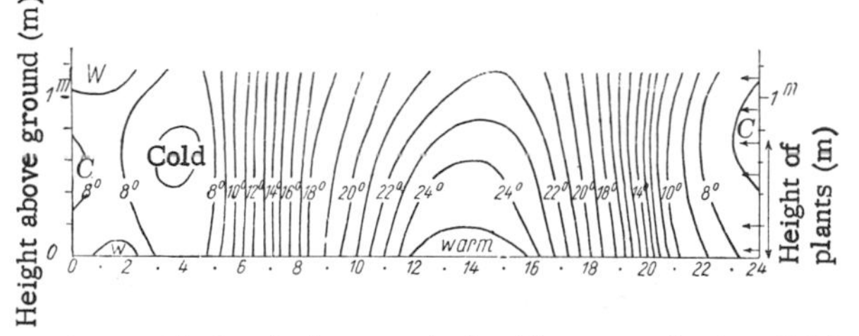

Fig. 156. Temperature variation during a typical midsummer day in a field of Jerusalem artichokes. (After A. Mäde)

Figure 156 shows the diurnal temperature variation in a Jerusalem artichoke (***Helianthus tuberosus***) field near Müncheberg (Brandenburg), after A. Mäde [578a]. Temperatures were recorded on a sunny day (4 August 1935) at heights marked by arrows on the right of the diagram, by six resistance thermometers free of radiation errors. The pattern of these temperatures resembles that of the mean temperature in August. The vertical temperature distribution at heights from 10 to 170 cm was investigated in Japan by R. Taguchi and H. Okumura [583] in a plantation of mulberry trees that were cut down every year and renewed by sprouting.

From the macroclimatologic point of view, German vineyards lie near the northern limit of the area in which vines can be cultivated. They are therefore by compulsion dependent on situations which have a microclimate that is especially sunny, warm, and free of frost. An introduction to living conditions on a vineyard terrace has been given by O. Linck [578] in a very readable, well-illustrated book which should be of interest to microclimatologists as well as botanists.

The microclimate of a "wine mountain," as it is called in German, is made up of many individual factors. To begin with, the term "mountain" indicates that the sunny slopes of hills are selected for the cultivation of grapes. The climate of terrace vineyard is therefore that of a slope, which will not be dealt with until Sec. 44. This type of climate is altered artificially by terracing. These flatter surfaces, on which the vines are planted, are easier to work and are bounded on the side toward the mountain by a stone terrace step, the microclimatic effect of which was discussed in Sec. 20. In many places the vineyards are divided by stone walls. These have been built through centuries by the laborers in the vineyard, collecting loose stones and building them into walls at the sides. They run down into the valley and form a shelter against the wind, hence providing warm spaces, which however do not impede the flow of colder air down into the valley. If the holes between the stones have not been blocked by fine accumulated dust, these stone walls have low thermal conductivity. They therefore become strongly heated during the day and act as sources of heat; they are correspondingly cool and moist deeper down, and in addition to promoting the growth of xerophytic surface flora they also support deep-rooted bushes and even trees by means of which their protective influence is increased.

The type of country in which the hill is situated is of equal influence. If its foot lies near a river or a lake shore, extra warmth is obtained through specular reflection, which was described on pp. 16f. If the hill is topped by a cold plateau, there will be an increased risk of night frosts through cold air flowing down at night (see Sec. 42). As a protection against this, the upper boundaries are often shut off by thick hedges or woods. According to R. Weise [588], the locations safest from frost in Franconia are surrounded by a crescent of slopes from SE through E round to W, and are open only toward the SW (see Sec. 54).

But the vines themselves, and the manner in which they are trained, also play a decisive role in creating the microclimate of the wine mountain, or, as it is called in the wine-producing country, the Wingerts or Wengerts. This aspect of vineyard climate, determined as it is by plants, is clearly the easiest to investigate, because this part of the hill is level. This is usually the case in the research reports that follow, but there are also exceptions to this rule.

The first instrumental measurements were made in 1928 by R. Kirchner [577] in the Palatinate wine district. They became known at a much later date, when K. Sonntag [582] began his research there. He recognized that a basic distinction had to be made between the climate of the rows of vines and that of the open lanes between them. Figure 157 shows the observed temperature distribution for midday (left) and at night (right) on and after a sunny September day of 1933. The sun is able to penetrate to the ground in the lane, which runs N-S, producing high surface temperatures and a large temperature gradient close to it, similar to the conditions in an open-planted vegetable field (see Fig. 156). In the rows of vines the highest temperatures are found below the outer active surface where the foliage gives protection from the wind, but these highest values are naturally much lower than in the lanes. At night, the lowest temperatures are found at the level of the radiating surfaces of the leaves (not at the ground in the lanes, which are shielded), which means that the dew formed is collected to the advantage of the plants. "Even outside the vineyard," K. Sonntag wrote, "an iron pole was dry from the ground up to the level of the stems, but was covered with water droplets above the level of the leaves." This double influence of the outer active surface of the vines and of the solid ground can be seen also in the temperature profiles measured by Y. Tsuboi, Y. Nakagawa, and I. Honda [585, 586] in Japanese vineyards.

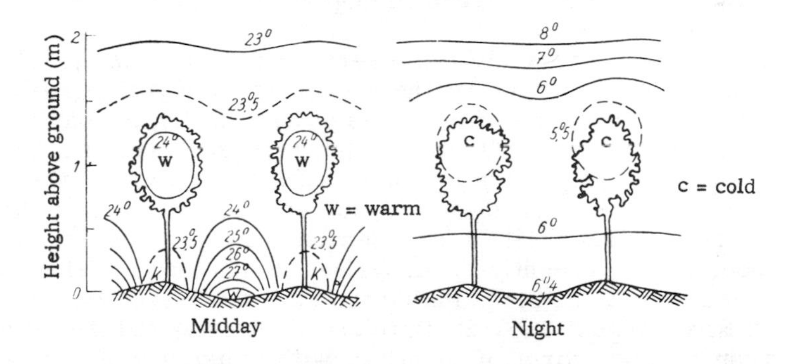

Fig. 157. Temperatures at midday (left) and at night (right) in a Palatinate vineyard on 17 September 1933. (After K. Sonntag)

R. Weise [588-590], O. Jancke [575], and H. Burckhardt [572, 573], in particular, advanced our knowledge of vineyard climate considerably. The influence of the height of the vine, with the distance between the rows constant, was investigated by Burckhardt in a level vineyard at Mussbach in the Weinstrasse district of Germany. The top of the foliage of the higher-trained Sylvaner grape reached 150 cm, while that of the lower was 90 cm above the ground and the distance between neighboring rows was 120 cm in both groups. Simultaneous readings with six psychrometers, including adjacent open ground, in August 1952 gave the following picture: the temperature decrease above the ground, such as that in Fig. 157, was found to exist in the N-S lanes between 10:00 and 14:00 only. In the higher growth, the temperatures were lower both by day and at night in comparison with the adjacent bare land, because of extensive shading. In the low growth there was hardly any difference between the lane temperatures and the bare land. The humidity gradient showed a lapse with height both by day and at night, greater in the low vines than over bare land, and greater in the high vines than in the low.

The crown area of the vines was subject to radiative heating and evaporation loss by day. In the higher vines the first of these was greater because of its greater leaf area, while in the lower vines the second factor was the larger. The temperature is therefore higher in the area of the grapes, in the taller vines, compared with the temperature at the same level in the lanes between them, and also higher than that of the shorter vines. However, since the grapes in the latter are closer to the ground, this balance out. Grapes of both taller and shorter vines enjoy the same kind of temperature in the warm hours at midday. This explains why the fear of the vintner that the yield of grapes will be reduced if they are trained too far from the heat-dispensing ground is not justified. Water-vapor pressure is, however, always higher over the vines that are trained higher because the mass of transpiring leaves is greater.

This result agrees with the observations of R. Weise [589] that in the Würzburg wine country the higher form of training (Frankish stem training) does not show a loss of heat during the day, in comparison with the lower form (Frankish head training). The first form, however, allows the cold night air to flow away more easily in the comparatively foliage-free space in the lower layer of air, and therefore runs less risk of frost damage than the second form. This has been proved recently by R. Weise [590] in the published results of measurements he made inside the vine shoots.

Then in August 1957 H. Burckhardt investigated the difference between an open form of planting, with a distance of 3 m between the rows of 2.1-m-high vines, which is desirable because of the advantages it offers in ease of working, and a normally trained vineyard with 1.2 m between rows of vines 1.1 m high. The microclimate of the lanes in the widely planted vines was more balanced, and the air more settled, than in the normal style, and gave the cold night air a better chance to flow away. However, the temperature was rather lower during the day in the area where the grapes were growing than in the normal vineyard, which is confirmed by the greater acidity that can be observed in grape juice from open vineyards. The relative humidity was also somewhat higher, and this increases the risk of infection by fungus diseases.

N. Weger [587] demonstrated the practical consequences of the considerable differences in the microclimate of two vineyards only 3 m apart in the Geisenheim area. The difference was apparent in the quality of the grapes and in the incidence of pests. Practical viticulture may derive great advantages from such measurements, provided they are made with the necessary instruments, and the requisite amount of time is devoted to them.

Frost danger and the method of dealing with it are discussed in Sec. 54.

It is only a short step from the microclimate of the vineyard to that of the orchard. O. Takechi and J. Kikuchi [584] investigated the microclimate of a lemon orchard in Japan. The height of the crown area was 2 m, which is low in comparison with German orchards. The microclimate of two stands of different densities was investigated. In the summer of 1953 the amount of light in the denser orchard was 63 percent of the lighter, the relative humidity 10 to 20 percent higher, and the fluctuation of the ground surface temperature was 8.3 against 10.4 deg, caused mainly by the maximum being lower. Two-hourly temperature profiles have been published for the range -30 to +200 cm for both of these orchards. R. Schröder [580, 581] investigated coffee plantations in the tropics, and pointed out the great influence on microclimate of artificially shaded crops and unshaded crops, and made the first measurements in South America. Orchards in Germany, however, bear a strong resemblance to thinly planted woods. It is therefore more appropriate if we now turn to the study of meteorologic problems in forests.

CHAPTER VI

PROBLEMS IN FOREST METEOROLOGY

When a forester is laying down the foundation of a new forest, the young plants are particularly sensitive to many dangers from weather, whether seeded, naturally or by design, or planted as seedlings. Late frosts, winter cold, dryness in spring, summer droughts, or persistent winds are the causes of most of the damage to young growth. The forester therefore has a special interest in all the microclimatologic questions that were raised in the last chapter. The tree nursery is subject to the same difficulties as an area used for farming or horticulture.

When the wood has grown to maturity, new problems and tasks arise. Possible weather damage due to high winds, rime, breakage under the weight of snow, destruction by lightning, and so forth will not be discussed in this context, but only the problems arising from forest climate, and forest microclimate in particular. The subject of discussion in Chapter VI is therefore the climate of the forest as a factor with which the forester has to reckon. The study of local climate has developed in recent times into a subordinate science to forestry.

In Secs. 33-36 the climate of a typical closed old stand will be studied, then the microclimate of its border areas (Sec. 37) or of clearings (Sec. 38) in which most of the young growth takes place. The influence of special measures taken when laying the foundations of a new forest will also be discussed, such as mixing of various types of trees, the building up of a stand (for example, various age classes), and care (for example, degree of thinning). The influence of all these factors on microclimate will be considered and tested, in order to understand the requirements of both silvics and forestry, and be able to improve them (Sec. 38). Only knowledge of these matters will furnish an approach to the age-old question of the influence of forests on general climate with any hope of success (Sec. 39).

33. Radiation in an Old Stand

An old stand of trees with its closed crown area differs from all the types of vegetation covers discussed in Chapter V, in that there exists in the trunk area a typical enclosed air space. The climate of the trunk area is a transitional climate between that of the open air separated from it by the dense crown area, and the climate of the forest floor below. This forest-floor climate, under the sheltering influence of the trunk area, is basically different from the climate of an area of bare ground.

In an old stand the outer active surface is situated at the top of the crown area (see p. 273). There, in a narrow vertical zone, radiation is received and emitted, the wind is allowed some degree of penetration, and the contrast between the outside climate and that of the trunk area is decreased by the turbulent mixing which, although strongly inhibited, takes place in the crown area. Microclimate of the crown area is therefore characterized by a great degree of instability. The contrast between it and the equilibrium in the trunk area is very noticeable, as is known by anyone who has experienced the striking difference on entering a wood from the blistering summer heat, a howling gale, or biting winter cold of the open country.

We shall now try to learn the characteristic properties of the microclimate of crown area, trunk area, and forest floor, beginning with the primary factor, radiation.

In Fig. 141 the decrease of short-wavelength radiation within a vegetation cover was given for a meadow. Similar measurements of light intensity within various types of vegetation have been made by A. Baumgartner [652], using selenium cells. When he took, as a unit of height, the height of the top of the plants, regardless of whether it was centimeters or meters, and expressed the light intensity as a percentage of the light outside, he discovered that

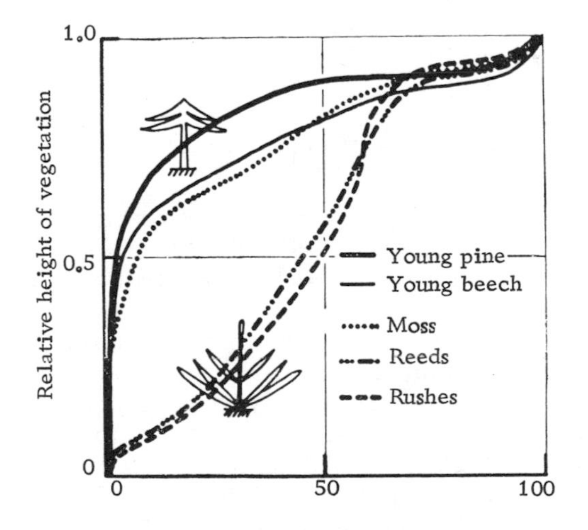

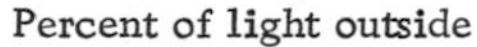

Fig. 158. Two different types of light diminution within vegetation cover. (After A. Baumgartner)

two different types of vegetation could be distinguished, as in Fig. 158. In one type the light-intensity curve is convex upward, and in the other it is linear or concave. This difference corresponds approximately to the distinction made earlier between plant communities that were mainly horizontal (Sec. 32) and those that were mainly vertical in structure (Sec. 31). Woods belong to the first group. The canopy of leaves or needles causes a sharp decrease in light intensity in the top section, which may be seen in Fig. 158 with the 2-3-m-tall young pines and beeches, and also recognizable in the moss cover, which is only 4 cm thick. Grasses, however, absorb large amounts of light only at lower levels, where stems are stronger and dry. There are, naturally, a number of intermediate stages between these two types.

Figure 159 gives the light curve, after E. Trapp [615], for a 31-m-high stand of red beeches 120 to 150 years old, intermingled with a few pines. It was situated at 1000 m above sea level on a 20° slope with a SE exposure, near Lunz in Austria. The full line gives the distribution of light on sunny days. When weather is murky the absolute light intensity is naturally lower in the wood, but its rate of decrease is slower, as shown by the broken curve, because

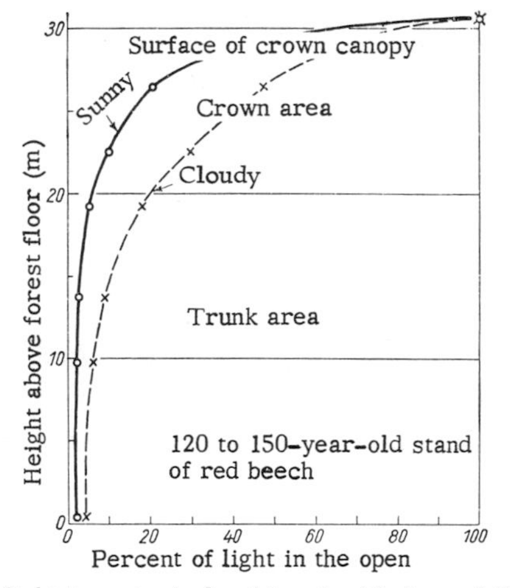

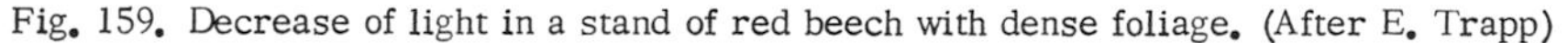
Fig. 159. Decrease of light in a stand of red beech with dense foliage. (After E. Trapp)

diffuse sky radiation then makes up a proportion of the global radiation bigger than on sunny days. Y. Harada [601] was able to show, in Japan, that there was a relatively greater proportion of light in a stand on foggy days than on days without fog, for similar reasons.

The decrease of short-wavelength radiation in a stand depends to a large extent on the type of tree, the character of the stand, its age, and its productivity. Many appropriate measurements are available from (arranged in chronologic order) R. Geiger and H. Amann [619], F. Lauscher and W. Schwabl [603], F. Sauberer and

E. Trapp [611], E. Trapp [615], W. Nägeli [608], G. Scheer [612], G. Sirén [613], D. H. Miller [605, 606], and A. Baumgartner [437, 591]. It is possible only to give limiting values, between which the light intensity will usually vary in the different kinds of stands of trees (Table 72).

Table 72

Light intensity (percent of that outside) in stands of trees.

Type of tree (old stand)	Without foliage	With foliage
Deciduous trees		
Red beech	26—66	2—40
Oak	43—69	3—35
Ash	39—80	8—60
Birch	—	20—30
Evergreen trees		
Silver fir	—	2—20
Spruce	—	4—40
Pine	—	22—40

Because of the great importance of light reaching the ground for plants growing there, several attempts have been made to deduce a relation between the age of the stand or its density and the light intensity at the forest floor. For example, G. Mitscherlich [607] used photographic exposure meters, in the Buntsandstein area of the Thuringian Forest, to measure the intensity of light at the floor of many fir stands in the province of Dietzhausen forest station, in its relation to age of stand and yield class. Figure 160 shows his results for 87 different stands. The young stand, still flooded with light at first, completes its growth in roughly 17 yr under the conditions prevailing there, and then permits only 10 percent of the external light to pass through. As it ages from then

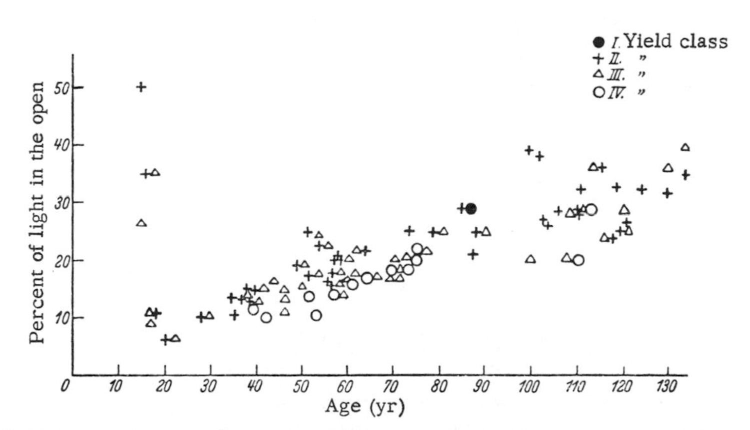

Fig. 160. Relation between the age of a stand of fir and the light penetrating to the forest floor. (After G. Mitscherlich)

onward, the stand becomes lighter again. The better-yielding classes have fewer but stronger trunks and therefore let more light penetrate than do the poorer-yielding types. Under the resulting climatic conditions, the forest floor remains dead when less than 16 percent of the outside light is able to penetrate; from 16 to 18 percent the first undemanding mosses appear; with 22 to 26 percent the berry plants are seen; and it is only with 30 percent that the first naturally seeded firs are observed.

The most comprehensive approach to the problem so far has been made by D. H. Miller [605, 606]. He laid down the following four rules to be adhered to in future series of measurements: (1) pure stands or mixed stands of as varied density as possible should be available; (2) the density of the stand must be measured as carefully as possible; (3) light measurements should cover the whole range of the spectrum; (4) every measurement must be sufficiently comprehensive to exclude any chance features of lighting or shading in the stand. It was Miller's experience that the most useful relations were obtained when he used the sum of all stem diameters per unit of surface area in assessing the light intensity at the forest floor. His unit of inches per acre has been converted to meters per hectare for use in Fig. 161. The nine different series of measurements from German and American sources, illustrated with different symbols in Fig. 161, led to a relatively satisfactory relation (for more details see [605]). Figure 161 also includes very light tree density of park land.

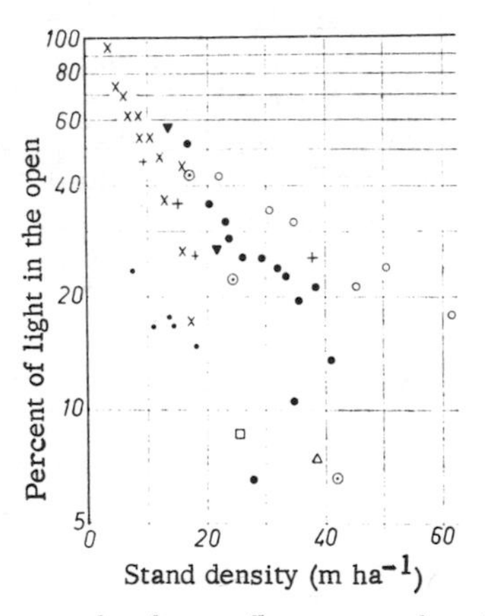

Fig. 161. Light penetration to the forest floor as a function of stand density. (After D. H. Miller)

It is well known that in tropical forests the amount of light penetrating to the ground is extremely small. The darkness of the tropical rain forest, the moist heat, and the odor of decay are the characteristics always emphasized. M. Gusinde and F. Lauscher [600] found, in a 30-m-high primeval tropical forest in the Congo, that there was only 1 percent of the outside light at 2 m above the ground; H. Slanar [614] recorded only 0.5 percent, and H. Eidmann [597] observed 0.4 percent in a mountain forest on Fernando Po. Figures at various heights in Panama, measured by W. C. Allee, are found in a book by P. W. Richards on the tropical rain forest [609]. According to this source the quantity of light as a percentage of the outer light was:

Position	Upper crown space	Tops of small trees	Trunk space	Forest floor
Height (m)	25—23	18—12	9—6	0
Light (percent)	25	6	5	1

Richards gives a figure of 0.5 to 1 percent for the light at the ground below the rain forest, exceeding 1 percent for short periods, but never more than 5 percent. Measurements made by G. C. Evans, using spectral filters, show that in the tropical forest 8 percent of the light was of short wavelength, in the range from 320 to 500 mμ, 22 percent in the green between 470 and 590 mμ, and 45 percent in the red beyond 600 mμ.

The preponderance of red in the internal radiation in forests is also encountered in mid-European woods. K. Egle [596] determined the illumination penetrating to the forest floor, expressed as a percentage of that incident on the foliage, for a number of wavelengths, obtaining the results shown in Table 73. It may be seen that, as the amount of foliage increases, the decrease in the intensity of the light that penetrates takes place more slowly in the long-wavelength than in the short-wavelength range; the light in the trunk zone will therefore contain a decreasing quantity of blue.

Table 73

Light intensity (percent of that on foliage) reaching forest floor.

Date	Wavelength (μ) and color					
	0.71 Red	0.65 Orange	0.57 Yellow	0.52 Green	0.45 Blue	0.36 Violet
12 March (buds still closed)	61	54	51	48	46	44
15 April	59	39	36	33	32	30
10 May	19	6	7	6	6	5
4 June	14	4	5	4	3	3

The difficulty of measuring the quantity of light in forests is largely due to its special structure under these conditions. At the forest floor light patches alternate with deep shadows, and the mosaic pattern of darkness and light changes with the movement of the sun. M. I. Sacharow [610] proved that the interior illumination of Russian forests increased as the wind grew stronger, which is understandable since the wind will change the disposition of the crowns of the trees. G. Sirén [613], on the other hand, was unable to prove this in the birch forests of Finland. It is difficult to establish representative values because of the great influence of the seasons, solar altitude, and weather. W. Nägeli [608] made a distinction between a basic illumination, corresponding to diffuse light in the wood, and the contribution due to direct sunlight, which is more or less great and of short duration. To understand the mosaic pattern of woodland light more fully, K. Brocks [593] investigated the differential illumination at surfaces inclined at different angles inside an oak and fir wood near Eberswalde. In each position he made 84 individual measurements of the illumination at a hemispherical surface 1 m above the forest floor. In comparison with the value outside the wood, measured simultaneously, the ratio of the lightest to the darkest area on the hemisphere was 10:1, in a stand of oaks without leaves it was 180:1, and after the trees were in leaf, 17:1.

Figure 162 illustrates a series of measurements made by H. Ellenberg [598] on two test areas 10 by 10 m, in two different types of woods. Light intensity was measured at the tops of the shoots of

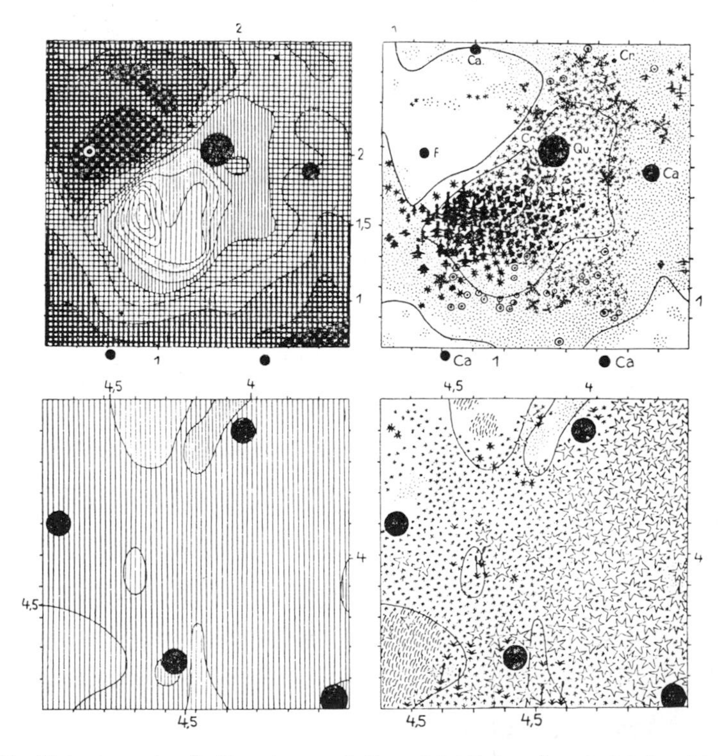

Fig. 162. Light intensity (left) and ground flora (right) in a forest of oak and hornbeam (above) and a beech forest (below). (After H. Ellenberg)

the flora growing on the floor of the wood, and the results are given in the left of Fig. 162. Black areas represent tree trunks. Lines of equal light intensity are drawn for every 0.5 percent, values for some of them being shown along the margins. The diagram on the top left is for a wood of oak and hornbeam while below it on the left is the diagram for a pure beech wood. The intensity of light in the latter is 4 to 5 percent of that in open country, and its distribution is remarkably even, while in the former it is noticeably darker, and differences of illumination in different places are very great because of the steplike build-up of the stand. The diagrams on the right show the flora on the surface of the test areas. Distribution is very closely correlated to the light factor, even if this is brought about in a very complex way, as may be seen from a glance at Fig. 162. The original work should be consulted to find the meaning of the symbols used to designate the flora.

Weakening of light intensity in woods means that twilight ends later in the morning and starts earlier in the evening. J. Deinhofer and F. Lauscher [594] were concerned with establishing the basis for legal definitions in this respect. With a cloudless sky the end of civil twilight (no longer possible to read a newspaper) was 16 min earlier in a deciduous forest, 20 min in a coniferous forest, and 28 min earlier in an old tall wood than in the open. With overcast skies the difference reached 45 min, and in one case when it was raining even 54 min. G. Scheer [612] estimated the seasonal variation of the intensity of global illumination in various tree-shaded areas of the Darmstadt Botanical Gardens by observing the time when birds started to sing in the early morning.

The exchange of long-wavelength radiation is determined, for the most part, by outgoing radiation from the upper active surface, in the case of an old stand with a closed crown area. This is made visible by the heavy deposition of dew in the upper part of the crown area after a night of outgoing radiation (see p. 325). In the trunk area there is little temperature difference, even in comparison with the ground below, so that the internal exchange of long-wavelength radiation is usually insignificant. If the trees are thinly planted, however, radiation may also take place from the ground below, according to the rules established for horizontal shielding influences in Sec. 5. This will be discussed further in connection with the microclimate of clearings in Sec. 38. If there is a snow cover in the wood, so that the temperature may not exceed 0°C, a stream of long-wavelength radiation will flow from the tree trunks, provided they are warmer, to the snow cover and contribute to melting it. This process was investigated, mainly by D. H. Miller [605], for the thinly wooded area of the Sierra Nevada in California.

Short- and long-wavelength radiation together determine the radiation balance. We are indebted to A. Baumgartner [591, 437] for the most complete sets of measurements of the radiation balance at various levels in a stand. The wood in question is not an old stand, but a young growth of firs 5 to 6 m high, 30 km southeast of Munich. The typical processes involved in arriving at the radiation balance may be seen in operation. These measurements will be referred to frequently in the subsequent discussion, and it is therefore appropriate that some general information about the measurement site should be given here. A full description is to be found in K. Mauerer [604].

The area was level, and the depth of penetration of weather factors was 60 to 90 cm. It had a 20-year-old natural growth of young firs, under which there was on the ground a 2-cm-deep layer of needles and brushwood. The trunk area was filled with dead underbrush to a height of about 2.5 m. The crown area, which began here, extended upward, in closed form, to about 4 m, above which there were numerous tops. This will be called the top space in the future, and it reached to a height of 6 m. The quantity of wood was 189 $m^3 ha^{-1}$, and there were 349 trees per 100 m^2 of surface in the crown area.

Climatic investigations were transferred to such a young and low stand because the costs in material (construction of observation station) and in labor are small in comparison with those in an old stand, and many of the more significant traits of forest climate can be recognized in the very early stages. Figure 163 shows how the instruments were set up. Cup anemometers *W* could be fitted at nearly three times the height of the stand without difficulty, thus obtaining a good picture of the wind field above the wood. Thermographs and hygrographs *T, H* were set up at five different levels, up to 10 m above the ground; radiation-balance meters *S*, as used by F. Albrecht and G. Hofmann [602] were fixed in three positions,

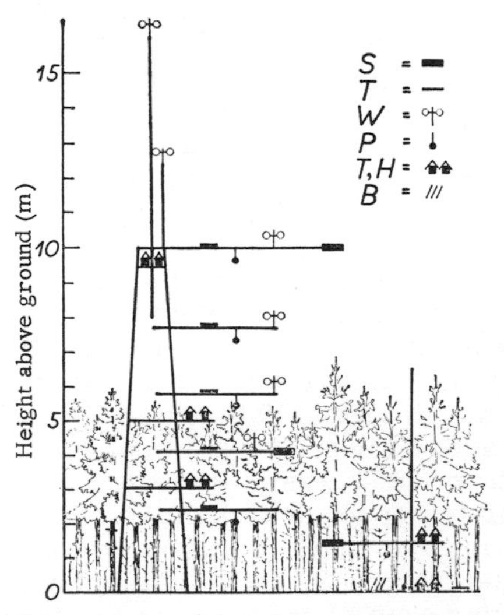

Fig. 163. Arrangement of instruments for studying stand climate in a growth of young fir near Munich. (After A. Baumgartner)

later increased to five, at the end of long horizontal poles so as to be completely exposed. Earth thermometers *B*, Leicke dew plates, and Piche evaporimeters *P* completed the instruments.

Figure 164 shows the typical variation of radiation balance for a bright day in the late summer of 1951. Above the stand (heavy line), the uniform, negative balance by night changes barely an hour after sunrise in the steeply rising day curve. Although the weather was cloudless until 11:00, and clouds appeared only on the southern horizon up to 13:00, the change in evaporation and, after 16:00, isolated cumulus clouds caused much unsteadiness in the radiation balance. One and a half hours before sunset, radiation equilibrium is again established. Below the crown area at 4.1 m (dotted line), radiation varies with random shadow effects from the tree tops and branches. At 2.4 m it is only when the sun is high that any radiation worth mentioning is able to penetrate inside the wood.

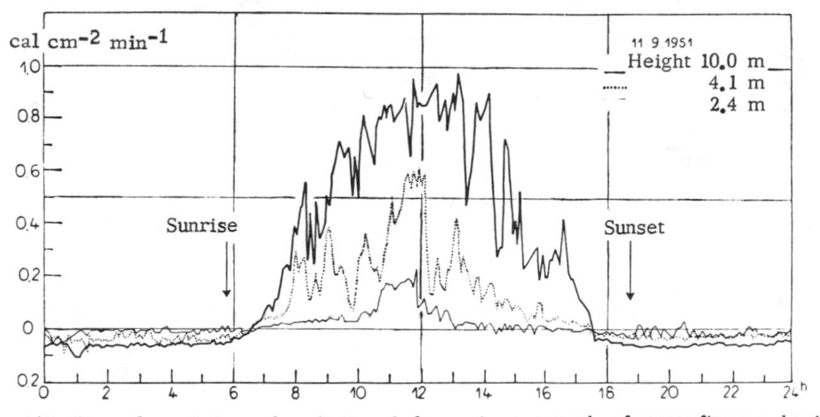

Fig. 164. Diurnal variation of radiation balance in a growth of young fir on a day in late summer. (After A. Baumgartner)

While Fig. 164 reflects the variations occurring on a single day, Fig. 165 is for the average values observed over a dry period in midsummer, from 29 June to 7 July 1952. When the sun's rays strike the tree tops tangentially at sunrise and sunset, the radiation-balance meter between the tree tops at 5 m records a remarkably low value, because of temporary screening, in comparison with the instrument in the open at 10 m. By night the radiation loss is reduced through partial shielding by the tree tops. Irregularities in the balance at the top of the crown area (4 m) and in the trunk area (3 m) reflect the random selection of the observation point, but the calculated daily mean values give a true picture of the situation at these levels.

The daily variation is shown, for the same dry period, in the form of isopleths in Fig. 166. The tops and not the crown area first start the absorption process. The height of the sun can be

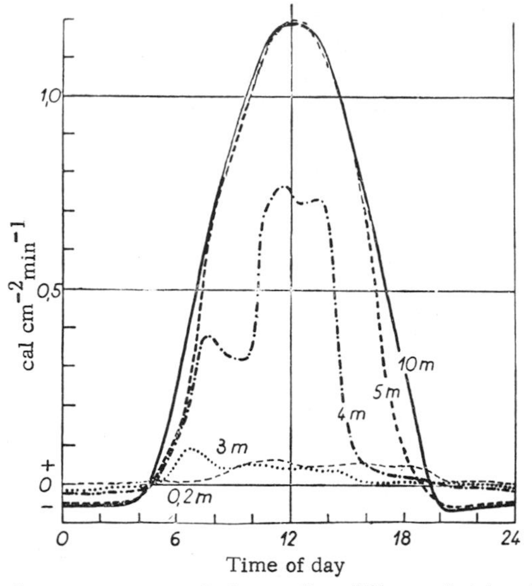

Fig. 165. Smoothed temperature variation at five different heights in a young fir plantation during a dry period in midsummer. (After A. Baumgartner)

noticed in the shallow dip in the crowded isopleths about the noon hour, when it reaches its maximum altitude. The heated tops and crown area radiate warmth downward in long waves to the cooler ground below. This flow of heat is kept up until midnight, until the radiation emitted from the tree-top zone from sunset onward has lowered the temperatures there too. Then the forest floor will also lose heat by radiation, part of it streaming toward the top zone, which is undergoing further cooling, and another part escaping through gaps toward the night sky.

The hourly average values of the radiation balance for five different heights can be found for 2, 4, 5, and 7 July 1952 in Baumgartner's publication of detailed figures [592]. If the averages are

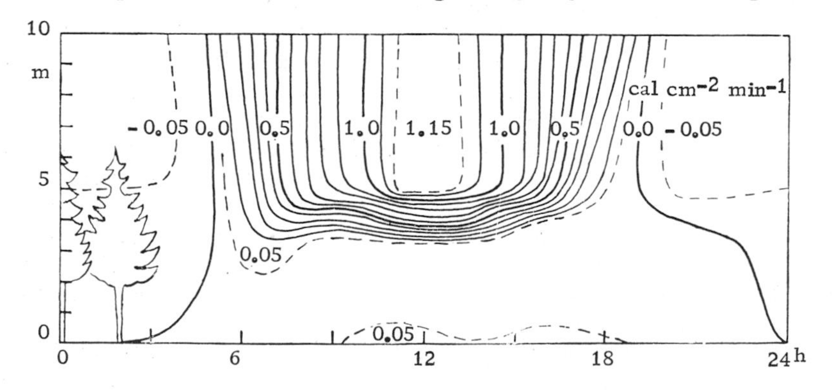

Fig. 166. Isopleths of radiation balance for the same time and place as in Fig. 165.

computed for the period, the results given in Table 74 are obtained, the term "night" meaning the period of negative radiation balance, and "day" that of positive radiation balance.

Table 74
Radiation balance at different heights in a stand of young firs.

Height of measurement (m)	Duration (hr)		Average balance (cal cm^{-2} min^{-1})		Daily balance	
					cal	
	Day	Night	Day	Night	cm^2 day^{-1}	%
10.0, above the forest	15.0	9.0	0.370	- 0.054	603	100
5.0, tree-top area	13.6	10.4	0.370	- 0.046	551	91
4.1, crown area	15.0	9.0	0.164	- 0.014	260	43
3.3, trunk area	17.0	7.0	0.013	- 0.011	29	5
0.2, on forest floor	21.0	3.0	0.020	- 0.005	44	7

In the radiating tree-top zone the period of negative balance is longest, but close to the ground it lasts only for the 3 hr before sunrise. The third position is uncertain in the value given for the mean balance, but the sequence of the figures indicates the reliability of the method of measurement. It is not possible that the positive balance, for the wood as a whole, at 0.2 m should be greater than at 3.3 m, as shown in the above figures and indicated by the 0.05 line at the forest floor in Fig. 166. But it is possible to understand that the accidental positioning of the instruments might mean that more "sun spots" were received below than in the trunk area where the higher instrument was located.

34. Heat Balance and Wind in an Old Stand

Knowledge of the radiation balance is prerequisite for understanding the heat and water budgets of an old stand. The main problems in the heat balance of the earth's surface without vegetation have already been dealt with in Secs. 25-28. The influence a vegetation cover might have on the heat balance was, however, left out of consideration for the time being. This section will deal with the tallest of all vegetation covers, the forest.

The influence of the forest depends on three factors. In the first place there has to be taken into account the quantity of heat required by plant metabolism. Secondly, the thermal capacity of trunk, branches, twigs, leaves, or needles has to be considered. Finally, we have to think of the part played by the air mass in the trunk area (p. 299), which forms a transitional zone between the forest floor and the free atmosphere.

When coal burns it releases the energy accumulated by the forest in earlier geologic history, obtained from sunlight and stored in the assimilation process (photosynthesis). In its green chloroplasts (the green coloring matter) the plant produces glucose from the carbon dioxide extracted from the air and water taken from the sap flow. The energy required to complete this chemical process is provided by solar radiation; photosynthesis therefore takes place only by day. It takes about 4000 cal to produce 1 g of glucose. We may therefore calculate the amount of energy used from the amount of material produced. According to A. Baumgartner [437], the energy requirement of a wood for purposes of photosynthesis is about 0.005 cal cm^{-2} min^{-1}, or roughly 1 percent of incoming solar radiation. D. H. Miller [605] reached a similar conclusion from examination of the growth of timber in the Sierra Nevada forests in California. The estimated annual increase of 0.01 g cm^{-2} is equivalent to a gain of energy of 50 cal cm^{-2} in the combustion of wood, which leads to an equally small requirement. The quantity of energy used up in photosynthesis has been determined in Japan recently by E. Inoue and his assistants [621-623] on the basis of aerodynamic measurements (carbon dioxide, temperature, moisture, and wind profiles) over wheat and rice fields. The values obtained, varying with position and time, had a maximum of 0.01 cal cm^{-2} min^{-1}, which is of the same order of magnitude. The amount of energy used varies during the day as radiation intensity increases; but since other factors also influence the rate of assimilation, such as the size of the openings (stomata), the maximum does not coincide with the noon maximum of radiation intensity; the stomata tend to close about noon. The daily variation shows one maximum in the late forenoon and a second during the afternoon.

Against the energy consumed in photosynthesis there is a gain of energy through photolysis to be placed in the balance. When plants breathe (respiration), sugar is broken down into carbon dioxide and water. In contrast to photosynthesis, photolysis takes place during both day and night and leads to a nocturnal loss of weight by the leaves. The gain of energy is of the same order as the energy consumed in photosynthesis, about 0.005 cal cm^{-2} min^{-1}. This, however, may amount to 10 percent of the total radiation exchange during the night, when it is much smaller. Heat released in respiration can lead to a temperature increase of a few tenths of a degree, and many believe they have observed this. Since respiration increases quickly when temperature rises, it is subject also to a diurnal variation with a maximum in the early afternoon, at the time of the temperature maximum.

The metabolism of the forest is therefore made up of the interplay of assimilation and respiration, energy consumed and energy recovered. For the young fir plantation near Munich, A. Baumgartner calculated the following values (mcal cm^{-2} min^{-1}):

Hour of day:	3—4	6—7	9—10	12—13	15—16	18—19	21—22
Photosynthesis:	0	— 10	— 14	— 12	— 13	— 10	0
Respiration:	5	6	7	8	8	7	7
Metabolism:	5	— 4	— 7	— 4	— 5	— 3	7

With the high degree of accuracy possible today in measuring the quantities involved in the heat balance, this amount of a few thousandths of a calorie may be neglected. Plant metabolism will therefore be left out of all further discussions.

The importance of the two other factors may be readily recognized from a consideration of Fig. 167, also for the young fir plantation near Munich. The diagram shows the diurnal variation of heat exchange for the fair-weather period from 29 June to 7 July 1952, evaluated by A. Baumgartner [437]. The dotted curve

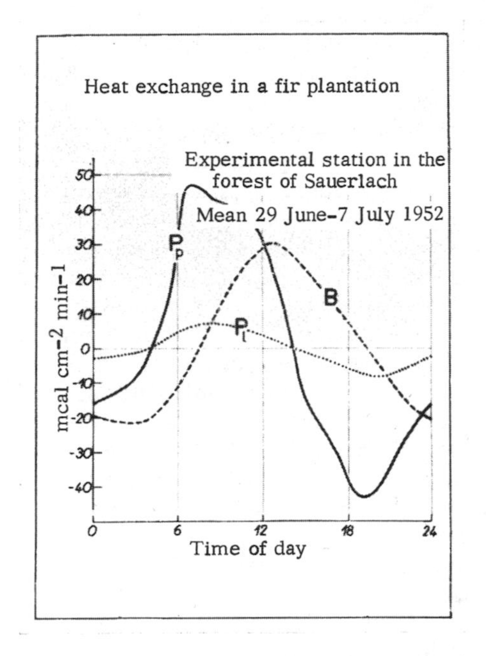

Fig. 167. Diurnal variation of the small quantitites of exchanged heat within a fir plantation and the ground below it. (After A. Baumgartner)

marked P_l shows the amount of heat exchange in the air mass within the wood. The maximum is ± 0.01 cal cm^{-2} min^{-1}, which is not surprising in view of the small thermal capacity of air. The quantity P_l may therefore practically always be neglected even in an old stand of considerable height. In contrast, the amount of heat P_p exchanged in the wood and the needles is great after sunrise, when the trees are heated, and at sunset, when they cool, and is even greater than the quantity of heat B exchanged at the forest floor. This depends on the thermal capacity of the substance, for which data were given on p. 258. While wood and air masses respond quickly to changes in radiation, the forest floor follows only indirectly the daily variation of the stand that shelters it. Hence the daily variation of B shows a phase difference of 4 to 6 hr in comparison with P_p and P_l.

In spite of the importance that the information in Fig. 167 has for our knowledge of the processes involved, we must not lose sight of the fact that a greatly exaggerated vertical scale has been chosen. If the heat exchange in the forest as a whole is investigated, P_l, P_p, and B become insignificant; it is more appropriate to include all that takes place in the stand as if it were a part of B. The value of B, after being corrected by P_l and P_p, can then be seen in its proper proportions in Fig. 120, which is for the same wood and the same period. The insignificance of B compared with S, V, and L is evident. The average daily energy balance also shows for that particular period of fair weather the following values for these factors: $S = 586$, $V = -386$, $L = -197$, while B is only -3 cal cm^{-2} day^{-1}. However, in contrast to the other factors, the factor B involves incoming and outgoing quantities of energy that are practically equal. A better idea of the proportions is therefore obtained by adding the flow of energy in both directions, regardless of sign, to obtain: $S = 644$, $V = 396$, $L = 209$, and $B = 67$ cal cm^{-2} day^{-1}. It may then be seen that B accounts for only 5 percent of the total exchange of 1316 cal cm^{-2} day^{-1}.

In order to understand the temperature and water-vapor distribution in the forest, it is essential to know the wind profile as well as the radiation budget in an old stand. The amount of eddy diffusion taking place, depending mainly on the wind strength, will determine whether the radiative heat being taken up or given out by the tree stand may remain in it or be transported elsewhere.

The braking effect of the wood on the wind and on air flow within the stand is not basically different from the influence of plant cover on the vertical wind profile, already discussed in Sec. 30. The height reduction d (described on p. 274) is naturally much greater for a wood than for a grain field or a plowed field. A special feature, which may often be observed with woods, is that air movement is less restricted in a trunk area free of branches, particularly when the wind can blow in through the open borders of the stand. The wind profile with three different wind speeds in a

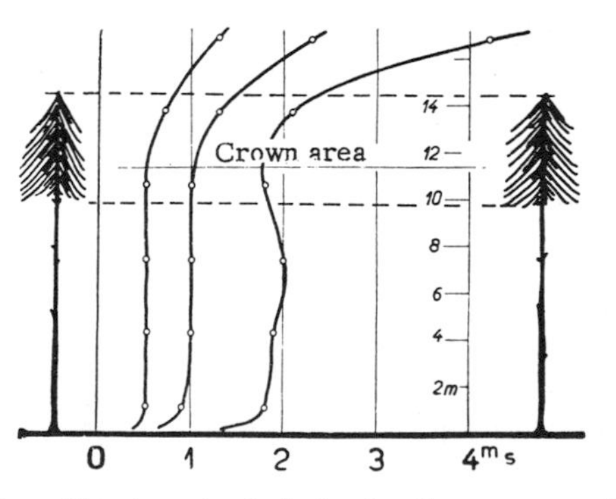

Fig. 168. Wind profiles in a stand of pine for three ranges of wind speed.

thin fir stand in the Bavarian forest of Wondreb (Upper Palatinate) is illustrated in Fig. 168, using measurements made by R. Geiger [618]. When winds are strong, the light current through the trunk area can be seen. H. Ungeheuer [626] was able to show, for a high beech wood on a slope in the Taunus hills, not only that the katabatic wind at night (see Sec. 43) was able to pass through the wood, but that its passage was even favored by its being protected from external disturbances by the ceiling of the forest crown.

Another peculiarity of the forest is the influence of its foliage on the wind field. R. Geiger and H. Amann [619] measured the wind profiles shown in Fig. 169 in a 24-m-high, 115-yr-old oak stand near Schweinfurt, which had beech saplings 40 to 50 yr old growing below. The diagram shows the wind profile before and after the trees were in leaf for similar winds outside (at a height of 28 m). Naturally the wind is able to penetrate more deeply into the wood before the leaves open. This is of practical importance

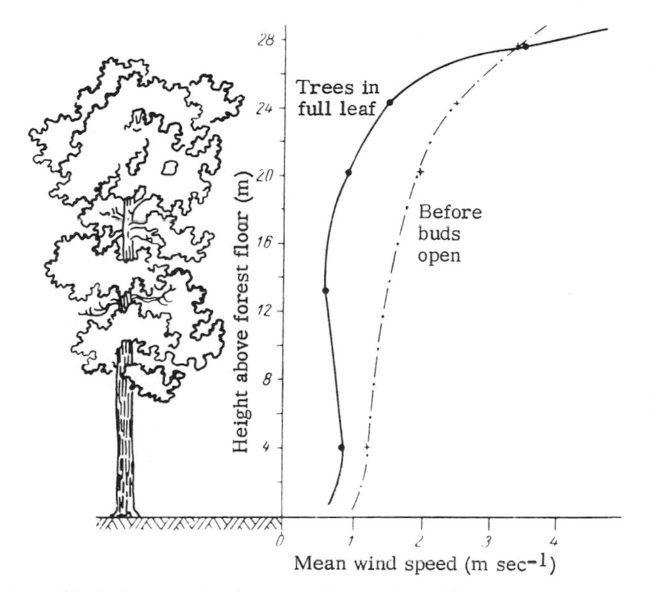

Fig. 169. Influence of foliage on the wind profile in an oak forest.

in attempting to destroy larvae by dusting with insecticide from an aircraft. Before the trees are in leaf, not only are surfaces lacking on which the poison dust might settle, but also the stronger winds carry it away more quickly downwind, and greater turbulence scatters the powder more widely than when the foliage is present later.

The contrast in conditions before and after breaking into leaf shows up most clearly in the frequency of hours with calm ($u < 0.7$ m sec^{-1}); in 206 hr of observation before, not a single hour of calm was recorded among the tree tops, whereas after leaves were present calm conditions were recorded in 10 percent of 494 hr of observation. At a height of 4 m above the forest floor, however, the number of hours of calm rose from 67 to 98 percent.

The extent to which wind can penetrate a wood depends on several factors. Thickets and growths of saplings bring the air to a complete halt. In old stands the type of woood and the degree to which the crown is closed are of importance. Figure 170, after R. Geiger [618], shows comparative wind profiles in two stands of the forest

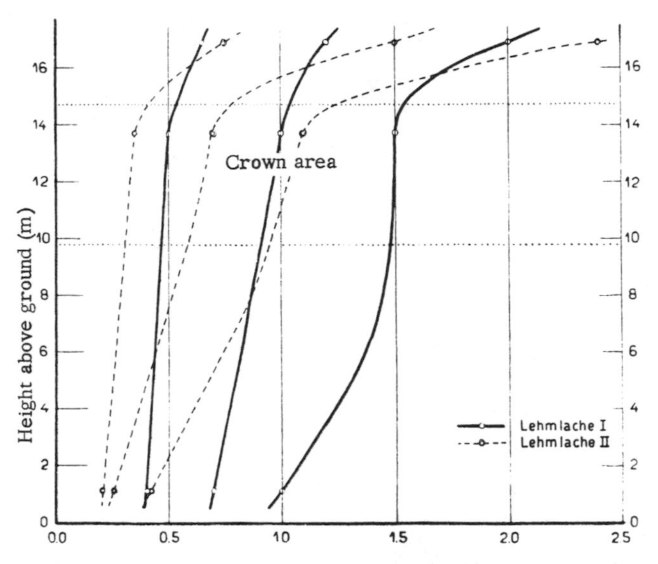

Fig. 170. Comparison of wind profiles in two stands of pine, one without (I) and the other with (II) an underbrush of fir.

of Wondreb, already mentioned, which had been developed on different lines, and which were only 86 m apart. The stand called "Lehmlache I" was a 65-yr-old pine forest with a uniform, loosely knit crown. In "Lehmlache II" the same kind of stand had an association of fir of different age groups below, so that crowns were found at all levels. This reduced the wind speed very effectively, as shown in Fig. 170, with the immediate result that in the trunk area it was cooler by day and warmer at night, while the crown area experienced more extreme temperatures than the thinner, better-ventilated wood.

G. Sirén [613] took anemometer readings three times daily at three different heights in a spruce and birch forest in north Finland at Siulionpalo (67° N). The comparison was made for five ranges of wind speed, and although wind speed inside the birch forest was somewhat less, with winds of over 2 m sec^{-1}, the differences were otherwise insignificant.

There is very little real information on the diurnal variation of wind speed within forests, but calm conditions are the distinctive feature of their climate. The normal type of cup anemometer used for continuous recording is often kept motionless by the friction of its support, and responds only to stronger isolated gusts which penetrate down from above after a surprisingly long time lag. This may best be observed in stormy weather, when individual tree trunks perfomr a circular oscillatory movement, continually altering the structure of the crown roof. This gives a good insight into the character of the forest in question, and is often more conclusive than measured wind profiles in deciding the degree of response to external winds.

An example of the daily wind variation is given in Fig. 171 for the plantation of young fir near Munich. In contrast to Fig. 166 and Figs. 175-177 which follow, the height scale is extended well beyond the top of the forest. The numbers on the isopleths give wind speeds (cm sec^{-1}) on the fine day selected for this experiment.

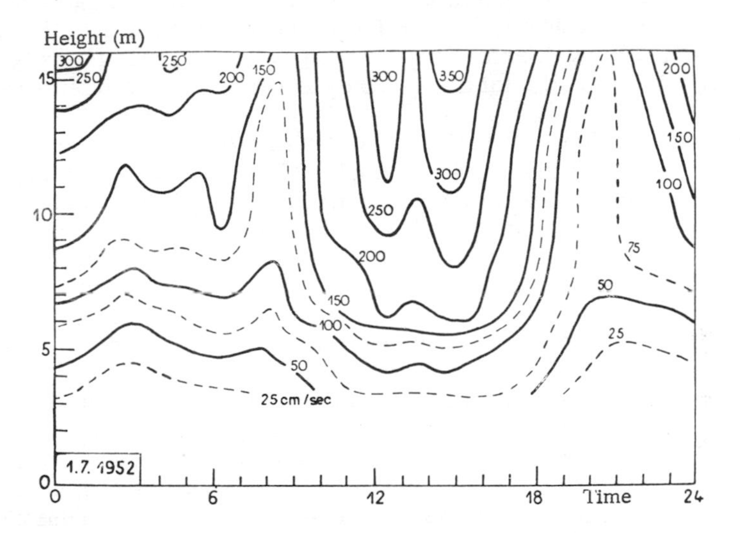

Fig. 171. Isopleths of wind speed in a fir plantation near Munich. (After A. Baumgartner)

The tree tops being at 5 m, it appears that winds of more than 0.5 m sec^{-1} are to be found in the crown area only between 10:00 and 18:00. The well-known diurnal variation of wind force in the open, with its maximum in the early afternoon, can be recognized clearly only above the stand. A second maximum about midnight is the result of a local down-slope wind, which usually blows over the upland plains in Bavaria in areas close to the Alps in this kind of summer-weather situation.

35. Air Temperature and Humidity in an Old Stand

Measurements of air temperatures in forests are obtained preferably by making recordings of vertical profiles. Table 75 summarizes the series of measurements available at the present time. The investigations of K. Göhre and R. Lützke extended over a number of cultivated surfaces, with and without a protective cover. N. v. Obolensky [700] made psychrometer measurements at four different heights in a pine and young oak plantation near Leningrad in the early summer of 1922. Reference should also be made to the work of K. Ermich [691-693] in low types of timber.

A distinction has been made in Table 75 between stands on level ground and those on hillsides, since the latter are subject to topographic influences (Chapter VII). In the column headed "method," the letters have the following meanings: *H* is for thermographs in screens shielded against radiation, though the results are not free from radiation errors; *HA* means that thermograph readings were reduced to true air temperature every 2 hr by taking readings with an aspiration psychrometer; *A* is for readings by aspiration psychrometer, *Th* by thermistors, which were housed in protective screens, but this did not necessarily exclude errors due to radiation.

In investigations by Geiger and Amann [619] in the Schweinfurt Forest of 115-yr-old oaks 24 m high with 40- to 50-yr-old saplings below, thermocouples were used in addition to thermograph recordings. Although subject to a noticeable radiation error, which became marked in the case of the elements placed above the crown area, the diurnal variation resulting from these observations on a sunny day, 18 August 1930, shown in Figs. 172 and 173, will be the subject of discussion, for the reason that these readings provide us with a vivid picture of a summer day in the forest.

Readings were started before sunrise, when temperature was at its lowest in the radiating crowns of oaks at a height of 23 m, and measurements were made at five heights at 30-sec intervals. At sunrise the first thermoelement to receive heat was the one above the level of the stand. At this hour there is a marked boundary surface in the crown area; within the stand it is still nighttime and the air is cool and moist; above it the morning sun is heating strongly and reducing the relative humidity. "The division is clearly perceptible to all who climb the observation ladder on a clear morning; thousands upon thousands of winged insects fill the space above the boundary surface, and this living cloud is sharply cut off at its lower edge" (R. Geiger [618], p. 852). Temperatures in the crown area react toward 08:00, slowly at first, because their rich coating of dew must first be evaporated. The great temperature instability in the crown area makes a contrast to the uniform and slowly increasing temperatures of the trunk area.

Table 75

Summary of air-temperature measurements in forests.

	Year of publication	Author [reference]	Stand investigated			Method of measurement
			Age (yr)	Type and character of wood	Height (m)	
On the level	1925–26	R. Geiger [618]	65	Old pine stand (with and without a fir underbrush)	14	*H*
	1931–32	R. Geiger and H. Amann [619]	115 45	Old oak stand with beech saplings	24	*H*
	1952–1956	A. Baumgartner [427]	20	Young growth of fir	6	*HA*
	1956	K. Göhre and R. Lützke [620]	90	(*a*) Mixed pine-beech stand	20	*A*
			81	(*b*) Old pine stand	17	*A*
			26	(*c*) Pine saplings	11	*A*
On slopes	1934	H. Ungeheuer [626]	136	Old stand of beeches on a NW slope	17	*H*
	1941	F. Sauberer and E. Trapp [624]	135	(*a*) Old beech stand on a SE slope	28	*H*
			90	(*b*) Mixed stand of beech, fir, and silver fir on a N slope	22	*H*
	1957	A. Baumgartner and G. Hofmann [616]	100	Old fir stand on a WSW slope	30	*Th*

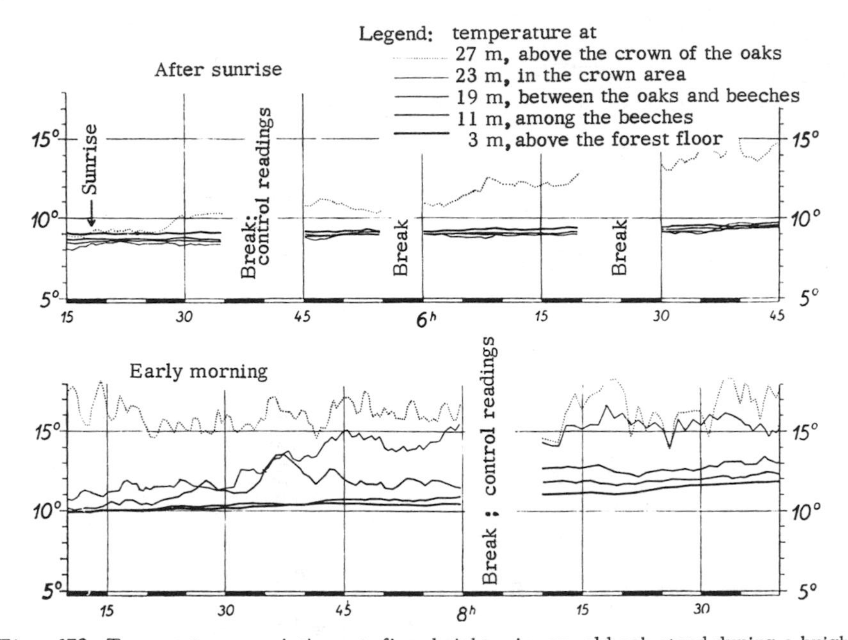

Fig. 172. Temperature variation at five heights in an old oak stand during a bright summer morning.

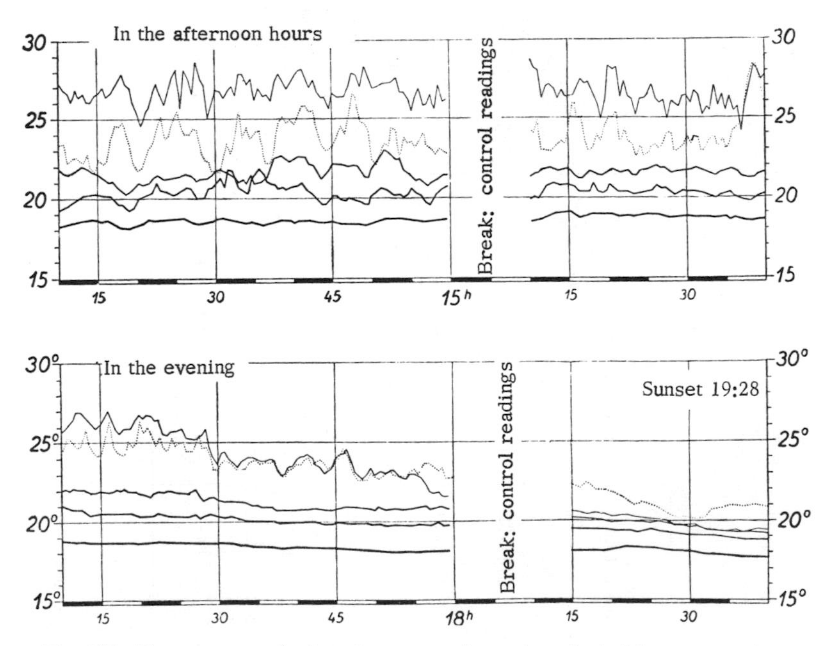

Fig. 173. Temperatures in the afternoon and evening of a bright summer day.

About noon (Fig. 173) a state of equilibrium is reached; temperature curves then run horizontally, being distinguished from each other by their degree of instability. Temperature decreases from the maximum in the crown area upward into the free atmosphere, and downward into the trunk area, as it should do in an incoming-radiation situation. The temperature decrease in the evening, in contrast to the increase in the morning, leads to an increasing stabilization in the stratification of the air, whereas morning heating produced increasing instability.

The interrelation between temperatures at different levels depends, naturally, on the type of tree, the age of the stand, and its structure. K. Göhre and R. Lützke [620] emphasized strongly that the vertical temperature gradients found in a stand may be markedly distorted by radiation errors affecting the instruments used. Since only those elements that are positioned in the crown area and above it are exposed to radiation both day and night, while others in the shelter of the stand are not so exposed, the diurnal temperature variation in the crown area is exaggerated. The readings give the erroneous impression of a much denser forest. Since this justifiable objection detracts from the value of earlier series of measurements, making them of little value, the discussion which follows will be based on two studies that are faultless in this respect.

Figures 174 and 175 represent two types of stand, which exhibit basic differences in their temperature variation, namely, a thin and a dense stand. Temperature isopleths in Fig. 174 are based on 24 vertical temperature profiles published by K. Göhre and R. Lützke [620], measured in a mixed pine-beech stand near Eberswalde, on 14 and 15 September 1953. The stand consisted of 82- to 100-yr-old pines of 20-m average height, intermingled with single birch and beech trees, and had an inter- and undergrowth of 10- to 60-yr-old, weak to strong beeches of an average height of 13 m. The measurements made on this clear day were at eight levels, extending a little beyond the crown of the firs.

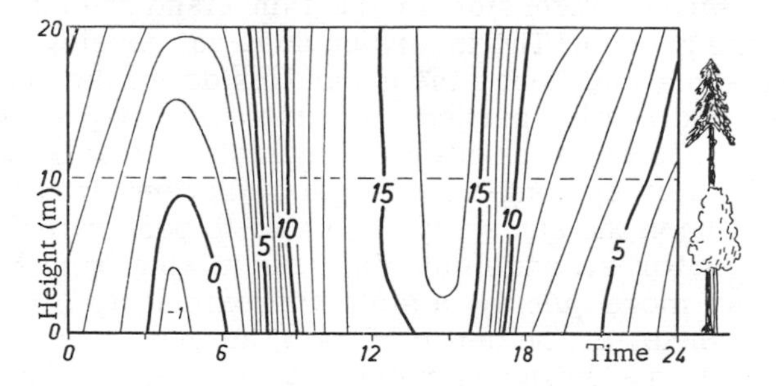

Fig. 174. Temperature variation during the day in a thin pine wood near Eberswalde. (After K. Göhre and R. Lützke)

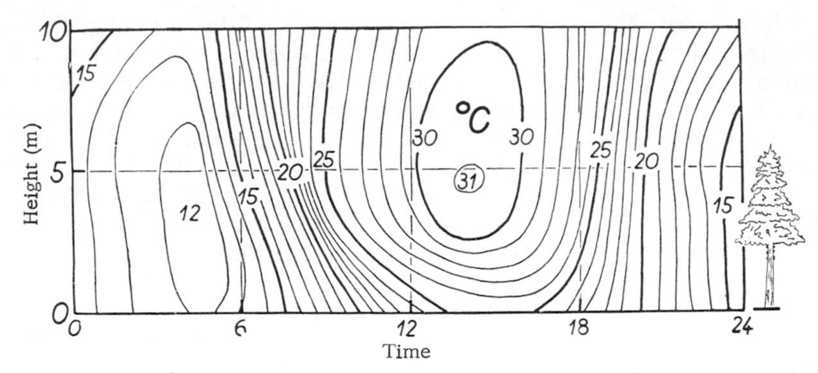

Fig. 175. Temperature variation during the day in a young fir plantation near Munich (compare Figs. 166 and 175-177).

The same method of representation is used in Fig. 175 for a dense stand, using the series of measurements, already consulted frequently, from A. Baumgartner [437] for the fir plantation near Munich. The readings used are averages of a 10-day period in summer (28 June to 7 July 1952), for six levels, extending to twice the height of the trees; this should be remembered when comparing the two diagrams.

The isotherms are crowded at the time of maximum heating a few hours after sunrise and during the cooling period before sunset, the former being more intense than the latter. In the less dense stand (Fig. 174), the whole air mass of the forest is affected at times, while in the denser (Fig. 175) heat penetrates only slowly from the crown down through to the forest floor, so that the bundle of isotherms is tilted downward from left to right. At the time when the temperature variation changes from plus to minus or vice versa, that is, at sunrise and after midday, the wide spread of the isotherms illustrates the change.

In the thin stand there is a difference between the tree tops and the forest floor of 1 deg at noon and 5 deg at night, while in the denser stand it is 6 deg at noon and 1 deg at night. Consequently, there is a strong inversion in the thin stand at night similar to that over bare ground. The inversion does not develop in the dense wood in the same way, since the ground inside it takes part in night cooling only indirectly, perhaps because cold-air pockets that develop in the crown area filter downward, or maybe because the ground loses heat by long-wavelength radiation toward the colder tree tops. At noon the sun's rays are able to penetrate farther into the less dense forest, and more vigorous mixing brings the heated air aloft down more quickly, a state approximating isothermy thus being established. In the denser forest, neither sunshine nor wind can penetrate to a significant extent; and so the air in the trunk area remains cool and moist, contrasting with the warmth in the zone of the tree tops.

Figures 174 and 175 should be taken only as indications of types of temperature distribution. The influence of radiation and wind varies with the weather situation, and it is possible for a single stand to exhibit different types of distribution on different occasions.

With such temperature profiles at the time of strongly positive radiation balance in summer, the mean day temperature must be higher in the crown area than in the interior of the forest or even at the ground. The daily mean temperatures for the Eberswalde pine forest from 14 to 15 September 1953 were:

Height (m):	0	5	10	15	20
Temperature (°C):	7.1	7.4	8.1	8.4	9.0
Relative humidity (%):	84.0	80.2	79.6	78.4	77.8

The high mean temperature in the crown area results from its warmth at night (Fig. 174). In the dense fir plantation, on the other hand, it is mainly due to the high midday temperatures and is therefore more strongly marked in spite of the stand's lack of height. This may be seen by a study of the figures for the dry period from 29 June to 7 July 1952 (Table 76).

Table 76

Temperature and relative humidity in a dense fir plantation.

Height of measurement (m)	Temperature (°C)		Average water-vapor pressure (mm-Hg)	Relative humidity (%)		
	Daily average	Daily fluctuation		Daily average	Daily fluctuation	Average for overcast days
10.0, above the forest	22.3	16.4	11.9	63	58	76
5.0, in tree-top area	21.6	19.4	11.2	63	62	80
3.0, in crown area	21.1	19.0	12.2	70	62	84
2.5, in trunk area	20.8	18.4	11.7	69	60	86
1.5, in area of dead branches	19.6	16.5	11.5	71	60	87
0.2, at forest floor	18.3*	14.0*	12.5	79	45*	90

The air at the ground is on the average 4 deg cooler, and the difference between day and night is 5.4 deg less than it is aloft. All of this applies to summer conditions. It may be expected that in winter, when the radiation balance is negative, the opposite will be found, with the air at ground level warmer than among the tree tops on the average; there are, however, no recent measurements to show this.

The practical importance of this heat distribution has been shown by E. Schimitschek [625] in his observation of the development of bark beetles (*Ips typographus*) in pine woods near Lunz in Austria. He established a link with temperature; the bark beetles being to swarm only at 20°C. Higher bark temperatures, and the more frequent and longer duration of periods of high temperature in the crown area, result in a quicker development of this forest pest there than at the ground. The primary attack on a closed fir stand starts in the crown area and spreads slowly downward.

The forest must transpire in order to live, and it therefore acts as a source of water vapor. Table 76 above gives also the vertical profile of water-vapor pressure in the Munich fir plantation. There is a first maximum with a vapor pressure of 12.5 mm-Hg at the ground, while in the area of dead branches above this there is a minimum. The second maximum, due to water vapor being given off by the tree crowns, appears to be lower because drier air from the outside is mixed with it from above. This process can be seen from the isopleths of water-vapor pressure in Fig. 176. The diagram is for the same dry period as in Figs. 166, 171, and 175.

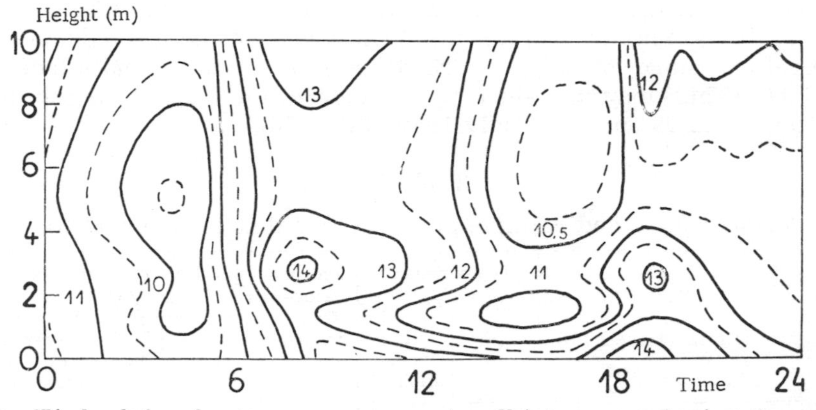

Fig. 176. Isopleths of water-vapor pressure (mm-Hg) in a young fir plantation near Munich.

In the late afternoon a minimum of water-vapor pressure is found in the region of the tree tops, perhaps accentuated by reduced stomatic openings and diminished transpiration. Another is located in the dead branch area, while the forest floor and the crown remain somewhat moister. The familiar double wave of vapor pressure can be followed from Fig. 176 into the trunk area. The night minimum is caused by a reduction in evaporation occasioned by lower temperatures, while the noon minimum is caused by water vapor being transported away rapidly by increased convective mixing (see pp. 40f). The main reason for its deep penetration, however, is the extreme dryness of the weather. This was observed within the forest by K. Göhre and R. Lützke, even before the trees were in leaf. Normally water-vapor pressure in the stand is constant with time during the day, although its distribution is unstable.

The distribution of relative humidity is much simpler. With the kind of temperatures found here, a difference of 1 mm-Hg in

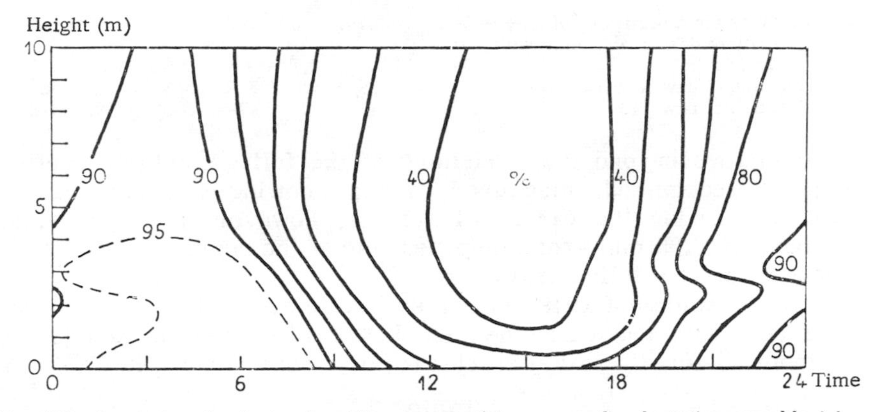

Fig. 177. Isopleths of relative humidity (percent) in a young fir plantation near Munich. (After A. Baumgartner)

water-vapor pressure is equivalent to a change of about 6 percent in relative humidity. The humidity field is therefore determined mainly by the temperature field. The configuration of the isopleths in Fig. 177 therefore bears a broad similarity to that of Fig. 175. The relative humidity is high during the day in the cooler trunk area, and low among the tree tops.

The daily average of relative humidity therefore decreases with height. This has already been indicated by the figures on p. 321 for the Eberswalde pine wood. In the fir plantation near Munich a similar result is found for overcast days, in the last column of Table 76, the period from 9 to 14 July 1952. Average figures for the dry period in midsummer, like those for vapor pressure, give two maxima: at the forest floor and in the crown area. This is a result of averaging for the whole day, because at night the crown area is often more humid (deposition of dew), while the air near the ground is always relatively the most humid during the day. The protective climate of the trunk area is characterized particularly by its small variation in relative humidity, as shown in the second-last column of the table.

36. Dew, Rain, and Snow in an Old Stand

A tree standing alone will shelter the ground below from outgoing radiation at night. Below a crown covered with dew there will be a dew-free patch of ground. A. Mäde [642] placed a number of dew-recording instruments under a beech tree on the experimental farm in Etzdorf near Halle. The lower rim of the canopy of foliage was 5.2 m from the tree trunk, and the instruments were situated in a radial line outward from the tree. Measurements in the months from May to July 1951, when the tree was in full foliage, gave the following quantities of dew, times of deposition, and duration of fall for the quarter year:

Distance from periphery of tree (m):	−3.2	0	2	4	6	8
Quantity of dew on all nights with dew (mm):	0.1	1.1	1.6	2.0	2.0	2.1
Duration of dew deposition (hr):	39	330	370	388	382	368
Duration of dew (hr):	47	420	479	500	499	473

Two meters beyond the periphery of the foliage no further effect can be recognized, measured values showing a purely random scatter. Under the circle of foliage, however, the very small amount of 0.05 mm arose only because at the beginning of May the leaves were not fully out.

In the autumn of 1949 Mäde installed dew recorders at a number of levels within and outside the shelter of the beech. Results for October, before the leaves started to fall, were as shown in Table 77.

Table 77

Dew measurements under an isolated beech.

Height above ground (m)	Quantity of dew (mm)		Duration of deposition of dew (hr)		Duration of dew (hr)	
	Inside	Outside	Inside	Outside	Inside	Outside
14.0, above the tree	0.95	0.95	94	94	135	135
10.5, under crown surface	0.42	0.90	92	118	118	155
5.0, in crown of beech	0.08	0.50	47	114	62	155
1.0, under the crown	0.04	1.42	13	200	17	264
0.1, under the crown	0.05	1.42	13	268	18	334

Radiation conditions were approximately constant at all levels outside the beech; the small divergence of the radiation flow, as established by H. Kraus [451], might bring a slight increase of dew with height, through the radiation contribution of the dewfall (see p. 254). However, the figures in Table 77 show a definite decrease with height in the open, which must be caused by the ventilation-humidity term, which increases with height as a result of saturation deficit and increasing wind speed, and therefore the quantity of dew deposited is less. The dew profile in the open is, of course, only a theoretical conception, since there are no surfaces there on which it can form.

In the crown area of the beech dew decreases very quickly both inward and downward, and in both intensity and duration. Close to the beech in the open air at 14 m the deposit of 0.95 mm of dew was a little greater than that at a height of 10 m. A probable reason for this result is the effect of the layer of cold air in the crown area of the radiating beech, an example of which has already been given in our discussion of the climate of vineyards (p. 295).

Turning from isolated trees to closed stands, the dew profiles measured by A. Baumgartner [437] in his young fir plantation and in a nearby afforestation area can be used. Figure 178 gives the measurements for all the nights at ten levels above the ground, using Leick dew plates, and joining up the values for each individual night. The difference between the two areas is quite striking. In the forest, maximum deposition of dew occurs a little below the zone of greatest negative radiation balance, while it is close to the ground in the open. When dewfall was slight, only the upper surfaces of needles and twigs were wet, the entire surface when it was a little heavier. It seems likely that some of the water vapor being precipitated in the crown originates from evaporation at lower levels in the stand.

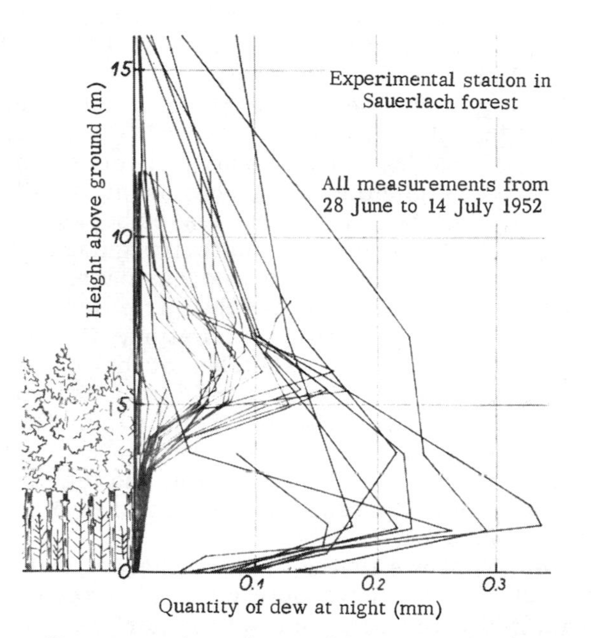

Fig. 178. Dew profiles measured simultaneously in a young fir plantation and on bare ground. (After A. Baumgartner)

In open country maximum dewfall is located 1 to 2 m above the ground. This type of maximum was also found in Mäde's observations, on a dew recorder at a height of 50 cm above ground, showing 1.44 mm in October, which is slightly more than that above or below that level. For the observations in Fig. 178 the average amount of dew per night was:

Height above ground (m):	0.5	1.0	1.5	2.5	4.0	5.5	8.0	10.0	12.0	16.0
Open land (mg cm^{-2}):	8	17	27	21	17	14	10	8	6	3
Young fir (mg cm^{-2}):	1/2	1/2	1/2	1	2	12	6	4	3	1

Dew found within a stand, in spite of the insignificance in quantity, may become large on some nights when there has been a heavy deposit, since large drops of dew may drip through from twigs in the crown area and release a veritable cascade, a shower of dew.

To study the distribution of rain in the forest, it is best to start again with an isolated tree. H. F. Linskens [641] measured the distribution of rainfall under an apple tree for a whole growing period. The tree selected was a bushy apple tree, 10 yr old, standing in loose association with others in an orchard, so as to keep wind effects as small as possible. When the tree had no foliage, rainfall was evenly spread under it, except for drops below the forks of branches. As the canopy of leaves developed, it held back the rain, and this was observed at first as a general reduction of the rain water arriving at the ground. It was only after the leaves had unfolded completely, and their regidity had decreased, that their ability to deflect precipitation showed an increase. Precipitation maxima then made an appearance below the outer rim of foliage, which acts as a gutter, sometimes as much as 160 percent of the precipitation in the open. The surface area of the sheltering canopy was estimated by making a leaf count and measuring the area of 50 sample leaves by planimeter, giving 54 m^2, while the area of ground below was 28 m^2. When the leaves began to fall, the distribution of precipitation became quite irregular below the tree, reaching uniformity again when the tree was bare.

H. F. Linskens [640] also studied the varied behavior of several different types of tree toward precipitation. Figure 179 gives a few examples. The top row has sketches of the five types: (*A*) red beech, (*B*) weeping apple, (*C*) pyramid oak, (*D*) maple, and (*E*) Lebanon cedar, to show the structure of the crown. The second row gives a bird's eye view (note the different scales used). The third row shows the distribution of winter rain, and the bottom row that of summer rain both expressed as percentages of the unaffected precipitation in the open.

The conifer (*E*) shows little difference between winter and summer. The thick crown allows only 60 to 90 percent of the rain to pass through, while the drip zone at the periphery receives 10 to 20 percent more than open land. Diagrams for the beech (*A*) show the fairly uniform distribution mentioned above in the winter picture. The weeping type (*B*) shows where water pours down from the tips of individual twigs. The umbrella type (*D*) even shows a distinct gutter effect when without foliage, but only a few irregularly situated dripping areas when in leaf. The first two types exhibit a marked channeling of water into the drip area below the outer rim of foliage. The pyramid oak (*C*) does not fall into either of these classifications, but in winter acts like a funnel, pouring all the water down the trunk, close to which the quantity of water reaches ten times the precipitation. Even when bearing summer foliage, the funnel effect still gives 110 percent in the trunk area, with another high value in the gutter area. H. F. Linskens describes this as a centripetal precipitation distribution, in contrast to the normal or centrifugal distribution.

From isolated trees the discussion will now pass to a closed tree stand. Others, in addition to the forester, have an interest in finding out how the forest utilizes the precipitation it receives. The

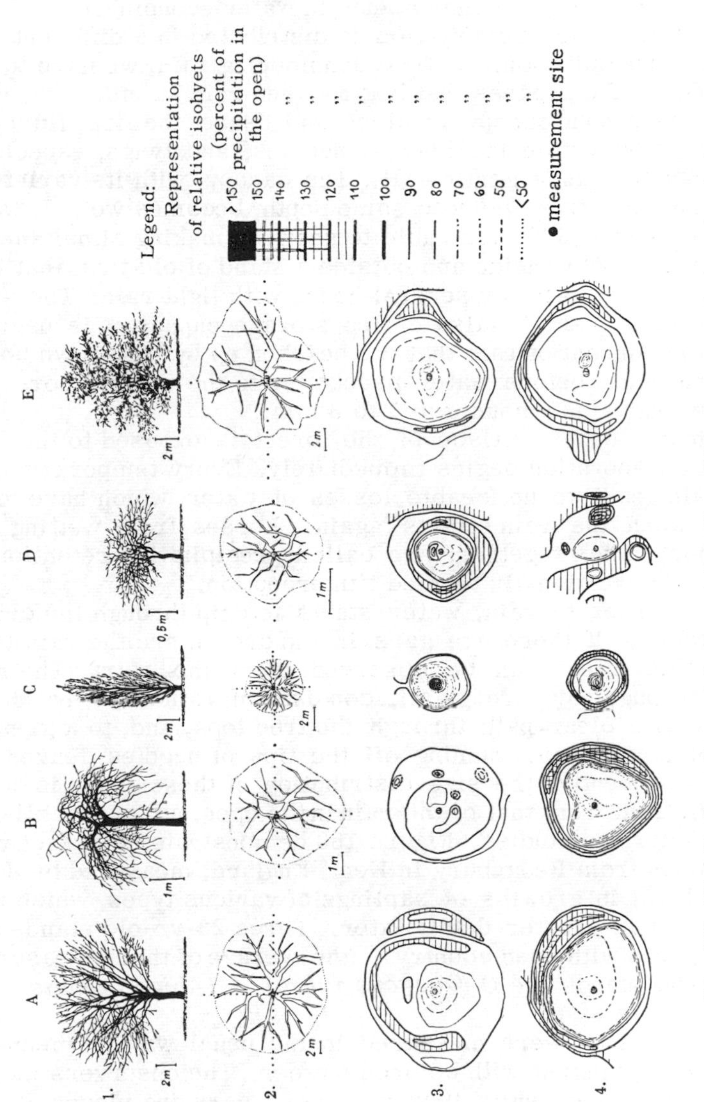

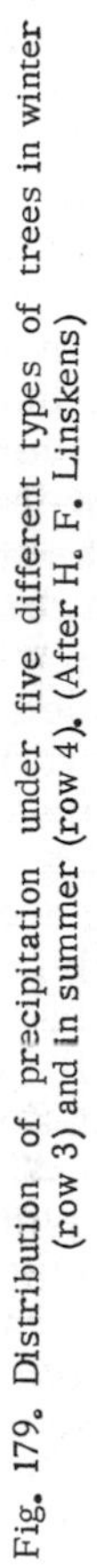

Fig. 179. Distribution of precipitation under five different types of trees in winter (row 3) and in summer (row 4). (After H. F. Linskens)

water economist must take into account the fact that less water reaches the ground in a wooded area than in the open country, and the contribution toward ground-water supply is therefore smaller than from open country. (It will be explained in Sec 39 that forests are nevertheless of great importance in water economy.)

Obviously, liquid precipitation is distributed in a different way from solid precipitation. In first considering rain, we have to get a clear idea of the process that begins when rain starts.

When the crown canopy is close and dense, the first flurry of water drops stays on the leaves, needles, and twigs, especially when it starts with a gentle fall. The canopy, with its variety of forms, arranged *en echelon* to some depth, becomes wet. J. Delfs and his assistants [628] were able to show, by making simultaneous measurements both inside and outside a stand of old firs, that this initial stage may last for several hours with light rain. The term "crown-wetting value" (also called storage capacity) is used to describe the amount of rain that can be taken up by the crown under these conditions, before water penetrates to the forest floor. The crown-wetting value amounts to 1 to 3 mm.

The moist upper surface of the forest is exposed to the free winds, and evaporation begins immediately. Every temporary letup in the rain leads to noticeable losses of water, which have to be replaced when the rain starts again. Losses from wetting and evaporation, taken together, are called precipitation retention in the forest, internationally termed "interception."

If it continues to rain, water starts to drip through the crown after a while. If there are gaps in the crown, and the rain falls heavily, this effect can be observed from the start. The rain dripping through, or throughfall, consists of raindrops that accidentally find a clear path through the tree tops, and, to a greater extent, of large drops running off the tips of needles, leaves, or twigs. Consequently, the size distribution of these drops is completely different from that of the rain in the open, a fact established in early American studies. Figure 180 demonstrates this fact with recent values from Bedgebury in Kent, England, measured by J. D. Ovington [643] in growths of saplings of various types, which will be discussed in greater detail later. Three 23-yr-old stands are here compared with open country F (the average of three measuring points). The stands are Q (*Quercus rubra*), A (*Abies grandis*), and L (*Larix eurolepsis*).

Raindrop sizes were measured in the usual way by means of filter paper sprinkled with colored powder. The observers moved about in the stand, while it was raining, carrying pieces of this paper 60 × 60 cm, so as to catch an entirely random selection of the drops falling through. There is a known connection between the size of the resulting stains and the size of the drops. Summation curves could therefore be drawn from Ovington's published frequency distributions of the diameters of the colored stains, to

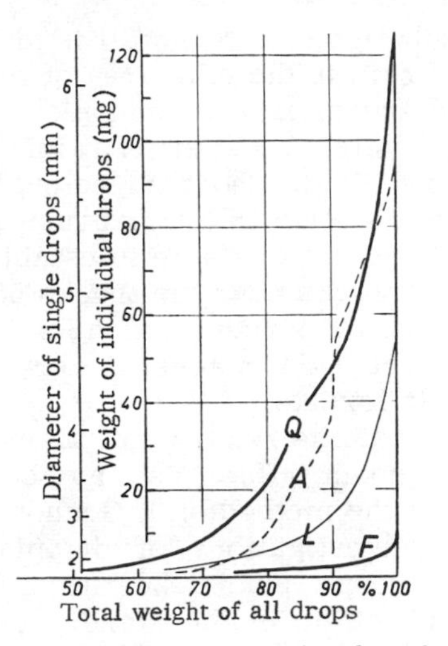

Fig. 180. Spectral distribution of drops in oak (*Q*), silver fir (*A*), and larch (*L*) woods compared with the open (*F*). (After J. D. Ovington)

express the share of each size of drop—beginning with the smallest—as a percentage of the total weight of rain.

The ordinate in Fig. 180 is therefore the weight of raindrops; the diameters of the drops are also shown to make the curves easier to visualize. These summation curves, which all begin at 0 percent for drops of diameter 0, show that in every case more than half the weight of rain comes from tiny droplets with diameters less than 1 mm. In the open air (*F*), as already known from meteorologic research, drops of more than 2 mm diameter are a rare occurrence; and in this investigation no raindrops were found in the open measuring more than 2.5 mm. The situation is different within the stand. It is not surprising that the forms of the three curves (*Q*, *A*, *L*) are not quite smooth, as is often found in this kind of statistical curve. In the stand of northern red oak (*Q*) drops of more than 6 mm diameter are found, since this is a deciduous tree. Such drops can never last long. They acquire such a high velocity of fall that they are broken up into smaller drops. But drops falling from the canopy of leaves have neither space nor time to reach break-up speed. This is why they reach the forest floor with such enormous sizes. As may be seen from Fig. 180, however, drops of over 5 mm diameter account for only about 5 percent of the total weight of rain; since the weight of a drop increases in proportion to the cube of the radius, their number must also be very small. Drops escape more readily from the needles of the other two types of trees (*A*, *L*) than from leaves, and therefore do not acquire a very large size. The large size of dripping raindrops in forests may become of importance with soils susceptible to erosion.

The spatial distribution of throughfall is always very irregular. The accumulation of rain in the drip area of the tree, noticeable in the case of isolated trees, is still conspicuous in a closed stand. The first to make a systematic study of rainfall distribution within forests was E. Hoppe [636], who carried out his investigations at the forestry research station of Mariabrunn in Austria, about the end of the last century. He was able to establish that, in a 60-yr-old stand of firs, there was near the trunks 55 percent, and at the borders of the crowns 76 percent, of the rain falling in the open. This uneven spread of rain makes it very difficult to measure rainfall distribution in forests.

It is quite clear that the readings of a single rain gage will be subject to chance. Random influences may be reduced if a position is selected, following the method of J. Grunow [634], that has been shown by previous investigations of dripping to have average characteristics. E. Hoppe placed 20 rain gages along two lines perpendicular to each other in the stand; C. L. Godske and H. S. Paulsen [633] selected 9 typical points. The best method, however, is that adopted by J. Delfs [627], who introduced the use of troughs 20 cm long and 5 cm wide, equivalent to 50 normal rain gages (1 lit of water withdrawn = 1 mm rain). Figure 181 shows such a trough in position in an old stand of firs in the Harz Mountains.

Fig. 181. Trough for measuring the amount of penetrating rain, and apparatus for collecting water flowing down the trunks in a stand of firs in the upper Harz Mountains. (After J. Delfs)

They may be used in winter as well, provided a chemical is added to melt the snow and oil to prevent evaporation as in rain totalizers. In young growths only, these troughs yield a rainfall value that is too high, since it is impossible to position them without overlapping the gutter area when there are many trunks.

After the rain has stopped, dripping continues inside the stand. It may last for 2 hr, a point that should not be overlooked when the amount of individual rainfalls is measured.

Besides interception in the crown area and the throughfall, there is still the flow down the trunks to consider. Like dripping rain, this begins only after complete wetting of the crown canopy. As may be seen from Fig. 181, which shows three examples, this is collected by a reception channel made of metal or plastic and fed into a container attached to the trunk. In coniferous woods the amount of trunk flow is small, but it cannot be neglected in deciduous forests.

The water economist is interested principally in the water taken up by the soil of the forest floor. This is the sum of the throughfall and the trunk flow. Interception and trunk flow together are usually called the interception loss or gross interception, which is equal to the difference between precipitation in the open and the quantity of rain dripping through. Since the trunk flow is often insignificant in coniferous forests, the interception loss is simply called interception. (E. Hornsmann [717] has suggested the term "Auffang," "capture," for interception.)

The way in which the forest behaves toward falling rain makes it clear that the rain intensity, duration, and variation with time are of vital importance to interception. When J. D. Ovington [643] summed up the results of his observations by saying that interception was "between 6 and 93 percent" he expressed the fact that practically any amount of interception was possible, depending on the character of the rain and of the stand. For a single stand, interception will be great in proportion to rainfall when the rain is light, of short duration, or frequently interrupted, and will be small for heavy showers or persistent rain over a wide area. This may be shown by a comparison of two readings, both from the same stand of firs in the Harz Mountains, which will be described later. With widespread rain of 50 hr duration yielding 70.5 mm, 75 percent passed through the crown area; while of a thundershower yielding 74.6 mm and lasting 3.5 hr, 98 percent got through.

In Japan, K. Takeda [650] attempted to sum the matter up in a series of comprehensive equations, which could then be compared with actual measurements. It is possible, however, to make some advance, on the simple assumption that once the wetting quantity b (mm) has been reached, the rain interception I (mm) will be proportional to the amount of rain N (mm). The equation in the form

$$I = b + cN$$

was developed by R. E. Horton [637] in 1919 and has been advantageously used. A comprehensive report of American studies was published in 1948 by J. Kittredge [638]. The equation was found to be approximately satisfied in more recent investigations from 1940 to 1946, by P. B. Rowe and T. M. Hendrix [647]. During this period they made 214 precipitation measurements in a 65- to 75-yr-old stand of Ponderosa pine near Bass Lake in California, obtaining for rain $I = 2.8 + 0.06\ N$ mm. Curve C in the lower half of Fig. 182 is based on this equation.

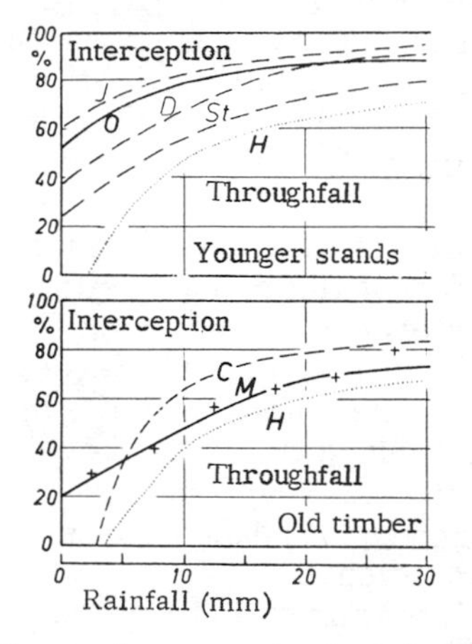

Fig. 182. Interception in fir stands, as a function of amount of precipitation (C for California pine)

A distinction must be made between the types of trees and ages of the stand in order to appreciate the results obtained up to date. In old fir stands, shown in the lower half of Fig. 182, the relation between interception and throughfall is shown as a function of the amount of precipitation.

The heavy curve M illustrates the very first set of measurements made by E. Hoppe in a 60-yr-old fir stand at Mariabrunn in Austria. The small crosses, which practically coincide with this line, are the results of the combined work of J. Delfs, W. Friedrich, H. Kiesekamp, and A. Wagenhoff [628], which was conducted in the Harz area between 1948 and 1953 at the instigation of E. Wiedemann, with the aim of determining the water economy of the whole forest, after the heavy losses it had sustained under military occupation. These figures were measured in an 80-yr-old 22-m-high fir stand. The dotted curve H is the result of readings made on the Hohenpeissenberg by J. Grunow [634] in an old stand on the steep southern slope, near the peak of this isolated mountain, which is almost 1000 m high; the difference in the results is therefore not surprising (the curve C is also for a NNE slope of 36 percent at 1020 m above sea level).

Younger stands allow more rain through to the forest floor, as shown in the upper half of Fig. 182. The curves are for measurements made in the Upper Harz Mountains; in 60-yr-old saplings 15 m high (St); a 30-yr-old 6-m thicket (D), and a 15-yr-old, not yet closed-up fir plantation (J). The average for the 4-yr period gave values of: $St = 29$ percent, $D = 24$ percent, $J = 12$ percent, in comparison with 37 percent for the old stand. The trunk flow was between 0.5 and 3 percent, compared with 0.8 percent in the old stand.

The line *O* is the result of C. L. Godske and H. S. Paulsen's research [633] in a 35- to 50-yr-old plantation of fir saplings at Os near Bergen. The value for interception here is somewhat lower, and is due to the thinner growth of Norwegian forests (60° N) compared with German stands of the same age (52° N). The dotted Hohenpeissenberg curve (*H*) is for a 40- to 60-yr-old stand of saplings on the north slope of the mountain.

The strong dependence of interception on the distribution of rain must mean that its average value is subject to great variations from year to year and even from month to month. Consequently it is extremely difficult to estimate an average value for the interception of a forest of greater extent, which will most likely consist of stands of various ages and perhaps also of various types of wood.

In the Harz investigation the absolute value of rain (mm) retained in the crown of the fir was determined for a number of individual cases. This increased with the amount of rainfall and with the age of the stand as follows:

Amount of rain (mm):	0	5	10	15	20	25
Thicket, 30 yr old (mm):	3	6	7	5	4	9
Saplings, 60 yr old (mm):	3	6	8	8	9	12
Old timber, 80 yr old (mm):	4	7	10	10	11	18

The effect due to water filtered out of driving mists by the forest can only be mentioned here. This process mainly concerns the forest edges and will be dealt with in Sec. 37.

In contrast to the needles of conifers, deciduous leaves collect water. In so far as they drip, the drops are very large, as we have already seen from the drop-distribution curve *Q* for oak in Fig. 180. It is clear, however, that the leaves are so disposed as to channel the water collected via twigs, branches, and trunk, to the advantage of the tree. The trunk flow is therefore of greater importance in deciduous than in coniferous forests. A comparison of rainfall density (quantity per unit time) in a forest in leaf and when leafless with rainfall density in the open will show the importance of the leaf structure of the forest.

In a typical hardwood forest of the southern slopes of the Appalachians consisting of oak, poplar, maple, acacia, birch, and so forth, G. R. Trimble and S. Weitzman [651] set up five recording rain gages on an 11° southern slope at 800 m above sea level within the stand, and in a clearing of 100 m^2. The crowns of this mixed 50-yr-old stand were between 10 and 23 m above the ground. When the stand was in full leaf (10 August to 7 October 1951 and 19 May to 6 July 1952, described as "summer" in subsequent references), 28 rain periods were investigated. When completely devoid of foliage (6 November 1951 to 28 April 1952, subsequently described as "winter"), after discarding cases when snow was present, 31 cases of rainfall were investigated. The average value of the five recorders in the stand was compared with the open-air figures.

In all rainfalls, 5-min periods with the highest precipitation density were chosen to be examined from the point of view of erosion just how much the forest might be able to reduce this rainfall density. Results are shown in Fig. 183 for the periods in leaf and without foliage. The abscissa is rain density in the open; the figures are easier to understand if one remembers that 5 mm in 5 min would be called pelting rain. The ordinate shows the amount that penetrates the crown of the forest, as a percentage of the rain density in the open.

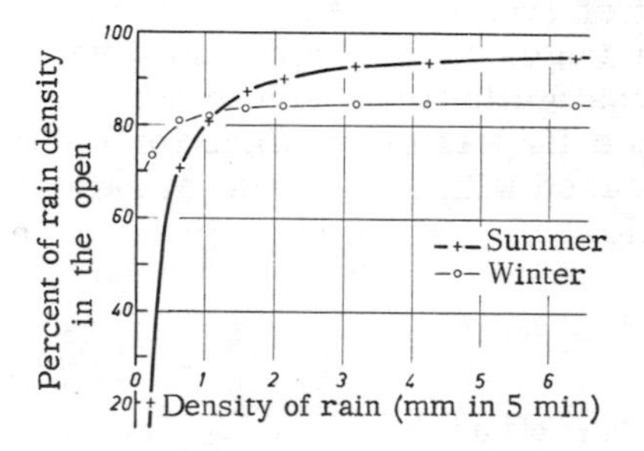

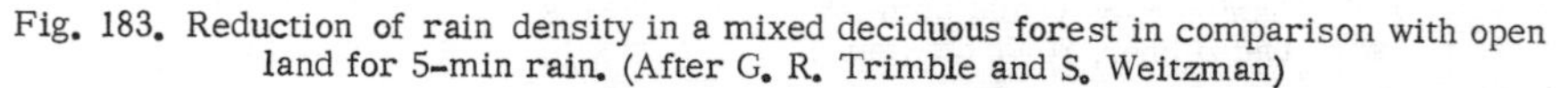
Fig. 183. Reduction of rain density in a mixed deciduous forest in comparison with open land for 5-min rain. (After G. R. Trimble and S. Weitzman)

Up to a density slightly above 1 mm in 5 min, the thick leaf canopy in summertime has, as expected, a strong reducing influence compared to the bare winter forest. The difference between summer and winter is naturally greatest with the lightest rain. But it is surprising to find that the proportions are reversed when rainfall is intense. Trimble and Weitzman ascribed this to the tendency of heavy rain to bend the leaves downward to such an extent that the rain falling on them can no longer flow along twigs and branches to the trunk, but pours straight onto the ground instead. In winter, the bare branches are no longer sheltered from the rain and therefore catch more than they do under the canopy of summer leaves in such heavy falls.

The increase of trunk flow in deciduous forests can be deduced from comparative measurements made from 1952 to 1958 by F. E. Eidmann [629] in the forest of Hilchenbach (Westphalia). The fir trees were 70 yr old and 25 to 28 m tall, and the beeches 95 yr old and 25 to 30 m in height. Both stands were growing adjacent to each other on a southern slope inclined at 25° to 28°, at 600 m elevation. Table 78 gave the distribution of the precipitation falling on the stands during the 6 yr as a percentage of the annual average,

Table 78

Distribution of precipitation (percent of annual average) falling on two stands of trees.

Type of tree	Season	Interception	Trunk flow	Throughfall
Fir	Summer	32.4	0.7	66.9
	Winter	26.0	0.7	73.3
Beech	Summer	16.4	16.6	67.0
	Winter	10.4	16.6	73.0

1216 mm. The rain not intercepted by the crown of the firs drops through. In the beech stand, in contrast, half the water caught by the leaves in summer is channeled to the trunk. In winter more water flows down the trunk than is retained in the crown (practically without foliage). In the deciduous forest, therefore, interception is less and the total water reaching the ground is more than in the evergreen forest. The proportion of water dropping through is about the same in both types, and is greater in winter than in summer for both.

Interception is proportionally great when amounts of precipitation are small, because of the quantity required to wet the crown, and decreases in percentage as the amount of rain increases. It follows that trunk flow begins only when wetting is complete, and that it will increase with the depth and duration of rainfall. Hoppe's measurements at Mariabrunn in Austria [636] demonstrated this result. In 1894 he observed the figures shown in the top row below in an 88-yr-old stand that contained 0.8 beech and 0.2 silver fir, and the lower row in 1895 in a pure beech stand (percent of total precipitation).

Amount of rain (mm):	0	5	10	15	20
1894 (percent):	8.4	14.5	15.7	20.2	
1895 (percent):	10.1	18.4	18.0	22.5	

F. E. Eidmann's mean values are in complete agreement with these results.

A wealth of information is contained in the survey of a large number of foreign trees in Europe, made at Bedgebury in Kent by J. D. Ovington [643], which has already been mentioned. In 13 different areas of young growth, each 1000 m^2 in area, with trees 22 to 23 yr old, rainfall was collected in ten rain gages, and the trunk flow in three containers attached to the trunks. Figure 184 shows the interception for seven different types of trees, as a function of rainfall depth. Table 79 gives details of the trees shown by initials in Fig. 184. The figures for throughfall are the lowest and highest yearly averages for the years 1949-1951. The amount of trunk flow is insignificant in all cases. The figures in the column headed "Drops" gives the weight percent of precipitation falling in larger drops than were observed simultaneously in the

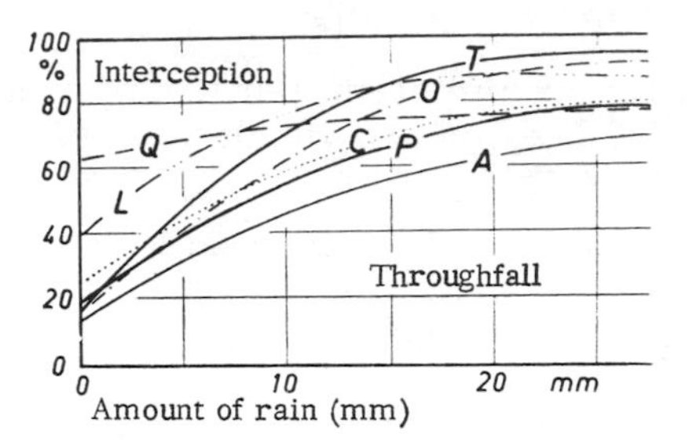

Fig. 184. Interception in various types of young stands in England. (After J. D. Ovington)

open; these figures serve to amplify Fig. 180, and at the same time show the great differences in the behavior of different types of trees.

Table 79

Interception of various types of trees; see Fig. 184.

Description of saplings			Quantities measured			
Type of tree (all 22-23 yr old)	Height of stand (m)	Height of crown canopy (m)	Through-fall (%)	Trunk flow (%)	Drops (%)	Snowfall (mm)
Deciduous						
Q *Quercus rubra*	7.3	2.4	68—71	0.3	68	13
Coniferous						
L *Larix eurolepsis*	14.6	3.0	70—90	0.1	45	7
T *Thuja plicata*	7.6	2.1	63—65	0.1	46	1
O *Picea omorica*	10.1	4.0	59—61	0.2	66	0
C *Chamaecyparis lawsonia*	8.8	1.8	56—57	0.1	48	—
P *Pinus nigra*	8.5	2.1	52—53	0.2	61	1
A *Abies grandis*	14.3	4.9	49	0.1	64	—

Little is known of the rain distribution within the tropical forest. In the subtropical primeval forests of Brazil (19° to 23°S, 41° to 45° W, 600 to 900 m), F. Freise [631] has observed rainfall over many years, making meticulous observations with a limited number of instruments. Of the total precipitation (100 percent), 20 percent evaporates in the crown space, 28 percent reaches the ground down the trunk, and 34 percent drops through, making a total of 82 percent. The remaining 18 percent disappears into the bark and hollow stems, and by further evaporation.

After rain lasting an hour, or several hours, ceases (not earlier and not after snow), forest smoke can be observed on mountain slopes over close stands, and also in the plain. This peculiar effect has been described and explained by F. Rossmann [646]. Small clouds or ragged wisps of fog cling about the tree tops just as the rain stops, and hang around for nearly an hour. In the rain-soaked crown area proper there is no fog, even though the air mass in it is close to saturation. Since the rain has a cooling effect, it is a little warmer in the crown area. If this warmer air is then mixed with the cooler air aloft, by vertical mixing, the small degree of cooling that results is sufficient to produce a slight condensation, lasting only a short time and not forming a coherent cloud. As soon as the quota of air near saturation in the crown area has been used up in the mixing process, the phenomenon will cease.

The snowfall distribution in the forest is important. It lies below the crown canopy, protected to a large extent from evaporation, thus providing a special form of winter water storage. It also melts more slowly in spring than snow in the open, by several weeks, thus retaining a significant quantity of water. Being more lightly borne by the wind than rain, snow responds more strongly to variations in configuration and density of the forest crown. Above all, the amount reaching the forest floor is determined by the type of snowfall.

Large flakes of wet snow cling easily to the crown. In winter measurements in the Harz Mountains [628], 80 percent of a fresh snowfall, 12 cm deep, with a 13.1-mm water content, was retained in the crown of a stand of old firs and saplings. A fir thicket trapped an entire fall of 10 cm of fresh snow in its top area. In catastrophes due to breakage under the snow load, it is nearly always a question of wet snow. When temperatures are low, however, powder snow penetrates easily into the woods.

In spite of all these facts, measurements show that, on the average, interception of winter precipitation is less than that of summer rain. The cause of this lies almost certainly in the reduction of evaporation at low temperatures, and the tendency for snow to accumulate and then crash down in clumps from twigs and branches.

It is certainly true that measurements of snow depth in woods and in the open show a smaller difference than is found with rainfall in summer. From 120 yr of observations at Prussian forest stations with instruments in paired exposures, J. Schubert [648] obtained from snow readings for each pair a snow-depth ratio open land: stand = 100:90, while for rain the ratio was 100:73. H. Hesselman [635] found almost the same snow depth in pine stands as in a felled area in Sweden.

This is confirmed by interception measurements. P. B. Rowe and T. M. Hendrix [647] subdivided the figures for the 6-yr series of measurements in California in a 70-yr-old pine stand, according to whether rain or snow fell. Through the crown fell 84 percent of the rain and 87 percent of the snow. In the Upper Harz, too, it was found that more precipitation reached the forest floor in winter than in summer, in spite of the single cases of wet snow mentioned earlier. The mean value for a 4-yr period in an old fir stand was 67 percent in winter against 60 percent in summer; in the fir saplings it was 73 percent against 69 percent. These are the figures for evergreen forests. This effect applies even more to the leafless deciduous forest in winter. The figures in Table 79 for average snow depth during an English winter with little snow gave a greater depth in the oak than in the coniferous stands. From estimates of the snow load on trees during snow breakage in Upper Silesia, W. Rosenfeld [645] deduced that in stands of silver fir and spruce 25 percent to 55 percent of the snow fell through to the ground, while it was 60 percent to 90 percent in a beech stand. G. R. Eitingen [630] observed that in a Russian birch forest 91 percent reached the ground in summer, and 100 percent in winter, with an annual precipitation of 545 mm.

Since with snow, just as with rain, the ground near the periphery of the tree crown is more deeply covered with snow sliding from the crown or by incident powder snow, large frost circles develop around individual tree trunks. Their development and influence on stands in the Bavarian Forests have been well described by G. Priehäusser [644].

37. Microclimate at the Stand Edges

When the forester wishes to bring up a new generation of trees, he often makes use of the special microclimatic conditions at the edge of old standing timber. Depending on the orientation of the forest edge, the ground and the air between the old trees may be sunny or shady, windy or still, warmer or colder, moister or drier. These qualities may be used to promote better growing in natural regeneration or artificial planting, depending on the requirements of the particular type of tree, and on the general climatic conditions. Knowledge of edge climate is therefore of great importance in forestry.

According to R. Geiger [656], the edge climate develops from two sources. First, it is a transitional climate. Influences of the nearby forest make themselves felt in the part called the outer edge (the strip of open land around the stand), as, for example, when cool forest air from the trunk area drifts over the open land on a summer day. In the inner edge, too, that is, in the belt of trees close to the open land, influences from the outside are at work, and their effect is greater when the border trees have little covering to provide a sheltering influence, particularly in the open border formed when felling begins. Secondly, the edge of a forest forms a high step in the topography, which takes up or wards off sunshine, wind, and rain. This second influence is usually more powerful than the first in modifying microclimate.

The proportion of sunlight and shade is the first factor determining local climate. As far as diffuse sky radiation is concerned, the orientation of the stand edge is immaterial, practically without influence. The greater the proportion of diffuse radiation to direct sunshine, therefore, as in cloudy weather and winter, the less will be the difference between different edges.

Figure 185 gives the number of hours of sunshine received by stand edges of all directions during the year. The curves, based on a long series of measurements in Karlsruhe from 1895 to 1934 by J. v. Kienle [745], show the influence of weather. The summer

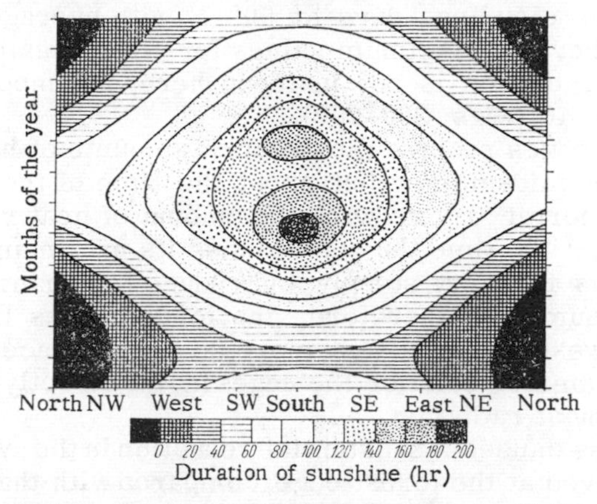

Fig. 185. Monthly sunshine duration at stand edges facing in all directions throughout the year.

maximum at the southern edge is split into two by the "European monsoon" in the month of June. In general, however, as far as radiation is concerned (but not the other elements) the symmetry of east and west edges and of spring and autumn is maintained. During winter, edges with a northern exposure are completely deprived of sun.

The intensity of the incoming radiation is of greater significance than the duration of sunlight. Figure 186 gives some information on

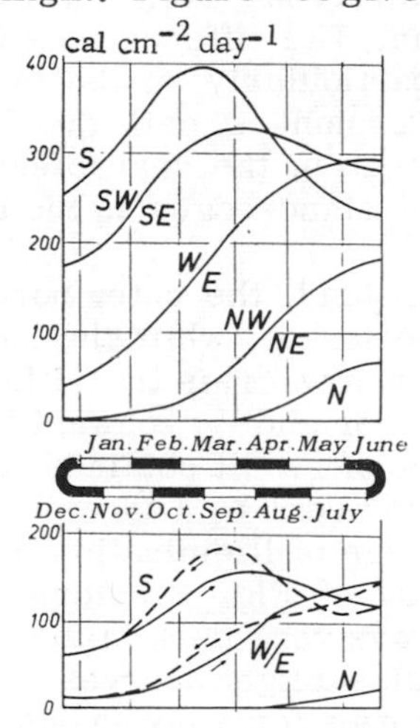

Fig. 186. Daily totals of direct solar radiation received at stand edges facing in various directions, on a sunny day (above), and on a normal day (below) throughout the year.

this subject. Above the scale, which divides the year into two halves, are shown the daily totals of incoming radiation for the eight most important cardinal points that the edge of the forest may face. The figures are based on the calculations of W. Kaempfert and A. Morgen [744] for cloudless days at 50° N with average atmospheric turbidity. They are maximum values for clear-weather situations. Diffuse sky radiation is not included here; its importance will be shown in Sec. 40 (Figs. 200-202).

Southern edges receive the greatest amounts of heat, naturally, though not in midsummer but about the time of the equinoxes. In winter the shorter day reduces the amount of heat received, while in summer, at the time the sun reaches its maximum of noon altitude, its rays fall very steeply and reduce the quantity of heat received. In summer, therefore, the forest edges that face more toward the west and east are more favorably placed, while at the time of the summer solstice the edges facing directly east and west receive the most radiation.

Cloudiness causes a substantial reduction in the average quantity of heat received at the forest edge, compared with that on cloudless days. Average values have been computed by J. Shubert on the basis of the Potsdam radiation records for the period 1907 to 1923, for S, E, W, and N edges. The daily variation in cloud cover favors the east-facing edges which are exposed to the morning sun, in comparison with the western edges which receive their quota of radiation during the afternoon period of greater cloudiness. The difference is small, however, and has been left out of Fig. 186. Nevertheless, the symmetry of the year about the sun's maximum altitude is thereby unbalanced, and the lower part of Fig. 186 therefore shows the difference by a solid curve for spring and a broken line for autumn. The difference between the two shows the brightness of May, particularly of the late summer—the Indian summer, which in German is call the "old wives' summer." There is good agreement in the characteristic variation between the same directions of stand edges in the upper and lower halves of Fig. 186.

At the edge of the stand, the outer border is made more dark by the nearness of the forest, while the inner border is lightened by the adjacent open country. This transition in radiation has been measured by F. Lauscher and W. Schwabl [665]. Figure 187 gives an example of the type of result obtained; brightness is measured on both sides of the stand edge by Lange photoelectric cells, and is expressed as a percentage of illumination in the open.

The firs (upper part of Fig. 187) form a very dense covering, and it is therefore always very dark in the inner border. The contrast is greater in full sunlight (curves 2 and 3) than in overcast conditions (1). The curves for a deciduous stand in the lower part of Fig. 187 are: wintertime bare of foliage (1) and (3), and summertime in leaf (2) and (4). Curve (1), for overcast skies, shows a

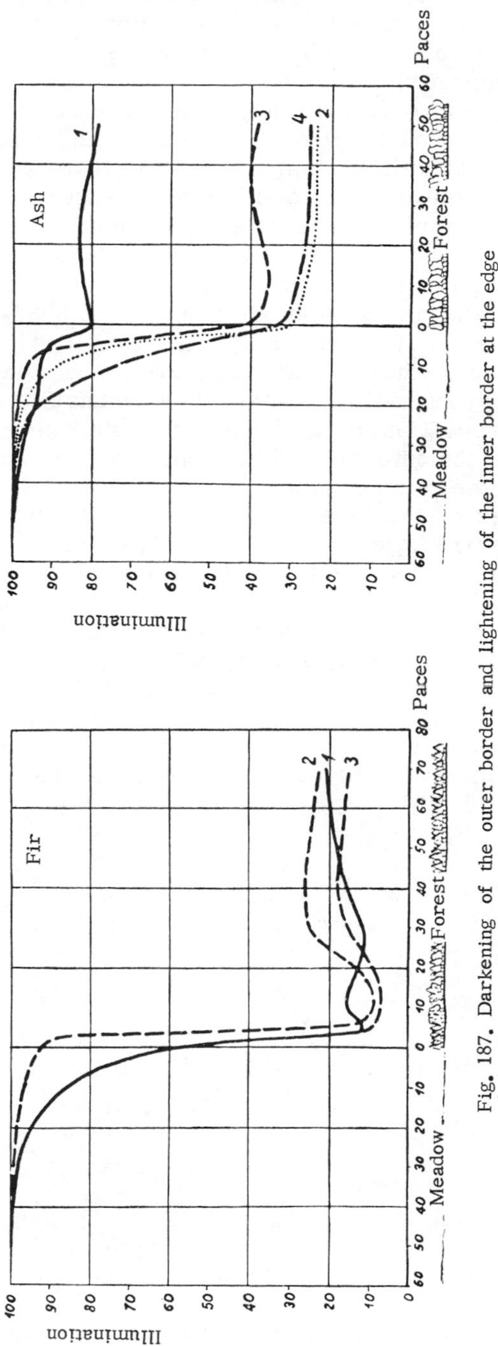

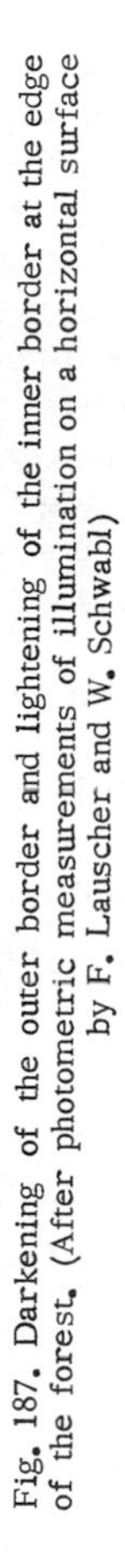
Fig. 187. Darkening of the outer border and lightening of the inner border at the edge of the forest. (After photometric measurements of illumination on a horizontal surface by F. Lauscher and W. Schwabl)

fairly uniform distribution of light, while curve (3) shows the effect of the shadow cast by branches. In the summer, curves (2) and (4), the difference between outside and inside is greater, but never reaches the amount produced by firs.

I. Dirmhirn drew attention [654] to the remarkable increase in the albedo of a meadow as the edge of a wood was approached (measured with respect to a walnut tree). The ratio of reflected to incident radiations, summed up over all wavelengths, is not altered by a change in the surface properties of the meadow as the walnut tree is approached, but its increase from 20 to 40 percent results from a change in the spectral structure of the radiation received. The filtering of light in the forest, described on p. 303, enriches the incident light in long-wavelength radiation as the stand is approached (even if the absolute amount is diminished). These long waves are reflected more strongly than the short-wavelength radiation, and therefore the proportion of radiation reflected decreases more slowly than does the intensity of incident radiation, hence leading to almost double the value of albedo.

What one edge of the forest gains in sunshine, the other edges lose by being overshadowed. The growth of young trees at the outer border depends on the length of time they spend in the shadow of the nearby stand, but they may also find this profitable in times of great heat and dryness, although it is unfavorable when heat is lacking. Figure 188, after R. Geiger [656], illustrates the width of shadow at 48° N (Munich) at the time the summer sun is on the turn. The abscissa is the direction in which the stand is facing, and the ordinate the time of day. The upper and lower horizontal dashed lines give the times of sunrise and sunset. The isopleths are for equal widths of shadow, expressed as a fraction of the height of the stand.

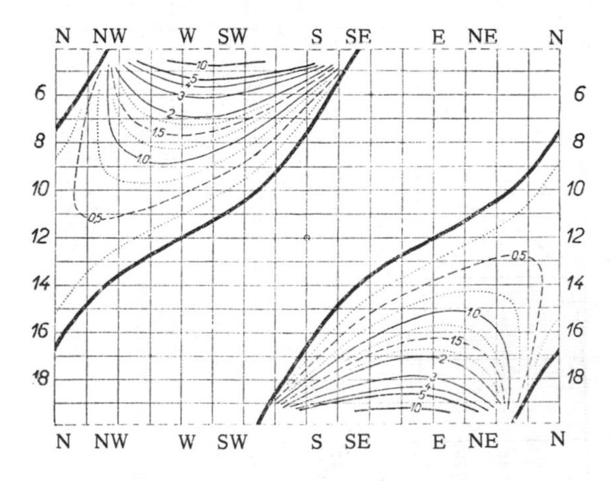

Fig. 188. Lines of equal shadow width at stand edges facing all directions, at the time of the summer solstice at 48°N.

The heavy zero lines join the times of day and the stand edges for which the sun's rays are tangential. The broad, clear band in the center, running from bottom left to top right, means that the edge is receiving direct sunlight. There are also clear areas in the top left-hand and bottom right-hand corners of the diagram. In midsummer the sun rises so far in the NE that it shines directly in the early morning on the stand edges facing as far around as NNW. In the same way, stands that face NNE are touched again by the setting sun in the early evening. To determine the time of day at which a 15-m-broad plantation will be completely in the shadow of a 20-m-high stand facing WSW, we find from Fig. 188 for a shadow width of 15:20 = 0.75 that on 21 June the plantation is in shadow from sunrise to 09:00. From this time onward the shadow narrows until shortly after 11:00 when the whole plantation comes into direct sunlight where it remains until sunset.

Figure 188 is valid only for the day of the summer solstice. For practical use, a great number of such diagrams must be produced (two more are given in [656]). R. J. van der Linde and J. P. M. Woudenberg [666] developed a simple means by which the required data can be obtained for any type of stand, at any time of the year, by the superposition of two sets of curves.

An astonishing example was found by J. Schubert [676]. For a north-facing edge the width of the shadow is independent of the time of day at the equinoxes, 21 March and 23 September. The long slanting shadows at sunrise and sunset extend the same perpendicular distance from the edge of the stand as the shorter noon shadows, which are in fact at right angles to it.

At night the negative radiation balance of the outer edge is improved, since part of the night sky is shielded by the nearby stand. Counterradiation from the forest is much more effective than that from the sky, since it is either at the same or a higher temperature than the ground in the outer border. This factor has already been shown in line C of Table 8 from F. Lauscher's figures [48]. In the following table line A gives the effective outgoing radiation as a percentage of the outgoing radiation from open country, using instead of the angle γ the distance D from the edge of the forest as a multiple of the height h of the stand. Since the part of the sky near

Distance D	0	$0.2h$	$0.4h$	$0.6h$	$0.8h$	h	$2h$	$3h$
A (Lauscher)	50	60	70	78	84	88	95	98
B (Bolz)	50	79	83	87	89	91	96	98

the zenith makes the least contribution toward counterradiation (p. 23), the protective influence of the forest decreases rapidly with distance from it. The decrease probably occurs much more quickly in actual fact. By more modern methods, H. M. Bolz [686] has measured the change in radiation balance with distance from a single tree 10 m high, his figures being shown in line *B* of the table. These figures were also confirmed by night observations. The difference from *A* depends only to a small extent on the fact that a single tree has a smaller shielding influence than the continuous forest edge and row *B* therefore approaches reality more closely than *A*.

Although the much-discussed sheltering influence appears to have little effect, it is nevertheless of great practical importance. The effect may be observed directly after individual nights with frost, when there is a strip free of hoarfrost along the edge of the forest, or the fresh green color of young spruce plants compared with the brown and limp shoots of those further away from the edge of the stand. The protective influence often controls the nature of growth in woodland clearings exposed to frost damage; outward from the edge of the stand the influence of open land gradually wins, and often a group of young plants may be seen growing under a single tree in the middle of other growths destroyed by frost.

While we are able to deduce radiation effects at the stand edges approximately from theoretical considerations, when it comes to problems of temperature, air humidity and soil moisture, precipitation, and wind we have to rely entirely on observations. In spite of the great importance to forest development of regeneration in the border areas, and in spite of the great care devoted to it by the forester in practice, there are virtually no measurements available. Since influences result from many factors, the orientation of the stand edge, its height, the type of trees and leaf cover, prevailing winds, soil characteristics, height above sea level, and general climate, researchers do not like to conduct investigations that are of doubtful general validity and applicability. Many basic questions can, however, be cleared up by using a method of investigation that is easy to put into practice in a country with extensive forests, which I suggested in a lecture at the Weather Bureau in Washington in 1950. The method is to cut an octagonal clearing at least 500 m in diameter in a uniform region of old timber with similar composition. This creates a set of inner edges facing the eight principal points of the compass, and subject to the same general climatic conditions (this is much more difficult to obtain at the outer edges of an octagonal wood). One central observation station, eight on the inner borders, and eight on the outer borders, recording all the important meteorologic elements, would be able to produce answers to many problems still outstanding today.

Occasionally it is possible to come across measurements that have been made for another purpose. W. Lüdi and H. Zoller [668] had four observation stations in operation for a time in 1943 and 1944 on the south side of the Hard Forest near Basel, while studying the influence of forests in airfield planning. The values given in Table 80 have been obtained from a graph of the temperature variation on a summer day. The hours selected were those nearest the maximum and minimum daily values; they are considerably later in the soil because of the time lag in penetration. The southern edge is substantially warmer, both by day and by night, than the forest or the open country. At a depth of 2 cm in the soil the difference between the stations at the edge and 35 m from the forest was 5 deg during the day, and was even greater at the surface of the ground. The variation of relative humidity was a reflection of that of temperature.

Table 80.

Air and soil temperatures (°C) at the edge of a forest.

Observation site	20 m in forest	At forest edge	35 m out of forest	100 m out of forest
Air temperature at 10 cm				
at 14:00	18.4	22.2	20.0	21.6
at 05:00	9.0	8.0	6.8	6.0
Difference	9.4	14.2	13.2	12.8
Temperature 10 cm in soil				
at 18:00	11.2	17.8	17.2	17.0
at 08:00	10.6	13.6	13.0	12.8
Difference	0.6	4.2	4.2	4.2

The effect of the heat of insolated stand edges has been demonstrated most strikingly by F. Lauscher [664] in describing the mass appearance of the timber pest *Ocneria monacha*, which is called "the nun" in Austria. If the spring has been characterized by a number of days rich in sunshine the larvae emerge from where the eggs were laid on the warm trunks in great numbers before the parasites that prey on them have come out of the ground. This temporary disturbance of biologic balance favors massive reproduction of pests. This phase advance of trunk temperature ahead of the temperature of the forest floor between 1927 and 1953 had a clear maximum in 1946, a catastrophic year for damage by the "nun."

In discussing wind influences on the stand edges, it is best to follow H. Pfeiffer [672] and distinguish between passive and active influences of the forest on the wind.

The passive forest influence arises from the fact that the edge of the forest is an obstacle to the flow of air. Even a single tree, a lone survivor, for example, influences the wind field in its surroundings. This was measured by M. Woelfle [679] with regard to symptoms of wearing away on a high old oak tree. On the windward edge of a stand the flow of air is brought to a halt. A wedge of stagnant air forms, or a lee eddy develops with a horizontal extent of about 1.5 times the height of the stand. Woelfle was able to record only 20 to 30 percent of the free wind behind the canopy of a dense fir stand. As wind speed increased further, this percentage decreased even more because the twigs arranged above each other then acted like a venetian blind. With an open stand, the wind can penetrate into it in gusts but is quickly slowed down by the tree trunks, and even more quickly by any undergrowth present. Streamlines are cramped together in the border zone above the forest.

When winds are strong there is very great turbulence above the rough forest surface and swaying tree tops. On the lee side of the wood there is an area sheltered from wind, which will be discussed more closely in Sec. 53. When the wind is blowing at an angle to the stand edge it acts as a steering line, and wind increases in the outer edge area. This strong cross wind presents a great danger in times of gale to unwary automobile drivers on roads leading out of the forest at right angles.

The influence of the forest becomes positive when the temperature difference built up between it and the open country produces a flow of air tending to counteract this development.

During the day, the air near open ground is heated while the air under the crown of the forest remains cool. When this happens, the cool air may flow out from the trunk area as a daytime forest breeze. L. Herr [659] and K. Dörffel [655] believe that this effect can be observed in the cooling and moistening of the air in the outer edge, coming from the wood. It is similar in origin to the sea breeze which blows during the day from the cool sea over the warm land. As early as 1920, A. Schmauss [675] called it a "sea breeze without a sea."

In contrast to the daytime forest breeze, a field breeze at night is not observed. The strong braking effect of the trees prevents its development. In mountainous wooded country, a night forest wind blows into the open country, representing the downflow of cooler air forming over the radiating crown area. H. G. Koch [663] has proved its existence by using balloons and has described it more fully. M. Woelfle [680] has rightly pointed out that this is a normal night cooling phenomenon, as described in Sec. 42, and that it is incorrect to apply the term forest wind to it.

As we may see from what has gone before, the passive influence of the forest on the wind field at its edge is much more effective than its active influence. The wind field for its own part is instrumental in controlling two other processes that take place at the forest edge, namely, the dissemination of seeds and distribution of dust. Along a country road at the western edge of a forest on a hot, dusty summer day, the filtering effect of the trees at the border can be seen in the white powder observed over everything. From measurements made by M. Rötschke [674] it appears that with winds perpendicular to the forest edge, in addition to the maximum at the edge itself, an increase in dust content is found in the inner border zone. For example, on 29 January 1935, with a wind of 2-3 m sec^{-1} in the open, the dust content in thousands of particles per liter in front of (-) and behind (+) the forest edge was:

Distance (m)	-100	-50	-25	+25	+50	+100
Dust content	10.1	10.2	10.3	14.0	11.8	11.5

Since in this particular case there was a thin snow cover with a temperature of -2°C, the filtering effect could be observed without interference from any secondary source of dust. The interior of the forest became more and more free of dust (cf. Sec. 39). If the wind blows at an angle to the forest edge, there is a marked increase in dust in the outer border zone, which results from the increase in wind mentioned earlier.

L. Kohlermann [72] measured in spring 1949 the distribution of seeds borne out into the surrounding fields from a 90-yr-old pure stand of pines, 23 m in height, in the forest of Dudenhofen in Hesse. The frequency of wind direction is given in the first line of the table below for the period of investigation. At the same time she determined the quantity of seeds released which, as we know from Sec. 9, is dependent on the weather at the time. The figures from the collecting boxes inside the forest are entered in the second line under the wind direction for which they are applicable, this having been observed at the same time. From these two lines of percentages it can be recognized that the release of seeds is low when moist winds from the SW, W, and NW are blowing, while it is comparatively great with dry winds from the N, and particularly from the NE. The collecting boxes at the edges of the forest were not set up in all eight principal directions. The distribution-rose for the seeds collected is shown in the third line for the eight main directions. It shows that turbulence and tangential winds at the forest edge make distribution much more uniform than might be imagined from statistics of wind direction.

Wind direction	S	SW	W	NW	N	NE	E	SE
Wind frequency (percent)	4	27	13	12	11	22	8	3
Seed dispersal (percent)	3	21	4	4	14	46	6	2
Quantity of seeds (percent)	10	15	10	10	13	17	15	10

H. Hesselman [660] investigated the distribution of seeds within the edge area of the stand by means of 262 collecting boxes buried flush with the ground, in the winter of 1936-37, in a 90-yr-old pine stand, 18 m tall, near Lund in Sweden. He found:

Distance from stand edge (m)	-37	-22	0	7-1/2	22	37
	Inner edge			Outer edge		
Number of seeds (m^{-2})	105	89	73	52	29	17
(percent)	144	122	100	71	40	23
Weights of seeds (g)	3.8	3.6	3.6	3.4	3.4	2.9

Distribution follows approximately the law of mass exchange, and a light seed is transported farther than a heavy seed. At the edge of the stand there was only 7 percent of hollow grains, but 19 percent at a distance of 37 m. The surprising distance to which even heavy fruits may be carried by turbulence is shown by the figures obtained by L. Kohlermann [72] for the distribution of 127,070 fruits of an ash. Although the mean rate of fall of the fruits was 1.4 m sec^{-1} and their points of release were 3.5 to 12.5 m above the ground, the following percentage distribution was found in nine circular areas surrounding the tree between 10 and 100 m: 26, 22, 15, 12, 10, 6, 4, 3, and 2 percent. She also investigated the distribution of the fruits of mountain maple, which has a rate of fall of 1.0 m sec^{-1}, up to a distance of 725 m.

Rain and, to a greater extent, snow because of its lightness, also enter the border area of the stand, reaching the ground in quantities that are determined to a large extent by the wind field. Snow accumulation in the zone of calm at the lee edge of the stand is a familiar sight (see Sec. 53). The only rain recordings I know of were made by A. Lammert and reported by O. Ziegler [681]. Depth of rain from May to August in the famous dry summer of 1947 on the east side of a 40-m-tall stand of poplars, which ran in a N-S direction, that is, in the shelter of the wind, was as follows:

Distance (m):	4	14	24	outside
Rainfall (mm):	8	33	76	105-110

Wind shadow therefore also means rain shadow in this case. Although measurements of this kind are easy to make, they are almost nonexistent, and the same applies to measurements of dew at stand edges.

In contrast to this lack, numerous investigations have been made of precipitation from fog, which arises from the filtering of droplets out of driving fog by the trees and is sometimes called "horizontal precipitation." The first investigation was made in 1906 by Marloth [669] on Table Mountain in Capetown. He set up a rain gage with a bundle of brushwood on top of it next to a normal gage, and collected ten times as much rain as in the latter. F. Linke [667] has shown, by a series of systematic measurements in the Taunus, Germany, from 1915 to 1919, that precipitation from fog is principally a feature of forest edges.

The first thing to establish, however, is that this phenomenon is by no means confined to the edge only. The east coast of the Japanese island of Hokkaido at 43°N is much plagued by dense sea fogs from the Pacific, similar in character to the Newfoundland fogs. From time immemorial there has been a belt of woods along the coast, which is said to have a protective influence against the fog. Under the leadership of T. Hori [661] a great number of modern theoretical and experimental investigations were made, in coniferous woods near Ochiishi in 1951 and in deciduous woods near Akkeshi in 1952, to test this protective influence and hence determine the best type of trees and size of belt. Turbulence above the tree tops resulted in some of the fog droplets falling into the wood and some outside it. The rate of fall caused more drops to fall inside than outside, and the difference between the two shows the amount of water extracted from the fog by the trees. This was determined experimentally by H. Ooura [671] by means of a well-planned arrangement of lattices and nets above the wood, and the careful limitation and exclusion of all drizzle. The order of magnitude is illustrated by the result that 0.5 mm of fog precipitation per hour was measured when the wind speed was 4 m sec^{-1} and the water content of the foggy air was 800 mg m^{-3}. That was six to ten times as much as that deposited on grassland under conditions otherwise identical. In all cases which fog is of climatological importance, the supplement due to fog in woodland areas must be taken into account, in addition to the distribution of precipitation mentioned in Sec. 36.

This is especially true for high positions reaching into the cloud level in mid-European climate. "Anyone who has observed," writes J. Grunow [634], "the way a mountain forest pours water down to the forest floor as after heavy rain, when the clouds are lying on it, who has seen how branches and the crowns of trees sway and bend under the heavy load of rime, and how whole areas of the forest can be broken asunder by the weight, or who has seen rime falling to the ground and lying like a snow cover, and even a sledge run being made where the ground was formerly clear of snow, will no longer doubt the great yield of fog precipitation." By comparing rainfall on the Hohenpeissenberg (989 m) with and without fog, Grunow calculated the supplement from fog to be 20 percent of the annual precipitation in a closed stand; in cases like this, the gain of water from driving fog compensates the interception loss to some extent.

At the stand edge fog precipitation makes up a substantially greater proportion. The fog collector of Ooura yielded 20 times as much water on the windward or weather side of the wood as a similar device on the lee side. On the Hohenpeissenberg, Grunow estimated fog precipitation to provide an average supplement of 57 percent to liquid precipitation. In a definitely foggy situation, from 25 to 29 April 1952, 18 mm fell in open country while there was 157 mm under the firs in the stand border. F. Linke [667] found over the years from 1915 to 1919, in fir woods at a height of 800 m in the Taunus, the percentage increase in the reading from a rain gage in the stand, compared with the one in the open, given in Table 81. More recent measurements by J. F. Nagel [670]

Table 81

Readings of rain gage in fir woods as percentage of reading in open.

Rain gage	Month		Summer	Winter	Year
	Least foggy (June)	Foggiest (November)			
Immediately on forest edge	104	301	131	184	157
A little farther into the stand	87	259	90	159	123
Average number of fog days	11	24	14	22	18

on Table Mountain near Capetown, gave for the year from 1 March 1954 to 28 February 1955 the following comparison: 3294 mm in a fog-precipitation gage of the Grunow type [657] against 1940 mm in a standard rain gage. Particulars will be given in Sec. 45 on the increase of the percentage precipitation due to fog on mountain slopes as a function of height above sea level.

When temperatures are below 0°C, the supercooled fog droplets freeze on contact with needles and branches to build dangerous accumulations of rime. The shape and quantity of these growths of ice have been studied and described by J. Rink [673] on the Schneekoppe, the highest peak in the Riesengebirge, and more recently on the Hohenpeissenberg by J. Grunow [658]. Measurements on the Feldberg (1493 m) in the Black Forest by K. Waibel [677] give an idea of the weight of the ice that may accumulate. The amount of ice on high-tension power cables was found to depend mainly on wind speed, but not on air temperature and any accompanying snow precipitation. The maximum hourly growth per meter of cable was 230 g, the highest daily total during two winters was 3.2 kg m^{-1}, and the greatest growth over an extended period was 32.3 kg m^{-1}. The trees at the edges of a forest may be crushed under the load of ice. Occasionally the damage becomes catastrophic, when heavy snow falls on the already overloaded branches, or when a gale snaps the top-heavy trees and tosses them about. M. Diem [653] has written on the stress to which overhead power cables are subjected.

38. Further Problems Concerning the Local Climate of Forests

The interests of forestry are not restricted to the microclimatology of the border areas which form the most favorable sites for regeneration, but are concerned with all areas where new growth occurs, and with all conditions that might promote the improvement of older growths. The science of local climatology is still in its infancy. In the following discussion little more can be done than to mention a few examples of the kind of research that is being undertaken, most of which is more random than systematic. The aim of all research must be the complete understanding of the heat and water budget of the locality in question (see Secs. 25-28). Only then will we have not only the necessary observational data on which to base our conclusions, but a true understanding of the processes involved, which is essential if we are to judge the applicability of the results to other localities. The great cost of such investigations, and the demands on trained personnel, mean that much of it must be postponed until the distant future. This does not mean that there are no rewarding tasks waiting to be tackled today. Only a short time ago, a list of the foresters' requirements was laid before forest climatologists by H. Gothe [696].

To rejuvenate the forest, the forester makes improvement cuttings or fellings in addition to using forest edge areas. The light necessary for the next generation of trees is provided by making circular or elliptical clearings by felling, in which new growth soon appears as a result of natural reproduction or direct seeding or planting. The young trees enjoy to a large measure the protection afforded by the neighboring stand and thus benefit from the woodland climate. They have wind protection, an even temperature, and relatively high humidity, all of which promote growth. Extension of the cuttings, desirable to give more space to the young trees, brings with it a risk of night frost, because of the still air in the open space and the increasing amount of outgoing radiation as the area widens. Only when the cutting becomes the size of a clearing is the wind able to exert its influence from above downward and thus once more reduce the danger of frosts.

It is therefore to be expected that in the transition from improvement cutting to a wide clearing there must be a definite size at which the climate of the microenvironment will become particularly extreme. This size depends on the ratio of the diameter D of the clearing (assumed circular) to the mean height H of the surrounding stand. The ratio $D:H$ is called the index of size of the clearing. B. Danckelmann [689] found, for example, as long ago as 1894 that clearings of index 1.25 in the Brandenburg Forest provided complete frost protection, that there was moderate damage with index 1.50, and from index 2 onward damage was considerable.

Since such consequential observations depend on the chance frequency of late frosts in the year in question, R. Geiger [694] undertook a systematic study in a mixed stand of pine and beech, averaging 26 m in height, near Eberswalde, by cutting seven circular clearings which differed only in diameter. Table 82 contains details of the dimensions of the clearings and of the measurements made in the centers. The screening angle h is the angle from the horizontal up to the tops of the trees in the surrounding stand, measured from the ground in the center of the clearing. According to F. Lauscher [48] outgoing radiation A in the middle of the circular clearing, as a percentage of the radiation from open land, can be calculated from the equation

$$A = 100\,(1 - \sin^{r+2} h)\,,$$

in which r is a function of the observed water-vapor pressure e (mm-Hg) that is given with reasonable accuracy by $r = 0.11 + 0.034\,e$. Values for A in Table 82 have been computed by this method and show that outgoing radiation from the largest clearing differed by only 13 percent from the radiation from the open land.

The midday temperatures in the table show how much warmer the aspiration psychrometer was at 10 cm above the central point of the clearing than in the surrounding stand. The temperature

Table 82

Measurements on forest clearings.

Measurement	Diameter D (m)						
	0	12	22	24	38	47	87
Size index $D:H$	0	0.46	0.85	0.93	1.47	1.82	3.36
Mean angle of shielding h	90°	72°	59°	58°	48°	40°	26°
Outgoing radiation A (percent of open land)	0	11	31	33	52	66	87
Rain (percent of open land)	—	87	—	—	105	—	102
Midday temperature excess (8 July 1940) of clearing over stand (deg)	0	0.7	2.0	2.0	5.2	5.4	4.1

excess increases at first, since the sun has better access to the clearing as its size increases, but beyond an index of 1.8 it again decreases, because wind influence then comes into play and begins to carry away the heated air close to the ground.

Night temperature minima, however, do not seem to indicate any critical size of clearing. Temperatures decreased in a fashion more nearly linear as the diameter of the clearing increased, as shown by Fig. 189, which gives the averages of cold nights in the spring and summer of 1940, and for the coldest night with late frosts.

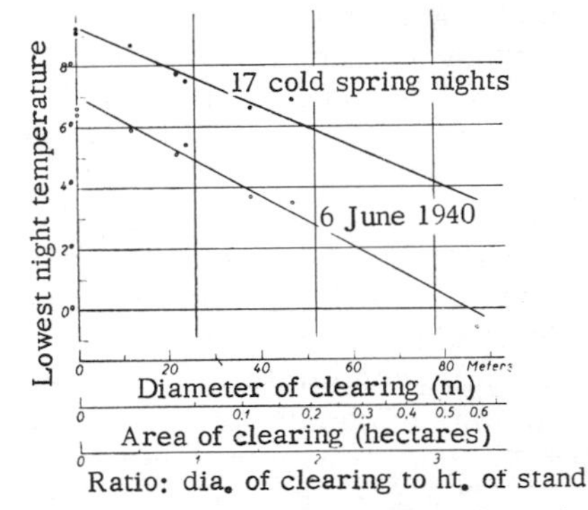

Fig. 189. The increase in frost danger as the size of the clearing increases.

Wind effects become of less importance since nights such as these are always associated with weather situations where winds are light. It may also be seen that outgoing radiation A is not the only influence at work, since the decrease in temperature is not proportional to the radiation values given in the table. In addition, there is a warming influence present in small clearings, or a cooling influence in larger clearings. The first may be explained by a mixing of cold air in the clearing with warmer air from the trunk area, and the second by a downflow of cooled air into the clearing from the crown area of the neighboring stand.

Microclimate within the clearing is by no means uniform; the most varied local differences are found there next to each other in the smallest of areas. Temperatures of the ground, the air, and the bark of trees in sunny and shaded areas were studied from random samples by H. Aichele [682] in a clearing cut to destroy the bark beetle in the central Black Forest. Further detailed and comprehensive investigations have been made by B. Slavík, J. Slavíková, and J. Jeník [701] in a forest area 35 km southwest of Prague.

A clearing of practically the same diameter as the height of the stand was cut in a mixed forest of 0.5 oak and 0.3 beech, with birch, larch, pine, and fir. The shape of the clearing is shown by the heavy dot-dash lines in Fig. 190. During the first 3 yr after cutting all environmental factors, and the development of vegetation and of the young trees, were measured. Figure 190 gives a selection of the observations of local climate.

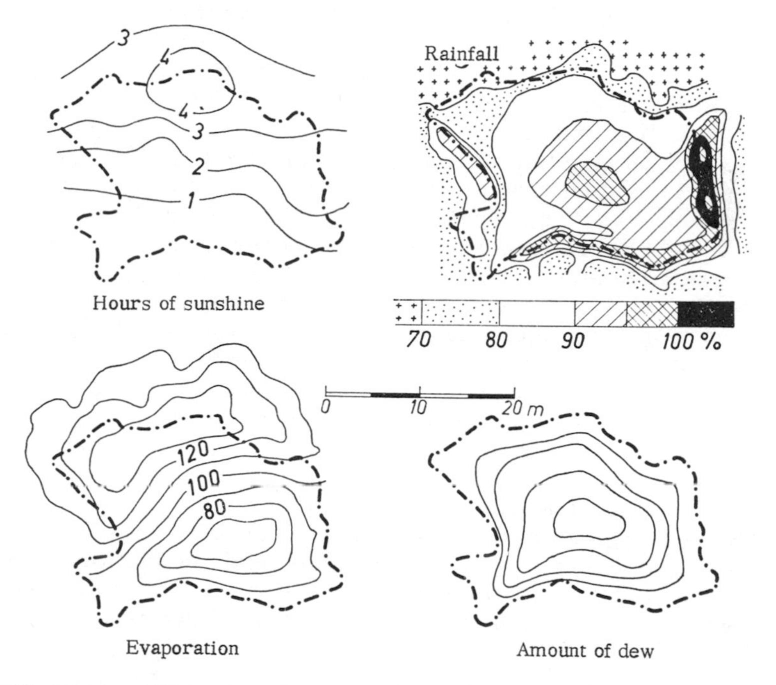

Fig. 190. Local conditions in a clearing in a mixed deciduous forest. (After B. Slavík, J. Slavíková, and J. Jeník)

Great variations are found in the duration of sunshine in the cleared area. The lines of equal sunshine duration are shown in the top left-hand chart, for 20 May 1953. The southwest corner receives only the early morning sun, and the southeast corner the evening sun. The north side of the clearing, being a "southern edge," is in the most favored position, of course. Daily temperature maxima correspond entirely to this distribution of thermal radiation.

The bottom right-hand diagram shows isopleths of equal dew deposit (without numerical values in the German summary). The maximum amount of dew at the center agrees with what would be expected from the nocturnal radiation balance of the clearing. Similar isopleths, running parallel to the edge of the stand, were found by H. M. Bolz [686] in direct radiation measurements at night in a Warnemünde park measuring 40 × 70 m. The night temperature minima behaved in a similar manner, and, since this is different from the distribution of maxima, it follows that there were some unexpected features in the distribution of diurnal temperature variation.

Amounts of precipitation in the two growing periods of 1953 and 1954 are shown on the top right. The prevailing west wind causes the rain to fall into the clearing at an angle, leading to an accumulation in the east, where rain dropping from the periphery of the trees augments it still further so that it is greater than in the open country (over 100 percent). The effect of the drip line at the periphery of the trees may be seen at other edges of the clearing, while in the calmer area near the center there still falls over 95 percent of the rainfall in the open.

The area in question is one that suffers from a shortage of rain (547 mm yr^{-1}), and special attention is devoted to the water budget for that reason. The final chart in Fig. 190, at the bottom left, gives lines of equal evaporation. The measurements were made by Piche evaporimeters at a height of 20 cm above the ground, and the figures give the mean values so obtained, expressed as percentages of the average evaporation in the old stand. The amount of evaporation is less in the southern part of the clearing than in the neighboring forest, because of the low night temperatures. This, however, applies only to the coldest layer of air at night, closest to the ground, up to a height of about 70 cm. In the northern part it is increased by the greater sunshine there. It is interesting to study the change that took place in soil moisture in 1953, the first of these years, shown in the N-S cross section of Fig. 191. While, from early summer to autumn, demands made by the forest on the water supply dried out the soil in the old stand, soil moisture was maintained in the clearing where the roots had been removed. Lack of interception also helped.

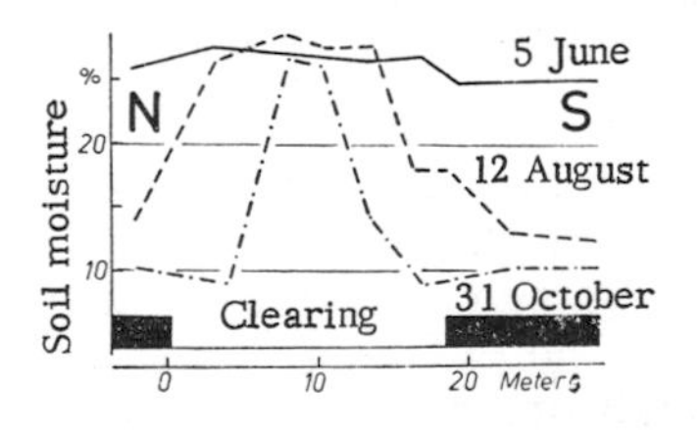

Fig. 191. Maintenance of soil moisture during 1953 in the clearing shown in Fig. 190.

A recent publication by J. Tomanek [703] describes microclimatic investigations in two circular clearings of 20-m and 40-m diameter, and two rectangular cuttings 20 × 60 m with their long axes pointing N-S and E-W respectively (temperature, precipitation, and evaporation extremes in the middle and at four points at the edges).

Observations are available from a tropical forest. M. Gusinde [600] carried out a series of observations in a cutting from 30 m to 50 m in diameter in the primeval forests of the Congo near Ituri from May to December 1934. These observations were analyzed by F. Lauscher. There was a surprising uniformity in the temperature variation in the clearing, and very high relative humidity. The daily fluctuation of temperature was 6 to 8 deg. The daily average relative humidity was 91 percent, the minimum was 70 percent, and in the "driest" month of May it was still 65 percent. Only on one occasion was 33 percent measured. Vapor pressure varied with changes in tropical climate between 11 and 23 mm-Hg.

Besides using clearings as nurseries for young trees, the forester often rejuvenates the growth by planting under the forest canopy. The first step is to cut down a sufficient number of single trees in the old stand to let enough light penetrate through to the forest floor. The shelter of the old timber provides protection from wind, in a similar way to a clearing, thus retaining the moisture and warmth of the old stand for the benefit of the young growth, and making enough radiation and space available for the development of natural seeding or planting. Thinning is carried out in areas or in strips. Another procedure is to build up a protective growth of fast-growing wood such as birch and larch, which lets a lot of light through, and which itself needs much light. This method is used when the trees to be cultivated are of a type sensitive to late frosts, but the protective screen of trees is naturally lower in height than in an old stand. A few series of measurements are available for both of these methods.

The procedure of making improvement cuttings has an effect similar to that of thinning on the climate of a stand. A. Ångström [684] was able to observe, over a period of several years, near Vindeln in north Sweden, the heat gain in the forest floor as the extent of improvement cutting increased. Ground temperatures were 2 to 3 deg higher in stands where this had been carried out extensively than in those that had not been improved, and in the spring the ground thawed 2 to 4 weeks earlier.

Comparisons between the screening effect of an old stand and a clearing were made in 1923-24 by C. v. Wrede [705] in a mixed stand of pine and fir, with single silver firs and beeches in the Altmühl valley in the Jura Mountains. Meteorologic shelters were set up 40 cm above the ground between young trees in a circular clearing 14 m in diameter and in a protective strip 50 to 60 m wide, in which the density of stock had been reduced to 0.43. Table 83 shows how a more moderate climate was produced in the clearing than in the strip because of a reduction of maximum temperature.

Table 83

Temperature differences (deg) in a clearing and a protective strip of trees.

Month 1923	Difference of absolute monthly extremes		Greatest diurnal temperature fluctuation in month	
	Clearing	Screen of old timber	Clearing	Screen of old timber
June	18.1	20.9	12.5	15.5
July	23.6	26.7	15.0	16.7
August	24.0	27.9	20.6	24.0

The night temperature minima were, however, lower in the clearing. The reason for this was that the wind speed at 1 m above the ground was 1.5 times as great in the screen of old wood as that in the group of young trees. The direction of the wind in the group was often opposite to that in the open country and in the screening stand. This is perhaps due to the creation of a large-scale eddy above the group, which is enclosed on all sides by the old stand, as shown in Fig. 192. H. Pfeiffer [672] has shown by smoke experiments how such eddies develop in the narrow central spaces surrounded by blocks of houses in the city, which are similar in general shape. This is of practical importance since houses standing on the right of Fig. 192 would have a wind blowing down the chimney, so that fires would not draw properly, and might even at times be blown into the

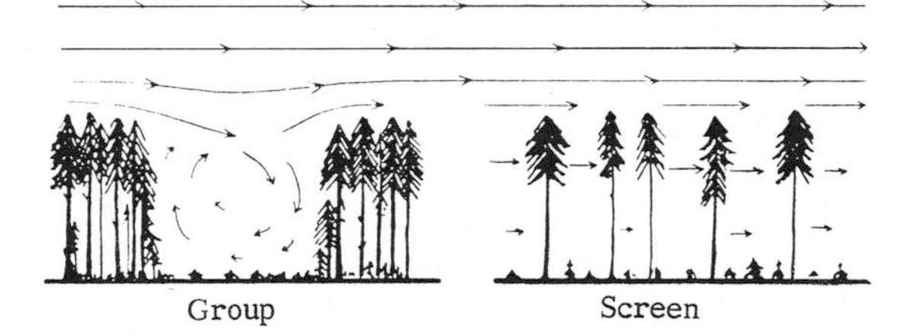

Fig. 192. Air flow in a regeneration area and under a screen of old trees.

room. Planting of trees in the open courtyard prevents the development of eddies and consequent damage. So little sun penetrated to the ground in the clearing investigated by Wrede (even if there were enough light) that heating of the air layer near the ground was due to the neighboring old stand, while in the strip screen it was partly due to heating from the ground which received direct sunshine at times, as may be seen in the analysis of vertical temperature profiles. It followed that the relative humidity was also 5 to 7 percent less in the screen than in the clearing.

A preliminary report on 3 yr of comparative measurements of the water budget in a forest screen (beeches) and cleared areas has been issued by F. Kortüm [698]. It was shown that the critical factors determining the mass of leaves produced were evaporation (which in turn depends on the amount of heat received in radiation, see Sec. 28) and the humidity of the air. This approach to the problem from the energy aspect proves to be very fruitful.

The changes in stand climate resulting from the establishment of an outer screen of birches was investigated by A. Baumgartner [685] in the Thiergarten forest in the principality of Thurn and Taxis near Regensburg. Impetus to make the investigation arose from the experience of the Forest Officer, A. Lindner, that spruce, fir, larch, and oak all grew much better on a southern slope below a screen of birch than in the open. Comparisons made in all kinds of weather showed without any doubt that the local forest economy, general climate, and soil conditions were the same in the two places (with one exception which will be mentioned shortly), and that the differences in growth could therefore be due only to indirect effects. The only element that differed was the diurnal temperature variation, illustrated in Fig. 193. The air in the lowest half meter shows greater temperature extremes in open country than in the outer screen. Above this level the situation is reversed. The curves in Fig. 193 are based on measurements in all kinds of weather, including a particularly wet year. It therefore follows that temperature extremes are greater at the ground on clear nights, and (Sec. 29) plant temperatures even more extreme than is indicated by these averages.

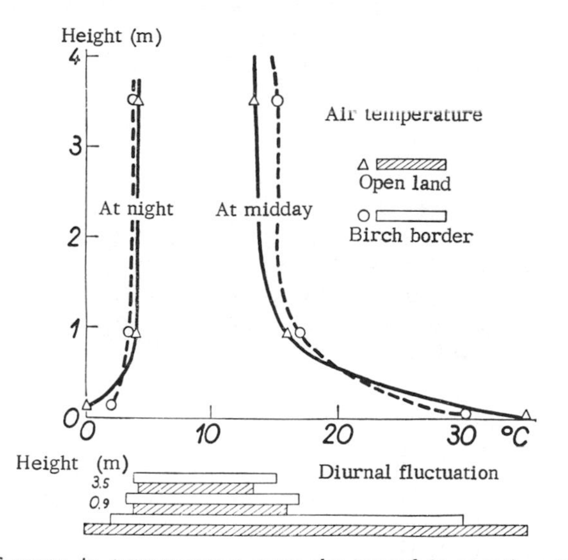

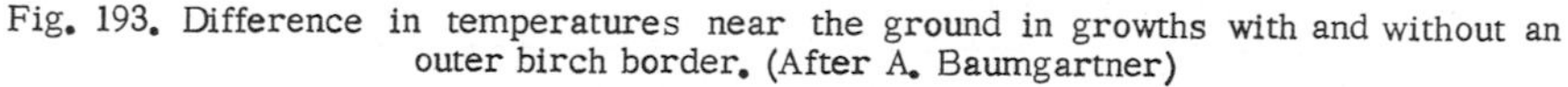

Fig. 193. Difference in temperatures near the ground in growths with and without an outer birch border. (After A. Baumgartner)

It is therefore correct to state that the outer screen provides effective protection at the time of late frosts. This was proved in 1927 by H. Amann [683] near Munich, and in 1952 by K. Göhre [695] near Eberswalde. Amann made his observations in May, in a

32-yr-old outer screen of birch 11 m high, designed to give frost protection to a 14-yr-old stand of young firs 0.8 to 1.5 m high, in the Anzing-Ebersberger forest, which is very susceptible to late frost damage. Temperature minima measured at a height of 25 cm in the birch screen and over nearby clear ground were as shown in Table 84.

Table 84

Temperature minima (°C) in a cleared area and a screen of birch trees.

Date	Felled area		At the edge of felled area			
	Point 1	Point 2	In the outer birch wood	In the interior: Point 1	In the interior: Point 2	Near to an old stand
11-12 May	-11.0	-10.9	-7.3	-6.2	-5.6	-5.2
14-15 May	- 8.0	- 7.7	-4.1	-2.6	-1.9	-2.0
15-16 May	- 3.8	- 3.5	-1.7	+0.4	+0.6	+1.0
25-26 May	- 2.9	- 2.1	+0.2	+1.5	+2.3	+2.5
Average of 11 May nights	- 4.1	- 3.7	-1.3	-0.2	+0.4	+0.5

The temperature gain on the average of the eleven coldest nights was 4 deg and reached 6 deg on one night. Near the edge of the open ground, an inflow of cold air can be recognized in the modification of the vertical temperature profile, giving a numerical proof that cold night air is entering the screen from the open surroundings.

When a stand border has been developed in a different way, its behavior may be quite different; this was shown by Göhre [695] in measurements made in an outer screen of birches that had developed from wind-borne seeds in an area left bare by a forest fire. The birches were 2 to 3 m high, and in their shelter pine seedlings 20 to 40 cm high were growing. Figure 194 shows the average temperature minima 10 cm above the ground at 15 observation points (M) in the period from 27 June to 10 July 1952. In this case the inner border of the birch screen is the warmest position; no cold air from outside is therefore flowing in under the screen.

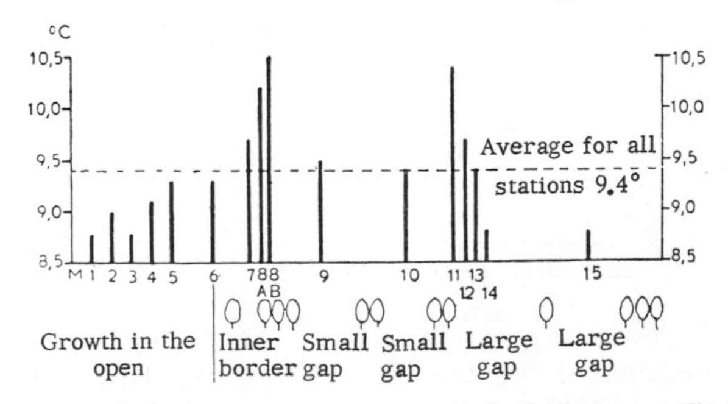

Fig. 194. Minimum temperatures in an outer wood of birch near Eberswalde. (After K. Göhre)

Gaps in the outer screen, as well as bare open ground outside, are areas of frost danger. Cold pools develop in these gaps through radiative heat losses, and because the air is still, not through advection. The pool of cold air is retained in the wood, and therefore an outer screen with gaps is only half a screen.

It is more difficult to assess the influence of measures adopted by forestry to develop older stands than to decide what is good at the nursery stage. The following chronologic summary will give some idea of the practical tasks involved in research in applied forest meteorology by means of instruments, although it is not possible to discuss the various works individually.

R. Geiger [618] made a comparison of two 65-yr-old stands of pines in northeast Bavaria, of which one had a uniform, loosely knit crown, and the other had a rich undergrowth of firs. Figure 170 was an example taken from this study. R. Geiger and H. Amann [619] made a similar investigation in two neighboring stands of old oaks with and without a subsidiary growth of beech in the Schweinfurt Forest (see Fig. 172). H. Burger [688] compared the climate in the trunk area of a fir plantation in the Canton of Bern in Switzerland with another of fir and spruce of mixed age groups. At Erdmannshausen to the south of Bremen ground temperature was measured in two mixed stands of pines and firs by Boos [687], and humidity, precipitation, and wind measurements were made in the air layer near the ground. J. Eggler [690] proved by means of random sampling on 16 different days that the rare stands of oak near Graz had a different slope climate from that of a nearby beech wood. B. Maran [699] measured air humidity during fair weather in mixed stands of pine and larch, with and without undergrowth, and in birch woods in central Bohemia. A. Sokolowski [702] compared four important groups of forests in Poland by means of systematic series of measurements of ground and air temperature, humidity, and wind over a period of 2 yr. R. Wagner [704] made observations in beech, oak, and willow forests in Hungary.

Increasingly detailed investigations into the local climate of forests lead finally to the desire to grasp fully the tremendous variability of all the climatic facts about forests. Figure 195 is an extract from the work of H. G. Koch [697] showing the temperature variation on a single day of fair weather in a closed forest consisting of many different components near Leipzig. The 7-km stretch is illustrated in section at the top of the diagram. Temperatures were measured in journeys by automobile.

Shortly before sunset, and about 2 hr before sunrise, the isotherms are crowded together and run in a mainly horizontal direction. This means that temperatures change sharply everywhere at the same time. The decrease in temperature at sunset and its increase in the morning are such large-scale meteorologic features that all differences in the stand are insignificant by comparison. When the heat balance approaches equilibrium, however,

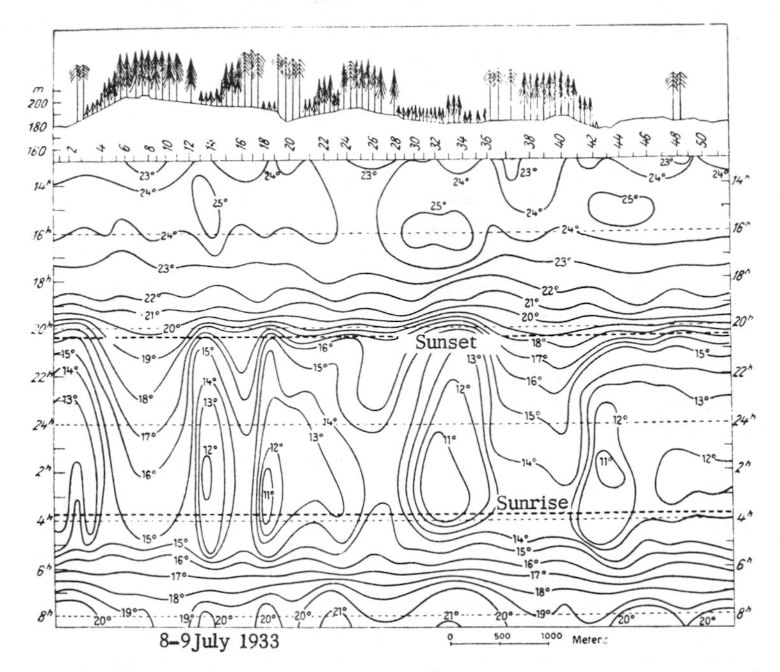

Fig. 195. Diurnal temperature variation within a closed forest area near Leipzig. (After H. G. Koch)

the local differences come into full effect. This is seen more clearly at night than during the day (Fig. 195), in the "islands" of isothermal pattern which are more sharply marked at night. The reason for this is the smaller movement of air by night and its thermally stable stratification. Temperature exceeds 25°C at noon in three places: these are—as seen from the sketch of the trees at the top—the clearings, the young growth, and the open land. These areas are very cold at night, falling below 11°C in several places. The isotherms also show very clearly how slowly the night cold penetrates into the old stand.

The influence of topography falls completely into the background in this example. Its effect will be discussed in Chapter VII. When forest and mountains are brought together it becomes much more difficult to evaluate local climate. This has been investigated by A. Baumgartner, G. Hofmann, H. G. Koch, D. H. Miller, J. Parker, G. Waldmann, J. N. Wolfe, and others. This subject will be dealt with in Sec. 47.

39. Remote Climatic Influences of the Forest

Sections 33 to 38 dealt with the influence of vegetation and climate on each other at a particular place. The question now is whether the sum total of all those influences of vegetation, which to begin with are tied to one locality, may not in the end have some effect on climate in the wider sense. For many hundreds of years, forests have had climatic influences ascribed to them by writers. Wooded countries, it is said, have a different climate from countries without woods. Felling and spoliation of forests worsen the climate, afforestation improves it.

There is much in history to show that highly developed human cultures, especially in the dry areas of the earth, went down at the same time as their forests were lost. The connections are so many-sided and capable of so many interpretations that great care must be exercised in coming to conclusions based on historical data. It is interesting to note that the development of forest meteorology in Germany had its origin in this very question. It was at first more a matter of forest politics than a discipline of the natural sciences. Its function was to provide proof that destruction of forests would be damaging to the people. The term "beneficial influence" began to be used when speaking of forests, "because Country and People will fare well as long as the forests stand." We will use the term "forest influence" without passing judgment on the value of the influence.

The beneficial effects of the forest have been described by E. Hornsmann [717] in his book *The Forest Helps All* (with a bibliography), which should appeal to every forest lover. It is known today that forests do indeed have a great influence on the nature of a country, but this is due only slightly to their effect on climate. In this section a survey will be made of the problem as a whole, especially with the aspects related to climate, to the extent that they are supported by definite results.

Deforestation of a country will have basically different results, depending on the general climate of the country concerned, the properties of its soil, and the type of vegetation. In steppes and in continental climates the consequences are more serious than in a humid, temperate climate, or in the region of the tropical rain forest. Mountainous areas are more sensitive to the loss of woods than flat areas. This is applicable even on a small scale. In the maritime climate of England, deforestation during World War II had practically no repercussions, while the large-scale cutting down of the Harz forests after 1945 resulted in serious damage. The importance of soil and vegetation has been discussed in many of its aspects by W. Wittich [730]. As he showed in his clear and critical exposition, it is only when the complicated relations are broken down into individual processes, and these processes are studied in typical limited areas, that advances can be made and the tendency avoided to transfer correct results incorrectly into other contexts.

The discussion of forest influences start with some generally known facts, which are not of a climatic nature. On slopes forests provide the best protection against soil erosion by water. In

Germany, according to J. H. Schultze [728], the critical angle of slope at which soil erosion occurs is between 1° and 7° on fields, on roads between 5° and 10°, and in woods between 20° and 30°. In mountain areas the forest helps control flash floods and offers protection against avalanches not only by "accomplishing this by splintering into nonexistence" (Hornsmann) but by binding the snow and holding it in avalanche release areas. Forests provide the surest protection against both wind erosion of loose soils and the accumulation of airborne soil. Changes caused in the wind field (discussed in Sec. 37 for forest edges), and planting of artificial windbreaks (see Sec. 53) are other ways in which the forest exerts influence without making any changes in the general climate.

Of the beneficial effects of forests, the central position is usually occupied by the water budget of the area. Many great systematic investigations have been made in this context. Switzerland led the way in 1900 by research in precipitation and runoff in two forest valleys, one of which (the Sperbelgraben) was completely wooded, while the other (the Rappengraben) had mostly undergone deforestation. The results of this experiment, extending over five decades, have been published by A. Engler [715] and H. Burger [710, 711]. In the Colorado mountains, C. G. Bates and A. J. Henry [707] selected two neighboring forest valleys 90 and 81 hectares in area, and investigated their climate and water budget from 1911 to 1919 before one of the valleys was clear-felled. Comparative measurements between the two were then continued until 1926. To this "Wagon Wheel Experiment" there were added the "Coweeta Experiment" [729] and several other large-scale investigations in the United States. In 1951 new experiments were started in the Canton of Freiburg in Switzerland, designed to be continued for 30 to 50 yr [713].

In Germany, investigations were carried out in the Upper Harz by J. Delfs, W. Friedrich, H. Kiesekamp, and A. Wagenhoff [628] during the period 1948-1953, already mentioned on pp. 332f. Figure 196 illustrates the water budget of the valleys of the Lange Bramke and the Winter, which are adjacent to each other, running ENE and

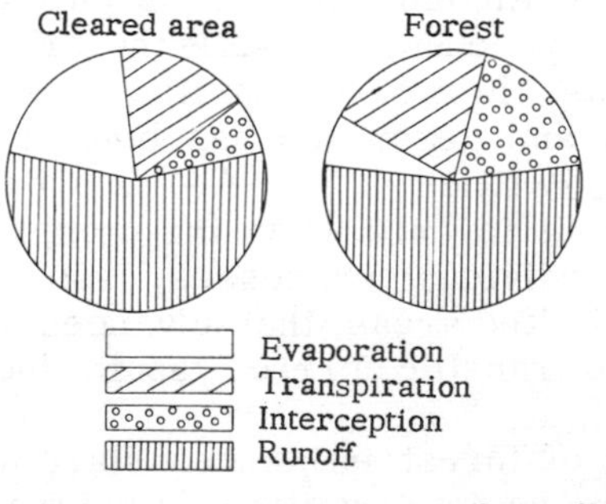

Fig. 196. Evaporation, transpiration, interception, and runoff in experimental plots in the Harz Mountains, averaged for the year. (After W. Friedrich)

NNE at the same height above sea level (600 to 700 m). The proportions of the various elements shown are for the 5-yr period 1949-1953. The Lange Bramke Valley (area 75 ha) was 97 percent cleared in 1948, while the Winter Valley (area 87 ha) had a stand of 82- to 122-yr-old firs, with a thicket and saplings on 20 ha. The slopes of the valleys were 13 percent and 15 percent respectively, while the hillsides in each had an average slope of 36 percent.

The interception loss (see Sec. 36) was substantially greater in the wooded valley than in the areas with surface vegetation and the reforested areas in the cleared valley, where the young firs reached a height of only 80 cm by the end of the experiment. Productive transpiration consumed more water in the forest than unproductive evaporation from the ground in the cleared area. Averaged over the 5 yr, in which period the amount of water in the soil and the loss of ground water balance each other, the bare area transferred a greater share to ground water than did the forest (57 against 54 percent). This is because the need of water is greater in the forest. The highest forest production is associated with greatest consumption of water (see ref. 718). In the large-scale Coweeta Experiment care was taken to retain the humus layer after deforestation, and the water recharge was increased to the extent of being equivalent to 430 to 560 mm of precipitation per year (Wittich).

The importance of forests to the water economy lies not in that they provide a supply of water, but rather in that they have a regulating influence on the water economy. It is only in extremely dry areas, such as the western United States, where it is important that as much as possible of the precipitation find its way into the groundwater supply, that the forest is thinned or built up of woods which are economical in their water requirements so that it would retain its good protective and regulating properties, but use as little water as possible. Forests of this special type have been practically nonexistent in the climatic conditions of Germany up to the present.

The regulating influence of the forest depends primarily on its ability to reduce surface runoff. A normal forest floor has a high water-absorption capacity, even when the quantity of water is great. Experiments with artificial sprinkling amounting to 100 mm, made in Switzerland by H. Burger [710], showed that compact soil under a willow tree took more than 3 hr to absorb the water; the soil under another willow took almost 2 hr, a plantation with gaps 20 min, while a good stand of fir, spruce, and beech needed only 2 min to absorb it. Investigations into mosses and lichens in woods, made by K. Mägdefrau and A. Wutz [721], showed that they were able to absorb between 3 and 10 mm of rain when the air was dry, depending on the type of moss. Laboratory experiments with 1 dm^2 of various mosses from Bavarian forests gave a quantity of 2.3 to 7.5 mm of water absorption for four fir stands, 8.6 mm in a pine stand, and as much as 14.7 mm for moss from a mixed stand of pine and fir.

The forest floor releases this water very slowly. The mosses mentioned above took 16 days. Therefore, depending on the water-holding capacity and permeability of the soil, new water always penetrated into lower depths, even in rainless periods. In the forest region recharge is more uniform. In winter a reserve of water is built up in the forest in the form of snow, protected from evaporation (see p. 337).

These processes bring about the further result that the high flood water in summer is more moderate in forest areas. In the two Harz valleys, for example (Fig. 196), rain on 7 July 1950 amounted to 16.4 mm in 37 min; this produced runoff of 200 lit km^{-2} of catchment area per second in the unwooded valley, whereas it was only 75 lit km^{-2} in the wooded valley. In winter, by contrast, the proportion of height and duration of high water in the two valleys may be interchanged. When the ground water is recharged in winter, floods may be higher in the forest if, for example, the remains of melting snow are carried off along with the rain, when the unwooded area has long since been snow-free. Suppression of surface runoff in the forest produces greater purity of water. Water from the bare valley in the Harz in 1950 contained 56.0 tons of suspended matter, and brought down 2.0 m^3 of pebbles in the stream per square kilometer of catchment area, against 18.6 tons and 0.05 m^3 for the wooded valley. During normal flow, both streams contained 5 to 10 mg/lit of suspended matter, which does not change much during flood in the forest valley, but rises to 550 mg/lit in the cleared valley.

This short review of nonclimatic beneficial effects of forests provides ample justification for the numerous European laws, dating back to the 14th century, which declared the forests of certain areas to be protected forests, forests for water conservation, or prohibited forests. Their economic exploitation was therefore subject to certain restrictions, which were intended to safeguard their maintenance. H. v. Pechmann [725] pointed out the necessity for protecting such forests by law, in conformity with present-day knowledge, basing his arguments on experience with Bavarian forests. Some forest laws are aimed at preserving these beneficial influences and are hence applicable to all forests.

Turning now to the climatic benefits arising from forests, we may begin with the oft-repeated question whether forests increase the amount of precipitation over a territory. Assertions of this nature are based on the established fact that the presence of a forest has a favorable influence on the water economy, and this was first attributed to increased precipitation. Moist air in the forest, the smoking phenomenon described on p. 336, the observed zone of high humidity in the air surrounding the forest, which has only just been established by H. Mrose [722], and many other observations may all have contributed to provide apparent support for such propositions.

Precipitation formation is, however, a process that takes place in the upper atmosphere, and it is very unlikely that the type of ground below will have any significant effect on it. A study of the African coast of the Mediterranean provides an impressive example of how little effect even a massive supply of water vapor from below may have. Although enormous quantities of water are transferred from the warm sea into the atmosphere by evaporation, the coasts remain arid desert because the general circulation of the atmosphere in these areas is unfavorable for precipitation.

Early attempts to establish some factual basis for argument by comparing precipitation amounts in wooded and unwooded country were inconclusive because there was a large number of other active influences present, such as height above sea level, proximity to the coast, type of soil, relative position to atmospheric pressure centers, and so forth. Measurements such as those made by J. Schubert [727] on the Letzlinger moors by means of a special network of observation stations could not be easily interpreted, because of extreme difficulty in assessing the influence of wind on precipitation. It is certain that the increase in annual precipitation of 5 to 6 percent, ascribed to forests, is too high. An attempt was also made to prove that there was a change in precipitation amounts at the time the island of Mauritius underwent deforestation from 1850 to 1880. The most interesting investigation was that made by H. F. Blanford [708] when a large area of southern central India was reforested as a result of a new forest law in 1875. By comparing precipitation figures before and after, he concluded that he had detected an increase in rainfall. A. Kaminsky [719] showed later, however, that this was due to a climatic variation, which had affected great areas of the country, but had left unaffected precisely the places Blanford had selected to set up his control observation stations, intended to verify the accuracy of his measurements before and after reforestation. This is an instructive example, showing how very difficult it is to arrive at useful conclusions from experiments covering such a wide field of observations.

The small-scale investigation made by H. Burckhardt [709] at the meteorologic station on the peak of the Erbeskopf (816 m) in the Hunsrück is illuminating. In 1949 the 20-ha beech forest of varied age groups was felled, and the depth of precipitation decreased to 62 to 85 percent of its previous value. Was this attributable to the forest? The answer is no, since we must assume likewise that some of the earlier rain might have been caused by the forest. It would be attributable to the forest if we take into consideration that the simultaneous proved increase of almost 40 percent in wind speed blew away the rain that formerly would have settled on the mountain. The soil, vegetation, subsoil, and springs all received less water after deforestation. This is the sense in which it is possible also to understand the greater quantities of rain falling on many prairie or steppe regions, the Ukraine for example, after afforestation, as shown by rain-gage measurements.

Toward the end of the 19th century, a new technique for tracing forest influences was brought into use in Europe. Twin forest observation stations were set up, consisting of meteorologic stations in pairs, one inside and one outside the forest, close to each other, so as to provide a good opportunity to study the differences between the climate of the trunk area and the open country. The initiator of this technique was E. Ebermayer in Bavaria [714], followed by H. E. Hamberg [716] in Sweden, A. Müttrich [723, 724] and J. Schubert [726] in Prussia, Lorenz-Liburnau [720] in Austria, and H. Burger [688, 711] in Switzerland.

It was by means of this technique that our first knowledge of precipitation conditions within and without the forest was gained, giving us some information about the amount of interception (already discussed in Sec. 36). The high humidity inside the forest was discovered; and the temperature variation in the forest floor and in the soil of open country, and the frequency and depth of frost penetration in winter frosts could be compared. It became possible to express numerically the moderation of temperature variation in the forest. Figure 197 shows examples

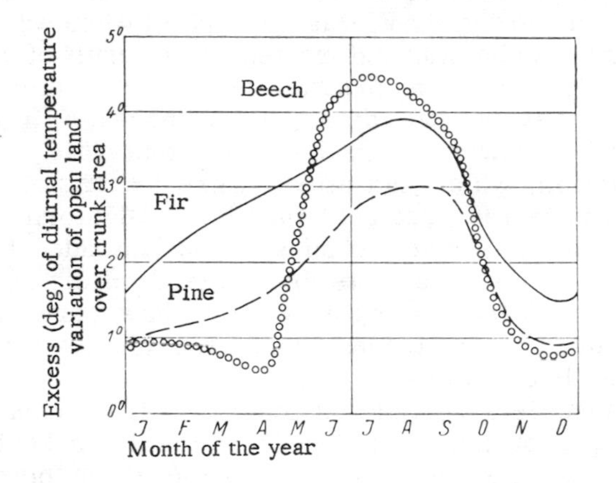

Fig. 197. The "characteristic curve of the type of trees" as an indicator of the temperature variations in the trunk-area climate. (After A. Müttrich)

of so-called "characteristic curves for tree type" constructed from observations made over 15 yr by A. Müttrich [723] with five paired stations in a fir stand, four in pine, and six in beech. These annual curves show by how many degrees the mean diurnal temperature variation is lower in the trunk area than at the same level outside the stand. The difference is naturally greatest in the summer. The annual variation is uniform in the coniferous forests, in comparison with beech forests where breaking into leaf in the spring and disappearance of foliage in the fall cause noticeable discontinuities.

Processes in the crown area were left out of consideration at that time. It was therefore not possible to reach any direct conclusions about the beneficial effects of the forest. This would be possible only by understanding completely the heat and water budgets of the forest and the adjacent open country, that is, by looking into nature's workshop and seeing how forest and open country utilize what nature has endowed them with. Nevertheless, in an indirect fashion, these investigations laid the foundation. Although, for example, the temperature variation is much greater in the crown area than in the trunk area (Sec. 35), it can never reach the value it has in the open. Wooded regions have a milder temperature, more similar to a maritime climate, than areas without forests.

In Sec. 37 we learned something about the ability of forest edges to filter out dust. Figure 198 shows the distribution of dust measured by M. Rötschke [674] during an automobile journey near Leipzig in the afternoon of 10 April 1935, using a Zeiss conimeter. Not only the edge, but also the whole area of the forest, which itself produces hardly any dust, shows a low dust content. A similar reduction was found in the smaller and more numerous condensation nuclei, in experiments made by H. Zenker [731] on the island of Usedom in 1953.

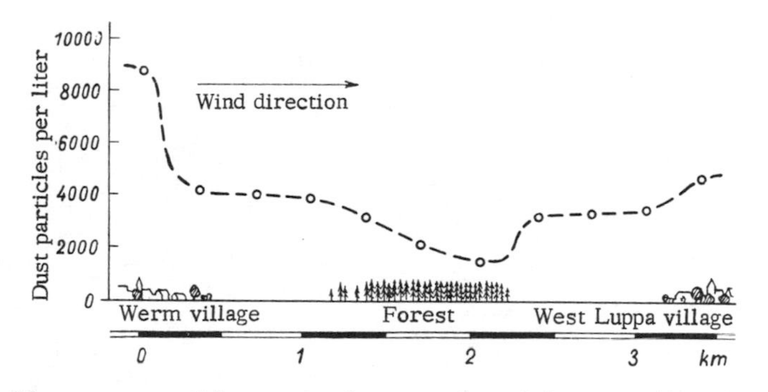

Fig. 198. Measurements of dust content in open and wooded country. (After M. Rötschke)

Twin observation stations do not record the difference in climate between forest and open country but the climatic contrast below and above a plant cover. Beetles at a height of 2 cm in summer grass will experience the same kind of properties of "trunk-area climate" as a human being would at a height of 2 m in the forest. In many respects, a far-reaching effect in climate is a question of whether the ground has some form of vegetation cover or not (sandy desert, gravelly waste). This is a more important distinction than that between a forest and the normal type of cultivated land which also has a plant cover, during the growing period at least.

At the present time it is certain that any increase in precipitation due to the presence of a forest is bound to be very small. It is also known that the presence of forests has a moderating influence on climatic extremes. However, the extent of this influence is even harder to assess. When it becomes a question of the whole economy of nature, as in this case, it is necessary to keep in mind and in sight, as A. Schmauss often said, that "everything fits in with everything else." This is particularly true when the living processes of plants are involved, since these are able to mask the physical relations as a result of their ability to make adaptations to circumstances. No one has yet tried to find out, for example, what reciprocal influence the ability of the forest to regulate water flow might have on climate.

There is no doubt that the influence of the forest is the greater the more inhospitable the nature of a territory is to the growth of trees. In the boundary zones between arid and humid climates, at the tree line in mountains and polar regions, and in regions of frequent storms or constant strong winds, the beneficial effects of forests may become very great. There is here a very great and rewarding field open for scientific investigation.

CHAPTER VII

THE INFLUENCE OF TOPOGRAPHY ON MICROCLIMATE

The discussion was introduced in Chapter II by considering the microclimate of a level surface without vegetation. Then in Chapters V and VI this presupposition was abandoned, and the mutual influences of vegetation and microclimate on each other were studied. Now the assumption that the ground is level will also be abandoned, and the influences exerted on microclimate by the topographic variations of the terrain will be investigated.

The first idea that comes to mind is that topographic influences must be much more noticeable during the day than at night. When the sun is shining, slopes of different inclination and orientation receive different amounts of heat, and the equalizing currents of air set in motion (such as up-slope or anabatic winds) determine the microclimate. The orientation of the slope is therefore of decisive importance. At night, the temperature distribution is regulated by the downward flow of air that has been cooled in contact with the ground. The orientation of the slope is of no importance, but differences of height become significant. The paragraphs that follow will therefore be arranged according to the time of day.

Consideration of topographic influences soon leads to new questions. When only level ground was considered, for example, microclimate was restricted to the lowest few meters of the atmosphere, and we ventured out of this shallow layer only occasionally, to obtain a better understanding of some process or other. When the land is one of strong relief, however, microclimate must be extended farther and farther into deeper layers. There is a continuous transition from the microclimate of a furrow in a plowed field to the climate of a long mountain valley, which is one of the subjects of study in general climatology. Since this is a textbook of microclimatology, the exposition will of necessity become all the more brief, the more closely we approach problems that are dealt with in textbooks on general climatology.

40. Insolation on Various Slopes

Sloping ground is described by its angle of inclination, or gradient, and the direction toward which it faces. These two quantities determine the situation or exposure of the slope. The connection between angle of inclination and gradient is as follows:

Slope:	0.1°	0.5°	1°	3°	7°	11°	27°
Gradient:	1:573	1:115	1:57	1:19	1:8	1:5	1:2

A west slope means a slope facing toward the west; a valley running N-S therefore has a west slope on its eastern side and an east slope on its western side (care must be taken when consulting references).

Slope climate or exposure climate is determined, in the first place, by the different amounts of direct solar radiation and heat received by an inclined surface as compared with a horizontal surface. This difference can have great significance; for example, a surface inclined at 20° facing toward the south, even allowing for the high degree of cloudiness in Germany, receives roughly twice as much radiation in January as a horizontal surface. The amount of radiation it enjoys is equivalent to a substantial displacement toward the equator.

The climate of a slope has therefore always been of significance. The German wine industry depends on it. In agriculture and horticulture it decides the quality of arable land and whether it will be possible to cultivate certain species of plant. The early strawberries that are on sale in Tokyo two months in advance of the main crop are grown on the steep terraces of Shizuoka, the exposure climate of which has been described by S. Suzuki [754]. In looking for sites for hospitals and sanatoriums, an attempt is made to find sunny slopes. Extreme slopes of 90°, that is, walls facing various directions, are of great importance in the architecture of dwelling houses, town planning, technology, and the cultivation of fruit on trellises. We are already familiar with some of the effects from the microclimate of stand borders in Sec. 37. This is of such importance that illumination engineering has already developed into a separate profession in German building technology.

Beginning with direct solar radiation, which is the thing first experienced by the observer, it is a simple mathematical operation to compute the quantity of radiation falling on a surface in any position and elevation, given the intensity and direction of solar radiation. K. Schütte [752] has devised a simple procedure for carrying this out. There are five factors involved: the geographic latitude, the declination of the sun (period of the year), the altitude of the sun (time of day), the angle of the slope, and the direction in which it faces. Sometimes it is desirable to have instantaneous values, or hourly average values, or diurnal or monthly totals of radiation intensity. Sometimes only duration of sunshine and times of sunrise and sunset are required; occasionally the values to be measured are for cloudless periods (with various degrees of atmospheric turbidity), and at times it is desired to correlate radiation received with the state of cloudiness. It is obvious that no set of tables can satisfy all requirements. The summary given in Table 85 will help the reader to consult the published tables most suited to his purpose; this is only a selection of the material available.

The astronomically possible sunshine duration and the actual amount recorded at Karlsruhe in a series of measurements extending over several years have been tabulated by J. v. Kienle [745]. The analysis is for the eight principal directions and for slopes inclined at 0, 15, 30, 45, 60, 75, and 90°, and also shows the hours of sunrise and sunset. Figure 185 gave the data for walls (90°) in graphic form. The summary in Table 85 is only for published measurements of radiation intensity. At one time people were satisfied with theoretical computations that excluded the effect of the atmosphere. By this means R. Gessler [737] computed the extraterrestrial radiation received by sloping surfaces on 17 selected days, for latitudes and slopes at intervals of 15° and for principal directions. M. R. Pers [750] made a similar theoretical analysis, using a general transmission coefficient of 0.8 for the atmosphere.

The works quoted in Table 85 are all based on actual measurements (over as many years as possible) and therefore come closer to reality. It may be readily discovered what a great difference there is if, as in Group A, only direct solar radiation is used as a basis and the total or horizontal radiation as measured is adopted to give a calculated value for the slope; or on the other hand whether the radiation-measuring instrument itself is tilted at the angle under consideration and thus the important effect of sky radiation (and reflected radiation from the ground) is also captured. Group B comprises the works that have used this new and highly welcome technique, which is much more valuable for practical purposes. Naturally the figures available in Group A are much more comprehensive and cover many aspects of the problem, while those in Group B are sparser, but are closer to reality.

The basic laws of irradiation of slopes are illustrated by Fig. 199. It is based on measurements of direct solar radiation in cloudless weather during the years 1930 to 1933 at Trier (49°45′N), made by W. Kaempfert [741]. The abscissa is the angle of slope in degrees, and the ordinate is local time. The figure consists of nine diagrams, for three directions of slope and three selected days, the figures printed on the isopleths being the quantity of radiative heat received, for cloudless skies, with a normal amount of atmospheric turbidity.

Since the gradient scale of each diagram begins with a slope of 0°, the left-hand margin of each shows the radiation received on the horizontal; this is therefore identical for each set of three in the same column (daily sunshine duration) and also for the isopleths

Table 85

Papers published on the intensity of solar radiation on various slopes. (For additional works, see text and bibliography.)

Author [reference] year of publication	Geographic latitude	Gradient and direction of slope	State of sky	Periods and quantities calculated	Method of representation
A. Calculated from measurements of direct solar radiation (sky radiation not included)					
J. Schubert [676] 1928	52°	0, 30, 90° N, E, S, W	(*a*) Cloudless (*b*) Partly cloudy	(*a*) 1-hr value at midday of each month (*b*) Daily totals for midday of each month	Tables (and graphs)
G. Perl [749] 1936	0, 15, 30 45, 60 75, 90°	0 and 90° N (S) and E (W)	Cloudless	Hourly and daily totals for the 4 days at the solstices and equinoxes	Tables (and graphs)
W. Kaempfert [741] 1942	50°	0, 10, 20, 30, 60, 90° N, E, S, W	Cloudless	Hourly totals for the 4 days at the solstices and the equinoxes	Tables (and graphs)
M. Nicolet and L. Bossy [748] 1950	51°	90 N, NE, E, SE, S	Cloudless	Half-hourly instantaneous values for the 5, 15, and 25 of each month, hourly and daily totals	Tables (and graphs)
A. Schedler [751] 1951	48°	10, 20, 30° N, NE, E, SE, S	(*a*) Cloudless (*b*) Partly cloudy	(*a*) Forenoon, afternoon, and daily totals for the middle of every month (*b*) Monthly totals	Tables (a few sketches)

W. Kaempfert and A. Morgen [744] 1952	50°	0, 10, 20 . . . . . . 80, 90° N, NE, E, SE, S	Cloudless	Annual variation	Graphical
A. Bögel [733] 1957	50° 60°	0, 10, 20, 30, 40° W (E)	Cloudless	Daily totals for 11 days with different solar declination	Tables (and graphs)
B. Global radiation measured on sloping surfaces					
J. F. Hand [740] 1947	42°	90° S and E	Partly cloudy	Ratios of radiation on a wall and on a horizontal surface. Hourly values for each week, 26 March 1945 to 3 June 1946	Tables and annual curves
J. C. Thams [755] 1956	46°	0° 25° S	(*a*) Cloudless (*b*) Partly cloudy	Monthly averages of the daily global-radiation totals	Tables
K. Gräfe [738] 1956	54°	0 and 90° N, E, S, W 45° S	Partly cloudy	Daily totals and averages of global radiation for individual days, for 4 yr of observation	Graphical
W. Collmann [32] 1958				Averages of daily global-radiation totals for 10-day and monthly periods	

of equal radiation intensity extending from the corresponding points on the margin. The right-hand margin of each diagram indicates the radiation received on a vertical wall, the upper limit being the time of sunrise and the lower the time of sunset. The north- and south-facing slopes naturally show a symmetric distribution of isopleths about the noon line, while the east slope gives an asymmetric distribution.

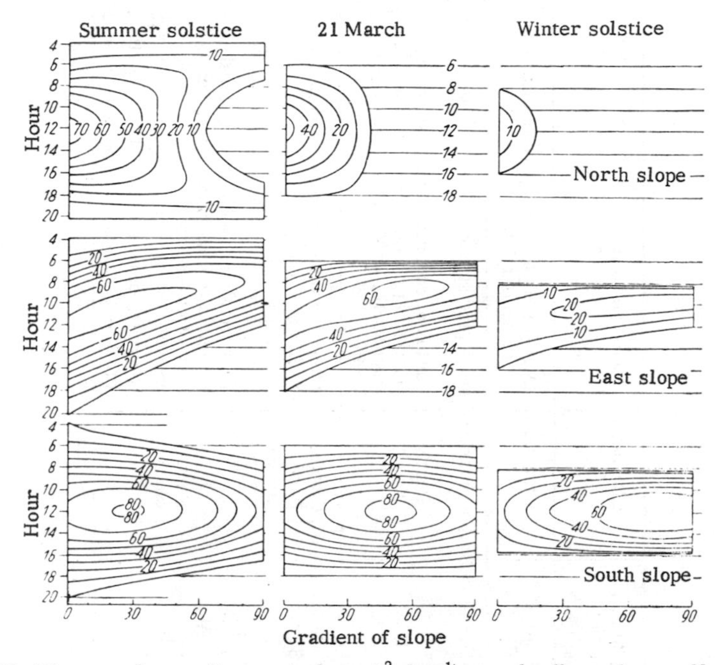

Fig. 199. Direct solar radiation (cal cm^{-2} hr^{-1}) on cloudless days on N, E, and S slopes of all inclinations on three selected days (50°N).

Considering first the lower row, for the south slope, on 21 March (center diagram) the sun rises exactly in the east and sets exactly in the west; the right-hand diagram shows that throughout the winter half-year the sun strikes southern slopes of all inclinations at the same instant, namely, at sunrise. The situation is different in the summer half-year; then the sun rises farther to the NE and some time elapses, after it appears over the horizon and begins to climb, before its rays can touch a slope inclined toward the south; and the steeper the slope, the longer this will take. The upper and lower boundaries in the left-hand diagram are therefore curved lines, compared with the straight boundaries in the center and right-hand diagrams. The intensity, which is always greatest at noon with respect to time, is maximum, with respect to slope on that slope which is perpendicular to the sun's rays. The maximum therefore moves from a relatively flat slope in summer (left) to a steep slope in winter (right). On 21 December a southern wall receives at noon an intensity of radiation which, at the summer equinox, is received on the flat surface by 09:00. This point, which is of practical importance, will be reconsidered in Sec. 41.

For the north-facing slope in the top row of diagrams in Fig. 199, it will be seen that at midsummer (left) the times of sunrise and sunset are the same for all main angles of slope. If the slope is very steep, the sun is no longer able to shine on it at noon, hence the "neck" cut out of the right-hand margin. Steep north slopes receive direct sunshine only in early morning and late evening. Similar to south slopes, the greatest intensity on north slopes with respect to time is at noon, but the maximum with respect to slope is always on the level at inclination 0°, in contrast to south slopes.

East slopes (center row) differ from both north and south slopes in that the place of greatest radiation intensity moves according to angle of slope and time of year. Here, too, the displacement from flatter slopes in summer to steeper slopes in winter may be recognized; sunrise is always at the same time, while sunset occurs earlier, the steeper the slope is.

Figure 199 is designed to allow differences in slopes to be studied in detail. From the practical point of view, however, cloudiness reduces the differences. For slopes important to agriculture and horticulture, Table 86 gives the amount of heat received per square centimeter in direct sunshine for each month of the year. The calculations were made by A. Schedler [751] and are based on radiation measurements for Vienna. Since cloudiness in our latitudes is distributed about noon in much the same way as the hours of sunshine, east and west slopes receive about the same quantity of heat, the former mostly in the forenoon, the latter in the afternoon.

Values of real practical application are obtained, however, only when diffuse sky radiation is also taken into account.

According to radiation measurements made at the Hamburg Meterological Observatory, reported by W. Collmann [32], a horizontal surface there receives annually a round sum of 34,000 cal cm^{-2} from direct sunshine, but 43,000 in diffuse radiation from the sky, and reflected radiation amounts to 14,000 cal cm^{-2}. In mid-European latitudes it therefore seems that more radiation is received indirectly than directly from the sun. It is true that sky radiation is greater near the sun than from distant parts of the sky, but in the discussion that follows it may be taken as being approximately uniform, that is, it is assumed to be independent of direction. This therefore represents a supplementary supply of radiation enjoyed by all slopes to the same extent. An impressive example of how diffuse radiation may even out differences due to orientation of slope was given in 1952 by J. Grunow [739]. He set

Table 86

Monthly totals (10 cal cm^{-2} mo^{-1}) of direct solar radiation, allowing for cloudiness, from radiation measurements at Vienna, 1930-1932. (After A. Schedler, 1951)

		January	February	March	April	May	June	July	August	September	October	November	December	Total
Level ground		107	179	380	585	740	819	893	788	569	294	107	82	5544
10° slope	N	52	116	296	510	690	774	827	703	462	205	58	31	4724
	NE (NW)	66	133	319	528	696	784	847	728	486	226	69	45	4928
	E (W)	104	178	375	577	731	812	877	778	555	281	102	74	5446
	SE (SW)	142	218	426	620	764	836	908	828	612	341	137	106	5938
	S	158	238	446	636	773	841	925	852	650	372	155	125	6171
20° slope	N	4	52	204	417	611	702	744	596	342	114	7	0	3793
	NE (NW)	36	92	256	462	633	725	774	644	402	165	41	21	4251
	E (W)	104	176	366	559	705	778	841	752	540	279	102	74	5274
	SE (SW)	175	253	466	646	764	825	909	846	649	382	165	134	6214
	S	203	289	501	677	790	847	927	884	723	442	193	160	6638

up Robitzsch radiation recorders on the N and S slopes of the Hohenpeissenberg, with their receiving surfaces parallel to the slopes, which were about 30°. According to Kaempfert's tables [741], the N slope should receive about 2 percent in December and about 73 percent in June of the radiation on the S slope. Actual measurements, which included diffuse radiation, showed values that varied between 32 percent (December) and 94 percent (July).

F. Volz [756] used two Linke star pyranometers at Mainz, in cloudless weather, setting one up on a horizontal surface to measure global radiation, and by placing the other at the inclination of various slopes measured the radiation incident on the slope in question. He thus obtained the amount of global radiation received by a spherical dome for the time of observation. Analyzing these results he found that the global radiation received by a slope could be arrived at with sufficient accuracy by using the figure for direct solar radiation as calculated for the particular slope and adding a supplement for the diffuse sky radiation, appropriate to the sun's elevation.

A good idea of the radiation incident on walls, at least for those that face N, E, S, and W, is given by the observations made at the Hamburg Meteorological Observatory over a period of years, using a Moll-Gorczinsky solarimeter. Figure 200 shows the annual variation of radiation received at a north-facing wall, from K. Gräfe [738], which shows the effect of diffuse sky radiation at its strongest. The lower diagrams (*a*) and (*b*) have had added to them the direct solar radiation *S* which is restricted to the half-year between the spring and autumn equinoxes, from W. Kaempfert and A. Morgen [744]. This curve was already shown in Fig. 186. The measured daily total values for 1952 (crosses), 1953 (solid circles), and 1954 (open circles) are entered separately and, with the exception of five days, all lie above the *S* line. The fact that this increased reception of radiation is due to reflection from the urban area lying to the north, in addition to diffuse radiation, may be seen from the daily values in winter, which are indicated by the meteorologic symbol used for snow cover. The high degree of reflection from snow cover (p. 205) raises the figure far above the normal maximum shown by the broken line. Diagram (*b*) shows the 3-yr average value of global radiation on the slope, and (*c*) the ratio of this value to that on a horizontal surface.

In the autumn the measurement points lie closer to the envelope curve than in the spring. This is because greater turbidity at that time of the year increases diffuse radiation at the expense of direct solar radiation. A south wall which, in contrast to a north wall, lies in direct sunshine, is therefore more favorably placed in the spring, as Fig. 201 shows. This figure gives the 3-yr average values of global radiation, as measured in Hamburg, for a S wall, an E (W) wall, and a 45° slope facing south, in addition to the N wall. The use of the same scale allows a direct comparison to

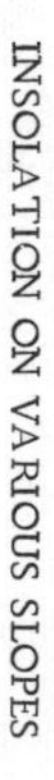

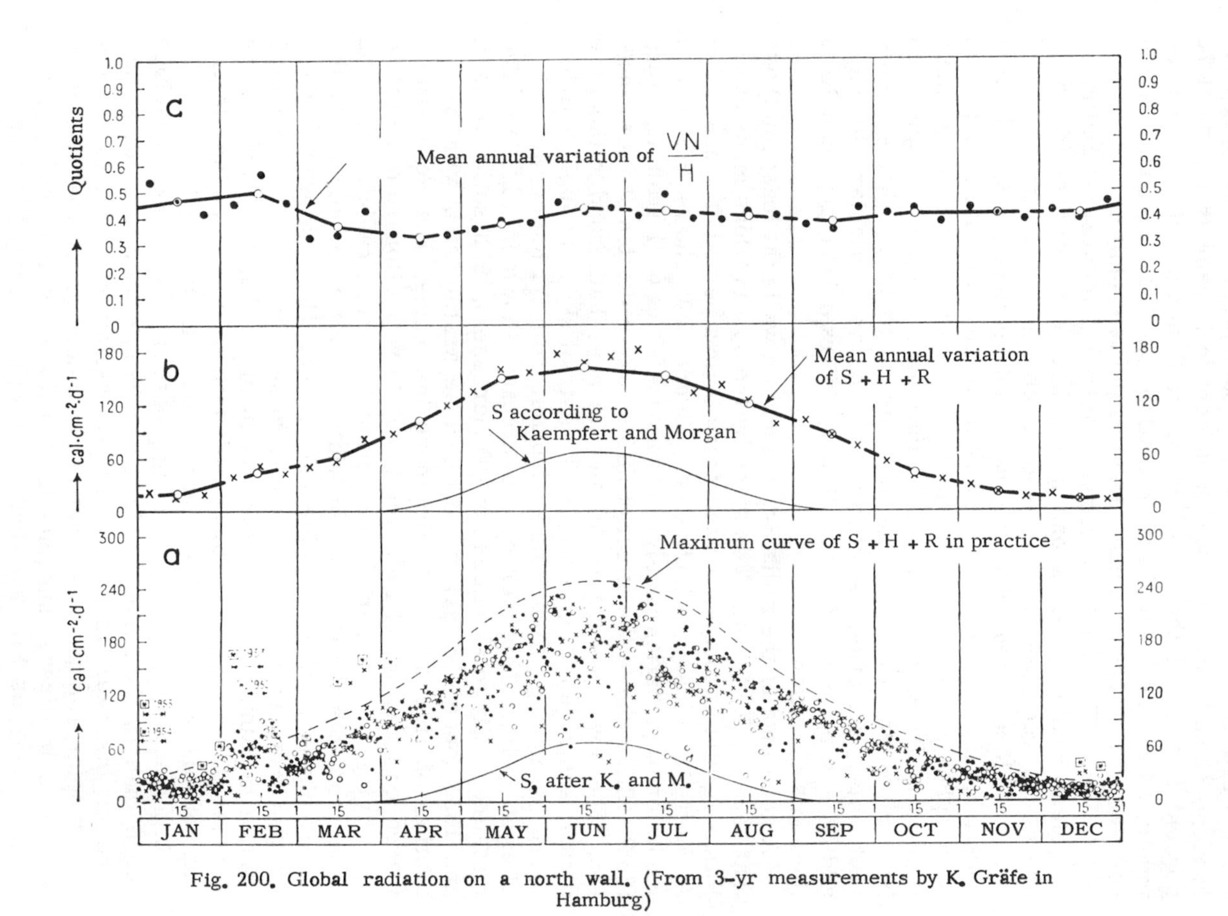

Fig. 200. Global radiation on a north wall. (From 3-yr measurements by K. Gräfe in Hamburg)

be made with the lower curves of Fig. 186, based on direct solar radiation, and taking cloudiness into account.

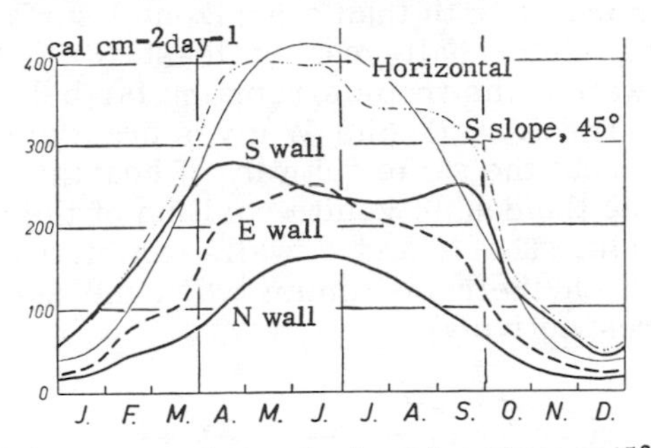

Fig. 201. Global radiation measured on a horizontal surface, a 45° slope facing south, and on three walls. (After K. Gräfe)

Now we shall discuss briefly some special investigations. Assuming that scattering of sunlight in the atmosphere took place according to Rayleigh's law, F. Möller [747] calculated the color of a W wall at 50° latitude at the equinox in cloudless weather, from measurements of the spectral distribution of intensity. The color changes from a yellowish hue in the morning, which does not owe its shade to the redness of morning light since the wall is in shadow, through hues of green to a midday blue. This, however, does not reach the chroma of sky blue, and after coming into the sunlight passes through white and bluish hues, to become toward evening a saturated chroma of golden yellow and orange-red. I. Dirmhirn [736] discusses the radiation of house walls from the point of view of home construction. W. Kaempfert [742] describes the duration and intensity of insolation throughout the year on south trellises, from the gardener's point of view. F. Volz [756] writes about the global radiation on a 15° slope facing south, I. C. Thams [755] a 25° S slope, and W. Kaempfert [743] the insolation of narrow streets.

Much valuable information has been given by G. Dupont and W. Schüepp about the incidence of sunlight on vertical walls in the tropics, based on radiation measurements in the Congo from 1951 to 1952. Both global radiation and diffuse radiation on a horizontal surface were measured. The difference between these two measurements, that is, the direct solar radiation, allowed the amount of direct sunlight on the wall to be calculated; to these values half the measured amount of diffuse radiation was added, since the vertical wall is exposed to only half the sky. The radiation on walls facing the eight principal directions was calculated for each month, for both clear and cloudy weather, by M. Decoster, W. Schüepp, and N. van der Elst [735] for Leopoldville at 4°S and by Decoster and Schüepp for Stanleyville at 1°N. The results are tabulated and summarized in graphical form.

Figure 202 shows global radiation on a horizontal surface and on walls in clear weather at the equator (Stanleyville). While a S wall receives more warmth than a horizontal surface in temperate latitudes, in the winter half-year at least (Fig. 201), the amount received by a wall in the tropics remains far below that falling on the horizontal. The E wall (the W gives practically the same results) receives about the same quantity of heat throughout the year, and only shows up the double zenith position of the sun, as does the horizontal surface. The N and S walls exhibit a marked seasonal variation and reach their maximum when the sun stands farthest north or south respectively.

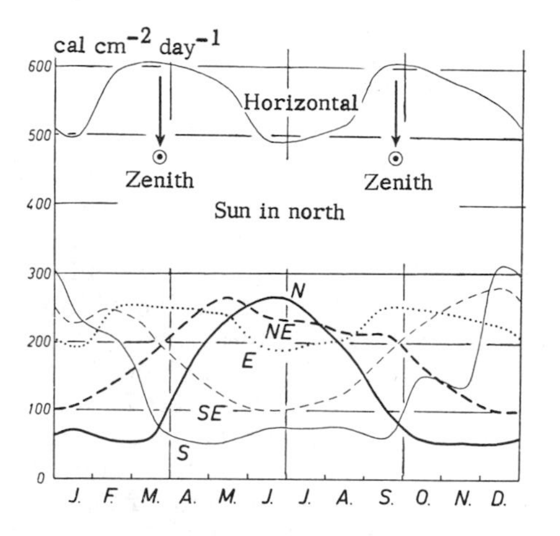

Fig. 202. Global radiation on a horizontal surface and on perpendicular walls near the equator. (After H. Decoster and W. Schüepp)

A comparison of Figs. 201 and 202 illustrates the difference between temperate latitudes and the tropics. In regions where there is an excess of heat and where the sun comes near the zenith, it is clear that the orientation of a slope does not have the practical importance it acquires in mid-European latitudes. In the arctic regions, the low altitude of the sun at any time leads to diffuse radiation taking over the leading role, and this balances out the difference due to varying slopes. On the other hand, where there is a general lack of heat, even a minute gain by virtue of a slope's direction can acquire great importance.

Whether some particular type of slope may be favorable or unfavorable for plants, animals, or man therefore depends entirely on general climatic conditions. Where rainfall is in plentiful supply but heat is somewhat lacking, as in Germany, southern slopes provide a better habitat, as the vine terraces show. Where there is plenty of heat but a shortage of water, the shady northern slopes are to be preferred. I was most impressed during my first flight over the Sinai Peninsula in October 1956 to observe the green shimmer of a skimpy vegetation cover on the north side of the mountains only. J. Walther wrote in his book on the laws governing the formation of deserts (Berlin, 1900): "While in polar lands all horizontal surfaces are empty of vegetation, because only the shrubs on the mountains can receive enough thermal radiation to meet the requirements of growth, a mountain slope in the desert is all the more bare the steeper it is and the more it is baked in the blistering sun. Only the northern slopes are able to support a richer vegetation, and shady valleys and sheltered hollows favor the settlement of plant communities."

41. The Effect of Differing Amounts of Sunshine on Microenvironment

In a mountain valley the difference in climate between the two opposite sides is a result of the different amounts of radiation they enjoy. However, such large-scale effects will be ignored at the moment, and attention will be turned to the consequences that are to be observed everywhere on a very small scale.

An antheap is a mountain in miniature, and the ants make use of the different amounts of radiation on different slopes in looking after the young of the colony (see p. 477). Figure 256 gives the temperature variations in a flat termite nest. The microclimate of clamps in which potatoes and similar products are stored in the open will be discussed on p. 483. Binding writes of the pillars of the Parthenon Temple: "The marble is brown and golden on the sunward sides as a result of the sunlight absorbed, and cool and bluish yellow and white on the sides away from the sun."

A lone tree trunk standing erect is circled by the sun in daytime. One half of the surface of the trunk is exposed to the sun at any time. K. Krenn [769] made use of measurements of total solar radiation intensity from Vienna at 202 m above sea level and from the top of the Kanzel Mountain in Carinthia, Austria, at 1474 m, to calculate the total radiation received by the sides of a tree trunk on a cloudless day. The trunk was considered circular in cross section, 1 cm in diameter, and divided into 16 sectors corresponding to points of the compass, each 1 cm in height. The total quantity of heat received by each of these areas was calculated for each hour during the day. The values arrived at for 1 April on the summit of the Kanzel are plotted by Krenn, as shown in Fig. 203.

The total amount of radiation received by each sector during each hour is represented by a vector length, laid out along the appropriate radius, the end of the vector being marked by the hour in question (for the scale see Fig. 205). Lines are then drawn connecting the points marked for the same hours, and the spaces between are alternately black and white to make the distribution more easily appreciated at a glance, and show the effect of the sun's movement from east to west. The envelope curve cuts off a length of radius representing the total of daily radiation received, and is symmetric about the N-S line since cloudless weather has been presupposed.

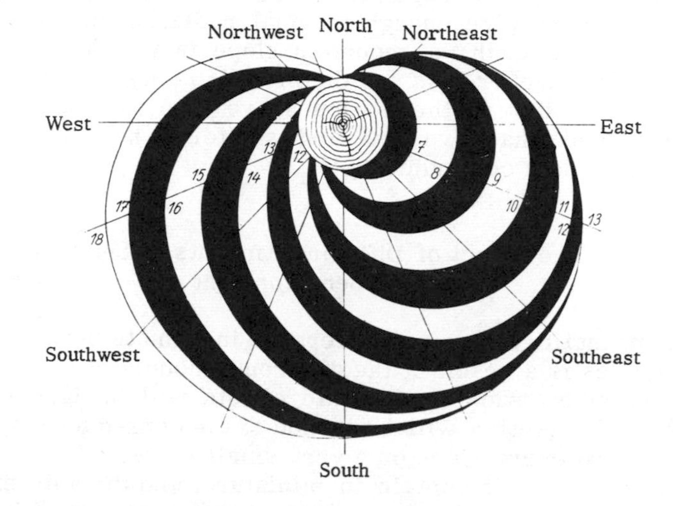

Fig. 203. Amount of heat per hour derived from direct solar radiation on a standing tree trunk. (After K. Krenn)

The temperature variation in the bark of the tree, in the living cambium below, and in the woody parts of the trunk corresponds to this pattern of radiation enjoyed by the trunk. The heat absorbed by the bark penetrates inward as if into a soil consisting of three layers. The density, thermal conductivity, and specific heat depend on the type of tree and its moisture content at the time. Numerical values for these quantities may be found in B. Koljo [768]. Temperature measurements have frequently been made in the trunks of standing trees. Figure 204 gives a somewhat simplified version of the daily temperature variation in a 50-yr-old stand of Sitka spruce on the Nystrup plantation in West Jutland, Denmark, on 20 July 1951, as measured by N. Haarlov and B. B. Petersen [763] with thermistors at a height of 1.3 m above the ground. The full lines give the temperatures below the bark for the exposures marked on them, the dotted curve is for temperature at a depth of 5 cm in the wood on the SW side, and the dashed line is the air temperature.

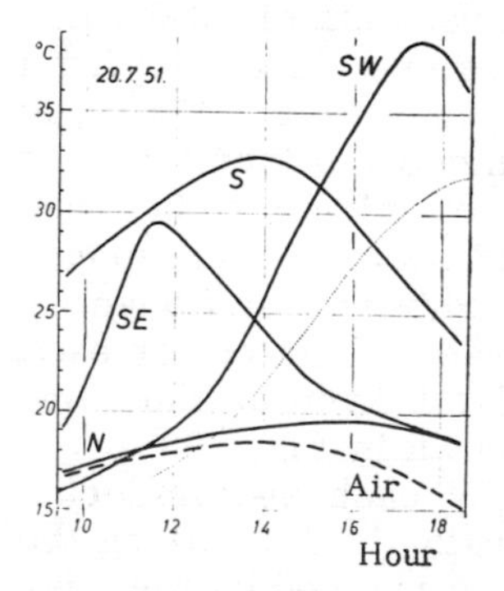

Fig. 204. Daily temperature variation in the bark of a Sitka spruce. (After N. Haarlov and B. B. Petersen)

The temperature maximum follows the sun from SE via S to SW. The parts which receive radiation later are drier and have already been warmed to some extent. Therefore the maxima increase as the day continues, in spite of the fact that radiation is symmetric about the noon line. The greatest maximum in the SW is 9 deg higher than the SE maximum, as seen from Fig. 204. Bark temperatures on the N side are a little above air temperature at first, because of absorption of diffuse radiation, but it is only toward evening that the increasing difference with respect to air temperature shows how the whole trunk has been warmed up, the N side also participating by means of conduction.

The differences are reduced when shading occurs by branches of neighboring trees, and cloudiness has a similar effect, in a way that is simple to understand. The highest bark temperatures measured by Haarlov and Petersen when the air temperature was 23°C were: 26°C in the N, 33°C in the SE, 38°C in the S, and 42°C in the SW. The curve of the root as it enters the trunk becomes several degrees warmer because of the favorable angle it presents to the direction of incident solar radiation. Further useful data on the temperature of trees are to be found in B. Primault [773], B. Koljo [768], E. Gerlach [762], and H. Aichele [757].

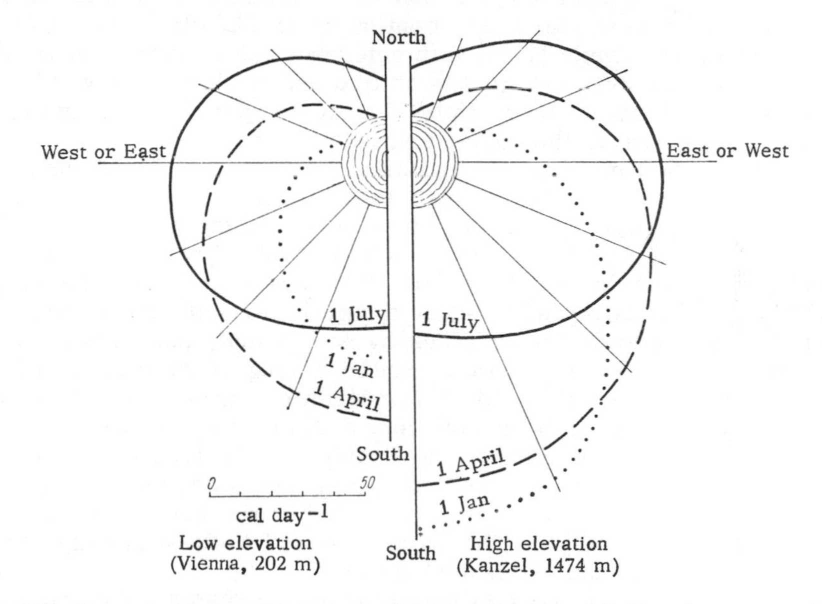

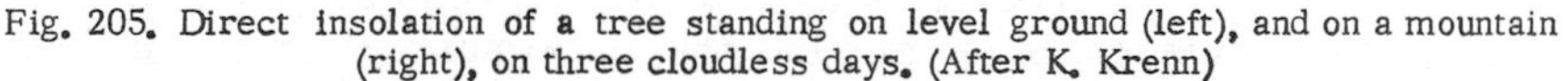
Fig. 205. Direct insolation of a tree standing on level ground (left), and on a mountain (right), on three cloudless days. (After K. Krenn)

In Fig. 205 the daily total radiation of the tree trunk on the Kanzel on 1 April has been repeated from Fig. 203 in a dashed line, but only its right half, because of the symmetry about the noon line.

Figure 205 permits the seasonal effect to be shown as well, by the addition of radiation curves for 1 January and 1 July, and the influence of elevation by similar curves for a tree in Vienna on the left of the diagram. The seasonal effect is as expected, on the N side. It must, however, be surprising at first to find that, on the south side in the higher position, the low winter sun provides the vertical trunk with more heat than at any other time of the year. On 1 January for the Kanzel this is more than double the value for Vienna. Even in Vienna the S side (but no other) receives more direct solar radiation in January than in July.

Great temperature contrasts are therefore experienced in winter between the cold air, which may be well below the freezing point, and the surface of the trunk on its southern side, which is strongly heated by the sun in clear winter weather. The layer of cambium cells below the bark, which in the red beech already contain considerable amounts of water in preparation for the approaching rise of sap, is unable to withstand the violent alternation of midday heat and night frost. It splits longitudinally with an audible report. The crack in the bark so formed may be hair-fine at first, widening later to form fissures in the bark. Figure 206 shows the damage caused to a red beech in a photograph taken by M. Seeholzer [777]. A relation between the widening of frost cracks and decreasing temperature and continuing frost action was demonstrated by H. Schulz [776], with references to more recent literature. By measurement he was able to establish that the width of a crack varied with the rhythm of daily temperature fluctuations, and was greatest in the poplar and least in the oak.

Trees suddenly exposed by felling are particularly susceptible, but fruit trees are also subject to this kind of damage. The risk can be reduced by taking measures to reduce the amount of radiation absorbed by the surface of the trunk. From thermocouple readings made by B. Primault [773] near Zurich in 1952-53, a straw-matting cover will reduce the diurnal temperature variation on the SW side of forest trees by 29 percent, and white paint by 35 percent. H. Aichele [758] painted a ring of whitewash on fruit trees and compared cambium temperatures under the white ring and under untreated bark both above and below. The temperature differences between the N and S sides of the tree were reduced from 8 deg (not whitewashed) to 4 deg (whitewashed) in January, and from 20 deg to 14 deg in February. In all cases it is better to apply the paint only on the parts exposed to frost damage, from SE to NW, not on the whole surface.

The trunk of a tree is usually oval in cross section, not circular. In stands on level ground the maximum diameter is preponderantly in the direction of the prevailing wind, and the smallest diameter perpendicular to this direction, as shown by G. Müller [772]. This is explained by more luxuriant growth in places sheltered from the wind and the firm anchorage against the wind.

The cross sections of tree trunks standing on slopes are no longer controlled by wind direction but by light received, which is least in the up-slope direction and greatest in the opposite direction, with the

Fig. 206. Cracks and peeling of bark of a red beech as a result of intense midday radiation at low air temperatures. (After M. Seeholzer)

result that the greatest diameter follows the line of steepest slope. Figure 207, for example, shows the results obtained in measuring 331 oaks in two test areas in the Palatinate Forest. The sketch on the left shows the area with contour lines drawn in, and on the right is the frequency distribution of the greatest (full lines) and smallest (broken lines) diameters arranged on a compass rose.

Hitherto only the trunk of the tree has been discussed. Now the whole tree will be considered.

G. Eisenhut [760] examined 5276 separate blossoms spread over the crown of an isolated 90-yr-old lime tree (*Tilia euchlora*) in Jesenwang with a uniform appearance, to establish at a particular instant how many were in full bloom (or withered) according to

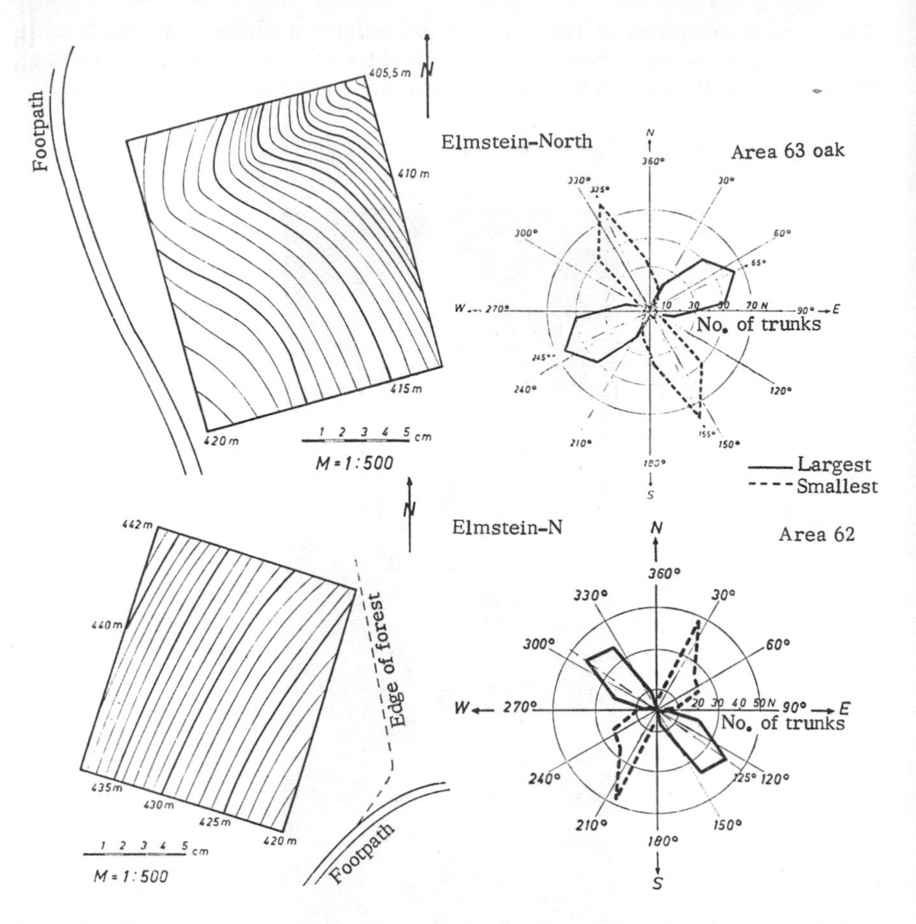

Fig. 207. The greatest trunk diameter of oaks, which need much light, in the Palatinate Forest, is in the direction of the slope. (After G. Müller)

position on the crown and exposure, and found the following percentages:

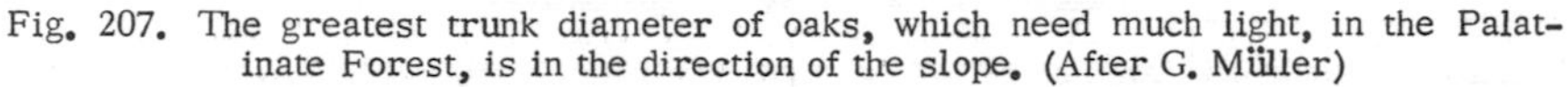

Exposure	S	W	E	N
Upper crown	38.7	—	—	—
Middle crown	30.2	22.1	16.9	2.9
Lower crown	14.3	19.5	21.9	5.7

Blossoming appears to depend on orientation. The tip of the crown, which receives sunlight all day, blossoms first. In the middle crown the directions follow in an order corresponding to that of maximum temperatures in Fig. 204. In the lower part of the crown the contrasts are more balanced. The sun at low altitude gives preference to the E in the morning and the W in the evening over the S; and the earlier blossoming in the N, for this part of the crown, is probably caused by its closeness to the ground. The number of blooms present also depended on exposure. In every 100 g of freshly weighed petals in the tip of the crown there were 296 blooms, in the middle crown 327 in the S, 130 in the N, and in the lower crown 251 in the S, and 115 in the N.

The daily sequence of blossoming of 181 male blooms on a 15-yr-old pine, standing alone near Eberswalde, was followed by A. Scamoni [774]. The flowers were on parts of branches at a height of 1.1 m above the ground, and the sequence in which they opened is shown graphically in Fig. 208.

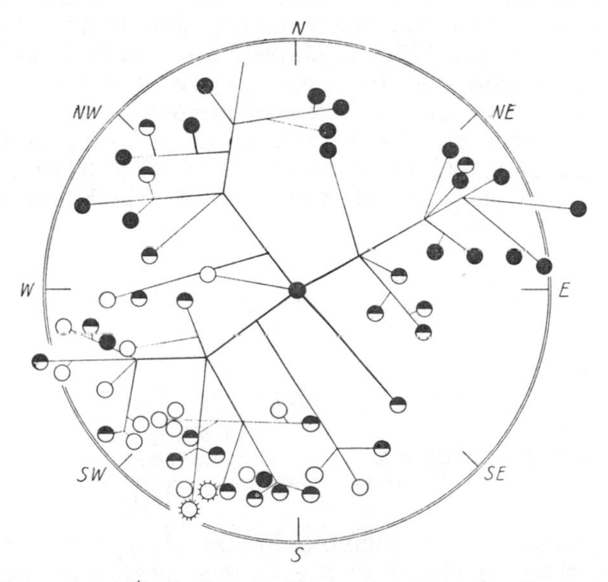

Time of blossoming: ☼ 15 May, ○ 16 May, forenoon, ◒ afternoon ● 17 May

Fig. 208. Time of blossoming of a pine tree, standing by itself, near Eberswalde. (After A. Scamoni)

Although there are many other influences at work, such as differences of shadow, density of flowers, supply of sap, and so forth, the effect of orientation is plainly to be seen. Prof. H. Walter of Stuttgart-Hohenheim wrote to me about this diagram: "I can give you a good example from Arizona. The enormous saguaro cactus always blooms first on the SW side. No buds develop on the NE side. The concentration of sap is also greatest in the cells on the SW side, and lowest on the NE side. The growth of *Echinocactus wislizeni* is restricted on its SW side by the greater dryness there, so that it gradually curves over toward the SW and finally topples."

The differences in the blossoming of fruit trees on their N and S sides was studied by N. Weger, W. Herbst, and C. F. Rudloff [779] over a period of 4 years in 18 species of pear tree at Geisenheim. The daily temperature variation in the air around a polygonum shrub and a dwarf fir was studied by P. Filzer [761] in the botanical gardens at Tübingen. The sequence in which tulips flowered in a circular bed with sloping edges in Munich, and its relation with temperature and humidity readings, is to be found in O. Härtel [764].

In agriculture, level fields are often modified by planting crops in drills pointing in various directions. Although the total amount of radiation received may be the same, it is nevertheless distributed differently over the small ridges and many benefits are thus acquired. Before the young plants had time to alter the microclimate, N. Weger [778] in Geisenheim and H. Lessmann [770] in Freiburg investigated the temperature at depths of 5 cm and 10 cm respectively in various experimental plots. Figure 209 shows Lessmann's results in the form of 2-deg isotherms for drills running N-S, of height 13 to 15 cm, 26 to 30 cm apart, with sides sloping at 45°, on 31 August 1948, which was a cloudless day.

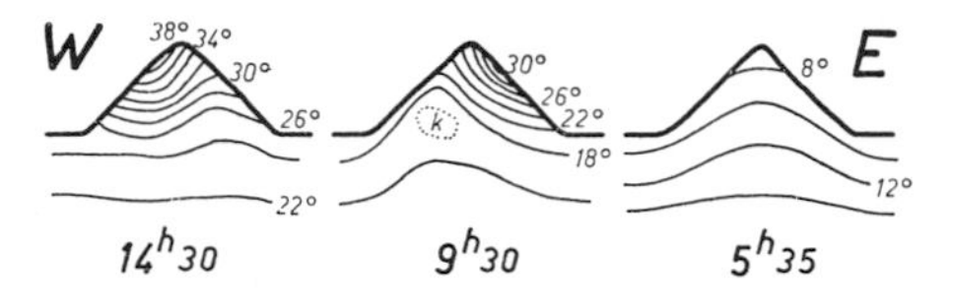

Fig. 209. Temperatures in unplanted drills, running N-S, on a sunny day in Autust. (After H. Lessmann)

Twenty minutes after sunrise (on the right of the diagram) the top of the ridge is naturally the coldest part, and is colder than the level ground; according to Weger it is also colder 5 cm under the surface by up to 2.8 deg. By 09:30 it is heated to 30°C under the surface of the sunny eastern slope (center diagram) while the west side is still cold from the night; heating of the western slope occurs in the early afternoon (left diagram) to over 39°C. It follows that during the day the plants enjoy a substantial gain of heat in N-S drills. Both investigations agree on this point; while for drills running E-W, Weger was still able to establish a gain, though a small one, over level ground, Lessmann was unable to do so.

N. Weger extended his temperature measurements at a depth of 5 cm to the ridge and furrow of the drills, and also to flat dams 100 cm wide at the base and 50 cm at the top, which ran E to W, and to the circular mounds of earth sometimes employed in vineyards, which are 18 cm high and 50 cm in diameter at the ground.

He was able to establish, particularly, a strong seasonal influence. In four periods of fine weather, the average daily gain of heat compared with level ground, computed in degree hours from the excess temperatures determined every 2 hr was as shown in Table 87.

Table 87

Average daily gain of heat (deg hr) in various earth configurations.

Periods (number of days of fine weather)	Conical mound	Dam	Drill			
			Wall		Furrow	
		E-W	N-S	E-W	N-S	E-W
21 July to 1 August (9)	50	30	16	14	0	0
29 August to 10 September (6)	39	32	14	10	1	2
19 to 25 September (4)	34	34	12	7	6	10
3 to 10 October (6)	26	30	6	5	4	2

The mounds therefore achieve the greatest gain. It is worth noting that the gain at the dam increased until autumn. In these figures no account is taken of the temperature deficit, compared with level ground, because this is mostly found in the valleys of the drills where nothing is growing. The young plants on the ridges respond to the extra heat by sprouting sooner and growing more quickly.

When the plants have grown, the radiation picture becomes quite different. W. Kaempfert [767] has studied the influence of the direction in which the rows of plants lie in relation to their "narrowness" E, that is, the ratio of the plant's height to its width (distance between rows, space width). Figure 210 considers the plants as vertical walls and portrays the annual variation of sunshine duration.

The time of day is read on the ordinate from the bottom up. The sunrise and sunset curves, the same for all plants, delimit the possible sunshine duration. The first diagram shows the conditions prevailing in rows running N-S. The second diagram is for rows turned through an angle of 22.5°, and so on, until the E-W rows are reached. The diagrams can then be continued by going back, but now reading the time scale from top to bottom.

The boundary case $E = \infty$ (dot-dash line) means that the rows of plants are in contact with each other (distance apart = 0). With rows running N-S the sun is then able to penetrate to the ground only for an instant at noon through the opening between the rows. This instant is independent of season. With NE-SW rows, however, this instant will be earlier in summer and later in winter. If the rows run ENE-WSW the sun can no longer penetrate the opening in winter and in spring and autumn only in the late afternoon from the

WSW; in midsummer, when the sun rises far in the NE, it can even penetrate a second time in the early morning.

The following directions should be sufficient to make it easy for the reader to use the diagram. If $E = 2$, that is, if the plants are twice as high as their perpendicular distance apart, the sun will shine into the rows only at the times covered by the white area in Fig. 210. If $E = 1$, then it will also shine in during the period covered by the dotted area.

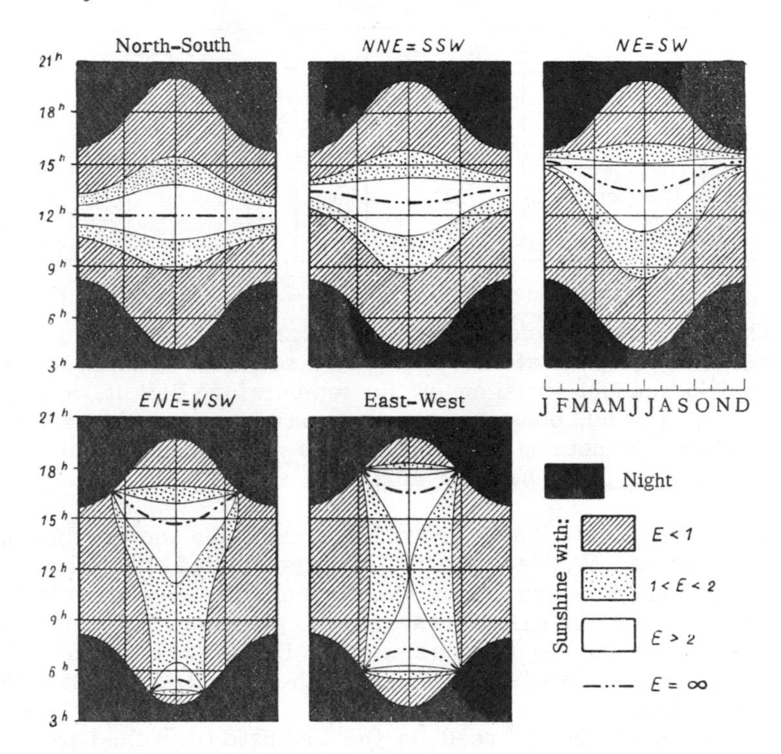

Fig. 210. Sunshine duration in rows of plants, for various directions of row, and various widths between plants. (After W. Kaempfert)

O. Härtel [765] was able to show from spectral-light measurements in barley fields on the island of Ischia that the spectral composition of light falling in rows parallel and perpendicular to the direction of the sun's rays at noon was very different, and would therefore be of importance to ecology. This work should be consulted for further information.

In a drought-threatened tree nursery the young firs were set in hollows, while the weed-covered topsoil was turned over and the divots placed around the plant hollows (Fig. 211). Only the plants marked B, with the divots on the eastern side of the hollow, came through well, since by this means they were placed on a slight W slope and were able to make use of the night dew and humidity for a few extra hours in the morning, before the welcome heat of the day was felt. The plants marked *A* on the E slope were soon dried by the morning sun and the afternoon shade was no substitute for the loss of moisture.

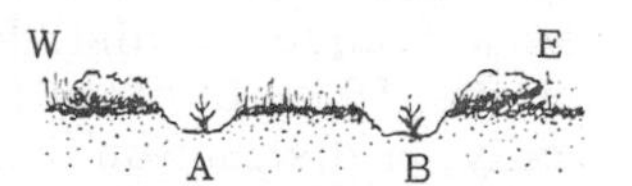

Fig. 211. When planting a forest, microclimate is determined by the position on which the divot is placed.

As a last example of the small-scale effect of sunshine, we shall take the compass plants, which have been studied in detail by H. Schanderl [775]. These plants have an inherited ability to place their leaves in any desired plane by rotating them. Figure 212 shows, for example, a wild lettuce (*Lactuca scariola*) that is protecting itself against the strong reflected radiation from the wall in front of which it is growing by placing its leaves in the "transverse compass position." All leaves are so positioned as to received a minimum of reflected radiation.

By studying a number of these plants, growing on a W slope of 30° inclination on dry limestone near Würzburg, H. Schanderl was

Fig. 212. The leaves of a wild lettuce in front of a wall are arranged in compass position. (Photo H. Schanderl)

able to establish the following. Four plants growing near each other had a total of 627 leaves. On the sunny 10 July 1931 he measured by compass the direction in which the broad axis of each of these leaves was directed. Arranging these in groups, that is, as a frequency distribution, he computed the percentage for each of the 16 main directions. Figure 213 shows these results graphically, a distinction being made between the range from the surface up to a height of 50 cm, and the leaves above 50 cm. The black

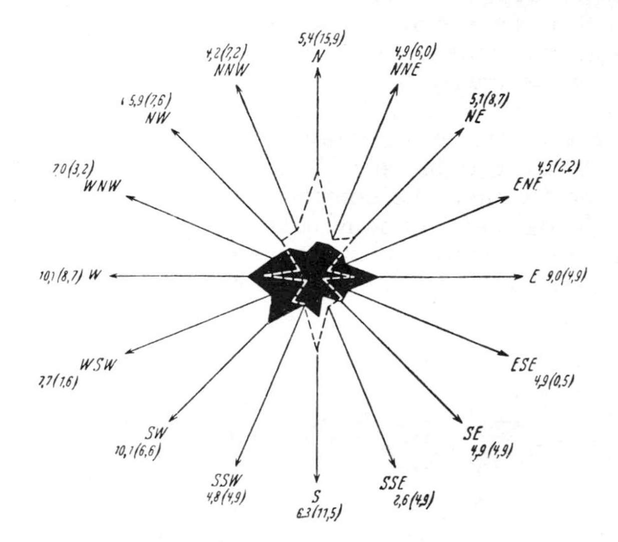

Fig. 213. The positions of wild lettuce leaves near the ground (black) and higher up (dotted), on a W slope. (After H. Schanderl)

distribution figure applies to the former, and the figure outlined by a broken line to the latter, for which the frequency values in brackets also apply. The leaves in the lower half meter of these lettuce plants are arranged mostly along an east-west direction, and those growing higher lie mainly north to south. The latter are therefore protecting themselves against direct radiation from the sun, while the former are guarding more against reflected radiation from the west slope. This is proof, which has been confirmed by Schanderl in other experiments, that the orientation of the leaves is directed in a way dependent on the radiation climate of the slope on which they are growing.

A remarkable special case among compass plants is provided by the gnomon or sundial plants, discovered by B. Huber [766]. *Aster linosyris* has narrow spear-shaped leaves which are normally held horizontal. In places where it is dry and the plants are exposed to strong radiation, they turn into a vertical position. On steep southwest slopes, however, the leaves on the downslope side are all "combed" forward in a southern direction and are all pointed fairly accurately in the direction of the midday sun.* At the time of strongest radiation the plants expose the minimum surface possible to the sun. In contrast to compass plants, gnomon plants on west slopes still point their leaves toward the south, with a 10° deviation, at the most, from the direction of the sun at noon.

42. Small-Scale Topographic Influences at Night (Cold-Air Currents, Frost Hollows)

R. Geiger and G. Fritzsche [277] investigated night temperature minima at a height of 10 cm above the ground during the spring and summer of 1939, in order to find the causes for the varied severity of frost damage to a growth of pines near Eberswalde. The effect of height was evident both on single nights with late frost and in the average of all observations, although the ground appeared to be level, and surveying disclosed only a gentle slope. Table 88 gives the temperature minima observed at five points within 100 m of each other. Differences of a few tenths of a meter give rise to substantial temperature differences. The lowest observation point always had the lowest temperature at night, and it fell below the freezing point on 17 nights during the growing period. It was warmer at the highest point of measurement, where the number of frost nights was only 12.

Table 88

Night minimum temperatures (°C) over nearly level ground.

Nights with frost, 1939	Elevation (m)				
	36.1	36.1	36.3	36.6	37.7
23-24 May	-7.6	-6.9	-5.4	-5.1	-3.7
2-3 June	-9.4	-7.9	-8.2	-6.7	-5.0
2-3 July	-2.1	-1.3	-1.1	0.0	+0.1
11-12 July	-2.5	-1.4	0.0	+1.6	+1.9
Mean of 30 coldest nights	-0.6	-0.4	+0.1	+0.7	+1.7

The impression is thus obtained that the air cooled in contact with the ground or the upper surface of plants increases in specific gravity and therefore starts to flow down to lower levels. It is therefore customary to speak of a nocturnal cold-air current, by

*In ancient times the gnomon had a *vertical* rod. It was the forerunner of the sundial, which had a rod *parallel to the axis of the earth*. The word "gnomon" is therefore used here with the more general meaning of "sun indicator," since plants in the "gnomon position" point toward the sun.

analogy to water which also flows in a current toward lower levels. However, while outflowing water is replaced by air without difficulty, because the empty space left does not have to be refilled with water, the outflowing air has to be replaced by a compensating air flow. Besides, the difference in density between warm and cold air, which differ in temperature by only a few degrees Celsius, is extremely small in comparison with the density difference between water and air.

M. Reiher [828] calculated the rate of flow of air from given differences in density, and arrived at values that were, for the most part, less than 1 m sec^{-1}. Small-scale air currents are therefore always weak, hardly detectable by an observer, and easily upset by obstructions. They must move as part of a closed circulation, and therefore can develop only after night cooling has progressed for a time. This movement of air is, however, of great practical importance, since it will continue throughout the night when outgoing radiation is maintained.

Figure 214 shows a fir plantation damaged by frost in the great Anzing-Ebersberg forest near Munich. The left-hand part of the diagram shows the distribution of the different age classes of trees in chart form, and the right-hand half shows the topography by height contours drawn at 10-cm intervals. To the eye it appears to be a level surface. Table 89 gives the night minima at 5 cm above the ground, measured by R. Geiger [786] at the nine observation states during the spring of 1925.

The most striking thing is the extremely high frequency of frost nights, comprising three-fourths of all nights in May and even half of all nights in June, hence creating a local climate that is known otherwise only from Scandinavia to Finland. Data from climatic stations of the regular Weather Service in Munich, which are also included in the table, therefore give a completely false picture, if taken to be valid for the vegetation of the area. These figures explain why young firs were unable to grow in this particular frosty area over the previous 30 years.

More detailed analysis shows that the temperature was not determined only by relative height in this case. The cold-air current at night was certainly directed toward the border of old trees to the north, but the protective effect of the high trees there, described in Sec. 37, reduces the severity of the frost. It is also difficult for the cold air to penetrate into the old stand, and so a cold-air dam develops, which will be the subject of later discussion.

An example from agriculture, taken from the fruit-growing area of southwest Germany, is illustrated in Fig. 215, by F. Winter [804]. The areas shaded in black show the extent of frost damage in May 1957 to the fruit on individual trees in a peach orchard near Jechtingen on the Kaiserstuhl. The figures on which the diagram is based were obtained by estimation and by random

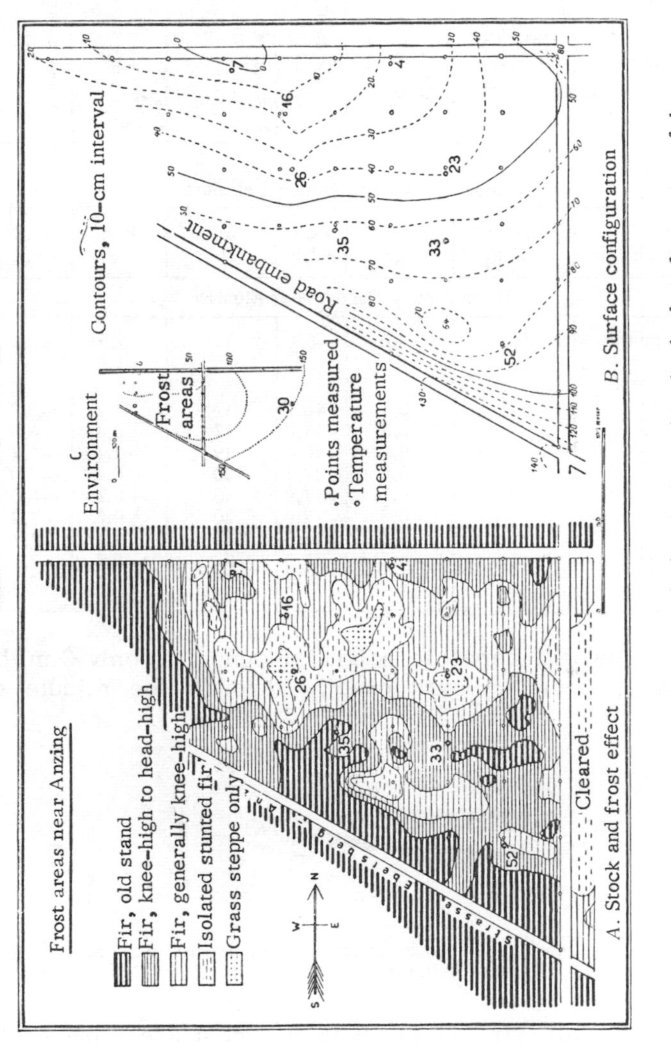

Fig. 214. Plant growth, terrain, and observation points in the late-frost areas of the Anzing Forest near Munich.

Table 89

Night minimum temperatures (°C) in Munich and in the Anzing Forest.

Station	Height of observation	May			June		
		Mean temperature	Coldest night, 3–4 May	Number of nights with frost	Mean temperature	Coldest night, 7–8 June	Number of nights with frost
A. For comparison (general climate)							
Munich City	8.4 m	8.8	+2.1	0	10.6	+8.2	0
Outside Munich	1.4 m	6.5	- 1.8	1	9.0	+4.2	0
B. In Anzing Forest near Munich							
Anzing Sauschütte	5 cm	1.6	- 8.4	12	4.5	-3.9	4
At the frost area:							
Point No. 30	5 cm	0.1	- 10.7	17	1.2	- 5.2	9
52	5 cm	- 0.3	- 11.0	17	0.4	- 7.9	12
35	5 cm	- 0.3	- 10.8	19	1.4	- 7.1	8
33	5 cm	- 0.6	- 12.4	20	0.4	- 7.0	12
4	5 cm	- 0.7	- 11.9	20	- 0.1	- 8.0	14
16	5 cm	- 0.8	- 12.8	20	0.3	- 7.2	13
7	5 cm	- 1.1	- 13.5	22	0.1	- 7.1	13
23	5 cm	- 1.5	- 13.5	22	- 0.2	- 7.1	15
26	5 cm	- 2.0	- 14.4	23	- 0.7	- 8.8	15

sampling. The difference in height amounted to only 3 m, but the frost-damage boundary was exactly through the middle of the orchard.

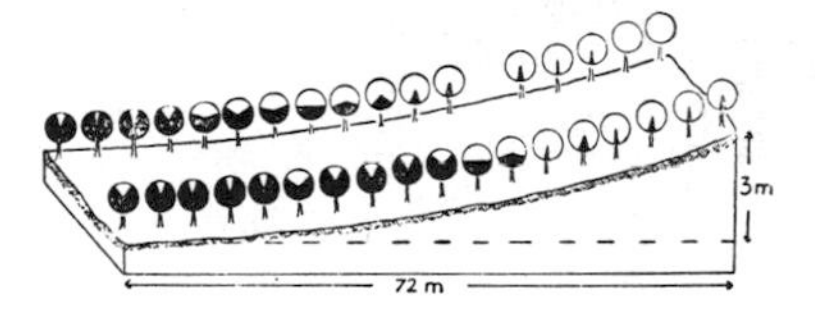

Fig. 215. Proportion of young peach fruits (black) frozen on a slope in May 1957. (After F. Winter)

A lake of cold air builds up in the most low-lying area, as a result of the flow of cold air. It is possible to visualize the level of its banks in Fig. 215. This phenomenon has a variety of names, because of the frequency of its appearance: cold island, frost hole, cold-air pool, frost hollow, and the like. The old-fashioned rule still applies: at night concave land surfaces are cold, convex surfaces warm.

The distribution of night temperature in broken terrain has been studied in a large number of carefully conducted investigations. Analysis of the numerous aspects of cooling phenomena has shown that the cold-air current just described is only one of five factors that cause the low temperatures observed in dish-shaped hollows. The other four factors are: (2) the increase in counter-radiation which results from the screening of the horizon in the hollow, leading to a change in radiation balance that reduces the frost hazard (a few figures were given earlier, on p. 23); (3) the reduction in turbulent exchange produced by the dish shape, which keeps the temperature low at night; (4) the heat supply from the soil of the slopes when the hollows are deep and narrow; and finally (5) shortening of daylight as the sun rises later and sets earlier.

The interplay of these five factors may vary according to the size and form of the depression in the ground. If the depression is small, it is useful to follow the method of H. M. Bolz [686] and to distinguish three types, shown in the sketches in Fig. 216. The

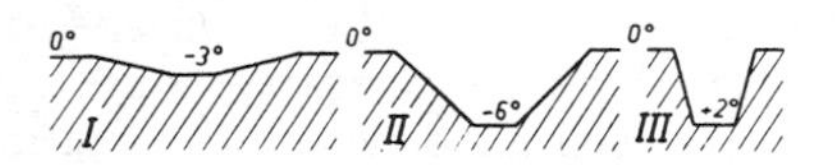

Fig. 216. The three types of small hollows. (After H. M. Bolz)

flat hollow (I) derives its low night temperature primarily from the flow of cold air; it is merely a cold-air pocket. The low temperatures in deeper hollows (II) are mainly due to a reduction in turbulence, while a narrow depression (III) remains comparatively warmer in spite of the absence of turbulence because of the heat supplied from the soil (side walls) and the strong screening of the horizon. The minimum temperatures given by Bolz illustrate the temperature characteristics of the three types of depression.

A few experiments made in open country give some insight into the combined influence of the five factors. K. Brocks [783] dug six 20-cm-deep trenches in Eberswalde, with the same base width but sides sloping at different angles. He recorded the night minimum temperatures at a height of 10 cm above the base and compared them with temperatures above level ground, finding:

Angle of slope:	0°	15°	30°	45°	60°	75°	90°
24 May 1937 (°C):	6.3	6.6	7.0	7.3	7.5	7.5	8.1
Average of 138 nights (°C):	6.23	6.23	6.27	6.34	6.44	6.59	6.67
With snow cover (°C):	-2.5	—	-4.4	-3.5	-2.7	-2.4	-2.4

When a shallow fresh snow cover cut off the supply of heat from the soil, the flow of cold air led to lower temperatures in the flat trenches than over level ground (the last line of the table,

24 March). Apart from this, the trenches were warmer than the air above level ground, because of screening against outgoing-radiation loss and conduction of heat from the soil from the open lateral walls; the steeper the sides, the warmer the trenches were. This is clearly shown in the May night in the first line, but is also true for the average of the other nights.

Both day maxima and night minima were recorded in Japan by E. Maruyama and Y. Yamamoto [796], in circular depressions of diameter 0.5, 1, 2, 3, and 5 m, and depths of 10, 30, 40, 70, and 100 cm, from April 1953 to February 1954. Temperatures were measured by shielded thermometers at a height of 10 cm above the soil. Maxima were higher throughout than over level ground, and the minima lower, but no close correlation with the dimensions of the depressions was apparent.

In experiments with models and in the open air, H. W. Georgii [787] artificially added dispersed particles to the air in depressions and, by measuring the decrease in the concentration of the particles, was able to compute the volume of air (y) exchanged per unit time. This was proved to be proportional to the austausch coefficient A and was related to the wind speed v by the equation $y = a + bv^p$, in which a was a quantity containing convective exchange (see pp. 38f), b depended on the dimensions of the depression, and the exponent p had a value of about 1.2. In further experiments [788] he measured the minima 5 cm above the floor (1.4 m^2 in area) of two 60-cm-deep holes with sides sloping at 30° and 90° respectively. He determined separately the influence of the heat flow from the soil, night radiation losses, and turbulent mixing by correlating his observations with each of these factors and establishing the causal connection and magnitude of the influence of each. Out of the 211 nights on which observations were made, the steep hole was warmer than the level ground 181 times, the flat hole only 118 times. The quantity of heat conducted from the sides was in exact agreement with the seasonal variation in ground temperature, and was therefore large in autumn, when the soil still retained its summer warmth, and small in the spring.

From these small-scale depressions attention will now be turned to features of larger dimensions. The example selected is the classical subject of study first made famous in Vienna by the work of Wilhelm Schmidt [801]: the Gstettneralm Sinkhole at an elevation of 1270 m near Lunz in Austria. Since 1928 special observations have been made there with ever-widening scope, and have been supplemented by laboratory experiments using a terrain model on the scale of 1:5000.

The sinkhole is a funnel-shaped depression in the limestone mountain, formed by faulting. Sepp Aigner [16], who has been a brilliant observer for many years, knew it as a 10-yr-old boy and has given a fine description of it; he writes: "I was allowed to go with the keeper to the grazing cattle, and we also climbed down into the Gstettner while doing so, and my companion told me that a dog froze in front of the shepherd's hut which once stood there, that when the weather was cloudless the cattle would always leave the bottom in the evening and climb up the slopes, not returning again until some time in the forenoon, that in the bottom of the sinkhole everything was white with hoarfrost in the morning, that pools of water often froze at night, and so on. Shepherds and hunters have known from time immemorial that the Gstettneralm Sinkhole was a cold hole. They also saw that dwarf pines grew at the bottom and firs started to appear only farther up."

Figure 217 shows a cross section of the sinkhole with the height slightly exaggerated. The figures show the temperatures measured before sunrise on 21 January 1930 at the points indicated on the slopes. On the upper slope, temperatures only 1 to 2 deg below zero were recorded. But below the edge of the WSW slope, over which the cold air can flow through the saddle and down into the

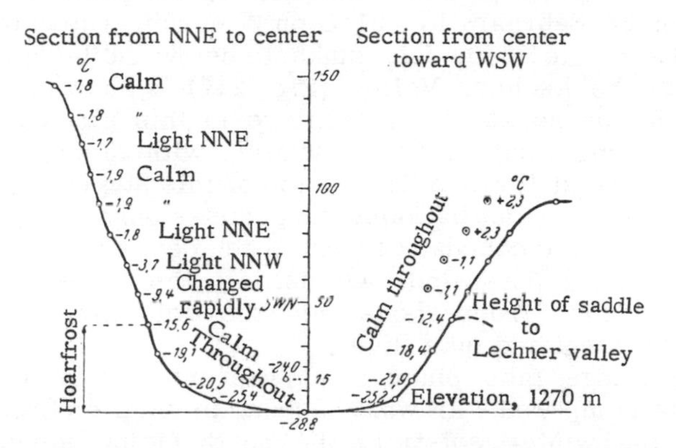

Fig. 217. Night temperatures in the Gstettneralm Sinkhole on 21 January 1930. (After Wilh. Schmidt)

Lechner Valley, temperature decreases very quickly in the deep cold-air basin, and reached - 28.8°C, on that particular day, at the bottom of the valley. The lowest temperatures recorded constitute a record for central Europe and are reached on many occasions because of the peculiar relief. The minimum thermometers, which were read 73 times between 1928 and 1942 in this isolated and barren landscape, gave temperatures lower than - 40°C 27 times, 8 times below - 50°C, and the absolute extreme recorded was - 52.6°C. Night frosts occur even in the middle of summer. Whereas vegetation elsewhere becomes poorer at higher altitudes in mountain regions, the reverse is true in the sinkhole. There are tall woods at the upper edge, followed lower down by stately firs, mixed with Alpine roses, then come the dwarf firs, and at the bottom of the sinkhole the only vegetation consists of a few hardy grasses and shrubs that are able to survive the winter under the snow. A similar reversal of the vegetation pattern has been described in more detail by J. Horvat [790] for the karst sinkholes of Yugoslavia.

Conditions prevailing on the Gstettner hill pastures have been analyzed in detail by F. Sauberer and I. Dirmhirn [799, 800] by means of radiation measurements, temperature records, surveying, and pilot-balloon ascents, to such an extent that a great deal is now known about the interaction of the five factors. For record temperatures to be reached, a low initial temperature is necessary; this is obtained when the season, elevation, topographic features, and weather situation are favorable. Here too the winter snow cover plays an important part by insulating the ground heat to some extent. This is shown by the fact that the lowest temperatures appear with a fresh snow cover, and do not sink further during periods of fair winter weather, but may even increase as the snow becomes denser because its thermal conductivity then increases (see p. 208). For example, temperature increased from - 30°C on 27 February to - 21°C on 3 March. The calmness of the air in the 70- to 80-m-deep sinkhole below the level of the saddle leading to the Lechner Valley (Fig. 217) is shown in all observations. The decrease of temperature in this stretch averages 10 deg in summer and 16 deg in winter, with 25 deg as an extreme value. The wind flows over the top of this stagnating pool of cold air, as shown by pilot balloons, and causes only a regular undulating motion in the boundary layer. When deer are in the basin they can be stalked downwind, against all the rules; only when the hunter descends from above into the cold air can the animals pick up the scent and take flight.

Many things take place in a different way in these large depressions compared with what happens in smaller features. At the time of the winter solstice, the north-facing slopes remain in shadow throughout the day, as proved by Lauscher and Dirmhirn from the model of the Gstettneralm, and these slopes therefore produce cold masses of air even during the day. The large-scale flow of cold air hence acquires new significance, since it was shown in Sec. 5 that counterradiation from the atmosphere came from a remarkably shallow atmospheric layer, 72 percent being from the lowest 87 m. It follows that in such deep sinkholes the amount of counterradiation will be greatly reduced when they are filled with cold air, and the radiation balance will therefore be worsened. The essential preliminary condition is that air cooled in contact with the ground should be able to flow in from a sufficiently wide area, that is, that the "frost catchment area," as it may be termed, is sufficiently large. If these conditions were not fulfilled on the Gstettneralm, it would not produce such record low temperatures, as many other sinkholes fail to do.

Finally, something must be said about water vapor. The cold bottom of the sinkhole is often filled with fog. This means that the temperature has fallen to dewpoint. R. Wagner [803] has described (with photographs) the fluctuating fogs of the sinkholes in the Bükk Mountains of Hungary in 1953–54 and compared them with temperature measurements. When a snow cover lay in the Gstettneralm, however, the excess water vapor sublimated directly onto the snow without forming fog and hence dried the air considerably. Quantitatively, the latent heat of sublimation thus set free was insignificant in comparison with the increased amount of radiation released through the dehydrated air; the heat gain was estimated to be only 6 percent of the loss (1 compared with 18 cal cm^{-2} in 6 hr).

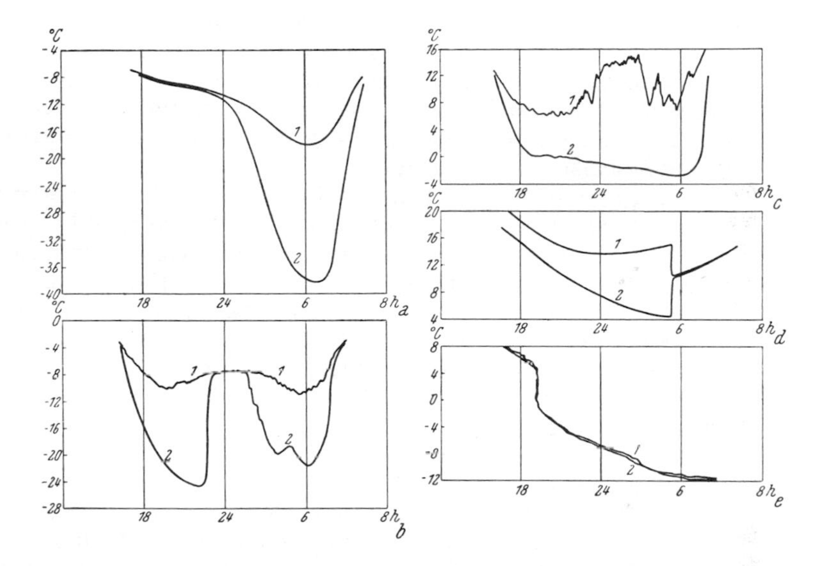

Fig. 218. Types of night temperature variation on a slope (1) and at the bottom (2) of the Gstettneralm Sinkhole. (After F. Sauberer and I. Dirmhirn)

Figure 218 illustrates a few of the night temperature recordings made on the slope (1), and at the bottom (2) of the sinkhole, with a height difference of 56 m. Curves (*a*) at the top left are for a normal, cloudless radiation night; (*b*) shows intervention by a foehn wind about 23:00, which had started in a slight heating on the slope about 20:00, but reached the bottom only at 23:00; by 02:00 the foehn effect had stopped and cooling began again. At (*c*) the foehn is unable to break through but moderates the temperature drop somewhat at the bottom. The curves (*d*) show a fresh influx of cold air at 05:00. This cools the slope, but the wind blowing warms up the bottom of the sinkhole. Curves (*e*) illustrate a cold air outbreak with a homogeneous air mass, with overcast skies and a strong wind.

The large flat dish-shaped depression forms a contrast to the deep sinkhole. The micrometeorologic processes taking place at night in this type were the subject of a study by E. Huss [791] in the Federsee area to the north of Lake Constance. Over a stretch of 5 to 10 km the moorland and the Federsee (1.5 km^2 in area) are surrounded by a row of hills about 50 m high. A cold pool of air develops here at night, with the isotherms following the height contours approximately.

Differences in relief and vegetation over a surface of such wide extent, and the ease with which the wind may intrude from higher levels, mean that a long period of the night must pass before the lake of cold air can acquire uniformity and stability. This has been demonstrated by E. Huss by means of numerous temperature measurements at several heights at four different points. A lamellar temperature structure, fluctuating patches of ground fog, a gliding movement of small limited bodies of air, and the appearance of natural oscillations resembling seiches in lakes can all be recognized from these records.

With the large number of effective factors that have been described in the foregoing paragraphs, it is difficult to make a correct forecast of the temperature distribution in broken terrain, and the effects of the ground surface, described in Chapter III, and of the plant cover, have to be taken into account as well. It is, nevertheless, of great practical importance to be able to assess the distribution of night minimum temperatures over an area, having in mind the danger of night frosts. Many countries have therefore engaged in special investigations to further aims of practical interest. Some of these will be referred to briefly, since at the same time they form valuable material for study.

The first method, which is independent of all hypotheses, is to plot all minimum temperatures observed during a particular night on a height contour map of the area. C. L. Godske [789] has published two-color maps (contours in black, isotherms in red) of this type for the 2- to 8-km-wide valley where the town of Bergen, Norway, is located for three spring nights and one autumn night. The maps are explained in the accompanying text which contains numerous other diagrams.

The second method, which is used much more frequently, is to plot the average frost danger, based on temperature measurements and observation of frost damage. The maps are then constructed to show three to five frost-danger zones, depending on the purposes for which they are designed. Maps of this type were drawn by H. Aichele [781] for the higher parts of the Baar in the source regions of the Neckar and the Danube, and for the western part of the Lake Constance area. Similar maps have been constructed by E. Kaps [793] for the Bendestorf valley near Harburg, by J. van Eimern [784] for Quickborn in Holstein, and by H. Kern [794] for the Danube moors near Ingolstadt, for which he measured the mean temperatures in May and June 1950 at 18 different observation points. Further details of frost maps designed for vineyards will be given in Sec. 54. The Davos Valley in Switzerland was mapped by W. Schüepp [802] in eight danger zones with the intention of protecting the potato crop. There are maps even for South Australia, for the artificially irrigated area around Morgan in the valley of the Murray River. These were produced by B. Mason [797] and show five frost zones in a two-color print. The effects of any single damaging frost, of course, will depend on the weather situation, wind, and cloud cover on that particular occasion. E. Franken [785] showed that in the area around the mouth of the Elbe River near Hamburg cold pools might occupy different areas at different times, and that the temperature differences between neighboring places might have different signs (+ or -) in different cases. This should not be left out of consideration when using frost-danger maps. These only summarize the average situation, and do not describe a particular one.

A third method is to map the actual damage caused by frost. G. Krühne [795] surveyed the damage done by May frosts in the area of Wernigerode in the Harz Mountains in 1953, and produced maps showing four different degrees of damage. G. Reichelt [798] constructed similar maps of damage to green crops for the same May in the Baar area mentioned above, and finally F. Winter [804] mapped the damage to walnut trees in May 1957 in the Rhine Valley west of the Black Forest at Offenburg and in the west of Lake Constance area.

43. Local Winds in the Mountains

In Sec. 37 a distinction was made between an active and a passive influence of the forest on the wind field. There are advantages in making a similar distinction in an investigation of the effects of the configuration of the land on air flow over it. We shall therefore speak of an active topographic effect when differences of temperature and air pressure, caused by the land, give rise to air currents. The most important of these are the mountain winds which vary periodically during the day, and the local flow of cold air described in Sec. 42 belongs also to this group. The influence is passive when topographic features, mountains, valleys, and slopes affect and modify the existing wind field. These two types of mountain wind will be the subject of this section, in which the general climatologic aspects of the problem will be treated briefly and the microclimatologic aspects in detail.

Mountain winds with a diurnal periodic fluctuation have been subdivided by A. Wagner [845] into three types: compensating winds, mountain and valley winds, and slope winds. The sphere of influence of the winds decreases in that order.

Compensating winds are brought into existence between level country and neighboring extensive mountain areas, through greater heating which causes the isobaric surfaces over the plain to bulge upward, thus creating a horizontal pressure gradient that produces a movement of the air mass above the plain toward the mountain. A return flow occurs at night with the reversal of the temperature difference. A snow cover on the mountains will increase the temperature contrast and thus favor the creation of compensating winds. These are large-scale air movements which can increase to gale force only in narrow passes linking mountains and plains. C. Troll [843] has studied their development in the tropical highlands of South America and Africa, and has described their influence on cloud formation, precipitation, and vegetation, and has also indicated that they are often inseparably associated with valley winds (to be described later), or with land and sea breezes, trade winds, and monsoons.

Let us now consider the other extreme, the slope winds, which are found as upslope winds during the day and downslope winds at night.

Nighttime downslope winds owe their existence to the type of cold-air flow described in Sec. 42. The term cold-air current is, however, usually restricted to small-scale air movements over almost level ground, which normally are not associated with a current in the opposite direction during the day. In mountainous terrain, by way of contrast, strong downslope winds are created at steep slopes and are always replaced by upslope winds during the day. These winds therefore possess some characteristics that distinguish them from the cold-air currents in the former sense, and these differences must now be explored.

F. Defant [809] used the double-theodolite pilot-balloon ascents of A. Riedel on the slopes of the north chain of mountains at Innsbruck to determine the wind components parallel to the slope as a function of height above the surface of the slope. He found the following vertical profile from measurements made on five undisturbed nights with a typically developed downslope of winds (the upslope wind will be discussed later):

Height above slope (m):	5	10	20	30	50	100
Downslope wind at night (m sec^{-1}):	1.0	1.5	2.3	2.4	1.9	0.2
Upslope wind at midday (m sec^{-1}):	2.3	2.9	3.7	3.9	3.4	2.4

The greatest speed was found to be at a height of 20 to 40 m, that is, close to the surface of the slope, although at some distance from its rough surface, and the speeds are far greater than those of cold air (p. 394).

The character of the air movement is also different. Over steeper slopes air flows in gusts; according to A. Defant [808], this begins when gradients reach about 1:100. A smaller-scale investigation of the movements of such air gusts was made by M. Reiher [828] on a steep 18° slope near Göttingen. He used platinum resistance thermometers to measure air temperature at heights of 10, 30, and 50 cm above the ground for three-fourths of the height of the slope. Figure 219 gives one example: the lower half of the diagram shows the temperature variation at the three measurement points, and the upper half gives the temperature stratification by the distribution of isotherms in the lowest half meter of the air over a period of only 5.5 min.

The increase of temperature with height is typical of the nocturnal negative radiation balance. It may be assumed that the distribution of isotherms over the time interval will reflect the spatial structure of the cold-air parcel passing the measurement point. This movement should therefore be thought of as taking place from right to left. At 19:04 a typical cold-air drop can be recognized easily from the arching upward of the isotherms. At 19:02 a tongue from this drop is lifting the warmer air off the ground, as shown by the crowding of the isotherms at 10 cm. Although at 19:05 warmer air is trying to break in from aloft (a gust of wind?) for a short time, the cold-air drop moves out only by 19:07. Such movements took place every 4 to 5 min; with an average wind speed of 1.4 m sec^{-1}, the horizontal extent of the cold-air drop was 300 to 400 m.

F. W. Nitze [827], who observed a similar process, gave the following vivid explanation: "It seems that the cold air dams up on the plateau of the slope. As long as its vertical dimensions remain small, it will only flow very slowly; friction in contact with the ground is very great. When the cold air has collected in sufficient quantity, however, it begins to slide down. In doing so it entrains the air behind it so that a strong cold-air current develops which persists until there is no longer enough cold air on the plateau to support it." H. Aichele [805] describes a case in which artificial fog was introduced over a valley near Lake Constance to combat frost, which performed a pulsating motion by first drifting down the valley on top of the cold air and then flowing back up again. In the dry fruit-growing area near Riverside, California, F. A. Brooks and H. B. Schultz [807] recorded the fluctuations of a nocturnal density current of this type and analyzed it theoretically.

In mountain areas these cold-air drops may assume the form of "air avalanches," to use the term of A. Schmauss [831]. He has described such avalanches in the Bavarian Alps. H. Scaëtta [830] has encountered them in such intensity in central Africa, on the Karisimbi (4000 m) northeast of Lake Kiwu, that at the same time on two nights the air avalanche almost swept away his tent. The fact that this is also a rhythmic pulsation has been shown by the careful observations made by J. Küttner [824] during a climbing tour in which he bivouacked at night among the rocks. He has

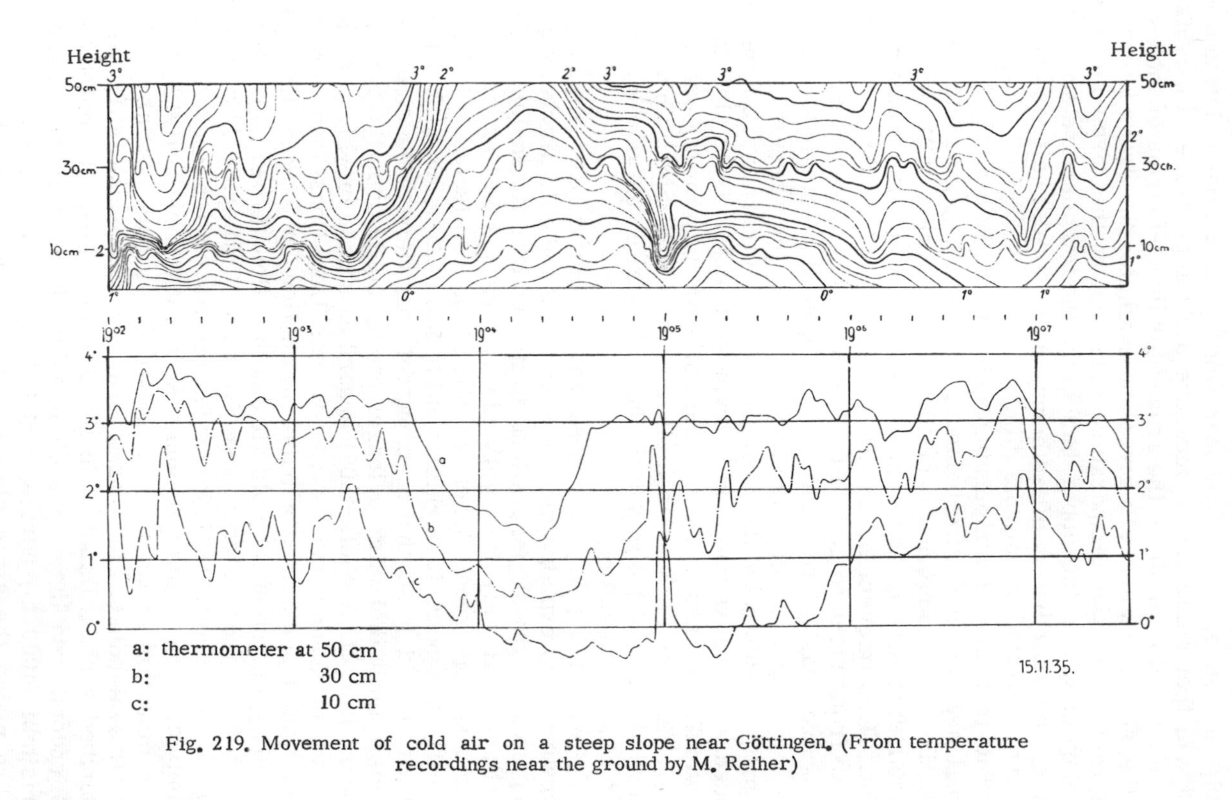

Fig. 219. Movement of cold air on a steep slope near Göttingen. (From temperature recordings near the ground by M. Reiher)

described it as follows: "I was able to follow the microstructure of cold-air drops during two nights that I had to spend in the upper part of the wall, when making a tour of the north wall of the inner Höllental peak (Wetterstein Mountains) on 22 to 24 July 1947. The point of observation was at 2700 m, about 100 m below the summit. A high-pressure weather situation with light winds prevailed. After sunset at 21:01 central European time, a weak air current developed which died down again within a minute. At 21:06 there was another air movement, and another at 21:11, followed by gusts at regular intervals of 5 min which persisted throughout the night, disturbing our sleep in a most unpleasant way. An occasional gust would be missing, but the next one would appear punctually on time. The greatest deviation was ± 1 min. The rhythm continued until 6 o'clock in the morning with an unbelievable regularity, always bringing a shower of cold air at h + 1, h + 6, etc. Soon after sunrise these periodic winds stopped, although the north wall still stood in unchanging shadow." The same 5-min period was kept up the following night. A second similar occurrence was later reported by A. Schmauss [833].

A. Schmauss [832] proved, during a forest fire that lasted from 4 to 6 October 1942 in the dwarf-fir region of the Karwendel Mountains (2000 m), that the fire had been carried down by the downslope wind at night into separated stands of dwarf fir a few hundred meters lower. In the prevailing good-weather situation, therefore, the night circulation was established over the whole area in spite of the local rising currents in the area of the fire.

Contrasting with the downslope winds at night are the winds blowing up the slope during the day. These are stronger becuase the radiation exchange is greater. Their direction is made obvious by the way the smoke blows from cottage chimneys and the direction in which airborne seeds are carried. The table on p. 404 indicates the profile of the upslope winds, calculated from the 11 pilot-balloon ascents in Defant's investigation [809]. Here too, the wind is strongest at 20 to 40 m above the ground. This means that glider pilots, who wish to use these winds as thermals, must fly with their wings, which have a span of 16 to 18 m, very close to the slope. R. Maletzke [826] reports that flying-corps students from the Technical College at Munich, touring the Alps by glider in 1958, were able to cover a distance of 30 km using the upslope wind from a single hillside, without losing altitude. Development of local slope winds is favored by the unstable temperature stratification of the atmosphere. H. G. Koch [823] was thus able to show on the stormy day of 28 June 1952, from the 14:00 synoptic observations of the meteorologic network, that winds were blowing up the slopes on both sides of the ridge of the Thuringian Forest.

Upslope winds were measured by J. Grunow [817] for the SE slope on which a ski jump had been built for the international championships held at Oberstdorf in the Allgäu from 28 January to

6 February 1950. They were found to be 2 to 3 m sec^{-1}, with a maximum of 3.6 m sec^{-1}. They were most strongly developed between 11:30 and 13:00, and it could be proved that they increased the length of jump. The ten best skiers made jumps of between 92 and 117 m with an upslope wind of 1.8 m sec^{-1}, and 106 to 128 m with 3.2 m sec^{-1}.

Figure 220 is a cross section of a mountain valley around noon. The winds flowing up both slopes, which are in sunshine, increase in vertical extent farther up the slope, in proportion to the areas shaded with fine dots. The rising air must be replaced continuously. This is achieved not only by the cross-valley flow shown in the sketch, but also by the wind flowing up the valley, which is indicated by the larger dots, intended to show a wind blowing perpendicular to the plane of the paper.

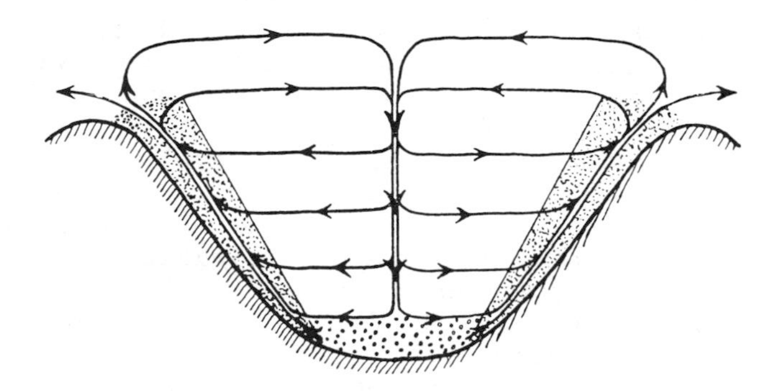

Fig. 220. The system of upslope and upvalley winds. (After A. Wagner)

If the directions of the arrows in Fig. 220 are reversed, we obtain a picture of the system of downslope and downvalley winds at night. In this case, however, the finely dotted areas have to be reversed so as to be narrow at the top, broadening out toward the bottom of the slope.

There is, therefore, a double system of periodic mountain winds; during the day there are upslope winds and upvalley winds, and at night downslope winds and downvalley winds, sometimes also called a valley-exit wind or a mountain wind because of its source. L. Prandtl, A. Defant, and F. Defant [810] have produced a complete theory of mountain winds. The slope winds are always initiated first, then the valley winds follow, and hence during the course of the day there is a change taking place with a phase difference, which is illustrated in Fig. 221 for a fine summer day.

Diagram *A* shows the position shortly after sunrise. The upslope winds have set in (light arrows), but, since the air in the valley is still colder from the night than the air in the plain outside, the downvalley wind is still blowing (black arrow) and is still being fed by the return flow from the upslope circulation. It soon dies out with further heating. For a time (*B*) the upslope winds occupy the picture alone. This is the time when the air in the valley heats most quickly. Toward midday (*C*) the wind up the valley sets in and feeds the upslope winds, and for its own part is supported by the return flow descending from greater heights in

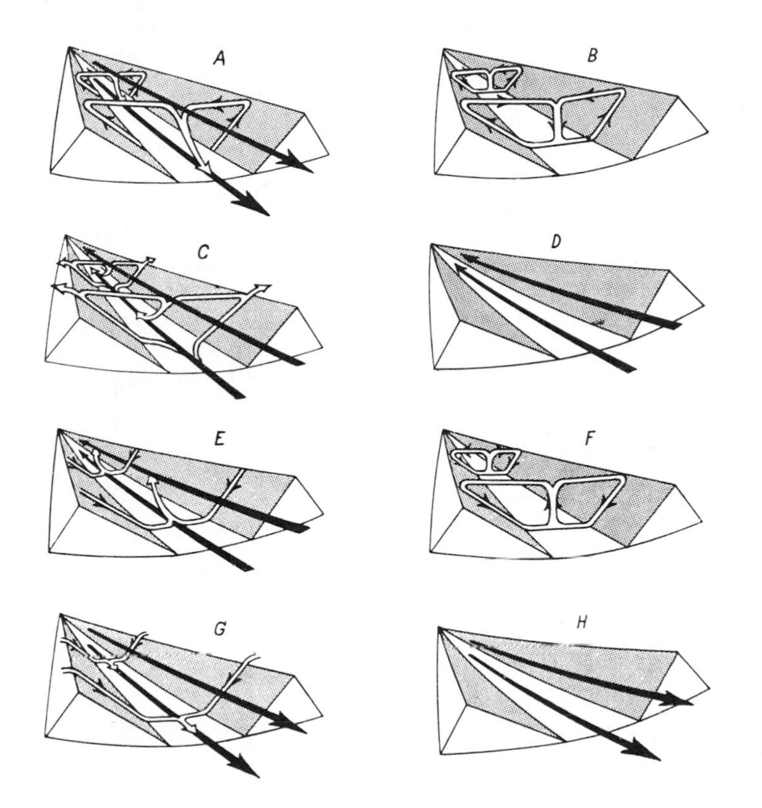

Fig. 221. The interplay of slope and valley winds during the course of a day. (After F. Defant)

the center of the valley (see Fig. 220). This is the hour at which chains of cumulus clouds appear above the slopes, while the sky remains clear where the air is sinking over the center of the valley. In the late afternoon (*D*) the upslope winds cease to blow; the upvalley wind continues to blow for only a short time. It has been shown by H. Berg [806] that in valleys running N-S in the Allgäu Alps an upslope wind continues to blow up the west slope when a downslope wind has set in over the east slope, which is in shadow. Such situations with different degrees of insolation must give rise to a cross-valley wind, according to the theory of T. A. Gleeson [816].

Sooner or later, as evening draws on, the downslope wind sets in. The change may often be observed from smoke trails, as shown in photographs taken by F. Winter [848] at the exit of the Heidelberg Valley. Diagrams *F*, *G*, and *H* of Fig. 221 corresponds to *B*, *C*, and *D* in the morning.

The valley winds which blow in both directions are on a larger scale, and as a rule are more strongly developed than slope winds. It is possible for upvalley winds to determine the shape in which trees grow, as shown by F. Runge [829] for the Zillertal and neighboring valleys. The upvalley wind in this case, even if its long-term effects were only slightly marked, was able to come into being against the nocturnal downvalley wind and even against the prevailing wind direction from the west.

Salzburg airport lies at the exit of the Salzach Valley. Surface winds recorded at a height of 18 m have been analyzed for the period 1949 to 1951 by E. Ekhart [813] to ascertain their daily variation. Figure 222 shows the results after smoothing. Wind

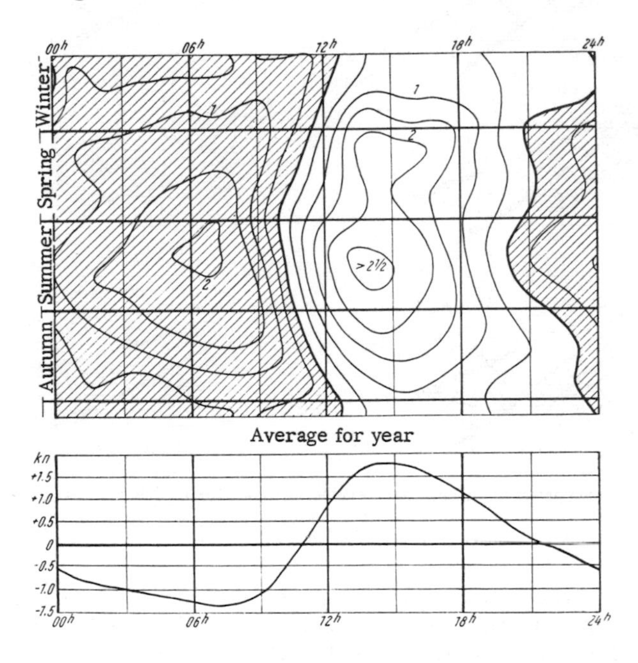

Fig. 222. Wind components in the direction of the valley at Salzburg airport. (After E. Ekhart)

components in the direction of the valley show a marked change of direction from downvalley (hatched) to upvalley wind (white). Toward noon, between 10:00 and 13:00, depending on the season, the upvalley wind sets in, reaching a speed of 2.5 knots (1.3 m sec^{-1}) in summer. The downvalley wind at night sets in at very different times in different seasons and reaches its maximum only during the late hours of the night. These are average values for all kinds of weather situations, therefore the speeds are small. Nevertheless, the periodicity of the wind system is so great that it shows up clearly in the analysis. A similar investigation has been carried out for Graz by R. Stöckl [841], using observations made at three different hours.

The downvalley wind has been given special attention since it is partly responsible for the night frost danger. One such wind that has been specially investigated is the "Wisperwind" which blows at night out of the Wisper Valley from the east into the warm valley of the Rhine with a speed of 3 to 4 m sec^{-1}. H. Schultz [837] showed that the speed increased linearly with the magnitude of the temperature reversal in the valley of the Wisper. In a similar way, R. Luft [825] showed the importance of the "Siebengebirge wind" for the climate of Bonn and Beuel on the Rhine. The downvalley wind at Braunlage health resort in the Harz has been investigated by L. Schulz [838].

Only rarely have valley winds been investigated both in vertical extent and in their relation to temperature and water-vapor pressure, therefore we are indebted to W. Schüepp [836] for having done this for the Davos Valley in Switzerland. Along with a few helpers he made measurements close to the ground on 20 winter mornings, and at the same time made use of the observations from the meteorologic station (Met. St. in Fig. 223), from Davos Observatory, and at the top of a 70-m church steeple. The temperature and wind profile thus derived is shown in the upper part of Fig. 223, the water-vapor distribution in the lower part.

The maximum speed of the downvalley wind is 2.5 m sec^{-1} at a height of 40 m above the bottom of the valley. Reduced to a speed of 1 m sec^{-1} by friction at the slopes, the wind flows perpendicular to the height contours as a downslope wind, according to simultaneous air-flow observations, while at higher levels it follows the general course of the valley as a downvalley wind. The temperature inversion amounted to about 9 deg. The maximum water-vapor pressure of 2.5 mm-Hg was found at the top of the slope, and, while the air at the cold valley floor is dry from the point of view of absolute humidity, relatively it is 70 to 90 percent saturated. The dot-dash line joining the water-vapor pressure maxima in the vertical section shows a downward drifting of moister air from the upper slope, along with the downvalley wind.

After World War II, it was intended to expand the Odenwald fruit-growing industry in a few valleys in the Erbach district, but a narrow limit was set to this extension by the risk of late frosts. The German Weather Service, the Agricultural College, and the Fruit Industry Advisory Center of the district joined their resources to carry out a microclimatologic survey, under F. Schnelle [834, 835]. The upper limit of the frost danger area for the slopes to be brought under cultivation was mapped on a scale of 1:10,000, based on carefully collected information from fruit farmers, on night temperature measurements over a period of several years,

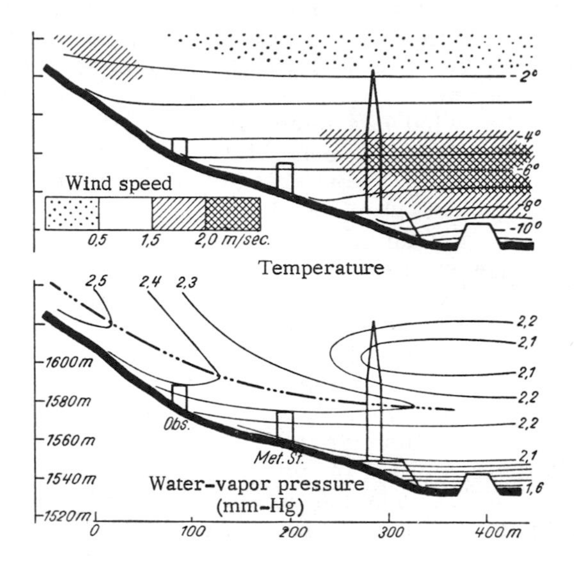

Fig. 223. Vertical section of the nocturnal downvalley wind in Davos Valley. (After W. Schüepp)

on observations of the frequent valley fogs (especially their upper limits), and on an exact study of the topography of the area with the eye of an experienced microclimatologist.

It was found, as shown in Fig. 224, that wherever a valley narrows the flow of cold air is dammed up. In the Mossau Valley, which is uniform in shape and about 8 km long, the upper limit of the cold air sinks very slowly, becomes horizontal, and is dammed up at all junctions of 10° to 20° slopes that are covered with tall trees. More detailed analysis of 20 such valley narrows indicated that only an opening of 400 to 500 m or more in the valley would allow the cold air to flow through. "Field observations showed repeatedly that cold air does not flow like water, but more like porridge or thick syrup." The cold air flowing down both slopes as downslope winds therefore forms a flat surface in the upper part of each section of the valley, extending up to 55 m vertically above the narrow. On nights with light winds the downvalley wind can be overcome by these cold-air dams, as was already shown on p. 394 for the small-scale and shallow cold-air flow described there.

It must be kept in mind that the periodic variation in the wind system can be disturbed or inhibited by outside influences. For example, H. G. Koch [822] was able to demonstrate, by means of aerologic ascents on the west side of the Tirso basin in Sardinia,

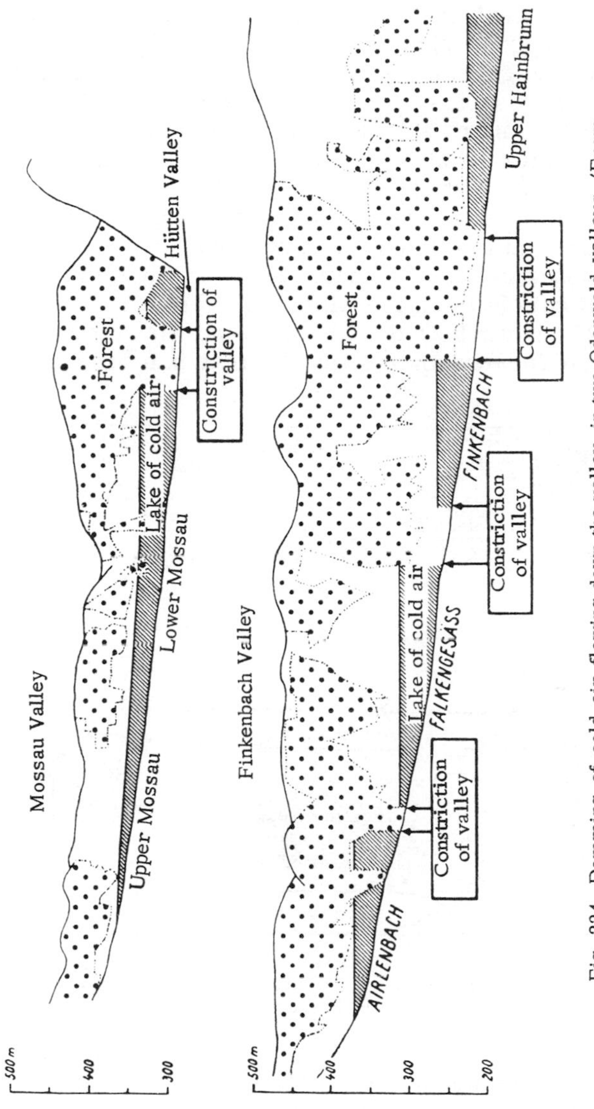

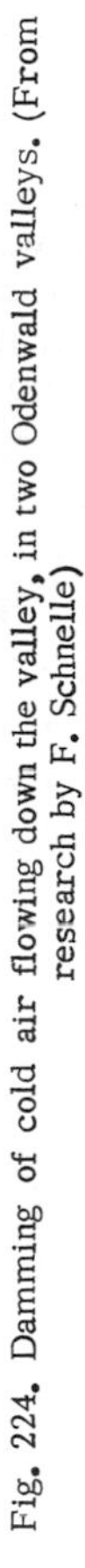
Fig. 224. Damming of cold air flowing down the valley, in two Odenwald valleys. (From research by F. Schnelle)

that a descending wind like a bora broke into the normal mountain wind system in the late afternoon, and that this was attributable to the sea breeze coming over the top of the mountains. E. M. Wilkins [847] described the development of a marked wind discontinuity in the plain of the Snake River in eastern Idaho, between the normal upvalley wind and a wind descending from a high mountain range close by (good photos!), which was of importance for the dispersal of waste gases from an atomic power plant situated there.

At this stage we must consider another cold wind of rare occurrence which is more strongly developed in great summer heat, namely, the glacier wind, or firn wind as it is sometimes called. It is included in the list of active mountain winds since it owes its origin to the contrast between the temperature of the glacier ice and the sunny ground in the surrounding areas. H. Tollner [842] was the first to describe it. E. Ekhart [812] and H. Hoinkes [818] have investigated it carefully on the Alpine glaciers.

Figure 225 shows how the glacier wind fits into the daily wind system. In the fine-weather situation that has been assumed, the

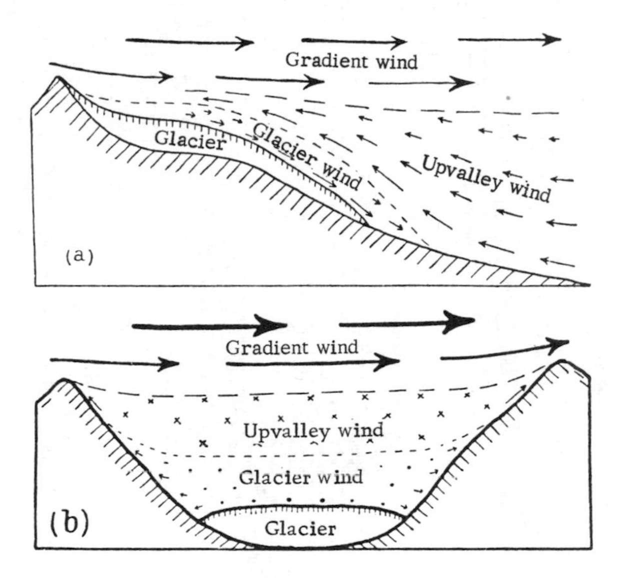

Fig. 225. Incorporation of glacier wind into the wind system on a fair-weather day. (After E. Eckhart)

gradient wind in the upper air layers is unable to extend its influence downward. Below it the upvalley wind occupies the whole mountain valley. The glacier wind, which is only a few tens of meters deep and has a maximum speed of 3 m sec^{-1}, which is reached at 2.5 m sec^{-1} above the surface of the glacier, slides down underneath the upvalley wind, in the opposite direction.

The glacier wind blows night and day in summer with approximately constant strength. There is a first maximum before sunrise, corresponding roughly to the maximum of cold-air flow, and a second maximum before sunset when the temperature contrast between the heated neighborhood and the glacier ice has reached its peak. In discussing the temperature of the air close to the ice, we follow Hoinkes in making a distinction between air that belongs to the glacier and air that has come from the outside. The latter, which has flowed in from the surroundings, has transferred only a little heat to the glacier and exhibits only a small increase of temperature with height (in contrast to the temperature decrease above the neighboring rock surface). The glacier air has become highly stabilized by long contact with the ice, has a steep temperature gradient (6 deg in 2 m), and short-term temperature fluctuations connected with horizontal wind variations, something like the cold-air gusts mentioned on p. 404. This is the motor that drives the glacier wind. The steep temperature gradient means there is a strong flow of heat from the surrounding air to the ice surface, and this assists the process of ablation, as was proved by Hoinkes [819]. The often-expressed opinion that the cold glacier wind has a "preserving" effect on the ice therefore does not apply.

The glacier wind is soon extinguished down in the valley. Nevertheless it has a refreshing effect on the mountaineer who approaches the glacier from below, and has an important influence on plant life in the neighborhood. According to H. Friedel [815], the Pasterze Glacier on the Grossglockner causes various types of wind damage such as wind tracks in grass, damages mosses, stunts trees, and so forth, and reduces the altitude limit of plants, *Elynelum*, for example, by about 500 m in its vicinity. During the summer of 1931, G. Schreckenthal-Schimitschek [917] measured air temperature at a height of 1 m and ground temperature at depths of 5 and 20 cm at distances of 3, 30, and 500 m from the front of the Mittelberg Glacier in Pitztal. The cooling effect was naturally greater on the ground temperature than on the air temperature. At a distance of 30 m ground temperatures in August and September no longer fell below the freezing point, and were 2-3 deg higher at the 5-cm depth, and 2-4 deg at the 20-cm depth, than they were at 3 m from the glacier terminus. The first sparse vegetation was encountered at 30 m distance.

The passive influence of topography on the wind field is a problem of flow. The results of laboratory research are, however, hardly applicable. The variety of topographic features and frictional effects at the surface require a direct approach in exploring nature by observation and measurement.

Large-scale effects can be studied from the observations of synoptic stations. Thus, F. Steinhauser [840] was able to prove, for the lower Carpathians, that winds which cross the mountain ridge at right angles are at times strengthened considerably on the lee side. Since the ridge runs with a relative height of 400 m from SW to NE, the storm frequency is greatest at the station lying immediately under the ridge on the NW side with winds coming from the SE, and at the stations on the SE of the ridge with winds from the NE. A picture is thus obtained in which the mountain range appears to act as a source of storms. In addition to being increased in speed, winds on the lee side are much more gusty, where, according to the lee-wave theory, air pressure is about 3.5 to 5 mb lower.

J. H. Field and R. Warden [814] used a 1:5000 scale model of the rock of Gibraltar in wind-tunnel experiments, to discover the influence of this isolated obstruction on the wind field as a function of wind direction, while investigating the causes of aircraft crashes when the airfield first came into use. Additional information on gustiness and vertical currents was obtained from 138 pilot-balloon ascents in the area. The practical result was a chart of the danger zones for aircraft for the four most frequent wind directions.

Investigation of passive wind influences on a microclimatologic scale has only just begun. This opens up a very important field for research since, after the radiation balance, the amount of ventilation is the most important factor determining local climate. Everyone is aware that river valleys also form wind lanes. The more pronounced the relief of an area is, the more strongly marked is the wind pressure on the windward slopes and the formation of eddies in the lee. All valleys are exposed to the risk, to a greater or lesser extent depending on their configuration, of becoming a region of stagnant air. All these features have an effect on temperature and humidity.

The manner in which trees grow may sometimes provide a point of departure in this study; details can be found in W. Weischet [846] and F. Runge [829]. In a short but careful study, J. van Eimern [811] described the influence of the steep slopes of the Harburg Mountains, some 60 to 100 m in height, and the Geest Ranges on the completely flat lower Elbe area. At a distance of 2 km from the steep edge there was a shift of up to 24° in the direction of winds coming from the mountain area, but this was due to increased friction in the mountains and not, as might be expected, to the direct effect of steering lines. The reduction in wind speed corresponded quantitatively to the pattern deduced from windbreak studies (see Sec. 53). The streamline chart that van Eimern [932] produced for a valley used for agricultural purposes, for the typical wind directions encountered there, will be discussed in Sec. 47.

E. Kaps [821] attempted to estimate the amount of ventilation from the shapes and sizes of valleys. As a measure of ventilation he took the time required for the air acquiring local characteristics to be replaced by new air. It was possible to formulate only a few general propositions. H. Kaiser [820] of the Essen weather bureau investigated flow over long, extended hill ranges, by making sections perpendicular to the hills, and setting up anemometers at 1.5 m above the ground. With winds of 3 to 6 m sec^{-1}, increases in wind of 50 to 65 percent were measured on the windward slopes of gradient 3° to 10°, with the zone of strong wind extending vertically up to 100 m above the ridge. On the lee side, no tendency for the air flow to detach itself from the ground was observed for slopes of 2° to 4° when winds were less than 4 m sec^{-1}. With wind speeds of 5 to 10 m sec^{-1}, however, there was a marked degree of protection in the lee, associated with a descending wind. The smallest wind speeds were observed at 200 to 400 m in the lee of the ridge, and the greatest (descending wind) at 500 to 600 m, and were to a large extent independent of the strength of the wind. The large- and small-scale topographic features led to the superposition of the corresponding wind fields. These results can therefore give only an idea of the kind of effect to expect.

A recent paper by H. R. Scultetus [839], describing the effect of topography, wind direction, and lapse rate on the distribution of wind speed and direction in a forest village near Cleves, may serve as an example for future research of this kind. The influence of cross-valley winds has been studied in Japan by M. Yoshino [849 to 851] in several valleys, using various wind-speed, direction, and gustiness recorders. It is not possible to do more here than recommend the works to be consulted.

44. The Climate of Various Slopes (Exposure Climate)

The different amounts of radiation received during the day on different slopes were the subject of Sec. 40, then in Sec. 41 the small-scale effects of these differences were illustrated by a number of examples. This discussion will now be broadened to include larger dimensions, and in doing so attention must be directed to those investigations in which effects due to the orientation of the slope have been studied in the most systematic way possible. The shape that best satisfies this preliminary condition is that of a regular mountain cone, sloping uniformly on all sides. It is possible to construct such a shape on a small scale, but it is unfortunately hardly ever found in nature on a large scale.

The radiation received by the slope, described in Sec. 40, however important it may be, is only one of a number of factors in the heat balance (see Sec. 25), and then only during the daytime. The actual character of the exposure climate is also influenced by many other factors.

Air movements created or modified by topographic features were discussed in Secs. 42 and 43. On a uniform mountain cone situated in an air stream, the greatest speeds are found on the mountain sides, the windward side will favor the formation of upslope winds, and the wind in the lee will be gusty, in both strength and direction. It follows that the frequency distribution of wind directions at any given place may constitute evidence of a varied wind climate.

The question must be raised whether it might not be possible for a special type of climate to develop near the ground on mountain slopes, and whether its temperature contrasts might not be eliminated by the orographic winds generated, particularly when the slopes are steep. This question was answered in the negative by R. Geiger [854] on the Hohenkarpfen in 1926.

The Hohenkarpfen is a conical mountain of almost ideal shape, situated on the edge of the hilly area of Württemberg. It has a relative height of 100 m, with a slope in the upper part of 30° and lower down a gentle one of 11°. Figure 226 shows the profile of the mountain, with vertical scale enlarged only twice. The air layer close to the slope has been enlarged 50 times so as to show the observation points at 25 and 100 cm. The diagram shows the highest and lowest daytime temperatures during the summer at the

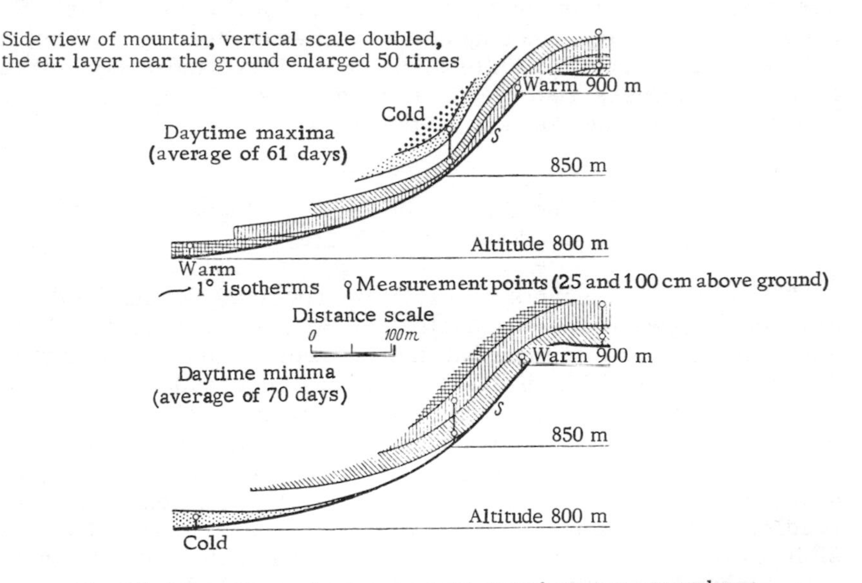

Fig. 226. Microclimate develops near the ground, even on steep slopes.

summit, the average for 8 points on the steep upper sides (25 cm high) arranged on the slopes in the 8 principal directions, averages for 16 stations at the point of inflection of the slope, and for 8 stations on the flatter slope below (25 cm).

Temperature stratification close to the ground can be recognized even on these steep slopes, but it is different from that of the plain. About midday (Fig. 226, top) in summer the strong upslope wind at the foot of the steeper slope draws the cooler air farther away from the slope, with the result that the lapse rate becomes particularly large in this position. The highest temperatures near the ground, therefore, are not found at the steep slope, but at the bottom of the valley and on the flat mountain plateau, which is protected from the wind by a growth of shrubs. At night (lower part of Fig. 226) the temperature increases with height at every observation point, because of the negative radiation balance, but the stratification of the 100-m layer also allows the drop of temperature with decreasing height above sea level to be recognized.

The climate of slopes facing in different directions is affected to a large extent by moisture conditions as well as radiation and wind.

In the first place, the precipitation distribution is not uniform. A distinction must be made between two different orders of magnitude in the influence of mountains on precipitation. The west side of the Harz Mountains is rich in rainfall, and the east side poor; this is due to the lifting of rain clouds by the prevailing westerly winds, and foehn-like drying on wind-protected slopes. This influence implies height differences which cause substantial temperature changes in the rising and descending air masses. The smaller the dimension, however, the more the local precipitation distribution is determined by the wind field. This distribution is often exactly opposite; more precipitation is found in the wind shadows than on the windward side, especially with brisk winds. The ordinary type of rain gages on the lee side of the Hohenkarpfen picked up 5 to 10 percent more precipitation than on the windward side. Snow may often be observed to pile up behind a fence or an undulation in the ground (see Fig. 271).

To the wind influence there is added the effect of the slope of the ground. By general agreement, precipitation is measured by means of gages set up so that the receiving surface is horizontal. If a gage parallel to the slope is used, of such a size that the horizontal projection of its receiving surface is equal to that of a standard gage, the amount collected is found by experinece to be different, and this will give a measure of the rainfall actually received by the slope, which of course is what interests us. The difference depends on the direction in which the slope faces, its gradient, and the angle at which the rain is falling; the latter is determined by the wind at the time, andby the type of precipitation. There is a special literature on this problem of measurement techniques.

We can best learn about the relation of these factors by studying J. Grunow's investigations on the Hohenpeissenberg [855]. Figure 227 shows the amount of precipitation received in a gage parallel to a 20° slope on the Hohenpeissenberg, compared with a horizontal gage, as a function of wind speed. When the wind is blowing on the slope (windward), the former always catches more than the latter; the difference is naturally greater with snow (solid line) which is more easily borne on the wind than rain (broken line). When the slope is in the lee, the "rain shadow" of the

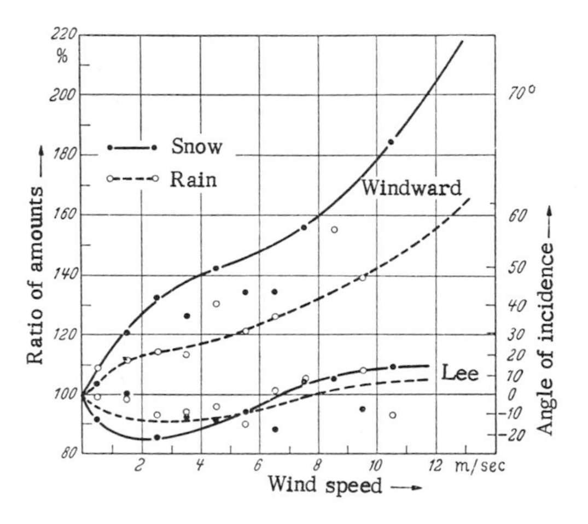

Fig. 227. Proportion of precipitation falling on a receiving surface parallel to the slope, and on a horizontal surface on the Hohenpeissenberg. (After J. Grunow)

mountain shows up clearly with moderate winds. Only when wind speeds fall below 7 m sec^{-1} does rain fall (in the lee eddy area) more on the sloping than on the horizontal reception surface.

The relation illustrated must depend on the angle at which the precipitation is falling, measured from the vertical. A scale on the right-hand edge of the diagram has therefore been introduced to show this angle. On the average, this was 20° for rain and 31° for snow on the windward slope, while it was 8° for both on the lee slope. With winds of more than 7 m sec^{-1} the angle changes, turning gradually against the direction of the general wind, as a lee eddy forms. On other slopes, different values for the proportions of rainfall collected and for the angle of rainfall are found, but the basic principles remain the same.

Nothing is known about the deposition of dew on different slopes, in particular about the periods of deposition, which must certainly be different.

When we come to consider soil moisture, on which the evaporation term of the heat-balance equation depends, both the amount of precipitation and the rate at which the soil dries, hence the type of soil, are of importance. We have already seen from Fig. 204 that radiation has an asymmetric effect in that, although it is evenly distributed on either side of the noon line, the radiative heat received in the forenoon is mainly used to dry the surfaces on which it falls, while in the afternoon most of it is used for heating the soil. The results that were found to be valid for the bark temperatures of a tree trunk apply here, in a similar way, to surface temperatures on differently oriented slopes. Since this process takes place every day it will, in general, affect the soil itself in the end. Even on a mountain cone with initially homogeneous soil, a different soil will gradually develop on slopes facing in different directions and therefore the temperature pattern in the soil will be different.

It is for these reasons (in the Northern Hemisphere) that the warmest slopes are not those facing south, but southwest, as a rule. This rule may be observed to apply even in very restricted fields of research; thus in 1878 E. Wollny [875] verified it using boxes filled with sifted soil, facing eight directions and tilted at 15°. Figure 228 shows the results of an experiment by A. Kerner

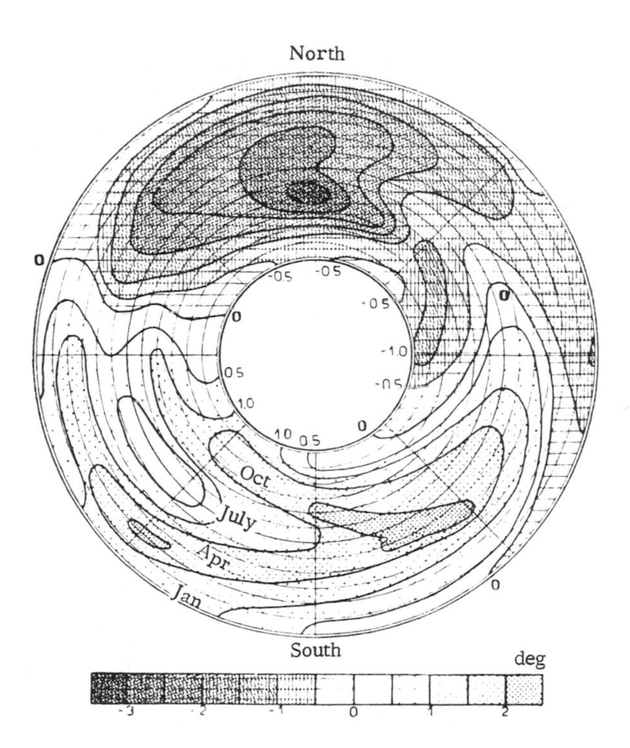

Fig. 228. Departure from the mean value of ground temperature at a depth of 70 cm, according to direction of slope and time of year. (From observations by A. Kerner near Innsbruck)

[862] on the Judenbühel near Innsbruck. This hill projects like a peak on the south slope of the Hungerberg plateau and today is called the "Spitzbühel" because of its shape. Soil temperatures

were measured at depths of 70 to 80 cm on its slopes from 1887 to 1890. The temperature differences are therefore small, but are sufficient to indicate the essential features of the different slope exposures.

The circular shape of Fig. 228 illustrates the directions of the slopes, and months are shown as concentric circles. The average soil temperature has been evaluated for all directions, for each month, and deviations from this mean are shown. Hatched areas are relatively cold and dotted areas are warm. The greatest temperature differences are, of course, found in summer (center of the circular band). The north slope is coldest, as expected. The warmest zone varies in position, however, with season. The southwestern position which we expected to be warmest from the previous discussion, occupies this position only from autumn to spring. The warmest area moves around to the southeast in summer. This is caused by the diurnal variation in cloudiness since, in the Alps, strongly developed cumulus build up by early afternoon, and occasionally produce thunderstorms. More radiation is therefore received by the ground in the forenoon when the cloud cover is less, showing once more how many factors are involved in the determination of slope climate.

G. H. Schwabe [870] used artificial soil mounds in experimental ecologic studies in South Chile. In this manner he studied the effect of the contrast between the comparatively low air temperatures caused by the Humboldt Current and the large amount of incoming radiation inherent to this latitude on the plant yield on various slopes. For this purpose he used earth mounds 1.2 to 1.5 m high standing in a horizontal circular area 10 to 15 m in diameter, the mounds having sides sloping about 35°, as estimated from the diagram. The surfaces of the mounds and the flat areas were sown uniformly, and later the harvest was collected in sectors.

The method is shown in Fig. 229 by two examples. In order to facilitate a comparison between these measurements from the Southern Hemisphere and the other diagrams, the sunny side, which in Chile is the north side, has been put at the bottom of the diagram to accomodate those who live north of the equator. The isopleths in the left-hand diagram give the percentages of beans, planted on 3 October 1955 that survived a night with late frost on 28 to 29 October on two mounds near Valdivia (40°S). The assessment was made for eight sectors of the mound and the flat circle, at positions shown by the small circles in Fig. 229. The differences, which are naturally greatest on the slopes, and were between 10.8 and 30.3 percent, were also observed in the surrounding flat area. The drier soil of the slopes acted as a heat reservoir for the night period and permitted most plants to survive in the area of the day maximum between the sunward side and the west.

The diagram on the right of Fig. 229 shows the yield of grass (kg/m^2) on eight mounds in a natural meadow near Valdivia. It had been unusually dry in spring and early summer, and the drying west winds had an unfavorable effect on the western slopes of the mounds, so that the region of highest yield was determined by humidity rather than by the warmth of the soil.

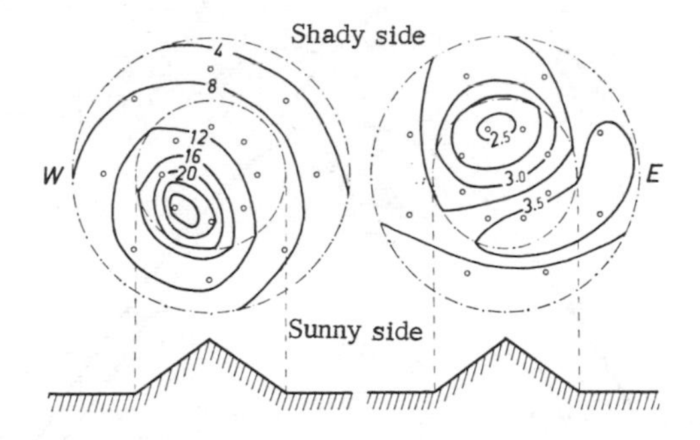

Fig. 229. Beans (percent) surviving a night frost, and the yield of pasture grass (kg m^{-2}) on artificial earth mounds 1.5 m high. (After G. H. Schwabe)

From similar latitudes (42°N), but from the arid Gobi Desert at Edsengol, a series of temperature measurements was made on the slopes of a sand dune by W. Haude [857]. He selected a dune 23 m high, in a position some 1400 m above sea level near the winter camp of the Sven Hedin Expedition to NW China in 1931-32, and took regular soil-temperature readings on the E, S, and W slopes, and on the top of the dune. Figure 230 shows the daily temperature variation from the averages of 12 almost cloudless days. The air temperature was read at the same time, in a shelter on the sand dune, showing that this was a period of daytime frost (max < 0°C). In spite of this, the temperature of the soil at a depth of 2 mm reached values of 22°C in this dry climate and on one occasion, 18 February 1932, rose as high as 32.8°C. The E slope is here warmer than the W because there was a flat pebbly area to the east, while there were other dunes to the west. There, too, in contrast to European climate, soil moisture, and hence evaporation, plays only a subordinate role in the heat balance.

A remarkable microclimatologic phenomenon in the eskers of southern and central Finland was investigated by V. Okko [866], as pointed out by C. Troll [873]. The debris carried off by melting glacier waters is piled up to a relative height of 25 m, and consists of coarse blocks and loose gravel with only a thin covering of fine sand on the surface, where plants find enough nourishment and moisture.

In some places on the crests of such eskers there are snow-free areas throughout winter (V. Okko located 28 places from 60° to 64°N). From these places warm air streams out, with a mean temperature of about + 3°C in midwinter, quite independent of the temperature of the surrounding air in which temperatures as low

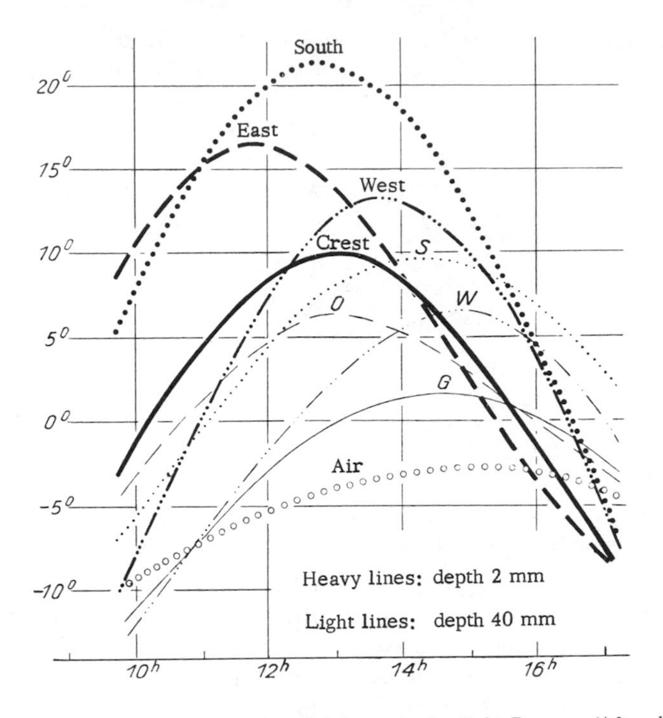

Fig. 230. Winter soil temperatures in sand dunes in the Gobi Desert. (After W. Haude)

as - 30°C have been measured. The current of air coming out of the ground is strong enough to blow out a match. At temperatures below - 20°C, a cloud of condensed water vapor 2 to 3 m high is produced. The circulation is fed from the foot of the esker where the coarse blocks are and air can enter freely from the outside. This winter circulation begins about the end of September and ends with gradually diminishing speeds in March and April. The explanation is to be found in the seasonal variation of the supply of heat in the ground, which, in the special conditions prevailing here, is delayed to an unusual extent.

During the summer, the slopes of the eskers between south and west undergo considerable heating. Slope temperatures up to 62°C were measured. Cold air flows out from the cold interior at this time, and a flow of air inward at the top may be observed. Until autumn more heat is stored in the esker than in the soil under a level surface. As soon as the first snow falls on the thin sand cover, the stored heat within the mass of rubble is protected against substantial loss. A rough computation suffices to show that this heat supply is able to provide the energy for the outward flow of air in winter, which keeps various areas free of snow. The groundwater level is usually a few meters down and the water has a temperature of from 5° to 6°C, thus also providing a barrier to thermal loss below. It is not surprising that the warm air flowing outward is always saturated, in contrast to the outside air. Vegetation near the snow-free areas is often loaded with a heavy deposit of rime.

Not many systematic investigations are available, covering all the sides of a mountain or a mountain peak. Geiger's measurements on the Hohenkarpfen in 1926 [854] indicated the influence of even sparse vegetation on a slope, and showed that it might be of the same order of magnitude as the influence of the slope itself. It is a pity that a Russian work by V. A. Smirnov [872] in 1935, which apparently had the same aims, was not accessible to me. A. Cagliolo [852] analyzed the records from four meteorologic shelters situated 30 cm above the ground on a flat hill 10 m in height, for the four main directions, in the province of Buenos Aires (39°S) during 1951. A similar investigation was carried out in Japan by T. Sakanoue [869]. F. K. Hartmann, J. v. Eimern, and G. Jahn [856] made a new and valuable study in 1954 of slope climate on a completely wooded hill in the Harz, which will now be studied in more detail.

The forest itself often provides us with a picture of the varying climate of different slopes. For example, T. Künkele [865] describes the Palatinate forest as follows: "Anyone who looks out from a single peak over this area, which from a geologic point of view is an apparently uniform terrain, deeply dissected by narrow valleys and steep mountain slopes, and directs his gaze in a NNE direction (that is, toward the slopes facing the sun and the wind), will have the impression of an almost uninterrupted dark-bluish sea of firs, interspersed with very few deciduous trees. Looking toward the SSW, however, it is surprising even to a forester to observe how completely the picture of the forest changes into the soft green shimmer of extensive areas of deciduous trees with a decreasing admixture of conifers on the winter sides. This difference shows up on the maps of the Forestry Department and of hiking clubs in which the green and yellow colors used to indicate foliage trees contrast with the other shades used for evergreens, reflecting the orography of the mountains. A site-quality map would provide a similar picture since the different slopes of one individual mountain (that is, the same geologic beds) always differ in site quality, often in the ratio 2:4." In a similar way J. Parker [867] has described forest distribution on the N and S slopes of the Palouse Range in Idaho, and has shown the climatic differences by temperature and soil-moisture records.

One of the three experimental areas used by F. K. Hartmann, J. v. Eimern, and G. Jahn, the Grosse Staufenberg near Zorge in the South Harz, was completely covered with a high beech wood on a deep and well-aerated soil. The stands, however, showed markedly different characteristics on different slopes. The question to be settled was whether this was due to microclimatic differences. During a summer and an autumn period measurements were made, near the ground, of radiation (Robitzsch actinograph), illumination (photoelectric cells), temperature, humidity, and evaporation (Piche) at 20 points in similar stands with an even crown canopy, around this hill with slopes inclined from 15° to 40° and a relative height of 200 m. There were clear and unequivocal signs of the orientation of the slope to be found, even in the strongly subdued features of the microclimate to be observed in the interior of the stands where only 5 to 15 percent of the external light penetrated, during both observation periods. Only a few points will be extracted from the wealth of data.

Figure 231 shows the distribution of day maxima, and Fig. 232 that of day minima of air temperature for four bright June days,

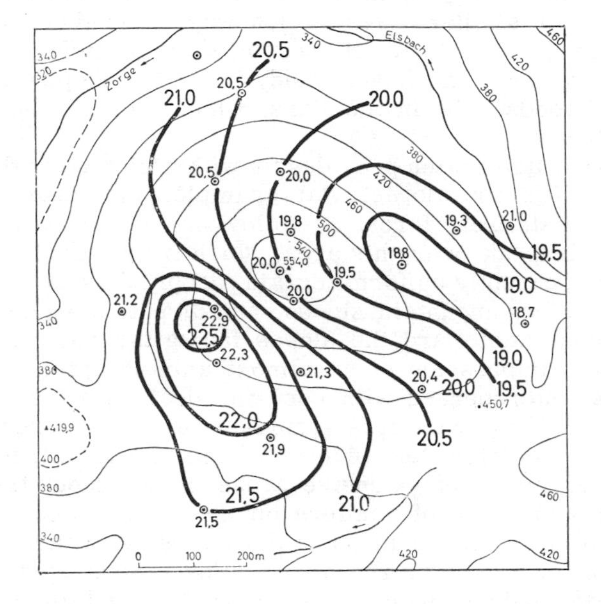

Fig. 231. Maximum temperatures in the climate near the forest floor on the Staufenberg, on four sunny June days in 1954. (From F. K. Hartmann, J. v. Eimern, and G. Jahn)

measured at a height of 40 cm above the forest floor, around the hill. This is similar to the distribution prevailing every day, but brings out the differences more clearly. The temperature maxima have, to a considerable extent, the same distribution as global radiation values, and the daily averages of soil and air temperature. The southwest slope is warmest, and its warmest area is in the center; the northeast slope is coldest, again in the middle, where the 19° isotherm extends toward the east under the influence of local topography. The Piche evaporation figure is also greatest in the middle of the SSW slope, with a maximum of 18 $cm^3\ d^{-1}$, and least in the NE with a minimum of 9 $cm^3\ d^{-1}$, while the air humidity is correspondingly lowest in the SSW. The north slope appears to be relatively warm in Fig. 231 because when the sun is high in summer it also shines on this slope; this was no longer the case during the autumn observation period.

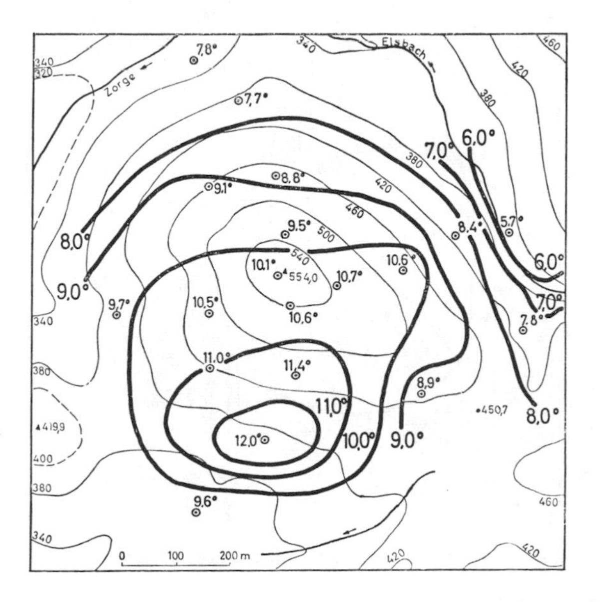

Fig. 232. Minimum temperatures for the same time as in Fig. 231.

Night minimum temperatures during the summer period (Fig. 232) are mainly determined by the flow of cold air from the north out of the Zorge Valley; this is dammed up by the Staufenberg and produces the lowest temperatures in the Elsbach Valley to the NE. The cold air is also able to penetrate into the closed stand. Its encroachment around the hillside makes the lower part of the southern slope the warmest part, in other seasons as well.

Although these comprehensive measurements are still insufficient to delimit microclimatic zones quantitatively, the authors have attempted, in a way that is worthy of imitation, to produce a chart of practical application although qualitative in nature. This chart is reproduced in Fig. 233. Observed values of incoming daytime radiation S, temperature T, and relative humidity F are divided into five steps, where 1 indicates the highest, and 5 the lowest, radiation, temperature, and humidity. The distribution thus arrived at was found to be most closely related to the type of stand. Only an indication can be given here of the rich supply of information available in the original work.

There are a number of other studies, which were not carried out in such a systematic way on a specially selected and most suitable area, but which nevertheless serve to resolve many problems of slope climate. A few of these, of interest to microclimatologists, are indicated in the following list.

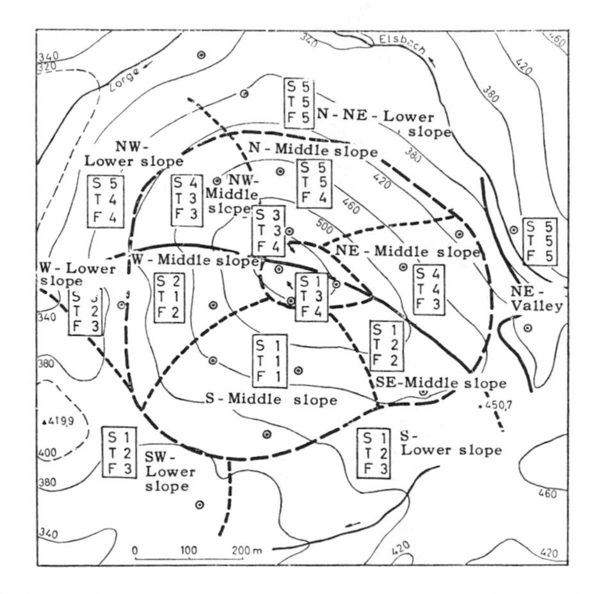

Fig. 233. Distribution of microclimate on the Staufenberg (explanation in text).

R. Wagner [874] set up ten observation stations along a N-S line across the Hosszubérc, which is a ridge with a relative height of 80 m, sloping 18° to 20°, running E-W, in the Bükk Mountains (48°N) in Hungary. He divided the area into local climate zones, but only observations made by random sampling have been published.

J. R. Held [861] compared the shelter records made on the N and S slopes of a long Alpine valley in Pinzgau, at a height of 1300 m above sea level. Differences were negligible during bad weather, amounted to 0.3 to 1.1 deg and 3 to 8 percent on sunny days, and reached a maximum (up to 3.8 deg and 26 percent) on foehn days. Comparison was made more difficult by the different locations of the stations: the cool and moist N station stood on a thickly wooded 30° slope, while the S station was in a meadow sloping at 20°, both on different geologic substrata. The daily average temperature was only 0.3 deg warmer on the sunny side. The differences must have been considerably greater near the ground.

I know only from references of the investigation made by J. E. Cantlon [853] into ground and air temperatures and humidity at heights of 5, 20, 100, and 200 cm at two stations each on the N and S slopes of Cucketunk Mountain in New Jersey, during three seasons.

In fine weather there was a marked decrease of temperature with height on the S slope, whereas there was always an increase over the N slope. Humidity gradients were also investigated and the plant communities on the two slopes were compared. R. E. Shanks and F. H. Norris [871] investigated the danger of late frosts on the N and S slopes of a valley running E-W near Knoxville, Tennessee, by recording extreme temperatures at a height of 30 cm at 14 stations. The S slope had the higher maxima, the N slope the lower minima; the daily average was 3 F deg warmer on the S slope, the minima only 1 to 2 F deg higher. The periods during which temperatures were below the freezing point showed a better correlation with observed frost damage than did minimum temperatures.

The W and E slopes of a valley in Cumberland, England, which ran N to S were investigated by W. E. Richardson [868]. The frequency of various large differences between shelter maxima and minima were counted for a series of observations extending over almost 2 yr, and their dependence on the characteristics of the air mass was established.

The great differences in soil moisture on different slopes is well known, but few numerical values have been published. Some figures for Idaho are to be found in the paper by J. Parker already referred to [867]. A German example is shown in Fig. 234, which

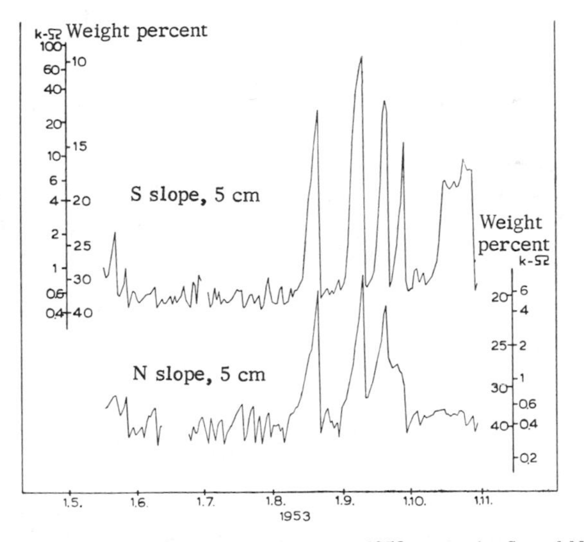

Fig. 234. Soil-moisture variation in summer 1953 on the S and N slopes of the Hohenpeissenberg. (After K. Heigel)

gives the change in soil moisture from May to October 1953 on the 20° S slope, and the N slope of the Hohenpeissenberg, measured at a depth of 5 cm below the grass-covered surface (after K. Heigel [860]). The kΩ scale gives the electrical resistance of the plaster block that is used to measure soil moisture and is dependent on its accidental properties. Comparison of the slopes is made possible by the two scales showing soil moisture in weight percent. The N slope is moister throughout; in the period of fair weather in the autumn, 20 percent is reached at the time of strongest drying, whereas a decrease to 10 percent is found on the S slope.

Heigel also investigated [858] the dependence of phenologic processes on exposure and height, on the slopes of the Hohenpeissenberg. Blossoms of the sweet cherry suffered a delay of 2 days for every 100 m of height in 1951; the difference between N and S slopes amounted to 5 to 7 days. The harvest of winter rye, on the other hand, responded primarily to height and was delayed when scattered widely (type of soil!) by 7 days per 100 m on the average. The dandelion (*Taraxacum officinale*) had an extremely sensitive reaction to sunshine duration on the slope, in the opening of its flowers. Comparing plants on the N and S slopes only, with equal gradients of 18° to 20°, the following dates for flowering were found in 1954:

Elevation (m):	820	860	900	940
South slope:	25 April	27 April	2 May	11 May
North slope:	9 May	17 May	(no vegetation)	

In 1953 and 1954 K. Kreeb [864] observed the development of 19 plant species in the grass cover of an oak and hornbeam forest on the SSW and NNE slopes of the Körschtal near Plochingen, Neckar. Because weather conditions differed in these two years, the unfolding of leaves and blossoming of each plant species occurred at different dates, but the time sequence was the same. The longitudinal growth of leaves was also measured on both slopes. This was, for example, on 19 March 1954, for *Milium effusum*, 8 cm on the SSW slope against 5 cm on the NNE, while on 26 March the lengths were 12 compared with 5 cm, and 17 and 10 cm on 9 April.

45. Mountain, Valley, and Slope

Section 44 was mainly concerned with the influence of the direction of a slope on the local climate; the effect of elevation or relative height will now be considered.

The best point of departure is the distribution of night temperatures. Let us consider a valley cutting into a high plateau, as shown in cross section in the sketches in Fig. 235. If the rules for the movement of cold air given in Sec. 42 applied here, the valley would contain an enormous lake of cold air, as shown in the top sketches. The effect of the greater dimensions here is that individual circulations are built up between the air that is cooling on the slope and the reservoir of warmer air above the valley floor, as shown in the lower sketches in Fig. 235. These circulations were observed directly and measured photogrammetrically by F. W. Nitze [827], using balloons fitted with small lamps. A lake of cold air therefore develops only near the bottom of the valley. Since a layer of cold air near the ground remains over the plateau above, an intermediate zone, known as the thermal belt, develops on the slope, where temperatures are higher at night. This corresponds to the night inversion above level ground.

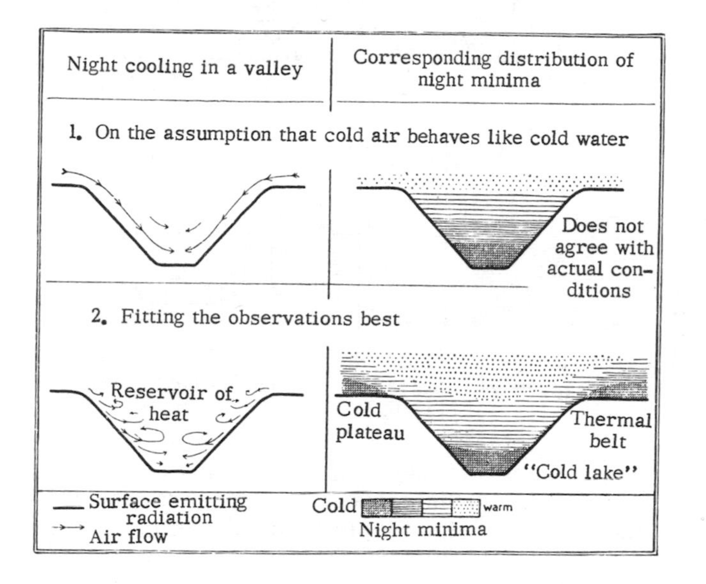

Fig. 235. Development of the thermal belt.

This vertical division into three zones can be recognized from temperatures recorded by F. D. Young [892] at five different heights, during the night of 27-28 December 1918 on the slopes of the San José Mountains in the Pomona Valley, California. The lowest curves for 0 and 8 m in Fig. 236 show the freezing temperatures in the lake of cold air at the bottom of the valley; they run almost horizontal in the undisturbed period before sunrise. At 15 m the warm thermal belt is being approached; the temperature distribution is much more varied. The warmest zone is at 68 m, above which (84 m) temperatures decrease again.

During the day, a division into three zones is maintained, but the temperature pattern is different. These laws of temperature distribution were the subject of a study by R. Geiger, M. Woelfle, and L. P. Seip [883] on the Grosse Arber (1447 m) in the Bavarian Forest in 1931 and 1932. A few years later (1935-1938), G. L. Hayes [885] set up four pairs of stations on an E-W ridge in the Priest River Experimental Forest in the Rocky Mountains. These were in the valley (700 m altitude) inside and outside the forest, on the hillside at 820 and 1160 m, and on the ridge at 1676 m, one station with a north and one with a south exposure. This investigation

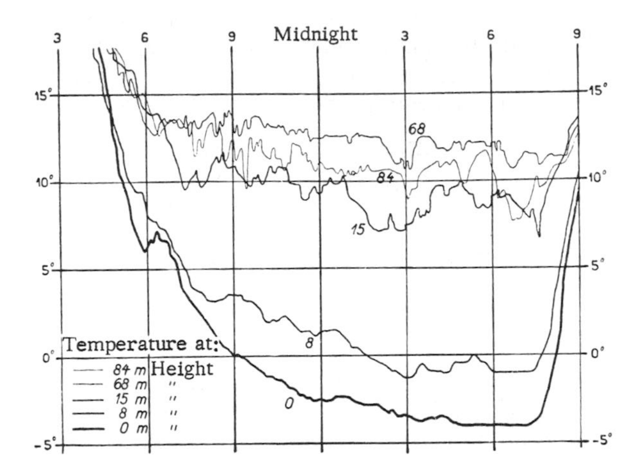

Fig. 236. Night temperature variation at five different heights in a valley in California. (After F. D. Young)

was designed primarily to determine the risk of forest fires as a function of position on the slope and exposure, but it contains a wealth of diagrams showing isopleths of air temperature, relative humidity, and wind speed for an average August day, as a function of height above sea level, and also shows the differences of these elements measured on the N and S slopes. Another investigation was carried out in the Bavarian Forest, on the Grosse Falkenstein (1308 m), by A. Baumgartner, G. Hofmann, G. Kleinlein, and G. Waldmann [876-879, 891, 616]. Also worth consulting is the series of measurements made by S. Morawetz [889] in the summer of 1951 at four stations differing in height by only 65 m, in West Steiermark, Austria, since they give a good picture of the slope zones and sites in the valley.

Figure 237 shows the arrangement of observation points in the investigation carried out on the Grosse Arber. Stations with meteorologic shelters were set up in the valley in the SW (Bodenmais, 665 m) and in the E (Seebachschleife, 645 m), on the hillsides at Kopfhäng (1008 m) in the SW, on a level area in the N at Mooshütten (946 m), and also on the peak (1447 m). Along the line of crosses 99 measuring points were set up for taking readings of night temperatures near the ground.

Figures 238 and 239 give the averages of the shelter readings for 25 clear days in May and June in the form of curves showing the diurnal air-temperature and relative-humidity variations. Since

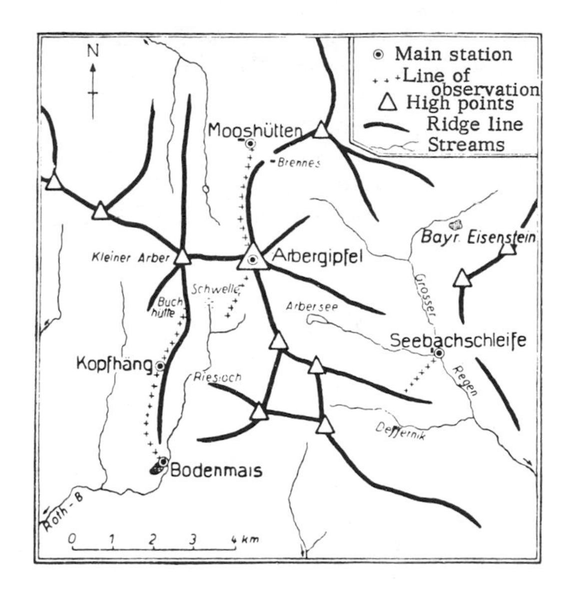

Fig. 237. Experimental setup on the Grosse Arber, 1931-1932.

the weather was fair and the year advancing, it was warmer at midnight than it was 24 hr earlier. The valley site, which is coldest and moistest at night, is warmest and driest during the day. The

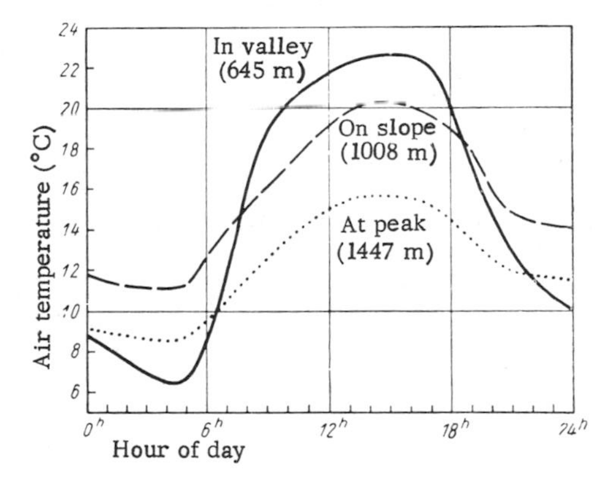

Fig. 238. Diurnal air-temperature variation on bright spring days on the Grosse Arber.

bottoms of valleys therefore enjoy a "continental climate." The hillside station is warmest at night, its daytime features depend on its relative height, and its relative humidity is about average. While the air enclosed in the valley undergoes a steep rise in temperature in the forenoon, the rate of temperature increase with time does not exceed a certain value at the station on the SW slope, since the upslope wind will otherwise be strengthened and force the establishment of an equilibrium. This is what causes the curious appearance of the lines during the forenoon temperature increase. The peak has the smallest diurnal variation; its night temperature is exceeded by that of the valley; it is driest at night and moistest during the warm part of the day, even though on these selected days there was none of the condensation frequently found on mountain tops.

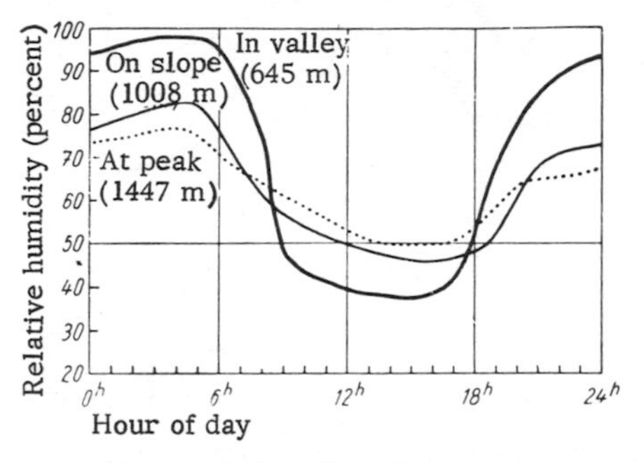

Fig. 239. Diurnal air-humidity variation for the same time and place as Fig. 238

The field of investigation on the Grosse Falkenstein in the Bavarian Forest is shown by the contour map in Fig. 243. The line of measurement followed a straight clear path on the WSW slope which was open enough for the instruments to be set up outside the stand, and yet narrow enough to inhibit the slope winds. Figure 240 shows the instrumental equipment. There were eight stations

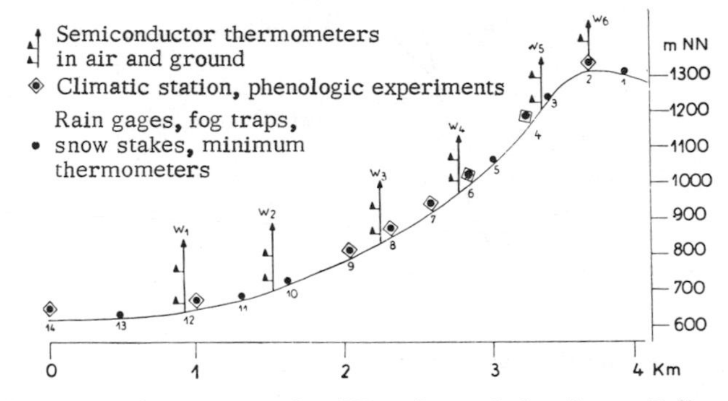

Fig. 240. Experimental setup on the WSW slope of the Grosse Falkenstein in 1955. (After A. Baumgartner and G. Hofmann)

with shelters, near which were small experimental phenologic plots, arranged down the hillside. Between these were points at which precipitation from clouds, precipitation from fog, depth of snow, and night minimum temperatures at two heights above the ground were measured. In the stand nearby semiconductor thermistors were set up in shelters to protect them from radiation, at six levels, on the forest floor, in the lower trunk area, in the crown area, and at the ends of long poles several meters above the tree tops (W_1 to W_6). All these electrical thermometers could be read from a central station on the peak. A report has been made by A. Baumgartner and G. Hofmann [616; see also Sec. 56] on this valuable method of making microclimatic measurements, which can be extended to include humidity and wind.

Table 90

Daily variation in meteorologic shelters on the west slope of the Grosse Falkenstein, in May 1955. (After A. Baumgartner)

Station No. (Fig. 240)	Elevation (m)	Hour of the day												Daily average
		2	4	6	8	10	12	14	16	18	20	22	24	
		Air temperature (°C)												
2	1,307	2.9	2.7	2.7	*3.6*	*5.2*	*6.6*	*7.0*	*6.5*	*5.2*	*4.0*	*3.4*	3.0	*4.4*
4	1,157	4.2	3.8	3.6	4.2	5.8	7.4	8.1	8.1	7.3	5.6	4.7	4.3	5.6
7	925	5.8	5.5	5.1	5.9	8.1	9.7	10.5	10.4	9.3	7.3	6.3	5.7	7.4
9	796	6.4	6.0	5.9	6.9	9.9	11.4	12.0	12.0	11.0	8.8	7.2	6.5	8.6
12	658	3.8	3.4	3.6	6.6	10.2	11.9	12.3	12.0	10.6	8.4	5.8	4.4	7.7
14	622	*1.9*	*1.5*	*2.4*	8.0	11.0	12.6	13.2	12.9	10.9	6.8	3.8	*2.4*	7.8
		Relative humidity (percent)												Water-vapor pressure (mm-Hg)
2	1,307	89	89	90	86	82	75	75	77	82	86	88	89	5.3
4	1,157	88	88	89	86	81	73	72	73	75	82	86	88	5.6
7	925	86	86	88	83	70	60	60	62	68	79	84	86	6.0
9	796	91	92	92	80	66	56	58	60	66	80	88	91	6.5
12	658	97	97	96	88	68	58	59	63	72	85	95	97	*6.6*
14	622	97	98	97	81	64	55	56	58	66	88	95	97	6.2

Table 90 gives the diurnal variation of air temperature and relative humidity and the mean water-vapor pressure (from the 07:00, 14:00, and 21:00 observations) for the month of May 1955 for six selected shelter stations, indicated in Fig. 240 by the numbers 2 (1307 m at the peak) to 14 (622 m in the valley). Heavy type is used to make the warmest and relatively driest zones of the vertical profile stand out clearly, and italic type to indicate the coldest and most humid zones. These results, published by A. Baumgartner [878], can be used to study the principal features of the transition of slope climate from the bottom of the valley to the peak.

The triple division of slope climate previously described is most conspicuous in situations with strong radiation; where there is precipitation or strong wind the normal temperature decrease is found as in the free atmosphere. In Fig. 241 the observations made on the Grosse Arber have been arranged according to the three most frequent air masses, for the warmest midday hours and the coldest hours at night. At midday the temperature always decreases with height; at night only with polar maritime air (*mP*), that is, when the temperature is comparatively low, precipitation is frequent, and winds are brisk. With a continental air mass (*c*), however, always associated in spring and summer with high temperatures and slack winds, an inversion and a warm thermal belt develop. This gives different temperatures on the W and E slopes of the Arber, but the temperature variation with height is similar. Although the temperature at the peak is 8 deg higher, the valley below the W slope is colder at night with *mP* than with *c* air. The regularity of the processes is made clear by the fact that the maritime air mass (*m*), shown between these other two also plays a role in the formation of inversions.

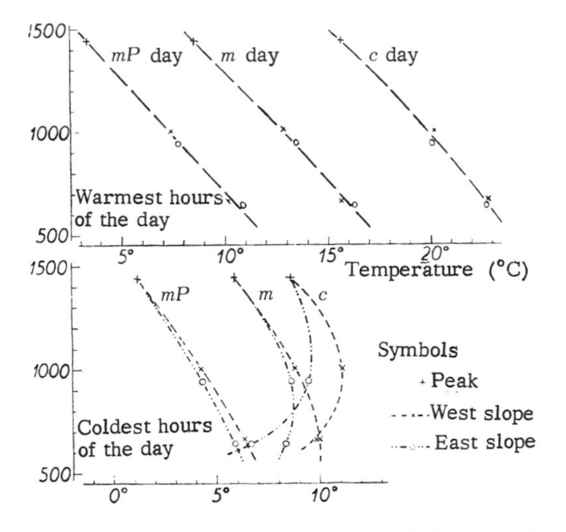

Fig. 241. Vertical temperature gradients on the Grosse Arber in spring as a function of time of day and air mass.

During the course of any individual night the warm thermal belt does not always remain at the same elevation on the hillside. In Japan H. Mano [888] established, by experiments on the slopes of the Bandai Mountains (1819 m) above Lake Inawashiro (514 m) in the autumn of 1954, that there were regular night-temperature fluctuations at a station set up at 830 m, which is the normal area of the thermal belt. These variations were of the order of 1 deg, and there was a time interval of 2 hr from one extreme to the other. By measuring the temperature and wind field in the free space above the mountain side up to a height of 1800 m, the cause of the fluctuation was established as a displacement of the boundary layer between the lower wind close to the hillside and the upper gradient wind. These regular displacements were initiated partly by thermal and partly by dynamic processes, and had the effect that the station at 830 m was sometimes inside and sometimes above the inversion layer (compare with the variation of the boundary surface mentioned on p. 400).

In spite of the many influences affecting the location of the thermal belt on a hillside, its position on any one hillside is remarkably constant when established by statistical averaging, a fact that lends it great practical importance. Figure 242 shows a side view of the section marked on Fig. 237 by a line of crosses running SE from Seebachschleife. On it, the positions of the 23 measurement points are shown by short vertical lines. Night minimum temperatures observed in the springs of 1931 and 1932 were used to establish the height above sea level of the highest temperature for each individual night. The frequency distribution thus arrived at is given on the right of Fig. 242, showing that there is a distinct preference for a height of 800 m. The second weak maximum at the bottom of the valley results from bad-weather situations in which temperature decreases regularly with height. This is the situation in spring, and from the Falkenstein results it should also apply to autumn. In summer, by contrast, the warm thermal belt usually occupies a lower position, according to A. Baumgartner [878].

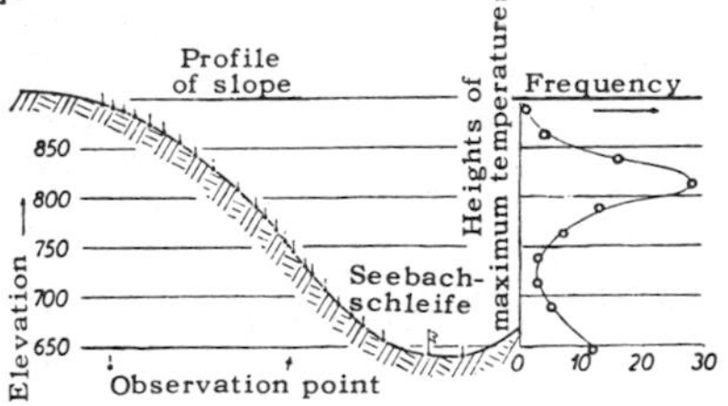

Fig. 242. Location of the warm zone on the slopes of the Grosse Arber in spring.

It is not surprising that the existence of such a warm belt was known from early times, long before there was any scientific knowledge of climatology. In Germany this area was preferred for the earliest villages, monasteries, and country houses. E. Bylund and A. Sundborg [880] have given some good examples from Swedish Lapland, where places 2 to 3 km apart horizontally and with a height difference of less than 100 m may show a temperature difference of as much as 8 to 9 deg.

The diagram in Fig. 243, taken from the Falkenstein experiment, shows the distribution of melting snow on 19 April 1955. The points at which snow-depth measurements were made were selected so as to be comparable and unaffected by the woods, in order to make clear the effect of position on the slope, the orientation of the slope, and terrain configuration. By making a comparison with the height contours it can be seen that the warm thermal zone at about 700 m is already clear of snow. Below this level snow depth increases, and above it increases even more, naturally, since winter accumulation was somewhat greater there. G. Waldmann [891] has published a number of such maps. This also provides an example of the close relation between the microclimate of an area and the snow-melt pattern, mentioned in Sec. 24 on p. 220.

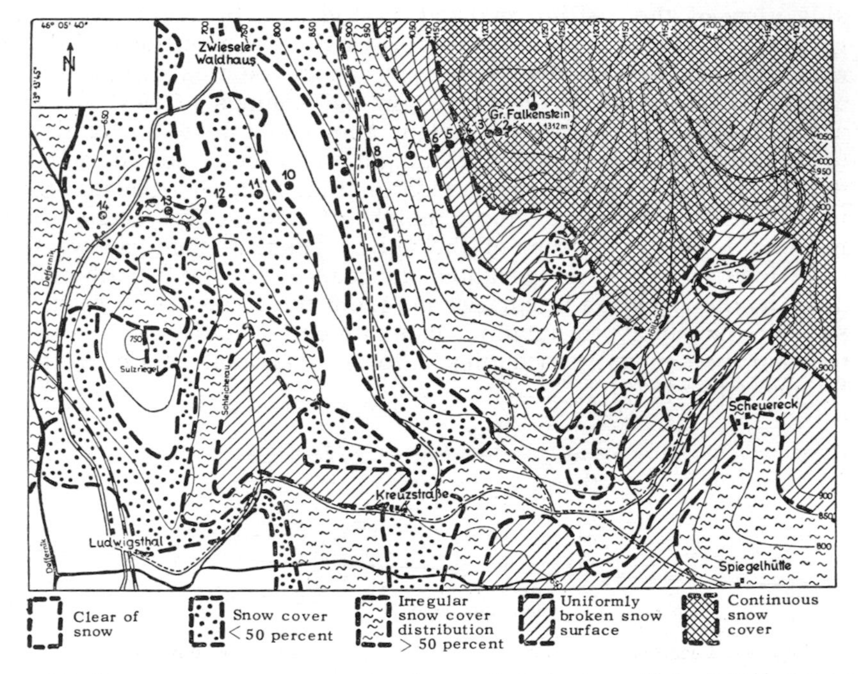

Fig. 243. Map of snow cover on the Grosse Falkenstein on 19 April 1955, showing early melting on the warm slope zone. (After G. Waldmann)

The development of plants also reflects these features. F. Schnelle [18] made use of German and British observations in his *Plant Phenology* to draw attention to the fact that vegetation first makes its appearance up to 20 days in advance, and trees blossom a few days earlier, on the thermal belt than in the valley 200 m below. From observations also made by F. Schnelle [890] in the German phenologic network from 1936 to 1939 the ears of winter rye first appeared:

At height of	150	200	300	500	700	1000 m
on	17	16	15	22	28 May	7 June

The most favorable position in this case was therefore 300 m. The thermal belt was found by B. Frenzel and H. Fischer [881] in their comprehensive phenologic observations to be at 1100 m above sea level in the Allgäu Alps.

The curve on the left of Fig. 244 gives the mean night minima over a 68-night period in May and June 1931 and 1932, for the area above the Seebachschleife (see Figs. 237 and 242). Although the period covers all kinds of weather, the warm thermal belt is easily recognized. On the right are three phenologic graphs, for which the time scale runs from right to left to make comparison easier. The similarity between the phenologic and the temperature curves is then quite evident.

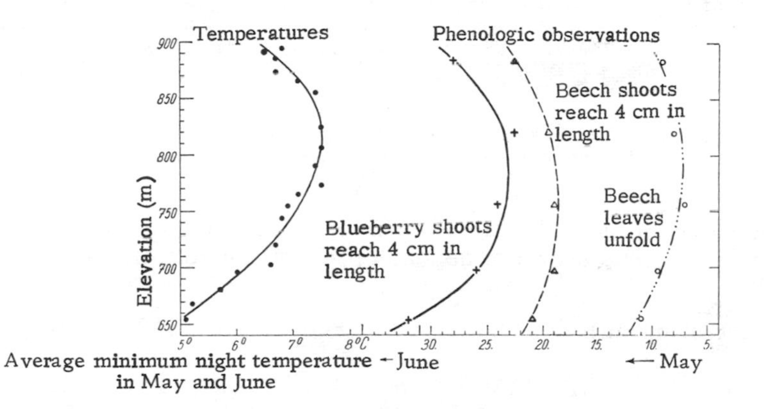

Fig. 244. The close relation between night temperatures (left) and plant growth on the Grosse Arber.

Figure 245 shows the development of a phenologic phase over a period of time on the Falkenstein (compare Fig. 243), similar to that shown in vertical section in Fig. 244. The first signs of green on the red beech (*Fagus silvatica*) appeared in the thermal belt before 1 May 1955. From that position the green area spread upward and downward and did not reach positions in the valley until 9 days later. A series of similar phenologic charts were published by A. Baumgartner, G. Kleinlein, and G. Waldmann [879], in which four species of trees and ground flora were observed continuously. The duration of the growing period in days in 1955, measured from the time when more than half of all twigs bore

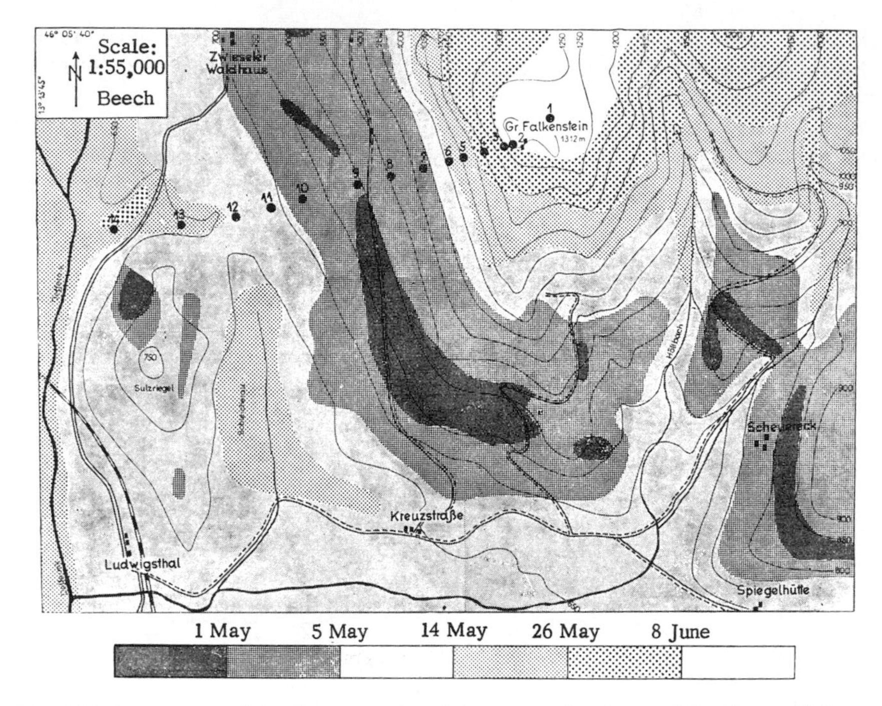

Fig. 245. Appearance of the first green beech leaves on the slopes of the Grosse Falkenstein. (From A. Baumgartner, G. Kleinlein, and G. Waldmann)

open leaves until the time when one tree had lost more than half its leaves, was shown to depend on elevation as follows:

Elevation (m):	620	700	800	900	1000	1100	1200
For beech (days):	139	157	*163*	161	135	127	121
For maple (days):	137	*146*	*146*	140	134	129	126

The method recommended by J. J. Higgins [886], of determining the state of phenologic development experimentally on garden peas (***Pisum sitavum***), by counting the nodes and measuring the length of the internodes, was used successfully on the phenologic experimental plot on the Falkenstein, already mentioned. Five sowings, between 12 May and 30 August, gave a maximum development on hillside positions from 800 to 850 m (thermal belt!) for the sowing on 18 June, that is, when the young plants were able to develop during both June and July.

In contrast to the warm belt on the hillside, valley sites are cold by night, and therefore exposed to frost danger; the air there is also still, hence hot and dust laden, and fogs are frequent.

The enrichment in dust content of the valley has been measured many times. From the point of view of industrial air pollution, attention may be directed to a few cases where it reached catastrophic proportions. The general weather situation in the early December days of 1930 favored the stagnation of air and the development of fog in the narrow valley of the Meuse near Liège, to such an extent that gases rich in fluorine discharged by the zinc and superphosphate factories in the area accumulated in a high concentration. Hundreds became ill with respiratory troubles and more than 60 died. That this was only an aggravation of a normal micrometeorologic state is shown by the fact that severe damage had already been caused in the same area in 1911. In Pennsylvania, smoke piling up under an inversion caused 20 deaths in the Donora area, and hundreds had to be evacuated, suffering from heart and asthma troubles. H. Koschmieder [887] therefore very correctly advised that positions such as these should be avoided by atomic power stations, and that they should be sited on higher areas, as suggested by M. Manig.

Fog is also a characteristic feature of valleys. For example, it was established by J. Grunow [884] that in the period 1950-1956 the Ammer basin had 200 hr less sunshine per year than the Hohenpeissenberg, and in December alone the difference was 32 hr. Above the warm thermal belt, however, fog is often the result of drifting clouds. Figure 246, after A. Baumgartner (877], shows the change in fog frequency with height on the western slopes of the

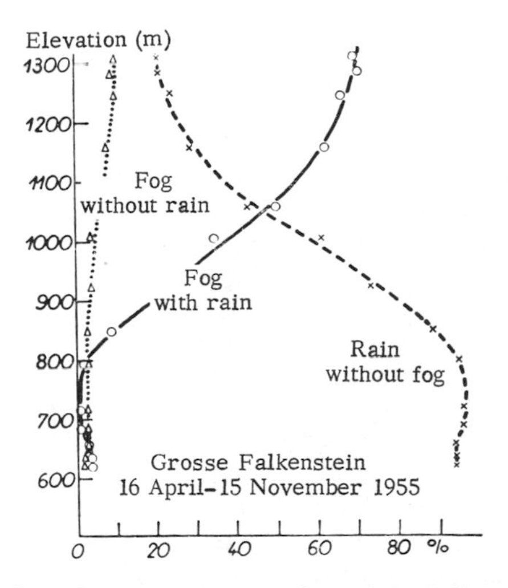

Fig. 246. Fog as a habitat factor on the west slope of the Falkenstein. (After A. Baumgartner)

Grosse Falkenstein. Three groups of days are differentiated in the period from 16 April to 15 November 1955, their totals adding up to 100 percent for each height. The fog days in which no rain falls (dotted curve) are rare, and increase slightly with height. When it is raining in the valley, the rain is nearly always associated with good visibility (rain without fog). The higher one goes up the slope, the more frequently is an association of rain and cloud encountered (fog and rain). Measurements of precipitation from fog, using Grunow's catchment nets [634], gave, in comparison with readings of ordinary rain gages, a supplement from about 1000 m upward, exceeding 100 percent at some places on the peak, and therefore doubling the amount of precipitation available to vegetation (see Sec. 37). This caused substantial differences in local conditions and provided an explanation for the moorland vegetation on the peak, which requires a high air humidity. Because of the limited extent of such areas in the peak region, the fog supplement was of no importance to the water economy.

46. Microclimate in the High Mountains

When a height of 1000 m is exceeded to a marked extent under European climatic conditions, the climate near the ground acquires new basic characteristics, which become increasingly conspicuous at high elevations. These special characteristics are not only of scientific interest, but also acquire increasing practical importance.

Increasing population pressure and more intensive economic exploitation have caused far-reaching and serious damage in the high mountains. Trees are particularly susceptible at the tree line, where they grow only slowly and are subjected to more severe and more frequent damage by weather if they are not handled and exploited properly. Under the influence of man, therefore, wasteland increases, large areas of forest are lost, and the increase in the number of catastrophes due to avalanches and flash floods, as well as the decrease in agricultural production in high elevations, can also be largely attributed to this influence. Apart from botanical and forestry problems the border areas can be reclaimed for the forest only if the local microclimatic conditions are fully understood, for it is on these that the imported young plants have to depend for growth. A large-scale investigation has been started in the Ötz Valley near Obergurgl (47°N, 11°E, 1940 m) by the Innsbruck office of the Austrian Forestry Section for flash-flood and avalanche control, under the supervision of R. Hampel, in preparation for extensive reforestation. Much more has been learned about local conditions at the timber line than heretofore, from the first results published, especially from H. Aulitzky [894-900].

The construction of new roads is being progressively extended to higher regions to meet the demands of tourist traffic. Visitors want to be able to admire wild mountain scenery without trouble, from good highways. The construction and maintenance of such roads is completely dependent on the climate near the ground, as is also the possibility of using them during glaze, snow, and unusual amounts of rain.

It is possible to understand the particular features of high-mountain microclimate properly only if one bears in mind at the same time the rapid change in the large-scale climate with elevation. No other mountain range in Europe has been so thoroughly investigated from the plains to the summits as the Eastern Alps have been by Austrian meteorologists. This mountain area has therefore been selected to provide data for Table 91, although the Swiss and German Alps are not substantially different. This table is a selection of data from various sources which will be referred to later. The figures given are of average conditions, and may at times lack precision because of interpolation. Therefore, these macroclimatic values should not be taken without further reflection as valid for any particular place at a given elevation. Nevertheless, they provide us with a good picture of mountain climate.

Direct solar radiation increases with height above sea level since the atmosphere, with its turbidity, scattering, and absorbing properties, has decreased in mass above the level in question. Diffuse sky radiation in cloudless conditions weakens with height, because the density of the air producing the scattering effect is less. In overcast conditions this increases with height, however, and the type of cloud is of importance. More details can be found in F. Sauberer and I. Dirmhirn [902, 916]. The table gives the daily total global radiation on a horizontal surface, from F. Sauberer, for an average day in the months richest and poorest in radiation and for the extreme cases of cloudless and overcast skies. From 200 to 3000 m above sea level, incoming radiation increases by 21 percent with clear sky, and by 160 percent with overcast skies, or by 1 percent and 4 percent, respectively, per 100 m. Short-wavelength radiation therefore increases substantially with height, radiation from clouds providing the main share.

The long-wavelength radiation budget, on the other hand, is virtually independent of height; this is because decreasing temperature reduces the amount of outgoing radiation from the ground, and the counterradiation from the atmosphere is also reduced because of its decreasing thickness. The gain in short-wavelength radiation is therefore retained, in the radiation term S of the heat balance.

Attention must be paid to the particularly high solar-radiation values to be found at high elevations under partly cloudy conditions, when the sun is shining through gaps in the clouds and there is added the amount due to cloud reflection. H. Turner [922], who made a detailed study of radiation at the Obergurgl station (1940 m) measured instantaneous values of as much as 2.25 cal cm^{-2} min^{-1},

Table 91
Changes in climatic conditions with height above sea level in the Eastern Alps (for basis, see text).

Elevation (m)	Mean daily global-radiation totals (cal $cm^{-2}d^{-1}$)				Mean air temperature (°C)				Annual number of—			
	Cloudless		Overcast									
	June	December	June	December	January	July	Year	Annual variation	Summer days	Frost-free days	Days of frost change	Days with frost
1	2	3	4	5	6	7	8	9	10	11	12	13
200	691	130	155	30	-1.4	19.5	9.0	20.9	48	272	67	93
400	708	136	168	32	-2.5	18.3	8.0	20.8	42	267	97	98
600	723	141	180	34	-3.5	17.1	7.1	20.6	37	250	78	115
800	735	146	192	36	-3.9	16.0	6.4	19.9	31	234	91	131
1,000	747	150	205	38	-3.9	14.8	5.7	18.7	15	226	86	139
1,200	759	154	220	40	-3.9	13.6	4.9	17.5	11	218	84	147
1,400	771	157	236	43	-4.1	12.4	4.0	16.5	7	211	81	154
1,600	782	160	253	47	-4.9	11.2	2.8	16.1	4	203	78	162
1,800	791	163	272	50	-6.1	9.9	1.6	16.0	2	190	76	175
2,000	799	166	293	54	-7.1	8.7	0.4	15.8	0	178	73	187
2,200	807	168	314	58	-8.2	7.2	-0.8	15.4	0	163	71	202
2,400	814	169	336	62	-9.2	5.9	-2.0	15.1	0	146	68	219
2,600	821	170	358	66	-10.3	4.6	-3.3	14.9	0	125	66	240
2,800	828	171	380	70	-11.3	3.2	-4.5	14.5	0	101	64	264
3,000	834	171	403	75	-12.4	1.8	-5.7	14.2	0	71	62	294

Table 91 (continued)

Elevation (m)	Annual number of days with—		Relative humidity (percent)	Annual precipitation (mm)	Relative snow frequency (percent)		Number of days with snowfall	Average quantity (cm d^{-1})	Total of new snow depths (cm)	Maximum depth of snow	
	Dry ground	Snow cover			Summer	Winter				Depth (cm)	Beginning on—
1	14	15	16	17	18	19	20	21	22	23	24
200	187	38	71	615	0	49	27	4.6	51	20	18 January
400	173	55	74	750	0	61	32	5.2	116	31	23 January
600	160	81	77	885	0	70	38	5.8	182	51	28 January
800	147	109	78	1,025	0	79	45	6.4	247	73	3 February
1,000	133	127	76	1,160	0	85	53	7.0	313	93	11 February
1,200	120	138	74	1,295	1	90	62	7.6	379	100	14 February
1,400	107	152	73	1,430	2	93	73	8.2	444	120	21 February
1,600	93	169	73	1,570	5	96	85	8.8	510	142	3 March
1,800	80	189	74	1,700	10	97	98	9.4	575	168	14 March
2,000	67	212	74	1,835	16	98	113	10.0	641	199	26 March
2,200	53	239	75	1,970	24	99	128	—	707	242	8 April
2,400	40	270	78	—	34	100	143	—	—	296	20 April
2,600	27	301	80	—	44	100	158	—	—	366	3 May
2,800	13	332	82	—	55	100	173	—	—	446	15 May
3,000	0	354	84	—	67	100	188	—	—	545	29 May

which is 112 percent of the solar constant. Figure 137 gave a similar example from a height of 2720 m in the mountains of Japan. In the presence of drifting clouds, therefore, radiation is subject to very great fluctuations with time. Turner observed variations in global radiation up to 7 times its value in 1 min, 11 times in 9 min, and as much as 15 times its value in 11 min, during the series of observations he made in the summer of 1953. Vegetation at high elevations therefore must be able to cope with these sharp variations. It is not surprising that Turner [923] was able to record changes of surface temperature of 10 deg within a few seconds, using very fine thermocouples.

To these temporal fluctuations are added local differences. At high elevations, the horizon is often screened by mountains which reflect strongly when covered with snow or glaciers, or the site in question may look down a valley. H. Turner [922] calculated for Obergurgl that there was a 10-percent loss of global radiation in cloudless weather, through restriction of the horizon, with the sun at its highest, increasing to 60 percent when the sun was at its lowest (December). This loss caused by the screening of the horizon offsets the gain resulting from higher elevations in cloudless weather. Only in the months of May and June was there a final gain.

The differences in slope exposure, described in Secs. 40 and 44, also gain in importance with elevation. On the S and N slopes of the Alps, the higher the elevation the more is plant life subjected to extreme radiation conditions, hence also local peculiarities. When the snows first melt in the spring, and when the first flowers appear, this effect can often be most impressive (see the fine photograph from Arosa in W. G. Kendrew's *Climatology*, Oxford, 1949, p. 280).

Stronger radiation produces higher surface temperatures. As long ago as 1900, J. Schubert [726] showed by means of paired observation points in forests (see p. 366) that the excess of soil temperature at a depth of 60 cm compared with air temperature at a height of 2 m was 0.75 deg on the plain, and 2 deg for stations at 1000 m. J. Maurer [914] found that the excess of soil temperature at a depth of 1.2 m over air temperature in Switzerland was:

Elevation (m):	600	1200	1800	2400	3000
Excess (deg):	0.5	1.3	2.0	2.5	2.9

If this shows up in climatologic statistics, it will be much more evident in the measurements made by modern techniques in microclimatology.

At a height of 2070 m in the Ötztal, at the tree line, H. Turner [923] was able to measure absolute surface temperatures that were substantially in excess of the highest known values for the European plain, in spite of the initially lower air temperature at this height (see Sec. 20). It is possible to understand the value of 80°C (highest estimated value, 84°C) measured by reliable methods several times in the hot July of 1957, if we consider the stronger radiation, the SW exposure and slope of 35°, a low albedo of 9 percent, and the very poor conductivity of a dark raw humus without vegetation (barren from excessive heating). In comparison with air temperature at 2 m, this is an excess of 50 deg; at the same time, the surface temperature on the NE slope was 57 deg lower. These hot, bare patches of ground can be repopulated only by plants that have an unusually high resistance to heat, in which leaf temperatures up to 44°C have been measured, and that roll up their limp leaves at times, or turn the light-colored lower side upward.

Average soil-temperature values taken from Turner's 4-yr series of measurements are of equal practical interest. Table 92 gives figures for the relatively coldest July 1954 and the relatively warmest July 1957, for a depth of 1 cm on a horizontal cleared area, where measurement is much simpler.

Table 92

Soil temperatures (°C) for a cold and a warm July.

Quantity	July 1954	July 1957
Monthly average	10.6	14.8
Warmest daily average	19.6	29.4
Absolute maximum	37.2	59.0
Absolute minimum	0.0	0.0
Average daily variation	21.0	27.4
Greatest daily variation	34.5	48.0

In September 1951, I. Dirmhirn [903] measured the surface temperatures of a gneiss slab 0.25 m^2 in area and 6 cm thick with an albedo of 33 percent, lying horizontally at 3050 m on the Hohen Sonnblick. They reached +29°C during the day and -4°C at night. The diurnal temperature variation in the surface of the stone, compared with that of the air, on days of fine weather, showed the following excesses, arranged according to wind speed:

Wind speed (km h^{-1}): (m sec^{-1}):	50 14	40 11	30 8	20 6	10 3
Excess (deg):	16.4	17.2	18.5	20.6	24.1

On clear September days (0-0.2 cloud cover) the average temperatures during the day were:

Hour of day:	2	4	6	8	10	12	14	16	18	20	22
In the air (shelter) (°C)	2.4	2.7	3.2	4.3	4.8	5.2	5.7	5.4	4.7	4.0	3.6
On the gneiss surface (°C)	-0.8	-1.0	0.0	9.0	19.6	24.0	22.7	13.0	6.6	0.0	-0.6

The neighboring rock was heated to a smaller extent because some of the heat was conducted away to lower layers. Very thin laminae of rock are therefore subjected to a high degree of weathering. Further data are available from H. Turner and I. Dirmhirn, and from H. Aulitzky [897].

In contrast, air (shelter) temperature decreases with height. Figures from F. Lauscher [910, 911] are given in columns 6 to 13 of Table 91. The winter temperature inversion (thermal belt of Sec. 45) can be recognized from the fact that the temperature ceases to decrease between 800 and 1200 m elevation in the January figures. It is evident, too, in the somewhat higher value for relative humidity at 800 m, column 16, from F. Lauscher [912], contains average daily relative humidity for each month. The great importance of this inversion layer in mountain climate has been pointed out by F. Steinhauser [920].

While air temperatures therefore decrease with height, and their annual variation becomes more maritime in character (column 9), the climate near the ground becomes more extreme and more continental in its characteristics under the influence of increased radiation. Measurements made by T. Asai at the summit of the Wutaischan in China, and reported by M. Schwind [918], show that the temperature gradient in the layer near the ground is substantially greater at a height of 2670 m than at 980 m at the foot of the mountain.

As a rule, wind speed increases with height. The layer near the ground in which extreme conditions develop will therefore, on the average, have a smaller vertical extent than over level ground. At the same time, however, local radiation conditions vary widely under the influence of slope exposure, and ventilation is also extraordinarily varied, since it depends on local topography and wind direction at the moment. There are always windswept ridges where there can hardly be any question of the development of surface climate proper, and there are also regions with light winds in the mountains. Wind recordings made by H. Aulitzky [898] from May to November 1953 over a steep slope in the Ötztal, facing WNW, near Obergurgl, gave only moderate speeds of 1 to 3 m sec^{-1} at 10 m above the slope, and 0 to 1.5 m sec^{-1} at 40 cm above Alpine rose bushes. During the 5-1/2 months the instantaneous value at 10 m never exceeded 10.2 m sec^{-1}, and normal daily maxima were between 2 and 9 m sec^{-1}. Analysis of wind directions showed that in 70 percent of all cases these were local slope and valley winds which were mentioned in Sec. 43 (see pp. 403ff, especially air avalanches in high mountains, p. 405). Only in 30 percent of the cases a gradient wind, corresponding to the pressure field, prevailed. The slope winds were steered by the thickly branched trees at the forest edges and were therefore able to exert a great microclimatologic influence. The predominantly N and W daytime winds left the southern edges in a wind shadow where the cushion of air was heated by the sun to temperatures fatal to young plants. In contrast, the shady northern edges were also well ventilated.

A further point to be considered in the assessment of mountain microclimate is the changes in the state of the ground with height. Precipitation distribution is extremely varied, depending on wind and exposure (see pp. 419ff). H. Friedel [905] has directed attention to some of the technical difficulties arising in making measurements, since a basic distinction must be drawn between precipitation from the atmosphere and the amount actually received. It is possible, for example, in heavy snowfall, for an increasing wind to blow away more snow from a pass than is deposited there.

The amount of precipitation usually increases with height. This is shown in the figures in column 17 of Table 91, taken from F. Hader [907] for the northeastern Alps for the period 1851 to 1950, in so far as sufficient observations are available for great heights. For this reason it becomes increasingly rare to find dry ground as height increases. F. Lauscher [913] provided a new basis for this argument, with the figures given in column 14. From these it is seen that at 1600 m the ground is dry only half as often as it is in the plain, and at 3000 m it is practically never dry. This is due to the factors V and B in the heat balance.

In addition, precipitation falls more and more frequently in the form of snow. Columns 15 and 18-23, from E. Ekhart [904], contain data on snowfall and snow-cover conditions. Columns 18 and 19 give the ratio of days with snowfall to days with precipitation, expressed as a percentage; column 21 gives the average snow depth in centimeters per day of snowfall; and column 23 gives the maximum depth of snow to be expected on the average throughout the year from the figures of F. Steinhauser [920] for locations below 1200 m. When these greatest snow depths are to be expected is shown in column 24, from H. Steinhäusser [919]. All these values serve to illustrate the increasing importance of snow with elevation.

As long as a plant is under the snow cover, it is protected from destructive weather influences. Parts projecting above the snow, however, are subjected to the extreme climatic conditions near the snow about which details were given in Sec. 24 (see pp. 216ff and Fig. 113). If the ground is bare of snow, as is the case on sunny slopes in winter, plants that are stimulated to transpire will suffer from the great lack of moisture, especially when the ground is frozen. W. Larcher [909] has reported the investigations of this phenomenon made in the Alpine garden on the Patscherkofel near Innsbruck. W. Tranquillini [921] published the first series of microclimatic observations at the mountain station near Obergurgl. Results for the snow-melting period in April and May 1955 are shown in Fig. 247. In addition to meteorologic data on global radiation, air temperature in the shelter, and the rapidly increasing temperature of the ground as the snow melts, there are also

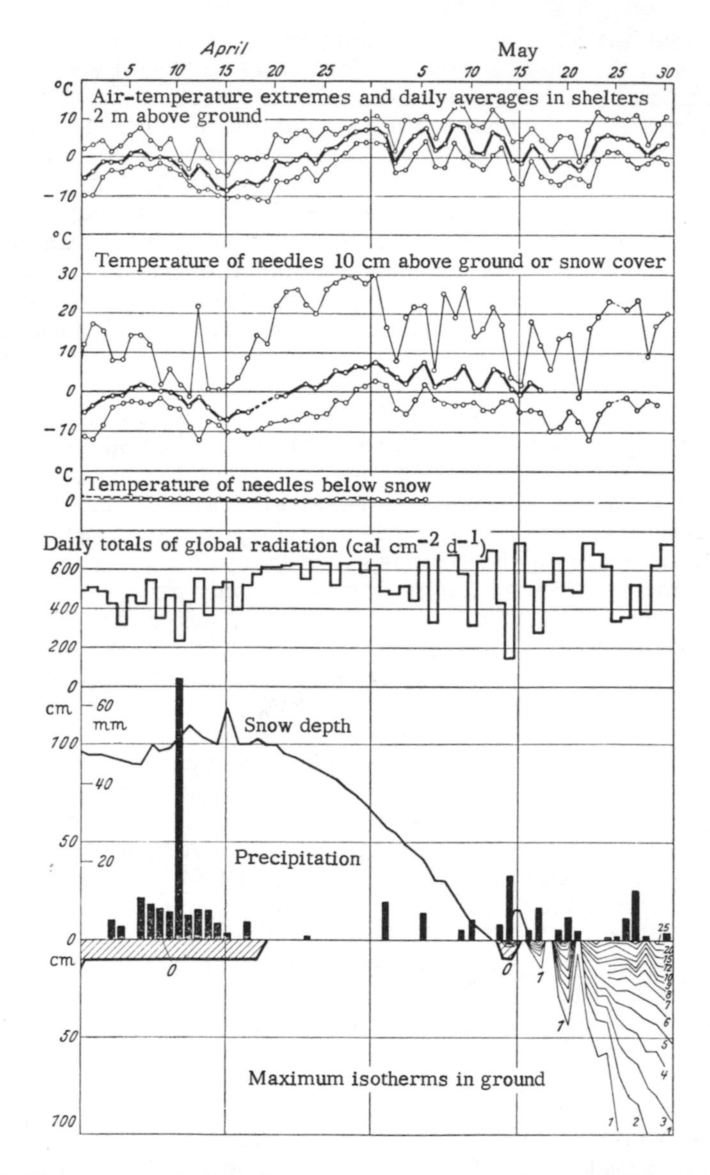

Fig. 247. Measurement of microclimate at the Alpine tree line near Obergurgl, during snow melting. (After W. Tranquillini)

records of the temperature variation in the needles of young *Pinus cembra* 1.5 m high. Within the snow cover the temperature is always 0°C; above the snow it may increase to 30°C, or about 20 deg higher than the shelter temperature, or it may fall to - 12°C. The greatest diurnal temperature variation measured in the pine needles amounted to 34 deg on an April day.

The way snow melts in mountain areas is closely related to topography, as already shown on p. 220; that is to say, even if the snow melts at different times in different years, the melting pattern is always the same. A photogrammetric survey of the melting snow was made in 1935 by H. Friedel [905] for an area of 32 km^2 in the eastern Alps near the Pasterze Glacier, and the analysis was supplemented by direct measurements on the ground. Division of the area into 620 sections enabled the effect of soil type and slope exposure on the melt pattern to be evaluated as a function of elevation. Figure 248 shows an extract from the results.

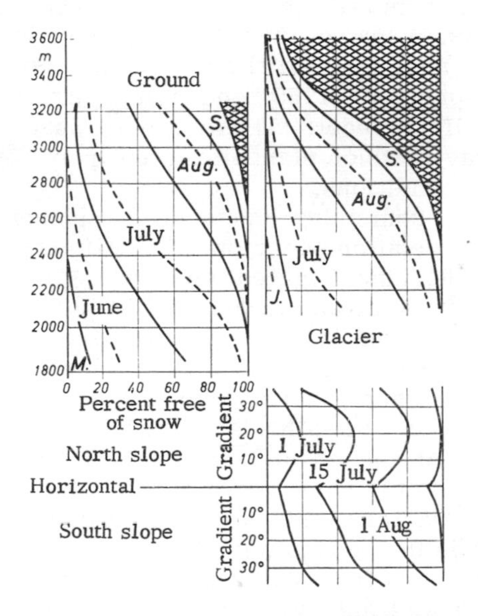

Fig. 248. The dependence of snow melting on the underlying surface in high mountains (above) and on slope exposure (below). (After H. Friedel)

Observations in the top left of the diagram are for snow on the ground, while those on the right are for snow on the glacier ice. The solid line gives the percentage of the ground free of snow at the height in question on the 1st of each month, and the broken line is for the 15th. It may be seen, for example, that at 2800 m on 1 August, 60 percent of the ground is snow-free, at 2500 m 80 percent, and at 2000 m 100 percent, while on the glacier at the same time only 35, 55, and about 90 percent of the snow has cleared. The date lines all join up with the 0- and 100-percent lines because the very first bare patches occur in certain high positions, and snow continues to lie for a very long time in the last hollows. The firn line lies where the upward extension of the melt pattern in September comes in contact with the first of the new snow falling in autumn (surfaces above this border are crosshatched).

The effect of slope exposure is shown in the lower part of Fig. 248. This applies to snow on the ground between 2400 and 3000 m. It follows from the surprising configuration of the date lines that the steep southern slopes thaw first, followed by the more gentle southern slopes and also northern slopes. These are followed, but with a marked delay, by level surfaces; this is because the snow depth is always greater in level areas from which the snow is unable to slide downward. Finally come the steep northern slopes sheltered from the sun.

Since the influence of relief is so strong, the high-mountain microclimate is a mosaic of vastly different conditions in the smallest of areas. A map of the patterns of melting snow in the experimental area near Obergurgl has been published by H. Aulitzky [899] for the spring of 1955, and a map of the vegetation of the same area by H. Friedel. These two maps give an impressive picture of the way in which plant life is entirely dependent on these rapidly changing conditions.

In conclusion, something must be said about the reciprocal influence of the vegetation cover on microclimate.

H. Desing [901] investigated an earth-slide area on a steep slope near Innsbruck at an elevation of 1300 to 1500 m, which resulted from soil erosion, in order to determine the effect on microclimate of the pioneer willow and alder plants introduced to hold the slide. Among others there were two comparative stations in similar positions in the slide area, which is called a "Blaike" there, one on a bare patch and one in alder bushes 2 m in height. The following results were recorded for the months of June to October 1950, in shelters 2 m above the ground: during the day the air temperature in the bushes was about 1 deg lower; the difference was least in the hours from 09:00 to 11:00, when the air trapped in the alder bushes was able to respond more quickly to heating by the sun. Relative humidity in the bushes was about 5 percent higher, therefore, than over the bare ground. Random sampling in August, however, showed extreme differences in soil temperatures. On 27 August 1950 the diurnal variation at the surface of the bare ground was 15.3 deg against 7.1 deg in the bushes. Desing draws the following conclusion: "A preliminary planting of protecting types of wood can also be of great importance, in such high positions, for the subsequent planting of productive timbers."

The air temperature was, on the average, still about 1 deg lower in the alder bushes at night than over bare ground. This was explained by its position near the bottom of a gradient where cold air flowed at night, while the station on bare ground was on an open slope. The alder bushes therefore had cold air streaming through at night, like the beech forest on the Staufenberg (see p. 427).

An investigation was carried out by H. Zöttl [924] on a west slope of 20° at a height of 1830 m in the Wetterstein Mountains. Three equally exposed stations were used to measure temperature in a pebble field, a belt of grass, and dwarf fir bushes. The plants were arranged in up-and-down strips so that differences in microclimate were due only to the influence of vegetation at the site in question. On the pebble field, pioneer types of grass had taken root over 30 percent of the surface. The grass itself (*Caricetum firmae*) covered 85 percent of the ground. The dwarf firs were an association of *Pinus montana prostrata* and *Erica carnea*. These three types of vegetation are characteristic of subalpine elevations. As an example, temperatures on 2 July 1949 (rounded off to whole degrees) were:

Height above the ground (cm)	Afternoon at 13:00			Early morning 04:30		
	Grass	Pebbles	Dwarf fir	Grass	Pebbles	Dwarf fir
+20	15	13	15	2	2	2
0	41	23	18	0	1	2
-20	10	8	7	11	8	8

The grass-covered surface had the most extreme microclimate, caused by poor conductivity in the uppermost layer, and the lack of air mobility, while incoming radiation was practically unhindered. The dwarf firs reduced the diurnal variation to an extraordinary extent, even at this high elevation. Figure 146 has already provided us with an example of this shielding effect.

The influence of the forest on climate, described in Chapter VI, is also effective on high mountains. This has been shown by H. Desing's observations (see above) at the stations in the fir stand. H. Aulitzky [895] has rightly pointed out that, as the tree line is approached and the forest becomes thinner and lighter, the protection it offers to the next generation of trees gradually decreases. At this stage many conditions are encountered that are similar to those described by D. H. Miller [605] for the forests in California, which are lighter from reasons of large-scale climate.

47. Terrain Climatology (Topoclimatology)

An attempt was made in Secs. 40-46 to deduce the influence of the configuration of the ground on the climate close to it. When the expression "terrain climatology" is used in the literature, it is generally taken to mean the climate in a particular place, which depends not only on the configuration of the ground, but also on the type of soil (Secs. 19-24) and its vegetation cover (Secs. 29-39). These factors, too, are considered features of the locality, since they are subject to only slow changes in the course of time.

Figure 249 is from the investigation made by R. Geiger, M. Woelfle, and L. P. Seip [883], already quoted on p. 431. The diagram shows the mean night minima for the months of May and June in 1931 and 1932 for all 99 stations on the Grosse Arber, as a function of height above sea level. Differences are naturally greater on clear nights (left diagram) than when all nights are taken together (right), yet each place preserves its temperature characteristics, as may be seen by comparing the pattern of dots on the left and right.

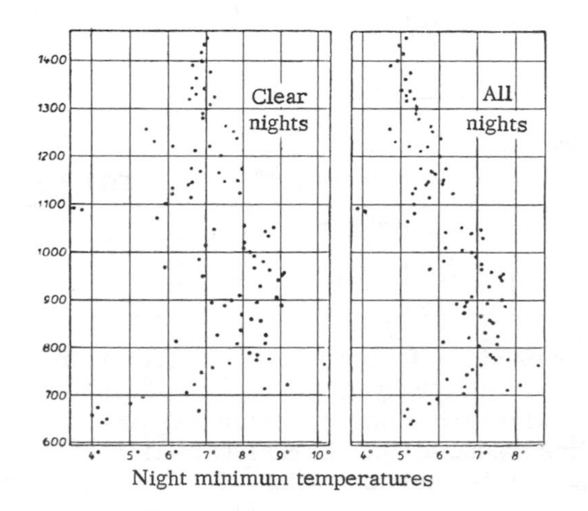

Fig. 249. Night temperatures in spring at 99 microclimatologic stations on the Grosse Arber, as a function of elevation

In the diagram on the right, it is possible to recognize the cold positions in the valley, the warm thermal belt about 800 to 900 m, and above it the decrease of temperature with height. The cloud of dots on the left has been scattered so widely by other local influences that, in the lowest 200 m of the mountain, night temperatures can be 3 deg colder or 3 deg warmer, on the average, than at the peak. The thermometers were not placed close to the ground but positioned at eye level. The two coldest stations just under 1100 m were in a very boggy meadow where cold air was prevented from flowing away. The warmest station, at 762 m, was on a gentle SW slope and, apart from being on the sunny side, was surrounded by forests and orchards. The climate of each individual position is therefore a combination of the effects of the topography, the nature of the ground, the vegetation, and many other things. It was therefore possible to demonstrate at an earlier stage (pp. 220 and 451) that microclimate was dependent on the locality.

Topographic maps show not only the configuration of the land but other characteristics associated with it, such as geologic strata, soil conditions, land use, and density of builidngs or vegetation on it. An attempt is also made in large-scale climatology to produce cartographic pictures, because they provide the means by which the most information can be presented in the most abbreviated form. Climate, however, is an abstract conception that can be deduced only from observations extending over long periods of years. There can, therefore, be only two basically different ways in which the problem of mapping of climate can be approached. Either individual elements can be mapped, thus producing temperature and precipitation maps, fog- or thunderstorm-frequency maps, and so on, or an attempt can be made to arrive at a single form of representation. Such a climatic map, in the narrower sense of the term, presupposes a classification of climate and consequently a division of existing climates into types (for example, a map of Köppen's climatic zones). While the first type of map provides a continuous picture, so that the value of an element can be extracted from it for any point in the area, the second type includes many places within a single type. Subdivision into a great number of types facilitates the use of this map for a particular area, just as a large number of lines of equal values (temperature, duration of sunshine) or of equal frequency makes a map of a single element easier to use. When vegetation is mapped, it is impossible to avoid some form of division into types, and no one will dispute the practical value of such maps.

The first type of climatic map to be considered is the normal type, on a fairly large scale. Large-scale climate maps are based on a network of observation stations 50 to 100 km apart, or even more. On this basis satisfactory maps can be prepared on a scale of 1:500,000 or smaller, the permissible limits depending on the density of the network and the relief. Maps contained in climatic atlases and geography books belong to this category.

A climatic map of a locality therefore implies a map on a fairly large scale, the construction of which is no longer dependent on the help of normal climatologic stations. Usually one thinks of a map on a scale of 1:25,000 to 1:10,000. At the same time a limit is drawn on the other side, because a drill in a plowed field has a very different kind of climate on its N and S sides, and this difference is due to topography. The charting of such fine distinctions, however, could no longer be described as a map of local climate, but must be termed a microclimatic map.

The concept of "terrain climate," for which C. W. Thornthwaite [957] has suggested the fitting name of "topoclimate," has therefore taken up an intermediate position between macroclimate and microclimate. This intermediate stage has long been found necessary. H. Scaëtta [951] suggested it should be termed "mesoclimate," but this term has found as little acceptance as the terminology proposed by R. Geiger and W. Schmidt [934]. W. Weischet [967], who divided macroclimate into regional and subregional climate, which H. Flohn [933] rightly questioned, called topoclimate in the above sense "local climate," a term that has been avoided here, since it is used in so many different senses in the international literature.

The most direct way of establishing the climate of a terrain, or topoclimate, is to increase the density of the network of observation stations. The usual method of setting up normal shelters with instruments at heights of about 2 m is used. The width of the meshes in this special network is substantially smaller, and therefore the duration of the period of observation at auxiliary stations must be decreased sharply, to keep the costs of material and personnel within reasonable limits. Results obtained at these temporary stations are added to data extending over many years from the ordinary stations.

For example, A. B. Tinn [958] analyzed the observations of four stations that were in operation for 2 years (1935 to 1937) and added the results to those of eight stations that had already been evaluated, in order to determine the influence of topography as a function of the prevailing weather situation in the area of Nottingham. F. Spinnangr [955] set up eight additional stations up to 7.7 km from the Fredriksberg station near Bergen, and carried out a series of observations from August 1941 to July 1942. The temperature and precipitation data thus obtained served as the basis for the information and expert opinion required by trade, industry, and building for the town of Bergen and its surroundings. An experiment made near the English spa of Bath by W. G. V. Balchin and N. Pye [926, 927], from October 1944 to December 1945, consisted of setting up 31 normal observation stations, as much as possible on the same type of soil, and always over grass so as to eliminate as much as possible any influence of the underlying surface and vegetation and to retain only the influence of the topography and the town. Day and night temperatures, wind conditions, air humidity, and the tendency for fog formation made it possible to classify the bioclimatic suitability of the area investigated for human habitation separately for summer and winter. Maps of the local rainfall distribution as a function of wind direction indicated the influence of topography. A similar investigation was carried out in the Thames valley near Reading by M. Parry [948] in 12 shelter stations recording from June 1951 to November 1952. Other studies based on similar large-scale climatic observations are those of F. Lauscher [946], on the extent to which the Viennese station on the Hohe Warte is representative of the southern area of the city, the lively descriptions of the climate of Tübingen by K. Daubert [930], or the investigation into frost danger in certain areas of northwest Argentina by J. J. Burgos, A. Cagliolo, and M. C. Santos [929], described as "a microclimatologic exploration." Most of these are descriptions rather than climatic maps.

Research into topoclimatology in the narrower sense of the term began only with the development of new methods of working, and the inclusion of the air layer near the ground. A pioneering investigation in this field was carried out by Wilhelm Schmidt [952, 953], after World War I, on the Lower Lunz Lake, 100 km southwest of

Vienna, using a special network of bioclimatologic stations. Thirteen stations were set up at elevations between 610 and 1530 m, in positions selected from the points of view of vegetation and geography. In addition to recorders screened from radiation, there were maximum and minimum thermometers at three heights above and below the soil surface, which were read once a week. The close relation between the conditions prevailing at a place and hence affecting plants and animals was thus established. The network was not dense enough to assess the topoclimate for mapping.

This classical investigation with its great wealth of data (see p. 398), unfortunately never published in full, was continued in a similar but extended form, although on a smaller scale, in the Neotoma Valley in Ohio. Microclimatic conditions affecting the plant communities found in the Neotoma Valley are described in a book by J. N. Wolfe, R. T. Wareham, and H. T. Scofield [968], which also provides a detailed analysis of the weather and climate in this part of Ohio. The study was based on observations from the normal meteorologic network, supplemented in part or throughout the period from 1939 to 1943 by the following additional stations: maximum and minimum temperatures were read at over 100 places once a week, at heights of 91 cm and between 10 and 30 cm above the ground, the temperature under the soil cover was read at 11 stations, and the 24-hr temperature variation was recorded at a few selected points. This comprehensive program was completed by 10 precipitation stations, and in 34 places evaporation was measured by Livingstone atmometers; light measurements were made by three different methods and air humidity was determined in four places by sling psychrometer. Phenologic observations were carried out throughout the year. It was found possible to express the different conditions of growth in the various places numerically and to describe them in detail.

New demands have been placed on the description of microclimate in the neighborhood of atomic power stations. The possibility of accidents, and the necessity for rapid dispersal of dangerous waste gases, make the ability of the atmosphere to scatter such products of prime importance, and this depends on topography and the weather situation. Measurements of vertical wind profiles, eddy diffusion, and temperature are of greatest use for this purpose, therefore, and also experimental investigations on air flow and diffusion by means of smoke clouds, or the addition of other easily detectable and noninjurious additives to the atmosphere. An example and model of such investigations is provided by the 584-page report by the U. S. Atomic Energy Commission [960] on measurements made in the Oak Ridge area in Tennessee.

Another investigation worth noting is that made by L. C. Bliss [928] in arctic tundras in northern Alaska and alpine tundras in the mountains of Wyoming. Eight microclimatic stations were employed to measure all the principal elements above and below the surface of the ground. Similarities and differences between climatic conditions in the two types of tundra could therefore be detected.

The procedures so far discussed have made it possible to describe and explain the climate of a particular area. Its mapping, however, requires the development of a new technique. K. Knoch [938-941], who foresaw the future requirements of agriculture and land planning for topoclimatic maps even before World War II, urged continuously that they should be prepared, and started off a large number of investigations. Above all, he laid down the lines for this new procedure.

The preparation of climatic maps assumes the existence of a new profession, that of a cartographer of topoclimate. His theoretical training is in climatology and microclimatology, followed by a period spent in the field, measuring, observing, listening, and testing. It requires long years of experience for a man gradually to feel himself part of the true nature of a countryside, and able to judge it rightly. J. van Eimern [931] has recently given us a valuable survey of his activities in this field.

The investigation begins systematically with the nearest ordinary stations in the climatologic network; "the local climate is embedded in the macroclimate" [K. Knoch]. The backbone of the investigation is formed by additional stations with standard shelters, set up for the duration of the survey. These are supplemented by a greater number of observation posts near the ground at which temperature, wind, humidity, and many other essential quantities are measured. It is indispensable to measure radiation on cloudless days and its relation to the albedo of the surface, the slope of the ground, and any restrictions of horizon. To map these elements, observations and measurements must be made at as many places as possible, following the observation techniques outlined in Secs. 55 and 56. H. Wagner and his associates [961, 962] have constructed annual radiation maps for the upper Vogtland area (Elster Mountains), based only on a 1:25,000 topographic map and Kaempfert and Morgen's nomogram ([744]; see p. 373); the fact that they were able to establish a good correlation between the mapped values on the one hand and zones of forest growth and phenologic phenomena on the other, is only an indication of the predominant importance of the radiation factor in this landscape of hills of medium height. The topoclimatologist must, however, establish this by observation and measurement in his own investigation area.

The basic information built up in the way just described is amplified by random sampling. These measurements are made at carefully selected times of day, in particular weather situations, at various times of the year, mostly on foot or by bicycle or, in investigations into city climate, by car (see p. 489). By returning repeatedly to certain fixed points, the variation with time can be taken into account. On occasion, it is possible to employ large numbers of less experienced observers; for example, G. M. Howe [936] used 22 students with sling psychrometers, making observations every quarter hour, to map temperature and humidity near a town in Wales for the duration of a single radiation day.

It requires a great deal of skill to collect information from local farmers, fruit growers, gardeners, foresters, and so forth, and to assess it. Such people have a store of practical experience that should not be left unexplored. Good psychologic understanding is required, and a fine ability to distinguish between true and false observations.

Observation of plants is of great importance. A good knowledge of botany, particularly of plant sociology, is needed. The theorem that the same plant community is found where the microclimate is the same, within a restricted area of countryside, is valid to only a limited extent in view of the nature and water content of soils. Nevertheless, it is a good point of departure when studying a new area. The picture provided by vegetation is of great service, particularly when selecting observation points. Trees damaged by frost or deformed by wind may show damage caused by weather, the existence of plants that are very demanding may indicate locations with a favorable climate.

Then there are phenologic observations. The value of these lies especially in the means they provide for making interpolations between the positions where meteoroglogic measurements have been made. W. Weischet [966] made lavish use of the results of 3 years of phenologic observations in his descriptions of the topoclimate of the lower Rhine basin. H. Aichele [925] demonstrated the close connection between the first appearance of apple blossoms and local microclimate in the area west of Lake Constance. Figure 245 illustrates another example, already described in more detail. The topoclimatologist, however, should never lose sight of the fact that he has to support his statements on meteorologic grounds, and plants are only aids and adjuncts; otherwise vegetation maps might be preferable.

The investigator of local climate must always study the local effects of various weather situations carefully. Where does fog occur? Where is dampness first noticeable on the ground? Where do the first cracks appear in dry periods? What is the pattern of dew, hoarfrost, glaze, or rime (see pp. 181ff). Where does moist vegetation first dry out? His diary must be full of such observations.

By these methods, the topoclimatologist builds up his material, using a 1:10,000 topographic map. There is no system of representation in general use at the present time. It is hardly possible to have a system with such elusive material as that describing abstract climate. It is only by working in the field, as K. Knoch repeatedly pointed out, that valid methods can gradually be worked out. There are two alternative ways open here: the preparation of several maps showing individual elements, or of a single map with contents and method of representation designed for a particular purpose. A few examples from the large number of previous investigations, which illustrate these two different alternatives, will now be described.

J. C. Thams [956] measured sunshine duration by heliograph at 71 different points in the Magadino plain, 34 km^2 in area, at the northern end of Lake Maggiore. This element, which depends on the screening of the horizon, was mapped on a scale of 1:150,000 for eight time intervals (year, month, the longest and the shortest day) and constructed lines of equal possible sunshine duration (isohels), supplemented by the actual sunshine duration in June. These maps were used to decide where in this plain, surrounded by high mountains, to plant tobacco, which needs considerable sunshine.

Frost-danger maps, based on the distribution of minimum night temperatures, provide an example of the mapping of a single element, which has been developed to a fairly advanced stage. This has already been discussed in some detail in Sec. 42. Precipitation maps which make it possible to detect the rain shadow in a particular area are another example. The dependence of precipitation distribution on wind direction in a small area leads to the preparation of micrometeorologic maps for individual cases or for particular wind directions. The most detailed analysis of this type was prepared by C. L. Godske [935] for the Hardanger Fjord area. The work of H. Kauf [937] in the Saale Valley near Jena can be placed alongside the investigation near Bath mentioned above. Another good example of a microclimatic map is that of J. van Eimern and E. Kaps [148] for the area of Harburg and Hamburg. The snow-cover maps mentioned in Sec. 46 should also be remembered.

If a climatologist has only one map at his disposal, the five-step method of mapping proposed by K. Knoch [939, 941] can be recommended. First, those areas are plotted where the climate is normal, shown by the nearest climatologic station. Depending on the purpose of the map, favorable areas (for example, rich in sunshine) and unfavorable areas (for example, with strong winds) are then plotted. Finally, the extremes among favorable and unfavorable situations (for example, frost hollows) are entered, normally only occupying small areas. The more experience the topoclimatologist has, the surer will his judgment be. Maps of this type with a five-step scale provide a valuable aid for experienced practical users.

One of the first maps produced was for the wine industry. N. Weger [964] used as the first basis for his study in 1947 of the area round Geisenheim the measured heat sums received by the slopes under study. Other factors important for the production of wind, such as frost danger, the risk of flooding, excessive slope steepness, and so forth, were included, using a scale of dots. Thus a four-step map was produced for the wine industry, ranging from "very good" to "unsuitable." He published [965] proposed methods for the detailed preparation of such maps on the basis of further experience, to help in deciding the possible yield of different types of vine, the limits of profit to be expected, and as an aid in making valuations when purchasing or exchanging vineyard properties.

A modern investigation to assist fruit growers was carried out in the Odenwald by F. Schnelle and his collaborators. The method and results have already been described on pp. 411ff.

A study by J. van Eimern [932] in the parish of Hütting, which is situated in a wide and dry valley formerly part of the channel of the Danube, was used to bring about better regulation of local climate. The valley is about 6 km long, open to the WSW, and forms part of the dry valley of Wellheim. This area was exposed to damage from wind as well as frost, and both these factors had to be taken into consideration when planning the investigation and the improvements to be made. A normal climatologic station formed the basis, and 17 additional fixed stations were added for the duration of spring 1957, and another 17 to 19 mobile stations, which were moved into different sites every 2 to 3 weeks, were set up. At all of these, maximum and minimum temperatures were read at a height of 50 cm, and the wind speed measured at 1.5 m. Three observers made continuous notes of wind speed and direction between the stations when making their rounds. Soil temperatures were measured at three points.

The first maps to be constructed were seven of the wind fields measured for seven typical wind directions, which also included lines of equal wind speed expressed as a percentage of the speed at a reference point. Frost danger was mapped by plotting the averages for 15 spring nights. In addition to the difference in temperature from that at a reference point, the risk of late frosts was divided into six zones by a quantitative assessment of the percentage frequency of - 2°C frosts (the limit for damage) occurring at a height of 50 cm after 15 May over 18 yr, to deduce which local observations were added to the frost statistics of neighboring climatologic stations. Maximum temperatures at 50 cm, which depend on slope exposure (sunshine) and are strongly affected by wind (ventilation) gave differences of only 3 deg, and were therefore of no significance in the analysis. The final result was a map of the proposed measures to be adopted for wind protection, arrived at by a careful evaluation of each individual measurement, giving the optimum wind shielding that could be adopted without increasing the risk of late frosts and occupying a minimum of space.

Climatic maps for the Harz Mountains, designed for use by foresters, have already been mentioned on pp. 425ff, and the method of mapping has been shown in Fig. 233. Further examples may be found in the bibliography [942-945, 947, 949, 950, 954, 959, and 963].

48. The Microclimate of Caves

Natural caves and the phenomena pertaining to them are described and explained by the science of speleology. Such caves belong properly to the interior of the earth. They lie underneath its surface, and climatic conditions found in them correspond, in many respects, to what has been said in Secs. 10 and 19-21 about soil climate.

Nevertheless, this subject is treated here as an appendix to topoclimate, since the one or many connections between the caves and the free atmosphere such as entrances, air shafts, and the like determine the particular kind of climate in the cave, and this depends to a large extent on its form and shape.

The most striking feature observed on first entering a mountain cave on a fine summer day is the rapid decrease in light intensity, and the way the air becomes cooler and moister. It may also soon be observed that a special type of animal and plant life has adapted itself to this unusual environment, as much as possible. It may even be that, under certain circumstances, an ice-covered floor and bizarre forms of ice on the walls and roof provide a remarkable contrast to the climate at the entrance, where vegetation is luxuriant.

It is of decisive importance for the climate of a cave whether it has one, two, or more entrances. A distinction is thus made, following R. Oedl [978], between caves with a single opening and transit caves. In the first, the air is at rest, and they are therefore sometimes called static caves; at best, turbulent mixing at the entrance may penetrate a short distance inside, or there may be an inward or outward flow of air as a result of temperature differences, which will be discussed later. In transit caves, on the other hand, there is a circulation of air, perhaps reaching considerable speeds in some cases, especially where there are narrow areas in the passages. It is not necessary to have "entrances" that can be used by people; wind shafts or cracks of sufficient width in rocks are enough. If the openings differ substantially in height above sea level the circulation is enhanced. Transit caves, open on many sides, are therefore also termed dynamic caves.

Static caves will be discussed first. Figure 250 shows a longitudinal section of a cave of this type at Jenin in Palestine. Measurements made by P. A. Buxton [971] at midday on 7 June 1931 show the typical temperature distribution in a cave with only one opening. At the point *A*, where daylight still penetrates and a man can stand upright, it is hot and relatively dry. The cave narrows after 7 m (*B*) and it was possible to penetrate farther only by crawling. In this position frogs and the larvae of water insects were found in puddles. Farther inside, the dry-bulb temperature decreased rapidly, and about 25 m from the entrance it became equal to the wet-bulb temperature, which was constant. From that point onward the saturated air maintained a constant temperature.

The change from open air to cave conditions, illustrated here by a single example, is best supported by the series of measurements

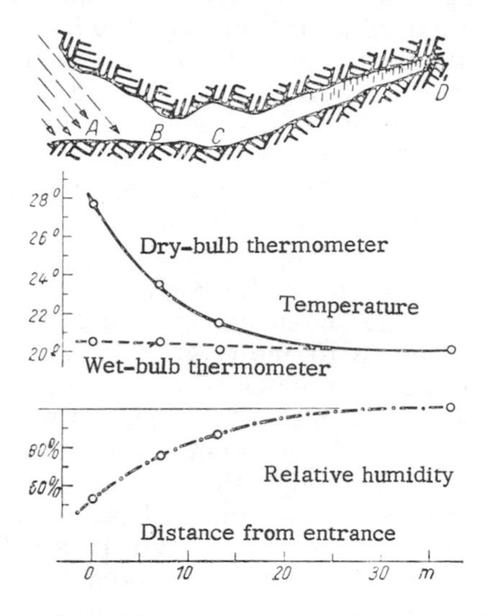

Fig. 250. Temperature and humidity measurements in a cave with one opening. (After P. A. Buxton)

made over a year in the stalactite cave of Baradla in Hungary by E. Dudich [972]. In this cave, which has its entrance on a southern slope, a tunnel 45 m long leads downward in steps. The values given in Table 93 were recorded over this stretch in 1928–1929. The maximum air humidity was 100 percent everywhere. The annual minimum values given show also the extent of its variation, which decreases just as quickly as temperature with distance from the entrance.

Table 93

Temperature and relative humidity in a cave.

Location	Entrance to cave		Steps leading downward			45 m from entrance
	In doorway	Behind door	10 steps	40 steps	68 steps	
Annual temperature (°C)						
Maximum	17.3	14.6	11.8	10.6	10.2	10.4
Minimum	1.8	4.8	6.6	7.1	7.8	8.8
Difference (deg)	15.5	9.8	5.2	3.5	2.4	1.6
Air humidity, minimum (%)	19	66	77	91	95	96

It might be expected, at first sight, that the same kind of annual mean temperature would be found within a cave as would be observed with deep borings. The configuration of the cave, however, exerts considerable influence.

If the cave slopes downward from the entrance, cold air flows downward inside it and is no longer affected by warmer and lighter air. Caves of this type are called sack caves and act as cold reservoirs. If the entrance itself lies at the cold floor of a valley (see Sec. 45), this effect is enhanced, as shown in an example described by H. Lautensach [976]. In extreme cases the shaft of the cave goes vertically downward. The "Eisbinge," at an elevation of 900 m in the Erzgebirge, explored by H. Mrose [977], is a cleft in the rock 1 m wide and 20 m deep, part of a mining works in the tin-bearing granite, which has collapsed. In winter 1.5 m of snow falls into the cave. The mean annual temperature of the rock is +4°C, which is conducive to melting. The cold air, however, for which there is no exit, is imprisoned there and in summer there is a layer of thin fog, about 5 to 10 m deep, in which there is only a small amount of mixing of the warm external air with the cold air trapped in the cave. Although the process of melting continues from the spring onward, by means of heat conduction through the rock, and reaches the bottom of the cave, it is not altogether completed when the first new snow falls in the autumn. The absence of any circulation makes the microclimate here substantially colder than it should be at this elevation.

Metorologic conditions in a vertical shaft were investigated by A. Baumgartner [970] in the "Hölloch" (hellhole) in the Allgäu, during the year 1949. This cave may be described as having a single entrance, since connections with the open air through cracks and joints in the rock have only a small effect, because of their narrowness. An almost circular vertical shaft, 71 m deep and about 8 m in diameter, led into the cave. Light intensity, measured by a selenium cell, decreased with depth in the following way, on 3 September 1949:

Depth in shaft (m):	0	5	10	15	20
Light intensity (%):	100	90	30	10	4

From 30 m down, there was only scattered light with an intensity of about 0.5 percent of the external light, which allowed objects to be recognized, even at the bottom of the shaft, after a long period of adaptation.

The air-temperature variation with depth on a bright summer day is shown in Fig. 251. The characteristics of this type of cave are easy to distinguish. Below 15 m temperatures remain between 4.5 and 5.5°C; high midday temperatures in the air outside are able to penetrate downward by mixing only for a few meters, and rapidly decrease in intensity. The relative humidity behaves inversely, reaching saturation point at a depth of 15 m. When weather conditions were suitable, that is, when the transport of air through cracks in the rock and the joints where water percolated was sufficient, the cold air in the cave flowed over the upper rim of the shaft. This intruding air was up to 20 deg colder than the environment and flowed down the mountainside at a speed of 0.5 m sec^{-1}, giving prospective visitors to the cave a foretaste of its microclimate.

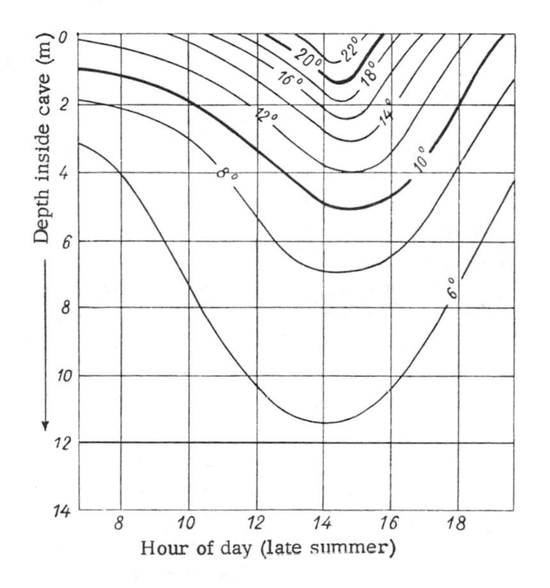

Fig. 251. Penetration of diurnal temperature fluctuation into a sack cave. (After A. Baumgartner)

The opposite thermal effect is obtained when a cave slopes upward from its single entrance. Cold air is able to flow out from the cave if it is cooler than the environment, and warm air from the outside is able to penetrate within. While there is often a sharp decrease in temperature at the entrance to a sack cave, especially in summer, the temperature may suddenly increase at the entrance to a rising cave, particularly in winter. Such caves are therefore preferred by insects and bats for hibernation. I have no knowledge of any actual measurements.

Conditions are quite different in caves with a number of openings. These caves are to a certain extent included in the atmospheric circulation. The main entrance is usually at a low level, and the links with upper levels are made through air shafts from which the percolation water that hollows out the caves is able to penetrate. A good example is the "Eisriesenwelt" in the Dachstein Mountains, where the climatic conditions have been described in some detail by R. Oedl [978], R. Saar [979], and others. The main entrance to this cave is at 1458 m on the steep downward slope from the Dachstein plateau, while the open wind ducts end on the plateau at heights of from 1600 to 1900 m. The altitude of the openings, the temperature differences inside and outside the cave, and the dynamic processes taking place in the outside atmosphere determine the kind of circulation inside the cave.

Caves with a number of openings are therefore ventilated by air currents, in contrast to static caves. The average speed of the air current in the tubular entrance gallery of the Eisriesenwelt, 13 m^2 in area, was found to be 4.0 m sec^{-1}. The maximum recorded was 10 m sec^{-1}, this upper limit apparently being due to the effect of friction on the walls. From this maximum it may be deduced that an average of 1.6×10^9 m^3 of air flow through in a year.

The direction of the air current is decided mainly by the difference in temperature between the air outside and inside, and the movement of air has, in turn, an effect on the temperature of the cave. In winter, from about December to March, cold air streams in through the lower entrance, cools the rock of the cave, and escapes upward. This cold air is comparatively dry; relative humidities down to 40 percent have been measured. The ice that is present in the lower part of the cave therefore evaporates strongly. Cave temperatures fall considerably below 0°C. When spring melting sets in outside, and melt water trickles inside, that is, in May and June, the amount of ice is greatest in the cave, which is still cold from the winter.

Warm summer periods are scarcely felt within the cave. From May to the middle of November its temperature ranges between - 1°C and + 1°C. The cold air flows out the lower entrance of the cave; for this reason the circulation from July to September is the reverse of that in winter. Summer air is sucked in from above and warms up the rock mass in the higher levels of the cave system. On cooling, the relative humdity of the air increases, and is always close to saturation level.

Continental weather, with very cold winters, which are able to penetrate into the cave, and warm summers, which go by leaving hardly a trace of their existence, therefore furthers cooling within the cave and promotes the formation of ice. The cold winter of 1928-29 caused the cave temperature in the Eisriesenwelt to fall by 1.5 deg everywhere.

When outside and inside temperatures do not differ much, which occurs mainly in the transitional seasons, but which may also be the case in summer and less often in winter when there are changes in weather, there can be a reversal of the direction of air circulation in the cave for a short time. W. Gressel [973] has rightly pointed out that dynamic processes in the atmosphere, the pressure gradient at the several openings of the cave, the enhancement or inhibition of local mountain winds (Sec. 43), frontal pressure differences, and so forth, may also play important roles.

The microclimate of caves is of scientific interest, but is also of practical importance when the caves are used for any purpose, as for dwellings, or when artificial caves are constructed.

M. Hell [974] proved that in the famous beer cellars of Kaltenhausen near Hallein, Salzburg, galleries 6 m long had been tunneled in the rear or mountain wall, leading to natural air vents. By this means, the air circulation in the cellars was ensured. When the temperature outside was 20°C there was a wind blowing inside with a temperature of 4°C. The climate of the cave gave the name to the whole district of Kaltenhausen. H. Lautensach [976] observed that people in Korea moved silkworm eggs into cold air shafts, similar to caves, when it suddenly became warm in spring, so that the caterpillars would not hatch out before the mulberry trees, their source of food, were in leaf. In the construction of tunnels, double-sided open shafts are cut to provide the conditions of temperature and air circulation suitable for railway travel. In World Wars I and II whole communities have saved their lives, property, and art treasures in artificial and natural caves. This leads us on to the question of the reciprocal effects of man and microclimate, which will be discussed in Chapter VIII.

To conclude, it is worth mentioning two rare and special types of cave formation. The first is the thermal cave, where the microclimate is entirely controlled by the flow of thermal water through it. Anyone who has been in the galleries of the famous hot springs at Gastein will remember well the extremely damp and warm climate there. The second type is the ice cave and the tunnels hewn out artificially in glaciers. The measurements made by W. Paulcke [422], H. Hess [975], and W. R. B. Battle and W. V. Lewis [969] should be referred to for details.

CHAPTER VIII

RELATIONS OF MAN AND ANIMALS TO MICROCLIMATE

When investigations are made of the dependence of the life (bios) of plants, animals, and man on atmospheric conditions and processes, the field of research is designated as bioclimatology in the widest sense of the term. It is further subdivided into three parts, biosynoptics, biometeorology, and bioclimatology in the more restricted sense.

Biosynoptics, according to H. Seilkopf [997], is the study of large-scale weather effects on biologic processes, as shown by surface and upper-air meteorologic charts. An influx of subtropical warm air at high levels, for example, may lead to an increase in the number of clinical illnesses, or to the unexpected appearance on the north German coast of butterflies from southeast Europe.

Biometeorology investigates the effect of individual weather phenomena on living processes. When lightning strikes a wood, it not only damages the tree it hits, but may bring about the death of surrounding trees (lightning clearing). A strong foehn wind not only increases the urge for bees and wasps to sting but may induce people with suicidal tendencies to kill themselves.

Bioclimatology, in the more restricted sense of the term, investigates the connections between climate and life. It determines, for example, how man adapts his style of house building to the various main types of climate on the earth, or how infantile paralysis is usually associated with certain periods of summer weather, or the way in which sunshine duration on grassland affects the milk yield of cattle.

In its broader sense, bioclimatology today has developed into a special science of such wide scope that it is not possible here to treat even its micrometeorologic and microclimatologic aspects in detail. In addition to making a general survey, only a few problems will be selected, which are of special interest to the microclimatologist and which have been investigated by him on a number of occasions.

The exposition will deal with plant life, animal life, and human beings in that order. Plants are rooted to the places where they grow. Bioclimatic problems therefore simply become questions of the local peculiarities. It is for this reason that the reciprocal effects of plant life and microclimate on each other were discussed in Chapters V and VI.

The argument is also restricted to the living conditions of healthy plants. It is well known that an ailing plant reacts more sensitively to atmospheric influences, just like a sick animal or man, than it does in a healthy state. It follows that plant protection, veterinary science, and medicine, which are particularly interested in the relations between the sick organism and weather and climate, must also be interested in microclimate. Unfortunately, these important practical problems are outside the scope of this book.

It follows, therefore, that phytopathology, or the study of plant illnesses, is omitted in dealing with plant life here. It is also concerned with the animal pests that damage plants. From the point of view of microclimate, therefore, phytopathology investigates the temperature and humidity conditions that favor the development of fungi and other sources of disease, or which type of climate will stimulate damage by aphids. A comprehensive survey of these problems from the meteorologic point of view is to be found in H. Schrödter [996]. Many new results are available in K. Unger [999], and reference should be made to the great joint work by H. J. Müller and others [990] treating damage by aphids, and many other references.

49. Animal Behavior

Plants are unable to escape from the microclimate of their environment, no matter how adverse the conditions may be. Animals, on the other hand, have freedom of movement, are able to avoid an unfavorable microclimate, and can search for suitable conditions. The present section will give a few examples of the behavior of animals.

Every animal has its own, usually rather restricted, sphere of movement. It is bound to a locality, and therefore subjected repeatedly to its conditions of microclimate and micrometeorology, as W. Kühnelt [986] put it. This applies most to lower and less mobile forms such as larvae, worms, and caterpillars. But even the swiftest and freest of creatures, like the birds, are restricted in a sense by having nests, in which they spend the night and bring up their young, and therefore subject to local conditions. The microclimate of animal habitat will be treated in Sec. 50.

A mobile animal first adjusts itself to the microclimate of its immediate environment, by reacting to local weather processes. Dancing swarms of midges avoid the changing directions of a freshening wind, as observed at times by F. Lauscher [987], by taking shelter from it in the lee of a hedge. According to W. Mosauer [989] the sand lizard (*Uma notata*) can stand the extreme temperatures of desert sand by raising the body a little above the heated ground when running fast. The African cricket, according to F. S. Bodenheimer [981], turns its body in the direction of the sun's rays about noon, so as to present the smallest possible cross-sectional area to it, but lies broadside to the sun in the cool of the morning (compare with compass plants, p. 391). Figure 252 is a photograph taken in the Tunisian desert near Kairouan in March 1930 by E. Kaiser [985]. The white spots are desert snails of the species *Leucochroa candidissima* which climb in hundreds up the branches of the jujube tree (*Zizyphus lotus*) to avoid the heat of the ground.

Fig. 252. Desert snails take refuge from the overheated layer of air near the ground. (After E. Kaiser)

In describing his experiences while trapping animals in the Himalayas for Hagenbeck's Zoo, H. Wiele [1000] tells of an automobile journey he made near Rawalpindi after a thunderstorm associated with a marked decrease in temperature. They had to get through a great swarm of locusts that had alighted exhausted and motionless from the cold. "When we had crossed through the dense core of the flight and came into the bright sunlight again, we found that the roadway with its smoothly rolled covering of light gray-blue granite gravel was so thickly covered with locusts that it looked like a thick loosely knit light-green carpet, from which there stood out sharply in pale gray the pattern of the shadows of the trees with their thousands of branches and twigs, because not a single insect was to be seen in the shadow pattern."

E. T. Nielsen [922] observed in Denmark that one type of cricket (*Tettigonia viridissima*) begins its song in the afternoon on low forms of growth, sitting on low plants or reeds, for example. At night it is heard singing higher up, in the trees. In order to find out whether different types of crickets were involved, or if it was a single kind that sang below, and climbed higher as darkness fell, he attached some test specimens to a few meters of thread. In fact, the crickets did climb spirally up the tree, when the air layer near the ground began to cool. Nielsen assumes that they seek the more comfortable temperatures of higher layers. T. Roos [994] demonstrated by research in the rivers of central Sweden that the fertilized females of insects living on the water flew in the opposite direction to the wind, which blew downstream (caused by surface drag? see p. 200), therefore opposing the "organic drift" which always acts in the direction of flow.

I have always been much impressed, while making observations in stands of tall pines during forest-meteorology investigations, by the regular accumulation of hordes of insects above the tree crowns in the first hours of the morning. Thousands upon thousands of hovering insects, gnats, and butterflies would collect there, such a mass of living creatures that it was hardly possible to believe they could exist in such numbers. This living cloud was cut off so sharply at the boundary surface of microclimate (see Fig. 172) that the observer climbing up the observation ladder seemed to be pushing his head through a barrier.

Sea gulls take the wind field and lapse rate in the layer close to the water into account when flying over the sea. Systematic observations were made by A. H. Woodcock, and the results have been illustrated by G. Neumann [991] in Fig. 253. When the water is

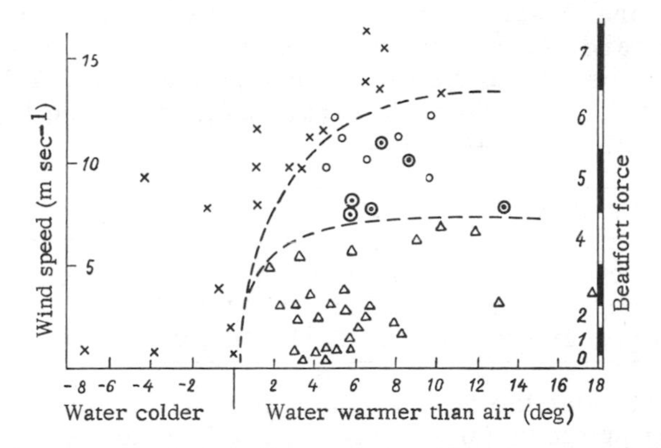

Fig. 253. The flight tactics of sea gulls are designed to take advantage of the wind and temperature structure. (After A. H. Woodcock)

warmer than the air, and wind speeds are light, a cellular circulation develops over the sea. The gulls make use of the upcurrents in these cells and climb by gliding upward in a circle. Such cases are indicated by small triangles in Fig. 253. When the water is colder than the air, on the other hand, or when the wind is stronger than Beaufort force 6 with the water warmer, there are no up-currents and the gulls climb in a straight line by flapping their wings (crosses in Fig. 253). Between these two possibilities there were two intermediate forms: a straight-line glide upward (circles), and an upward glide partly in a straight line and partly in a circle (circles with dots). Each type of flight therefore corresponds to a state of microclimate over the sea, with well-marked boundaries.

The Japanese lark (*Alauda arvensis japonica*) usually flies in a horizontal circle about 60 m in diameter at a definite height. From flight-path measurements by double theodolite on a 50-m base, S. Suzuki and his colleagues [998] established that this height depended on the surface temperature. It was 80 to 100 m with temperatures of 24° to 28°C, and averaged only 40 to 70 m at 16°C.

Small, light insects, easily borne on the wind, exhibit a density of distribution in the atmosphere similar in many respects to that of the inorganic particles and suspended matter described in Sec. 17. C. G. Johnson [984] investigated the number of aphids in the air near Bedford, England, up to heights of 600 m, by towing nets. From 151 measurements at six different heights, the density was found to be between 1.8 and 0.0008 aphids per cubic meter. The decrease with height h was found to follow an approximate index law $D = h^{-b}$, in which D is the density relative to that at a height of 1 m, h (m) is the height, and the index b varies to a marked degree with the lapse rate. With an adiabatic structure (see p. 38), b was approximately 1, and the density was therefore inversely proportional to height. With a smaller lapse rate, the density of aphids aloft was always very low. The greater the decrease of temperature with height, and the more unstable the air was, the higher were the aphids carried upward. After conversion of units, the value of the index b was found to be 1.04 to 0.80 for a lapse rate of 1.0 to 1.2 deg/100 m and 0.55 for 1.4 deg/100 m. Using these figures there is a decrease in density with height as follows:

Height (m)	1	10	50	100	500	1000
for 1.0 deg/100 m	1	0.091	0.017	0.008	0.002	0.001
for 1.2 deg/100 m	1	0.158	0.044	0.025	0.007	0.004
for 1.4 deg/100 m	1	0.282	0.116	0.079	0.033	0.022

For these reasons the decrease in aphids with height was slower in spring and summer (mean value from May to August, $b = 0.78 \pm 0.21$) than in autumn (mean of September and October, 1.25 ± 0.31). Two maxima were found in the diurnal distribution, one in the late forenoon, and one toward evening, the reasons for which are given on p. 126.

Locusts fly in close swarms, but here too the density within the swarm is a function of the lapse rate and varies, according to the observations of R. C. Rainey [993a], between 0.001 and 10 locusts per cubic meter in Kenya and Somaliland. The swarms therefore have a vertical spread between a few meters and several kilometers. Rainey's excellent air photographs show swarms sometimes looking like a thick fog lying on the ground, and sometimes like a veil of cumulus shape.

A consequence of the way animals adjust themselves to the circumstances in which they live is that, even in a small area, their distribution both by number and by species will be adjusted to microclimatic differences.

An example of this kind of adjustment is taken from E. Schimitschek [995] on the development of bark beetles (*Ips typographus*) in a tree trunk lying on the ground. Figure 254 shows a cross section of a fir log lying in a NW to SE direction. In sector 1 on the SW side, where the effect of the sun is greatest, bark temperatures of 50°C were recorded, and no beetle eggs were found to have been laid. Eggs laid in both of the sectors marked 2 died. Larvae developed from eggs laid in sectors 3, but these dried up later. It was only in sector 4, which is narrow on the sunny side and broad on the shady side, that conditions suitable for normal development were found. Where the trunk was in contact with the ground (5) and became very damp in wet weather the mortality rate rose again to 75 to 92 percent.

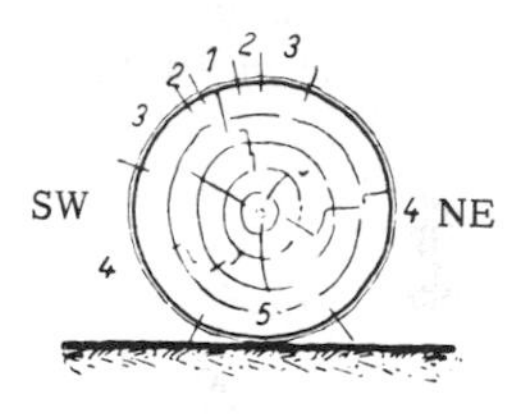

Fig. 254. Development of bark beetles around a tree trunk (explanation in text).

The damage caused by pests to grain in storage was investigated by E. Bernfus [980] with a view of developing protective measures. When the humidity of the grain increases to 12-13 percent, the mites that cause the damage begin to develop, and the optimum development is reached with 17-18 percent humidity. Wheat was stacked 12 sacks high on top of a wooden base, in a large storage barn on level ground, shut off from the air outside by sliding doors. The decrease of humidity with height observed in the open air in summer and autumn (Sec. 15) also applied to the stored grain. The number of mites per kilogram of wheat amounted to 35, 9, 7, and 4 for the average of three sacks, from the bottom up. The lowest three sacks gave the figures 46, 38, and 21. Totally different results were obtained in the warm and moist spring of 1953. Air temperature in the barn was measured at 22°C, and the surface of loose grain stored was 15°C. The warm and moist weather led to condensation. Here the incidence was on the average 36 mites per kilogram of grain near the surface, 5 in the middle, and 3 in the layer near the ground. Sampling within the heaped-up grain showed that the number of beetles was greatest in the proximity of the iron pillars used to support the roof. This was thought by Bernfus to be due to condensation on the iron, which enabled the beetles to maintain equilibrium in their requirements for water.

According to H. Franz [982], the orthoptera found in the moist bottom of a valley near Parndorf, southeast of Vienna, are of a different variety from those inhabiting the drier slopes or the meadows at higher levels.

W. O. Pruitt [993] established the difference of microclimate in five different types of stand, differing in type of soil and density of growth, by making measurements of soil and air temperature, and showed that these differences accounted for the different distribution of shrews (*Blarina brevicanda*).

The unusual temperature stratification in the sinkhole at Gstettneralm near Lunz was described in Sec. 42, and the adaptation of vegetation to these different conditions was mentioned on p. 399. Similar adaptation by animals was the subject of an investigation by E. Schimitschek [995].

A particular butterfly (*Evetria turionana* Hb.) is frequently found up to the level of the saddle, the importance of which for temperature stratification was emphasized on p. 419. Inside the sinkhole, 10 m below the saddle, it is very rare, and 30 m farther down it occurs only sporadically, as most of the caterpillars die. A true bark beetle, *Pityogenes conjunctus* Reitt., was found in the dying and dead branches of dwarf firs in the sinkhole. Describing conditions in the cold bottom of the sinkhole Schimitschek writes, "On 19 July we found nothing but eggs in all the hatching areas examined, with a few larvae that had just emerged in a few exceptional cases only. In addition to these freshly laid broods there were also some others that had been abandoned or still held a few young beetles, and others with larvae three-quarters grown. On 23 September the larvae of the July broods were about half-grown; some of them had died. Eggs laid fairly late in this area do not reach the young larval stage, but die off. Generations here are of 2-yr duration, in the most favorable cases 1.5 yr. On the higher slopes of the sinkhole perfect 1-yr generations can be found. The frequency of the incidence of *Pityogenes conjunctus* increases from below upward! The number of eggs per brood is greater at the upper edge of the sinkhole than at the bottom."

It therefore appears that the distribution of animal life depends not only on the general large-scale climatic conditions, but also on microclimate. According to H. Grimm [983], animal geography is compelled to take the problems of microclimate into account. The limits of the area in which any particular species is found is a boundary zone in which the creatures are found to exist in pockets. In much the same way as plants, animals are able to exist in an unfavorable general climate only if they can find places where the microclimate is favorable. The egg and the larva of the hookworm, for example, which are used to tropical temperatures, find themselves at home in the conditions prevailing in tunnels and mine shafts, where they create risks for miners. In underground heating systems in Paris, the rat flea that carries plague is able to thrive, although it is a visitor from warmer countries. E. Martini and E. Teubner [988] proved by laboratory studies and field observations that the true malaria mosquito (*Anopheles*) needs microclimatic conditions different from those favorable to other mosquitos. This has an immediate effect on the malaria risk to which men are exposed in the tropics. The microclimate of tropical dwellings (including adjacent buildings such as stables) decides which kinds of insects shall be either able or unable to live along with human beings. Many examples can be found to illustrate this point.

50. Animal Dwellings

All the higher animal forms have homes in which they can bring up a new generation. The requisite microclimatic conditions are taken into account with the sureness of instinct. Modern zoology is making an ever-widening study of the ecology of breeding places, often as a joint effort of zoologist and microclimatologist, a most desirable form of cooperation. The time is past when the British biologist P. A. Buxton [1001] could say to the meteorologist: "You study climate in your well-ventilated shelter where, at the best, only an earwig can live, but what we want to know is what is the climate in a bird's nest or in a rathole."

The example of an American zoologist [1017] will serve to illustrate the fine techniques that modern methods of measurement can adapt for this purpose. In the desert near Salt Lake City there lives a type of rabbit that is unable to find water, and can therefore obtain its supplies of liquid only from its scanty nourishment. To investigate these unusual conditions of life it was necessary to establish the humidity conditions of its environment, and therefore also of its burrow. To do this a rabbit was caught in a trap in front of its burrow, and a recording device was tied to its tail. This device was smaller in cross section than the rabbit and could record temperature and humidity for a few hours. On being freed the animal rushed back into the protection of its burrow, and thus the conditions in the interior, and usually also in the environment, were recorded. The device was recovered by killing the animal and digging it out. By this method it was proved that the animal was able to maintain its fluid balance without ever drinking a drop of water, because of the surprisingly high relative humidity of its burrow, where it remained for long periods.

I have always been amazed at the way rabbits on the North Sea island of Sylt arranged their burrows. One had its entrance, for example, on the upper slope of a sand dune, so that when there was heavy rain it collected below, and none came through the opening. An overhanging growth of heather roofed over the entrance and protected it from rain and dripping water. It had a southerly exposure, so that the entrance was sunny, and shielded from north winds. A very large bush to the west provided additional protection from the stormy west winds.

H. Löhrl [1010] showed that one type of swallow (*Nyctalus noctula* Schreb.) exhibited considerable skill in selecting, out of the wide area in which it lived, a place to nest in winter with the warmest microclimate. Those observed in Munich first selected (as often) a place in the town, where it is warmer in winter than in the country, the favored position being the corner of a house pointing inward into the warmest part in the center of the town, and with an exposure to the southeast. About 12 m above the street, and therefore above the cold layer of surface air, the birds hollowed out holes about 50 cm deep in the walls behind the gutter of the roof, which therefore protected them from rain. The interior of the house was heated, and in addition one of the main pipes of the steam central heating ran close to the nest. The following temperatures were measured simultaneously on one occasion: - 14°C at the meteorologic station, - 5°C on the roof of the house, and almost 0°C in the nesting area.

Figure 255 shows photographs of a termite hill in Arnhem Land in northern Australia, after R. Hesse [1006]. Its asymmetric shape,

Fig. 255. Termite nest orientated in compass direction in northern Australia. (After R. Hesse)

which is made plain by the views from the east on the left and from the north on the right (its exposure to the midday sun), has most probably been developed as a protection against strong solar radiation. These constructions pointing N-S are the counterpart of the compass plants described on p. 391. The earth galleries also constructed by termites over their lines of communication are thought by W. Kühnelt [1008] to be another protection against the dangers of microclimate, this time evaporation losses.

The flat nests, rather similar to those built by ants in Europe, of one type of termite (*Eutermes exitiosus*) were investigated by F. G. Holdaway and F. J. Gay [1007]. Figure 256 shows the temperatures measured on a winter day, 9 August 1933, in a circular nest 38 cm high and about 80 to 90 cm in diameter, near the Australian capital of Canberra (35°S). When air temperatures were below 9°C the walls of the nest were heated extensively by radiation, up to 25°C on the N side (see Secs. 41 and 44). The termites are, however, able to maintain temperatures higher than this in the interior of the nest, that is, always more than 25°C. This is brought about by the production of heat in the process of living (metabolism). The authors were able to establish, by numerous comparative measurements, that inhabited termite nests were on the average 8.1 to 10.3 deg warmer than those that had been abandoned. In such cases there is added to the selection of favorable natural conditions, a modification of microclimate controlled by instinct.

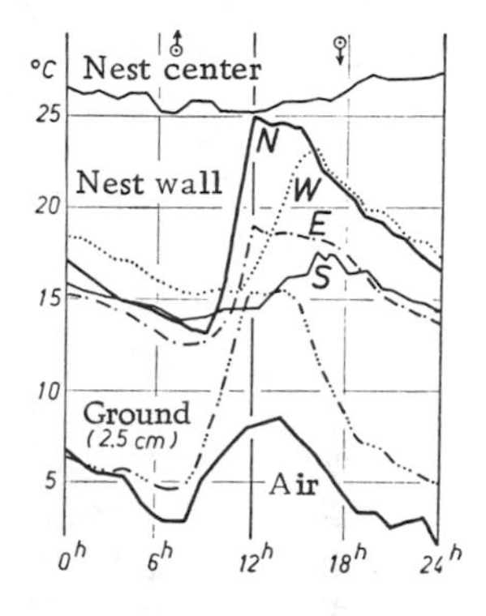

Fig. 256. Temperature variation on a winter day in a flat termite nest near Canberra, in 35°S latitude. (After F. G. Holdaway and F. J. Gay)

The high dome-shaped nests built by ants in the European climate, which is poor in sunshine, apparently are intended to exploit to full advantage the small amount of solar radiation in forests. Since the materials used (pine needles, straw) and the loose method of construction make thermal conductivity poor, differences of exposure climate within short distances are not evened out. G. Wellenstein [1015a] investigated in September 1927 a nest of red forest ants, 80 cm high and more than 12 m in circumference, in a stand of young firs on a steep slope, not far from Trier. The nest was 3 to 4 deg warmer on the shady side, 25 cm under the surface, than the air in the environment, and below the sunny side it was 5 to 9 deg warmer than the shady side.

A. Steiner [1014] describes the construction and operation of a nest in the following way (abbreviated): "A dome-shaped nest with numerous air spaces in the interior is built of earth and parts of plants in a place with a southern exposure, sheltered from the wind. The shape varies from that of a hemisphere to a cone, depending on the prevailing radiation and precipitation conditions, and serves to capture heat. The relatively greatest degree of efficiency in this respect is obtained when the elevation of the sun is low; it has been calculated for a latitude of 47°N that a hemispherical dome picks up twice the radiation incident on a horizontal surface at noon on 21 December, 1.25 times as much at the equinoxes, and 1.05 times the quantity of 21 June. The increased quantity of heat picked up is protected from loss by various means, principally by the thick roof of the dome, constructed of plant materials with poor conductivity, by the heat-insulating air chambers in the interior, and by closing the openings at night. By these means an average temperature is obtained in the interior of the nest, at a depth of 30 cm, that often varies between 23° and 29°C, and is 10 deg higher than the corresponding ground temperature. The varying temperatures from place to place in the upper parts of the dome are utilized to the optimum extent by tireless shifting of the eggs, which are also protected from overheating by being transported down into deeper parts of the nest.''

According to K. Gösswald [1005], the zones in which the three types of red forest ants are distributed are also determined by microclimatic conditions. One type is able to stand the cool moist conditions in deciduous forests, and builds its high nests there. Another type searches out light, dry, and well-heated evergreen forests, and there lays down colonies of nests which are often nearly flat in form. K. Palat [1012] states that the ant *Formica rufa pratensis,* living at the forest edges, uses its completely flat type of nest in the warm climate of southern exposure only in the spring. In summer the unimpeded solar radiation is so strong that the ants use only the shady side between 08:00 and 15:00.

An extraordinary degree of microclimate regulation has been achieved by the Australian megapode *Leipoa ocellata* (mound bird, jungle fowl), which has been called the ''thermometer bird'' in consequence. H. J. Frith [1004] has shown that the temperature of its nest is maintained at 34.5°C, with a fluctuation of only 1 to 2 deg with air temperatures that vary between 16° and 49°C during the year. Nests are roughly circular in shape, about 5 m in diameter, and about 20 eggs are laid over a period of a week, to hatch out in 6 months. During the cold spring the birds collect large quantities of leaves which are then covered with about 50 cm of sand and left to ferment and produce heat, thus raising the temperature of the nest. During the summer the sun's heat is used, with daily adjustments to the thickness of sand, involving 2 to 3 hr of work, depending on the amount of radiation and the air temperature.

What these highly developed animals do by instinct, men do deliberately for their domestic animals. Early forms of stall were designed mainly as a protection against theft and severe weather, but modern breeding requirements have given great significance to the controlled climate of stalls. P. Lehmann [1009] and W. Zorn and G. Freidt [1016] drew attention to the importance of this factor some time ago. Modern selective breeding is designed for specific purposes so that the genetic factors can become fully effective only in an environment that has a comfortable microclimate. M. Cena and P. Courvoisier [1002] have made a new survey of this problem and its relevant literature.

Sweating and excretion by animals make it necessary to have good ventilation in spite of all modern hygienic installations. The improvement in the chemical content of the air that is thus achieved usually brings about a deterioration in its physical characteristics, and this in turn is best counteracted by shutting off the outside air in the way most suitable for the animals in question. The construction of stalls must try to strike a balance between these opposing requirements.

Even an open shed offers some protection against weather, as A. Raeuber [1013] was able to show recently with a stall for calves that was open toward the south. Temperature and humidity were approximately the same as in the open, but there was protection against rain and wind, although the latter depended very much on the wind direction.

A closed shed gives still air and an even microclimate, where much depends on the body heat of the animals. By making comparative measurements in a shed when empty and fully occupied, A. Mehner and A. Linz [1011] showed that the temperature fluctuation in the former case was one-half and in the latter one-eighth that of the open air. They are of the opinion that the correlation between open-air and stall temperatures is a measure of the goodness of the shed. Figure 257 gives an example of the balanced type of climate prevailing in a shed, from measurements made by H. Wächtershäuser [1015] in a byre measuring 27 × 11 × 4 m in which there were stalls for 35 to 40 milch cows. The average diurnal temperature variation for the cold period from October to April for the years 1949-1952 was 3 to 4 deg, hardly ever more than 6 deg inside (top diagram), while outside the values were widely scattered, following the general climatic pattern. The lower part of Fig. 257 shows the frequency of various mean temperatures over the same period. In the byre, the daily average never fell below + 8°C, even when outside temperatures were as low as - 8°C.

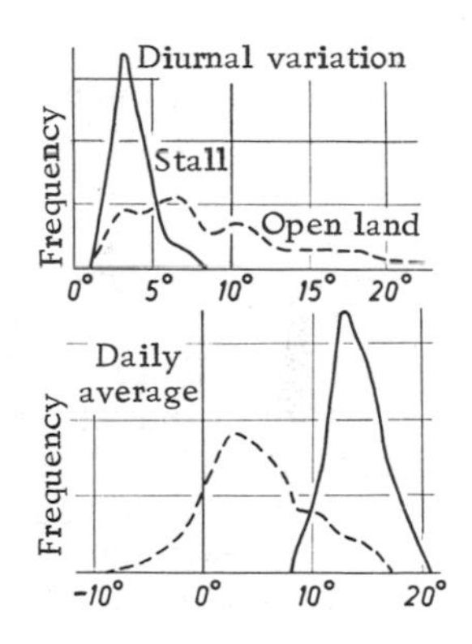

Fig. 257. Average temperatures for four winters in a byre for 40 milch cows, compared with open land. (After H. Wächtershäuser)

A particularly good indicator of bioclimatic conditions in the housing of animals, according to M. Cena and P. Courvoisier [1002], is the amount of cooling from various surfaces, which they measured with frigorimeters. The skin temperature of the animals and the long-wavelength radiation exchange with the walls are of importance. The ultimate aim of modern stabling is a stall with controlled artificial climate, similar to air-conditioned houses. H. Dahmen [1003] has reported on one such attempt. It is not possible to go into further detail here concerning this rapidly developing special study.

51. Man and Microclimate

In much the same way as animals, man reacts unconsciously in his first responses to favorable or unfavorable microclimatic conditions. When the first warm days of spring arrive, mothers know how to take their baby carriages to the warmest and most sheltered places, richest in radiation, without going into any theoretical calculations. In the summertime you can come across real artists in finding the most comfortable places to lie in both on the beach and on mountain pastures. I once read that the homeless in London, wandering at night on Victoria Embankment, would get to know the temperature of almost every wall, and chose the outer walls of hotel kitchens as favorite places to rest against.

The first factor of great importance is the unconscious or unintentional effect of human civilization on microclimate. Where man begins to achieve dominance, the infinite variety of undisturbed creation is upset. Plant and animal worlds are made poorer, the network of water supply is regulated, moors, swamps, and pools are eliminated, irregularities in type and form of the ground are smoothed out, even forests take on a cultivated appearance. Human intervention builds up to produce a total effect far beyond its geographic framework.

Men are also forever creating new kinds of microclimate. Every building constructed displaces the original climate of its site, creating a warm, sunny, and dry climate with a southern exposure on the one hand, and a shady, cold, and damp northern climate on the other. Industrial works are shrouded with thick haze which alters the whole radiation economy. Snow melts in city streets in autumn when the surrounding countryside already is covered with white. A rail or road embankment dams up the cold night air and creates new frost-danger zones. The air quivers in the heat rising from the southern slopes of coal heaps. These thousand and one small effects can offset each other only partially, as experience shows. Replacement of the original order of creation by the all-encompassing economic order ordained by man makes changes in the general climate of a country as well, and the measures adopted to further the advancement of industry are, in a long-term view, by no means always such that they also favor the advancement of life. History has taught us more than once that increasing urbanization has brought in its train great damage to the heat and water budgets of a land.

The risk of bringing about a deterioration of climate by human intervention depends on the type of climate, and is greatest where plant life is fighting for its existence because of a shortage of water or heat. It is possible for very limited intervention into the water balance to produce far-reaching results in the bordering regions of arid zones. Great "regeneration areas" have been laid out around the town of Broken Hill in Australia, the great mining center for silver, lead, and zinc, situated in the middle of arid desert land. Wise foresight indicating that the original sparse vegetation in the area would have to be removed by the inhabitants, a comparatively luxuriant type of vegetation was grown in surrounding fenced-in and irrigated areas, improving the climate of the town, providing recreation spaces, and offering protection against dust storms in summer.

At high mountain elevations, and at the arctic tree line, the lack of heat makes plant life particularly sensitive to human intervention, some of the consequences of which have already been mentioned in Sec. 46. Critical situations may also be created by the absence of calm periods, a characteristic of many coastal and highland climates. This question will be raised again in Sec. 53 when we discuss wind protection.

Knowledge of the problems of microclimate has therefore acquired a general importance today, extending far beyond its original sphere of reference. People are more aware today of the dangers that may result from disturbances of the natural order of growth than in previous times. The person who would upset microclimate through ignorance or business interests must be converted into a conscious and responsible controller of microclimate.

Conscious direction of microclimatic conditions extends to the control of the plants and animals that provide nourishment for men, and to their own lives, homes, and work.

Nearly every chapter of this book has shown, as far as plants are concerned, especially Chapters V and VI, how man can improve the living conditions of plants by influencing microclimate. Sections 19 and 20 included measures that might be adopted to alter soil conditions and thus improve plants. Artificial protection against frost and wind will be the special subjects of Secs. 53 and 54. Irrigation should also be included in this category. This has a double purpose: to further the optimum growth of plants by giving them the supplies of water they need, and to increase air humidity during the hottest hours of the day in a dry period; this has the effect of reducing, as far as possible, the passive transpiration of the plant, which is its method of protecting itself against high temperatures, but which makes unnecessary demands on energy. The first of these aims is achieved by watering at night, the second by watering during the day with as fine a spray as possible. The period of watering, its duration, the quantity and temperature of the water must all be adjusted to suit the type and age of the fruit to be protected, and the weather at the time. Irrigation techniques must therefore go hand in hand with microclimatology (Fig. 86 has already shown the effect of watering on soil temperature). A survey of this matter has been given by S. Uhlig in [1051]. Microclimatic measurements in tobacco fields were published by M. Diem [1027, 1028] and R. Trappenberg [1050], for irrigated meadows in the Black Forest by G. Reichelt [1045], and for vineyards by H. Burckhardt [1022]. C. W. Thornthwaite [1048, 1049] is of the opinion that in no other way can man modify climate so much as by altering the water balance.

In this context two problems of microclimate have a direct influence on plant growth: the climate of greenhouses and of clamps.

A new and comprehensive exposition of greenhouse climate, and the numerous possibilities of influencing it, was published by J. Seeman [1046], including 62 references to other works published up to 1957. In greenhouses, radiation from the sun and sky enters partly through the glass roof and partly through the glass side walls. The fact that the space within is enclosed prevents the plants from exchanging the heat received freely with the atmosphere. Heat is given off rather in the form of a direct exchange of long-wavelength radiation with the glass roof, which in spite of all its cooling maintains a comparatively high radiation temperature. The customary method of glazing with clear glass (window panes) or with gardening clear glass, ground on one side, has only a small effect on incoming short-wavelength radiation, according to laboratory experiments. The amounts lost are 7 to 8 percent by reflection and 8 to 15 percent by absorption for the whole range of solar radiation (320-2800 mμ), but only 2 to 3 percent for visible radiation. However, unavoidable soiling of the glass must be taken into account in practice, for which A. Niemann [1042] has recently given figures. It is possible to keep this loss down to 5 percent, but in industrial areas it may exceed 50 percent and even 70 percent in extreme cases. Nowadays, plastic materials with properties similar to those of glass are also used.

The microclimate of greenhouses is also influenced by the way they are built and used. Those built in an E-W direction have been shown to be superior to those lying N-S in the winter and in the transitional seasons of European climate. Steep glass roofs are also better in winter than flat ones. The position and number of the parts of the framework that cast shadows are also of importance. The gardening glass, with its ability to scatter light, is preferable to window glass, since it reduces the sharp contrast of light and shadow from the roof struts. Double glazing should be considered for purposes of heat insulation in winter.

Practical measures employed to control climate are: shading by movable mats, coir materials, wooden blinds, and so forth, to cut down incoming radiation; ventilation, for which there are very many different methods; artificial heating of the air or soil, or direct heating by radiation, and finally covering to prevent loss of heat. There is a movement toward automatic control of greenhouse climate, which can be adjusted to any special plant requirements, by controlling both temperature and the amount of light available, so that plants will not breathe so much, with the lower temperatures at night, but will have optimum temperatures for assimilation in sunny conditions (the Luvatherm process). It is understandable that the study of hothouse climate is now a separate branch of science.

Winter supplies of potatoes and turnips, and more recently of fruit as well, are stored in the open in clamps. The most favorable storage temperature in the inside of the hill-shaped clamp is 2° to 4°C. The first temperature measurements in the study of the climate of clamps were published by W. Kreutz [1036] in 1948. Diurnal variations and small irregularities of temperature are eliminated by a covering of straw heaped up with loose earth, which has poorer thermal conduction than the soil itself. To this is added the layer of winter snow, which may, however, increase the risk of frost on the south side of the clamp by melting earlier. The decisive factors are the site selected, the time of storage, and the weather when the stack is built, the ventilation of the clamp by providing channels or windows or perhaps by building the clamp in a position exposed to winds, and the changing of its covering at the right time following the trend of winter weather. In the German Federal Republic, according to W. Kreutz [1038], temperatures are measured continuously in 20 model clamps, extending from Bremen to the foothills of the Alps, and advice is broadcast to farmers over six radio stations on what action to take with their clamps. Details of clamp microclimate are to be found in F. Hummel [1032], H. Kern [1034], and W. Kreutz [1036-1038] for Germany, and in E. M. Crook and D. J. Watson [1024] for Britain.

The way man looks after animals, as far as microclimatology is concerned, consists mainly of housing domestic animals. The climate of stalls has already been discussed on pp. 478ff, but modern meteorology is now exploring new paths, in extending heat-balance investigations to include the heat budget of the animal's body. In Australia, for example, where sheep occupy such an important position in the national economy, C. H. B. Priestley [1044] calculated the amount of solar and sky radiation, and short-wavelength radiation reflected from the ground, incident on a sheep standing under the tropical sun, its long-wavelength radiation budget, and its heat exchange with the surrounding air by means of conduction and eddy diffusion. All these factors were calculated for an ideal sheep in the form of a horizontal circular cylinder. Two layers were distinguished in the calculations, the fleece with its poor thermal conductivity, and the actual body of the sheep. The investigation made it possible to evaluate the external physical heat exchanges in terms of measured meteorologic conditions, and hence to distinguish them from internal physiologic processes.

Man only only is concerned with suitable microclimatic conditions for plants and animals, but has a direct interest in making conditions suitable for himself. This leads us into the realm of cryptoclimate. This term is used to describe the climate of completely or partially enclosed spaces, especially spaces of the kind used by man and animal. C. E. P. Brooks and G. J. Evans [1020], who defined the term, have published an annoted bibliography of 282 references describing the climate of dwelling houses, schools, theaters, hospitals, libraries, museums, factories, warehouses, offices, laboratories, cellars, ships, tunnels, tents, and many other places; there are also references to the climate of caves, which has already been mentioned (Sec. 48), stall climate (Sec. 50), and conditions inside clamps. The relations between climate (and microclimate) and architecture are discussed in a book by J. E. Aronin [1018]. It is possible to give only a few of the basic ideas and some explanatory examples from this range of bioclimatic problems, the scope of which is rapidly increasing to gigantic proportions. These will be limited to the living habits, the housing, and the work of men.

The first type of microclimate to be considered is that prevailing in our means of transport. I. Dirmhirn [1029] investigated the conditions to which travelers were exposed in 13 types of streetcars in Vienna, in winter and summer. The excess temperature in winter of the interior above the air outside depended on radiation, wind, inflow of fresh air through windows and doors, and the number of passengers. It varied from 2.9 deg in trailers with open platforms to 10.3 deg in powered cars of American make. The excess was 3.5 deg in empty cars, 6.0 deg in those half filled, 7.7 deg in fully occupied cars, and 12.7 deg when they were overcrowed. There was a marked temperature stratification within the car; on one occasion, 0°C was measured at the floor, 6.0°C at 60 cm, and 8.5°C at a height of 1.7 m. The relative humidity fell, as was to be expected, as the excess of temperature increased.

Measurements were made inside an automobile by E. King [1035] at Tübingen in May 1956, using a sedan of volume 2.1 m^3, driven by a rear engine. After the first set of measurements had been made, its color was changed from black to white. When the car was left standing in the palace yard at Tübingen, with two people inside it, the excess of temperature compared with a shelter nearby, and the actual temperature of the air in the interior were as given in Table 94. Although in the second case the weather was warmer and richer in radiation, there was less heating inside. In both cases, however, the temperature at 12:00 inside the automobile was close to or even higher than blood temperature (compare with the number of babies rescued by police from parked cars during the October carnival in Munich). The temperature of the sunny side of the automobile was 2.5 to 3.5 deg higher, and that of the shady side about the same amount lower, than the air temperature inside.

Table 94

Temperature conditions in an automobile.

Color	Measurement	Time of day 08:00	10:00	12:00
Black (9 May)	Global radiation (cal cm^{-2} min^{-1})	0.13	0.34	0.41
	Temperature excess (deg)	3.0	10.6	19.4
	Automobile temperature (°C)	15.6	28.0	40.6
White (29 May)	Global radiation (cal cm^{-2} min^{-1})	0.19	0.40	0.45
	Temperature excess (deg)	1.2	3.0	12.0
	Automobile temperature (°C)	21.0	26.6	37.4

With the automobile in motion, the excess of interior temperature over that of the air above the street, measured simultaneously, other conditions being similar, is given in Table 95. It is seen that the influence of color remains considerable. Nowadays the air-conditioning of automobiles is being developed; see H. L. von Cube [1025].

Table 95

Excess (deg) of temperature in a moving car over that of the outside air.

Speed (km hr^{-1})		0	20	40	60	80
Black sedan	Closed	14.0	12.5	12.1	11.9	11.3
	Window open	9.2	7.9	6.8	6.1	5.4
White sedan	Closed	7.3	6.6	6.0	5.7	5.6
	Window open	4.4	3.9	3.5	2.6	2.5

I know of only a few measurements made in trains and aircraft. The main point of interest here is the effect on goods that are sensitive to weather conditions. A. Cagliolo [1023] measured maximum and minimum temperatures in three types of cars designed for transport of fruit in Argentina; W. L. Porter [1043] in Arizona recorded temperatures up to 67°C under the roofs of covered freight cars when temperatures outside were 46°C.

Men know how to reduce the adverse effects of climate on their buildings, and how to create an appropriate climate in their dwellings. In a comprehensive report on the climate of houses, H. Landsberg [1040] showed that man can feel comfortable only with a temperature between 18° and 32°C but is able to extend his habitation into parts of the earth where temperatures range from - 76°C to + 63°C.

The climate of a room depends in the first instance on the situation of the house and on the position of the room inside the house. The direction in which it faces, its height above sea level, the arrangement of windows, and the immediate neighborhood decide the atmospheric conditions to which the dwelling is exposed. The way in which a room is able to cope with these conditions depends on the building material, the strength of the walls, the size of the room, its doors and windows, and so forth. It depends, as W. Leistner [1041] once expressed it, on the "invasion" of outdoor conditions into interior climate.

The exchange of air between a room and its environment takes place mainly through gaps in doors and windows, but also occurs through pores in the material of the walls. H. W. Georgii [1030] was able to separate these two ventilation factors by enriching the air in a sealed room with carbon dioxide, the penetration of which through walls has been known since 1858 from the work of Max von Pettenkofer, and then carrying out the same experiment with an artificial aerosol (a sprayed $CaCl_2$ solution) which can escape only through the gaps. In each case the quantity measured was the rate of decrease of concentration with time, which was then expressed as a function of wind speed and temperature difference between the outside and inside, with both doors and windows closed. The results refer to a second-floor laboratory, of volume 32 m^3, in the Frankfurt Meteorological Institute, with two outside walls of brick, two doors, and one window.

The surprising result was found that there was no discernible connection between the temperature difference of the exterior and interior and the amount of self-ventilation occurring. By contrast, the influence of the wind blowing outside was strong, as may be seen from Fig. 258. In calm conditions, the CO_2 readings gave an exchange of air amounting to 2.3 lit sec^{-1}, which increased very quickly as wind speeds rose. The carbon dioxide was involved in both methods of self-ventilation, whereas the aerosol could escape only through cracks, and therefore indicated a much lower rate of air exchange. The aerosol in this concentration undergoes through sedimentation, coagulation, and diffusion at the walls a certain amount of alteration with time, which, when allowed for, gives the corrected curve (dashed line). The average of 35 series of measurements gave the result that 0.46 of the air volume in the room was exchanged every hour by autoventilation (the ventilation coefficient). The value 1, that is, an exchange of air equivalent to the total volume of the laboratory in an hour, is just reached with winds of 6 to 7 m sec^{-1}. This is in good agreement with the results obtained by other investigators.

It is worth noting that the ventilation coefficient of a cellar in the same building was more than double this value (1.15). This was found to result from the suction effect from the warm staircase. As a consequence, there was a marked dependence of the

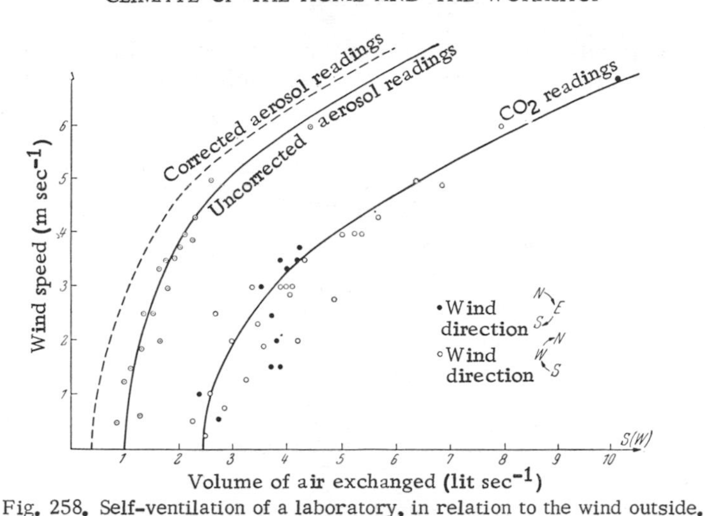

Fig. 258. Self-ventilation of a laboratory, in relation to the wind outside. (After H. W. Georgii)

effect, in this case, on the temperature difference between exterior and interior.

Measurements made in Innsbruck by R. Giner and V. F. Hess [1031], over a whole year, established that the dust content of a closed room (measured by Owen's method) was on the average only 58 percent, and the content of condensation nuclei (measured by the Aitken-Lüdeling method) was only 31 percent of the values on the balcony. After ventilation, this state became re-established in 3 to 4 hr. The situation with regard to dust is rather different where people are active (98 percent in schoolrooms), or some form of work is being done (the rag-pulping room of a paper mill has 800 to 900 percent). The "dead inside air" (C. Dorno) in a room, according to measurements made by F. Dessauer, W. Graffunder, and J. Laub [1026] in Frankfurt, has a much higher ion count than the "live outside air," amounting roughly to the quantity found only with summer thunderstorms, and the ion concentration is much more variable.

It should be remembered that great contrasts are also possible within a room. This has been proved in the clock room of the Geophysical Observatory on the Collmberg near Leipzig, which has no windows, and which is maintained at constant temperature. Figure 259, from D. Stranz [1047], gives a cross section of the room. It appears, from measurements made in February 1941, that a double air circulation developed, with a boundary surface at a height of about 1.5 m. In spite of the stationary state prevailing, in which the same amounts of heat were given out and taken up, there were differences of up to 6 deg within the room.

Air conditioning makes it possible to exercise complete control of the climate within a room. Sometimes such control is indispensable, as in deep mines, or in industrial storage rooms containing goods sensitive to temperature or humidity; sometimes it is desirable for comfort, and then brings a risk of enfeebling through living too much in a "hothouse."

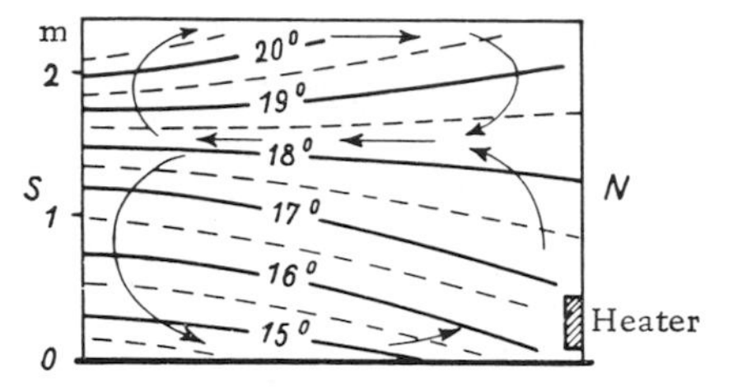

Fig. 259. Temperature stratification in a room without windows, in which temperature is kept constant. (After D. Stranz)

The climate of a room depends on the climate of the house as a whole. The effect of sunshine on houses has been studied in detail by architects and meteorologists, for details of which one might well consult the research reports of O. Völckers, F. Tonne, and A. Becker-Freyseng [1052; see also 1151 and 1197]. Wind stresses and protection against wind have been studied recently on models by M. Jensen [1033] and by H. Blenk and H. Trienes [1072]. R. E. Lacy [1039] has even measured the rain distribution around a house. Questions such as these have an added importance for hospitals and convalescent homes, in which connection it is worth referring to the survey by H. Landsberg [1040, with bibliography]. City climate, which in turn affects house climate, will be discussed in Sec. 52.

A great many tasks of practical importance in bioclimatology arise from problems of working conditions and storage and transportation of goods. The microclimates of schoolrooms, hospitals, libraries, factories, storage cellars, air-raid shelters, and tents have been measured [1019, 1020]. The first treatment of climate of living quarters and holds in ships sailing through different climatic zones was given in the report [1021] of a conference in Hamburg in 1956. It is not possible to go into these individual problems in any more detail here.

52. City Climate

The total transformation of natural landscape into houses, streets, squares, great public buildings, skyscrapers, and industrial installations has brought about changes of climate in the region of large cities. If a large city is defined as one with more than 100,000 inhabitants, there were 879 of them in the world in 1951, containing 160 million population; in the German Federal

Republic there were 47 with 13.6 million inhabitants or 28 percent of the total population. City climate (also called polisclimatology in German) is therefore of considerable practical importance.

A. Kratzer [1060] published a monograph on city climate in 1956 (second edition) containing a bibliography of 563 references. H. E. Landsberg [1061] published another survey, taking special account of American experience. There are also available descriptions of individual towns, from which much can be learned, for example, A. Sundborg's monograph on Uppsala [1069], published in 1951, and the comprehensive studies of Vienna [1067] and Linz [1062] by teams of investigators.

The method of investigation formerly used to establish the nature of city climate, by making simultaneous observations inside and outside the confines of the city, makes it difficult to eliminate the additional influences due to topography and vegetation, even when the series of observations can be made homogeneous with respect to measurement techniques and simultaneity. A special meteorologic study was begun in 1929 when Wilh. Schmidt in Vienna, at the same time as A. Peppler in Karlsruhe, began to record temperatures in the area of the city by automobile. Care was taken to allow for the variation with time of the elements measured, by making loop circuits, returning several times to the same place in the area being investigated. The method (Sec. 56) has been improved and extended in the meantime, and has led to reliable investigations into the microclimate of large cities in all parts of the world.

The basic reason for the differences found in city climate is the alteration of the heat and water budgets. This is caused by natural ground becoming largely replaced by stone, from which precipitated water is quickly lost, and because the roughness of the surface has been increased by the presence of buildings. In addition, heat is supplied by domestic and industrial fires, and, finally, city air is rich in dust stirred up by traffic, and in exhaust fumes from vehicles, from fires, and from industrial works. A. Kratzer [1060] estimated from the coal consumption of large towns in Germany that, on the average, the supply of heat was about 40 cal cm^{-2} per day. From radiation measurements made in Hamburg, the sun and sky supply an average of 34 cal cm^{-2} on a December day, and 50 cal cm^{-2} on a January day. In winter, therefore, the artificial supply of heat is about the same as that obtained from radiation, and is not to be neglected in summer either.

The most striking impression of the extent of atmosphere pollution in a city is obtained when it is approached on a clear winter day, especially from a higher level. A dark-gray, sometimes colored, cap of smoke, often piled up like a dome, lies over the city. This alters the whole radiation balance of the city area. Measurements within the smoke cap often allow three layers to be distinguished: the first of these lies above the surface of the streets between the houses and contains most of the stirred-up dust and exhaust gases emitted by vehicles; a second layer, at a height of about 20 m, is fed by gases coming out of the chimneys of houses; and a third layer, about 50 to 60 m above the ground, is caused by smoke from industrial stacks.

Figure 260 shows measurements made with a Zeiss conimeter in the lowest dust layer in Leipzig by A. Löbner [1063] when an

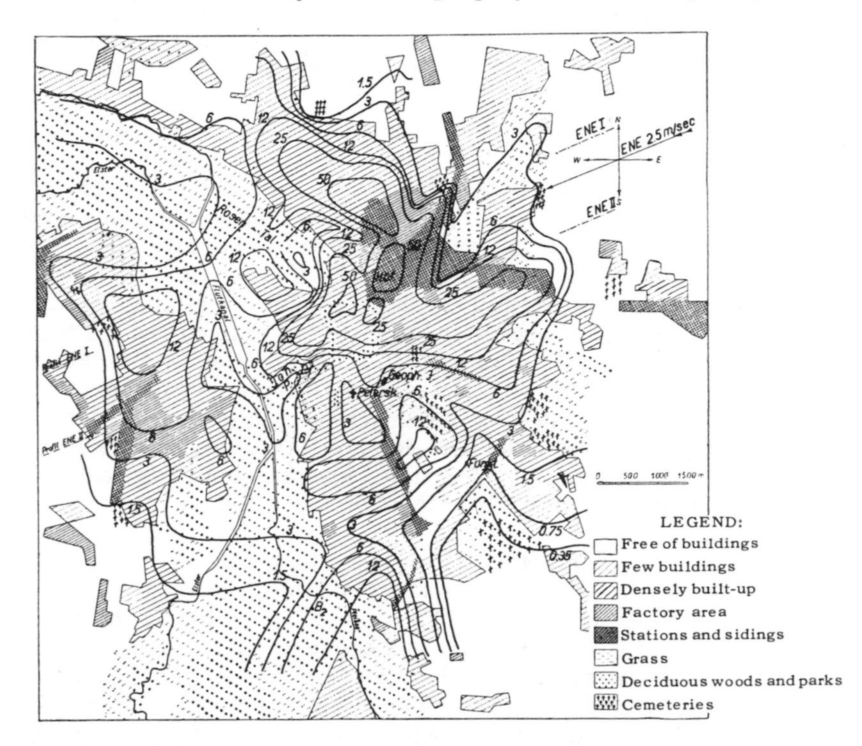

Fig. 260. Distribution of dust in Leipzig with an ENE wind. (After A. Löbner)

ENE wind was blowing. The figures on the isopleths of equal dust content give the number of dust particles per liter of air, when multiplied by 100. Air blowing over areas not built-up is still comparatively clean, but increases its dust content immediately on entering Leipzig. The increase is very great, and occurs suddenly in the area of the railroad station in the northeast of the city. The green areas of Rosental (upper left of diagram) filter the dust out again almost as quickly. In passing through the industrial area to the west the increase in dust is insignificant, probably because the high factory chimneys feed the uppermost dust layer, and not the lowest, in which these measurements were made.

More dangerous than dust are chemically active solid and gaseous products. Their composition depends on the type of industry present, about which some information may be found in H. E. Landsberg [1061]. The special case of Los Angeles smog has already been mentioned on p. 133. The principal contaminant of this type is sulfur dioxide (SO_2), which may cause damage to plants, and which, in particular, may drive coniferous trees out of a city. In transient conditions of high concentration it may also be injurious to human health. New results have been published for measurements made in Vienna in 1956-57 by A. Bangerl and F. Steinhauser [1054; see also 1066]. Figure 261 shows the mean distribution of sulfur dioxide in the town and surroundings of Linz in

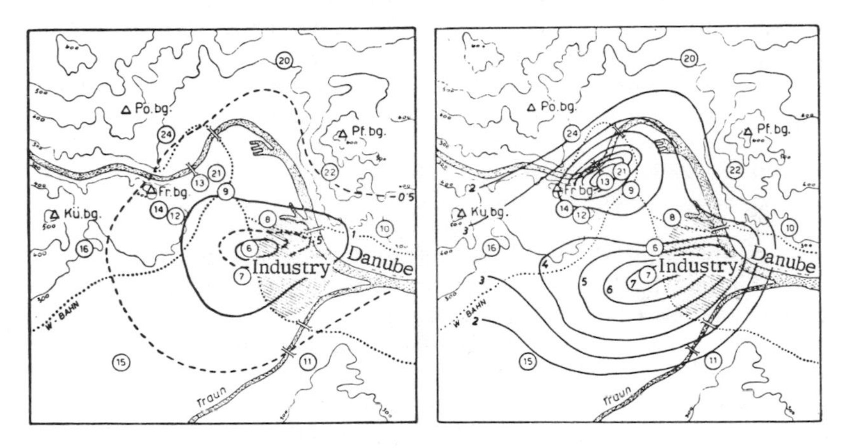

Fig. 261. Distribution of sulfur dioxide in Linz in summer (left) and winter (right). (After E. Weiss and J. W. Frenzel)

summer (left) and in winter (right), from E. Weiss and J. W. Frenzel [1071]. The figures in small circles give the points at which measurements were made, the sulfur dioxide being absorbed by cotton cloth impregnated with a solution of potassium carbonate and glycerine in water. The figures on the lines of equal concentration give the number of milligrams of sulfur per 100 cm^2 of absorbent cloth per 100 hr. Being average values, they are low in comparison with the values possible on individual occasions, and also low compared with other cities.

In summer, the maximum sulfur content in Linz is found in the industrial area in the southeast of the town. In winter, when consumption of domestic coal is higher, there is a second maximum in the heart of the city, and concentrations are higher on the whole. In addition to this variation over the year, which was well marked everywhere, there was a weekly variation which could be explained by the living habits of people. In Hamburg, E. Effenberger and A. Linder [1057] used modern statistical methods to evaluate a series of measurements made over a 3-yr period (1954-1957) in which the coarse particles suspended in the air were picked up on tape by the impact method, evaluated optically, and converted into a particle count. They established a mean count of 41 cm^{-3} on Sunday, 49 on Monday, and 50 to 53 on other weekdays. The diurnal variation has already been treated in Fig. 65. It should be mentioned in passing that the normal electric field is disturbed by towns. Less mobile large ions are favored by contaminants in preference to small ions, and this increases the potential gradient to about double its value (about 80 V m^{-1} in open country, up to 200 V m^{-1} in cities).

Since some suspended particles are hygroscopic, they are particularly active as condensation nuclei and therefore increase the frequency of fogs in urban areas. The annual fog frequency therefore increases as the town grows larger, which may be seen clearly from the following mean values, taken from a publication by B. Hrudička [1059], for 20-yr periods for Prague:

1800–1820	1820–1840	1840–1860	1860–1880	1880–1900	1900–1920
83	80	87	79	158	217

Beyond this point there is no further increase, because of regular improvements in the construction of fireplaces, and because steam locomotives have been replaced by electric ones.

An excellent regional study of the fog frequency in urban areas is provided in the *Climatological Atlas of the British Isles* (London, 1952). It contains a chart of the Thames Valley, showing lines of equal fog frequency, drawn for the 09:00 observations, for the winter period from 1 September 1936 to 31 March 1937. In the region of the Thames estuary fog was present in 5 to 10 percent of all observations, whereas in the City the figure rose to over 40 percent.

The modification of the radiation balance caused by air pollution is one of the most important factors in determining the climate of a city. If atmospheric turbidity is measured in the middle of a city, it is found to be about 1 unit greater than on the outskirts; if the blue of the sky is measured, it is displaced 2 to 3 units farther toward white on the Linke scale. During the day, between 10 and 40 percent of the global radiation fails to reach the ground surface, depending on the altitude of the sun and the degree of turbidity, because it is absorbed and scattered by the layers of mist and haze. A considerable amount of long-wavelength radiation is emitted by the mist, providing a good protection against nocturnal radiation losses. (The theory, put forward in many books, that the roughness of the surface in cities gives protection against radiation loss because "most surfaces in the city are able to emit radiation only toward restricted areas of the sky" is based on a fallacy; see p. 272).

Taken over the year as a whole, a city is warmer than the surrounding countryside, the difference being of the order of 1 deg.

When winds are light, the difference may become considerable. Figure 262 shows the arrangement of isotherms in Karlsruhe on a midsummer evening, drawn up from temperatures measured on automobile trips by A. Peppler [from 1060]. Cooling is slowed down in the evening where there are large masses of stone buildings, because heat has been stored up during the day, and ventilation is impeded. On 23 July it was up to 7 deg warmer in the city than on the outskirts. The cooler air outside is able to force its way inward only slowly; it may even be that a freshening wind will reverse the process and cause warm air to flow out of the town, as shown by H. Berg [1055] for Cologne one night in June.

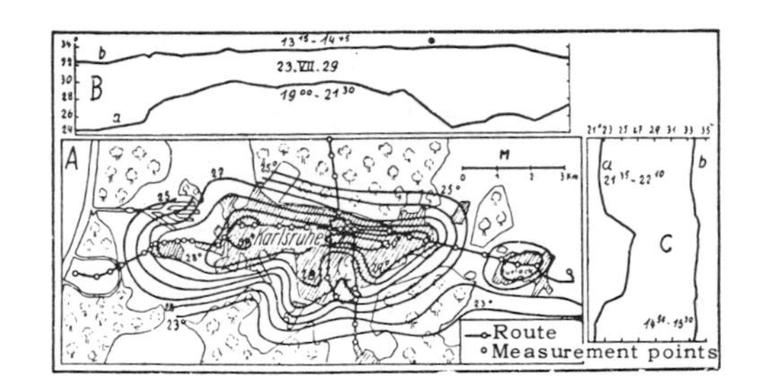

Fig. 262. Distribution of temperature in Karlsruhe on a hot summer evening. (After A. Peppler)

An observant city dweller may be able to recognize temperature differences from the behavior of vegetation. Earlier sprouting in spring, for example, was shown for Hamburg by E. Franken [1058], who mapped the time of appearance of forsythia blossoms in May 1955, with the assistance of 270 volunteers. The end of the growing period (leaf fall) also usually occurs earlier in towns. Often a town will be surrounded by snow while it is melting in the streets (not only because of traffic). From a number of older series of observations, it seems probable that precipitation in towns is more often in the form of rain instead of snow than is the case over country areas, and that the frequency of snow will consequently be less in towns, but this has by no means been proved for towns in general. It would seem, rather, that the numerous condensation nuclei available might cause falls of light snow to be released from low-lying clouds from time to time in winter, and that such falls would be restricted to the built-up area.

In winter weather situations there may also be convective influences at work, due to the heat of the town, similar to those often observed to release shower precipitation in midsummer. D. Stranz [1068] was able, by means of aerologic ascents over Göteborg on 1 March 1949, to make it quantitatively probable that the heat of the town was just sufficient to have triggered off precipitation from a belt of cloud moving over south Sweden. A similar case has been reported from Mannheim. Figure 263 gives the lines of equal precipitation (mm) for a summer cloudburst over the city of Munich. A shower coming from the west on 25 July 1929 produced a first maximum over the suburb of Pasing, and the principal maximum in the northeast of the city. Distributions of this type arise more frequently than can be attributed to chance. Using Bavarian statistics for hail, from 1884 to 1940, A. Weinländer [1070] was able to demonstrate that there was clear evidence of a similar increase on the eastern edges of Munich as well.

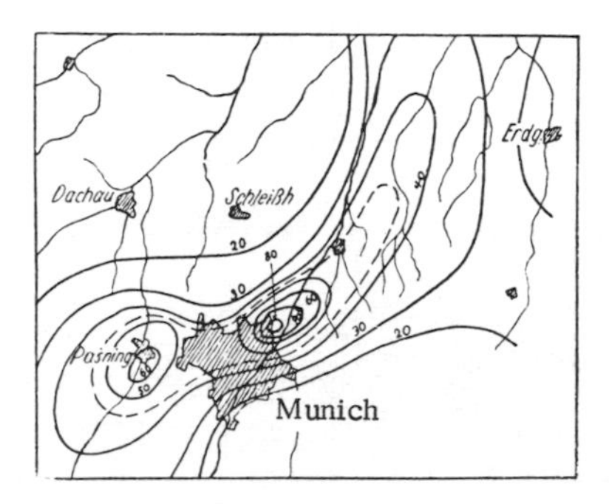

Fig. 263. A cloudburst is triggered over the city of Munich. (After J. Haeuser)

In general, it may be said of the climate of a city that its topographic situation (see Chapter VII) is the deciding factor in determining the extent to which the influences described above will be effective. It is only natural that a town in a sheltered situation in a valley, where winds are light, will show greater climatic differences inside and outside than will another situated on a plateau exposed to the winds. Situations on mountain slopes or on coasts will favor certain aspects of city climate. In extending cities or founding new ones, the known laws of city climatology should be taken into account from now on, much more than they have been in the past.

53. Artificial Protection against Wind

A distinction is made between damage to vegetation caused by storms and by winds blowing continuously for long periods. Storm damage is caused by the high wind speeds. Gales flatten grain standing in fields, and in the forests may uproot and blow down single trees, groups of trees, or whole stands. If the trees are firmly anchored and the ground is hard enough, the tree trunks break if the gusts are strong enough, or they may develop a resonant swaying motion under varying wind impulses before finally breaking. These are isolated weather phenomena which will not be discussed further.

Winds that are moderate to strong but that blow continuously, as in coastal areas, on high plateaus, or on mountain tops, have an injurious effect on the water economy of plants, by placing too heavy demands on it; they also adversely affect the water content and other properties of the soil. With increasing intensity of cultivation, especially in years of drought, soil erosion may occur, although this has only rarely been observed in Germany. Artificial wind protection binds the soil, promotes the growth of vegetation, and increases agricultural yield. Protection is provided by planting copses and rows of trees, or by erecting windbreaks of nonliving materials, such as fences of wooden slats, reeds, and the like.

The bibliography on this subject has expanded enormously in the last 10 years. Papers have been published in many countries, with many and varied climates, from different points of view, and employing different methods of research. Only the most important results will be mentioned in what follows, and it is therefore appropriate at this stage to make a short review of the more important works in the last 20 years (arranged chronologically). In doing so, the research technique will be described at times, to clarify the results obtained.

Modern research in this field was begun in Switzerland by W. Nägeli [1098-1101]. In 1941 he started making systematic observations of wind speeds in large forest strips and natural wind-breaking hedges of all sizes and types, most of them on level ground. Later he also included screens made of reeds, and extended the measurements up to a height of 9 m above the ground. Several examples will be quoted from his work.

R. J. van der Linde and I. P. M. Woudenberg [666] published in 1946 details of a practical method of determining the extent of shielding in the lee of a protective hedge, which has already been described (p. 343). Later they made nine separate investigations, each lasting several days, in the flat Dutch countryside near Oldebroek in northern Gelderland, measuring air temperature, humidity, and wind inside and outside the windbreak, and published the results in 1950 [1096]. The windbreak consisted of hedges of oak underbrush, 66 to 72 percent *Quercus robur*, several 100 m long, irregularly arranged, mostly running from NW to SE. It was planted on low earthen dikes 5 to 14 m in breadth, was cut every 8 to 9 yr, and grew to about 5 m in height. No other investigation offers such an exhaustive analysis as this of temperature changes in the area of influence of a windbreak.

In 1952 W. Kreutz [1093] published a book in Germany on wind protection, in which he collected all previous literature and the results of his own work in a form suited to the practical requirements of agriculture. A joint study was carried out in Fukuoka, Japan, by eight professors of Kyushu University under the leadership of K. Sato [1104], including:

(1) Wind-tunnel investigations of various windbreak models;

(2) Measurements in a level experimental plot 60 by 40 m, over which the natural variation of wind speed did not exceed 5 percent; this was brought about by leveling, and by keeping the grass on the plot at a height of 10 cm. One or more shelterbelts, 20 to 23 m long, were erected on this plot, consisting of the tops of Japanese cypresses (*Chamaecyparis obtusa*) buried 0.5 m in the ground, and cut off at a height of 2 m to give an even upper surface. The following points were then investigated, using anemometers at four distances from the windbreak, and, at five heights up to 6 m, aspiration psychrometers and artificial smoke trails: (a) The effect of one to three shelterbelts for comparison with wind-tunnel measurements; (b) the effect of the density of the shelterbelt (distance between trees within the belt); (c) the influence of the number of belts set up one behind the other as well as of (d) the distance between these; then (e) the effect of gaps in the shelterbelt; and finally (f) the effect of various shelterbelt shapes.

(3) Measurements in two shelterbelts of Japanese cypress 10 to 13 m high, with 1 row and 16 rows of trees respectively, in the rough terrain on Mount Aso, in the shelter of which highland rice was growing. Twelve rows of instruments were arranged perpendicular to the principal wind direction for measuring wind, temperature, and humidity.

Another Japanese research group, under S. Tanaka, T. Tanizawa, and others [1111], has published since 1953 the results of various measurements made in front of and behind hedges for wind protection; nine such reports have appeared by 1957.

In 1954 M. Jensen [1089] continued the classical experiments of the Danish Health Society. These comprehensive publications summarize the basic aerodynamic laws pertinent to wind-protection problems, and give numerous results of a new assessment of investigations made by C. Nökkentved, some in wind tunnels, some on experimental plots (two parcels of land with different degrees of protection, between two unprotected areas), and some in W-E sections through the Jutland Peninsula. In addition to wind, soil and air temperatures, the water and heat budgets (one area heated electrically), and the crop yield were measured.

The aerodynamic studies of wind protection published in 1956 by H. Blenk and H. Trienes [1072] contain detailed analyses of wind-tunnel experiments on models in the Brunswick Technical College (wind-speed profiles and soil evaporation). By making comparative measurements in the open, the question was examined whether results obtained with models could be applied in practice (effect of roughness of the ground, turbulence, and the shape of the obstacle). The results obtained in model experiments (p. 488) on the shelter effect of protective hedges on a house standing alone are also included.

In Scotland, J. M. Caborn [1074] published in 1957 the results of measurements on the wind field and the microclimate of shelterbelts in the hilly area near Edinburgh, together with some wind-tunnel experiments. This bulletin is a summary of the wind-protection problems for the forester.

Four regional investigations appeared in Germany in the same year. J. van Eimern [1079] investigated the variability of the wind field with the height of measurement, duration of measurement, speed in the open, wind direction, and the season of the year. Measurements were made near Hamburg at 17 points in the lee of a double row of trees (distances from 6 to 430 m) at heights of 1.5 and 3.0 m. The trees were maples 12 m high, running N-S, the spaces between the trees being shielded by two rows of shrubs running close by. Readings were made in summer when the trees were in leaf and in winter when they were bare. An investigation by E. Franken and E. Kaps [1082] into protection of farmland near the coast in the Emsland area contains a description of methods for precise analysis of shelterbelts and for computing the surface effects of various shelterbelt systems. G. Casperson [1075] measured the heat budget, with special emphasis on soil, air, and plant temperatures, in a sheltered and an open field near Potsdam. U. Eskuche [1081] measured local conditions of plant growth and yield in a sheltered position near Herbertingen in the Danube Valley.

In 1959 H. Kaiser [1091] finally published the results of an analytic investigation into the problem of wind protection, based on the analogy between the diffusion of smoke from chimneys and the decreased wind in the lee of a windbreak. The shelter effect was interpreted as the diffusion of negative impulses in the turbulent boundary layer. Diffusion equations and boundary-layer theory made mathematical treatment possible, and there was satisfactory agreement with many observations in wind tunnels and in the open.

After this short review of the more recent literature, we shall begin our discussion of the results obtained with the modification of the wind field, assuming that the ground is level. The effect of shelterbelts on microclimate will be discussed later.

Shelter from wind is generally provided by long narrow strips set up perpendicular to the direction of the prevailing wind. The length, breadth, height, and material may vary within the widest limits. Trees with and without undergrowth, hedges and rows of bushes, often growing on low earth and stone walls, sunflowers or other tall plants, lath trellises, straw matting, fences of woven reeds, wire netting with a small mesh, and finally board fences or walls are used as material. Figure 264 shows the braking effect on the wind within a shelterbelt 75 m wide consisting of a mixed stand 20 m in height for three wind-speed groups. The results were obtained by W. Nägeli [1099], using a close echelon of wind recorders in the Rhone Valley. The double effect of any loose shelterbelt can be recognized: the trunks, branches, leaves, or needles reduce the wind speed by friction, but at the sime time the cross section of the stream narrows as the air enters the shelterbelt and this increases the speed. It therefore appears, particularly with stronger winds, that the braking effect is delayed, and there are even cases in which an increase of wind has been measured in the entry zone. Rows of single trees therefore sometimes produce a kind of "funnel" effect, showing perhaps in the way grain is blown down by the side of a road lined with trees, especially when there are no low branches or undergrowth. These differences, however, are reduced after a

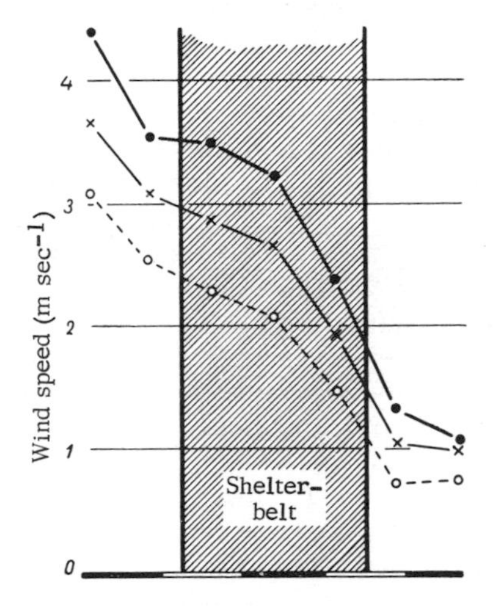

Fig. 264. Braking of wind in a shelterbelt of trees 20 m high. (After W. Nägeli)

distance, as proved by H. Blenk and H. Trienes [1072], and the shelter provided by the row of trees is the same as behind a hedge of the same penetrability. In the third of the Japanese investigations on Mount Aso, mentioned above [1104], the greatest wind speed was recorded inside the 16 rows of trees of the shelterbelt because the first branches of the cypress trees were several meters above the ground; the greatest shelter effect was produced 50 m behind the belt.

The extent to which a single shelterbelt is able to reduce the wind speed, its effect in depth, depends on wind direction, and the height, breadth, and nature of the belt. In all cases, the protective effect is a function of distance from the shelterbelt. It is substantially greater downwind than upwind. All investigations have shown, fortunately, that the effect in depth is proportional to the height h of the shelterbelt, to a very close approximation, and also proportional to the wind speed v. Distances x in front of the shelterbelt (-) or behind it (+) are therefore expressed in units of h, that is, the quantity used is $\pm\ x/h$ = multiples of the height of the belt. Wind speeds are expressed as percentages of the wind speed in the open. Figure 265 uses these quantities as coordinates to illustrate the shelter effect of various types of shelterbelts, for winds blowing at right angles to them. They are based on measurements made in the open by W. Nägeli [1099] at 12 different shelterbelts in Switzerland, with anemometers at a height of 1.4 m. All later researches confirm the basic laws thus established.

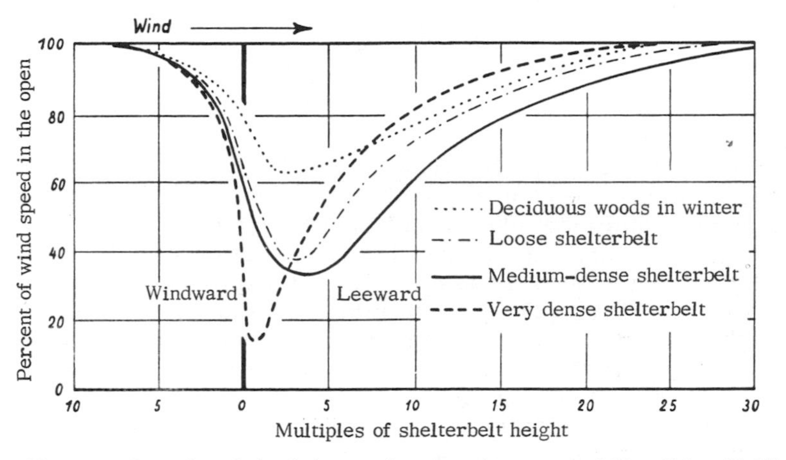

Fig. 265. The effect of a shelterbelt as a function of its penetrability. (After W. Nägeli)

As a rule of thumb it may be taken that a 10-percent reduction in wind speed is obtained for values of x between $-3h$ and $+20h$. The denser the obstruction, the greater is the amount of shelter immediately behind it, but the effect in depth is less. In the Japanese experiment described in 2(b) above [1104], the single row of cypresses standing 0.5 m apart was thinned systematically, and the mean speeds were observed as percentages of the wind in the open (Table 96). According to these figures (column $5h$) the reduction in wind speed generally increases as the density of the belt increases; but the distance upwind and to the lee of the belt extends relatively farther, as shown by the heavy-type figures, the lowest in each column, when the hedge is more penetrable. In these circumstances there is no formation of a lee eddy, but the flow of air is weakened by the obstruction. According to M. Jensen [1089], the most favorable practical results are obtained with a penetrability of 35 to 40 percent; according to H. Blenk and H. Trienes [1072], it is 40 percent to 50 percent. It has been shown by wind-tunnel experiments that the shape and arrangement of the empty spaces is without significance. When penetrability is variable, it is better if it increases from below upward than in the reverse direction. Shelterbelts without foliage have, according to Jensen, 60 percent of the effect they have in summer when in leaf; detailed comparative measurements in summer and winter were made by J. van

Table 96

Wind speed (percent of that in open) as shelterbelt is thinned.

Number removed from each six plants in a row	Distance from shelterbelt					
	-2.5h	0.5h	2.5h	5h	7.5h	10h
0	93	75	56	29	53	74
2	91	72	60	38	60	69
4	83	90	78	69	78	76
5	97	96	98	93	96	95
6	100	100	100	100	100	100

Eimern [1079] in a double row of maples. He found also that a change in wind direction of ± 45° from the perpendicular did not produce any significant change in the shelter effect.

It follows from the above that the shelter effect of a dense forest does not extend as far as that of a narrow penetrable strip of trees (see Nägeli [1101]). The minimum width is determined by the degree of closeness of form desired, and particularly by the biologic requirements of the plants, which have to be observed in setting up and maintaining the shelterbelt.

All the results mentioned up to now have been based on wind measurements made with instruments set up at 1 to 1.5 m above the ground, that is, high enough to be free of any chance effects due to the ground surface, but low enough to measure the conditions to which the plants to be protected would be exposed. Figure 266 gives an example of the measurements made by W. Nägeli [1100] at nine levels up to 8.8 m above the ground, on both sides of a reed screen 2.2 m in height in the Fur Valley near Zurich. Similar measurements have also been made by the Japanese [111].

The lines of equal wind speed (isotachs) in the upper diagram are for a reed screen with a penetrability of 45 to 55 percent, the lower for 15 to 20 percent penetrability. The effect on the windward side of the obstruction can be observed in the upper diagram, in the way the isotachs bend upward. Above the obstruction, where the streamlines are crowded, the isotachs also form into a bundle; the zone of increased speeds above the screen can be followed up to more than 9 m, and is displaced slowly toward the lee as height increases. This effect higher up in the diagram is accentuated not only because intermediate isotachs have been inserted, but also because the wind is still only increasing slowly with height, and individual isotachs therefore become markedly deformed. Behind the obstruction the isotachs fan out, and the line for 5 m sec^{-1} begins to sink again only at a distance of 15h.

The lower diagram shows the more powerful shelter effect in the neighborhood of the denser reed screen. The isotachs rise more steeply, are crowded more tightly above the obstruction, and there is a closed isotach of 1 m sec^{-1} in the lee eddy. However, in conformity with the basic rule, the isotachs bend down again toward the ground more quickly, showing that the denser screen has a smaller effect at a distance.

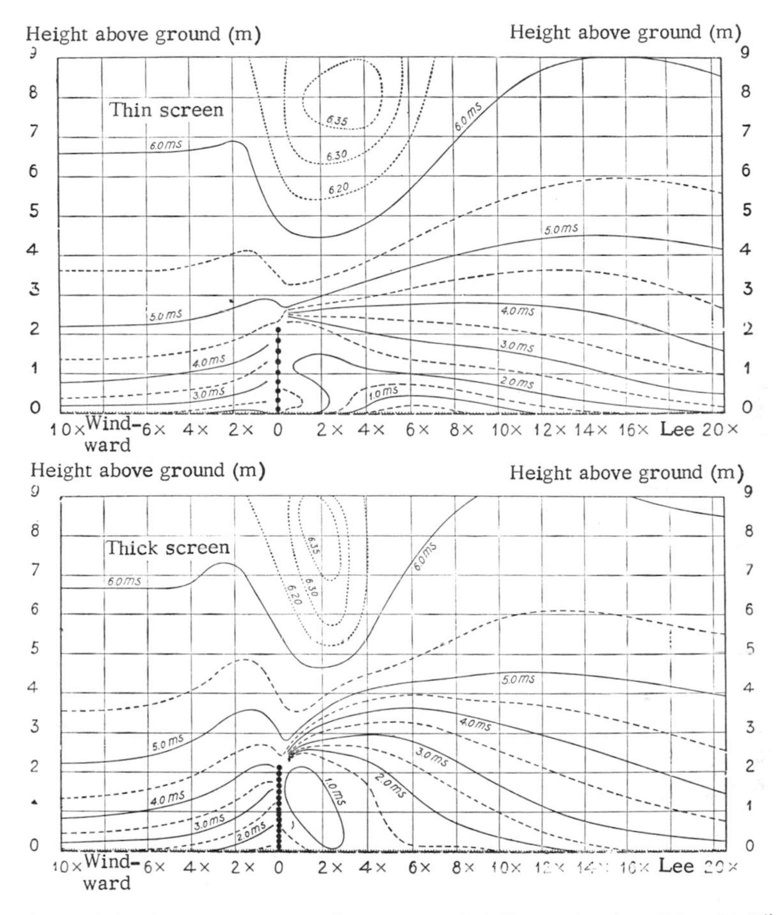

Fig. 266. Wind field aroung two reed screens of different density. (After W. Nägeli)

Results given so far are all based on mean values over fairly long periods. It is important to know, for practical purposes, just how long observations have to be continued in order to obtain a representative mean. This question was the subject of experiments made by J. van Eimern [1079] with the double row of trees already mentioned. The results illustrated in Fig. 267 are for measurements made at a height of 3 m. They are summaries of the summer values at distances of 15, 48, and 100 m with winds perpendicular to the direction of the trees. The abscissa is the wind speed in the open, and the ordinate is the wind speed in the shelter, expressed as a percentage of the former. Each dot corresponds to an hourly average, and the symbols used indicate the time of day. The points shown below the hatched line on the left lie below the limits of the accuracy of measurement.

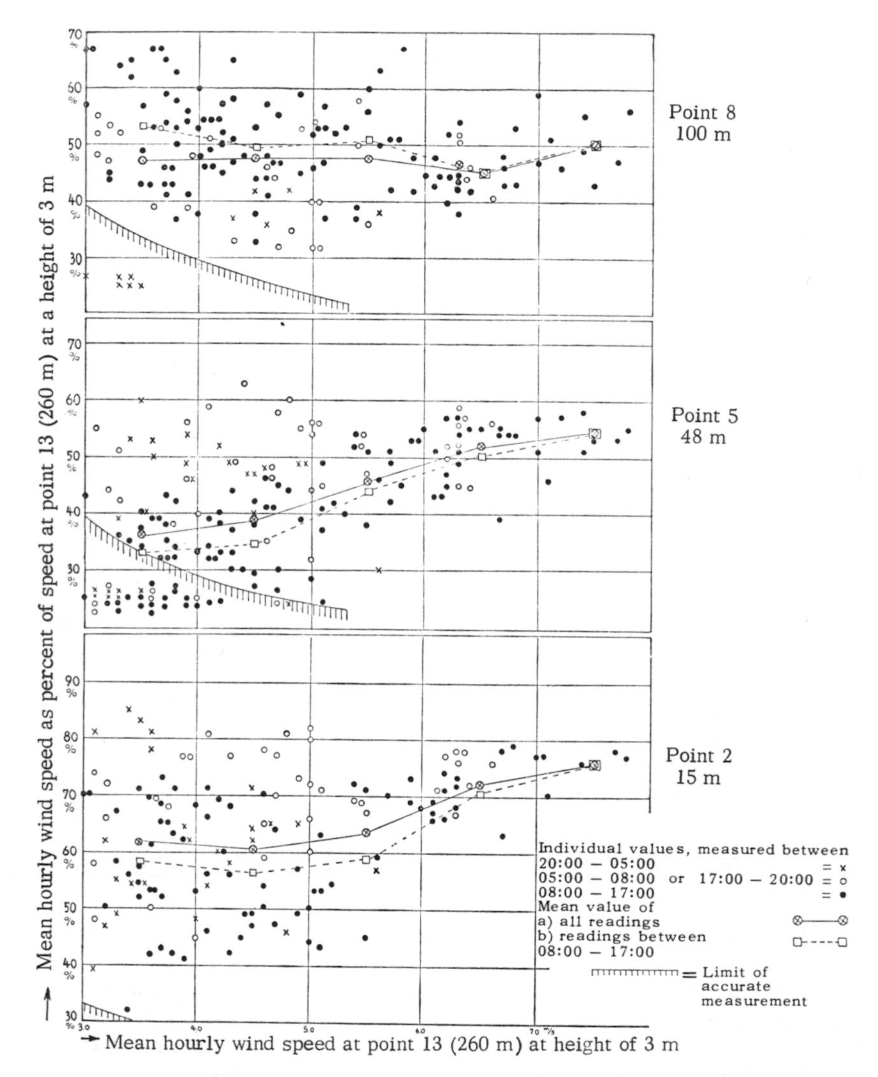

Fig. 267. Wind measurements at three points behind a double row of trees in foliage. (After J. van Eimern)

The scatter of these hourly averages seems extraordinary (more marked because only part of the ordinate scale is shown). It is naturally greater with weak than with strong winds. Similar measurements made in winter gave only half the amount of scatter. Analysis of the results in this case showed that, for a given wind direction, at least 20 hr of recording were required to collect data capable of being evaluated. The broken line, joining daylight values only, gives a relatively lower speed at a distance of 15 m and a relatively higher speed at 100 m than the average of all readings. The row of trees therefore appears to behave as if it were denser during the day and less dense by night. The reason for this lies in the more unstable structure of the air by day, with more stable stratification at night. More recent studies by H. R. Scultetus [1107, 1108] confirm this effect of temperature stratification and hence show that the shelter effect will also depend on the type of air mass.

Wider tracts of land are shielded by a number of shelterbelts in echelon. Figure 268, which is based on W. Nägeli's measurements [1099], illustrates how the protective influence of the second of two belts of trees, 20 to 25 m in height, is already effective at a distance of 400 m (= 18h) from the first, where the wind speed has again risen to 80 percent of its value in the open. H. Kaiser [1090] called this the "back coupling" of shelterbelts. It is natural for the practical man to ask what is the most effective distance apart, but it is not possible to answer this question in general terms. There are too many factors at work, such as the nature of the shelterbelt, the frequency distribution of wind directions, local features, and the topography of the area; it would also be impossible to devise experimental areas such that the local results would lead to general conclusions.

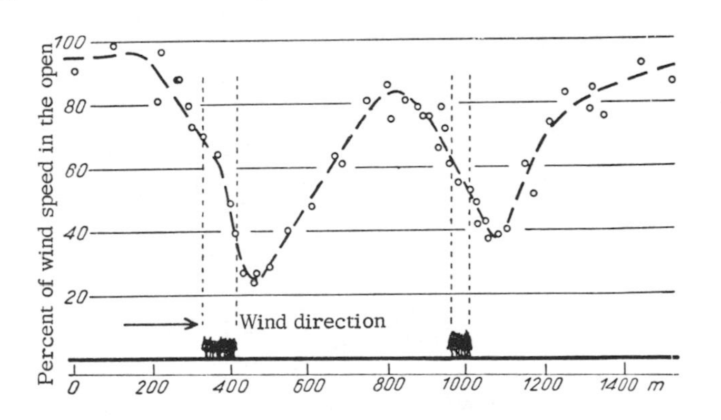

Fig. 268. The effect of shelterbelts arranged behind one another. (After W. Nägeli)

The features that could be generalized here were best summarized by H. Kaiser [1090] in 1959. Experience shows that in passing from the sea, where friction is small, the wind field becomes adjusted to the greater friction over land, after crossing a coastal transition zone, and a similar kind of modification takes place where there is a change from unprotected to protected areas. Shelterbelts established at equal distances one behind the other, thus increasing the roughness of the land (see Sec. 30), have been shown by experinece to have a substantially more favorable effect than might be expected from the individual experiments described above; this is found to be the case also, when not too great preference is given to the arrangement that places the shelterbelt and principal wind direction perpendicular to each other.

A much more favorable arrangement is that of a network of shelterbelts perpendicular to each other, in which the orientation of the belts is of fundamental importance only when the ground is not level. In this case the distance between the belts can be increased greatly: the shelter effect is more uniform and is generally least near the center of the network. The corners provide positions where it is particularly suitable to make the openings necessary for access. The hope that it might be possible to use rows of high trees as "wind-steering lines," to guide the wind into paths that would be convenient for the protective hedges, ignored the turbulent character of the wind field, and therefore cannot be realized.

An arrangement proposed by M. Woelfle [1115] in 1938, shown in Fig. 269, proved its value in the Upper Rhone; see S. Uhlig [1114].

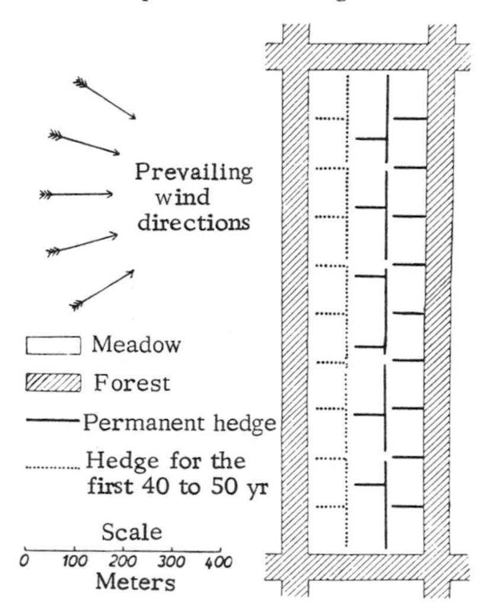

Fig. 269. Wind protection on the Woelfle plan.

The hatched strip shows the network consisting of a mixed stand 50 m wide and 15 m high. The area inside is subdivided by hedges 5 m high that perform a secondary function by fencing off cattle, giving protection to birds, and providing a source of hazel nuts and a place to build up stones cleared from arable land. The hedges shown by dotted lines could be taken away when the trees of the mixed stand had grown to a sufficient height. Establishment of shelterbelts in an unknown territory requires a thorough preliminary study of the areas. The natural distribution of wind speeds and directions must be analyzed as functions of the general wind direction, a survey must be made of its microclimatic features (see Chapter VII, especially Sec. 47), and the effects of the shelterbelts on the microclimate, discussed below, must also be taken into account. W. Kreutz [1093] gives two examples of this kind of planning, and the most instructive example is given in J. van Eimern's investigation [932] already quoted on p. 461.

The next discussion deals with the effect of shelterbelts on microclimate.

The first general remark to be made is that the simplified rule established for the wind field, namely, that low shelterbelts short distances apart have the same effect as high ones at correspondingly greater distances, must no longer be thought valid for microclimate. Further, a warning must be given against speaking of favorable or unfavorable effects, since such opinions depend entirely on local conditions. For example, a weak wind is usually advantageous for the water balance of the soil, but not in swampy marshland, and not when ripe but still moist crops must dry quickly at harvest time.

The radiation balance is modified by the shadows thrown by the shelterbelt. More can be learned about this from Sec. 41 and in R. J. van der Linde and J. P. M. Woudenberg [666]. Radiation by day is treated by I. Dirmhirn [1077] and S. Sato [1105]. The former computed, for Vienna, the amount of global radiation, subdivided into I and H, taking into account the observed duration of sunshine received by the ground between distances of 0 and $10h$ from shelterbelts, over the four seasons, for shelterbelts running W to E and S to N, and for directions between these. The differences are less in summer than in spring and autumn because of the high altitude of the sun. For example, the total daily global radiation in spring, expressed as a percentage of the undisturbed radiation, is shown in Table 97 (a much abbreviated extract). The loss of short-wavelength radiation is therefore restricted to the proximity of the shelterbelt. At night, outgoing radiation is reduced near the shelterbelt, for reasons given in Sec. 5, p. 24, and Sec. 37, p. 343.

The indirect effects of the modification of the wind field are worthy of note. Since a limit is placed on the peak values of wind speed, the danger of soil erosion is reduced sharply, as was proved statistically by M. Jensen [1089] and J. van Eimern [1079]. An incidental result of employing a single windbreak is the undesirable increase in wind speed which always occurs around the ends of the obstruction. Figure 270 gives one example, for the shelterbelt

Table 97

Total daily global radiation (percent of undisturbed radiation) near shelterbelts.

Direction of shelterbelts	Side of belt	Distance in units of height				
		0	0.2	0.5	1	2
West-East	N	27	33	39	48	97
	S	81	85	90	95	98
Southwest-Northeast	NW	37	45	60	79	92
	SE	71	77	81	92	97
South-North	Both	53	60	72	84	94

shown in Fig. 268. The fields at the edge of the belt experienced winds over 20 percent stronger. The same applies to gaps (funnel effect). There are many Japanese figures [1104] available for these two effects. Protection against wind should therefore be a community undertaking. An example of the rules set down by an association can be found in a silviculture periodical, *Die Holzzucht 10* (1956), 29f.

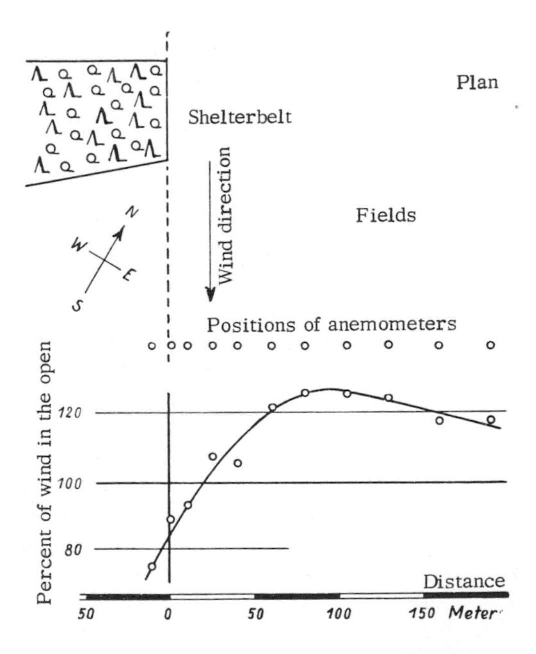

Fig. 270. Side effects of a shelterbelt. (After W. Nägeli)

The effect of wind protection on the dissemination of weed seeds, spores that might cause a risk of infection, and so forth, has been investigated by K. Illner [1088] and H. Schrödter [1106]. It has been proved, convincingly, by J. D. Rüsch [1103] that the air in the sphere of influence of the shelterbelt contains rather less carbon dioxide than it does outside.

A much-debated indirect effect of the reduction in wind is the increase in the risk of night frosts. It is clear that in sloping areas the cold-air dam (Sec. 42) retained behind the upper side of the shelterbelt will produce lower temperatures than on its other side. This has been proved by Y. Daigo and E. Maruyama [1076] for a 6° slope with a hedge of Japanese cedars, 1.2 m in height. Care must therefore be taken to allow spaces for the cold air to flow downward when planning wind protection in such situations. Dutch [1096] and Danish [1089] studies in the plain gave the following results: when there is a calm at night, temperatures are higher in the shelterbelt, because outgoing radiation is reduced and soil temperatures have been higher during the day. With brisk winds the temperature is the same inside and outside. Only with extremely weak winds—Jensen says 0.5 to 2 m sec^{-1}—is there increased danger of night frost in the area of the shelterbelt, particularly in the first weeks of spring, a time of year when the stronger heating during the day is still ineffective in the shelter area. But then the danger is less for the areas immediately adjacent to the shelterbelt.

On the other hand, it appears from all these considerations that during the day, especially in clear and dry weather, soil and air temperatures are substantially higher in the shelter areas. The difference in the uppermost soil layer is about 2 deg on the average, and 0.5 to 1 deg in the air near the ground. Observations made by G. Casperson [1075] on this very subject showed that excess temperatures of 10 deg were found on days of strong radiation in the shelter of a 3-m-high hedge of hawthorn near Potsdam. The following extreme example of a day in May illustrates the process well. The diurnal temperature variations (deg), differences in which depended mainly on the change in the daytime maxima, were:

Height (cm)	In the ground			In the air		
	-10	- 5	- 2	5	75	140
In the open	4.1	5.9	8.7	26.9	22.9	22.0
8 m behind hedge	7.2	11.4	15.5	32.9	26.0	24.2

This gain in heat, in conjunction with reduction of excessive evaporation, is certainly the cause of the improvement in growth and hence the increased crop yield in areas protected from the wind.

As far as the water economy is concerned, the lack of air movement favors atmospheric precipitation. This is the reason for the "increase" in precipitation in the field of influence of a shelterbelt. The essential points have already been discussed in Sec. 39, particularly on p. 365. When precipitation is in the form of snow, this feature is particularly marked. The familiar snow fences seen on country roads allow snow to pile up in their shelter, thus keeping it off the roadway. Observations and measurements of this process are to be found in [1079, 1092, 1093, 1099]. T. Müller [1097] mapped the funnel effect (see p. 498) by observing the way in which snow was swept away in and behind a gap in a hedge.

Similar processes can be observed at work in drifting sand, where it is possible to follow the effects for a considerably longer time than with snow. Figure 271 shows the results obtained by H. Kaiser [1091] with piles of sterile industrial sand. The wind-speed profile in the lee of the obstruction was measured before the sand started to drift, for winds perpendicular to it, at a height of 1.4 m in (*a*), 38 cm in (*b*), and 45 cm in (*c*). The sand accumulation has been exaggerated four times. The figure shows accumulation behind: (*a*) an impenetrable brushwood barrier after 7 yr; (*b*) a penetrable brushwood fence after 2 yr, and (*c*) a snow fence with a clearance at the ground, after 6 mo. The shape of the sand deposits follows the laws given on p. 500.

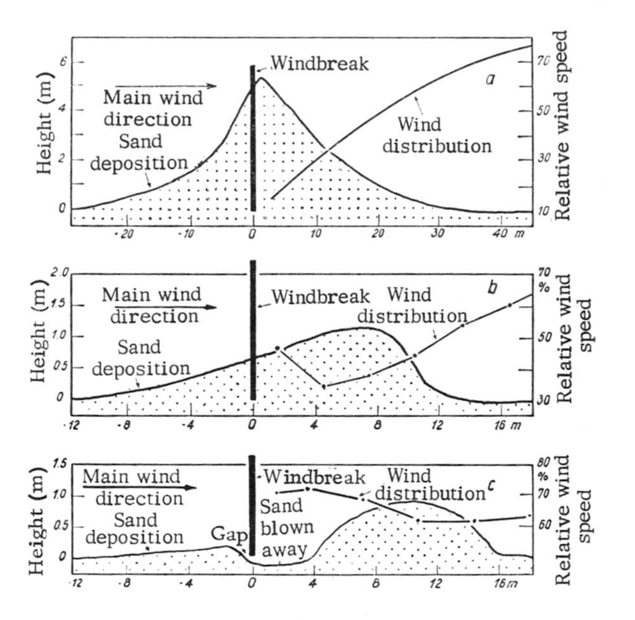

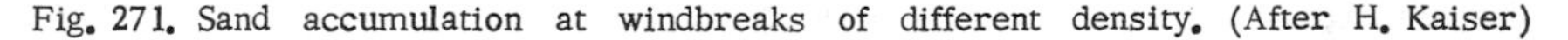
Fig. 271. Sand accumulation at windbreaks of different density. (After H. Kaiser)

Measurements made by L. Steubing [1110] using Leicke dew plates always gave greater quantities of dew in places sheltered from the wind. The difference was not greatest on clear nights with weak winds and deposition of dew, but on nights with strong air exchange. The greater amount of dew therefore results from the stillness of the air behind the obstruction and also to some extent from the higher air humidity there. L. Steubing found a maximum increase of up to three times the value in the open at a distance of 2 to 3*h* behind the obstruction. J. van Eimern [1078] also established that there was a delay in the disappearance of dew after sunrise, and found the greatest amount of dew midway between the hedges, where night temperatures were lowest.

The degree of wind protection has often been shown by measuring evaporation instead of wind; the close relation between wind and evaporation under similar radiation conditions (see Sec. 28) always led to the same results qualitatively. The reduction in evaporation is often considered to be the most valuable result of artificial protection against wind. It should not be overlooked, however, as M. Jensen [1089] pointed out, that the prevention of heat loss will increase evaporation, since this will become greater as the temperature rises and the mass of vegetation becomes greater. This productive evaporation has the effect that the water budget, as measured by Jensen, was not substantially different inside and outside the area protected by the hedges.

The problem of protection against wind is associated with many biologic problems, which cannot be dealt with here. In areas exposed to strong winds the plants to be used for the establishment of shelterbelts, the width of the belt, the distances between the rows and plant stands, and the suitable mixture of trees must be properly selected. Later comes the problem of regenerating the belt. A good guide to all these practical problems was published in 1959 by H. H. Hilf [1087]. Crop yield, which has been measured frequently, is so strongly affected by general climate, local conditions, and the type of plant that it can be evaluated properly only by an experienced agriculturalist.

Wind protection therefore becomes a problem of modifying the landscape. Geographers very rightly declare that this can in no way be regarded merely as a wind problem. The works of C. Troll [1113] and W. Hartke [1086] should be consulted for information on this subject. In Europe there are only certain regions where hedge landscapes could develop, namely, the coastal areas with a maritime climate in Denmark, Britain, France, and Portugal, the mountain regions in the Alps, and stony plowland where the farmers pile up stones turned up in plowing at the edges of fields, which are then populated by shrubs. The ever-increasing removal of the original natural landscape in Germany keeps the problem of wind protection, to which almost too much attention is paid, always before our eyes. It is to be hoped that attention to individual research will not cause people to lose sight of the countryside as a whole.

54. Artificial Protection against Frost

Damaging frosts that occur during the growing period are called late frosts when they occur in the spring and early frosts when they occur in the autumn. Night frosts in summer are frequent in high latitudes in Europe, but occur also in some places in the latitude of Germany, as shown by Table 89.

It is useful to make a distinction between advection frosts and radiation frosts. Advection frosts have their origin in the advection of cold air, are therefore dependent on the general weather situation, and are associated with more or less strong winds. A. Vaupel [1133] proposed that two stages should be distinguished according to the mass, the vertical depth, and the path of the cold air, and according to wind speed and structure. Radiation frosts are caused by radiative cooling by night when winds are still and skies are clear, and therefore affect the air layer near the ground. Damaging frosts are generally due to a combination of both factors: the advection of cold air reduces temperatures in general to near the danger level, and outgoing radiation then causes the plants to die. The basic condition for such frosts are outlined in Secs. 5, 14, 19-21, 29, 42, 43, and 47.

The most effective weapon that men possess against frost damage is preventive action. "The best time to protect an orchard against frost is when the orchard is being established," as W. J. Humphreys said in 1914. It is incomprehensible that even today the most fundamental laws of microclimatology are disregarded time and again, when new orchards are laid out at great cost in notable frost hollows.

Preventive measures should be extended into the surroundings of the area to be protected. The site selected should not lie in the zone of a cold-air current (Sec. 42); this can be shut off, as shown earlier, by hedges at higher levels or by forest belts, and in all cases the damming up of cold air should be prevented by providing means for the cold air to flow away downward. The whole area from which cold air flows at night, the so-called frost-catchment area, may be modified by the cultivation of wasteland, or drainage of grassland. Advice on this subject, illustrated for the practical farmer, has been given by H. Burckhardt [1118] and F. Schnelle [1125]. The function of the outer forest in providing frost protection was described in Sec. 38.

Information should be collected on frost danger as a function of topography when protective measures are being prepared. The best method of doing this is to map the danger areas; good examples are to be found in W. Baier's [1116] report for the wine country in the north of Württemberg and Baden, and in A. Vaupel [1133] for the wine industry of the Rhineland-Palatinate. The method of mapping frost danger has already been dealt with in Sec. 42. The geloscope, by which the advent of night frosts can be recorded, was described in Sec. 21. Alarm systems are also used, so that frost-protection plans can be put into operation at the right time. A thermometer set up in the plot to be protected (at a height depending on the type of plant) sets off an alarm bell on reaching a preset temperature.

Experience has shown that it is necessary to have a switching-off device so that the bell does not ring all night, but there is a risk of forgetting to reactivate the warning system in the morning in preparation for the following night.

To combat damaging frosts effectively requires a substantial amount of money, a state of constant readiness of the labor force in time of danger, and long hours of night work. This expenditure is economically justifiable only when it is a question of protecting valuable plants such as vines or expensive fruit, or when damaging frosts are rare, as perhaps in the California orange groves, which are roughly at the same latitude as North Africa. If there are a number of late night frosts in a single spring, as is often the case in Germany, the entire results depend on the night when protection has been least successful.

Frost protection may be achieved either by reducing the amount of heat lost by radiation (protection against outgoing radiation) or by supplying heat. The second method is effective and expensive; the first is practical on a small scale, but doubtful in large businesses.

Prevention of radiation is achieved by covering individual plants or by covering the whole area with a pall of smoke which will reduce the amount of radiation.

In small market gardens effective protection can be provided by solid coverings or by straw matting over plots with early plants. This method has been used, and proved effective, for decades. On a larger scale, frost covers have been successfully tried on vines [1126, 1130]. The covers are cone-shaped and must satisfy three requirements: they must be of a poorly conducting material, since radiation from their outer surface will cool them considerably, and this fall of temperature must as long as possible be kept from affecting the inside surface and the airspace below; secondly, the airspace inside the cone must not be too small, so that it will be possible for a separate microclimate to develop; and finally, it has proved useful to have an opening in the cone immediately above the ground. A gardener who covered plants with empty cans of the same size as the plants ignored all three of these rules and so condemned his "protected" plants to death by frost.

Modern techniques provide many substances that will produce artificial fog for frost protection. Radiation can be reduced to about half by very dense smoke clouds. Nevertheless, smoke clouds have not proved their worth in frost protection, and the reasons have been explained by J. Drimmel [1121] in a detailed study. To be effective, the smoke cloud must, of course, be spread over the area long before the freezing point is reached, if possible before the radiation balance becomes negative, or about 1.5 hr before sunset. To hinder advection of cold air from the neighborhood, the whole of the frost-catchment area, which usually means a wide surrounding area, must also be covered with the smoke.

This must be a coherent cloud of uniform and sufficient density, which is often difficult to achieve. It is not the maximum value of the reduction in radiation, often mentioned in writings on the subject, but the average value over the whole area that determines its success. A wind increase, even of little strength, will displace the cloud. For all these reasons, smoke clouds are used nowadays only in conjunction with other methods of heating as an additional help. The smoke cloud is, however, particularly suitable for use in the early morning after a night frost to protect the parts of the plants stiffened by frost from the strong rays of the sun. Experience shows that damage is less if the plants are allowed to recover slowly from the night damage, that is, if they are allowed to thaw slowly.

Plants can be protected completely from radiative cooling if they are covered with water. This is done in the United States with cranberries, which are grown in fields surrounded by earthen walls for this reason. The protective influence of small bodies of water on plants close to them or at a higher level is often spoken of, but it is too small quantitatively to be of practical importance. J. van Eimern and E. L. Loewel [1123] have been able to prove the existence of a "tube of slightly warmer air" over the ditches in the marshy land near Hamburg. The recommendation that the empty ditches be filled before a frost night cannot be justified since the dried-up and weed-filled ditches could then become an additional source of frost (grass frost). J. Hanyu and K. Tsugawa [1124] showed that the watering of Japanese rice fields for frost protection gave a warming effect only up to 60 or 70 cm; flooding with water cannot by itself give frost protection, but can at the most be used as an adjunct to other measures.

There are today three different methods of providing frost protection by supplying heat. It can be done by ventilation, the cold air near the ground being mixed with warmer air from aloft, by heating with burners over a definite area, or by utilizing the latent heat of freezing of water.

Figure 272 shows a propeller with three blades, 6 m in diameter, mounted on a tubular steel pole 9 m high above apple trees in a fruit-growing experimental station at Buxtehude. The axis of the propeller can be adjusted to any inclination, and the propeller can be set automatically to traverse horizontally through any required angle; power for a speed of 170 rpm is provided by a 10-hp electric motor. When the strong night inversion is established in the calm conditions on a night with late frost (Sec. 14) the propeller blows the air aloft, which is considerably warmer, down into the cold layer close to the ground. J. van Eimern [1122] has published a comprehensive report, with a bibliography, on the results obtained so far by this method, including his own research.

In America, propellers that blow horizontally are used; motors of up to 210 hp, giving speeds up to 900 rpm, have been tested.

Helicopters flying just above the tree tops have also been used with success as described by R. T. Small [1131]. In England, propellers blowing vertically downward have been used; these bring down the warm air more effectively, but have a limited horizontal range.

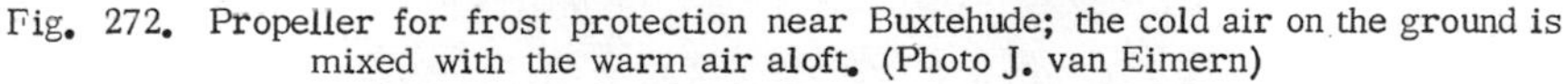

Fig. 272. Propeller for frost protection near Buxtehude; the cold air on the ground is mixed with the warm air aloft. (Photo J. van Eimern)

The two methods have been combined in Germany and Australia by having the propeller blow down at a slant, the most suitable angle depending on the height of the plants to be protected. Plants growing close to the ground, such as strawberries, can be protected more effectively than high fruit trees, if we remember the undisturbed temperature profile on a frosty night (see p. 96). The decisive question of how large an area can be protected by a wind machine can hardly be answered in general terms, because it depends on the structure of the garden. To give some idea, it might be mentioned that J. van Eimern, in experiments among apple trees, found that the propeller shown in Fig. 272 had a continuous effect of 0.5 deg over 1.2 ha and 1 deg over 0.8 ha, when inclined at 45° and traversing around 180°. For strawberry plants near the ground, it is estimated that there is an additional gain of at least 1 deg. Many small propellers are always more effective than a few big ones.

The cost is very high, as with all really effective measures. Compared with the heating processes next to be described, the advantages are that a minimum of personnel is needed (for turning the propellers on) and full effectiveness is reached within 1 or 2 min. There is no need to be afraid of the cooling caused by evaporation, as all the dew present is blown away in 5 to 10 min. There is perhaps an adverse effect in the boundary zone, where the wind has become too weak to produce heating by mixing, but still strong enough to affect the plants in the otherwise calm air of the night.

Frost protection by heating has been described in detail by O. W. Kessler and W. Kaempfert [1126] in their fundamental work published in 1940. Heat is provided either by oil stoves—the preferred means in the United States—or by briquettes which are either arranged in small heaps on the ground or burned in stoves. Figure 273 shows a vineyard in the Saar where seven rows of briquette stoves, arranged 4.5 m apart in the lowest row, can be seen burning during the frosty night of 2-3 May 1935. Farther up the slope the distance between burners can be increased because the smoke is blowing upward. Sometimes wood, brushwood, and other waste material is burned. In Japan, as reported by Y. Tsuboi and others [1132], the oil stoves are sometimes placed not on the ground in vineyards, but at a height of 90 cm, so that a cold layer of air, which is not dangerous, is left near the ground and all the heat goes to the benefit of the grapes.

Fig. 273. Ovens for burning briquettes are lit in a vineyard when night frost sets in. (After O. W. Kessler and W. Kaempfert)

O. Dinkelacker and H. Aichele [1119] in 1958 produced an increase in temperature of 2 to 3 deg by using two to three briquette stoves per 100 m^2 at a cost of 5 DM per 100 m^2 per night. With oil stoves, which are much easier to light, and which can be adjusted to suit the temperature variation during the night, an increase of about 3 deg is obtained with two to five stoves per 100 m^2 at a cost of 8 or 9 DM.

Frost protection by heating is more dependent on terrain and the vagaries of the weather than are the other two methods. Figure 274 illustrates the results of measurements by O. W. Kessler with 28 thermometers inside and 17 outside a heated experimental field near Oppenheim, on the evening of 26 April 1929. The initial temperatures of the thermometers, which were set up at 50 cm above the ground, were between 6.0 and 6.5°C before heating was started

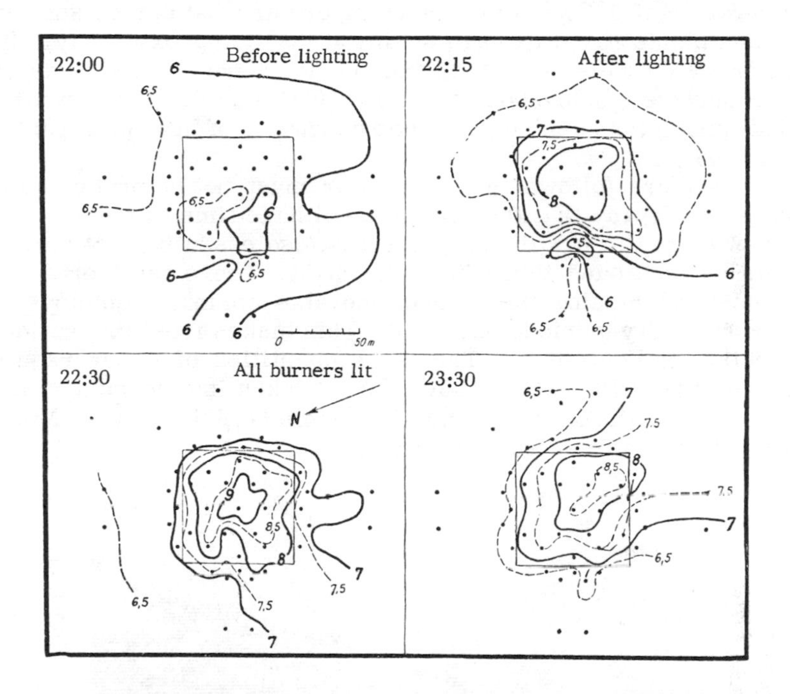

Fig. 274. Isotherms at 50 cm above the ground, in an attempt to provide frost protection by oil burners. (After O. W. Kessler)

(diagram at top left). After the first stoves were lit at 22:15, the temperature rose to 8°C in the middle of the field, while at the southern edge, where the stoves had not yet been lit, there was a decrease to 5°C (cold air drawn in by the first fires?). A quarter of an hour later, when all the stoves had been lit (below left), the warm zones are well distributed and show an increase of 2 to 3 deg in the center. The temperature increase was kept up in this stand, as shown by the distribution of the isotherms an hour later (below right). The only difference is that the warm center has been displaced a little to the side.

A method of protecting sensitive plants against frost that at first sight seems surprising is to cover them with a layer of ice by spraying with water. Every gram of water that freezes liberates the 80 cal of heat that had to be supplied to change it from the solid to the liquid state. This latent heat is made use of in the spraying process. When temperatures are below 0°C the leaves become covered with a layer consisting of ice inside but of spray water on top. The upper surface of the layer gives off heat by radiation, by heat exchange with the surroundings, and by evaporation. At the boundary between ice and water, however, the freezing process causes latent heat to be liberated, maintaining the boundary at a temperature of 0°C, and this is a temperature that can be sustained by most plants (frost damage begins at - 2°C approximately). If no further water were supplied, and the ice were allowed to dry, its upper surface would soon fall below 0°C, and the plants would be killed quickly, since ice is a good conductor of heat (see Table 10 and p. 209).

It therefore follows that the water must not be broken up into such a fine spray that the droplets will have time to freeze during their passage through the air, and it is also essential that spraying should be continued throughout the night. There can therefore be no question of using the kind of movable sprinkler employed for watering in dry periods (Sec. 51). This makes frost protection by sprinkling rather costly. The large quantities of water required for the long cold nights must also be taken into account. Figure 275, from O. Dinkelacker and H. Aichele [1120], shows a Moselle vineyard being protected by spraying.

Fig. 275. Sprinkler arrangement for frost protection in a Moselle vineyard. (After O. Dinkelacker and H. Aichele)

The stream of water from a rotating sprinkler moves around in a circle, while that from a swinging sprinkler moves to and fro. Interruption of the water supply to a single plant must never last long enough for the film of water to freeze completely. This may be avoided by making the time of interruption small or by having a heavy fall of spray. Table 98 gives the relation between the minimum density of spray necessary for complete frost protection, the period of interruption with a rotating sprinkler, and the observed temperatures found by H. von Pogrell [1129] in 20 experiments in the open air in late winter nights with frost and mainly weak winds.

Table 98

Minimum temperatures (°C) for different spray densities and interruption times.

Spray density ($mm\ hr^{-1}$)	Interruption time (min)			
	3	2	1	0.5
< 1.5	-2.5 to -3	-3	—	—
1.8-3.0	-4.5	-4 to -5	-5 to -6	—
3.4-4.1	—	—	-6.5	-7.5
11.3	—	—	—	9

Naturally, the temperature that can be counteracted will depend on the spray density and the period of interruption, ranging from a low density with long interruptions to high density and short interruptions. The greater the quantity of water, the more latent heat will be set free, and therefore the lower the temperatures that can be coped with successfully. (See A. Morgen [1127] for the latest arguments against the widespread but erroneous opinion that the quantity of water is unimportant.)

The valuable computations and graphs (Fig. 276) of A. Niemann [1128] give a quantitative representation of the heat budget of the watered parts of the plants. These computations could be carried out only on the basis of certain assumptions, and the diagram is not of general validity as a guide for frost spraying in practice. The assumptions were that outgoing radiation was 0.1 cal cm^{-2} min^{-1} during the night, the relative humidity was 90 percent, and the air flow over the parts of the plant was subject to certain provisos. The solid lines relate to nights with an air temperature of -5°C, and the broken lines to nights with -10°C. The lower curves give the loss of heat by conduction and exchange with the surrounding air (L), while the upper curves give the heat loss including that due to evaporation ($L + V$). The abscissa is the heat-transfer coefficient α_L, which increases rapidly as wind speed increases, as discussed in Sec. 28. As temperature decreases and wind increases, the loss of heat naturally increases, and this is shown in Fig. 267.

The scale on the right-hand edge of the diagram sets off this heat loss against the gain due to the latent heat of freezing for a certain spray density (mm hr^{-1}), assuming that all the water sprayed is received by the plants (which is not possible in practice). The zero of this scale is lowered by 0.1 cal cm^{-2} min^{-1}, equal to the radiation loss, which must be replaced irrespective of wind and air temperature. It may be deduced from the graphs that, with the given assumptions, a spray density of 3 mm hr^{-1} will provide effective frost protection and will not damage the soil (soaking). This agrees in order of magnitude with the figures of Pogrell's investigation in the open, mentioned above, which did not take into account the influence of wind. In practice, sensitive plants were effectively protected from any frost damage down to air temperatures of - 12°C in their alarming-looking armor of ice.

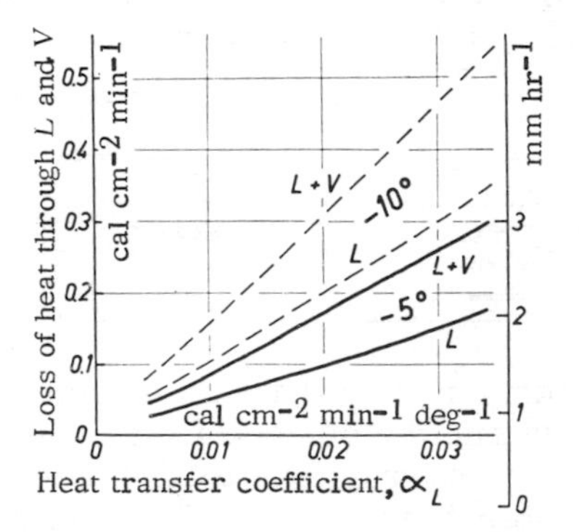

Fig. 276. Heat budget of plants sprayed with water for frost protection (From the calculations of A. Niemann)

Damage often occurs in the border areas where the spray is barely able to reach. This is not only because the quantity of spray is insufficient, or because the interruptions in it are too long, but also because the wetted plants are more liable to freeze. In some cases damage was observed in these areas, when plants within and outside the area were untouched.

It is important to start and cease watering at the right time. The first drops cool the plants immediately to the wet-bulb temperature of a psychrometer. When the air is relatively dry and motionless, cooling can be far below the air temperature at the moment. If spraying is begun too late, after the freezing point is reached, the plants may be killed before the latent heat released becomes effective. Many of the ill effects in early days are attributable to this mistake. The opinion is often heard that it should be stopped only after sunrise when all ice has melted. If, however, the air temperature is only a few degrees above the freezing point, and is expected to stay there, or higher, then the ice will remain at the melting point, that is, at 0°C; drying and cooling below this level is no longer possible. Extra heat from the sun makes melting easier.

Spraying with water as a means of frost protection is increasingly coming into use. It is effective to much lower temperatures than other methods and does not lose its efficacy with light winds that would blow away artificial fogs or heated air. The topography of the area to be protected, which often causes great difficulties with other methods, is of no importance provided the spray is able to reach the plants. The great expense of the installations, although offset to some extent by their usefulness in dry periods, is such that the method is justifiable economically only with very valuable plants.

Two other rarely used methods of frost spraying are worth mentioning briefly. The method that uses the wetting of the soil below the plants is based on the fact that moist soil has a better heat budget in times of frost than dry soil (Sec. 21). Before the expected frost has set in, therefore, water is poured over the ground. The surface of the ground must, however, be dry when temperatures reach the freezing point. The method is therefore full of risks and can be employed with some chance of success only on light sandy soils. The other method uses a mobile sprinkler to form a barrier against the movement of cold air at night (Sec. 42). This method has been tested successfully by H. Burckhardt [1117] in a Palatinate vineyard. An essential condition is that the configuration of the area should be such that the cold air breaking through from above can be brought to a halt at a comparatively narrow breach. More about these two methods, with full particulars, can be found in H. Burckhardt [1118].

CHAPTER IX

HINTS ON MEASUREMENT TECHNIQUES USED IN MICROCLIMATOLOGIC AND MICROMETEOROLOGIC INVESTIGATIONS

BY GUSTAV HOFMANN

55. Measurements and Quantitative Representation in General

In many cases a general knowledge of the measurement techniques used in meteorology [1137, 1140, 1136] and physics and its branches [1134, 1141] will be adequate for the selection of suitable instruments, although many special types of instruments have been developed for taking measurements near the ground, descriptions of which can be found only in individual publications. When the quantities to be measured are to be correlated with the observations of the official network, the instructions issued to observers in the Weather Service [1145, 1144, 1147] should be consulted for guidance. To a limited extent, descriptive literature and instructions issued by instrument manufacturers provide some assistance.

In deciding which instruments to use in measuring a particular quantity, the selection will depend, in most cases, on which one appears to be most suitable, but often on the availability of instruments, the capabilities of a workshop for constructing instruments, the finances available, and last but not least the cost involved in the future analysis of the results. These points must also be considered in judging descriptions of instruments in writings on the subject; for example, if a millivolt recorder is available a thermoelectric recording device may offer a less expensive solution to the problem of measurement than mercury thermometers. If a completely new assembly of instruments has to be prepared, it is usually cheaper in the end to buy existing instruments than to develop one's own instruments of a similar type.

Depending on the purpose of the investigation, it is useful to make a distinction between microclimatologic and micrometeorologic measurements, although in practice there are many intermediate stages.

Microclimatologic measurements should establish characteristic values of the meteorologic parameters for a particular place. Such values may be means, extremes, frequencies, duration or time of occurrence of certain values, and so on. The measurements made by the network of official climatologic stations can be regarded as being of this type. In Germany these stations are about 20 to 30 km apart so that the data for each will represent an average area of roughly 500 km^2. It is the task of topoclimatology to establish the peculiarities of the climate between these macroclimatologic stations (Sec. 47). In an investigation into the climate of a locality, if it is required to obtain valid data equivalent to those of macroclimatologic stations [1139] it will be necessary, in general, to use the same methods of measurement, that is, they will have to extend over a long period in order to balance out any special features of the weather at particular times. Because of the cost, this will be possible only to a limited extent. There have been, in fact, only a few investigations of this kind carried out up to the present time that permit good spatial interpolation to be made, because they employ the same measurement techniques. In general, we have to be content with the measurement of fewer meteorologic parameters, values of less reliability, or indirect methods.

In contrast, for practical purposes only a few parameters, perhaps even only one, are needed and the degree of accuracy required is not that which would be expected for climatologic data. One such case in point is frost danger, which can nearly always be deduced from measurements of minimum temperatures at appropriate heights for the crop in question. Measurements need be made on only a few nights to discover all the essential points of difference ([1146, p. 127]; cf. pp. 402f).

It is precisely for this reason, the urge to carry out local investigations for the most varied purposes (frost protection, wind protection, clearing of land, town planning, and so forth), that so many different methods of approach have developed [1146, p. 120], leading to the danger that it might not be possible to make comparisons, or to link up with each other the results of much research carried out at great expense, however valuable the data might be for the particular problem at hand. It will require a good deal of work to establish the right balance between solving the practical problem and establishing results that could be applied to the macroclimatic network. The more the methods of measurement used in local research differ from those of the national network, the more difficult it will be to use them to interpolate between the observations of stations working over long periods of time. In this sense, for example, temperatures measured at different heights constitute different climatic data, so that W. Kreutz [1146, p. 179] recommended the use of a shelter at 50 cm and another at the usual height of 2 m, and H. Burckhardt [1135] showed that, in making mobile measurements, attention had to be paid to the height of the sensing element on the fixed measuring device. Particular care must be taken in making comparisons and coming to conclusions based on similarity of land forms. Later results may lead to different conclusions, as for example the remark [1146, p. 11], "even in slightly undulating country, it is hardly possible to estimate correctly the different winds blowing over different parts of the area" (see also p. 416).

As a rule, it is necessary to make a great many measurements to obtain good meteorologic data, so that recording devices have to be used in many cases. Even with individual readings a great many data are needed as a basis. This means that, in every case, micrometeorologic research involves a great deal of work in analyzing and evaluating. Since every occasion on which an observation is made has to be included with equal weight in the computation of a climatologic value, it follows that the volume of work can be reduced only by curtailing the number of elements observed. Recording instruments, which are very valuable in themselves, yielding a great number of readings for relatively small outlay, unfortunately require more work later, when it is a question of evaluation, since they contain much more than it is possible to use. The observations left unused are valueless, and the work spent on getting them is wasted. Before embarking on a micrometeorologic investigation, therefore, it should first be considered carefully whether it is going to be possible to analyze the data expected in a way appropriate to the object of the investigation. It is much better to make a well-thought-out limitation of the elements to be measured than to be faced later with mountains of data that will long be left untouched, waiting for evaluation. It is, unfortunately, much easier nowadays to get money for expensive and complicated instruments than for the uninspiring work, requiring great patience, of reducing the primary data to values that can be used in proper scientific study.

Micrometeorologic research is differentiated from microclimatologic by the method used to analyze the recorded data. The primary data are used here to derive combinations of values of meteorologic parameters that will serve to establish or to prove meteorologic laws of general validity (that is, independent of locality). This system of analysis, which resembles that of physical science, differs however from that of the laboratory physicist mainly because the meteorologist is seldom able to verify the relation between the elements in question by direct experiment. He must, so to speak, wait patiently until nature decides to perform the required experiment for him. Since the point of time at which this will happen is not known in advance, the individual parameters must be recorded continuously, so that the valid association of values that will explain the connection can be extracted retrospectively from a great number of combinations. The fact that only a small fraction of the records is used in analysis does not mean there is a waste of data. This is in fact necessary, because meteorologic parameters near the ground appear to be coupled with each other because of superimposed control (diurnal and yearly periodicity, weather situation). Certain associations of values will therefore appear more frequently than others, whose usefulness in explaining the relation being investigated is nevertheless just as great, and which in many cases are particularly suitable for doing so, being values found in the boundary zones of the area.

The division of investigations in the air layer near the ground into microclimatologic and micrometeorologic is naturally to some extent deliberate since, as already said, there are many transitional forms and much research can serve both purposes. As far as choice of instruments is concerned, however, care is taken in microclimatologic investigations that the results are comparable with other similar investigations elsewhere. To a great extent, the usual types of instrument are employed, such as might be used at official stations, and development of special instruments is avoided in so far as it is possible to achieve the aim of the investigation without doing so. There is a good deal of sense in the often-expressed wish for a "standard instrument kit" where such research is concerned. This does not apply to micrometeorologic work, however, where the demands placed on the sensing elements are so varied that it is hardly possible to get satisfactory instruments. In this field any attempt at standardization would place barriers in the way of the continuous efforts made to develop instruments suited to the advancing aims of research.

In every case, the question of accuracy is important. It is proper that attempts should be made to obtain values that are as accurate as possible. However, the cost increases very rapidly as accuracy of measurement increases, and it is therefore important to consider just how significant single readings may be within the framework of the problem as a whole. If the establishment of the required quantities is based on formulas that contain the measured value, then the situation may be made clear by an assessment of the error involved, and thus great expense can be avoided.

A distinction has to be made between the error involved in reading the scale and the total error of the instrument. The latter error is usually much greater, since it includes the inaccuracies due to the lack of constancy of the connection between the sensing element and the indicator, to aging, and to variable external influences. It is unfortunate that these two types of error are often confused, not only in research papers but also in the literature supplied by the firms producing the instruments, so that they do not fulfill expectations if they are not read sufficiently critically. It is no problem today, even at moderate expense, to devise a method of reading temperatures that will make it possible to detect differences of 0.01 deg and even less at the sensor. The temperature of the sensor can, however, by no means be read to an accuracy of 0.01 deg, since uncontrollable variations in the members connecting it to the indicator cause changes in the readings. Obviously, this can be avoided to a considerable extent by careful construction, at a corresponding expense. Even if it is then possible—to keep to the example of temperature—to read the temperature of the sensory element with the desired degree of accuracy, it does not follow that this will be a true value of the quantity it is required to measure, the air temperature, for example. This applies in a similar way to the measurement of other quantities. Two further errors found particularly in micrometeorologic measurements will be indicated briefly, again using the measurement of air temperature as an example.

The first of these concerns the disturbance of natural conditions caused by introduction of the measuring instrument. The temperature of the sensory element is determined not only by that of the air, which has to be measured, but by other factors as well, such as radiation and ventilation. When a bare sensor is exposed, the error, that is, the difference between its temperature and that of the air, may amount to several degrees. If an attempt is made to reduce this error by shielding and by artificial ventilation, the natural conditions are disturbed, particularly when there is a marked temperature stratification close to the ground. It follows that a compromise solution must be sought that will capture the true state as effectively as possible, and this will depend to a great extent on the actual situation. If an error of 0.5 deg which cannot be controlled, is to be expected, there is little point in aspiring to an instrument that will show the temperature of the sensory element to 0.01 deg.

The second type of error, often called the "sampling error," depends on the limitation of the number of readings. Many meteorologic parameters exhibit short-term fluctuations which, by analogy with the same phenomenon in wind, are often referred to as gustiness or unsteadiness. If this irregular variation in air temperature amounts on the average to ± 0.3 deg, then an instrument with little time lag will give readings that may vary on the average by ± 0.3 deg from the average of a long series of measurements, or that may show this amount of spread on either side of the long-term temperature variation (Fig. 13). An instrument with an accuracy of 0.01 deg will then give readings with a chance dependence on the time the measurement is made. Only about ten readings will give a mean value that is accurate to about 0.1 deg. If it is required to find a mean value accurate to 0.01 deg it is necessary—assuming that the average is constant—to take about 1000 single readings.

The possible expense of evaluation then places limitations on the accuracy possible. Many types of instruments balance out the short-term fluctuations to a certain extent, because of the time lag from which they suffer; the mean value they give depends on the magnitude of the time lag. It would be worth trying to produce a coefficient of lag in the instrument that would be equivalent to the interval in making readings. This is generally impossible for technical reasons; it is a complete waste of effort to strive for instruments completely free of lag—"showing the true value"—since then the readings do not give the mean value for the measurement interval, but chance values. Similar arguments may be used in discussing the size of the sensor. It is not always an advantage to try to achieve point measurements.

The usefulness of an investigation does not depend, however, only on the quality of the instruments used and the expenditure on analysis. In many cases, the further utilization of published data is made difficult because some information is lacking, and, although this may seem unimportant to the author, it is not so to the reader. A full description of the instruments and the method of analysis is essential, but equally necessary is a complete description of the place where the measurements were made, especially the name of the place, or, better still, its latitude and longitude. The description of the site should also give the characteristics of the locality, especially the height of the measurement point above mean sea level and relative to its surroundings. When heights of measurement are given, there should be some indication of the position of the zero of height. No less important is information about the gradient and direction of surface slope at the measurement point, and the type and height of vegetation growing in the neighborhood as well. In the mountains, and particularly in forests, data should be given about any obstruction of the horizon, as this is important for insolation and wind conditions. An example of this kind of description has been taken from F. Lauscher's [1138] excellent work on the climatologic station at Lunz. The method is best shown by the top left-hand diagram in Fig. 277. The outside circle represents the horizon and the central point the zenith. Between horizon and zenith, circles for altitudes of 30° and 60° are drawn at equal distances. The paths of the sun on 21 June, at the equinoxes, and on 21 December are then drawn in. Hour points are marked by dots on the sun's paths and some of them are joined by dotted lines. Then the projections above the horizon are plotted at the appropriate compass bearing. A rough idea may also be given by the mean angle of obstruction of the horizon, which in the four stations shown in Fig. 277 is 4°, 17°, 30°, and 49°. The simple instruments that can be used, in addition to theodolites, are indicated on pp. 536f. In giving weather descriptions the standards used in the meteorlogic services should be adhered to as much as possible. Comprehensive descriptions can be made successfully by employing the international codes used in synoptic meteorology

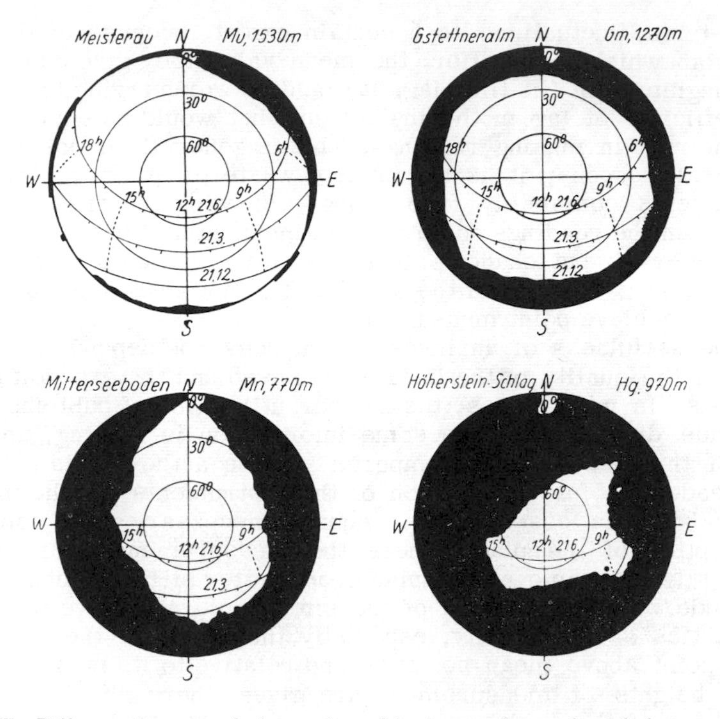

Fig. 277. Different degrees of obstruction of horizon at four selected observation stations near Lunz. (After F. Lauscher)

[1143], thus conveying a maximum of information in the smallest possible space. In giving the time it should be made clear whether a mean zone time is being used (Central European Time, for example) or whether it is mean local time, which is more suitable for climatology.

Particular care must be taken when describing the quantities measured and the relations deduced from them. Every measurement is, in principle, a comparison between the quantity to be measured and another, namely, the unit. Only the numerical value and the name of the unit together make the data complete. The numerical value by itself is often of little value, for the very reason that in meteorology many different units are in common use. At best, the reader has to waste time searching for the most probable unit. This objection applies also to general descriptions of units, such as "cgs." Units should be described completely, not only when giving single results, but also in tables, diagrams, and especially in formulas. Much confusion can be caused here by mixing equations involving the various elements with equations of numerical values. It must be said, unfortunately, that it would be easy to fill many pages with a bibliography of works in which failure to observe these points makes it difficult, if not impossible at times, to use the data that have been collected at such great expense. It would be most desirable if a simple proposition could be observed everywhere, which was expressed by the committee on units and formulas thus: "The symbols used in physical equations should, as a rule, signify the physical 'quantities,' that is, they should be numbers with names attached. It is perhaps more appropriate to regard them as a symbolic product of the numerical value (the number measured) and the unit in question, according to the equation: physical quantity = numerical value times unit. If, in the symbols used, the numerical value, or the unit and the numerical value alternately, are left to be understood, then some method must be adopted to eliminate any confusion, such as the addition of supplementary 'numerical equations.' In operating with equations of physical quantities, the symbols for the units as well as the numbers and the letters used should be multiplied or divided by each other."

One of many possible examples in which a symbolic equation of physical quantities can be used as a basis for calculation, as well as a numerical equation, was provided by the discussion of the Sverdrup method in Sec. 25. Bowen's ratio is given on p. 230 as

$$\frac{L}{V} = \frac{c_p}{r_w} \frac{\dfrac{d\Theta}{dx}}{\dfrac{dq}{dx}}.$$

In this symbolic equation for L/V all the elements involved can be given in any desired unit. The calculated value of L/V will always be the same, regardless of whether $d\Theta/dx$ is measured in degrees per 100 m or per centimeter, or dq/dx in grams per kilogram per 100 m or per centimeter. This no longer applies to the difference form of the equation:

$$\frac{L}{V} = 0.49 \frac{\Delta t}{\Delta e}.$$

This numerical equation will lead to the correct value only when Δt is the difference of the numerical values of the temperatures in Celsius degrees and Δe is the numerical difference of the vapor pressures measured in millimeters-of-mercury. If other units are used, such as Fahrenheit degrees, millibars, atmospheres, and the like, the numerical equation will give false values of L/V. It follows that, with this form of equation, the units must be stated for which it is valid. In the above example, these were stated just before the numerical equation was given. It is easy to see that, if this information had been omitted, the whole equation would make little sense to an eventual user. Unfortunately, as was already said, too little attention is paid to this point in many publications. It may also be seen from this example that there is an advantage in expressing relations in the form of a symbolic equation.

A preference for this type of equation has also been expressed by U. Stille [1142]: "The method of formulating symbolic equations, which has been used, consciously or unconsciously, in mechanics from time immemorial and is being extended ever further today in physics and technology, allows us to describe physical laws in a general form that is altogether independent of any choice of units that may be made later for any special reason." Nevertheless, it is not possible to exclude the use of numerical equations, especially as a means of expressing empirical relations. In such cases it is necessary to state, immediately before or after the equation in what units the quantities involved must be measured, so that the equation will be satisfied by the numerical values obtained.

56. Measurement of Individual Elements

The apparently simple measurement of air temperature is beset with difficulties in the region close to the ground, because the sensor of the instrument is both in contact with the air (and therefore in a position to conduct heat) and exposed to radiation. The radiation error is directly proportional to the radiation balance S (see p. 13) of the sensor, and inversely proportional to the heat-transfer coefficient a_L (see p. 254), which increases approximately as the square root of the wind speed v. Shielding from radiation by meteorologic shelters and similar arrangements certainly reduces S, but at the same time v, and hence a_L, are also reduced. In addition, the air flowing over the sensor will be warmed by the radiation shield, so that while the radiation error may be reduced it will not be eliminated [1137, p. 5; 1150, 1172]. For this reason, particularly in climatologic work, the instrument used to measure temperature, including the shielding against radiation, has been standardized so that, even if correct absolute values are not obtained, their comparability will be assured. It is also for reasons of comparison that the same rules for making measurements are still observed today, although it is technically possible to obtain better values. To conclude from this that meteorology has a different conception of temperature from that of physics would be going too far. Other methods of reducing S are to paint the sensor white, and to use platinum-wire hard-glass thermometers [1146, p. 107], which at least absorb only part of the short-wavelength radiation. Open stretched wires [1180] have a comparatively high value of a_L because of their small diameter, and thus their radiation error is reduced. R. Weise

[1146, p. 107] noted, however, that thermometers of this type, with platinum wire, broke from the slightest cause and it took a great deal of work to stretch the wire again, to calibrate and recalibrate. Finally, sling thermometers must be mentioned [1137, p. 16], although they can be used only for making single readings. A combination of artificial ventilation and shielding, which has proved its value over many years, is the aspiration psychrometer, which also allows the air humidity to be measured. The large model of the Assmann aspiration psychrometer [1137, pp. 15 and 221] is particularly recommended. The smaller model can be used only where its lighter weight is a distinct advantage, because it is less precise and more easily disturbed in its action. There are also aspiration psychrometers with electrical sensors [1159] which can be used for making continuous records.

Figure 278 is an illustration of one such fully electrical psychrometer. The aspirator head can be recognized on top. The outer cover, which rotates along with the motor, prevents penetration of rain and, even more, the accretion of any solid precipitation. Air is sucked into the instrument, as in the Assmann psychrometer, by two radiation-shielded tubes, but here these are arranged horizontally. In these protected tubes the two platinum-wire hard-glass thermometers are located, the electrical leads from

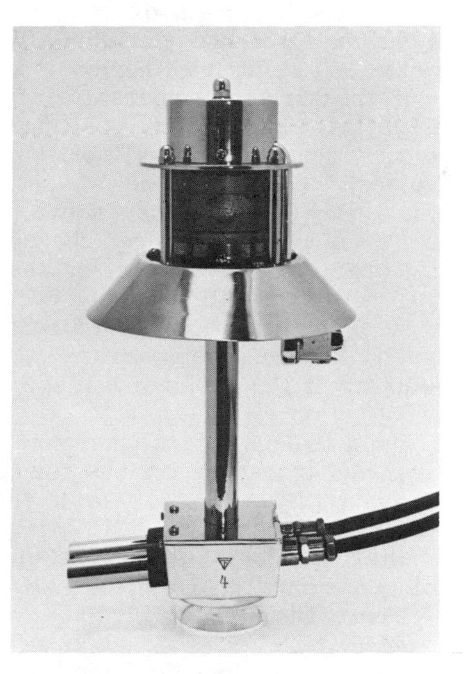

Fig. 278. Electric psychrometer after E. Frankenberger. (Photo from Th. Friedrichs, Hamburg)

which are seen on the right of the photograph. One of the thermometers is supplied, at the far side, with water from the container below, so that it gives the wet-bulb temperature. The suction tube in the center of the instrument ensures that there is sufficient distance between the level at which air enters the instrument and that at which it leaves. In the space nearest the ground, artificial ventilation may upset the natural temperature stratification. J. Bartels [1149] therefore proposed that, when measurements are made near the ground, the instrument should be moved forward at a constant height, so as to be provided with a continual new supply of undisturbed air. In addition to devices for reducing the radiation errors, there are proposals for methods of compensating errors [1188]. K. Brocks [228] and R. G. Fleagle [232] have described methods of measuring lapse rates optically.

Except for readings made immediately under the surface, measurements of soil temperature are free of radiation errors. The lag of the thermometer scarcely comes into consideration. On the other hand, the lack of homogeneity in the soil may cause the measured value to be unrepresentative of the required layer, especially if the soil has been disturbed at the precise position where the instrument was introduced. The simultaneous use of several thermometers at the same depth [13, p. 24] can, however, be only justified for very special requirements. In measuring water temperatures, electrical methods can be used with success [1201], permitting readings to be made immediately and continuously, in addition to bucket and reversing thermometers [1137, p. 20].

Measurement of the surface temperature of the ground presents special difficulties, since in this position shielding against radiation and the provision of artificial ventilation lead to completely unusable values. When the ground is bare, a method similar to that suggested by J. Bartels, mentioned above, can be employed, by drawing a flat sensing element over the surface, which will lead to useful readings. To cover a glass thermometer with a thin layer of soil, however, will scarcely provide raw data for discussion. It is possible that the method already tested in technology, of measuring surface temperatures by means of the characteristic radiation of the surface, will soon provide a better basis [1179, 1176].

In addition to liquid thermometers (mercury, alcohol, and so forth), which are manufactured in various forms and can be inserted into the ground [1137, p. 17; 1145], thermographs with Bourdon tubes or bimetallic elements [1137, p. 35] may be used. Light portable shelter [1178, 1190] may be used as well as the standard meteorologic shelter [1137, p. 3; 1145] employed in the national network. There are also many other forms of protection against radiation [1146, pp. 129, 437; 1139, 1172]. According to J. van Eimern [1146, p. 126], bicycle trips for making measurements have proved the value of meteorographs. Automobiles used for this purpose have been fitted on a number of occasions with remote-reading thermometers so that temperatures could be read inside the vehicle [1146, p. 146; 1156]. Long-term mean temperatures can be obtained also by measuring the optical rotation in ampoules filled with sugar solution. Although this method [1189] does not lead to the customary arithmetic mean temperature, because the speed of the reaction depends exponentially on the temperature, it can be applied successfully in investigations for bioclimatologic purposes. It has the particular advantage of making possible simultaneous measurements at a great number of points.

Electrical methods of measuring temperature are gradually gaining ground in comparison with nonelectrical methods, since they allow both recording and remote reading. When temperatures are measured by thermocouples, a method particularly suited to the measurement of temperature differences, it is advantageous to have a high degree of constancy in the reference temperature (the passive junction), particularly when temperature recordings have to be made. When working in the open, the reference junction can be buried about 20 to 50 cm in the ground, thus making use of the small temperature fluctuation at that depth. It is just as important to make a sufficient number of readings of the fixed reference temperature in this case, as it is when using a metal body with a themal lag [1158], or a bath of ice and water [1162], where care must be taken to ensure that electrical insulation is good [1181], and that electrical influences are eliminated. It is possible to compensate for fluctuations in the reference temperature, by using connections with resistance that varies with temperature [1167]. When simple thermocouples are linked directly to an intermittent contact recorder, or other type of recorder, it is hardly possible to obtain an accuracy of 0.1 deg. In many cases, therefore, amplifiers [1181] or thermopiles [1199] are used. These latter have recently been produced [1148] by the electrical deposition of copper on constantan wire, in a relatively simple way. The extent to which semiconductor thermocouples will soon be used in the production of practical measuring instruments cannot be foreseen at present.

When electric resistance is used to measure temperature, platinum-wire hard-glass thermometers have the advantage of a good degree of constancy with time. Effects due to the variable resistance of connecting wires, as temperature changes, can to a great extent be eliminated by the use of a three-core cable [1204]. While cross-coil instruments in general allow an accuracy of only 0.3 deg to be attained, it is possible to achieve an accuracy of 0.1 deg or better by means of the compensation recorder, though it is expensive. Attention must be paid to self-heating of the sensing element by passage of the electric current when measuring air temperatures. Systems of temperature measurement by metal-wire thermometers are necessary in various forms in industry, so that in many cases instruments can be found that are suitable for meteorologic purposes and technically well developed. In contrast to these, measurement of temperature by thermistors (thermally sensitive resistors) is still in the laboratory stage [1172, 1166]. Thermistors have the advantage over metal-wire thermometers that the relative change of resistance is about 10 times greater (-2 to -6×10^{-2} deg^{-1} against about 4×10^{-3} deg^{-1} for platinum). Where requirements of accuracy are similar, therefore, thermistors can be employed together with fairly robust indicator systems. Since, among other types, there are some with high resistance, it follows that the resistance of the leads, plugs, and switches has hardly any importance. The manufacturing tolerances, which are considerable at times, can be compensated by ballast resistances [1172]. Owing to the high relative changes in resistance, the resistance decreases approximately exponentially as temperature increases. This can be allowed for by using suitable potentiometers. However, the exponential characteristic and the high manufacturing tolerance are probably the reasons why there are hardly any industrially produced recording instruments that employ thermistors.

The importance of temperature measurement in meteorologic technique is not restricted to its importance as a parameter. A whole series of methods of measuring other elements depend on temperature [1169]. One example is the psychrometer, which enables the air humidity to be determined from the readings of a dry- and a wet-bulb thermometer [1137, p. 208]. In addition to the Assmann aspiration psychrometer, which was mentioned earlier, other types, including some that are not ventilated, may be used when the degree of accuracy required is lower [1186]. The errors are tolerable when the thermometers used are very thin [1153]. Attention is also directed to the thermocouple psychrometer described by J. L. Monteith and P. C. Owen [1184], which measures relative humidity by means of condensation on a thermocouple junction cooled by the Peltier effect, and which can be used just below saturation level. Methods that use the deposition of dew on a polished plate, now seldom used in meteorology, are becoming more popular because of substantial technical improvements (heating by eddy currents, determination of dew deposit by photoelectric cells) [13, p. 176]. In addition to these, measuring elements with a hygroscopic surface, which may be heated [1175] or cooled [111, 1141], have come into use recently. In spite of many errors, the human hair has not lost its importance as an instrument for measuring humidity. Its time lag can be reduced substantially by proper treatment. Besides the large hair hygrometer [1137, p. 235] of the Koppe type, there are smaller versions that can be used to advantage in micrometeorologic research [1153]. Much of its continued popularity is due to the convenience and simplicity of the use of hair in recorders (hygrographs). Relative humidity is a quantity frequently used, although it provides no measure of the amount of moisture in the air, but only its degree of saturation.

From van't Hoff's law, relative humidity is closely connected with the osmotic pressure of a moist body, such as the soil. The wide range of possible osmotic pressures, extending over several powers of 10, has led to the introduction of the pF value, which is equal to the common logarithm of the osmotic pressure expressed as the height of a column of water in millimeters. Tensiometers [13, p. 50; 1171] are used to measure low osmotic pressures (pF < 4); greater values can be obtained by aid of the methods used to measure relative air humidity.

In addition to osmotic pressure, which indicates the strength with which the water is bound to the soil, we have the water content of the soil. The relation between these two quantities is approximately constant only when the characteristics of the soil remain the same. Nevertheless, there are a number of methods for determining the water content of soil that are in fact methods of osmotic-pressure measurement [1168]. The most reliable values are still obtained by taking sample borings [332], even though sufficiently accurate values are obtainable only by making a number of simultaneous borings, because of the great inhomogeneity of soil. For meteorologic purposes the water content per unit volume ($g\,cm^{-3}$) is more appropriate than the ratio of the weight of water to that of dry soil (percent) and so borings are made *in situ* [1182]. The large number of tests that have to be made in assessing the site quality of soil has led to the development of rapid methods of measurement [1146, p. 188]. H. Aichele [1146, p. 86] has pointed out, however, that measurement of the water content by the calcium carbide method (the C-M apparatus) saves hardly any time in comparison with drying, when a very large number of tests are made. Apart from measurements in the upper layer (20 cm), the new type of soil-moisture measuring instruments, which use radioactive substances [1170], can be employed successfully.

The main instrument used to measure horizontal wind speeds is the cup anemometer [1137, p. 349]. When fitted with contact points, this can be used in a simple way in conjunction with chronographs [1137, p. 357], totalizers [1203], or telephone counter assemblies to give mean wind speeds. To make evaluation simpler when using chronographs, every tenth contact can be indicated [1146, p. 152]. If current is supplied through resistance and condenser, there is no risk of batteries being run down should the anemometer remain stationary on a contact. If the instruments are of light construction [13, p. 140] and contacts are transmitted without mechanical action on the axis [13, pp. 134 and 136], it is possible to devise types that are set in motion by very light winds. Average values for short periods can be found by measuring the interval between successive contacts by means of a stopwatch. Fan anemometers [1137, p. 348], which are set in motion by light winds, can be used in a similar way, but differ from cup anemometers in that it is necessary to point them in the direction of the wind. This also applies to the pressure-tube type of anemometer making use of the stream effect [1207], which looks like a Pitot tube from the outside, but through which some of the air passes. This type, like the cup anemometer, gives instantaneous values when used with a dynamo and recording device [1206, 1207].

The run of the wind can be shown, both with cup and fan anemometers, by revolution counters geared mechanically to the anemometer axis. It is convenient only to use types that give the run of wind up to 10^6 m, as with other types it may, in certain circumstances, be necessary to record too often [1146, p. 138]. When it is required to set up an installation to run for a substantial period, advanced types are available which, in addition to the run of wind, record instantaneous wind speeds (gust recorders) and direction, and yet other types for only one or two of these parameters [1206, 1207, 1202].

Thermal anemometers can be employed to measure low wind speeds, which would be badly indicated by cup anemometers, or not at all. Such instruments depend on the relation between ventilation and heat transfer. In addition to the katathermometer [1137, p. 381] and the hot-wire anemometer of Albrecht [1137, p. 390], many instruments of this type are described in the bibliography [1172, 1164, 1152], including some that measure wind direction on this basis [1165].

As shown by mention of the katathermometer among wind-measuring instruments, thermal anemometers are closely related to the instruments used to measure the factors involved in the cooling process, such as perhaps the frigorimeter [1137, p. 52] or frigorigraphs [1187].

Although there are whole ranges of tried and tested instruments and methods of measuring horizontal wind speeds, reliable methods of measuring vertical winds still present serious difficulties. This can be understood if we remember that the average vertical component of the wind is small in comparison with its horizontal component. In spite of many attempts [1137, pp. 346, 382; 1160; 13, p. 256], there are still no satisfactory instruments.

In considering instruments used for measuring radiation, a distinction has to be made between the calorimetric type, dependent on heating effects and, in principle at least, suitable for the whole range of radiation, and the photometric type, which responds to only a comparatively small range of the spectrum.

Calorimetric radiation-measuring instruments are too numerous to be even listed here. They can be divided, according to the principle on which they operate, into (*a*) those which measure the temperature increase of the sensing element or of some body joined to it by a thermal conductor, or its phase change, (*b*) others which, in a state of stationary equilibrium, measure the temperature difference between the sensor and a body that is unaffected or differently affected by the radiation, and (*c*) those which compare the effects of radiative heating with the effect of an (electric) heater. According to the nature of the quantity to be measured, a distinction is made between pyrheliometers, which measure direct solar radiation I_0 (see p. 12), do not require calibration (absolute instruments), and are generally available only to large observatories, and actinometers which are used for most measurements of direct solar radiation, and are related by calibration to absolute instruments. Frequently used instruments of this type are the Michelson-Marten (*b**) bimetallic actinometer [1137, p. 141], and the silver disk [*a*; 1137, p. 138]. In micrometeorologic work, pyranometers are more often required, that is, instruments that measure the amount of short-wavelength radiation (from 0.3 to 3 μ approx., see p. 8) from a hemisphere, in other words, the global radiation $I + H$, and short-wavelength reflected radiation R (see p. 12). Besides pyranometers with only one black sensor surface, such as the Moll-Gorczynski solarimeter [*b*; 1154], there are others with black and white radiation-reception surfaces, such as the Linke star pyranometer [*b*; 1177], and Robitzsch's [*b*; 1191] bimetallic pyranograph (radiation recorder). With shielding against solar radiation I, the sky radiation H can be measured with suitable pyranometers. By combining two pyranometers, one directed downward and the other upward, or by using one instrument in both directions in turn, the short-wavelength radiation balance $I + H - R$ of the surface below the instrument is obtained. The quotient $R/(I + H)$ is then the (short-wavelength) reflection index or albedo of energy. In addition to the instruments used for measuring radiation incident on a flat surface, there are types that measure radiation incident on a spherical surface. The best-known representative of this group is the spherical pyranometer of Bellani (*a*), which is now available in a well-devised model [1146, p. 182]. The long-wavelength (about 6 to 60 μ, see p. 8) radiation received from a hemisphere, that is, the counterradiation G of the atmosphere, and the radiation emitted by the ground $\sigma T_B{}^4$ (see p. 12) is measured by infrared pyranometers the best-known of which is Ångström's pyrgeometer [*c*; 1137, p. 172]. The combination of two instruments can be used here, too, to measure radiation balance, in this case the long-wavelength balance $G - \sigma T_B{}^4$. Among infrared pyranometers, the types with an opening of restricted angle, with which the (apparent) surface temperature of a body can be measured (see p. 530),

*The letters, *a, b, c* against instruments denote the principle on which they operate, as described above.

have a certain claim to be mentioned. The total short- and long-wavelength radiation received from a hemisphere, that is, $I + H + G$ or $R + \sigma T_B^4$ or the effective radiation $I + H + G - \sigma T_L^4$ can be measured by effective pyranometers. Schulze's [*b*; 1194] radiation balance meter, and Georgi's [*b*; 1161] universal radiation meter should be noted in this group. The latter instrument also allows the quantities mentioned earlier to be measured by a very simple method of changing covers, and therefore appears suitable for all measurements made in a particular direction. The combination of the two instruments, mentioned above, gives the radiation balance of the ground $S_B = I + H - R + G - \sigma T_B^4$ (see p. 13), that is, the radiation factor S of the heat-balance equation. Together with two pyranometers, this assembly of instruments allows the radiation exchange to be split up into the influx of short- and long-wavelength radiation from above and from below ($I + H$, G, R, σT_B^4). If it is required to measure only the radiation balance, then simpler instruments, such as the radiation-balance meter, can be used. These, and other instruments belonging to this group, are described in detail in [1205].

Photometric radiation-measuring instruments pick up only the particular part of short-wavelength radiation appropriate to their spectral sensitivity. By the use of suitable filters, this can be made to correspond to the range of spectral light sensitivity of the human eye, on which the definition of photometric units is based [1142]. However, even in this case the values measured in luxes (lx) can be converted into the corresponding energy units, such as cal $cm^{-2} min^{-1}$, only if the spectral composition is sufficiently well known. As a rough point of departure, 1 cal $cm^{-2} min^{-1} \cong 70$ klx can be used for global radiation (daylight). In spite of this defect, photometric radiation-measuring instruments are well liked, especially for measurements in particular directions, because they are simple to use, and also because for many types of studies (assimilation, and the like) the caloric effect of the radiation loses importance in comparison with its photometric effect [1193]. The flux of light can be reduced by reduction filters [1208] to an amount that will not damage the element when making measurements with photoelectric cells. Color filters allow interesting ranges of the spectrum to be filtered out. The quotient of reflected to incident radiation, measured simultaneously, gives the light albedo, which is not always equal to the albedo of energy mentioned above (p. 535), but which can be taken as approximately equal in most circumstances.

The true duration of sunshine is obtained from autographic sunshine recorders [1137, p. 185]. When making topoclimatologic investigations, simple devices such as the semicircular pinhole camera of Becker-Freyseng [1151] can be used; the local sunshine meter of Morgen [1185], or the Tonne [1197] horizontoscope, can be employed to give basic data for the study of the effect of surface slope or of the restriction of the horizon on insolation.

The Tonne horizontoscope illustrated in Fig. 279 consists of a curved plexiglass surface on which surrounding bodies, especially obstructions of the horizon, are reflected. On looking vertically downward into the instrument, which has been oriented by compass, a direct picture, like those in Fig. 277, is obtained by tracing the outline of the obstructions on a piece of transparent paper brought into position from below. If this tracing is then laid over a state-of-sun diagram (as in Fig. 277), it is a simple matter to discover the times at which the measurement point will be touched by the sun's rays. Other diagrams allow the total sunshine to be determined.

In measuring precipitation (rain, snow), rain gages with a reception surface of 100 cm^2 can be used, as well as the usual Hellman rain gages with 200-cm^2 surface area [1137, p. 276], but somewhat different values may be obtained. Rainfall can be measured by means of recording gages [1137, p. 279]. In areas that are not easily accessible, totalizers [1137, p. 276] can be set up to collect rain over long periods (rainfall collectors). On sloping surfaces (see p. 419) a receiving surface can be set parallel to the slope for some purposes. A measure of the precipitation from fog (Sec. 37, p. 348) is obtained if additional gages fitted with devices for capturing fog droplets are used [658]. To measure dew, dew plates [1173; 1155; 1146, p. 185]

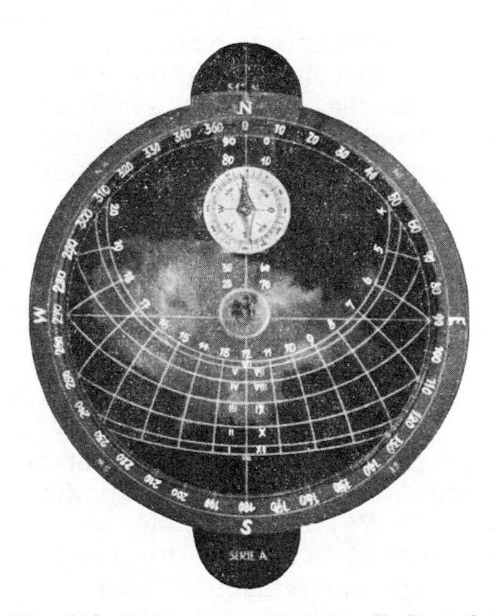

Fig. 279. Horizontoscope. (After H. Tonne)

and dew recorders [1183] can be used. These, like fog devices, give relative values [484], so that a form of standardization is very much needed. There is a simple apparatus [1174] which is useful for measuring the duration of wetting from rain, fog, or dew, an item of interest for many purposes. These are also useful in providing some idea of the type and duration of rainfall [1190], or simply to indicate whether or not there has been any precipitation during the night [1200]. The points of interest with snow are its depth and density. Besides instruments projecting from the snow [1137, p. 286], there have been developed recently methods of measurement employing radioactive substances which can be used to good advantage [1157].

In spite of many defects, it is generally possible to devise a sufficiently adequate method of measuring precipitation under most circumstances, but we are still very far from reaching the same position (see Sec. 28, p. 249) in the measurement of evaporation [490]. The main reason for this is that the precipitation falling on a given surface—apart from fog and dew—is not basically dependent on the small-scale structure of the surface. In contrast to this, evaporation is fundamentally dependent on the nature of the evaporating surface, on its heat budget, and on the water supply. Introduction of the concept of potential evaporation (see p. 252) leads to a quantity that makes some sense meteorologically, and that is largely free from influences of the underlying surface. For its measurement, lysimeters [1192, 1196] can be used to ascertain the actual evaporation. When installed suitably and subjected to close scrutiny, these may yield values that are usable. Values given by atmometers or evaporimeters [1137, p. 226; 1198], for which the expression "evaporation requirement of the air" is often used, are very difficult to compare with one another, because of the vastly different heat budgets of the sensors, and the results do not correspond to the potential evaporation of the ground. They bear roughly the same ratio to the actual evaporation as the rate of cooling measured by the katathermometer or similar instrument does to the flow of heat from the air to the ground surface [485]. Even the values of evaporation measured by small lysimeters and by a few plants standing apart from one another, cannot be applied to communities of plants because of the totally different heat budgets [1163].

57. Combined Measurement and Computation Procedures

The expression 'combined measurement procedure' in this context does not mean an assembly of individual instruments such as that proposed by W. Kreutz [1146, p. 178] as basic equipment for agrometeorologic stations, for example, but a combination of various single instruments, including a carrier when required, designed for some particular purpose. The mobile stations for topoclimatologic investigations fall into this category. A vehicle equipped with a wealth of recording instruments for study of the plant climate has been described by U. Berger-Landefeldt [1209]. A Volkswagen station wagon was fitted out as a mobile laboratory, in which photographic work could also be carried out. When in operation it was mounted on jacks, so as to be steady. One of the best mobile laboratories was the "Neustadt" [1146, p. 134], which consisted of a VW station wagon and trailer. This can also be used as a carrier for the equipment used on field trips. In contrast, R. Geiger [1214] looks on the use of the automobile mainly as a fast means of transport and of carrying working equipment. Most of the instrument assemblies that make up a car load are arranged outside the vehicle during the investigation. The mobile station "Nordrhein-Westfalen" [1146, p. 148] used for measurement in wind-protection investigations, has a trailer that is towed into the research area, and then the instruments are distributed over the terrain. The various ways in which an automobile or trailer can be used are largely determined by the nature of the investigation. In every case, the first consideration is mobility and the ease with which the instrument assemblies can be put into operation.

Fig. 280. Research ship and observation buoy, making stratification measurements in Heligoland Bight. (After K. Brocks)

A small floating instrument carrier for making measurements above water has been described by W. Kreutz [1146, p. 63]. Many new difficulties arise when observations are made at sea. K. Brocks [1210] used a buoy with a mast almost 10 m high to make stratification measurements (Fig. 280). The instruments fitted on the mast are cup anemometers and electric aspiration psychrometers; the recording apparatus is housed in the research vessel. The use of instruments separated from the ship involves considerable complications in procedure, which must be accepted to obtain readings that will be as undisturbed as possible, and these difficulties were evidently overcome by Brocks.

Stratification observations are carried out more frequently over land. The arrangement of instruments to read temperature, humidity, and wind on a mast over 100 m high has been described by A. C. Best and his colleagues [111]. H. Kraus [1172] used two short masts with instruments to measure these quantities, and also the radiation flux. By using several instruments to measure the same quantity at each level, average values over a space are obtained (see Fig. 281).

Fig. 281. A simple mast for stratification measurements. (After H. Kraus)

Stratification measurements are needed especially for heat-balance investigations. A survey of the methods employed for this purpose is obtained by consulting the list of works given in Table 59. Considering the four terms in the energy-balance equation of the earth's surface, the radiation balance S is mostly determined by means of radiation-balance meters, in which group the combination of two effective pyranometers should be included. The flow of heat B out of the ground may be established from the changes taking place in the heat content of the soil (see Sec. 25). In individual cases, instruments to measure the flow of heat in the soil may be used [13, p. 58]. For the two remaining terms, the flux of sensible heat L and the flux of latent heat of evaporation V to the ground, the corresponding heat-exchange currents at some height above the surface are taken, although this implies leaving advective and convective influences out of consideration [467]. To determine the austausch coefficients (see Sec. 25, p. 229) to be used in these expressions for eddy flux of heat, to be inserted in the equations, either wind stratification or the heat balance is used. In the first case relations are derived, such as those given by M. H. Halstead [447], who applies the same values for the quotients of the austausch coefficients for the various properties as for the similar molecular transport coefficients. In the second case, the Sverdrup method (described on p. 230) is used. H. H. Lettau [13, p. 305] has described a procedure that also takes into account the effect of stratification on the austausch coefficient for the flux of sensible heat. All these methods presuppose a knowledge of the temperature and humidity stratifications. Independent of this knowledge, and also of the theoretical model used to describe transport by eddy diffusion in the atmosphere, are the instruments that measure eddy flux directly by time integration of the product of the variation of mass flux and the value of the property in question at the corresponding time.

Instruments of this type may also be regarded as analog computers, which evaluate immediately the data received from the sensing elements. A similar type of calculating machine, into which measurements are fed in the form of photographically recorded curves [1217], is linked to analog computers and simulators [1215], which calculate, for a given selected time sequence of individual meteorologic parameters, by means of fixed relations built into the machine, the other parameters and in particular the terms of the heat-balance equation. Assemblies such as this permit experimentation with various types of meteorologic models, and make it possible to verify the extent to which the models reflect the true state of affairs.

In comparison with analog computers, digital computers (the type that deals with numbers, like ordinary calculating machines) are gaining in importance; it is only by using such fast-working machines that we can ever hope to deal with the rapidly increasing flow of observational data. Evaluation of microclimatologic observations will also be affected by the efforts of the Weather Service to evaluate the great mass of available material more comprehensively and faster by punched-card and punched-tape systems [1211]. Automatic weather stations [1218] and digital recorders [1212] will yield more and better data, and will also simplify and accelerate the evaluation of raw data, which still remains a wearisome task today. Developments in this field are still undergoing evolution, but we can expect that the coming years may bring with them many fundamental changes.

BIBLIOGRAPHY

SECTION 1

[1] *Alissov, B. P., Drosdov, O. A.*, und *Rubinstein, E. S.*, Lehrbuch d. Klimatologie (bes. V., VI., VIII u. IX. Kap.). * Berlin 1956.

[2] *Baum, W. A.*, und *Court, A.*, Research status and needs in microclimatology. * Transact. Amer. Geophys. Union **30**, 488—493, 1949.

[2a] *Brooks, F. A.*, An introduction to physical microclimatology. * Univ. of California, Davis Cal. 1959.

[3] *Brunt, D.*, Some factors in microclimatology. * Quart. J. **71**, 1—10, 1945.

[4] *Franklin, T. B.*, Climates in miniature: a study of microclimate and environment. * Faber & Faber, London 1955.

[5] *Geiger, R.*, Das Stationsnetz z. Untersuchung d. bodennahen Luftschicht. * Deutsch. Met. Jahrb. f. Bayern 1923—1927.

[6] —, Mikroklima u. Pflanzenklima. * Handb. d. Klimatologie, herausgeg. v. *W. Köppen* und *R. Geiger*, Teil I D, Borntraeger, Berlin 1930.

[7] —, Microclimatology. * Compendium of Met., Amer. Met. Soc. 993—1003, Boston 1951.

[8] *Grimm, H.*, Justus von Liebig u. d. Mikroklimatologie. * Z. f. angew. Met. **48**, 30, 1931.

[9] *Kraus, G.*, Boden u. Klima auf kleinstem Raum. * Fischer, Jena 1911.

[10] *Landsberg, H. E.*, Trends in climatology. * Science **128**, 749—758, 1958.

[11] —, und *Woodrow, C. J.*, Applied climatology. * Compendium of Met., Amer. Met. Soc. 976—992, Boston 1951.

[12] *Lettau, H.*, Synthetische Klimatologie. * Ber. DWD-US Zone **6**, Nr. 38, 127—136, 1952.

[13] *Lettau, H. H.* und *Davidson, B.*, Exploring the atmospheres first mile, vol. I u. II. * Pergamon Press, London 1957.

[14] *Peerlkamp, P. K.*, Bodenmeteorolog. onderzoekingen te Wageningen. * Med. Wageningen, Landbouwhoogeschool **47**, Nr. 3, 1944 (96 S.).

[15] *Ramdas, L. A.*, Microclimatological investigations in India. * Arch. f. Met. (B) **3**, 149—167, 1951.

[16] *Ruttner, F., Sepp Aigner* †. * Wetter u. Leben **10**, 96—97, 1958.

[17] *Sapozhnikova, S. A.*, Mikroklimat i mestnyii klimat. * Leningrad 1950 (nach MAB **3**, 692, 1952).

[18] *Schnelle, F.*, Pflanzen-Phänologie. * Probleme der Bioklimatologie **3**, Akad. Verl. Ges. Leipzig 1955.

[19] *Sutton, O. G.*, Micrometeorology. * McGraw Hill Book Co, New York 1953.

[20] —, The development of meteorology as an exact science. * Quart. J. **80**, 328—338, 1954.

[21] *Thornthwaite, C. W.*, A charter for climatology. * WMO-Bulletin **2**, 40—46, 1953.
[22] *Wolfe, J. N.*, The possible role of microclimate. * Ohio J. of science **51**, 134—138, 1951.

SECTIONS 2-5

[23] *Angström, A.*, Über die Gegenstrahlung der Atmosphäre. * Met. Z. **33**, 529—538, 1916.
[24] —, The albedo of various surfaces of ground. * Geograf. Ann. **7**, 323—342, 1925.
[25] *Baumgartner A.*, Das Eindringen des Lichtes in den Boden. * Forstw. C. **72**, 172—184, 1953.
[26] *Bolz, H. M.*, Über die Wirkung der Temperaturstrahlung des atmosphärischen Ozons am Erdboden. * Z. f. Met. **2**, 225—228, 1948.
[27] —, Die Abhängigkeit der infraroten Gegenstrahlung von der Bewölkung. * Z. f. Met. **3**, 201—203, 1949.
[28] —, und *Falckenberg, G.*, Neubestimmung der Konstanten der Angströmschen Strahlungsformel. * Z. f. Met. **3**, 97—100, 1949.
[29] —, und *Fritz, H.*, Tabellen und Diagramme zur Berechnung der Gegenstrahlung und Ausstrahlung. * Z. f. Met. **4**, 314—317, 1950.
[30] *Brooks, F. A.*, Atmospheric radiation and its reflection from the ground. * J. Met. **9**, 41—52, 1952.
[31] *Büttner, K.*, und *Sutter, E.*, Die Abkühlungsgröße in den Dünen etc. * Strahlentherapie **54**, 156—173, 1935.
[32] *Collmann, W.*, Diagramme zum Strahlungsklima Europas. * Ber. DWD 6, Nr. 42, 1958.
[33] *Czepa, O.*, und *Reuter, H.*, Über den Betrag der effektiven Ausstrahlung in Bodennähe bei klarem Himmel. * Arch. f. Met. (B) **2**, 250—258, 1950.
[34] *Dirmhirn, I.*, Einiges über die Reflexion der Sonnen- u. Himmelsstrahlung an verschied. Oberflächen. * Wetter u. Leben **5**, 86—94, 1953.
[35] —, Zur spektralen Verteilung der Reflexion natürlicher Medien. * Ebenda **9**, 41—46, 1957.
[36] *Dubois, P.*, Nächtliche effektive Ausstrahlung. * Gerl. B. **22**, 41—99, 1929.
[37] *Falckenberg, G.*, Absorptionskonstanten einiger met. wichtiger Körper für infrarote Wellen. * Met. Z. **45**, 334—337, 1928.
[38] —, Die Konstanten der Angströmschen Formel zur Berechnung der infraroten Eigenstrahlung d. Atmosph. aus dem Zenit. * Z. f. Met. **8**, 216—222, 1954.
[39] *Fleischer, R.*, und *Gräfe, K.*, Die Ultrarot-Strahlungsströme aus Registrierungen des Strahlungsbilanzmessers nach Schulze. * Ann. d. Met. **7**, 87—95, 1955/56.
[40] *Fritz, S.*, The albedo of the ground and atmosphere. * Bull. Am. Met. Soc. **29**, 303—312, 1948.
[41] —, und *Rigby, M.*, Bibliography on albedo. * MAB **8**, 949—999, 1957.
[42] *Gates, D. M.*, und *Tantraporn, W.*, The reflectivity of deciduous trees and herbaceous plants in the infrared to 25 microns. * Science **95**, 613—616, 1952.
[43] *Hinzpeter, H.*, Die effektive Ausstrahlung u. ihre Abhängigkeit v. d. Absorptionseigenschaften im Fenster der Wasserdampfbanden. * Z. f. Met. **11**, 321—329, 1957.

[44] *Hofmann, G.*, Zur Darstellung der spektralen Verteilung der Strahlungsenergie. * Arch. f. Met. (B) **6**, 274—279, 1955.
[45] *Houghton, J. T.*, The emissivity of the earth's surface. * Quart. J. **84**, 448—450, 1958.
[46] *Köhn, M.*, Zur Kenntnis des Lichthaushaltes dünner Pulverschichten, insbesondere v. Böden. * Naturw. **34**, 89—90, 1947.
[47] *Lauscher, F.*, Bericht über Messungen der nächtl. Ausstrahlung auf der Stolzalpe. * Met. Z. **45**, 371—375, 1928.
[48] —, Wärmeausstrahlung u. Horizonteinengung. * Sitz-B. Wien. Akad. **143**, 503—519, 1934.
[49] —, Strahlungs- u. Wärmehaushalt. * Ber. DWD **4**, Nr. 22, 21—29, 1956.
[50] *Linke, F.*, Die nächtl. effektive Ausstrahlung unter verschiedenen Zenitdistanzen. * Met. Z. **48**, 25—31, 1931.
[51] *Lönnquist, O.*, Synthetic formulae for estimating effective radiation to a cloudless sky etc. * Ark. f. Geofysik **2**, 245—294, Stockholm 1954.
[52] *Penndorf, R.*, Luminous reflectance (visual albedo) of natural objects. * Bull. Am. Met. Soc. **37**, 142—144, 1956.
[53] *Philipps, H.*, Zur Theorie der Wärmestrahlung in Bodennähe. * Gerl. B. **56**, 229—319, 1940.
[54] *Sauberer, F.*, Bemerkungen zu Fragen der praktischen Strahlungskunde. * Arch. f. Met. (B) **1**, 54—62, 1949.
[55] —, Das Licht im Boden. * Wetter u. Leben **3**, 40—44, 1951.
[56] —, Registrierungen der nächtlichen Ausstrahlung. * Arch. f. Met. (B) **2**, 347—359, 1951.
[57] —, Beiträge zur Kenntnis des Strahlungsklimas von Wien. * Wetter u. Leben **4**, 187—192, 1952.
[58] —, und *Dirmhirn, I.*, Untersuchungen über die Strahlungsverhältnisse auf den Alpengletschern. * Arch. f. Met. (B) **3**, 256—269, 1951.
[59] *Schnaidt, F.*, Zur Absorption infraroter Strahlung in dünnen Luftschichten. * Met. Z. **54**, 234—242, 1937.

SECTION 6

[60] *Albrecht, F.*, Meßgeräte des Wärmehaushalts an der Erdoberfläche als Mittel d. bioklimat. Forschung. * Met. Z. **54**, 471—475, 1937.
[61] *Becker, F.*, Die Erdbodentemperaturen als Indikator der Versickerung. * Met. Z. **54**, 372—377, 1937.
[62] *Diem, M.*, Bodenatmung. * Gerl. B. **51**, 146—166, 1937.
[63] *Hofmann, G.*, Die Thermodynamik der Taubildung. * Ber. DWD **3**, Nr. 18, 1955.
[64] *Lettau, H.*, Improved models of thermal diffusion in the soil. * Transact. Amer. Geophys. Union **35**, 121—132, 1954.
[65] *Schubert, J.*, Grundlagen der allgemeinen u. forstlichen Klimakunde. * Z. f. F. u. Jagdw. **62**, 689—705, 1930; **64**, 715—734, 1932; **72**, 257—273, 1940.

SECTIONS 7-9

[66] *Berger-Landefeldt, U.*, *Kiendl, J.*, und *Danneberg, H.*, Betrachtungen zur Temp.- u. Dampfdruckunruhe über Pflanzenbeständen. * Met. Rundsch. **10**, 11—20, 1957.

[67] *Defant, A.*, Über die Abkühlung der untersten staubbeladenen Luftschichten. * Ann. d. Hydr. **47**, 93—105, 1919.
[68] *Diem, M.*, Messungen der Staubausbreitung aus den Schloten einer Indudustrieanlage am Niederrhein. * Mitt. d. Vereinig. d. Großkesselbesitzer **42**, 1—23, 1956.
[69] *Firbas, F.*, und *Rempe, H.*, Über die Bedeutung der Sinkgeschwindigkeit für d. Verbreit. des Blütenstaubes durch d. Wind. * Biokl. B. **3**, 49—53, 1936.
[70] *Haude, W.*, Temperatur u. Austausch der bodennahen Luft über einer Wüste. * Beitr. Phys. d. fr. Atm. **21**, 129—142, 1934.
[71] *Katheder, F.*, Auflösung einer Bodennebeldecke durch ein startendes Flugzeug. * Z. f. angew. Met. **54**, 61—63, 1937.
[72] *Kohlermann, L.*, Untersuchungen über d. Windverbreitung der Früchte u. Samen d. mitteleurop. Waldbäume. * Forstw. C. **69**, 606—624, 1950.
[73] *Kramer, M. P.*, *Assur, A.*, und *Rigby, M.*, Selective annotated bibliography on turbulent diffusion and micrometeorological turbulence. * MAB **4**, 186—249, 1953.
[74] *Kreutz, W.*, und *Walter, W.*, Der Wind als Träger von Zementstaub und dessen Ablagerung auf Boden und Pflanze. * Gartenbauwissensch. **3**, 151—164, 1956.
[75] *Lettau, H.*, Atmosphärische Turbulenz. * Akad. Verl. Ges. Leipzig 1939.
[76] —, Über die Zeit- u. Höhenabhängigkeit d. Austauschkoeff. im Tagesgang innerhalb d. Bodenschicht. * Gerl. B. **57**, 171—192, 1941.
[77] *Meetham, A. R.*, Atmospheric pollution. * Pergamon Press London 1956.
[78] *Priestley, C. H. B.*, Free and forced convection in the atmosphere near the ground. * Quart. J. **81**, 139—143, 1955 u. **82**, 242—244, 1956.
[79] *Ramdas, L. A.*, und *Malurkar, S. L.*, Surface convection and variation of temp. near a hot surface. * Indian J. of Physics **7**, 1—13, 1932.
[80] *Rombakis, S.*, Über die Verbreitung von Pflanzensamen u. Sporen durch turbulente Luftströmungen. * Z. f. Met. **1**, 359—363, 1947.
[81] *Schmauß, A.*, Die nächtliche Abkühlung der untersten Luftschichten. * Ann. d. Hydr. **47**, 235—236, 1919.
[82] *Schmidt, W.*, Der Massenaustausch in freier Luft u. verwandte Erscheinungen. * H. Grand, Hamburg 1925.
[83] *Sutton, O. G.*, Atmospheric turbulence. * 2. Aufl. Methuen & Co., London 1955.
[84] *Swinbank, W. C.*, Turbulent transfer in the lower atmosphere. * Proc. Canberra Symposion 1956, 35—37, UNESCO Paris 1958.
[85] *Trappenberg, R.*, Theoret. u. experimentelle Untersuchungen zur Staubverteilung einer Rauchfahne. * Diss. T. H. Karlsruhe 1956.
[86] U. S. Weather Bureau, Meteorology and atomic energy * Washington 1955.
[87] *Wilkins, E. M.*, Computation of instantaneous source diffusion coefficients from smoke puff observations. * J. Met. **15**, 175—179, 1958.

SECTION 10

[88] *Albrecht, F.*, Ergebnisse von Dr. Haudes Beobachtungen etc. * Rep. Scient. Exped. to the NW Prov. China under the leadership Dr. Sven Hedin **9**, Met. 2, Stockholm 1941.
[89] *Batta, E.*, Tägliche Normalwerte der Bodentemp. u. die Bestimmung einer Aussaattemperatur. * Idöjárás **59**, 351—358, 1955.

[90] *Chang, Jen-hu,* World patterns of monthly soil temperature distribution. * Ann. Assoc. Amer. Geograph. **47**, 241—249, 1957.

[91] —, Global distribution of the annual range in soil temperature. * Transact. Amer. Geophys. Union **38**, 718—723, 1957.

[92] —, Ground temperature (2 Bde.). * Blue Hill Met. Observ. Milton Mass. 1958.

[93] *Dirmhirn, I.,* Registrierungen d. Temp. d. Bodenoberfläche an d. Zentralanstalt f. Met. u. Geodynamik Wien. * Jahrb. Zentralanstalt f. Met. u. Geodyn. Wien N. F. **87**, D 45—51, 1950, Wien 1951.

[94] —, Tagesschwankung d. Bodentemp., Sonnenscheindauer u. Bewölkung. * Wetter u. Leben **3**, 216—219, 1951.

[95] *Hausmann, G.,* Unperiodische Schwankungen der Erdbodentemp. in 1 m bis 12 m Tiefe. * Z. f. Met. **4**, 363—372, 1950.

[96] *Hautfenne, M.,* Une année d'enregistrement continu de la température du sol. * Mém. Inst. Royal Mét. (Belgien) **50**, 1952.

[97] *Herr, L.,* Bodentemperaturen unter besonderer Berücksichtigung der äußeren met. Faktoren. * Diss. Leipzig 1936.

[98] *Homén, Th.,* Der tägliche Wärmeumsatz im Boden u. die Wärmestrahlung zwischen Himmel u. Erde. * Leipzig 1897.

[99] *Katić, P.,* The soil temperatures at Novi Sad. * Ann. Scient. Work Fac. of agriculture Novi Sad **1**, 1957.

[100] *Kullenberg, B.,* Biological observations during the solar eclipse in southern Sweden etc. * Oikos **6**, 51—60, 1955.

[101] *Lehmann, P.,* Raumeinteilung der klimagebundenen Lithosphäre. * Ber. DWD-US Zone **7**, Nr. 42, 274—276, 1952.

[102] *Leyst, E.,* Über die Bodentemperaturen in Pawlowsk. * Rep. f. Met. **13**, Nr. 7, 1—311, Petersburg 1890.

[103] —, Untersuchungen über die Bodentemp. in Königsberg. * Schr. d. phys.-ökonom. Ges. Königsberg **33**, 1—67, 1892.

[104] *McCulloch, J. S. G.,* Soil temperatures near Nairobi 1954—55. * Quart. J. **85**, 51—56, 1959.

[105] *Müller, W.,* Großräumige Änderungen der Temperatur des Erdbodens seit 1911. * Met. Rundschau **11**, 145—150, 1958.

[106] *Paulsen, H. S.,* On radiation, illumination and met. conditions in S. Norway during the total solar eclipse of june 30, 1954. * Arbok Univ. Bergen 1955, Nr. 7.

[107] *Schmidt, A.,* Theoretische Verwertung der Königsberger Bodentemperaturbeobachtungen. * Schr. d. phys.-ökonom. Ges. Königsberg **32**, 97—168, 1891.

[108] *Toperczer, M.,* Die Bodentemperaturen in Wien 1911—1944. * Jahrb. d. Zentralanstalt f. Met. u. Geodyn. 1946, Anhang 6, Wien 1947.

[109] *Unger, K.,* Bearbeitung der Bodentemperaturen von Quedlinburg. * Angew. Met. **1**, 85—90, 1951.

[110] *Zikeev, N. T.,* A selective annotated bibliography on soil temperature. * MAB **2**, 207—232, 1951.

SECTIONS 11-12

[111] *Best, A. C., Knighting, E., Pedlow, R. H.,* und *Stormonth, K.,* Temperature and humidity gradients in the first 100 m over SE-England. * Geophys. Mem. **89**, London 1952.

[112] *Brocks, K.*, Über den tägl. u. jährl. Gang der Höhenabhängigkeit der Temp. in den untersten 300 m d. Atmosphäre u. ihren Zusammenhang mit d. Konvektion. * Ber. DWD-US Zone **1**, Nr. 5, 1948.

[113] —, Die Höhenabhängigkeit der Lufttemperatur in der nächtlichen Inversion. * Met. Rundschau **2**, 159—167, 1949.

[114] —, Temperatur u. Austausch in der untersten Atmosphäre. * Ber. DWD-US Zone **2**, Nr. 12, 166—170, 1950.

[115] *Flower, W. D.*, An investigation into the variation of the lapse rate of temperature in the atmosphere near the ground at Ismailia, Egypt. * Geophys. Mem. **71**, London 1937.

[116] *Henning, H.*, Pico-aerologische Untersuchungen über Temperatur- und Windverhältnisse d. bodennahen Luftschicht bis 10 m Höhe in Lindenberg. * Abh. Met. D. DDR **6**, Nr. 42, 1—66, 1957.

[117] *Johnson, N. K.*, A study of the vertical gradient of temperature in the atmosphere near the ground. * Geophys. Mem. **46**, London 1929.

[118] —, und *Heywood, G. S. P.*, An investigation of the lapse rate of temperature in the lowest 100 m of the atmosphere. * Geophys. Mem. **77**, London 1938.

[119] *Mal, S.*, *Desai, B. N.*, und *Sircar, S. P.*, An investigation into the variation of the lapse rate of temperature in the atmosphere near the ground at Drigh Road, Karachi. * Mem. India Met. Dep. **29**, Part 1, Calcutta 1942.

[120] *Reuter, H.*, Die Modifikation einer Luftmasse durch die nächtliche Abkühlung der Erdoberfläche. * Arch. f. Met. (A) **1**, 252—263, 1948.

[121] *Rink, J.*, Über das Verhalten der mittleren vertikalen Temperaturgradienten d. bodenn. Luftschicht (1—76 m) u. seine Abhängigkeit von speziellen Witterungsfaktoren u. Wetterlagen. * Abh. Met. D. DDR **3**, Nr. 18, 1—43, 1953.

SECTION 13

[122] *Battan, L. J.*, Energy of dust devil. * J. Met. **15**, 235—237, 1958.

[123] *Baum, W. A.*, Note on the theory of super-autoconvective lapse rates near the ground. * J. Met. **8**, 196—198, 1951.

[124] *Best, A. C.*, Transfer of heat and momentum in the lowest layers of the atmosphere. * Geophys. Mem. **65**, London 1935.

[125] *Czepa, O.*, Über die Energieleitung durch langwellige Strahlung in der bodennahen Luftschicht. * Z. f. Met. **5**, 292—300, 1951.

[126] *Deacon, E. L.*, Radiative heat transfer in the air near the ground. * Australian J. Scient. Res. (A) **3**, 274—283, 1950.

[127] *De Mastus, H. L.*, Pressure disturbances in the vicinity of dust devils. * Bull. Am. Met. Soc. **35**, 497—498, 1954.

[128] *Flower, W. D.*, Sand devils. * Met. Office London Profess. Notes **71**, 1936.

[129] *Franssila, M.*, Mikroklimatische Temperaturmessungen in Sodankylä. * Mitt. Met. Z. Anst. Helsinki **26**, 1—29, 1945.

[130] *Hartmann, K.*, Beobachtung einer Kleintrombe. * Z. f. Met. **8**, 189—191, 1954.

[131] *Ives, R. L.*, Behaviour of dust devils. * Bull. Am. Met. Soc. **28**, 168—174, 1947.

[132] *Klauser, L.*, Beobachtung einiger Kleintromben bei Potsdam. * Z. f. Met. **4**, 187—188, 1950.

[133] *Kyriazopoulos, B. D.*, Micrometeorological phenomenon in Byzantine decoration. * Publ. Met. Inst. Univ. Thessaloniki **4**, 1955.

[134] *Linke, F.*, und *Möller, F.*, Langwellige Strahlungsströme in der Atmosphäre und die Strahlungsbilanz. * Handb. d. Geophysik **8**, 651—721, Gebr. Borntraeger, Berlin 1943.

[135] *Möller, F.*, Strahlungsvorgänge in Bodennähe. * Z. f. Met. **9**, 47—53, 1955.

[136] —, Strahlung in der unteren Atmosphäre. * Handb. der Physik (herausgegeben von S. Flügge) **48**, Geophysik II, 155—253, J. Springer 1957.

[137] *Priestley, C. H. B.*, Heat conduction and temperature profiles in air and soil. * J. Austral. Inst. Agricult. Science **25**, 94—107, 1959.

[138] *Rossi, V.*, Über mikroklimatische Temp.- u. Feucht.-beobachtungen mit Thermoelementpsychrometern. * Soc. Scient. Fennica **6**, Nr. 25, 1933.

[139] *Schlichting, H.*, Kleintrombe. * Ann. d. Hydr. **62**, 347—348, 1934.

[140] *Tait, G. W. C.*, The vertical temperature gradient in the lower atmosphere under daylight conditions. * Quart. J. **75**, 287—292, 1949.

[141] *Thornthwaite, C. W.*, Micrometeorology of the surface layer of the atmosphere. * Publ. Climat. **1** bis **5**, 1948—52.

SECTION 14

[142] *Albani, F.*, Investigaciones sobre la distribución vertical de las temp. minimas en las capas de aire próximas al suelo. * Arch. Fitotécnico del Uruguay **4**, 361—376, 1951.

[143] *Brooks, D. L.*, A tabular method for the computation of temp. change by infrared radiation in the free atmosphere. * J. Met. **7**, 313—321, 1950.

[144] *De Quervain, F.*, und *Gschwind, M.*, Die nutzbaren Gesteine der Schweiz. * H. Huber, Bern 1934.

[145] *Dimitz, L.*, Hüttenminimum oder Erdbodenminimum? * Wetter u. Leben **1**, 321—326, 1949.

[146] —, Untersuchungen über die Frostdauer in 2 m und 5 cm über dem Erdboden. * Ebenda **2**, 58—61, 1950.

[147] *Drimmel, J.*, Über theoret. Formeln zur Berechnung der nächtlichen Abkühlung der Erdoberfläche und über ihre Anwendungsmöglichkeiten. * Arch. f. Met. (B) **5**, 18—40, 1953.

[148] *Eimern, J. van*, und *Kaps, E.*, Lokalklimatische Untersuchungen im Raum der Harzburger Berge u. d. benachbart. Elbniederung. * Landwirtsch.-Verl. Hiltrup 1954.

[149] *Falckenberg, G.*, Der Einfluß der Wellenlängentransformation auf das Klima der bodenn. Luftschichten u. d. Temp. der freien Atmosphäre. * Met. Z. **48**, 341—346, 1931.

[150] —, Der nächtliche Wärmehaushalt bodennaher Luftschichten. * Met. Z. **49**, 369—371, 1932.

[151] —, Experimentelles zur Absorption dünner Luftschichten für infrarote Strahlung. * Met. Z. **53**, 172—175, 1936.

[152] —, und *Stoecker, E.*, Bodeninversion u. atmosphärische Energieleitung durch Strahlung. * Beitr. Phys. d. fr. Atm. **13**, 246—269, 1927.

[153] *Fleagle, R. G.*, A theory of fog formation. * J. Marine Res. **12**, 43—50, 1953.

[154] —, The temperature distribution near a cold surface. * J. Met. **13**, 160—165, 1956.

[155] *Hader, F.*, Kann der Erdbodenabstand der Thermometerhütte verkleinert werden? * Wetter u. Leben **6**, 27—31, 1954.

[156] *Heyer, E.*, Über Frostwechselzahlen in Luft u. Boden. * Gerl. B. **52**, 68—122, 1938.

[157] *Lake, J. V.*, The temperature profile above bare soil on clear nights. * Quart. J. **82**, 187—197, 530—531, 1956.

[158] *Ramanathan, K. R.*, und *Ramdas, L. A.*, Derivation of Angströms formula for atmosph. radiation and some general considerations regarding nocturnal cooling of air-layers near the ground. * Proc. Ind. Acad. Sciences **1**, 822—829, 1935.

[159] *Ramdas, L. A.*, und *Atmanathan, S.*, The vertical distribution of air temperature near the ground during night. * Gerl B. **3**, 49—53, 1936

[160] —, *Kalamkar, R. J.*, und *Gadre, K. M.*, Agricultural met. studies in microclimatology. * Indian J. Agric. Sc. **4**, 451—467, 1934, und **5**, 1—11, 1935.

[161] *Raschke, K.*, Über das nächtliche Temperaturminimum über nacktem Boden in Poona. * Met. Rundsch. **10**, 1—11, 1957.

[162] *Reuter, H.*, Die Modifikation einer Luftmasse durch die nächtliche Abkühlung der Erdoberfläche. * Arch. f. Met. (A) **1**, 252—263, 1948.

[163] *Schwalbe, G.*, Über die Temperaturminima in 5 cm über dem Erdboden. * Met. Z. **39**, 41—46, 1922.

[164] *Siegel, S.*, Messungen des nächtlichen Gefüges in der bodennahen Luftschicht. * Gerl. B. **47**, 369—399, 1936.

[165] *Sverdrup, H. U.*, Austausch u. Stabilität in der untersten Luftschicht. * Met. Z. **53**, 10—15, 1936.

[166] *Szakály, J.*, Temperature minimum above bare soil during the night. * Idöjárás **61**, 158—160, 1957.

[167] *Troll, C.*, Die Frostwechselhäufigkeit in den Luft- u. Bodenklimaten der Erde. * Met. Z. **60**, 161—171, 1943.

[168] *Witterstein, F.*, Die Differenz zwischen Hütten- u. Erdbodenminimumtemp. nach heiteren und trüben Nächten in Geisenheim. * Met. Rundschau **2**, 172—174, 1949.

SECTION 15
(See also the references to Secs. 11-14 and 25-26)

[169] *Berger-Landefeldt, U.*, *Kiendl, J.*, und *Danneberg, H.*, Beoabachtungen des Temp.- u. Feuchtigkeitsgeschehens über Pflanzenbeständen. * Met. Rundschau **9**, 120—130, 1956.

[170] *Büdel, A.*, Die Feuchtemessung in der bodennahen Luftschicht. * Z. f. angew. Met. **48**, 289—293, 1931.

[171] *Büttner, K. J. K.*, Die Aufnahme von Wasserdampf durch die menschliche Haut, Pflanze u. Erdboden. * Arch. f. Met. (B), **9**, 80—85, 1958.

[172] *Diem, M.*, Tagesgang der relativen Feuchtigkeit in der bodennahen Luftschicht im Gras u. über nacktem Humus. * Arch. f. Met. (B) **2**, 441—447, 1950.

[173] *Lehmann, P.*, und *Schanderl, H.*, Tau und Reif. * R. f. W. Wiss. Abh. **9**, Nr. 4, 1942.

[174] *Ramdas, L. A.*, The variation with height of the water vapour content of the air layers near the ground at Poona. * Biokl. B. **5**, 30—34, 1938.
[175] —, und *Katti, M. S.*, Prelim. studies on soil moisture in relation to moisture in the surface layers of the atmosphere during the clear season at Poona. * Indian J. Agric. Sc. **4**, 923—937, 1934.
[176] *Thornthwaite, C. W.*, Summary of observations made at O'Neill, Nebraska. * Publ. Climat. **6**, 149—238, 1953.
[177] *Vieser, W.*, Temperatur- u. Feuchtigkeitsverhältnisse in bodennahen Luftschichten. * Beitr. z. naturkundl. Forsch. in Südwestdeutschland **10**, 3—34, 1951.
[178] *Vowinckel, E.*, Temperatur u. Feuchtigkeit der bodennahen Luftschicht in Pretoria. * Met. Rundsch. **4**, 22—23, 1951.

SECTION 16

[179] *Carruthers, N.*, Variations in wind velocity near the ground. * Quart. J. **69**, 289—301, 1943.
[180] *Davidson, B.*, und *Barad, M. L.*, Some comments on the Deacon wind profile. * Transact. Amer. Geophys. Union **37**, 168—176, 1956.
[181] *Deacon, E. L.*, Vertical profiles of mean wind in the surface layers of the atmosphere. * Geophys. Mem. **11**, Nr. 91, London 1953.
[182] —, Gust variation with height up to 150 m. * Quart. J. **81**, 562—573, 1955.
[183] *Hallenbeck, C.*, Night-temperature studies in the Roswell fruit district. * M. W. Rev. **46**, 364—373, 1918.
[184] *Halstead, M. H.*, The relationship between wind structure and turbulence near the ground. * Publ. Climat. **4**, Nr. 3, 1951.
[185] *Hellmann, G.*, Über die Bewegung der Luft in den untersten Schichten der Atmosphäre. * Met. Z. **32**, 1—16, 1915 u. Sitz-B. Berlin Akad. 1919, 404—416.
[186] *Heywood, G. S. P.*, Wind structure near the ground and its relation to temperature gradient. * Quart. J. **57**, 433—455, 1931.
[187] *McAdie, A. G.*, Studies in frost protection — effect of mixing the air. * M. W. Rev. **40**, 122—123, 779, 1912.
[188] *Paeschke, W.*, Experimentelle Untersuchungen zum Rauhigkeits- und Stabilitätsproblem in der bodennahen Luftschicht. * Beitr. Phys. d. fr. Atm. **24**, 163—189, 1937.
[189] *Prandtl, L.*, Führer durch die Strömungslehre, 5. Aufl. * Fr. Vieweg, Braunschweig 1957.
[190] *Sutton, O. G.*, Note on the variation of the wind with height. * Quart. J. **58**, 74—76, 1932.
[191] *Thornthwaite, C. W.*, und *Halstead, M.*, Note on the variation of wind with height in the layer near the ground. * Transact. Amer. Geophys. Union **23**, 249—255, 1942.
[192] *Trappenberg, R.*, Ein Beitrag zu den Windverhältnissen in den ersten 100 m der Atmosphäre. * Ber. DWD **8**, Nr. 57, 1959.
[193] *Young, F. D.*, Notes on the 1922 freeze in southern California. * M. W. Rev. **51**, 581—585, 1923.

SECTION 17

[194] *Auer, R.*, Über den täglichen Gang des Ozongehalts der bodennahen Luftschicht. * Gerl. B. **54**, 137—145, 1939.

[195] *Becker, F.*, Messung des Emanationsgehalts der Luft in Frankfurt a. M. und am Taunusobservatorium. * Gerl. B. **42**, 365—384, 1934.

[196] *Buch, K.*, Kohlensäure in Atmosphäre u. Meer. * Ann. d. Hydr. **70**, 193—205, 1942.

[197] *Demon, L.*, *De Felice, P.*, *Gondet, H.*, *Pontier, L.* u. *Kast, Y.*, Recherches effectuées par la section de physique du centre de recherches sahariennes en 1954, 1955 et 1956. * J. des recherches du centre nat. de la rech. scientif., Lab. de Bellevue **38**, 30—63, 1957.

[198] *Effenberger, E. F.*, Kern- u. Staubuntersuchungen am Collmberg. * Veröffentlicht. Geoph. I. Leipzig **12**, 305—359, 1940.

[199] —, Meßmethoden zur Bestimmung des CO_2-Gehalts der Atmosphäre etc. * Ann. d. Met. **4**, 417—427, 1951.

[200] —, Das Kohlenoxyd u. dessen Bedeutung. * Mediz.-met. Hefte Nr. 12, Hamburg 1957 (128 S.).

[201] *Ehmert, A.* und *H.*, Über den Tagesgang des bodennahen Ozons. * Ber. DWD-US Zone **1**, Nr. 11, 58—62, 1949.

[202] *Fett, W.*, Der atmosphärische Staub. * VEB Deutsch. Verl. d. Wiss. Berlin 1958.

[203] *v. Gehren, R.*, Die Bodenverwehungen in Niedersachsen 1947—51. * Veröff. Niedersächs. Amt f. Landesplanung u. Statistik, Reihe G, Bd. 6, Hannover 1954.

[204] *Glückauf, E.*, The composition of atmospheric air. * Compend. of Met. (Amer. Met. Soc.) 3—10, Boston 1951.

[205] *Goldschmidt, H.*, Messung der atmosphärischen Trübung mit einem Scheinwerfer. * Met. Z. **55**, 170—174, 1938.

[206] *Haude, W.*, Ergebnisse der allgemeinen met. Beobachtungen etc. * Rep. Scient. Exped. to the NW China under the leadership Dr. Sven Hedin IX Met. **1**, Stockholm 1940.

[207] *Huber, B.*, Über die vertikale Reichweite vegetationsbedingter Tagesschwankungen im CO_2-Gehalt der Atmosphäre. * Forstw. C. **71**, 372—380, 1952.

[208] —, Der Einfluß der Vegetation auf die Schwankungen des CO_2-Gehalts der Atmosphäre. * Arch. f. Met. (B) **4**, 154—167, 1952.

[209] —, und *Pommer J.*, Zur Frage eines jahreszeitlichen Ganges im CO_2-Gehalt der Atmosphäre. * Angew. Botanik **28**, 53—62, 1954.

[210] *Iizuka, I.*, On carbon dioxide in the peach orchard. * J. Agr. Met. Japan **11**, 84—86, 1955.

[211] *Inoue, E.*, *Tani, N.*, *Imai, K.*, und *Isobe, S.*, The aerodynamic measurement of photosynthesis over the wheat field. * J. Agr. Met. Japan **13**, 121—125, 1958.

[212] *Mühleisen, R.*, Atmosphärische Elektrizität. * Hand. d. Physik (herausg. v. S. Flügge) **48** (Geophysik II) 541—607, Springer 1957.

[213] *Niemann, A.*, Die natürlichen u. künstlichen Aerosole, ihre Bedeutung für den Gartenbau und ihre Probleme. * Z. f. Aerosol-Forsch. u. -Therapie **5**, 341—351, 1956.

[214] *Paetzold, H. K.*, Neue USA-Forschungsergebnisse über die Erdatmosphäre. * Physikal. Blätter **13**, 395—403, 1957.

[215] *Priebsch, J.*, Die Höhenverteilung radioaktiver Stoffe in der freien Luft. * Met. Z. **49**, 80—81, 1932.

[216] *Siedentopf, H.*, Zur Optik des atmosphärischen Dunstes. * Z. f. Met. **1**, 417—422, 1947.

[217] *Sindowski, K. H.*, Korngrößen- und Konformen-Auslese beim Sandtransport durch Wind. * Geolog. Jahrb. Hannover **71**, 517—525, 1956.

[218] *Stepanova, N. A.*, A selective annotated bibliography on carbon dioxide in the atmosphere. * MAB **3**, 137—170, 1952.

[219] *Teichert, F.*, Vergleichende Messung des Ozongehaltes der Luft am Erdboden und in 80 m Höhe. * Z. f. Met. **9**, 21—27, 1955.

[220] Vereinigung der Großkesselbesitzer, Luftverunreinigung durch Rauchgase aus Dampfkesselanlagen. * Mitt. d. Vereinig. etc. Essen 1955.

[221] *Wentzel, K. F.* (ed.), Wald-Rauchschaden. * Allg. F. **13**, 597—616, 1958.

[222] *Zimmermann, G.*, Kleinklimatische Ozonmessungen in Bad Kissingen. * Mitt. DWD **1**, Nr. 1, 1—13, 1953.

SECTION 18

[223] *Ashmore, S. E.*, North Wales road mirage. * Weather **10**, 336—342, 1955.

[224] *Aujeszky, L.*, Kleinklima u. Schallklima. * Forsch. u. Fortschr. **14**, 413—415, 1938.

[225] *Brocks, K.*, Die terrestrische Refraktion, ein Grenzgebiet der Meteorologie und Geodäsie. * Ann. d. Met. **1**, 329—336, 1948.

[226] —, Über vertikale Luftdichtezunahme in Bodennähe. * Met. Rundsch. **2**, 227—229, 1949.

[227] —, Die Lichtstrahlkrümmung in Bodennähe. * Deutsche Hydrograph. Z. **3**, 241—248, 1950.

[228] —, Eine räumlich integrierende optische Methode für die Messung vertikaler Temp. u. Wasserdampfgradienten in der untersten Atmosphäre. * Arch. f. Met. (A) **6**, 370—402, 1954.

[229] *Ertel, H.*, Ein Problem der meteorologischen Akustik. * Sitz-Ber. D. Akad. d. Wiss. Berlin (Kl. Mathem., Phys. u. Techn.) 1955, Nr. 2.

[230] *Fényi, J.*, Über Luftspiegelungen in Ungarn. * Met. Z. **19**, 507—509, 1902.

[231] *Findeisen, W.*, Über Beobachtungen von Luftspiegelungen auf dem Neuwerker Watt. * Ann. d. Hydr. **62**, 423—426, 1934.

[232] *Fleagle, R. G.*, The optical measurement of lapse rate. * Bull. Am. Met. Soc. **31**, 51—55, 1950.

[233] *Hamilton, R. A.*, The determination of temperature gradients over the Greenland ice sheet by optical methods. * Proc. Royal Soc. Edinburgh (A) **64**, 381—397, 1956/57.

[234] *Heybrock, W.*, Luftspiegelungen in Marokko. * Met. Rundsch. **6**, 24—25, 1953.

[235] *Hrudička, B.*, Zu den optischen u. akustischen Eigenschaften des Klimas einer Großstadt. * Gerl. B. **53**, 337—344, 1938.

[236] *Ives, R. L.*, Meteorological conditions accompanying mirages in the salt lake desert. * J. Franklin Inst. (USA) **245**, 457—473, 1948.

[237] *Marquardt, W.*, Eine seltene bodennahe Haloerscheinung. * Z. f. Met. **13**, 133—134, 1959.

[238] *Meyer, R.*, Atmosphärische Strahlenbrechung. * Linke-Möller, Hdb. d. Geophysik VIII, Borntraeger Berlin 1956 (Kap. 13).

[239] *Pernter, J. M.*, und *Exner, F. M.*, Meteorologische Optik. * W. Braunmüller, Wien u. Leipzig 1922.

[240] *Portig, W.*, Halo im Eisnebel. * Met. Z. **59**, 207—208, 1942.

[241] *Ramdas, L. A.*, und *Malurkar, S. L.*, Theory of extremely high lapse rates of temperature very near the ground. * Indian J. Physics **6**, 495—508, 1932.

[242] *Schiele, W. E.*, Zur Theorie der Luftspiegelungen. * Veröff. Geoph. I. Leipzig **7**, Nr. 3, 1935.
[243] *Scott, R. F.*, Letzte Fahrt I. * Brockhaus Leipzig 1913.
[244] *Wegener, A.*, Optik der Atmosphäre. * Müller-Pouillets Lehrb. d. Physik **5** (1), 11. Aufl. 1928.

SECTION 19

[245] *Baden, W.*, und *Eggelsmann, R.*, Ein Beitrag zur Hydrologie der Moore. * Moor u. Torf **4**, Beilage 3, 1952.
[246] *Bender, K.*, Die Frühjahrsfröste an der Unterelbe u. ihre Bekämpfung. * Z. f. angew. Met. **56**, 273—289, 1939.
[247] *Duin, R. H. A. van*, Influence of tilth on soil- and airtemperature. * Netherlands J. Agric. Science **2**, 229—241, 1954.
[248] *Grasnick, C. H.*, Die Temperaturänderung der Troposphäre über Lindenberg im Bereich von Antizyklonen durch Strahlungs- und Turbulenzvorgänge. * Abh. Met. D. DDR **43**, 1—102, 1957.
[249] *Homén, T.*, Der tägliche Wärmeumsatz im Boden u. die Wärmestrahlung zwischen Himmel u. Erde. * Leipzig 1897.
[250] *Johnson, N. K.*, und *Davies, E. L.*, Some measurements of temperature near the surface in various kinds of soils. * Quart. J. **53**, 45—59, 1927.
[251] *Kern, H.*, Die Temperaturverhältnisse in Niedermoorboden im Gegensatz zu Mineralboden. * Landw. Jahrb. f. Bayern **29**, 587—602, 1952.
[252] *Kreutz, W.*, Der Jahresgang der Temperatur in verschiedenen Böden unter gleichen Witterungsverhältnissen. * Z. f. angew. Met. **60**, 65—76, 1943.
[253] *Morgen, A.*, Zur künstl. Wärmesteuerung im Wurzelraum der Pflanze. * Met. Rundsch. **10**, 135—139, 1957.
[254] *Nidetzky, L.*, Temperaturverhältnisse in hydroponischen Versuchsbeeten. * Wetter u. Leben **6**, 151—154, 1954.
[255] *Olsson, A.*, Undersökning över vältuingens iuverkan på marktemp. och på lufttemp. närmast markytan. * Lantbruksakad. Tidskr. Stockholm **92**, 220—241, 1953.
[256] *Pessi, Y.*, Studies on the effect of the admixture of mineral soil upon the thermal conditions of cultivated peat land. * Publ. Finnish State Agric. Res. Board **147**, 1956 (89 S.).
[257] —, On the effect of rolling upon the barley and oat crop yield and upon the thermal conditions of cultivated peat land. * Ebenda **151**, 1956 (23 S.).
[258] —, On the thermal conditions in mineral and peat soil at Pelsonsuo in 1955—1956. * Ebenda **159**, 1957 (31. S.)
[259] *Philipps, H.*, Zur Theorie der Wärmestrahlung in Bodennähe. * Gerl. B. **56**, 229—319, 1940.
[260] *Reuter, H.*, Zur Theorie der nächtlichen Abkühlung der bodennahen Schicht u. Ausbildung der Bodeninversion. * Sitz-B. Wien. Akad. **155**, 333—358, 1947.
[261] *Sauberer, F.*, Messungen des nächtlichen Strahlungshaushaltes der Erdoberfläche. * Met. Z. **53**, 296—302, 1936.
[262] *Schmidt, W.*, Über kleinklimatische Forschungen. * Met. Z. **48**, 487—491, 1931.
[263] *Szász, G.*, Die Einwirkung von Bodenverschiedenheiten auf das Bestandsklima der Wintergerste. * Agrokemia es Talajtau **5**, 471—484, 1956.

[264] *Vries, D. A. de*, Het Warmtegeleidingsvermogen van grond. * Med. Landbouwhogeschool Wageningen **52**, 1—73, 1952.

[265] *Wächtershäuser, H.*, Harmonische Analyse des mittl. tägl. Temperaturgangs in extremen Böden etc. * Met. Rundsch. **4**, 23—24, 1951.

[266] *Winter, G.*, Antibiose u. Symbiose als Elemente der Mikrobenentwicklung im Boden u. Wurzelbereich. * Naturw. Rundschau 1951, 116—123.

[267] *Yakuwa, R.*, Über die Bodentemperaturen in den verschiedenen Bodenarten in Hokkaido. * Geophys. Mag. Tokio **14**, 1—12, 1945.

[268] —, On the effect of soil dressing upon the temperature of peat soil. * J. Agr. Met. Japan **8**, 92—96, 1953.

SECTION 20

[269] *Aderikhin, P. G.*, Ob uteplenii pochv putem izmeneniia ikh tsveta. * Met. i. Gidrologiia **8**, 28—30, 1952.

[270] *Andó, M.*, Beitrag zur Bodentemperatur des Flugsandes. * Acta Geographica **1**, 1—7, Szeged 1955.

[271] *Dirmhirn, I.*, Studie über die Oberflächentemperatur fester Körper. * Wetter u. Leben **10**, 136—144, 1958.

[272] *Dorno, C.*, Über die Erwärmung von Holz unter verschiedenen Anstrichen. * Gerl. B. **32**, 15—24, 1931.

[273] *Dufton, A. F.*, und *Beckett, H. E.*, Terrestrial temperatures. * Met. Mag. **67**, 252—253, 1932.

[274] *Ehrenberg, W. W.*, Künstliche Geländefärbung als Beispiel für physikalische Katalyse. * Arch. f. Met. (B) **4**, 470—482, 1953.

[275] *Firbas, F.*, Über die Bedeutung des thermischen Verhaltens der Laubstreu für die Frühjahrsvegetation des sommergrünen Laubwaldes. * Beih. z. Botan. Centralbl. **44**, Abt. II, 179—198, 1927.

[276] *Fuß, F.*, Veränderungen des Bodenklimas durch Bodenbedeckung bei ausdauernden Kulturen. * Angew. Met. **3**, 74—76, 1958.

[277] *Geiger, R.*, und *Fritzsche, G.*, Spätfrost u. Vollumbruch. * Forstarchiv **16**, 141--156, 1940.

[278] *Hadas, A.*, Soil temperatures at the evaporation station Lydda airport, Israel, in 1951—52. * Israel Met. Service, A Met. Notes **9**, 1—14, 1954.

[279] *Huber, B.*, Der Wärmehaushalt der Pflanzen. * Datterer, Freising—München 1935.

[280] *Keil, K.*, Frostbekämpfung im hohen Norden. * Met. Rundsch. **1**, 40—41, 1947.

[281] *Kiss, A.*, Temperaturextreme auf dem Sande von Üllés. * Acta Geographica **1**, 9—13, Szeged 1955.

[282] *Landsberg, H. E.*, Interaction of soil and weather. * Proc. Soil Science Soc. Amer. **22**, 491—495, 1958.

[283] *Neubauer, H. F.*, Notizen über die Temperatur der Bodenoberfläche in Afghanistan. * Wetter u. Leben **4**, 165—168, 1952.

[284] *Neumann, J.*, Some microclimatological measurements in a potato field. * Israel Met. Service, Ser. C, Miscell. Pap. 6, 1953.

[285] *Ramanathan, K. R.*, On temperatures of exposed rails at Agra. * India Met. Dep. Scient. Notes **1**, Nr. 4, 1929.

[286] *Ramdas, L. A.*, und *Dravid, R. K.*, Soil temperatures. * Current Science **3**, 266—267, 1934.

[287] *Ramin, v.*, Maßnahmen gegen das Verwehen der jungen Zuckerrüben. * Zuckerrübenbau **17**, 66, 1935.

[288] *Regula, H.*, Die Wetterverhältnisse während der Expedition und die Ergebnisse der met. Messungen. * Ergeb. d. Antarkt. Exped. 1938/39 **2**, 16—40, Hamburg 1954.

[290] *Sato, S.*, Studies on the methods to lower the high temp. of paddy field in the warm districts in Japan — mulching with rice-straw and grass. * J. Agr. Met. Japan **11**, 39—40, 1955.

[291] —, Heat economy in clear-water-plot and in carbon-black powdered plot. Ebenda **13**, 30—32, 1957.

[292] —, und *Funahashi, Y.*, Studies on the methods of lowering the field-temp. in the common-cultivation of rice in warm districts of Japan. * Ebenda **13**, 89—92, 1958.

[293] *Sauberer, F.*, Über die Strahlungsbilanz verschiedener Oberflächen und deren Messung. * Wetter u. Leben **8**, 12—26, 1956.

[294] *Schanderl, H.*, und *Weger, N.*, Studien über das Mikroklima vor verschiedenfarbigen Mauerflächen und der Einfluß auf Wachstum und Ertrag der Tomaten. * Biokl. B. **7**, 134—142, 1940.

[295] *Schropp, K.*, Die Temperaturen technischer Oberflächen unter dem Einfluß der Sonnenbestrahlung und der nächtl. Ausstrahlung. * Gesundheits-Ing. 1931, 729—736.

[296] *Thornthwaite, C. W.*, Estimating soil tractionability from climatic data. * Publ. Climat. **7**, Nr. 3. 1954.

[297] *Vaartaja, O.*, High surface soil temperatures. * Oikos **1**, 6—28, 1949.

[298] *Vaupel, A.*, Mikroklima und Pflanzentemperaturen auf trocken-heißen Standorten. * Flora **145**, 497—541, 1958.

[299] *Vries, D. A. de*, und *Wit, C. T. de*, Die thermischen Eigenschaften der Moorböden und die Beeinflussung der Nachtfrostgefahr dieser Böden durch eine Sanddecke. * Met. Rundsch. **7**, 41—45, 1954.

[300] *Weger, N.*, Beiträge zur Frage der Beeinflussung des Bestandsklimas, des Bodenklimas und der Pflanzenentwicklung durch Spaliermauern und Bodenbedeckung. * Ber. DWD-US Zone **4**, Nr. 28, 1951.

[301] —, Höhere Tomaten- u. Gurkenerträge durch Abdecken des Bodens mit Glas. * Z. f. Acker- u. Pflanzenbau **97**, 115—128, 1953.

[302] *Yakuwa, R.*, und *Yamabuki, F.*, Studies on the raising of the temperature of irrigation water. * Bull. Experim. Farm Fac. of Agric., Hokkaido Univ. **10**, 28—35, 1952.

SECTION 21

[303] *Bracht, J.*, Über die Wärmeleitfähigkeit des Erdbodens u. des Schnees u. den Wärmeumsatz im Erdboden. * Veröff. Geoph. I. Leipzig **14**, 145—225, 1949.

[304] *Brooks, F. A.*, und *Rhoades, D. G.*, Daytime partition of irradation and the evaporation chilling of the ground. * Transact. Americ. Geophys. Union **35**, 145—152, 1954.

[305] *Dücker, A.*, Der Bodenfrost im Straßenbau. * E. Schmidt, Berlin und Detmold 1947.

[306] *Fleischmann, R.*, Vom Auffrieren des Bodens. * Biokl. B. **2**, 88—90, 1935.

[307] *Franssila, M.*, Mikroklimatische Untersuchungen des Wärmehaushalts. * Mitt. Met. Zentralanstalt Helsinki **20**, 1—103, 1936.

[308] *Fukuda, H.*, Über Eisfilamente im Boden. * J. College of Agric. Tokyo **13**, 453—481, 1936.
[309] *Gilbert, T.*, Das bewegliche Luftvolumen des Erdbodens. * Arch. f. Met. (A) **8**, 397—410, 1955.
[310] *Herdmenger, J.*, Flugzeug u. Vorgeschichte. * Orion **4**, 474—475, 1949.
[311] *Keränen, J.*, Wärme- und Temperaturverhältnisse der obersten Bodenschichten. * Einführung in d. Geophys. II, Springer 1929.
[312] —, On frost formation in soil. * Fennica **73**, Nr. 1, 1—14, 1951.
[313] *Kretschmer, G.*, Messungen der vertikalen Volumänderung von Ackerböden. * Wiss. Z. Fr. Schiller Univ. Jena **4**, 639—645, 1954/55.
[314] —, Die Ursache für Eisschichtenbildung in Böden. * Ebenda **7**, 273—277, 1957/58.
[315] *Kreutz, W.*, Das Eindringen des Frostes in Böden unter gleichen und verschiedenen Witterungsbedingungen während des sehr kalten Winters 1939/40. * R. f. W. Wiss. Abh. **9**, Nr. 2, 1942.
[316] —, Bodenfrost. * Umschau in Wiss. u. Techn. 1950, Heft 2.
[317] —, Volumänderung der Bodenoberfläche in Abhängigkeit vom Wetter. * Met. Rundsch. **6**, 138—140, 1953.
[318] *Lehmann, P.*, Einfachgerät für den Frostspurennachweis. * VDI-Zeitschr. **99**, 415—416, 1957.
[319] *Maruyama, E.*, Effects of moisture on the heat conductivity, diffusivity, specific heat and apparent specific volume of the loam and the mixture of loam and clay. * J. Agr. Met. Japan **12**, 125—127, 1957.
[320] *Mayer, H.*, Beobachtungen über die Wärmeleitfähigkeit. * Synopt. Bearbeit. d. Frankfurter Wetterd., Linke-Sonderheft 1933, S. 67.
[320a] *Monteith, J. L.*, Visible microclimate. * Weather **13**, 121—124, 1958.
[321] *Pfau, R.*, Statistische Bearbeitung von Bodenfeuchtigkeitswerten. * Ber. DWD **7**, Nr. 46, 1958.
[322] —, Ein Beitrag zur Bodenwasserbewegung. * Met. Rundsch. **11**, 116—120, 1958.
[323] *Ramdas, L. A.*, und *Dravid, R. K.*, Soil temperatures. * Current Science **3**, 266—267, 1934.
[324] *Rettig, H.*, Beitrag zum Problem der Wasserbewegung im Boden. * Met. Rundsch. **9**, 182—184, 1956.
[325] —, Darstellung u. Verwertung von Ergebnissen der Bodenfeuchtigkeitsmessungen etc. * Ebenda **11**, 47—51, 1958.
[326] *Ruckli, R.*, Der Frost im Baugrund. * Springer, Wien 1950.
[327] *Schaible, L.*, Frost- und Tauschäden an Verkehrswegen und deren Bekämpfung. * W. Ernst, Berlin 1957.
[328] *Schmid, J.*, Der Bodenfrost als morphologischer Faktor. * A. Hüthig, Heidelberg 1955.
[329] *Slanar, H.*, s. (432).
[329a] *Stelzer, F.*, Das Auffrieren des Bodens. * Wetter u. Leben **11**, 1—8, 1959.
[330] *Uhlig, S.*, Die Untersuchung und Darstellung der Bodenfeuchte. * Ber. DWD-US Zone **4**, Nr. 30, 1951.
[331] —, Die Darstellung der Wasserverhältnisse im Boden. * Naturwiss. Rundschau 1951, 10—15.
[332] —, Acht Jahre Bodenfeuchte-Bestimmungen des Deutschen Wetterdiensts. * Met. Rundsch. **10**, 163—170, 1957.
[333] *Unger, K.*, Bodenfeuchtigkeitsbestimmungen unter einer Standardfläche und unter versch. Pflanzenbeständen zur Charakterisierung der Dürrewirkung. * Abh. Met. D. DDR **3**, Nr. 19, 23—27, 1953.

[334] *Unglaube, E.*, Ergebnisse der Bodenfeuchtigkeitsmessungen in Geisenheim. * Ber. DWD-US Zone **6**, Nr. 38, 1952.

[335] *Vries, D. A. de*, Some results of field determinations of the moisture content of soil from thermal conductivity measurements. * Netherl. J. Agric. Science **1**, 115—121, 1953.

[336] *Weger, N.*, Die Wasserbewegung u. die Wassergehaltsbestimmung in gefrorenem Boden. * Met. Rundsch. **7**, 45—47, 1954.

SECTION 22

[337] *Czepa, O.*, Über die spektrale Lichtdurchlässigkeit von Binnengewässern. * Wetter u. Leben **6**, 122—128, 1954.

[338] *Dirmhirn, I.*, Neuere Strahlungsmessungen in den Lunzer Seen. * Wetter u. Leben **3**, 258—260, 1951.

[339] *Frey, H.*, Der Frühlingseinzug am Züricher See. * Naturforsch. Ges. Zürich **133**, 1—48, 1931.

[340] *Herzog, J.*, Thermische Untersuchungen in Waldteichen. * Veröff. Geoph. I. Leipzig **8**, Nr. 2, 1936.

[341] *Höhne, W.*, Experimentelle u. mikroklimatische Untersuchungen an Kleingewässern. * Abh. Met. D. DDR **4**, Nr. 26, 1954.

[342] *Keil, K.*, Verfahren zur Verhütung des Zufrierens von Häfen. * Met. Rundsch. **1**, 312, 1948.

[343] *Lauscher, F.*, Sonnen- u. Himmelsstrahlung im Meer u. in Gewässern. * Linke-Möller, Hdb. d. Geophysik VIII, Kap. 12, Borntraeger, Berlin 1955.

[344] *Mahringer, W.*, Über die spektrale Durchlässigkeit des Traunsees. * Wetter u. Leben **10**, 24—27, 1958.

[345] *Pesta, O.*, Alpine Hochgebirgstümpel u. ihre Tierwelt. * Naturwiss. Rundschau **8**, 65—68, 1955.

[346] *Pichler, W.*, Der Almtümpel als Lebensstätte. * Biokl. B. **6**, 85—89, 1939.

[347] *Rathschüler, E.*, Der Einfluß eines Wasserfalles auf die Luftfeuchtigkeit der Umgebung. * Arch. f. Met. (B) **1**, 108—114, 1948.

[348] *Sauberer, F.*, Über das Licht im Neusiedlersee. * Wetter u. Leben **4**, 12—15, 1952.

[349] —, und *Ruttner, F.*, Die Strahlungsverhältnisse der Binnengewässer. * Probl. d. kosm. Physik **21**, Akad. Verl. Ges. Leipzig 1941.

[350] *Schanderl, H.*, Studien über die Körpertemperatur submerser Wasserpflanzen. * Ber. D. Bot. G. **68**, 28—34, 1955.

[351] *Schmidt, W.*, Absorption der Sonnenstrahlung im Wasser. * Sitz-B. Wien. Akad. **117**, 237—253, 1908.

[352] —, Über Boden- u. Wassertemperaturen. * Met. Z. **44**, 406—411, 1927.

[353] *Sverdrup, H. U.*, *Johnson M. W.*, und *Fleming, R. H.*, The oceans, their physics, chemistry and general biology. * New York 1946.

[354] *Volk, O. H.*, Ein neuer für botanische Zwecke geeigneter Lichtmesser. * Ber. D. Bot. G. **52**, 195—202, 1934.

[355] *Willer, A.*, Kleinklimatische Untersuchungen im Phragmites-Gelege. * Verh. Internat. Ver. f. theoret. u. angew. Limnologie **10**, 566—574, 1950.

SECTION 23

[356] *Brocks, K.*, Wasserdampfschichtung über dem Meer und „Rauhigkeit" der Meeresoberfläche. * Arch. f. Met. (A) **8**, 354—383, 1955.

[357] —, Atmosphärische Temperaturschichtung und Austauschprobleme über dem Meer. * Ber. DWD **4**, Nr. 22, 10—15, 1956.

[358] *Bruch, H.*, Die vertikale Verteilung von Windgeschwindigkeit u. Temperatur in den untersten Metern über der Wasseroberfläche. * Veröff. Inst. f. Meereskde. Berlin, Heft 38, 1940.

[359] *Conrad, V.*, Oberflächentemperaturen in Alpenseen. * Gerl. B. **46**, 44—61, 1935.

[360] —, Zum Wasserklima einiger alpiner Seen Österreichs. * Beih. z. Jahrb. d. Zentralanst. f. Met. u. Geodyn. Wien 1930, Wien 1936.

[361] *Deacon, E. L.*, *Sheppard, P. A.*, und *Webb, E. K.*, Wind profiles over the sea and the drag at the sea surface. * Austral. J. Physics **9**, 511—541, 1956.

[362] *Dirmhirn, I.*, Über die Strahlungsvorgänge in Fließgewässern. * Wetter u. Leben, Sonderheft **2**, 52—63, 1953.

[363] *Eckel, O.*, Über die interdiurne Veränderlichkeit der Fluß- und Seeoberflächentemperaturen. * Wetter u. Leben **3**, 203—212, 1951.

[364] —, Mittel und Extremtemperaturen des Hallstättersees. * Ebenda **4**, 87—93, 1952.

[365] —, Über Mittel- u. Extremtemp. des Lunzer Untersees. * Ebenda, Sonderheft **1**, 20—25, 1952.

[366] —, Zur Thermik der Fließgewässer: Über die Änderung der Wassertemperatur entlang des Flußlaufs. * Ebenda Sonderheft **2**, 41—47, 1953.

[367] —, und *Reuter, H.*, Zur Berechnung des sommerlichen Wärmeumsatzes in Flußläufen. * Geograf. Ann. **32**, 188—209, 1950.

[368] *Hay, J. S.*, Some observations of air flow over the sea. * Quart. J. **81**, 307—319, 1955.

[369] *Koizumi, M.*, A note on the diurnal variation of air temperature on the open sea. * Pap. in Met. and Geophys. Tokyo **7**, 322—326, 1956.

[370] *Kuhlbrodt, E.*, und *Reger, J.*, Wissenschaftl. Ergebnisse der Deutsch. Atlantischen Expedition auf dem Meteor 1925—27, Bd. **14**, Berlin 1938.

[371] *Lettau, H.*, Windprofil, innere Reibung u. Energieumsatz in den unteren 500 m über dem Meer. * Beitr. Phys. d. fr. Atm. **30**, 78—96, 1957.

[372] *Montgomery, R. B.*, Observations of vertical humidity distribution above the ocean surface and their relation to evaporation. * M. I. T. and Woods Hole Ocean. Inst. Pap. **7**, Nr. 4, 1940.

[373] *Neumann, G.*, Wind stress on water surfaces. * Bull. Am. Met. Soc. **37**, 211—217, 1956.

[374] *Peppler, W.*, Beitrag zur Kenntnis der Oberflächentemperatur des Bodensees. * Z. f. angew. Met. **44**, 250—256, 1927; **45**, 14—20, 99—105, 1928.

[375] *Reiter, E. R.*, Der mitführende Einfluß einer Flußoberfläche auf die darüberliegenden Luftschichten. * Arch. f. Met. (A) **8**, 384—396, 1955.

[376] *Roll, U.*, Zur Frage des tägl. Temperaturgangs u. des Wärmeaustauschs in den unteren Luftschichten über dem Meere. * Aus d. Arch. d. Seewarte **59**, Nr. 9, 1939.

[377] —, Das Windfeld über den Meereswellen. * Naturwiss. **35**, 230—234, 1948.

[378] —, Wassernahes Windprofil u. Wellen auf dem Wattenmeer. * Ann. d. Met. **1**, 139—151, 1948.

[379] —, Gedanken über den Zusammenhang der vertikalen Profile von Windgeschwindigkeit u. Temperatur in der wassernahen Luftschicht. * Ebenda **3**, 1—9, 1950.

[380] —, Temperaturmessungen nahe der Wasseroberfläche. * D. Hydrograph. Z. **5**, 141—143, 1952.

[381] *Schmitz, W.*, Grundlagen der Untersuchung der Temperaturverhältnisse in den Fließgewässern. * Ber. Limn. Flußstat. Freudenthal **6**, 29—50, 1954.

[382] *Sverdrup, H. U.*, The humidity gradient over the sea surface. * J. Met. **3**, 1—8, 1946.
[383] *Takahashi, T.*, Micro-meteorological observations and studies over the sea. * Met. Notes, Met. Res. Inst. Kyoto Univ. Ser. 2, Nr. 12, 1958.
[384] *Wahl, E.*, Temperaturmessungen in der Nordsee im Sommer 1948. * Ann. d. Met. **2**, 65—71, 1949.
[385] —, Strahlungseinflüsse bei der Wassertemperaturmessung an Bord von Schiffen. * Ebenda **3**, 92—102, 1950.
[386] *Wegner, K. O.*, Windprofilmessungen über Flußoberflächen bei schwachem Wind. * Arb. d. Met. Inst. Univ. Köln 1956.
[387] *Wüst, G.*, Temperatur- u. Dampfdruckgefälle in den untersten Metern über der Meeresoberfläche. * Met. Z. **54**, 4—9, 1937.

SECTION 24

[388] *Abels, H.*, Beobachtungen der tägl. Periode der Temp. im Schnee u. Bestimmung des Wärmeleitungsvermögens des Schnees als Funktion seiner Dichtigkeit. * Rep. f. Met. **16**, Nr. 1, 1892.
[389] *Ambach, W.*, Über den nächtlichen Wärmeumsatz der gefrorenen Gletscheroberfläche. * Arch. f. Met. (A) **8**, 411—426, 1955.
[390] —, Über die Strahlungsdurchlässigkeit des Gletschereises. * Sitz-B. Wien. Akad. **164**, 483—494, 1955.
[391] —, Ein Strahlungsempfänger mit kugelförmiger Empfängerfläche zur Bestimmung des Extinktionskoeffizienten in Gletschern. * Arch. f. Met. (B) **8**, 433—441, 1958.
[392] —, Ein Beitrag zur Kenntnis der Lichtstreuung im Gletschereis. * Ebenda **9**, 441—463, 1959.
[393] —, und *Mocker, H.*, Messungen der Strahlungsextinktion eines kugelförm. Empf. in der oberflächennahen Eisschicht eines Gletschers u. im Altschnee. * Ebenda **10**, 84—99, 1959.
[394] *Band, G.*, Strahlung u. Temperaturmessung über Schnee. * La Mét. 1957, 363—369.
[395] *Berg, H.*, Temperaturmessungen in der schneenahen Luftschicht. * La Mét. 1957, 357—361.
[396] *Bührer, W.*, Über den Einfluß der Schneedecke auf die Temperatur der Erdoberfläche. * Met. Z. **19**, 205—211, 1902.
[397] *Dirmhirn, I.*, Über neuere Strahlungsmessungen auf den Ostalpengletschern. * La Mét. 1957, 345—351.
[398] *Eckel, O.*, und *Thams, C.*, Untersuchungen über Dichte, Temperatur und Strahlungsverhältnisse der Schneedecke in Davos. * Geologie d. Schweiz, Hydrol. **3**, 275—340, 1939.
[399] *Friedrich, W.*, Schneerollen. * Wetter u. Leben **5**, 82—83, 1953.
[400] *Fukutomi, T.*, Effect of ground temperature upon the thickness of snow cover. * Low Temp. Sc. Sapporo Japan **9**, 145—148, 1953.
[401] *Gabran, O.*, Die Luftdurchlässigkeit einer Schneedecke u. deren Einfluß auf die Überwinterung d. Pflanzen. * Met. Z. **56**, 354—356, 1939.
[402] *Georgi, J.*, Die bodennahe Luftschicht über dem grönländischen Eis. * Veröff. D. Wiss. Inst. Kopenhagen **1**, 1—27, 1943.
[403] *Götz, F. W. P.*, Das Strahlungsklima von Arosa. * Springer, Berlin 1926.
[404] *Gressel, W.*, Über das Auftreten von Schneerollen u. Schneewalzen in Niederösterreich. * Met. Rundsch. **6**, 94—96, 1953 (s. auch Universum Wien **13**, 139—141, 1958).

[405] *Grunow, J.*, Zum Wasserhaushalt einer Schneedecke etc. * Ber. DWD-US Zone **6**, Nr. 38, 385—393, 1952.

[406] *Hoinkes, H.*, Zur Mikrometeorologie der eisnahen Luftschicht. * Arch. f. Met. (B) **4**, 451—458, 1953.

[407] *Hoinkes, H.*, Über Messungen der Ablation und des Wärmeumsatzes auf Alpengletschern etc. * Publ. Assoc. Internat. d'Hydrolog. **39**, 442—448, 1954.

[408] —, Über den Zusammenhang zwischen Temperaturgradient und Temperaturunruhe. * Ber. DWD **4**, Nr. 22, 16—18, 1956.

[409] —, Zur Bestimmung der Jahresgrenzen in mehrjährigen Schneeansammlungen. * Arch. f. Met. (B) **8**, 56—60, 1957.

[410] International Union of Geodesy and Geophysics, Internat. Classification for snow. * Techn. Mem. Nat. Res. Council of Canada, Comm. Soil and Snow Mech. **31**, 1—11, 1954.

[411] *Keränen, J.*, Über die Temperaturen des Bodens und der Schneedecke in Sodankylä. * Helsinki 1920.

[412] *Köhn, M.*, Über den Einfluß einer Schneedecke auf die Bodentemperaturen. * Wetter u. Klima **1**, 303—306, 1948.

[413] *Kreeb, K.*, Die Schneeschmelze als phänologischer Faktor. * Met. Rundschau **7**, 48—49, 1954.

[414] *Liljequist, G. H.*, Radiation and wind and temperature profiles over an antarctic snowfield. * Proc. Toronto Met. Conf. 1953, Am. Met. Soc. & R. Met. Soc. 78—87, 1954.

[415] *Löhle, F.*, Absorptionsmessungen an Neuschnee u. Firnschnee. * Gerl. B. **59**, 283—298, 1943.

[416] *Loewe, F.*, Etudes de glaciologie en terre Adélie 1951— 1952. * Expéd. polaires françaises (*P. E. Victor*) IX, Paris 1956.

[417] *Michaelis, P.*, Ökologische Studien an der alpinen Baumgrenze V. * Jahrb. f. wiss. Botanik **80**, 337—362, 1934.

[418] *Monteith, J. L.*, The effect of grass-length on snow melting. * Weather **11**, 8—9, 1956.

[419] *Müller, H. G.*, Zur Wärmebilanz der Schneedecke. * Met. Rundsch. **6**, 140—143, 1953.

[420] *Niemann, A.*, Die Schutzwirkung einer Schneedecke, dargestellt am farbigen Frostschadenbild. * Photograph. u. Wissensch. **5**, 27—28, 1956.

[421] *Nyberg, A.*, Temperature measurements in an air layer very close to a snow surface. * Geograf. Ann. **20**, 234—275, 1938.

[422] *Paulcke, W.*, Praktische Schnee- u. Lawinenkunde (Verständl. Wissensch. **38**). Springer, Berlin 1938.

[423] *Pichler, F.*, Über den Kohlensäure- u. Sauerstoffgehalt der Luft unter einer Schneedecke. * Wetter u. Leben **1**, 15, 1948.

[424] *Pluvinage, P.*, und *Taylor, G.*, La témperature de l'air dans les premiers mètres au-dessus de l'Inlandsis Groenlandais. * Rapp. Scient. Expéd. Polaires Franç. S IV **1**, 1956.

[425] *Portman, D. J.*, Air and soil temperature distributions near a melting snow cover. * Publ. Climat. **5**, Nr. 1, 4—8, 1952.

[426] *Reuter, H.*, Über die Theorie des Wärmehaushalts einer Schneedecke. * Arch. f. Met. (A) **1**, 62—92, 1948.

[427] —, Zur Theorie des Wärmehaushalts strahlungsdurchlässiger Medien. * Tellus **1**, Nr. 3, 6—14, 1949.

[428] *Roller, M.*, Über die Auswirkung mikroklimatischer Faktoren auf das Abschmelzen der Winterschneedecke. * Wetter u. Leben **5**, 31—33, 1953.

[429] *Roßmann, F.*, Beobachtungen über Schneerauchen u. Seerauchen. * Z. f. angew. Met. **51**, 309—317, 1934.

[430] *Sauberer, F.*, Versuche über spektrale Messungen der Strahlungseigenschaften von Schnee u. Eis mit Photoelementen. * Met. Z. **55**, 250—255, 1938.

[431] —, Die spektrale Durchlässigkeit des Eises. * Wetter u. Leben **2**, 193—197, 1950.

[432] *Slanar, H.*, Schneeabschmelzung im bewachsenen Gelände. * Met. Z. **59**, 413—416, 1942.

[433] *Takahasi, Y.*, *Soma, S.*, u. *Nemoto, S.*, Observations and a theory of temperature profile in a surface of snow cooling through nocturnal radiation. * Seppyo, Tokyo **18**, 43—47, 1956.

[434] *Thuronyi, G.*, *Zikeev, N. T.*, und *Rigby, M.*, A selective annotated bibliography on the micrometeorology of snow cover. * MAB **7**, 873—921, 1009—1025, 1956.

SECTIONS 25-26

[435] *Albrecht, F.*, Untersuchungen über den Wärmehaushalt der Erdoberfläche in verschiedenen Klimagebieten. * R. f. W. Wiss. Abh. **8**, Nr. 2, 1940.

[436] = [88].

[437] *Baumgartner, A.*, Untersuchungen über den Wärme- u. Wasserhaushalt eines jungen Waldes. * Ber. DWD **5**, Nr. 28, 1956.

[438] *Bryson, R. A.*, Preliminary estimates of the surface heat budget, summer, clear days at Point Barrow, Alaska. * Madison Dep. Met. Wisconsin 1956.

[439] *Budyko, M. I.*, Atlas teplovogo balansa. * Leningrad 1955 (s. Ref. *H. Flohn*, Erdk. **12**, 233—237, 1958).

[440] —, The heat balance of the earths surface (translated by *N. A. Stepanova*). * US Dep. of Commerce Washington 1958.

[441] *Deacon, E. L.*, und *Swinbank, W. C.*, Comparison between momentum and water vapour transfer. * Proc. Canberra Symposion 1956, 38—41, UNESCO, Paris 1958.

[442] *Fleischer, R.*, Der Jahresgang der Strahlungsbilanz u. ihrer Komponenten. * Ann. d. Met. **6**, 357—364, 1953/54.

[443] —, Registrierung der Infrarotstrahlungsströme der Atmosphäre u. des Erdbodens. * Ebenda **8**, 115—123, 1957.

[444] *Frankenberger, E.*, Ergebnisse von Wärmehaushaltsmessungen. * Met. Rundsch. **7**, 81—85, 1954.

[445] —, Über vertikale Temperatur-, Feuchte- und Windgradienten in den untersten 7 Dekametern der Atmosphäre, den Vertikalaustausch u. den Wärmehaushalt an Wiesenboden bei Quickborn/Holstein 1953/54. * Ber. DWD **3**, Nr. 20, 1955.

[446] —, Der Austauschkoeffizient über Land. * Beitr. Phys. d. fr. Atm. **30**, 170—176, 1958.

[447] *Halstead, M. H.*, The fluxes of momentum, heat and water vapour in micrometeorology. * Publ. Climat. **7**, 326—361, 1954.

[448] *Hoinkes, H.*, Wärmeumsatz u. Ablation auf Alpengletschern II. * Geograf. Ann. **35**, 116—140, 1953.

[449] *Hoinkes, H.*, und *Untersteiner, N.*, Wärmeumsatz u. Ablation auf Alpengletschern I. * Ebenda **34**, 99—158, 1953; **35**, 116—140, 1953.

[450] *Horney, G.*, Vorschläge für die Durchführung von Messungen des Wärme- und Wasserhaushalts von Kulturpflanzenbeständen. * Met. Rundsch. **4**, 100—103, 1951.

[451] *Kraus, H.*, Untersuchungen über den nächtlichen Energietransport und Energiehaushalt in der bodennahen Luftschicht bei der Bildung von Strahlungsnebeln. * Ber. DWD **7**, Nr. 48, 1958.
[452] *Lauscher, F.*, Strahlungs- und Wärmehaushalt. * Ber. DWD **4**, Nr. 22, 21—29, 1956.
[453] *Mather, J. R.*, und *Thornthwaite, C. W.*, Microclimatic investigations at Point Barrow, Alaska, 1956. * Publ. Climat. **9**, Nr. 1, 1956.
[454] *Miller, D. H.*, The influence of snow cover on local climate in Greenland. * J. Met. **13**, 112—120, 1956.
[455] *Niederdorfer, E.*, Messungen des Wärmeumsatzes über schneebedecktem Boden. * Met. Z. **50**, 201—208, 1933.
[456] *Raethjen, P.*, Wärmehaushalt der Atmosphäre (Abriß d. Met. II). * Geophys. Einzelschr., Geophys. Inst. d. Univ. Hamburg 1950.
[457] *Raschke, K.*, Über die physikalischen Beziehungen zwischen Wärmeübergangszahl, Strahlungsaustausch, Temperatur und Transpiration eines Blattes. * Planta **48**, 200—238, 1956.
[458] *Rider, N. E.*, und *Robinson, G. D.*, A study of the transfer of heat and water vapour above a surface of short grass. * Quart J. **77**, 375—401, 1951.
[459] *Schulze, R.*, Einige Meßergebnisse zum Energieumsatz am Erdboden. * Arch. f. Met. (B) **9**, 254—271, 1958.
[460] *Seo, T.*, A microclimatological study of thermal exchange at the earths surface. * Res. Rep. Kochi Univ. Japan **6**, Nr. 18, 1957; **7**, Nr. 10 und 21, 1958.
[461] *Sverdrup, H. U.*, The eddy conductivity of the air over a smooth snow field. * Geofysiske Publ. **11**, Nr. 7, 1—69, 1936.
[462] *Untersteiner, N.*, Glazial-meteorologische Untersuchungen im Karakorum. * Arch. f. Met. (B) **8**, 1—30, 137—171, 1957.

SECTION 27

[463] *Berg, H.*, Die Bedeutung des Insel- u. Küstenklimas für die Klimatherapie. * Geofisica pura e appl. **21**, 15 S., 1952.
[464] —, Mikroklimatische Beobachtungen am Rande einer Wasserfläche. * Ann. d. Met. **5**, 227—235, 1952.
[465] *Craig, R. A.*, Measurements of temperature and humidity in the lowest 1000 feet of the atmosphere over Massachusetts bay. * Mass. Inst. Technology, Pap. phys., ocean. and met. **10**, Nr. 1, 1946.
[466] *Geiger, R.*, Über selbständige und unselbständige Mikroklimate. * Met. Z. **46**, 539—544, 1929.
[467] *Hofmann, G.*, Wärmehaushalt u. Advektion. * Arch. f. Met. (A) **11**, 474—502, 1960.
[468] *Kaiser, H.*, Über den Strahlungstyp u. den Windtyp des Mikroklimas. * Met. Rundsch. **11**, 162—164, 1958.
[469] *Knochenhauer, W.*, Inwieweit sind die Temperatur- u. Feuchtigkeitsmessungen unserer Flughäfen repräsentativ? * Erfahr. Ber. d. D. Flugwetterd., 9. Folge Nr. 2, 1934.
[470] *Landeck, J.*, und *Uhlig, S.*, Kondensationserscheinungen über Brachland. * Met. Rundsch. **5**, 107, 1952.
[471] *Mäde, A.*, Über die Methodik der meteorologischen Geländevermessung. * Sitz-B. Deutsche Akad. d. Landwirtsch. Wiss. Berlin **5**, Nr. 5, 1—25, 1956.
[472] *Möller, F.*, Zur Bestimmung der Gebietsverdunstung aus der Advektion. * Ber. DWD-US Zone **6**, Nr. 35, 172—173, 1952.

[473] —, und *de Bary, E.*, Der Wärme- u. Wasserdampfhaushalt der freien Atmosphäre. * Arch. f. Met. (A) **4**, 142—155, 1951.

[474] *Nyberg, A.*, und *Raab, L.*, A remark on the energy transport from a warm river surface into cold air. * Tellus **5**, 529—532, 1953.

[475] —, An experimental study of the field variation of the eddy conductivity. * Ebenda **8**, 472—479, 1956.

[476] *Philip, J. R.*, The theory of local advection I. * J. Met. **16**, 535—547, 1959.

[477] *Runge, H.*, Entstehung von Bodennebel durch Auspuffgase. * Z. f. angew. Met. **2**, 289—300, 1956.

[478] *Täumer, F.*, Eine Methode zur Bestimmung der Gebietsverdunstung auf meteorologischer Grundlage. * Veröff. Inst. f. Agrarmet. Leipzig **1**, Nr. 1, 1955.

[479] *Woodcock, A. H.*, und *Stommel, H.*, Temperatures observed near the surface of a fresh-water pond at night. * J. Met. **4**, 102—103, 1947.

[480] *Zenker, H.*, Lokalklimatische Studien in Heringsdorf/Usedom. * Angew. Met. **2**, 289—300, 1956.

[481] *Zoltán, D.*, Kritérium a függö mikroklima jelenlétének megállapitásához. * Idöjárás **5**, 287—291, 1956.

SECTION 28

[482] *Brogmus, W.*, Zur Theorie der Verdunstung der natürlichen Erdoberfläche. * Einzelveröff. Seewetteramt Hamburg **21**, 1959.

[483] *Grundy, F.*, The use of cetyl alcohol to reduce reservoir evaporation. * J. Inst. of Water Eng. London **11**, 429—437, 1957.

[484] *Hofmann, G.*, Die Thermodynamik der Taubildung. * Ber. DWD. **3**, Nr. 18, 1955.

[485] —, Verdunstung u. Tau als Glieder des Wärmehaushalts. * Planta **47**, 303—322, 1956.

[486] *Keller, R.*, Das Schema des Wasserkreislaufes berechnet für das Deutsche Bundesgebiet. * Geograph. Taschenbuch, herausg. v. E. Meynen 1951/52. 203—205.

[487] —, Das Wasserdargebot in der Bundesrepublik Deutschland. * Forsch. z. Deutsch. Landeskde. **103**, 1—81, 1958.

[488] —, und *Clodius, S.*, Schema des Wasserkreislaufs für das Gebiet der Bundesrepublik Deutschland u. das Jahr 1951. * Der Große Herder, 5. Aufl. **9**, 906, 1956.

[489] *Mansfield, W. W.*, Reduction of evaporation of stored water. * Proc. Canberra Symposium 1956, 61—64, UNESCO Paris 1958.

[490] *McIlroy, I. C.*, The measurement of natural evaporation. * J. Austral. Inst. Agricult. Science **23**, 4—17, 1957.

[491] *Meinardus, W.*, Über den Kreislauf des Wassers. * Festrede Univ. Göttingen, Dietrichsche Univ. Buchdr. 1928.

[492] *Monteith, J. L.*, Editorial note. * Weather, London **12**, 225, 1957 (vgl. S. 203—210).

[493] *Penman, H. L.*, Evaporation: an introductory survey. * Netherlands J. of Agric. Science **4**, 9—29, 1956.

[494] *Reichel, E.*, Der Stand des Verdunstungsproblems. * Ber. DWD-US Zone **5**, Nr. 35, 155—172, 1952.

[495] *Unger, K.*, Zur Abschätzung von Verdunstungsunterschieden verschiedener Pflanzenbestände am natürlichen Standort. * Angew. Met. **2**, 1—14, 1954.

[496] *Wüst, G.*, Die Kreisläufe des Wassers auf der Erde. * Schrift. Naturwiss. V. f. Schleswig-Holstein **25**, 185—195, 1951.

SECTION 29

[497] *Angerer, E. v.*, Landschaftsphotographien in ultrarotem u. ultraviolettem Licht. * Naturw. **18**, 361—364, 1930.

[498] *Berger-Landefeldt, U.*, Temperaturbeobachtungen um ein Blatt. * Ber. D. Bot. G. **71**, 21—33, 1958.

[499] *Büdel, A.*, Das Mikroklima der Blüten in Bodennähe. * Z. f. Bienenforschung **4**, 131—140, 1958.

[500] *Casperson, G.*, Untersuchungen über das thermische Verhalten der Pflanzen unter dem Einfluß von Wind u. Windschutz. * Z. f. Botanik **45**, 433—473, 1957.

[501] *Egle, K.*, Zur Kenntnis des Lichtfeldes u. der Blattfarbstoffe. * Planta **26**, 546—583, 1937.

[502] *Falckenberg, G.*, Absorptionskonstanten einiger meteorologisch wichtiger Körper für infrarote Wellen. * Met. Z. **45**, 334—337, 1928.

[503] *Huber, B.*, Der Wärmehaushalt der Pflanzen. * Verl. Datterer, Freising-München 1935.

[504] —, Mikroklimatologische u. Pflanzentemperaturregistrierungen mit dem Multithermograph von Hartmann u. Braun. * Jahrb. f. wiss. Bot. **84**, 671—709, 1937.

[505] *Hummel, K.*, Über Temperaturen in Winterknospen bei Frostwitterung. * Met. Rundsch. **1**, 147—150, 1947.

[506] *Keßler, O. W.*, und *Schanderl, H.*, Pflanzen unter dem Einfluß verschiedener Strahlungsintensitäten. * Strahlentherapie **39**, 283—302, 1931.

[507] *Kunii, K.*, The tree temperature of Yoshino-Zakura. * J. Met. Res. Tokyo **4**, 1035—1038, 1953.

[508] *Lange, O. L.*, Hitze- u. Trockenresistenz der Flechten in Beziehung zu ihrer Verbreitung. * Flora **140**, 39—97, 1953.

[509] —, Einige Messungen zum Wärmehaushalt poikilohydrer Flechten und Moose. * Arch. f. Met. (B) **5**, 182—190, 1953.

[510] *Mäde, A.*, Temperaturuntersuchungen an Obstbäumen. * Gerl. B. **59**, 201—213, 1942.

[511] *Michaelis, G.* und *P.*, Über die winterlichen Temperaturen der pflanzlichen Organe, insbesondere der Fichte. * Beih. z. Bot. Centralbl. **52**, 333—377, 1934.

[512] *Nakagawa, Y.*, Studies on the plant temperature. * J. Agr. Met. Japan **13**, 17—21, 1957.

[513] *Raschke, K.*, Mikrometeorologisch gemessene Energieumsätze eines Alocasiablattes. * Arch. f. Met. (B) **7**, 240—268, 1956.

[514] —, Über die physikalischen Beziehungen zwischen Wärmeübergangszahl, Strahlungsaustausch, Temperatur u. Transpiration eines Blattes. * Planta **48**, 200—238, 1956.

[515] —, Über den Einfluß der Diffusionswiderstände auf die Transpiration und die Temperatur eines Blattes. * Flora **146**, 546—578, 1958.

[516] *Rohmeder, E.*, und *Eisenhut, G.*, Untersuchungen über das Mikroklima in Bestäubungsschutzbeuteln. * Silvae Genetica **8**, 1—36, 1959.

[517] *Sauberer, F.*, Zur Kenntnis der Strahlungsverhältnisse in Pflanzenbeständen. * Biokl. B. **4**, 145—155, 1937.

[518] —, Über die Strahlungseigenschaften der Pflanzen im Infrarot. * Wetter und Leben **1**, 231—234, 1948.
[519] —, und *Härtel, O.*, Pflanze u. Strahlung. * Akad. Verl. Ges. Leipzig 1959.
[520] *Schanderl, H.*, und *Kaempfert, W.*, Über die Strahlungsdurchlässigkeit von Blättern u. Blattgeweben. * Planta **18**, 700—750, 1933.
[521] *Seybold, A.*, Über den Lichtfaktor photophysiologischer Prozesse. * Jahrb. f. wiss. Bot. **82**, 741—795, 1936.
[522] *Steubing, L.*, und *Casperson, G.*, Pflanzentemperatur u. Taubeschlag. * Angew. Met. **3**, 219—224, 1958.
[523] *Suzuki, S.*, The nocturnal cooling of plant leaves and hoarfrost deposited thereon. * Geophys. Mag. Tokyo **25**, 219—235, 1954.
[524] *Takasu, K.*, Leaf temperatures under natural environments. * Mem. College of Science Kyoto (B) **20**, 179—187, 1953.
[525] *Tranquillini, W.*, Über den Einfluß von Übertemperaturen der Blätter bei Dauereinschluß in Küvetten auf die ökologische CO_2-Assimilationsmessung. * Ber. D. Bot. G. **67**, 191—204, 1954.
[526] *Ullrich, H.*, und *Mäde, A.*, Untersuchungen über den Temperaturverlauf beim Gefrieren von Blättern u. Vergleichsobjekten. * Planta **31**, 251—262, 1940.
[527] *Weger, N.*, Über Tütentemperaturen. * Biokl. B. **5**, 16—19, 1938.
[528] —, *Herbst, W.*, und *Rudloff, C. F.*, Witterung u. Phänologie der Blühphase des Birnbaums. * R. f. W. Wiss. Abh. **7**, Nr. 1, 1940.

SECTION 30

[529] *Angström, A.*, The albedo of various surfaces of ground. * Geograf. Ann. **7**, 323—342, 1925.
[530] *Bartels, J.*, Verdunstung, Bodenfeuchtigkeit u. Sickerwasser. * Z. f. F. u. Jagdw. **65**, 204—219, 1933.
[531] *Filzer, P.*, Untersuchungen über den Wasserumsatz künstlicher Pflanzenbestände. * Planta **30**, 205—223, 1939.
[532] *Friedrich, W.*, Messung der Verdunstung vom Erdboden. * Deutsche Forsch. **21**, 40—61, 1934.
[533] —, Über die Verdunstung vom Erdboden. * Gas- und Wasserfach **91**, Heft 24, 1950.
[534] *Göhre, K.*, Der Wasserhaushalt im Boden. * Z. f. Met. **3**, 13—19, 1949.
[535] —, Lysimeterversuche über den Wasserhaushalt von unbewachsenem und mit jungen Eichen bewachsenem Boden. * Arch. f. Forstwesen **3**, 645—674, 1954.
[536] *Kanitscheider, R.*, Temperaturmessungen in einem Bestand von Legföhren. * Biokl. B. **4**, 22—25, 1937.
[537] *Köstler, J. N.*, Waldbau. * Verlag P. Parey, Berlin 1950.
[538] *Paeschke, W.*, Mikroklimatische Untersuchungen innerhalb und dicht über verschiedenartigem Bestand. * Biokl. B. **4**, 155—163, 1937.
[539] *Stocker, O.*, Klimamessungen auf kleinstem Raum an Wiesen-, Wald- und Heidepflanzen. * Ber. D. Bot. G. **41**, 145—150, 1923.
[540] *Unger, K.*, Agrarmeteorologische Studien I. * Abh. Met. D. DDR **3**, Nr. 19, 1—22, 1953.
[541] *Walter, H.*, Grundlagen des Pflanzenlebens, 2. Aufl. * E. Ulmer, Stuttgart 1947.

SECTION 31

[542] *Biebl, R.*, Bodentemperaturen unter verschiedenen Pflanzengesellschaften. * Sitz-B. Wien. Akad. **160**, 71—90, 1951.

[543] *Daigo, Y.*, und *Maruyama, E.*, Air temperature in observation field and in wheat field. * J. Met. Res. Tokyo **3**, 106—108, 1951.

[544] *Delany, M. J.*, Studies on the microclimate of calluna heathland. * J. Animal Ecology Cambridge **22**, 227—239, 1953.

[545] *Gadre, K. M.*, Microclimatic survey of a sugarcane field. * Ind. J. Met. and Geophys. **2**, 142—150, 1951.

[546] *Heigel, K.*, Egebnisse von Verdunstungsmessungen mit Piche-Evaporimetern, ihre Abhängigkeit von einigen met. Faktoren u. von verschiedenen Standorten. * Met. Rundsch. **10**, 101—107, 1957.

[547] *Knapp, R.*, und *Linskens, H. F.*, Beobachtungen über den Einfluß einiger mediterraner Pflanzengesellschaften auf Mikroklima u. Bodenfeuchtigkeit. * Angew. Botanik **27**, 48—69, 1953.

[548] —, *Lieth, H.*, und *Wolf, F.*, Untersuchungen über die Bodenfeuchtigkeit in verschiedenen Pflanzengesellschaften nach neueren Methoden. * Ber. D. Bot. G. **65**, 113—132, 1952.

[549] *Kullenberg, B.*, Quelques observations microclimatologiques en Côte-d'Ivoire et Guinée Française. * Bull. de l'I.F.A.N. **17**, 755—768, 1955.

[550] *Kumai, M.*, und *Chiba, T.*, On the heat energy balance of the paddy field. * J. Agr. Met. Japan **8**, 117—119, 1953.

[551] *Mäde, A.*, Die Agrarmeteorologie in der Pflanzenzüchtung. * R. f. W. Wiss. Abh. **9**, Nr. 6, 1—48, 1942.

[552] *Monteith, J. L.*, The heat balance of soil beneath crops. * Proc. Canberra Symposium 1956, 123—128, UNESCO Paris 1958.

[553] *Müller-Stoll, W. R.*, und *Freitag, H.*, Beiträge zur bestandsklimatischen Analyse von Wiesengesellschaften. * Angew. Met. **3**, 16—30, 1957.

[554] *Norman, J. T.*, *Kemp, A. W.*, und *Tayler, J. E.*, Winter temperatures in long and short grass. * Met. Mag. **86**, 148—152, 1957.

[555] *Primault, B.*, La température d'un pré. * Ann. Schweiz. Met. Zentralanst. **91**, 1954, Anh. 7, 1956.

[556] *Ramdas, L. A.*, *Kalamkar, R. J.*, und *Gadre, K. M.*, Agricultural studies in microclimatology. * Ind. J. Agric. Science **4**, 451—467, 1934 und **5**, 1—11, 1935.

[557] *Sato, S.*, On the simple methods of temperature observations in the paddy field. * J. Agr. Met. Japan **9**, 96—98, 1954.

[558] *Satone, H.*, Electrical measurement of the temperature in the paddy field and thermal diffusibilities of its soil. * J. Agr. Met. Japan **6**, 45—48, 1951.

[559] *Schrödter, H.*, Bodenfeuchtigkeitsmessungen in einer Himbeerparzelle mit unterschiedlich behandelten Teilstücken. * Angew. Met. **1**, 92—94, 1951.

[560] *Specht, R. L.*, Micro-environment (soil) of a natural plant community. * Proc. Canberra Symposium 1956, 152—155, UNESCO Paris 1958.

[561] *Suzuki, S.*, Microclimatical studies of rice paddy field. * J. Agr. Met. Japan **1**, 1—7, 1943.

[562] —, The conduction of heat in a paddy field soil. * Ebenda **7**, 11—12, 1951.

[563] *Szász, G.*, Das Bestandsklima der Wintergerste. * Debrecen Met. Univ. Inst. Nr. 13, 1956.

[564] *Tamm, E.*, Bodentemperaturen unter verschiedenen Pflanzenbeständen. * Z. f. Pflanzenernährung, Düngung, Bodenkde. **47**, 29—34, 1949.

[565] —, und *Funke, H.*, Pflanzenklimatische Temperaturmessungen in einem Maisbestand. * Z. f. Acker- u. Pflanzenbau **100**, 199—210, 1955.

[566] *Tsuboi, Y.*, und *Nakagawa, Y.*, Study on the relation between the microclimatic characteristics and the growth of rice plant. * J. Agr. Met. Japan **9**, 59—62, 1954.

[567] *Wagner, R.*, Adatok a kopancsi rizföldek éghajlatához. * Idöjárás **61**, 266—277, 1957.

[568] *Waterhouse, F. L.*, Humidity and temperature in grass microclimates with reference to insolation. * Nature London **166**, 232—233, 1950.

[569] —, Microclimatological profiles in grass cover in relation to biological problems. * Quart. J. **81**, 63—71, 1955.

[570] *Yabuki, K.*, Absorptivity of paddy field and its effects for water temperature. * J. Agr. Met. Japan **9**, 121—124, 1954.

SECTION 32

[571] *Broadbent, L.*, The microclimate of the potato crop. * Quart. J. **76**, 439—454, 1950.

[572] *Burckhardt, H.*, Zur Abhängigkeit des Bestandsklimas in Weinbergen von der Erziehungsform der Reben. * Met. Rundsch. **11**, 41—47, 1958.

[573] —, Der Umweltfaktor Klima im Weinbau. * Wein-Wissensch. 1958, 59—65.

[574] *Hirst, J. M.*, *Long, I. F.*, und *Penman, H. L.*, Micrometeorology in the potato crop. * Proc. Toronto Met. Conf. 1953, 233—237, 1954 (London, R. Met. Soc.).

[575] *Jancke, O.*, Wärmemessungen im Weinberg etc. * Angew. Met. **2**, 19—23, 1954.

[576] *Justyák, J.*, Die Resultate unserer 5jährigen Untersuchungen über das Bestandsklima der Tomaten. * Veröff. Met. Inst. Debrecen 1957.

[577] *Kirchner, R.*, Beobachtungen über das Mikroklima der Weinberge. * Mitt. Pfälz. Ver. f. Naturkde. u. Naturschutz **5**, 93—101, 1936.

[578] *Linck, O.*, Der Weinberg als Lebensraum. * Hohenlohesche Buchh. Öhringen 1954.

[578a] *Mäde, A.*, Widerstandselektrische Temperaturmessungen in einem Topinamburbestand. * R. f. W. Wiss. Abh. **2**, Nr. 6, 1936.

[579] *Raeuber, A.*, Untersuchungen zur Witterungsabhängigkeit der Krautfäule der Kartoffel im Hinblick auf einen Phytophtora-Warndienst. * Abh. Met. D. DDR **6**, Nr. 40, 1—38, 1957.

[580] *Schröder, R.*, Resultados obtenidos de una investigacion del microclima en un cafetal. * Boletin informativo (Bibl. centro nac. de invest. de cafe), Columbia **2**, 33—43, 1951.

[581] —, Die klimatischen Bedingungen für den Kaffeeanbau auf der Erde etc. * Peterm. Mitt. **100**, 122—136, 1956.

[582] *Sonntag, K.*, Bericht über die Arbeiten des Kalmit-Observatoriums. * D. Met. Jahrb. f. Bayern 1934, Anhang D.

[583] *Taguchi, R.*, und *Okumura, H.*, Study on the microclimate in the mulberry field, with special reference to the vertical distribution of air temperature among the trees. * J. Agr. Met. Japan **8**, 81—83, 1953.

[584] *Takechi, O.*, und *Kikuchi, J.*, Study on the micrometeorology in the citrus-orchards different only in the tree density. * Ebenda **9**, 115—118, 1954.

[585] *Tsuboi, Y.*, und *Honda, I.*, Micro-meteorological characteristics in the vineyard (2). * Ebenda **10**, 37—41, 1954.

[586] —, und *Nakagawa, Y.*, dasselbe (1). * Ebenda **8**, 77—80, 1953.

[587] *Weger, N.*, Mikroklimatische Studien in Weinbergen. * Biokl. B. **6**, 169—179, 1939 u. **10**, 76—84, 1943.

[588] *Weise, R.*, Über die Rebe als Klima-Kriterium. * Ber. DWD-US Zone **12**, 121—123, 1950.

[589] —, Wettkundliches bei Rebenerziehungsversuchen. * Weinberg u. Keller **1**, 85—90, 1954.

[590] —, Wie beeinflußt die Erziehungsform die Temperaturen im Rebinnern? * Ebenda **3**, 332—338, 383—390, 1956.

SECTION 33

[591] *Baumgartner, A.*, Die Strahlungsbilanz in einer Fichtendickung. * Forstw. C. **71**, 337—349, 1952.

[592] —, Beobachtungswerte u. weitere Studien zum Wärme- u. Wasserhaushalt eines jungen Waldes. * Wiss. Mitt. Met. Inst. d. Univ. München 4, 1957.

[593] *Brocks, K.*, Die räumliche Verteilung der Beleuchtungsstärke im Walde. * Z. f. F. u. Jagdw. **71**, 47—53, 1939.

[594] *Deinhofer, J.*, und *Lauscher, F.*, Dämmerungshelligkeit. * Met. Z. **56**, 153—159, 1939.

[595] *Dordick, I. L.*, und *Thuronyi, G.*, Selective annotated bibliography on climate of the forest. * MAB **8**, 515—539, 1957.

[596] *Egle, K.*, Zur Kenntnis des Lichtfeldes u. der Blattfarbstoffe. * Planta **26**, 546—583, 1937.

[597] *Eidmann, H.*, Meine Forschungsreise nach Spanisch-Guinea. * D. Biologe **10**, 1—13, 1941.

[598] *Ellenberg, H.*, Über Zusammensetzung, Standort und Stoffproduktion bodenfeuchter Eichen- und Buchen-Mischwaldgesellsch. NW Deutschlands. * Mitt. flor.-soz. Arb. Gem. Niedersachsen **5**, Hannover 1939.

[599] *Ermich, K.*, The light conditions in the patches of mountain pine of Tatra. * Fragmenta florist. et geobot. **3**, 69—77, 1957.

[600] *Gusinde, M.*, und *Lauscher, F.*, Meteorologische Beobachtungen im Kongo-Urwald. * Sitz-B. Wien. Akad. **150**, 281—347, 1941.

[601] *Harada, Y.*, A study of sunlight forests on foggy days. * Bull. Forest Exp. Science Tokyo **64**, 170—181, 1953.

[602] *Hofmann, G.*, Ein Strahlungsbilanzmesser für forstmeteorologische Untersuchungen. * Forstw. C. **71**, 330—337, 1952.

[603] *Lauscher, F.*, und *Schwabl, W.*, Untersuchung über die Helligkeit im Wald und am Waldrand. * Biokl. B. **1**, 60—65, 1934.

[604] *Mauerer, K.*, Der Versuchsbestand. * Forstw. C. **71**, 324—330, 1952.

[605] *Miller, D. H.*, Snow cover and climate in the Sierra Nevada California. * Univ. Calif. Publ. Geography **11**, 1—218, Berkeley and Los Angeles 1955.

[606] —, Transmission of insolation through pine forest canopy, as it affects the melting of snow. * Mitt. Schweiz. Vers. Anst. f. d. Forstl. Versuchswesen **35**, 59—79, 1959.

[607] *Mitscherlich, G.*, Das Forstamt Dietzhausen. * Z. f. F. u. Jagdw. **72**, 149—188, 1940.

[608] *Nägeli, W.*, Lichtmessungen im Freiland und im geschlossenen Altholzbestand. * Mitt. Schweiz. Vers. Anst. f. d. Forstl. Versuchswesen **21**, 250—306, 1940.

[609] *Richards, P. W.*, The tropical rain forest (hier: Kap. 7: microclimates 158—190). * Cambridge Univ. Press England 1952.

[610] *Sacharow, M. I.*, Influence of wind upon illumination in a forest. * Akad. Nauk SSSR **67**, 913—916, 1949 (MAB **8**, 528, 1957).

[611] *Sauberer, F.*, und *Trapp, E.*, Helligkeitsmessungen in einem Flaumeichenbuschwald. * Biokl. B. **4**, 28—32, 1937.

[612] *Scheer, G.*, Über Änderungen der Globalbeleuchtungsstärke durch Belaubung und Horizonteinengung. * Wetter u. Leben **5**, 65—71, 1953.

[613] *Sirén, G.*, The development of spruce forest on raw humus sites in northern Finland and its ecology (408 S.). * Acta Forest. Fennica **62**, Helsinki 1955.

[614] *Slanar, H.*, Das Klima des östlichen Kongo-Urwalds. * Mitt. Geograph. Ges. Wien 1945.

[615] *Trapp, E.*, Untersuchung über die Verteilung der Helligkeit in einem Buchenbestand. * Biokl. B. **5**, 153—158, 1938.

SECTIONS 34-35

[616] *Baumgartner, A.*, und *Hofmann, G.*, Elektrische Fernmessung der Luft- und Bodentemperatur in einem Bergwald. * Arch. f. Met. (B) **8**, 215—230, 1957.

[617] *Evans, G. C.*, Ecological studies on the rain forest of S. Nigeria II. * J. of Ecology **27**, 436—482, 1939.

[618] *Geiger, R.*, Untersuchungen über das Bestandsklima. * Forstw. C. **47**, 629—644, 848—854, 1925; **48**, 337—349, 495—505, 523—532, 749—758, 1926.

[619] —, und *Amann H.*, Forstmeteorologische Messungen in einem Eichenbestand. * Ebenda **53**, 237—250, 341—351, 705—714, 809—819, 1931; **54**, 371—383, 1932.

[620] *Göhre, K.*, und *Lützke, R.*, Der Einfluß von Bestandsdichte und -struktur auf das Kleinklima im Walde. * Arch. f. Forstwesen **5**, 487—572, 1956.

[621] *Inoue, E.*, An aerodynamic measurement of photosynthesis over a paddy field. * Proc. 7. Jap. Nation. Congreß f. appl. mech. 1957, 211—214.

[622] —, Energy budget over fields of waving plants. * J. Agr. Met. Japan **14**, 6—8, 1958.

[623] —, *Tani, N.*, *Imai, K.*, und *Isobe, S.*, The aerodynamic measurement of photosynthesis over a nurserey of rice plants. * Ebenda **14**, 45—53, 1958.

[624] *Sauberer, F.*, und *Trapp, E.*, Temperatur- und Feuchtemessungen in Bergwäldern. * Centralbl. f. d. ges. Forstw. **67**, 233—244, 257—276, 1941.

[625] *Schimitschek, E.*, Bioklimatische Beoabachtungen und Studien bei Borkenkäferauftreten. * Wetter u. Leben **1**, 97—104, 1948.

[626] *Ungeheuer, H.*, Mikroklima in einem Buchenhochwald am Hang. * Biokl. B. **1**, 75—88, 1934.

SECTION 36

[627] *Delfs, J.*, Die Niederschlagszurückhaltung im Walde. * Mitt. d. Arbeitskreises „Wald u. Wasser" **2**, Koblenz 1955.

[628] —, *Friedrich, W.*, *Kiesekamp, H.*, und *Wagenhoff, A.*, Der Einfluß des Waldes u. des Kahlschlages auf den Abflußvorgang, den Wasserhaushalt u. den Bodenabtrag. * Mitt. Niedersächs. Landesforstverwaltung **3**, mit Tab.band, Hannover 1958.

[629] *Eidmann, F. E.*, Die Interception in Buchen- u. Fichtenbeständen. * UGGI, Symp. of Hannov.-Münden I, 5—25, 1959.

[630] *Eitingen, G. R.*, Interception of precipitation by the canopy of forests. * Les i Step, Moskau **8**, 7—16, 1951 (MAB **8**, 530, 1957).

[631] *Freise, F.*, Das Binnenklima von Urwäldern im subtropischen Brasilien. * Peterm. Mitt. **82**, 301—304, 346—348, 1936.

[632] *Gérard, P.*, Une année d'observations microclimatiques en forêt secondaire à Bambesa (Uélés). * Compt. Rend. 11. Congr. Un. Intern. des instituts de recherches forest. 206—209, Florenz 1954.

[633] *Godske, C. L.*, und *Paulsen, H. S.*, Investigations carried trough at the station of forest met. at Os II. * Univ. Bergen Årb. 1949, Naturw. Rekke **8**, 1—39, Bergen 1950.

[634] *Grunow, J.*, Der Niederschlag im Bergwald. * Forstw. C. **74**, 21—36, 1955.

[635] *Hesselman, H.*, Einige Beobachtungen über die Beziehung zwischen der Samenverbreitung von Fichte u. Kiefer u. die Besamung der Kahlhiebe. * Meddel. Fran. Stat. Skogförsöksanst. Stockholm **27**, 145—182, 1934; **31**, 1—64, 1938.

[636] *Hoppe, E.*, Regenmessungen unter Baumkronen. * Mitt. a. d. Forstl. Versuchswesen Österreichs **21**, Wien 1896.

[637] *Horton, R. E.*, Rainfall interception. * M. W. Rev. **47**, 603—623, 1919.

[638] *Kittredge, J.*, Forest influences. * McGraw-Hill Book Co., New York 1948.

[639] —, Influences of forests on snow in the Ponderosa-sugar pine fir zone of the central Sierra Nevada. * Hilgardia, Berkeley, Cal. **22**, 1—96, 1953.

[640] *Linskens, H. F.*, Niederschlagsmessungen unter verschiedenen Baumkronentypen im belaubten u. unbelaubten Zustand. * Ber. D. Bot. G. **64**, 215—221, 1951.

[641] —, Niederschlagsverteilung unter einem Apfelbaum im Laufe einer Vegetationsperiode. * Ann. d. Met. **5**, 30—34, 1952.

[642] *Mäde, A.*, Zur Methodik der Taumessung. * Wiss. Z. d. Martin-Luther-Univ. Halle-Wittenberg **5**, 483—512, 1956.

[643] *Ovington, J. D.*, A comparison of rainfall in different woodlands. * Forestry London **27**, 41—53, 1954.

[644] *Priehäußer, G.*, Bodenfrost, Bodenentwicklung u. Flachwurzeligkeit der Fichte. * Forstw. C. **61**, 329—342, 381—389, 1939.

[645] *Rosenfeld, W.*, Erforschung der Bruchkatastrophen in den ostschlesischen Beskiden in der Zeit v. 1875—1942. * Forstw. C. 1944, 1—31.

[646] *Roßmann, F.*, Das Rauchen der Wälder nach Regen und die Unterscheidung von Wasserdampf und Wasserrauch. * Wetter u. Leben **4**, 56—57, 1952.

[647] *Rowe, P. B.*, und *Hendrix, T. M.*, Interception of rain and snow by second-growth ponderosa pine. * Transact. Amer. Geophys. Union **32**, 903—908, 1951.

[648] *Schubert, J.*, Niederschlag, Verdunstung, Bodenfeuchtigkeit, Schneedecke in Waldbeständen und im Freien. * Met. Z. **34**, 145—153, 1917.

[649] *Slavík, B.*, Das Durchdringen der Regenniederschläge durch den Laubbestand. * Verh. d. 2. ganzstaatl. bioklimatolog. Konferenz 3. bis 5. Nov. 1958, 377—380, Tschechoslowak. Akad. d. Wiss. 1960.

[650] *Takeda, K.*, Rainfall in forest. * J. Met. Soc. Japan **29**, 199—212, 1951.

[651] *Trimble, G. R.*, und *Weitzman, S.*, Effect of a hardwood forest canopy on rainfall intensities. * Transact. Amer. Geophys. Union **35**, 226—234, 1954.

SECTION 37

[652] *Baumgartner, A.*, Licht und Naturverjüngung am Nordrand eines Waldbestandes. * Forstw. C. **74**, 59—64, 1955.

[653] *Diem, M.*, Höchstlasten der Nebelfrostablagerungen an Hochspannungsleitungen im Gebirge. * Arch. f. Met. (B) **7**, 84—95, 1955.

[654] *Dirmhirn, I.*, Zur spektralen Verteilung der Reflexion natürlicher Medien. * Wetter u. Leben **9**, 41—46, 1957.

[655] *Dörffel, K.*, Die physikalische Arbeitsweise des Gallenkampschen Verdunstungsmessers etc. * Veröff. Geoph. I. Leipzig **6**, Nr. 9, 1935.

[656] *Geiger, R.*, Die Beschattung am Bestandsrand. * Forstw. C. **57**, 789—794, 1935 u. **58**, 262—266, 1936.

[657] *Grunow, J.*, Nebelniederschlag. * Ber. DWD-US Zone **7**, Nr. 42, 30—34, 1952.

[658] —, Kritische Nebelfroststudien. * Arch. f. Met. (B) **4**, 389—419, 1953.

[659] *Herr, L.*, Bodentemperaturen unter besonderer Berücksichtigung der äußeren met. Faktoren. * Diss. Leipzig 1936.

[660] *Hesselman, H.*, Einige Beobachtungen über die Beziehung zwischen der Samenverbreitung von Fichte u. Kiefer u. die Besamung der Kahlhiebe. * Meddel. Fran. Stat. Skogförsöksanst. **27**, 145—182, 1934 u. **31**, 1—64, 1938.

[661] *Hori, T.*, Studies on fog (399 S.). * Tanne Trading Co. Sapporo, Hokkaido, Japan 1953.

[662] *Koch, H. G.*, Der Wald-Feldwind, eine mikro-aerologische Studie. * Beitr. Phys. d. fr. Atm. **22**, 71—75, 1934.

[663] —, Der Waldwind. * Forstw. C. **64**, 97—111, 1942.

[664] *Lauscher, F.*, Die Rolle mikroklimatischer Faktoren beim Massenauftreten von Waldschädlingen. * Wetter u. Leben **5**, 195—200, 1953.

[665] —, und *Schwabl, W.*, Untersuchungen über die Helligkeit im Wald und am Waldrand. * Biokl. B. **1**, 60—65, 1934.

[666] *Linde, R. J. van der*, und *Woudenberg, J. P. M.*, A method for determining the daily variations in width of a shadow in connection with the time of the year and the orientation of the overshadowing object. * Med. en Verh. (A) Nr. 52, Kon. Nederland. Met. Inst. Nr. 102, 1946.

[667] *Linke, F.*, Niederschlagsmessung unter Bäumen. * Met. Z. **33**, 140—141, 1916 u. **38**, 277, 1921.

[668] *Lüdi, W.*, und *Zoller, H.*, Über den Einfluß der Waldnähe auf das Lokalklima. * Ber. Geobotan. Forsch. Inst. Rübel Zürich f. d. Jahr 1948, 85—108, 1949.

[669] *Marloth*, Über die Wassermengen, welche Sträucher u. Bäume aus treibendem Nebel u. Wolken auffangen. * Met. Z. **23**, 547—553, 1906.

[670] *Nagel, J. F.*, Fog precipitation on Table Mountain. * Quart. J. **82**, 452—460. 1956.

[671] *Ooura, H.*, The capture of fog particles by the forest. * J. Met. Res. Tokyo **4**, Suppl., 239—259, 1952.

[672] *Pfeiffer, H.*, Kleinaerologische Untersuchungen am Collmberg. * Veröff. Geoph. I. Leipzig **11**, Nr. 5, 1938.

[673] *Rink, J.*, Die Schmelzwassermengen der Nebelfrostablagerungen * R. f. W. Wiss. Abh. **5**, Nr. 7, 1938.

[674] *Rötschke, M.*, Untersuchungen über die Meteorologie der Staubatmosphäre. * Veröff. Geoph. I. Leipzig **11**, 1—78, 1937.

[675] *Schmauß, A.*, Seewinde ohne See. * Met. Z. **37**, 154—155, 1920.

[676] *Schubert, J.*, Die Sonnenstrahlung im mittleren Norddeutschland nach den Messungen in Potsdam. * Met. Z. **45**, 1—16, 1928.

[677] *Waibel, K.*, Die meteorologischen Bedingungen für Nebelfrostablagerungen an Hochspannungsleitungen im Gebirge. * Arch. f. Met. (B) **7**, 74—83, 1955.

[678] *Woelfle, M.*, Wald u. Windschutz. * Forstw. C. **57**, 349—362, 1935; **58**, 325—338, 429—448, 1936.

[679] —, Windverhältnisse im Walde. * Ebenda **61**, 65—75, 461—475, 1939; **64**, 169—182, 1942.

[680] —, Bemerkungen zu „Der Waldwind" von *H. G. Koch*. * Forstw. C. 1944, 131—136.

[681] *Ziegler, O.*, Die Bedeutung des Windes u. der Thermik für die Verbreitung der Insekten etc. * Z. f. Pflanzenbau u. -schutz **1**, 241—266, 1903.

SECTION 38

[682] *Aichele, H.*, Beitrag zum Mikroklima eines Käferkahlschlages. * Arch. d. Wiss. Ges. f. Land- u.. Forstw. Freiburg **1**, 43—49, 1949.

[683] *Amann, H.*, Birkenvorwald als Schutz gegen Spätfröste. * Forstw. C. **52**, 493—502, 581—592, 1930.

[684] *Angström, A.*, Jordtemperatur i bestand av olika täthet. * Medd. Stat. Met. Hydr. Anst. Stockholm **29**, 187—218, 1936.

[685] *Baumgartner, A.*, Über die Unterschiede in den klimat. Wuchsbedingungen einer freien u. einer birkenüberstellten Wiederaufforstungsfläche. * Forstw. C. **75**, 223—239, 1956.

[686] *Bolz, H. M.*, Der Einfluß der infraroten Strahlung auf das Mikroklima. * Abh. Met. D. DDR **1**, Nr. 7, 1—59, 1951.

[687] *Boos*, Untersuchungen über das Bestandsinnenklima im Pr. Forstamt Erdmannshausen. * Mitt. a. Forstwirtsch. u. Forstwiss. **10**, 254—259, 1939.

[688] *Burger, H.*, Waldklimafragen I—III. * Mitt. Schweiz. Centr. Anst. f. d. forstl. Versuchswesen **17**, 92—149, 1932; **18**, 7—54, 153—192, 1933.

[689] *Danckelmann, B.*, Spätfrostbeschädigungen im märkischen Wald. * Z. f. F. u. Jagdw. **30**, 389—411, 1898.

[690] *Eggler, J.*, Kleinklimatische Untersuchungen in den Flaumeichenbeständen bei Graz. * Biokl. B. **9**, 94—110, 1942.

[691] *Ermich, K.*, Microclimatic investigations in plant communities of the Puszcza Niepolomicka. * Act. Soc. Bot. Poloniae **22**, 483—559, 1953.

[692] —, Ecological investigations in two forest communities. * Fragmenta Floristica et Geobot. **2**, 28—71, 1956.

[693] —, Thermic conditions in the patches of mountain pine on Hala Gasienicowa in the Tatra mountains. * Ebenda **5**, 99—115, 1959.

[694] *Geiger, R.*, Das Standortklima in Altholznähe. * Mitt. H. Göring Akad. d. D. Forstwiss. **1**, 148—172, 1941.

[695] *Göhre, K.*, Kleinklimatische Untersuchungen auf einer Kiefernkultur unter Birkenvorwald. * Arch. f. Forstwesen **3**, 441—474, 1954.

[696] *Gothe, H.*, Forstliche Klima-Aufnahmen — eine notwendige Ergänzung forstlicher Standortsaufnahmen. * Allg. Forstz. **9**, 316—318, 1954.

[697] *Koch, H. G.*, Temperaturverhältnisse u. Windsystem eines geschlossenen Waldgebietes. * Veröff. Geoph. I. Leipzig **3**, Nr. 3, 1934.

[698] *Kortüm, F.*, Der Energie- u. Stoffumsatz durch Verdunstung u. Transpiration unter Bestandesschirm u. auf Schlagflächen. * Verh. d. 2. ganzstaatl. Bioklimatolog. Konferenz 3. bis 5. Nov. 1958, 349—355, Tschechoslowak. Akad. d. Wiss. 1960.

[699] *Maran, B.*, Influence of undergrowth on the relative humidity of the atmospheric surface layers. * Tschechoslow. Akad. Prag, Sbornik **21**, 5—12, 1949 (MAB **8**, 1473, 1957).

[700] *Obolensky, N. v.*, Effect of arborous vegetation on the temperature of soil and the temperature and humidity of the air. * J. Geophys. and Met. **3**, 113—139, Moskau 1926.

[701] *Slavík, B.*, *Slavíková, J.*, und *Jeník, J.*, Ökologie der gruppenweisen Verjüngung eines Mischbestandes. * Rozpravy Tschechoslow. Akad. **67**, 2, 1957.

[702] *Sokolowski, A.*, Zur mikroklimatischen Charakteristik der wichtigsten Waldassoziationen des Forstreviers Ruda bei Pulawy. * Act. Soc. Bot. Poloniae **26**, 373—412, 1957.

[703] *Tomanek, J.*, Mikroklimatische Verhältnisse im Lochhiebe. * Verh. d. 2. ganzstaatl. Bioklimatolog. Konferenz 3. bis 5. November 1958, 297—313, Tschechoslowak. Akad. d. Wiss. 1960.

[704] *Wagner, R.*, Von dem Klima des Waldes. * Idöjárás **61**, 117—125, 1957.

[705] *Wrede, C. v.*, Die Bestandsklimate u. ihr Einfluß auf die Biologie der Verjüngung unter Schirm u. in der Gruppe. * Forstw. C. **47**, 441—451, 491—505, 570—582, 1925.

[706] *Woelfle, M.*, Waldbau u. Forstmeteorologie. 2. Aufl. * Bayer. Landwirtsch. Verl. München 1950.

SECTION 39

[707] *Bates, C. G.*, und *Henry, A. J.*, Forest and streamflow experiment at Wagon Wheal Gap, Colo. * MWRev. Suppl. Nr. **30**, 1928.

[708] *Blanford, H. F.*, On the influence of Indian forests on the rainfall. * J. Asiat. Soc. of Bengal **56**, II, 1, 1887.

[709] *Burckhardt, H.*, Lokale Klimaänderungen auf einem Berggipfel durch Kahlhieb. * Angew. Met. **1**, 150—154, 1952.

[710] *Burger, H.*, Einfluß des Waldes auf den Stand der Gewässer II—V. * Mitt. d. Schweiz. Anst. f. d. forstl. Vers.w. **18**, 311—416, 1934; **23**, 167—222, 1943; **24**, 133—218, 1944; **31**, 7—58, 1954/55.

[711] —, Waldklimafragen IV. * Ebenda **27**, 17—75, 1951.

[712] —, Wald u. Wasser in der Schweiz. * Allg. Forstz. **9**, 14—17, 1954.

[713] —, Der Wasserhaushalt in der Valle di Melera. * Mitt. d. Schweiz. Anst. f. d. forstl. Vers.w. **31**, 493—555, 1954/55.

[714] *Ebermayer, E.*, Die physikalischen Einwirkungen des Waldes auf Luft u. Boden. * Aschaffenburg 1873.

[715] *Engler, A.*, Einfluß des Waldes auf den Stand der Gewässer I. * Mitt. d. Schweiz. Anst. f. d. forstl. Vers.w. **12**, 1—626, 1919.

[716] *Hamberg, H. E.*, De l'influence des forêts sur le climat de la Suède (5 Teile). * Stockholm 1885—1896.

[717] *Hornsmann, E.*, Allen hift der Wald. * Bayer. Landwirtsch. Verl. München 1958.

[718] *Huber, B,*. Was wissen wir vom Wasserverbrauch des Waldes? * Forstw. C. **72**, 257—264, 1953.

[719] *Kaminsky, A.*, Beitrag zur Frage über den Einfluß der Aufforstung der Waldlichtungen in Indien auf die Niederschläge. * Nachr. d. Geophys. Centr. I. Leningrad **4**·

[720] *Lorenz-Liburnau, v.*, Resultate forstlich-meteorologischer Beobachtungen. * Mitt. a. d. forstl. Vers.w. Österr. **12** u. **13**, Wien 1890.

[721] *Mägdefrau, K.*, und *Wutz, A.*, Die Wasserkapazität der Moos- u. Flechtendecke des Waldes. * Forstw. C. **70**, 103—117, 1951.

[722] *Mrose, H.*, Der Einfluß des Waldes auf die Luftfeuchtigkeit. * Angew. Met. **2**, 281—286, 1956.

[723] *Müttrich, A.*, Über den Einfluß des Waldes auf die periodischen Veränderungen der Lufttemperatur. * Z. f. F. u. Jagdw. **22**, 385—400, 449—458, 513—526, 1890.

[724] —, Bericht über die Untersuchung der Einwirkung des Waldes auf die Menge der Niederschläge. * Neumann, Neudamm 1903.

[725] *Pechmann, H. v.*, Gedanken zur Schutzwaldfrage. * Allg. Forstz. **4**, 419—421, 429—430, 1949.

[726] *Schubert, J.*, Der jährliche Gang der Luft- u. Bodentemperatur im Freien und in Waldungen. * J. Springer, Berlin 1900.

[727] —, Über den Einfluß des Waldes auf die Niederschläge im Gebiet der Letzlinger Heide. * Z. f. F. u. Jagdw. **69**, 604—615, 1937.

[728] *Schultze, J. H.*, Neuere theoretische u. praktische Ergebnisse der Bodenerosionsforschung in Deutschland. * Forsch. u. Fortschr. **27**, 12—18, 1953.

[729] US Forest Service, Watershed management research Coweeta Experimental Forest. * US Dep. Agric. SE-Forest Exp. St. Ashville, N-Carolina 1948.

[730] *Wittich, W.*, Der Einfluß des Waldes auf die Wasserwirtschaft des Landes. * Allg. Forstz. **7**, 433—438, 1952; **8**, 144—148, 436—438, 1953.

[731] *Zenker, H.*, Waldeinfluß auf Kondensationskerne u. Lufthygiene. * Z. f. Met. **8**, 150—159, 1954.

SECTION 40

[732] *Becker, C. F.*, Solar radiation availability on surfaces in the US as affected by season, orientation, latitude, altitude and cloudiness. * J. Solar Energy Sc. a. Engin. Phönix **1**, 13—21, 1957.

[733] *Bögel, A.*, Die direkte Sonnenstrahlung auf Westhänge. * Z. f. Met. **11**, 70—83, 1957.

[734] *Decoster, M.*, und *Schüepp, W.*, Le rayonnement sur des plans verticaux à Stanleyville. * Serv. Mét. Congo Belge **15**, Léopoldville 1956.

[735] — —, und *Elst, N. van der*, Le rayonnement sur des plans verticaux à Léopoldville. * Ebenda **7**, 1955.

[736] *Dirmhirn, I.*, Himmelsstrahlung im Bauwesen. * Wetter u. Leben **7**, 200—201, 1955.

[737] *Geßler, R.*, Die Stärke der unmittelbaren Sonnenbestrahlung der Erde in ihrer Abhängigkeit von der Auslage unter verschiedenen Breiten und zu verschiedenen Jahreszeiten. * Abh. Pr. Met. I. **8**, Nr. 1, 1925.
[738] *Gräfe, K.*, Strahlungsempfang vertikaler ebener Flächen; Globalstrahlung von Hamburg. * Ber. DWD **5**, Nr. 29, 1—15, 1956.
[739] *Grunow, J.*, Beiträge zum Hangklima. * Ber. DWD-US Zone **5**, Nr. 35, 293—298, 1952.
[740] *Hand, J. F.*, Preliminary measurements of solar energy received on vertical surfaces. * Transact. Amer. Geophys. Union **28**, 705—712, 1947.
[741] *Kaempfert, W.*, Sonnenstrahlung auf Ebene, Wand u. Hang. * R. f. W. Wiss. Abh. **9**, Nr. 3, 1942.
[742] —, Zur Besonnung südseitiger Spaliermauern. * Gartenbauwiss. **17**, 531—542, 1943.
[743] —, Zur Frage der Besonnung enger Straßen. * Met. Rundsch. **2**, 222—227, 1949.
[744] —, und *Morgen, A.*, Die Besonnung. * Z. f. Met. **6**, 138—146, 1952.
[745] *Kienle, J. v.*, Die tatsächliche und die astronomisch mögliche Sonnenscheindauer auf verschieden exponierte Flächen. * D. Met. Jahrb. f. Baden 1933, Anhang.
[746] *Kimball, H. H.*, und *Hand, J. F.*, Daylight illumination on horizontal, vertical and sloping surfaces. * MWRev. **50**, 615—628, 1922.
[747] *Möller, F.*, Über die Farbe der Sicht und des Tageshimmels. * Arch. f. Met. (B) **5**, 1—17, 1953.
[748] *Nicolet, M.*, und *Bossy, L.*, Ensoleillement et orientation en Belgique II. * Mém. Inst. Roy. Mét. **36**, 1950.
[749] *Perl, G.*, Die Komponenten der Intensität der Sonnenstrahlung in verschiedenen geographischen Breiten. * Met. Z. **53**, 467—472, 1936.
[750] *Pers, M. R.*, Calcul du flux d'insolation sur une façade on pente. * La Mét. **11**, 429—435, 1935.
[751] *Schedler, A.*, Die Bestrahlung geneigter Flächen durch die Sonne. * Jahrb. Zentralanst. f. Met. u. Geodyn. Wien, N. F. **87**, D 52—64, 1950, Wien 1951.
[752] *Schütte, K.*, Die Berechnung der Sonnenhöhen für beliebig geneigte Ebenen. * Ann. d. Hydr. **71**, 325—328, 1943.
[753] *Schulze, R.*, Strahlungsempfang geneigter ebener Flächen. * Arch. f. Met. (B) **6**, 128—138, 1954.
[754] *Suzuki, S.*, Early strawberries and their cultivation on slopes. * Agric. and Horticulture Japan **16**, 1185—1188, 1941.
[755] *Thams, J. C.*, Die Globalstrahlung eines Südhanges von 25° Neigung. * Arch. f. Met. (B) **7**, 190—196, 1956.
[756] *Volz, F.*, Globalstrahlung auf geneigte Hänge. * Met. Rundsch. **11**, 132—135, 1958 (s. **12**, 135—136, 1959).

SECTION 41

[757] *Aichele, H.*, Der Temperaturgang rings um eine Esche. * Allg. Forst- u. J. Zeitung **121**, 119—121, 1950.
[758] —, Untersuchungen über die Frostschutzwirkung eines Kalkanstrichs an Obstbäumen. * Ber. DWD-US Zone **5**, Nr. 32, 70—73, 1952.
[759] *Bruckmayer, F.*, Bautechnik u. meteorologische Werte. * Wetter u. Leben **7**, 186—193, 1955.
[760] *Eisenhut, G.*, Blühen, Fruchten u. Keimen in der Gattung Tilia. * Diss. Univ. München 1957.

[761] *Filzer, P.*, Das Mikroklima von Bestandsrändern u. Baumkronen u. seine physiologischen Rückwirkungen. * Jahrb. f. wiss. Bot. **86**, 228—314, 1938.

[762] *Gerlach, E.*, Untersuchung über die Wärmeverhältnisse der Bäume. * Diss. Univ. Leipzig 1929.

[763] *Haarløv, N.*, und *Petersen, B. B.*, Measurement of temperature in bark and wood of Sitka spruce. * Kopenhagen 1952.

[764] *Härtel, O.*, Mikroklima u. Wachstum in Tulpenbeeten. * Biokl. B. **6**, 134—137, 1939.

[765] —, Über spektrale Veränderungen des Lichtfeldes durch die Vegetation. * Wetter u. Leben **6**, 118—122, 1954.

[766] *Huber, B.*, Aster linosyris, ein neuer Typus der Kompaßpflanzen (Gnomonpflanzen). * Flora **29**, 113–119, 1934.

[767] *Kaempfert, W.*, Ein Phasendiagramm der Besonnung. * Met. Rundsch. **4**, 141—144, 1951.

[768] *Koljo, B.*, Einiges über Wärmephänomene der Hölzer u. Bäume. * Forstw. C. **69**, 538—551, 1950.

[769] *Krenn, K.*, Die Bestrahlungsverhältnisse stehender u. liegender Stämme. * Wien. Allg. F. u. Jagdz. **51**, 50—51, 53—54, 1933.

[770] *Leßmann, H.*, Temperaturverhältnisse in Häufelreihen. * Jahr. Ber. d. Bad. Landeswetterd. 1950, 35—45.

[771] *Morgen, A.*, Einfluß der Pfahlneigung auf den Strahlungsgenuß der Rebe. * Der Weinbau **5**, 149—152, 1951.

[772] *Müller, G.*, Untersuchungen über die Querschnittsformen der Baumschäfte. * Forstw. C. **77**, 41—59, 1958.

[773] *Primault, B.*, L'influence de l'insolation sur la température du cambium des arbres frutiers. * Rev. Romande Agric., Vitc. Arboric. **10**, 26—28, 1954.

[774] *Scamoni, A.*, Über Eintritt u. Verlauf der männlichen Kiefernblüte. * Z. f. F. u. Jagdw. **70**, 289—315, 1938.

[775] *Schanderl, H.*, Der derzeitige Stand des Kompaßpflanzenproblems. * Biokl. B. **4**, 49—54, 1937.

[776] *Schulz, H.*, Untersuchungen an Frostrissen im Frühjahr 1956. * Forstw. C. **76**, 14—24, 1957.

[777] *Seeholzer, M.*, Rindenschäle u. Rindenriß an Rotbuche im Winter 1928/29. * Forstw. C. **57**, 237—246, 1935.

[778] *Weger, N.*, Bodentemperaturen in Beeten verschiedener Form u. Richtung. * Met. Rundsch. **2**, 291—295, 1949.

[779] —, *Herbst, W.*, und *Rudloff, C. F.*, Witterung u. Phänologie der Blühphase des Birnbaums. * R. f. W. Wiss. Abh. **7**, Nr. 1, 1940.

SECTION 42

[780] *Aichele, H.*, Frostgefährdete Gebiete in der Baar, eine kleinklimatische Geländekartierung. * Erdk. **5**, 70—73, 1951.

[781] —, Lokalklimatische Froststudien am westlichen Bodensee. * Met. Rundschau **6**, 126—130, 1953.

[782] *Aigner, S.*, Die Temperaturminima im Gstettnerboden bei Lunz am See, Niederösterreich. * Wetter u. Leben, Sonderheft 1, 34—37, 1952.

[783] *Brocks, K.*, Nächtliche Temperaturminima in Furchen mit verschiedenem Böschungswinkel. * Met. Z. **56**, 378—383, 1939.

[784] *Eimern, J. van*, Kleinklimatische Geländeaufnahme in Quickborn/Holstein. Ann. d. Met. **4**, 259—269, 1951.

[785] *Franken, E.*, Über eine Abhängigkeit der Temperaturverteilung in Strahlungsnächten von Geländeformung u. Windrichtung. * Met. Rundsch. **12**, 25—31, 1959.

[786] *Geiger, R.*, Spätfröste auf den Frostflächen bei München. * Forstw. C. **48**, 279—293, 1926.

[787] *Georgii, H. W.*, Untersuchung über die Durchmischung bodennaher Luft in Mulden. * Diss. Univ. Frankfurt 1955.

[788] —, Über die nächtliche Abkühlung der Luft in kleinen Gruben. * Met. Rundsch. **9**, 99—102, 1956.

[789] *Godske, C. L.*, On the minimum temperatures in the Bergen valley. * Bergens Mus. Arbok 1943, naturw. rekke **11**, 1—27, 1944.

[790] *Horvat, J.*, Die Vegetation der Karstdolinen. * Geografski Glasnik **14—15**, 1—25, Zagreb 1953.

[791] *Huss, E.*, Kleinraummeteorologische Studien im Federseegebiet in Strahlungsnächten. * Arch. f. Met. (B) **6**, 329—352, 1955.

[792] *Junghans, H.*, Temperaturmessungen aus einem Frostloch des Tharandter Waldes. * Angew. Met. **3**, 230—234, 1959.

[793] *Kaps, E.*, Die Frostgefährdung im Bendestorfer Tal. * Ber. DWD-US Zone **7**, Nr. 42, 258—263, 1952.

[794] *Kern, H.*, Kleinklimakartierung zur Erfassung der bodennahen nächtlichen Tiefsttemperaturen im Donaumoos. * Mitt. f. Moor- u. Torfwirtsch. **1**, 53—58, 1951.

[795] *Krühne, G.*, Karte der Maifrostschäden 1953 in Wernigerode/Harz. * Z. f. Met. **8**, 180—182, 1954.

[796] *Maruyama, E.*, und *Yamamoto, Y.*, Air temperature in depressed grounds. * J. Agr. Met. Japan **12**, 121—124, 1957.

[797] *Mason, B.*, An example of climatic control of land utilization. * Proc. Canberra Symposium 1956, 188—194, UNESCO Paris 1958.

[798] *Reichelt, G.*, Über Spätfrostschäden im Grünland in Abhängigkeit vom Relief, am Beispiel der Baar. * Wetter u. Leben **6**, 1—6, 1954.

[799] *Sauberer, F.*, und *Dirmhirn, I.*, Über die Entstehung der extremen Temperaturminima in der Doline Gstettner Alm. * Arch. f. Met. (B) **5**, 307—326, 1953.

[800] — —, Weitere Untersuchungen über die Kaltluftansammlungen in der Doline etc. * Wetter u. Leben **8**, 187—196, 1956.

[801] *Schmidt, W.*, Die tiefsten Minimumtemperaturen in Mitteleuropa. * Naturw. **18**, 367—369, 1930.

[802] *Schüepp, W.*, Frostverteilung und Kartoffelanbau in den Alpen auf Grund von Untersuchungen in der Landschaft Davos. * Schweiz. Landw. Monatshefte 1948, 37—59.

[803] *Wagner, R.*, Fluktuierende Dolinen-Nebel. * Idöjárás 1954, 289—298.

[804] *Winter, F.*, Das Spätfrostproblem im Rahmen der Neuordnung des südwestdeutschen Obstbaus. * Gartenbauwissensch. **23**, 342—362, 1958.

SECTION 43

[805] *Aichele, H.*, Kaltluftpulsationen. * Met. Rundsch. **6**, 53—54, 1953.

[806] *Berg, H.*, Beobachtungen des Berg- u. Talwindes in den Allgäuer Alpen. * Ber. DWD-US Zone **6**, Nr. 38, 105—109, 1952.

[807] *Brooks, F. A.*, und *Schultz, H. B.*, Observations and interpretation of nocturnal density currents. * Proc. Canberra Symposium 1956, 272—277, UNESCO Paris 1958.

[808] *Defant, A.*, Der Abfluß schwerer Luftmassen auf geneigtem Boden. * Sitz-B. Berlin. Akad. **18**, 624—635, 1933.
[809] *Defant, F.*, Zur Theorie der Hangwinde, nebst Bemerkungen zur Theorie der Berg- u. Talwinde. * Arch. f. Met. (A) **1**, 421—450, 1949.
[810] —, Local winds. * Compend. of Met. (Am. Met. Soc.) 655—672, Boston 1951.
[811] *Eimern, J. van*, Über eine Windbeeinflussung durch die Randhöhen des Elbtals bei Hamburg. * Met. Rundsch. **8**, 97—99, 1955.
[812] *Ekhart, E.*, Neuere Untersuchungen zur Aerologie der Talwinde. * Beitr. Phys. d. fr. Atm. **21**, 245—268, 1934.
[813] —, Über den täglichen Gang des Windes im Gebirge. * Arch. f. Met. (B) **4**, 431—450, 1953.
[814] *Field, J. H.*, und *Warden, R.*, A survey of the air currents in the bay of Gibraltar. * Geophys. Mem. **59**, London 1933.
[815] *Friedel, H.*, Wirkungen der Gletscherwinde auf die Ufervegetation der Pasterze. * Biokl. B. **3**, 21—25, 1936.
[816] *Gleeson, T. A.*, On the theory of cross-valley winds arising from differential heating of the slopes. * J. Met. **8**, 398—405, 1951.
[817] *Grunow, J.*, Der Wetterdienst bei der internationalen Skiflugwoche vom 28. 2. bis 6. 3. 1950 in Oberstdorf. * Met. Rundsch. **4**, 62—64, 1951.
[818] *Hoinkes, H.*, Beiträge zur Kenntnis des Gletscherwindes. * Arch. f. Met. (B) **6**, 36—53, 1954.
[819] —, Der Einfluß des Gletscherwindes auf die Ablation. * Z. f. Gletscherkde. u. Glazialgeologie **3**, 18—23, 1954.
[820] *Kaiser, H.*, Über die Strömungsverhältnisse im Bergland. * Met. Rundsch. **7**, 214—217, 1954.
[821] *Kaps, E.*, Zur Frage der Durchlüftung von Tälern im Mittelgebirge. * Met. Rundsch. **8**, 61—65, 1955.
[822] *Koch, H. G.*, Ein lokaler Tagesfallwind in Mittelsardinien. * Veröff. Geophys. I. Leipzig **15**, 40—60, 1949.
[823] —, Hanggewitter, Hangerwärmung u. Hangströmung am Thüringer Wald. * Peterm. Mitt. **99**, 107—109, 1955.
[824] *Küttner, J.*, Periodische Luftlawinen. * Met. Rundsch. **2**, 183—184, 1949.
[825] *Luft, R.*, Das Klima von Bonn-Beuel. * Z. f. angew. Met. **55**, 155—158, 191—197, 234—239, 1938.
[826] *Maletzke, R.*, siehe Alpenwandersegelflug 1958, Mitt.heft I/1958 der Akaflieg München.
[827] *Nitze, F. W.*, Untersuchung der nächtlichen Zirkulationsströmung am Berghang durch stereophotogrammetrisch vermessene Ballonbahnen. * Biokl. B. **3**, 125—127, 1936.
[828] *Reiher, M.*, Nächtlicher Kaltluftfluß an Hindernissen. * Biokl. B. **3**, 152—163, 1936.
[829] *Runge, F.*, Windgeformte Bäume in den Tälern der Zillertaler Alpen bzw. Allgäuer Alpen). * Met. Rundsch. **11**, 28—30, 1958 (bzw. **12**, 98—99, 1959).
[830] *Scaëtta, H.*, Les avalanches d'air dans les Alpes et dans les hautes montagnes de l'Afrique centrale. * Ciel et Terre **51**, 79—80, 1935.
[831] *Schmauß, A.*, Luftlawinen in Alpentälern. * D. Met. Jahrb. f. Bayern 1926, Anhang F.
[832] —, Absinken einer Inversion. * Z. f. angew. Met. **59**, 260—263, 1942.
[833] —, Über Luftlawinen. * Ber. DWD-US Zone **4**, Nr. 31, 14—16, 1952.
[834] *Schnelle, F.*, Kleinklimatische Geländeaufnahme am Beispiel der Frostschäden im Obstbau. * Ber. DWD-US Zone **2**, Nr. 12, 99—104, 1950.

[835] —, Ein Hilfsmittel zur Feststellung der Höhe von Frostlagen in Mittelgebirgstälern. * Met. Rundsch. **9**, 180—182, 1956.

[836] *Schüepp, W.*, Untersuchungen über den winterlichen Kaltluftsee in Davos. * Verh. Schweiz. Naturf. Ges. 1945, 127—128.

[837] *Schultz, H.*, Über Klimaeigentümlichkeiten im unteren Rheingau, unter besonderer Berücksichtigung des Wisperwindes. * Frankfurter Geogr. Hefte **7**, Nr. 1, 1933.

[838] *Schulz, L.*, Lokalklimatische Untersuchungen im Oberharz. * Biokl. B. **3**, 25—29, 1936.

[839] *Scultetus, H. R.*, Geländeausformung u. Bewindung in Abhängigkeit von der Austauschgröße. * Met. Rundsch. **12**, 73—80, 1959.

[840] *Steinhauser, F.*, Über die Windverstärkung an Gebirgszügen. * Arch. f. Met. (B) **2**, 39—64, 1950.

[841] *Stöckl, R.*, Der Talauf- und Talabwind von Graz. * Wetter u. Leben **5**, 169—171, 1953.

[842] *Tollner, H.*, Gletscherwinde in den Ostalpen. * Met. Z. **48**, 414—421, 1931.

[843] *Troll, C.*, Die Lokalwinde der Tropengebirge und ihr Einfluß auf Niederschlag und Vegetation. * Bonner Geogr. Abh. **9**, 124—182, 1952.

[844] *Voigts, H.*, Experimentelle Untersuchungen über den Kaltluftfluß bei verschiedenen Neigungen und verschiedenen Hindernissen. * Met. Rundsch. **4**, 185—188, 1951.

[845] *Wagner, A.*, Theorie und Beobachtungen der periodischen Gebirgswinde. * Gerl. B. **52**, 408—449, 1938.

[846] *Weischet, W.*, Die Baumneigung als Hilfsmittel zur geographischen Bestimmung der klimatischen Windverhältnisse. * Erdk. **5**, 221—227, 1951.

[847] *Wilkins, E. M.*, A discontinuity surface produced by topographic winds over the upper Snake River Plain, Idaho. * Bull. Am. Met. Soc. **36**· 397—408, 1955.

[848] *Winter, F.*, Schornsteinrauch veranschaulicht das abendliche Einsetzen des Hangabwindes. * Met. Rundsch. **9**, 224, 1956.

[849] *Yoshino, M.*, The structure of surface winds crossing over a small valley. * J. Met. Soc. Japan **35**, 184—195, 1957.

[850] —, Local characteristics of surface winds in a small valley. * Science Rep. Tokyo Kyoiku Daigaku **5**, 129—151, 1957.

[851] —, Winds in a V-shaped small valley. * J. Agr. Met. Japan **13**, 129—134, 1958.

SECTION 44

[852] *Cagliolo, A.*, Estudio microclimatico de pendientes en el sudeste de la provincia de Buenos Aires. * Meteoros **1**, 134—149, 1951.

[853] *Cantlon, J. E.*, Vegetation and microclimates on north and south slopes of Cucketunk Mountain, New Jersey. * Ecolog. Monogr. Durham N. C. **23**, 241—270, 1953.

[854] *Geiger, R.*, Messung des Expositionsklimas. * Forstw. C. **49**, 665—675, 853—859, 914—923, 1927; **50**, 73—85, 437—448, 633—644, 1928; **51**, 37—51, 305—315, 637—656, 1929.

[855] *Grunow, J.*, Niederschlagsmessungen am Hang. * Met. Rundsch. **6**, 85—91, 1953.

[856] *Hartmann, F. K., Eimern, J. van,* und *Jahn, G.*, Untersuchungen reliefbedingter kleinklimatischer Fragen in Geländequerschnitten der hochmontanen Stufe des Mittel- u. Südwestharzes. * Ber. DWD **7**, Nr. 50, 1959.

[857] *Haude, W.*, Ergebnisse der allgemeinen met. Beobachtungen etc. * Rep. Scient. Exp. to the NW Prov. China (Sven Hedin) IX, Met. 1, Stockholm 1940.
[858] *Heigel, K.*, Exposition u. Höhenlage in ihrer Wirkung auf die Pflanzenentwicklung. * Met. Rundsch. **8**, 146—148, 1955.
[859] —, Die Bodenfeuchte in Abhängigkeit von Exposition, Bodenart und Bewuchs. * Wetter u. Leben **9**, 104—108, 1957.
[860] —, Ergebnisse von Bodenfeuchtemessungen mit Gipsscheibenelektroden. * Met. Rundsch. **11**, 92—96, 1958.
[861] *Held, J. R.*, Temperatur- u. relative Feuchtigkeit auf Sonn- u. Schattenseite in einem Alpenlängstal. * Met. Z. **58**, 398—404, 1941.
[862] *Kerner, A.*, Die Änderung der Bodentemperatur mit der Exposition. * Sitz-B. Wien. Akad. **100**, 704—729, 1891.
[863] *Kiss, A.*, Adatok a futóhomok mikroklimájához. * Idöjárás **59**, 235—238, 1955.
[864] *Kreeb, K.*, Phänologisch-pflanzensoziologische Untersuchungen in einem Eichen-Hainbuchenwald im Neckargebiet. * Ber. D. Bot. G. **69**, 361—374, 1956.
[865] *Künkele, T.*, und *Geiger, R.*, Hangrichtung (Exposition) u. Pflanzenklima. * Forstw. C. **47**, 597—606, 1925.
[866] *Okko, V.*, On the thermal behaviour of some Finnish eskers. * Fennia, Helsinki **81**, Nr. 5, 1957.
[867] *Parker, J.*, Environment and forest distribution of the Palouse range in northern Idaho. * Ecology **33**, 451—461, 1952.
[868] *Richardson, W. E.*, Temperature differences in the South Tyne Valley, with special reference to the effects of air mass. * Quart. J. **82**, 342—348, 1956.
[869] *Sakanoue, T.*, Microclimatic investigation of waste heap. * J. Agr. Met. Japan **15**, 59—63, 1959.
[870] *Schwabe, G. H.*, Der künstliche Erdkegel als Gegenstand der experimentellen Ökologie. * Arch. f. Met. (B) **8**, 108—127, 1957.
[871] *Shanks, R. E.*, und *Norris, F. H.*, Microclimatic variation in a small valley in eastern Tennessee. * Ecology **31**, 532—539, 1950.
[872] *Smirnov, V. A.*, siehe MAB **6**, 355, 1955.
[873] *Troll, C.*, Unterirdische Jahreszeitenwinde in finnischen Äsern. * Erdk. **13**, 150—152, 1959.
[874] *Wagner, R.*, Die geographische Anordnung der Mikroklimate auf dem Hosszubérc Berg im Bükkgebirge. * Acta Geograph. **1**, 27—43, Szeged 1955.
[875] *Wollny, E.*, Untersuchungen über den Einfluß der Exposition auf die Erwärmung des Bodens. * Forsch. a. d. Geb. d. Agrik. Physik **1**, 263—294, 1878.

SECTION 45

[876] *Baumgartner, A.*, Die Regenmengen als Standortsfaktor am Großen Falkenstein. * Forstw. C. **77**, 230—237, 1958.
[877] —, Nebel u. Nebelniederschlag als Standortsfaktoren etc. * Ebenda **77**, 257—272, 1958.
[878] —, Die Lufttemperatur als Standortsfaktor etc. * Ebenda im Druck.

[879] —, *Kleinlein, G.*, und *Waldmann, G.*, Forstlich-phänologische Beobachtungen u. Experimente am Großen Falkenstein. * Ebenda **75**, 290—303, 1956.

[880] *Bylund, E.*, und *Sundborg, A.*, Lokalklimatische Einflüsse auf die Platzwahl der Siedlungen etc. * Ymer Stockholm **1**, 1—30, 1952.

[881] *Frenzel, B.*, und *Fischer, H.*, Beobachtungen zur Phänologie eines Alpentales. * Arch. f. Met. (B) **8**, 231—256, 1957.

[882] *Geiger, R.*, The modification of microclimate by vegetation in open and in hilly country. * Proc. Canberra Symposium 1956, 255—258, UNESCO Paris 1958.

[883] —, *Woelfle, M.*, und *Seip, L. P.*, Höhenlage u. Spätfrostgefährdung. * * Forstw. C. **55**, 579—592, 737—746, 1933; **56**, 141—151, 221—230, 253—260, 357—364, 465—484, 1934.

[884] *Grunow, J.*, Die Abschirmung des Sonnenscheins durch Talnebel im Alpenvorland. * Wetter u. Leben **9**, 99—104, 1957.

[885] *Hayes, G. L.*, Influence of altitude and aspect on daily variations in factors of forest-fire danger. * US Dep. of Agric., Circular 591, Washington 1941.

[886] *Higgins, J. J.*, Instructions for making phenological observations of garden peas. * Publ. Climat. **5**, Nr. 2, 1952.

[887] *Koschmieder, H.*, Die meteorologischen Probleme der radioaktiven Luftverseuchung. * Beitr. Phys. d. fr. Atm. **30**, 119—122, 1958.

[888] *Mano, H.*, A study of the sudden nocturnal temperature rises in the valley and on the basin. * Geophys. Mag. Tokyo **27**, 169—204, 1956.

[889] *Morawetz, S.*, Kleinklimatische Beobachtungen in der Weststeiermark bei St. Stefan ob Stainz. * Angew. Met. **1**, 146—150, 1952.

[890] *Schnelle, F.*, Studien zur Phänologie Mitteleuropas. * Ber. DWD-US Zone **1**, Nr. 2, 1948.

[891] *Waldmann, G.*, Schnee- und Bodenfrost als Standortsfaktoren am Großen Falkenstein. * Forstw. C. **78**, 98—108, 1959.

[892] *Young, F. D.*, Nocturnal temperature inversion in Oregon and California. * MWRev. **49**, 138—148, 1921.

[893] *Zenker, H.*, Lokalklimatische Besonderheiten an Tuberkulose-Heilstätten im Raume Bad Berka etc. * Angew. Met. **3**, 173—187, 1958.

SECTION 46

[894] *Aulitzky, H.*, Forstmeteorologische Untersuchungen an der Wald- und Baumgrenze in den Zentralalpen. * Arch. f. Met. (B) **4**, 294—310, 1952.

[895] —, Waldkrone, Kleinklima u. Aufforstung. * Centralbl. f. d. ges. Forstw. **73**, 7—12, 1954.

[896] —, Über mikroklimatische Untersuchungen an der oberen Waldgrenze zum Zwecke der Lawinenvorbeugung. * Wetter u. Leben **6**, 93—99, 1954.

[897] —, Die Bedeutung meteorologischer u. kleinklimatischer Unterlagen für Aufforstungen im Hochgebirge. * Ebenda **7**, 241—252, 1955.

[898] —, Über die lokalen Windverhältnisse einer zentralalpinen Hochgebirgs-Hangstation. * Arch. f. Met. (B) **6**, 353—373, 1955.

[899] —, Waldbaulich-ökologische Fragen an der Waldgrenze. * Centralbl. f. d. ges. Forstw. **75**, 18—33, 1958.

[900] —, Hinweise für eine naturnahe Waldwirtschaft im Bereich der Waldgrenze. * Allg. Forstzeitung Wien **69**, 4—6, 1958.

[901] *Desing, H.*, Klimatische Untersuchungen auf einer großen Blaike. * Wetter u. Leben **5**, 46—52, 1953.
[902] *Dirmhirn, I.*, Untersuchungen der Himmelsstrahlung in den Ostalpen mit besonderer Berücksichtigung ihrer Höhenabhängigkeit. * Arch. f. Met. (B) **2**, 301—346, 1951.
[903] —, Oberflächentemperaturen der Gesteine im Hochgebirge. * Ebenda **4**, 43—50, 1952.
[904] *Ekhart, E.*, Zur Kenntnis der Schneeverhältnisse der Ostalpen. * Gerl. B. **56**, 321—358, 1940.
[905] *Friedel, H.*, Gesetze der Niederschlagsverteilung im Hochgebirge. * Wetter u. Leben **4**, 73—86, 1952.
[906] *Geiger, R.*, Probleme der Mikrometeorologie des Hochgebirges. * Wetter u. Leben **5**, 21—28, 1953.
[907] *Hader, F.*, Nordostalpine Seehöhenmittel der Niederschlagsmenge. * Arch. f. Met. (B) **5**, 331—343, 1954.
[908] *Jaag, O.*, Untersuchungen über die Vegetation u. Biologie der Algen des nackten Gesteins in den Alpen, im Jura und im schweizerischen Mittelland. * Beitr. z. Kryptogamenflora d. Schweiz **9**, Nr. 3, 1—560, Bern 1945.
[909] *Larcher, W.*, Frosttrocknis an der Waldgrenze und in der alpinen Zwergstrauchheide. * Veröff. Ferdinandeum Innsbruck **37**, 49—81, 1957.
[910] *Lauscher, F.*, Neue klimatische Normalwerte für Österreich. * Beih. z. Jahrb. d. Zentralanst. f. Met. u. Geodyn. 1932, 1—13, Wien 1938.
[911] —, Langjährige Durchschnittswerte für Frost u. Frostwechsel in Österreich. * Ebenda 1946, D 18—30, Wien 1947.
[912] —, Normalwerte der relativen Feuchtigkeit in Österreich. * Wetter und Leben **1**, 289—297, 1949.
[913] *Lauscher, F.*, Durchschnittliche Häufigkeiten der Erdbodenzustände in verschiedenen Höhenlagen der Ostalpenländer. * Ebenda **6**, 47—50, 1954.
[914] *Maurer, J.*, Bodentemperatur u. Sonnenstrahlung in den Schweizer Alpen. * Met.-Z. **33**, 193—199, 1916.
[915] *Sauberer, F.*, Zur Abschätzung der Globalstrahlung in verschiedenen Höhenstufen der Ostalpen. * Wetter u. Leben **7**, 22—29, 1955.
[916] —, und *Dirmhirn, I.*, Das Strahlungsklima. * Klimatographie von Österreich 13—102, Wien 1958.
[917] *Schreckenthal-Schimitschek, G.*, Klima, Boden u. Holzarten an der Wald- u. Baumgrenze in einzelnen Gebieten Tirols. * Univ. Verl. Wagner, Innsbruck 1934.
[918] *Schwind, M.*, Mikroklimatische Beobachtungen am Wutaischan. * Erdk. **6**, 44—45, 1952.
[919] *Steinhäußer, H.*, Normalhöhen zur Kennzeichnung der Schneedeckenverhältnisse. * Met. Rundsch. **3**, 32—34, 1950.
[920] *Steinhauser, F.*, Die Schneehöhen in den Ostalpen und die Bedeutung der winterlichen Temperaturinversion. * Arch. f. Met. (B) **1**, 63—74, 1949.
[921]. *Tranquillini, W.*, Standortsklima, Wasserbilanz u. CO_2-Gaswechsel junger Zirben an der alpinen Waldgrenze. * Planta **49**, 612—661, 1957.
[922] *Turner, H.*, Über das Licht- u. Strahlungsklima einer Hanglage der Ötztaler Alpen etc. * Arch. f. Met. (B) **8**, 273—325, 1958.
[923] —, Maximaltemperaturen oberflächennaher Bodenschichten an der alpinen Waldgrenze. * Wetter u. Leben **10**, 1—12, 1958.
[924] *Zöttl, H.*, Untersuchungen über das Mikroklima subalpiner Pflanzengesellschaften. * Ber. Geobot. Forsch. Inst. Rübel für 1952, 79—103, Zürich 1953.

SECTION 47

[925] *Aichele, H.*, Der Beginn der Apfelblüte 1953 am westlichen Bodensee als Hilfsmittel der kleinklimatischen Geländekartierung. * Met. Rundsch. **6**, 204—206, 1953.

[926] *Balchin, W. G. V.*, und *Pye, N.*, A microclimatological investigation of Bath and the surrounding district. * Quart. J. **73**, 297—323, 1947.

[927] — —, Local rainfall variations in Bath etc. * Ebenda **74**, 361—378, 1948.

[928] *Bliss, L. C.*, A comparison of plant development in microenvironments of Arctic and Alpine tundras. * Ecolog. Monogr. Durham N. C. **26**, 303—337, 1956.

[929] *Burgos, J. J.*, *Cagliolo, A.*, und *Santos, M. C.*, Exploracion microclimatica en la selva Tucumano-Oranense. * Meteoros **1**, 314—341, 1951.

[930] *Daubert, K.*, Klima u. Wetter in Tübingen. * Tübinger Blätter 1957, 10—24.

[931] *Eimern, J. van*, Zur Methodik der Geländeaufnahme. * Mitt. DWD **2**, Nr. 14, 125—131, 1955.

[932] —, Geländeklimaaufnahmen für landwirtschaftliche Zwecke. * Bayer. Landw. Jahrb. **35**, 193—210, 1958.

[933] *Flohn, H.*, Zur Frage der Einteilung der Klimazonen. * Erdk. **11**, 161—175, 1957.

[934] *Geiger, R.*, und *Schmidt, W.*, Einheitliche Bezeichnungen in kleinklimatischer und mikroklimatischer Forschung. * Biokl. B. **1**, 153—156, 1934.

[935] *Godske, C. L.*, Studies in local meteorology and representativeness I. * Univ. Bergen Arbok 1952, naturw. rekke **10**, 1—100.

[936] *Howe, G. M.*, Observations on local climatic conditions in the Aberystwyth area. * Met. Mag. **82**, 270—274, 1953.

[937] *Kauf, H.*, Die Einwirkung der Orographie des mittleren Saaletales auf die Niederschlagsverteilung II. * Mitt. Thüring. L. W. W. **10**, 35—62, 1950.

[938] *Knoch, K.*, Weltklimatologie u. Heimatklimakunde. * Met. Z. **59**, 245—249, 1942.

[939] —, Die Geländeklimatologie, ein wichtiger Zweig der angewandten Klimatologie. * Ber. z. Deutsch. Landeskde. **7**, 115—123, 1949.

[940] —, Über das Wesen einer Landesklimaaufnahme. * Z. f. Met. **5**, 173—177, 1951.

[941] —, Plan einer Landesklimaaufnahme. * Ber. DWD-US Zone **5**, Nr. 32, 106—108, 1952.

[942] *Koch, H. G.*, Bestandstemperaturen eines bewaldeten Seitentals bei Jena. * Mitt. Thüring. L.W.W. **7**, 69—98, 1948.

[943] *Kreutz, W.*, Lokalklimatische Studie im oberen Vogelsberg. * Ber. DWD-US Zone **7**, Nr. 42, 171—176, 1952.

[944] —, und *Schubach, K.*, Lokalklimatische Geländekartierung der südlichen Bergstraße etc. * Mitt. DWD-US Zone **13**, 1—11, 1952.

[945] — —, Das Klima der Gemarkung Espenschied u. Vorschläge zur Klimaverbesserung durch Windschutz. * Arch. f. Raumforsch. (Hessen) 1958, 27—71.

[946] *Lauscher, F.*, Flachland- u. Hügelklima im Gebiet von Wien. * Wetter u. Leben **3**, 99—102, 1951.

[947] *Parker, J.*, Environment and vegetation of Tomer's Butte in the forest-grassland transition zone of north Idaho. * The Amer. Midland Naturalist **51**, 539—552, 1954.

[948] *Parry, M.*, Local temperature variations in the Reading area. * Quart. J. **82**, 45—57, 532—534, 1956.

[949] *Penzar, I.*, Mikroklimatologische Untersuchungen des geophysikal. Inst. in der Umgebung von Krizevci im Jahre 1953. * Schrift. d. Geophys. Inst. Univ. Zagreb, 3. Ser. **7**, 1956.

[950] *Sauberer, F.*, Kleinklima-Untersuchungen im Breitenfurter Becken. * Wetter u. Leben **4**, 122—127, 1952.

[951] *Scaëtta, H.*, Terminologie climatique, bioclimatique et microclimatique. * La Mét. **11**, 342—347, 1935.

[952] *Schmidt, W.*, Bioklimatische Untersuchungen im Lunzer Gebiet. * Naturw. **17**, 176—179, 1929.

[953] —, Observations on local climatology in Austrian mountains. * Quart. J. **60**, 345—352, 1934.

[954] *Schöne, V.*, Geländeklimatische Untersuchungen im Forschungsraum Huy-Hakel der D. Akad. d. Landwirtsch.wissenschaften. * Angew. Met. **3**, 129—135, 1958.

[955] *Spinnangr, F.*, Temperature and precipitation in and around Bergen. * Bergens Mus. Arbok 1942, naturw. rekke **9**, 1—30, 1943.

[956] *Thams, J. C.*, Zur Bestimmung der Sonnenscheindauer in einem stark kupierten Gelände. * Arch. f. Met. (B) **6**, 417—430, 1955.

[957] *Thornthwaite, C. W.*, Topoclimatology. * The Johns Hopkins Univ., Laboratory of Climat. Seabrook (manuscr.) 1953.

[958] *Tinn, A. B.*, Local temperature variations in the Nottingham district. * Quart. J. **64**, 391—405, 1938.

[959] *Uhlig, S.*, Beispiel einer kleinklimatologischen Geländeuntersuchung. * Z. f. Met. **8**, 66—75, 1954.

[960] US Weather Bureau, Meteorological survey of the Oak Ridge area, final report 1948—1952. * US Atomic Energ. Comm. 1953.

[961] *Wagner, H.*, und *Dinger, H. J.*, Die Besonnung im oberen Vogtland und ihre Bedeutung für das Pflanzenwachstum. * Angew. Met. **2**, 122—125, 1955.

[962] — —, und *Hunger, W.*, Über Besonnungsverhältnisse im Raum des Elsterer Kessels in ihrer Bedeutung für forstliche Standortsfragen. * Ebenda **2**, 365—369, 1957.

[963] *Wagner, R.*, Mikroklimatérségek és térképezésük. * Különnyomat a földrajzi közlemények 1956, 201—216.

[964] *Weger, N.*, Die vorläufigen Ergebnisse der bei Geisenheim begonnenen kleinklimatischen Geländeaufnahme. * Met. Rundsch. **1**, 422—423, 1948.

[965] —, Zur Methodik der Kleinklimakartierung im Weinbau. * Mitt. DWD **2**, Nr. 14, 132—133, 1955.

[966] *Weischet, W.*, Die Geländeklimate der niederrheinischen Bucht und ihrer Rahmenlandschaften. * Münchener Geograph. Hefte **8**, 1—169, 1955.

[967] —, Die räumliche Differenzierung klimatologischer Betrachtungsweisen, ein Vorschlag zur Gliederung der Klimatologie und ihrer Nomenklatur. * Erdk. **10**, 109—122, 1956.

[968] *Wolfe, J. N.*, *Wareham, R. T.*, und *Scofield, H. T.*, Microclimates and macroclimate of Neotoma, a small valley in central Ohio. * Bull. Ohio Biolog. Survey **8**, Nr. 1, 1—267, 1949.

SECTION 48
(See also reference 1020)

[969] *Battle, W. R. B.*, und *Lewis, W. V.*, Temperature observations in Bergschrunds and their relationship to cirque erosion. * J. Geology **59**, 537—545, 1951.

[970] *Baumgartner, A.*, Meteorologische Beobachtungen am Hölloch. * Im Druck.

[971] *Buxton, P. A.*, Climate in caves and similar places in Palestine. * J. Animal Ecology **1**, 152—159, 1932.

[972] *Dudich, E.*, Biologie der Aggteleker Tropfsteinhöhle „Baradla" in Ungarn. * Speläolog. Monogr. **13**, Wien 1932.

[973] *Gressel, W.*, Über die Bewetterung der alpinen Höhlen. * Met. Rundsch. **11**, 54—57, 1958.

[974] *Hell, M.*, Die kalten Keller von Kaltenhausen in Salzburg. * Forsch. u. Fortschr. **10**, 336, 1934.

[975] *Heß, H.*, L. Handl's Temperaturmessungen des Eises und der Luft in den Stollen des Marmolata-Gletschers etc. * Z. f. Gletscherkde. **27**, 168—171, 1940.

[976] *Lautensach, H.*, Unterirdischer Kaltluftstau in Korea. * Peterm. Mitt. **85**, 353—355, 1939.

[977] *Mrose, H.*, Eine seltsame Höhlenvereisung. * Z. f. angew. Met. **56**, 350—353, 1939.

[978] *Oedl, R.*, Über Höhlenmeteorologie mit besonderer Rücksicht auf die große Eishöhle im Tennengebirge. * Met. Z. **40**, 33—37, 1923.

[979] *Saar, R.*, Meteorologisch-physikalische Beobachtungen in den Dachsteinrieseneishöhlen, Oberösterreich. * Wetter u. Leben **7**, 213—219, 1955.

SECTION 49

[980] *Bernfus, E.*, Probleme des Vorratsschutzes aus eigener Erfahrung. * Mitt. d. Vers. Stat. f. d. Gärungsgewerbe, Wien 1957. Nr. 11/12.

[981] *Bodenheimer, F. S.*, Studien zur Epidemiologie etc. der afrikanischen Wanderheuschrecke. * Z. f. angew. Entomol. **15**, 435—557, 1929.

[982] *Franz, H.*, Über die Bedeutung des Mikroklimas für die Faunenzusammensetzung auf kleinem Raum. * Z. f. Morph. u. Ökol. d. Tiere **22**, 587—628, 1931.

[983] *Grimm, H.*, Kleintierwelt, Kleinklima u. Mikroklima. * Z. f. angew. Met. **54**, 25—31, 1937.

[984] *Johnson, C. G.*, The vertical distribution of aphids in the air and the temperature lapse rate. * Quart. J. **83**, 194—201, 1957 u. **85**, 173—174, 1959.

[985] *Kaiser, E.*, Ideen zu einer Biogeographie der Sahara. * Peterm. Mitt. **98**, 86—100, 1954.

[986] *Kühnelt, W.*, Die Bedeutung des Klimas für die Tierwelt. * Biokl. B. **1**, 120—125, 1934.

[987] *Lauscher, F.*, Mückentanz u. Windschutz. * Biokl. B. **6**, 186, 1939.

[988] *Martini, E.*, und *Teubner, E.*, Über das Verhalten von Stechmücken bei verschiedener Temperatur u. Luftfeuchtigkeit. * Beih. z. Arch. f. Schiffs- u. Tropenhyg. **37**, 1933.

[989] *Mosauer, W.*, The toleration of solar heat in desert reptiles. * Ecology **17**, 56—66, 1936.

[990] *Müller, H. J.*, *Unger, K.*, *Neitzel, K.*, *Raeuber, A.*, *Moericke, V.*, und *Seemann, J.*, Der Blattlausbefallsflug in Abhängigkeit von Flugpopulation und witterungsbedingter Agilität in Kartoffelabbau- u. Hochzuchtlagen. * Biolog. Zentralbl. **78**, 341—383, 1959.

[991] *Neumann, G.*, Bemerkungen zur Zellularkonvektion im Meer und in der Atmosphäre und die Beurteilung des statischen Gleichgewichts. * Ann. d. Met. **1**, 235—244, 1948.

[992] *Nielsen, E. T.*, Zur Ökologie der Laubheuschrecken. * Saertryk af Ent. Medd. **20**, 121—164, 1938.
[993] *Pruitt, W. O.*, An analysis of some physical factors affecting the local distribution of the shorttail shrew etc. * Mus. of Zoology, Univ. of Michigan, Misc. Publ. **79**, 1—39, 1953.
[993a] *Rainey, R. C.*, Some observations on flying locusts and atmospheric turbulence in eastern Africa. * Quart. J. **84**, 334—354, 1958.
[994] *Roos, T.*, Studies on upstream migration in adult stream-dwelling insects I. * Rep. Inst. Freshwater Res. Drottningholm **38**, 167—193, 1957.
[995] *Schimitschek, E.*, Standortsklima u. Kleinklima in ihrer Beziehung zum Entwicklungsablauf und zur Mortalität von Insekten. * Z. f. angew. Entomol. **18**, 460—491, 1931.
[996] *Schrödter, H.*, Agrarmeteorologische Beiträge zu phytopathologischen Fragen. * Abh. Met. D. DDR **2**, Nr. 15, 1—83, 1952.
[997] *Seilkopf, H.*, Biosynoptik. * Mediz.-met. Hefte (Ann. d. Met.) **8**, 74—83, 1953.
[998] *Suzuki, S.*, *Tanioka, K.*, *Uchimura, S.*, und *Marumoto, T.*, The hovering height of skylarks. * J. Agr. Met. Japan **7**, 149—151, 1952.
[999] *Unger, K.*, und *Müller, H. J.*, Über die Wirkung geländeklimatisch unterschiedlicher Standorte auf den Blattlausbefallsflug. * Der Züchter **24**, 337—345, 1954.
[1000] *Wiele, H.*, Für Hagenbeck im Himalaja. * D. Buchwerkst. Dresden 1925.

SECTION 50

[1001] *Buxton, P. A.*, Insects of Samoa. * British Mus. **9**, Fasc. 1, 1930.
[1002] *Cena, M.*, und *Courvoisier, P.*, Untersuchungen über die physikalischen Faktoren des Stallklimas unter besonderer Berücksichtigung der Abkühlungsgröße. * Schweiz. Arch. f. Tierheilkde. **91**, 303—336, 459—468, 1949.
[1003] *Dahmen, H.*, Untersuchungen im klimatisierten Stall. * Gesundh.-Ing. **67**, 61—64, 1944.
[1004] *Frith, H. J.*, Wie regelt der Thermometervogel die Temperatur seines Nesthügels? * Umschau **56**, 238—239, 1956.
[1005] *Gößwald, K.*, Rassenstudien an der großen Waldameise etc. * Z. f. angewandte Entomol. **28**, 62—124, 1941.
[1006] *Hesse, R.*, Tiergeographie auf ökologischer Grundlage. * G. Fischer, Jena 1924.
[1007] *Holdaway, F. G.*, und *Gay, F. J.*, Temperature studies of the habitat of Entermes exitiosus etc. * Austral. J. Sci. Res., B **1**, 464—493, 1948.
[1008] *Kühnelt, W.*, Der Einfluß des Klimas auf den Wasserhaushalt der Tiere. * Biokl. B. **3**, 11—15, 1936.
[1009] *Lehmann, P.*, Das Sonderklima des Stalls. * Forsch. d. Landwirtsch. **6**, 642—647, 1931.
[1010] *Löhrl, H.*, Der Winterschlaf von Nyctalus noctula Schreb. auf Grund von Beobachtungen am Winterschlafplatz. * Z. f. Morph. u. Ökol. d. Tiere **32**, 47—66, 1936.
[1011] *Mehner, A.*, und *Linz, A.*, Untersuchung über den Verlauf der Stalltemperatur. * Forschungsdienst **8**, 525—543, 1939.
[1012] *Palat, K.*, Massenreaktion der roten Waldameise auf Sonnenbestrahlung. * Wetter u. Leben **2**, 110—113, 1950.
[1013] *Raeuber, A.*, Meteorologische Vergleichsmessungen zwischen Schuppenstall u. Freiland in Groß-Lüsewitz. * Angew. Met. **2**, 217—222, 1956.

[1014] *Steiner, A.*, Neuere Ergebnisse über den sozialen Wärmehaushalt der einheimischen Hautflügler. * Naturw. **18**, 595—600, 1930.

[1015] *Wächtershäuser, H.*, Beitrag zum Stallklima. * Ber. DWD-US Zone **7**, Nr. 42, 382—384, 1952.

[1015a] *Wellenstein, G.*, Beiträge zur Physiologie der roten Waldameise etc. * Z. f. angew. Entomol. **14**, 1—68, 1929.

[1016] *Zorn, W.*, und *Freidt, G.*, Der Einfluß von Wetter u. Klima auf unsere landwirtschaftlichen Nutztiere. * Z. f. Züchtkde. **14**, Nr. 1, 1939.

[1017] Zitat verloren und z. Z. nicht zu ermitteln.

SECTION 51

[1018] *Aronin, J. E.*, Climate and architecture. * Reinhold Publ. Corp. New York, 304 S., 1953.

[1019] *Brooks, C. E. P.*, und *Evans, G. T.*, Annotated bibliography on industrial meteorology. * MAB **5**, 1187—1230, 1954.

[1020] —, und *Evans, G. J.*, Annotated bibliography on the climate of enclosed spaces (cryptoclimates). * MAB **7**, 211—264, 1956.

[1021] *Bullig, H. J.*, und *Höller, E.*, Laderaum-Meteorologie. * DWD, Seewetteramt, Einzelveröff. **9**, 1—43, 1956.

[1022] *Burckhardt, H.*, Künstliche Beregnung im Weinbau. * Weinberg u. Keller **2**, 154—158, 1955.

[1023] *Cagliolo, A.*, Marcado gradiente térmico en vagones de ferrocarril. * Meteoros **4**, 395—398, 1954.

[1024] *Crook, E. M.*, und *Watson, D. J.*, Studies on the storage of potatoes. * J. Agr. Sci. Cambridge **40**, 199—232, 1950.

[1025] *Cube, H. L. v.*, Klima im Auto. * Umschau **57**, 571—572, 615, 1957.

[1026] *Dessauer, F.*, *Graffunder, W.* und *Laub, J.*, Beobachtungen über Ionenschwankungen im Freien und in geschlossenen Räumen. * Ann. d. Met. **7**, 173—185, 1956.

[1027] *Diem, M.*, Das Mikroklima in einem künstlich beregneten Tabakbestand. * Arch. f. Met. (B) **5**, 215—233, 1953.

[1028] —, und *Trappenberg, R.*, Die Beeinflussung des Mikroklimas und der Bodenstruktur durch künstliche Beregnung. * Ebenda **8**, 382—406, 1958.

[1029] *Dirmhirn, I.*, Über das Klima in den Wiener Straßenbahnwagen. * Wetter u. Leben **4**, 158—162, 1952.

[1030] *Georgii, H. W.*, Untersuchung über den Luftaustausch zwischen Wohnräumen u. Außenluft. * Arch f. Met. (B) **5**, 191—214, 1953.

[1031] *Giner, R.*, und *Heß, V. F.*, Studie über die Verteilung der Aerosole in der Luft von Innsbruck u. Umgebung. * Gerl. B. **50**, 22—43, 1937.

[1032] *Hummel, F.*, Mietenklima u. Windeinfluß. * Ber. DWD-US Zone **5**, Nr. 32, 44—47, 1952.

[1033] *Jensen, M.*, The model-law for phenomena in natural wind. * Ingeniøren (Dänemark) **2**, 121—128, 1958.

[1034] *Kern, H.*, Mietentemperaturmessungen auf Niedermoorboden. * Ber. DWD-US Zone **6**, Nr. 38, 186—189, 1952.

[1035] *King, E.*, Medizin-meteorologische Einflüsse auf den Straßenverkehr. * Z. f. Verkehrssicherheit **4**, 116—136, 1958 (siehe auch Wetter u. Leben **8**, 213—219, 1956).

[1036] *Kreutz, W.*, Beitrag zum Mietenklima. * Met. Rundsch. **1**, 348—351, 1948.

[1037] —, Merkblatt zur Mietenbehandlung. * Mitt. DWD-US Zone 7, 1950.

[1038] —, Der Mietenklimadienst. * Mitt. DWD **2**, Nr. 14, 168—170, 1955.
[1039] *Lacy, R. E.*, Distribution of rainfall round a house. * Met. Mag. **80**, 184—189, 1951.
[1040] *Landsberg, H.*, Bioclimatology of housing. * Met. Monographs **2**, 81—98, 1954.
[1041] *Leistner, W.*, Beziehungen zwischen Raumklima u. Außenklima. * Ann. d. Met. **2**, 53—56, 1949.
[1042] *Niemann, A.*, Über die Strahlungsbeeinflussung durch verschmutzte Gewächshausscheiben. * Ann. d. Met. **8**, 344—352, 1959.
[1043] *Porter, W. L.*, Occurence of high temperatures in standing boxcars. * Siehe MAB **4**, 263, 1954.
[1044] *Priestley, C. H. B.*, The heat balance of sheep standing in the sun. * Austral. J. Agric. Res. **8**, 271—280, 1957.
[1045] *Reichelt, G.*, Zum Einfluß der Bewässerung auf das Mikroklima von Mittelgebirgswiesen. * Arch. f. Met. (B) **6**, 374—399, 1955.
[1046] *Seemann, J.*, Klima u. Klimasteuerung im Gewächshaus. * Bayer. Landwirtsch. Verl., München 1957.
[1047] *Stranz, D.*, Über die Wärmeverteilung u. Luftzirkulation in einem „temperaturkonstanten" Raum. * Gerl. B. **59**, 214—223, 1942.
[1048] *Thornthwaite, C. W.*, Climate and scientific irrigation in New Jersey. * Publ. Climat. **6**, Nr. 1, 1953.
[1049] —, Modification of rural microclimates. * W. L. Thomas, Man's role in changing the face of the earth, 567—583, Chicago 1956.
[1050] *Trappenberg, R.*, Untersuchungen über die mikroklimatischen Wirkungen künstlicher Beregnung im Tabakbestand. * Arch. f. Met. (B) **4**, 65—84, 1952.
[1051] *Uhlig, S.*, Siehe Agrarmet. Literaturbericht des DWD **1**, Nr. 5, 16—28, 1955.
[1052] *Völckers, O.*, *Tonne F.* und *Becker-Freyseng, A.*, Licht und Sonne im Wohnungsbau. * Ber. d. Deutsch. Forsch. Gem. Bauen u. Wohnen, Stuttgart 1955 (51 S.).
[1053] *Witte, K.*, Klimatologische, pflanzenphysiologische u. technische Probleme der Beregnung. * Verl. Wasser u. Boden, Hamburg 1954.

SECTION 52
(Includes only items not listed in the bibliography in A. Kratzer [1060])

[1054] *Bangerl, A.*, und *Steinhauser, F.*, Die Verteilung des SO_2-Gehaltes der Luft im Stadtgebiet von Wien. * Arch. f. Met. (B) **10**, 132—153, 1959.
[1055] *Berg, H.*, Nächtlicher Temperaturanstieg am Rande einer Großstadt. * Wetter u. Leben **11**, 9—11, 1959.
[1056] *Brooks, C. E. P.*, Selective annotated bibliography on urban climates. * MAB **3**, 734—773, 1952.
[1057] *Effenberger, E.*, und *Linder, A.*, Statistische Untersuchungen über den Wochengang (bzw. Tagesgang) des Staubgehaltes in einer Großstadt. * Ann. d. Met. **8**, 149—162 (bzw. 235—255), 1958.
[1058] *Franken, E.*, Der Beginn der Forsythienblüte in Hamburg. * Met. Rundschau **8**, 113—114, 1955.
[1059] *Hrudička, B.*, Zu den optischen u. akustischen Eigenschaften des Klimas einer Großstadt. * Gerl. B. **53**, 337—344, 1938.
[1060] *Kratzer, A.*, Das Stadtklima, 2. Aufl. * Verl. Vieweg, Braunschweig 1956.
[1061] *Landsberg, H. E.*, The climate of towns. * W. L. Thomas, Man's role in changing the face of the earth, 584—606, Chicago 1956.

[1062] *Lauscher, F., Roller, M., Wacha, G., Grammer, M., Weiß, E.*, und *Frenzel, J. W.*, Witterung u. Klima von Linz. * Wetter u. Leben, Sonderh. VI, Wien 1959.

[1063] *Löbner, A.*, Horizontale u. vertikale Staubverteilung in einer Großstadt. * Veröff. Geoph. I. Leipzig **7**, Nr. 2, 1935.

[1064] *Sauberer, F.*, Der Baum in der Großstadt. * Wetter u. Leben **7**, 77—78, 1955.

[1065] *Siebert, A.*, Landschaft u. Großstadt. * Umschaudienst d. Akad. f. Raumforschung u. Landesplanung **8**, Nr. 1/2 u. 3/4, 1958.

[1066] *Steinhauser, F.*, Messungen der Luftverschmutzung in Wien. * Arch. f. Met. (B) **10**, 200—209, 1960.

[1067] —, *Eckel, O.*, und *Sauberer, F.*, Klima u. Bioklima von Wien. * Wetter u. Leben, Sonderh. III, 1955; V, 1957 u. VII, 1959.

[1068] *Stranz, D.*, Auslösung von Schneefall über einer Großstadt. * Z. f. Met. **3**, 359—360, 1949.

[1069] *Sundborg, A.*, Climatological studies in Uppsala. * Geographica **22**, Uppsala 1951.

[1070] *Weinländer, A.*, Großstadt u. Klima. * Z. f. angew. Met. **59**, 390—393, 1942.

[1071] *Weiß, E.*, und *Frenzel, J. W.*, Untersuchungen von Luftverunreinigungen durch Rauch- u. Industriegase im Raume von Linz. * Wetter u. Leben **8**, 131—147, 1956.

SECTION 53

[1072] *Blenk, H.*, und *Trienes, H.*, Strömungstechnische Beiträge zum Windschutz. * Grundlagen d. Landtechn. **8**, I u. II, VDI-Verl., Düsseldorf 1956.

[1073] *Bringmann, K.*, und *Kaiser, H.*, Maisstreifen als Windschutz. * Z. f. Acker- u. Pflanzenbau **99**, 321—334, 1955.

[1074] *Caborn, J. M.*, Shelterbelts and microclimate. * Forestry Comm. Bull. 29, Edinburgh 1957 (135 S.).

[1075] *Casperson, G.*, Untersuchungen über den Einfluß von Windschutzanlagen auf den standörtlichen Wärmehaushalt. * Angew. Met. **2**, 339—351, 1957.

[1076] *Daigo, Y.*, und *Maruyama, E.*, Experimental studies of windbreaks for protection from frost damage. * Mem. Industr. Met. **20**, 1—7, 1956.

[1077] *Dirmhirn, I.*, Zur Strahlungsminderung an Windschutzstreifen. * Wetter u. Leben **5**, 208—213, 1953.

[1078] *Eimern, J. van*, Beeinflussung meteorologischer Größen durch ein engmaschiges Heckensystem. * Ann. d. Met. **6**, 213—219, 1953/54.

[1079] —, Über die Veränderlichkeit der Windschutzwirkung einer Doppelbaumreihe bei verschiedenen meteorologischen Bedingungen. * Ber. DWD **5**, Nr. 32, 1—21, 1957.

[1080] —, *Franken, E.*, und *Harries, H.*, Ergebnisse von Windschutzuntersuchungen in Hamburg-Garstedt 1952. * Landwirtsch. Verl., Hiltrup 1954.

[1081] *Eskuche, U.*, Über Windschutzuntersuchungen an der Donau bei Herbertingen. * Veröff. d. Landesst. f. Naturschutz u. Landschaftspflege Baden-Württemberg 25, 1957.

[1082] *Franken, E.*, und *Kaps, E.*, Windschutzuntersuchung Emsland 1955. * Ber. DWD **5**, Nr. 33, 1—37, 1957.

[1083] *Geiger, R.*, Der künstliche Windschutz als meteorologisches Problem. * Erdk. **5**, 106—114, 1951.

[1084] *Hanke, E.*, und *Kaiser, H.*, Untersuchungen über den Einfluß eines künstlichen Windschutzstreifens auf den Ertrag von Zuckerrüben. * Z. f. Acker- u. Pflanzenbau **102**, 81—100, 1956.

[1085] — —, Unters. über d. Auswirk. v. Windschutzstreifen auf d. Ertrag etc. * Ebenda **103**, 90—110, 1957.

[1086] *Hartke, W.*, Die Heckenlandschaft. * Erdk. **5**, 132—152, 1951.

[1087] *Hilf, H. H.*, Wirksamer Windschutz. * Die Holzzucht **13**, 33—43, 1959.

[1088] *Illner, K.*, Über den Einfluß von Windschutzpflanzungen auf die Unkrautverbreitung. * Angew. Met. **2**, 370—373, 1957.

[1089] *Jensen, M.*, Shelter effect (264 S.). * The Danish Techn. Press, Kopenhagen 1954.

[1090] *Kaiser, H.*, Beiträge zum Problem der Luftströmung in Windschutzsystemen. * Met. Rundsch. **12**, 80—87, 1959.

[1091] —, Die Strömung an Windschutzstreifen. * Ber. DWD **7**, Nr. 53, 1959.

[1092] —, Schneeverwehungen an Windschutzanlagen, eine Gefahr für Felder u. Wege? * Umschau **60**, 33—36, 1960.

[1093] *Kreutz, W.*, Der Windschutz (167 S.). * Ardey Verl., Dortmund 1952.

[1094] —, und *Walter, W.*, Der Strömungsverlauf sowie die Erosionsvorgänge u. Schneeablagerungen an künstlichen Windschirmen nach Unters. im Windkanal. * Ber. DWD **4**, Nr. 24, 1956.

[1095] — —, Wind- u. Strömungsverhältnisse im Bereich von Windschutzanlagen etc. * Z. f. Acker- u. Pflanzenbau **101**, 279—290, 1956.

[1096] *Linde, R. J. van der*, und *Woudenberg, J. P. M.*, On the microclimatic properties of sheltered areas (151 S.) * K. Nederl. Met. Inst., Nr. 102, 1950.

[1097] *Müller, T.*, Versuche über die Windschutzwirkung von Hecken auf der Schwäbischen Alb. * Umschaudienst **6**, Nr. 1/2, 1956.

[1098] *Nägeli, W.*, Über die Bedeutung von Windschutzstreifen zum Schutze landwirtschaftlicher Kulturen. * Schweiz. Z. f. Forstw. **11**, 265—280, 1941.

[1099] —, Untersuchungen über die Windverhältnisse im Bereich von Windschutzstreifen. * Mitt. d. Schweiz. Anst. f. d. forstl. Versuchswesen **23**, 221—276, 1943 und **24**, 657—737, 1946.

[1100] —, Untersuchungen über die Windverhältnisse im Bereich von Schilfrohrwänden. * Ebenda **29**, 213—266, 1953.

[1101] —, Die Windbremsung durch einen größeren Waldkomplex. * Ber. 11. Kongr. Intern. Verb. Forstl. Forsch. Anst. Rom 1953, 240—246, Florenz 1954.

[1102] *Olbrich, A.*, Windschutzpflanzungen. * M. & H. Schaper, Hannover 1949.

[1103] *Rüsch, J. D.*, Der CO_2-Gehalt bodennaher Luftschichten unter Einfluß des Windschutzes. * Z. f. Pflanzenernähr., Düngung, Bodenkde. **71**, 113—132, 1955.

[1104] *Sato, K.*, *Tamachi, M.*, *Terada, K.*, *Watanabe, Y.*, *Katoh, T.*, *Takata, Y.*, *Sakanoue, T.*, und *Iwasaki, M.*, Studies on windbreaks (201 S.). * Nippon-Gakujutsu-Shiukokai, Tokyo 1952.

[1105] *Sato, S.*, Calculations of the received solar radiation in the shade of windbreak at Miyazaki City. * J. Agr. Met. Japan **11**, 12—14, 1955.

[1106] *Schrödter, H.*, Untersuchungen über die Wirkung einer Windschutzpflanzung auf den Sporenflug und das Auftreten der Alternaria-Schwärze an Kohlsamenträgern. * Angew. Met. **1**, 154—158, 1952.

[1107] *Scultetus, H. R.*, Windschutz immer noch ein Problem etc. * Met. Rundschau **11**, 23—28, 1958.

[1108] —, Bewindung eines Geländes und vertikaler Temperaturgradient. * Ebenda **12**, 1—10, 1959.

[1109] —, Geländeausformung u. Bewindung in Abhängigkeit von der Austauschgröße. * Ebenda **12**, 73—80, 1959.
[1110] *Steubing, L.*, Der Tau u. seine Beeinflussung durch Windschutzanlagen. * * Biolog. Zentralbl. **71**, 282—313, 1952.
[1111] *Tanaka, S.*, *Tanizawa, T.*, *Sano, H.*, *Kakinuwa, S.*, und *Kodera, S.*, Studies on the wind in front and back of the shelter-hedges (1—9). * J. Agr. Met. Japan **8**, 61—63, 1953; **9**, 66—68, 1954; **10**, 30—32, 1954; **11**, 49—52, 91—94, 97—99, 1955; **12**, 9—12, 73—78, 1956; **13**, 7—8, 1957.
[1112] *Tani, N.*, On the wind tunnel test of the model shelter-hedge. * Bull. Nat. Inst. Agric. Sci. Japan (A) **6**, 1—80, 1958.
[1113] *Troll, C.*, Heckenlandschaften im maritimen Grünlandgürtel und im Gäuland Mitteleuropas. * Erdk. **5**, 152—157, 1951.
[1114] *Uhlig, S.*, Windschutzanlagen auf der Hohen Rhön. * Mitt. DWD **1**, Nr. 9, 1954.
[1115] *Woelfle, M.*, Windschutzanlagen. * Forstw. C. **60**, 52—63, 73—86, 1938.

SECTION 54
(For earlier references on propeller methods see ref. 1122, and on other methods see ref. 1126)

[1116] *Baier, W.*, Frostbekämpfung im Weinbau. * Ber. DWD **2**, Nr. 15, 1955.
[1117] *Burckhardt, H.*, Frostschutz durch Abriegelung der Kaltluftzufuhr. * Umschau **54**, 276—278, 1954.
[1118] —, Vorbeugender Frostschutz u. Frostschutz durch Beregnung. * Mitt. DWD Nr. 16, 26—53, 1956.
[1119] *Dinkelacker, O.*, und *Aichele, H.*, Versuche u. Erfahrungen über die Spätfrostbekämpfung in Württemberg. * Techn. Mitt. d. Instr.wesens d. DWD, N. F. **4**, 9—20, 1958.
[1120] — —, Frostbekämpfung im Wein- u. Obstbau. * Umschau **59**, 241—244, 1959.
[1121] *Drimmel, J.*, Der Frostschutz durch Lufttrübung — rechnerisch betrachtet. * Arch. f. Met. (B) **6**, 443—451, 1955.
[1122] *Eimern, J. van*, Frostschutz mittels Propeller. * Mitt. DWD **2**, Nr. 12, 1955.
[1123] —, und *Loewel, E. L.*, Haben die Wassergräben in der Marsch des Alten Landes eine Bedeutung für den Frostschutz? * Mitt. d. Obstbauversuchsrings des Alten Landes **8**, Nr. 10, 1953.
[1124] *Hanyu, J.*, und *Tsugawa, K.*, On the effect of frost protection of paddy rice by irrigation. * J. Agr. Met. Japan **10**, 125—127, 1955.
[1125] *Hilkenbäumer, F.*, *Schnelle, F.*, und *Breuer, W.*, Bestands- und Ertragssicherung im Obstbau durch Frostschadenverhütung. * Verl. Neumann, Radebeul 1951.
[1126] *Keßler, O. W.*, und *Kaempfert, W.*, Die Frostschadenverhütung. * R. f. W. Wiss. Abh. **6**, Nr. 2, 1940.
[1127] *Morgen, A.*, Zur Frage des Wasserbedarfs bei der Frostschutzberegnung. * Met. Rundsch. **10**, 133—135, 1957.
[1128] *Niemann, A.*, Neue Untersuchungen zur Physik der Frostberegnung und deren Bedeutung für die Bemessung und Bedienung der Anlagen. * Akt. d. Intern. Tag. f. Frostberegnung, Bozen 1957.
[1129] *Pogrell, H. v.*, Grundlegende Fragen der direkten Frostschutzberegnung. * Verl. L. Leopold, Bonn 1958.
[1130] *Schmidt, W.*, Meteorologische Feldversuche über Frostabwehrmittel. * Anh. z. Jahrb. d. Zentralanst. f. Met. u. Geodyn. 1927, Wien 1929.
[1131] *Small, R. T.*, The use of wind machines and helicopter flights for frost protection. * Bull. Am. Met. Soc. **30**, 79—85, 1949.

[1132] *Tsuboi, Y., Honda, I., Hatagoshi, K.*, und *Yamato, M.*, On experiments of protection against frost damage by oil-burning in vineyard. * J. Agr. Met. Japan **10**, 109—112, 1955.

[1133] *Vaupel, A.*, Advektivfrost u. Strahlungsfrost. * Mitt. DWD **3**, Nr. 17, 1959.

SECTION 55
(See also references 13, 111, 172, 228, 232, 332, 437, 484, 485, 490, and 658)

[1134] *Bongards, H.*, Feuchtigkeitsmessung. * Verl. Oldenbourg, München 1926.

[1135] *Burckhardt, H.*, Probleme u. Möglichkeiten der Kartierung der Frostgefährdung. * Met. Rundsch. **9**, 92—98, 1956.

[1136] *Grundmann, W.*, Meteorologische Meßgeräte am Erdboden. * Linkes Met. Taschenbuch, N. Ausg. III, 272—347, Akad. Verl. Ges., Leipzig 1957.

[1137] *Kleinschmidt, E.*, Handbuch der meteorologischen Instrumente und ihrer Auswertung. * Springer, Berlin 1935.

[1138] *Lauscher, F.*, Grundlagen des Strahlungsklimas der Lunzer Kleinklimastationen. * Beih. z. Jahrb. d. Zentralanst. f. Met. u. Geodyn. Wien 1931, Wien 1937.

[1139] *Mäde, A.*, Temperaturbeobachtungen an vereinfachten Klimastationen. * Angew. Met. **1**, 53—56, 1951.

[1140] *Middleton, W. E. K.*, und *Spilhaus, A. F.*, Meteorological instruments, 3. ed. * Univ. of Toronto Press 1953.

[1141] *Spencer-Gregory, H.*, und *Rourke, E.*, Hygrometry. * Crosby Lockwood and Son Ltd., London 1957.

[1142] *Stille, U.*, Messen u. Rechnen in der Physik. * Vieweg, Braunschweig 1955.

[1143] Internationale Wetterschlüssel, Deutsche Ausgabe, Bad Kissingen 1948 (mit Ergänzungen).

[1144] Anleitung für die Beobachter an den Niederschlagsmeßstellen des Deutschen Wetterdienstes, Bad Kissingen, 1950.

[1145] Anleitung für die Beobachter an den Wetterbeobachtungsstellen des Deutschen Wetterdienstes. Ausg. f. d. Klimadienst, 6. Aufl., Bad Kissingen 1951.

[1146] Die Agrarmeteorologen-Tagung in Frankfurt/M. vom 14. bis 17. 3. 1955. * Mitt. DWD **2**, Nr. 14, 1955.

[1147] Handbook of meteorological instruments. Part I: Instruments for surface observations. * Air Ministry, Met. Office, London 1956.

SECTION 56

[1148] *Albrecht, F.*, Einige neue Meßgeräte für Ausstrahlung u. Globalstrahlung. * Ann. d. Met. **5**, 97—121, 1952.

[1149] *Bartels, J.*, Temperaturmessung in Bodennähe u. Aspiration. * Met. Z. **47**, 76—77, 1930.

[1150] *Bauer, W.*, und *Buschner, R.*, Beitrag zur Messung der Lufttemperatur mit verschiedenen Formen des Strahlungsschutzes. * Ber. DWD **3**, Nr. 19, 1955.

[1151] *Becker-Freyseng, A.*, Besonnung von Grundstücken. * Bauzeitung **56**, 211—213, 1951.

[1152] *Deacon, E. L.*, und *Samuel, D. R.*, A linear temperature compensated hot wire anemometer. * J. Scient. Instr. **34**, 24—26, 1957.

[1153] *Diem, M.*, Feuchtemessung mit Hilfe thermoelektrischer Psychrometer. * Arch. f. Met. (B) **5**, 59—65, 1954.

[1154] *Drummond, A. J.*, On the measurement of sky radiation. * Arch. f. Met. (B) **7**, 413—436, 1956.

[1155] *Duvdevani, S.*, An optical method of dew estimation. * Quart. J. **73**, 282—296, 1947.

[1156] *Fimpel, H.*, Das elektrische Psychrometer für den Forschungswagen. * Wiss. Mitt. Met. Inst. Univ. München **3**, 10—12, 1956.

[1157] *Fischmeister, V.*, Die Bestimmung des Wasserwertes einer Schneedecke mit radioaktiven Stoffen. * Österr. Wasserwirtsch. **8**, 86—93, 1956.

[1158] *Forster, H.*, Die zweite Lötstelle und ihre Temperatur bei thermoelektrischen Temperaturmessungen mit Kupfer-Konstantan-Elementen. * Met. Z. **59**, 298—301, 1942.

[1159] *Frankenberger, E.*, Untersuchungen über den Vertikalaustausch in den unteren Dekametern der Atmosphäre. * Ann. d. Met. **4**, 358—374, 1951.

[1160] —, Über vertikale Luftbewegungen in der untersten Atmosphäre. * Ebenda **5**, 368—372, 1952.

[1161] *Georgi, J.*, Meteorologischer Universal-Strahlungsmesser. * Met. Rundsch. **9**, 89—92, 1956.

[1162] *Halstead, M. H.*, Reference temperature compensator. * Publ. Climat. **4**, Nr. 2, 5, 1951.

[1163] *Hesse, W.*, Ergebnisse von Pflanzentranspirationsmessungen mit Kleinlysimetern in Zusammenhang mit met. Einflüssen. * Angew. Met. **2**, 65—82, 1954.

[1164] *Höhne, W.*, Über die Weiterentwicklung des thermoelektrischen Feinwindmessers. * Z. f. Met. **8**, 243—247, 1954.

[1165] —, Windrichtungsschreiber für schwache Luftströmungen. * Ebenda **9**, 135—143, 1955.

[1166] *Höhne, W.*, Theoretische Betrachtungen über die Verwendbarkeit von Halbleiter-Widerständen in der Mikrometeorologie u. -klimatologie. * Ebenda **11**, 143—156, 1957.

[1167] —, *Mäde, A.*, und *Schmidt, M.*, Kompensation des Einflusses der Vergleichstemperatur bei thermoelektrischen Messungen. * Z. f. Met. **10**, 131—136, 1956.

[1168] *Höschele, K.*, Untersuchungen zur Methode der elektrischen Bodenfeuchtemessung. * Diss. Landw. Hochsch. Hohenheim 1957.

[1169] *Hofmann, G.*, Die Temperaturmessung als Basis meteorologischer Meßverfahren. * Wiss. Mitt. Met. Inst. Univ. München **3**, 13—29, 1956.

[1170] *Holmes, J. W.*, Measuring soil water content and evaporation by the neutron scattering method. * Netherl. J. Agric. Sc. **4**, 30—34, 1956.

[1171] *Kausch, W.*, Saugkraft u. Wassernachleitung im Boden als physiologische Faktoren unter besonderer Berücksichtigung des Tensiometers. * Planta **45**, 217—263, 1955.

[1172] *Kraus, H.*, Untersuchungen u. Entwicklungsarbeiten mit Thermistoren. * Wiss. Mitt. Met. Inst. Univ. München **3**, 30—57, 1956.

[1173] *Leick, E.*, Grundsätzliches zur Taumessungsfrage. * Die Kulturpflanze **1**, 53—78, 1953.

[1174] *Liebster, G.*, und *Eimern, J. van*, Hilfsinstrumente zur Bestimmung der Spritztermine bei der Schorfbekämpfung. * Der Erwerbsobstbau **1**, 70—74, 1959.

[1175] *Lieneweg, F.*, Absolute u. relative Feuchtebestimmung mit dem Lithiumchloridfeuchtemesser. * Siemens-Zeitschr. **29**, 212—218, 1955.

[1176] —, und *Schaller, A.*, Ardonox, ein neues Ardometer. * Ebenda **28**, 67—73, 1954.

[1177] *Linke, F.*, Ein Meßapparat für Sonnen- und Himmelsstrahlung für bioklimatische Stationen. * Biokl. B. **1**, 171—172, 1934.

[1178] —, Eine transportable Thermometerhütte für lokalklimatologische und mikroklimatologische Untersuchungen. * Ebenda **5**, 110, 1938.

[1179] *Lorenz, D.*, Experimentelle Untersuchungen mit Strahlungsmeßgeräten zur Messung der Oberflächentemperaturen und der Gegenstrahlung aus kleinen Raumwinkeln. * Dipl. Arb. Univ. Mainz 1957.

[1180] *Mäde, A.*, Ein Schutzkasten für das Platinwiderstandsthermometer des Reichswetterdienstes. * Met. Z. **55**, 415—417, 1938.

[1181] —, Zur Methodik mikroklimatischer Temperaturmessungen. * Angew. Met. **1**, 215—219, 1952.

[1182] —, Zur Methodik der Bodenfeuchtigkeitsmessungen. * Ber. DWD-US Zone **6**, Nr. 32, 195—197, 1952.

[1183] —, Zur Methodik der Taumessung. * Wiss. Z. Univ. Halle **5**, 483—512, 1956.

[1184] *Monteith, J. L.*, und *Owen, P. C.*, A thermocouple method for measuring relative humidity in the range 95—100 %. * J. Sci. Instr. **35**, 443—446, 1958.

[1185] *Morgen, A.*, Der Trierer Geländebesonnungsmesser. * Ber. DWD-US Zone **7**, Nr. 42, 342—343, 1952.

[1186] *Penman, H. L.*, und *Long, I. F.*, A portable thermistor bridge for micrometeorology among growing crops. * J. Sci. Instr. **26**, 77—80, 1949.

[1187] *Pfleiderer, H.*, Kritische Betrachtungen über die Abkühlungsgröße. * Ber. DWD-US Zone **6**, Nr. 38, 267—270, 1952.

[1188] *Raschke, K.*, Die Kompensation des Strahlungsfehlers thermoelektrischer Meßfühler. * Arch. f. Met. (B) **5**, 447—455, 1954.

[1189] *Schmitz, W.*, und *Volkert, E.*, Die Messung von Mitteltemperaturen auf reaktionskinetischer Grundlage mit dem Kreispolarimeter und ihre Anwendung in Klimatologie u. Bioökologie, speziell in Forst- u. Gewässerkunde. * Zeiß-Mitt. **1**, 300—337, 1959.

[1190] *Schnelle, F.*, und *Breuer, W.*, Meteorologische Meßgeräte und Voraussetzungen für den Schorfwarndienst. * Ber. DWD **6**, Nr. 41, 1958.

[1191] *Schöne, W.*, Bemerkungen zur Registrierung der Globalstrahlung mit dem Bimetallpyranographen nach Robitzsch. * Z. f. Met. **11**, 11—14, 1957.

[1192] *Schubach, K.*, Wasserhaushaltsuntersuchungen an verschiedenen Bodenarten unter besonderer Berücksichtigung der Verdunstung. * Ber. DWD-US Zone **7**, Nr. 40, 1952.

[1193] *Schulze, L.*, Ein Vorschlag zur Verbesserung und Vereinfachung der Lichtmeßtechnik bei ökologischen Versuchen. * Arch. f. Met. (B) **7**, 223—239, 1956.

[1194] *Schulze, R.*, Über ein neues Strahlungsmeßgerät mit ultrarotdurchlässiger Windschutzhaube am Met. Observ. Hamburg. * Geof. pura e appl. **24**, 3—10, 1953.

[1195] *Sonntag, D.*, Hinweise für die Praxis der Hygrometereichung u. -messung. * Z. f. Met. **12**, 36—38, 1958.

[1196] *Thornthwaite, C. W.*, und *Mather, J. R.*, The role of evapotranspiration in climate. * Arch. f. Met. (B) **3**, 16—39, 1951.

[1197] *Tonne, F.*, Besser bauen mit Besonnungs- und Tageslichtplanung. * K. Hofmann-Verl., Schorndorf b. Stuttgart 1954.

[1198] *Uhlig, S.*, Bestimmung des Verdunstungsanspruchs der Luft mit Hilfe von Piche-Evaporimetern. * Mitt. DWD **2**, Nr. 13, 1955.

[1199] *Unger, K.*, Eine Thermobatterie mit kompensierter Vergleichstemperatur für mikrometeorologische Temperaturmessungen. * Arch. f. Met. (B) **8**, 378—381, 1958.
[1200] *Weise, R.*, Über ein einfaches Hilfsmittel zum Taunachweis und seine praktischen Anwendungsmöglichkeiten. * Ann. d. Met. **5**, 378—381, 1952.
[1201] *Wilhelm, F.*, Vorläufiger Bericht über die Temperatur- und Sauerstoffaufnahmen im Schliersee 1956. * Gewässer u. Abwässer 19, 40—65, 1958,
[1202] *Woelfle, F.*, Ein einfaches vollelektrisches Registriergerät für Windrichtung und mittlere Windgeschwindigkeit. * Met. Rundsch. **5**, 133—134, 1952.
[1203] Ein kleiner Windsummenschreiber. * Sonderheft Techn. Mitt. Instr. Abt. MANWD, Hamburg 1950.
[1204] Elektrische u. wärmetechnische Messungen. * Hartmann u. Braun AG., Frankfurt/M., 9. Aufl., 1959.
[1205] Bericht über die Vergleichsversuche an Strahlungsmeßgeräten beim Met. Observ. Hamburg z. H. d. Radiation Comm. d. IAM, IUGG., I u. II, Hamburg 1955 u. 1956.
[1206] Druckschrift der Firma R. Fueß, Berlin-Steglitz.
[1207] Druckschrift der Firma W. Lambrecht, Göttingen.
[1208] Druckschrift der Firma B. Lange, Berlin.

SECTION 57
(See also references 13, 111, 447, and 467)

[1209] *Berger-Landefeldt, U.*, Der Pflanzenklima-Meßwagen. * Geof. pura e appl. **30**, 195—204, 1955.
[1210] *Brocks, K.*, Ein neues Gerät für störungsfreie meteorologische Messungen auf dem Meer. * Arch. f. Met. (A) **11**, 227—239, 1959.
[1211] *Dammann, W.*, Vor einer Neuordnung des Beobachtungs- und Arbeitssystems der praktischen Klimatologie. * Arch. f. Met. (B) **7**, 1—10, 1956.
[1212] *Emschermann, H. H.*, Die Darstellung von Meßwerten in Zahlenform. * Arch. f. techn. Messen, J 071—5, 1956.
[1213] *Frankenberger, E.*, Ein Meßgerät für vertikal gerichtete atmosphärische Wärmeströme. * Techn. Mitt. Instr.wesen d. DWD, N. F. **4**, 21—28, 1958.
[1214] *Geiger, R.*, Der Forschungswagen für mikrometeorologische Untersuchungen. * Wiss. Mitt. Met. Inst. Univ. München **3**, 1—9, 1956.
[1215] *Halstead, M. H., Richman, R. L., Covey, W., und Merryman, J. D.*, A preliminary report on the design of a computer for micrometeorology. * J. Met. **14**, 308—324, 1957.
[1216] *Swinbank, W. C.*, The measurement of the vertical transfer of heat. * J. Met. **8**, 135—145, 1951.
[1217] *Taylor, R. J.*, und *Webb, E. K.*, A mechanical computer for micrometeorological research. * Techn. Pap. Nr. 6, CSIRO Melbourne 1955.
[1218] *Woelfle, F.*, Automatische Wetterstationen. * Met. Rundsch. **11**, 60—67, 1958.

ABBREVIATIONS USED IN THE BIBLIOGRAPHY

Abh. Met. D. DDR. *Abhandlungen des Meteorologischen und Hydrologischen Dienstes der Deutschen Demokratischen Republik* (Akademie Verlag, Berlin, 1950 –).

Abh. Pr. Met. I. *Abhandlungen des Preussischen Meteorologischen Instituts* (Berlin, 1901 – 1935).

Allg. F. *Allgemeine Forstzeitschrift* (Bayerischer Landwirtschaftsverlag, Munich, 1946 –).

Angew. Met. *Angewandte Meteorologie,* supplements to *Zeitschrift für Meteorologie* (Berlin, 1951 –).

Ann. d. Hydr. *Annalen der Hydrographie und maritimen Meteorologie* (Deutsche Seewarte, 1873 – 1944).

Ann. d. Met. *Annalen der Meteorologie* (Meteorologisches Amt für Norswestdeutschland, 1948 – 1951, and Seewetteramt, 1952 –).

Arch. f. Met. (A). *Archiv für Meteorologie, Geophysik und Bioklimatologie, ser. A: Meteologie und Geophysik* (Springer, Vienna, 1949 –).

Arch. f. Met. (B). *Archiv für Meteorologie, Geophysik und Bioklimatologie, ser. B: Allgemeine und biologische Klimatologie* (Springer, Vienna, 1949 –).

Ark. f. Mat. *Arkiv för Matematik, Astronomi och Fysik* (Stockholm).

Beitr. Phys. d. fr. Atm. *Beiträge zur Physik der freien Atmosphäre* (Akademische Verlagsgesellschaft, Leipzig and Frankfurt am Main).

Ber. D. Bot. G. *Berichte der Deutschen Botanischen Gesellschaft* (Fischer, Jena).

Ber. DWD. *Berichte des Deutschen Wetterdienstes* (Bad Kissingen, Frankfurt am Main, and Offenbach, 1953 –).

Ber. DWD-US Zone. *Berichte des Deutschen Wetterdienstes in der US-Zone* (Bad Kissingen, 1947 – 1952).

Biokl. B. *Bioklimatische Beiblätter der Meteorologischen Zeitschrift* (Vieweg, Brunswick, 1934 – 1942).

Bull. Am. Met. Soc. *Bulletin of the American Meteorological Society* (Boston).

Erdk. *Erdkunde: Archiv für wissenschaftliche Geographie* (Dümmler, Bonn, 1947 –).

Forstw. C. *Forstwissenschaftliches Centralblatt* (Parey, Berlin).

Geograf. Ann. *Geografiska Annaler* (Stockholm).

Geophys. Mem. *Geophysical Memoirs* (Meteorological Office, London).

Gerl. B. *Gerlands Beiträge zur Geophysik* (Akademische Verlagsgesellschaft, Leipzig and Frankfurt am Main).

J. Agr. Met. Japan. *Journal of Agricultural Meteorology* (Tokyo).

Jahrb. f. wiss. Bot. *Jahrbücher für wissenschaftliche Botanik* (Borntraeger, Berlin).

J. Met. *Journal of Meteorology* (American Meteorological Society, Boston).

La Mét. *La Météorologie* (Paris).

MAB. *Meteorological Abstracts and Bibliography* (American Meteorological Society, Boston, 1950 –).

Meteoros. *Meteoros: Revista trimestrial de meteorologia y geofisica del Servicio Meteorologico Nacional* (Buenos Aires, 1951 – 1955).

Met. Mag. *The Meteorological Magazine* (London).

Met. Rundsch. *Meteorologische Rundschau* (Springer, Heidelberg, 1947 –).

Met. Z. *Meteorologische Zeitschrift* (Vieweg, Brunswick, 1866 – 1944).

Mitt. DWD. *Mitteilungen des Deutschen Wetterdiensts* (Bad Kissingen, Frankfurt am Main, and Offenbach, 1953 –).

Mitt. DWD-US Zone. *Mitteilungen des Deutsches Wetterdiensts in der US-Zone* (Bad Kissingen, 1948 – 1952).

M. W. Rev. *Monthly Weather Review* (U. S. Department of Agriculture, Washington).

Naturw. *Die Naturwissenschaften* (Springer, Berlin).

Planta. *Planta: Archiv für wissenschaftliche Botanik* (Springer, Berlin).

Publ. Climat. *Publications in Climatology* (The Laboratory of Climatology, Seabrook and Centerton, N. J., 1948 –).

Quart. J. *The Quarterly Journal of the Royal Meteorological Society* (London).

R. f. W. Wiss. Abh. *Wissenschaftliche Abhandlungen* (Reichsamt für Wetterdienst, Berlin, 1935 – 1942).

Sitz-B. Berlin Akad. *Sitzungsberichte der Preussischen Akademie der Wissenschaften zu Berlin.*

Sitz-B. Wien Akad. *Sitzungsberichte der Akademie der Wissenschaft in Wien. Mathematisch-naturwissenschaftliche Klasse.*

Tät-B. Pr. Met. I. *Tätigskeitbericht des Preussischen Meteorologischen Instituts* (Berlin, 1893 – 1933).

Tellus. *Tellus: A Quarterly Journal of Geophysics* (Stockholm, 1949 –).

Thar. Forstl. Jahrb. *Tharandter Forstliches Jahrbuch* (Parey, Berlin, 1842 – 1942).

Veröff. Geoph. I. Leipzig. *Veröffentlichungen des Geophysikalischen Instituts der Universität Leipzig,* ser. 2.

Wetter. *Das Wetter: Monatschrift für Witterungskunde* (Salle, Berlin, 1884 – 1927).

Wetter u. Klima. *Wetter und Klima: Monatschrift für angewandte Meteorologie* (Haug, Tübingen, 1948 – 1949).

Wetter u. Leben. *Wetter und Leben: Zeitschrift für praktische Bioklimatologie* (Österreichische Gesellschaft für Meteorologie, Vienna, 1948 –).

Z. f. angew. Met. *Zeitschrift für angewandte Meteorologie* (Akademische Verlagsgesellschaft, Leipzig, 1928 – 1944).

Z. f. F. u. Jagdw. *Zeitschrift für Forst- und Jagdwesen* (Springer, Berlin, 1869 – 1942).

Z. f. Met. *Zeitschrift für Meteorologie* (Meteorologischer und Hydrologischer Dienst der Deutschen Demokratischen Republik, Berlin, 1946 –).

SYMBOLS

a	Absolute humidity (g m^{-3})
a	Thermal diffusivity (cm^2 sec^{-1})
A	Austausch coefficient (g cm^{-1} sec^{-1})
B	Flow of heat from soil to surface of the ground (cal cm^{-2} min^{-1})
c	Specific heat (cal g^{-1} deg^{-1})
d	day
D	Penetrability (percent)
e	Water-vapor pressure (mm-Hg)
E	East (international symbol)
G	Counterradiation from atmosphere (cal cm^{-2} min^{-1})
hr	hour
H	Diffuse sky radiation (cal cm^{-2} min^{-1})
I_0	Direct solar radiation on a surface perpendicular to the rays (cal cm^{-2} min^{-1})
I	Direct solar radiation on a horizontal surface (cal cm^{-2} min^{-1})
k	Solar constant (cal cm^{-2} min^{-1})
K	A/ρ, eddy diffusivity (cm^2 sec^{-1})
L	Flow of heat from the air to the surface of the ground (cal cm^{-2} min^{-1})
q	Specific humidity of air (grams of water vapor per gram of moist air; the Weather Services mostly use grams per kilogram as a unit, which is 1000 times the numerical value)
Q	Quantity of heat (cal)
r_w	Latent heat of vaporization of water (cal g^{-1})
R	Reflected radiation from a natural surface (cal cm^{-2} min^{-1}) or reflection index (albedo) (percent)
S	Radiation or radiation balance (cal cm^{-2} min^{-1})
$-S_L$	Effective outgoing radiation by night (cal cm^{-2} min^{-1}) = negative radiation balance of the measuring instrument for air temperature
S_B	Radiation balance of the surface of the ground, called radiation loss when negative (cal cm^{-2} min^{-1})
t	Temperature (°C); data from papers using the F scale have been converted into °C, unless specifically stated otherwise. Also other units in the British and American systems (feet, inches, etc.) have been converted into the metric system
T	Absolute temperature (°K)
t_B, T_B	Temperature of the surface of the ground
t_L, T_L	Air temperature
u	Wind speed (cm sec^{-1} or m sec^{-1})
$-V$	Evaporation (cal cm^{-2} min^{-1}); the numerical values are approximately equal to those in mm hr^{-1}
W	Flow of heat from water to the surface of the water (cal cm^{-2} min^{-1})

x	Height above ground (cm)
z	Time (unit stated when used)
$-a_L$	Heat transport coefficient (cal cm^{-2} min^{-1} deg^{-1})
γ	Vertical temperature gradient (deg/100 m), negative when temperature decreases with height, positive with inversions
Θ	Potential temperature (°K)
κ	Coefficient of refraction (see Sec. 18)
λ	Thermal conductivity (cal cm^{-1} sec^{-1} deg^{-1})
λ	Wavelength (μ)
μ	Micron = 10^{-3} mm
ν	Extinction coefficient (cm^{-1}), see Sec. 24
ρ	Density of air or homogeneous substance (g cm^{-3})
ρ_m	Density of natural soil (g cm^{-3})
$(\rho c)_m$	Heat capacity by volume of natural soil (cal cm^{-3} deg^{-1})
σ	Stefan-Boltzmann constant, = 8.26×10^{-11} cal cm^{-2} min^{-1} deg^{-4}

INDEX